INTRODUCING PHYSICAL GEOGRAPHY

Updated and Upgraded

This classic photo of the earth from space was taken by astronauts returning from a lunar voyage in December 1972. At this time of year, the south polar region is tilted toward the sun and much of Antarctica is visible.

Africa and South America receive the lion's share of the sunlight striking the earth. These continents lie near the equator, where they meet the strong, vertical rays of the sun. North America and Europe, on the other hand, are short-changed because they intercept the sun's rays at a low angle. At the lower rim of the earth's disk, the stark, white continent of Antarctica rotates slowly but remains sunlit, experiencing 24-hour daylight.

As the earth days go by, you often gaze at your planet, marking the changes. Away from the equator, huge swirls of clouds form, dissolve, and reform in its atmosphere, making their way eastward across the disk. First one portion of a continent, and then another, is obscured. These swirls mark the passage of weather fronts and storms that are moved eastward by persistent westerly winds above the surface.

Near the earth's equator in Africa you notice a band of patchy, persistent cloudiness bracketed by reddish-brown areas of earth that are normally clear. The clouds result when warm, moist air is heated by excess solar energy and rises. As the air rises, it cools, and the moisture it contains condenses, forming clouds and producing rain. The lush, green landscape that is occasionally visible underneath the cloud belt seems to thrive on the warm temperature and abundant rainfall it receives. The reddish-brown areas are vast deserts. These receive the air that rises over the equator and becomes depleted of moisture. As the air descends toward the desert, it warms. Showing the colors of rock and soil, the hot, dry deserts are barren of plants.

During the remainder of your six-month tour of duty at the lunar base, you watch the slow changes of the planet with the season. By June, the earth has changed its position in the sun. Antarctica has moved below the southern rim. The sunlit northern rim of the disk now includes northern Canada, Siberia, and the Arctic Ocean. In previous months, you watched the green wave of vegetation sweep northward, up and across North America, Europe, and eastern Asia, following the warming temperatures of spring. The band of tropical cloudiness has also moved northward, bringing a green wave in Africa along with it as well. Clearly, each region of the earth has its unique climate, responding to the rhythms of the season in different ways. You remember fondly the days of June on earth—the smell of warm soil mixed with the scent of flowers, the hot sidewalks and streets at noon, the long evenings and warm nights alive with quiet rustlings. As the end of your tour draws near, you count the days eagerly, thinking of your return to your true home, planet Earth.

It's sixteen hundred hours, June 18, Houston time, aboard the Space Station Enterprise, *in low earth orbit. Your duty watch is just over. Being an environmental specialist in orbit around the earth is not the hardest job on the space station, but it isn't the easiest either. Today's shift required changing the filter on the nutrient pump for the algae, and in the weightlessness of orbit, it was a ticklish, delicate job.*

It's been quite a trip so far. You recall the events of the last few weeks—the arrival of the new crew on the lunar shuttle at Moon Base Alpha's spaceport, the preparations for the departure of your crew, the relaunch of the lunar shuttle, the slow days in transit to earth orbit, and, at long last, the docking of the lunar craft at the space station. But the most memorable part of the return trip was approaching the earth ever closer. With each day and hour, it grew larger and larger. Continents became regions of forests and deserts, mountain ranges, and inland seas. At last the planet became a sphere, its edges falling away and out of view as the lunar shuttle neared the space station. Now, in low earth orbit, finer detail is visible—systems of rivers, agricultural patterns, large cities.

A Landsat satellite photo of deforestation in the Rondonia region, Brazil. In this false-color image, vegetation appears red and forest clearings appear blue.

the ocean's edge. As you have surveyed the world's landforms from your orbital vantage point, you have noted how many seem to be related to the great structural features of the earth's surface—its great upwarps and downwarps creating mountain zones and inland seas, its folded and faulted crust producing mountains, valleys, and plains.

It's nine hundred hours, June 26, Houston time, aboard the Space Shuttle Discovery. The tension is building strongly now. Gravity is starting to return, and the earth shuttle is beginning to feel the slowing of friction with the atmosphere. The communications blackout period is about to begin. Strapped firmly into your seat, you contemplate the bumpy ride ahead with a racing heart. Fortunately, you are just a passenger on this segment and can leave the tricky duties of navigating the shuttle to a safe landing to the flight crew and their computers. You'll be landing at Edwards Spaceport, in the California desert, since the two other spaceports are experiencing winds and rain. For now, there is nothing much to see out of the nearby viewport—it is night over the Pacific. The minutes pass slowly—too slowly. The attitude-control jets fire from

Still drawn to the majesty and serenity of the earth turning below you, you float gently to the viewport for yet another few minutes of observation. Beneath you stretches Amazonia, its lush green rainforests visible through gaps in the clouds. The verdant pattern is interrupted by patches and streaks of recently cleared land, some with hazy smoke plumes at their edges. The rainforest is but one of the many types of vegetation covers you have noted from space, ranging from spruce forests to tropical grasslands. The vegetation patterns have fit the cloud patterns well and have also marked the seasonal changes of solar illumination you noted from your lunar base.

Some hours later, your view is of China. Finally, the clouds have cleared, and a disaster is in view—the Yangtze River is in flood. The river is now a wide, shining ribbon, filling its broad valley from edge to edge. A few tiny green islands are visible, but you know that from this distance they are actually ranges of hills or diked cities. Where the Yangtze meets the sea, a plume of sediment-laden river water curls lazily across the ocean waters. In this view, the work of rivers is especially evident—moving sediment from highlands to lower lands and the ocean, under the power of rains gen-

erated ultimately by unequal solar heating of the globe.

The following day, your view is of the San Francisco Bay area. The San Andreas fault is clearly visible on the east side of the peninsula as a line marked by valleys and linear lakes. Crossing the Golden Gate, it heads northwest, near

The San Francisco Bay area, viewed by the Landsat satellite from earth orbit.

The Space Shuttle runway at Edwards Air Force Base, California, as photographed from the Space Shuttle *Challenger*.

time to time, as the shuttle moves through its slowing maneuvers.

Soon the line of the horizon begins to redden. The sun rises quickly, spectacularly, moving rapidly skyward. Below, the California coast approaches, the white stripe of its beaches gleaming in the sun. The light brown scars of recent debris avalanches on the steep marine cliffs dissect the gray-green of the chaparral. Off to the right is Los Angeles. The pink bubble of polluted air over the city is plainly visible at this early hour.

Still descending, you cross the Tehachapi Mountains, which are now very close, and drop toward the Mojave Desert. A patch of stripes flashes by— lines of sand dunes silhouetted in the early light. The landing gear descends, shaking your craft with a thud as it locks in place. Your grip on the handrest tightens as your body tenses in anticipation. The shuttle flares out over the flat, salty soil of the dry lake. Bang! The wheels hit the ground hard. Touchdown! The brakes and drogue chute press you hard against the seat. You're home at last!

THE PHYSICAL GEOGRAPHY OF OUR PLANET

Our opening essay touches on a number of subjects that are part of physical geography, ranging from earth–sun relationships to evolution of landforms. These are a few of the important topics in physical geography that we will develop further in this book. However, there are some essential ideas and themes that need to be discussed first. We begin with the "spheres"—the four great realms of earth—mentioned in the essay. Next is an exploration of scale—in space and in time. Finally, we examine the earth as the human habitat and introduce physical geography as the science that studies it.

The Four Great Realms

At its most basic level, physical geography is concerned with the atmosphere, the lithosphere, the hydrosphere, and the biosphere and their interactions (Figure P.1). The **atmosphere** is a gaseous layer that surrounds the earth. It receives and returns flows of heat and moisture from and to the surface. The atmosphere also supplies vital elements—carbon, hydrogen, oxygen, and nitrogen—needed to sustain the life-forms on the lands.

The solid earth, or **lithosphere**, forms the stable platform for the life layer. The solid rock of the lithosphere bears a shallow layer of soil in which nutrient elements become available to organisms. The lithosphere is sculpted into landforms. These relief features, like mountains, hills, and plains, provide varied habitats for plants, animals, and humans.

The earth's four great realms are the atmosphere, lithosphere, hydrosphere, and biosphere.

Water, in all its forms, constitutes the **hydrosphere**. The main mass of the hydrosphere lies in the world's oceans, but water also occurs in the atmosphere as gaseous vapor, liquid droplets, and solid ice crystals. Water is found in or atop the uppermost layers of the lithosphere, forming ground water reservoirs, lakes, streams, and rivers. Water is essential to life and required for the survival of organisms.

The **biosphere** encompasses all living organisms of the earth. Life-forms on earth utilize the gases of the atmosphere, the water of the hydrosphere, and the nutrients of the lithosphere, and so the biosphere is dependent on all three of the other great realms.

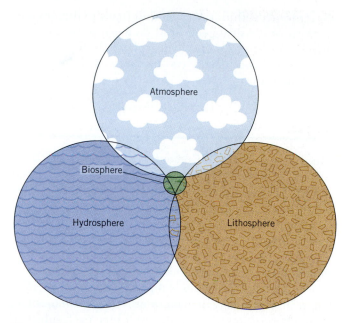

Figure P.1   The earth realms, shown as intersecting circles.

The Life Layer

The **life layer** is the shallow surface zone that contains most of the biosphere (Figure P.2). It includes the surface of the lands and the upper 100 m (about 300 ft) or so of the ocean. On land, the life layer is the zone of interaction among the biosphere, lithosphere, and atmosphere. The hydrosphere is represented on land by rain, snow, still water in ponds and lakes, or running water in rivers. In the ocean, the life layer is the zone of interactions among the hydrosphere, biosphere, and atmosphere. The lithosphere is represented by nutrients dissolved in the upper layer of sea water.

Throughout our exploration of physical geography, we will often refer to the life layer and the four realms

Figure P.2   As this sketch shows, the life layer is the layer of the earth's surface that supports nearly all of the earth's life. This includes the land and ocean surfaces and the atmosphere in contact with them.

that interact within it. We will find that the themes and subjects of physical geography concern processes that involve the atmosphere, hydrosphere, lithosphere, or biosphere—and usually two or three of these realms simultaneously.

SCALE

Think again about your recent trip home to earth from the lunar base. While making your trip, you noted aspects of several themes in physical geography. These include earth–sun relationships, global atmospheric circulation, precipitation and weather systems, world climates, natural vegetation, soils, the earth's surface structure, and landforms as well as the agents that produce them.

But what is interesting is that you observed them in about that order as you made your return from moon to earth. This relates to the **scale** with which each subject is concerned. Earth–sun relationships come first, because the sun is the energy source that powers most of the important processes that occur within the life layer. To study these processes, we need to consider the planet and its global energy balance as a whole.

The sun's energy is not distributed evenly across the earth's surface. This unequal solar heating produces currents of air and water, just as a current of hot air rises from a candle flame, and these currents constitute the global atmospheric circulation. When we move still closer to the earth, the cloud patterns of weather systems become obvious. By watching the earth over time, we see the regular movements of weather systems. Taking these movements along with solar control of surface temperature, we have the basis for climates of the world.

The nature of the earth's cover of plants, and the soil layers beneath them, are strongly related to climate. Local factors are also important in determining the exact patterns of vegetation and soils. Is the region a mountain or a valley? Is it being farmed or do sheep graze there? This specific information is not useful when we consider the planet as a whole, but it is important to collect when we consider the nature of the land's vegetation cover and soils at a middle scale. We look at a still closer scale to see features such as a sand dune on a beach or a rocky cliff cut away by a river. These individual landforms are produced by local action of wind or water.

The point here is that the processes acting in nature have a wide range of scales in time and space. Some processes—for example, the global circulation of the atmosphere—are best understood by considering the earth as a whole. Still other examples are concerned with very local phenomena, such as sand grains being blown along a beach. Time scales vary widely, too. It may take millions of years to produce a chain of mountains like the Himalayas, but a fracturing of the earth's crust and the resulting earthquake may last only a few minutes.

If you look at the table of contents for this book, you will note that our study of physical geography moves from the coarse scale to the fine, from global phenomena to local phenomena. There will be exceptions to this overall plan, but understanding the focus of this organization will be helpful in learning about the systems and processes of physical geography.

PHYSICAL GEOGRAPHY AND THE HUMAN HABITAT

The lands of the earth are the habitat of the human species, as well as other forms of terrestrial life. Since physical geography is the study of the earth's surface and the processes that act on it, physical geography is also the study of the **human habitat**. This personal involvement helps make physical geography interesting and rewarding to learn. Everyday concerns like weather systems are a good example. Since the weather is always changing, physical geography will help you understand how and why it changes.

> **Because physical geography focuses on the life layer, it is also the study of the human habitat.**

Still other aspects of physical geography help explain your local landscape. The shapes of hills and valleys normally change too slowly to be noticed on a daily basis. But if you watch soil being washed down a slope during a sudden rain shower, you can extend this phenomenon to the land around you and come to understand how years of rainfall slowly carve the landscape into unique landforms.

If you have the opportunity to travel, studying physical geography will help you understand more about the landscapes you visit. While considering a Peace Corps assignment to the country of Mali, on the African continent, you might expect to find a tropical jungle. But your study of physical geography will tell you that much of Mali looks like our American southwest desert!

Physical geography also encompasses a concern for our environment. For example, global climate change is a widespread concern today. Is the earth getting warmer? Is it getting cooler? Will there be another ice age? Scientists offer us conflicting opinions. To understand their conclusions, you need some background about the processes and systems that influence global climate. While our survey of physical geography focuses on natural processes, it will often explain environmental problems. We will encourage you to sift the data and the opinions drawn from current studies.

Humans are now the dominant species on the planet. Nearly every part of the earth has felt our impact in some way. As our population continues to grow, human impacts on natural systems will continue to increase. Each of us is charged with the responsibility to treat the earth well and respect it. Understanding the processes that shape the human habitat as they are described by physical geography will help us carry out this mission.

THE PROLOGUE IN REVIEW

Physical geography is a branch of science that investigates the earth's surface, and how and why it changes. It focuses on the life layer in which the four great realms—atmosphere, hydrosphere, lithosphere, and biosphere—interact. The processes of interaction between the realms can be examined at different scales. For example, earth–sun relationships control global heating and in this way influence global wind systems. Vegetation and soils are strongly related to climate, which varies at a regional scale. And individual landforms, such as a sand dune, are produced by processes acting at local scales. Since physical geography focuses on the land surface, it includes the study of the human habitat—the local environment in which we live and interact with natural processes.

KEY TERMS

physical geography
atmosphere
lithosphere

hydrosphere
biosphere
life layer

scale
human habitat

REVIEW QUESTIONS

1. What are the four great realms of earth? Name and describe each realm.
2. What is the life layer? Where is it located?
3. Provide two examples of processes that operate at global, middle, and local scales.
4. What relation does physical geography have to the human habitat?

A DEEPER VIEW OF *INTRODUCING EARTH*

Imagine that you are planning a mountain backpacking trip—for example, along the Appalachian or Pacific Coast Trails. Which aspects of physical geography might be useful to know about? Why?

Chapter 1

Our Place in the Sun

Stonehenge, England

PUTTING YOU IN THE PICTURE

At last the sky is lightening. Perhaps it won't be much longer now, you think, until the first rays of sun become visible on the horizon next to the Slaughter Stone. Exhaling slowly, you watch the condensation from your breath float away on the still air. A shiver overwhelms you and you pull your jacket more tightly around you.

Although the morning is cold, the sky is mostly clear. June weather in the southwest corner of England is notoriously changeable, but it looks like the viewing conditions will be good. The showers that sprang up in the night are now gone, leaving only a dampness in the soft soil underfoot and a few small gray tufts of cloud overhead. According to the information you noted yesterday on your visit to the Greenwich Observatory, the summer solstice will occur at 3:28 P.M. this afternoon. So, today will be the longest day of the year—almost $16\frac{1}{2}$ hours at this latitude of more than 50 degrees north. It will also be the day of the year on which the sun rises at its most northerly position from due east. You check your watch. Sunrise was forecast at Greenwich for 4:51 A.M., daylight saving time, but since you are nearly 100 miles west of Greenwich, it will occur a few minutes later—8.8 minutes to be exact. Still about 10 minutes to go.

You pause for a moment to survey your surroundings. Not far away is a trilithon, its two immense slabs arching upright to nearly three times your height. Crowning the two slabs is a massive horizontal lintel, chiseled from the same brown sandstone as the uprights. This trilithon is one of five, arranged in a horseshoe. Three are still standing, and two are represented only by a single, still-erect stone. Beyond the trilithons by four or five yards are the remaining vertical slabs of the sarsen circle, arranged in a ring of about 100 feet in diameter that encloses the trilithons. Several of the sarsen slabs are still topped by stone lintels as well. Stonehenge was quite an accomplishment, you think to yourself, considering that it was built 3000 to 4000 years ago using only the most primitive tools and machines.

The sky is now quite light. The low voices of the crowd are suddenly quiet in anticipation. A ruby gleam brightens the sky, and then the first solar rays are visible. Slowly, the red solar disk emerges. Beyond the sun, the sky is washed with red and pink, fading to deep blue. Small, puffy clouds above glow a spectacular rosy pink. What a sight!

From your location near the center of the ruin, the red solar ball is visible just to the left of the immense Slaughter Stone that marks the entrance to the ancient ruin. You visualize the missing twin slab to the Slaughter Stone, trying to imagine how the scene must have looked to the builders of the ancient monument. The sun would have been centered perfectly between the two huge monoliths. The long shadow of the Slaughter Stone to your right would have been joined by a second shadow on your left, providing a broad sunlit path extending from the entrance to the center of the structure. Quite amazing, you think, that the ancient peoples that built Stonehenge were so fascinated with the paths of the sun and the moon in the sky that they marked them by hauling huge stones hundreds of miles to this hallowed place. Suddenly, a wave of reverence for earth and nature hits you, and you realize that the scene has touched something primitive deep inside you that springs from the common heritage of humanity.

Humans have observed the movements of the sun since the dawn of civilization. These movements stem from two motions of the earth with respect to the sun. From day to day, the earth's rotation on its axis produces the cycle of daylight and darkness. From month to month, the revolution of the earth around the sun produces changes in the length of daylight that create the cycle of the seasons. These two great cycles continue, day after day, year after year, regulating the processes of earthly life. They are truly an appropriate starting point for our study of the physical geography of our planet.

Chapter Outline

The Shape of the Earth
Earth Rotation
  Direction of Rotation
  Environmental Effects of Earth Rotation
The Geographic Grid
  Parallels and Meridians
  Latitude and Longitude
  Map Projections
  Polar Projection
  Mercator Projection
  Goode Projection

Global Time
  Standard Time
  World Time Zones
  International Date Line
  Daylight Saving Time
The Earth's Revolution Around the Sun
  Tilt of the Earth's Axis
  Solstice and Equinox
  Equinox Conditions
  Solstice Conditions
Eye on the Environment • Earth–Sun Cycles and Ice Ages

Chapter 1

OUR PLACE IN THE SUN

Some people are "day people"—preferring the day for working and playing—while others are "night people"—preferring the night for their most important activities. How about you? Either way, both types of people react to the fact that the earth rotates on its axis, completing a full rotation with respect to the sun in a 24-hour period. This rotation determines our daily time system. As we will see in this chapter, the plan of rotation on an axis also leads naturally to the system of latitude and longitude—our method for locating places on the surface of the earth.

Some people like hot summers on the beach, while others like cold winters on the ski slopes. Which season do you prefer? The seasons are produced by the revolution of the earth around the sun in a cycle that repeats about every 365 days. As the earth revolves about the sun, the axis of the earth's rotation remains pointed toward the same place in the heavens, no matter where the earth is located. This means that in one part of the annual cycle the northern hemisphere is tilted toward the sun, while the southern hemisphere is tilted away. This produces summer conditions in the northern hemisphere. Later in the cycle, the situation is reversed, providing winter for the northern hemisphere and summer for the southern hemisphere.

The two eternal cycles of rotation and revolution are fundamental rhythms that not only influence our daily life, but are also essential to our study of physical geography. We begin our study of physical geography by devoting a chapter to the motions of the earth and sun and the implications of these motions for timekeeping, earth location, and the seasons.

THE SHAPE OF THE EARTH

Our planet is a rotating body that revolves around the sun. Its shape is very close to that of a sphere, as we all learn early in school. Pictures taken from space by astronauts and by orbiting satellites also show us that the earth is a round body. We learn this fact so early in life that it seems quite unremarkable.

Many of our ancestors, however, were not aware of our planet's spherical shape. To sailors of the Mediterranean Sea in ancient times, the shape and breadth of the earth's oceans and lands were unknown. On their ships and out of sight of land, the sea surface looked perfectly flat and bounded by a circular horizon. Given this view, many sailors concluded that the earth had the form of a flat disk and that their ships would fall off if they traveled to its edge.

High in a jet airplane at cruising altitude, you could come to the same conclusion. Peering out the window, you see below a vast expanse of earth or ocean fading into a distant and level horizon. However, at sunset, you might notice an interesting phenomenon (Figure 1.1).

Figure 1.1 Sunset over the Pacific Ocean, seen from a highflying aircraft. As the setting sun bathes these clouds in its last light, the ocean surface below is in shadow. For an observer at the surface, the sun has already left the sky.

While the sun is still visible to you, illuminating the clouds at your altitude with an orange glow, the surface below you is in shadow. This means that the sun is below the horizon to an observer at the surface but above the horizon to an air traveler. This only happens when the curvature of the earth blocks the sun for the surface observer but not for you, the air traveler.

The earth's spherical shape is an effect of the force of gravity acting to pull all portions of the earth toward the earth's center with equal force. Although rock may seem strong and rigid to the inhabitants of the earth's surface, rocks are plastic and can flow slowly at the temperatures and pressures encountered deep within the earth. (Chapter 11 will provide more information about this topic.) Thus, the earth responds to the pull of gravity over time by assuming the shape of a sphere.

Actually, the earth is not perfectly spherical, because the outward force of the earth's rotation causes the earth to bulge slightly at the equator. This effect produces a shape in which the earth's equatorial diameter, about 12,756 km (7926 mi), is very slightly larger than the polar diameter, about 12,714 km (7900 mi). The difference is very small—about three-tenths of a percent. The earth's true shape is described as an *oblate ellipsoid*.

Scientists still study the earth's shape and attempt to measure it as precisely as possible. This is important since the information is needed by satellite navigation systems for aircraft, ocean vessels, and ground vehicles seeking to determine their exact location. The more precisely the earth's shape is known, the more accurately location can be determined.

EARTH ROTATION

Another fact about our planet that we learn early in life is that it spins slowly, making a full turn with respect to the sun every day. We use the term **rotation** to describe this motion. One complete rotation with respect to the sun defines the *solar day.* By convention, the solar day is divided into exactly 24 hours.

The earth rotates on its *axis,* an imaginary straight line through its center. The intersections of the axis of rotation and the earth's surface are defined as the poles. To distinguish between the two poles, one is called the *north pole* and the other, the *south pole.*

Direction of Rotation

The direction of earth rotation can be determined by using one of the following guidelines (Figure 1.2):

- Imagine yourself looking down on the north pole of the earth. From this position, the earth is turning in a counterclockwise direction (Figure 1.2*a*).
- Imagine yourself off in space, viewing the planet much as you would view a globe in a library, with the

(a)   (b)

Figure 1.2 The direction of rotation of the earth can be thought of as (*a*) counterclockwise at the north pole, or (*b*) from left to right (eastward) at the equator.

north pole on top. The earth is rotating from left to right, or in an eastward direction (Figure 1.2*b*).

The direction of the earth's rotation is counter-clockwise, when we look down on the north pole from above.

The earth's rotation is important for three reasons. First, the axis of rotation serves as a reference in setting up the geographic grid of latitude and longitude, which we will discuss later in the chapter. Second, it provides the day as a convenient measure of the passage of time. Third, it influences physical and life processes on earth, as we will now describe.

Environmental Effects of Earth Rotation

The effects of the earth's rotation are of great importance to us and our environment. The first—and perhaps most obvious—effect of rotation is that it imposes a daily, or *diurnal*, rhythm in daylight, air temperature, air humidity, and air motion. Plants and animals respond to this diurnal rhythm. Green plants store energy during the day and consume some of it at night. Among animals, some prefer the day, others the night, for food gathering. The daily cycle of incoming solar energy and the corresponding cycle of fluctuating air temperature will be topics for analysis in Chapters 2 and 3.

Second, flow paths of both air and water are turned consistently in one direction because of the earth's rotation. When viewed from the starting point, flows are turned toward the right in the northern hemisphere and toward the left in the southern hemisphere. This phenomenon is called the *Coriolis effect*. It is of great importance in studying the earth's systems of winds and ocean currents. We will investigate both the Coriolis effect and its influence on winds and currents in Chapter 5.

A third physical effect of the earth's rotation is the movement of the tides. The moon exerts a gravitational attraction on the earth, while at the same time the earth is turning with respect to the moon. These forces induce a rhythmic rise and fall of the ocean surface known as the *tide*. The tide in turn causes water currents of alternating direction to flow in the shallow, salty waters of the coastal zone. The ocean tide may have no importance for a grain farmer in Kansas, but for the clam digger or charter boat captain on Cape Cod, the tidal cycle is a clock regulating daily activities. The ebb and flow of tidal currents is a life-giving pulse for many plants and animals that live in coastal saltwater environments. The tide and its currents are discussed further in Chapter 17.

THE GEOGRAPHIC GRID

The geographic grid provides a system for locating places on the earth's surface. Because the earth's surface is curved, and not flat, we cannot divide it into a rectangular grid, like a sheet of graph paper. Instead, we divide it using imaginary circles set on the surface that are perpendicular to the axis of rotation in one direction and parallel to the axis of rotation in the other direction.

Parallels and Meridians

Imagine a point on the earth's surface. As the earth rotates, the point traces out a path in space, following an *arc*—that is, a curved line that forms a portion of a circle. With the completion of one rotation, the arc forms a full circle. This is known as a parallel of latitude, or a **parallel** (Figure 1.3*a*). Parallels cut the globe much as you might slice an onion to produce onion rings—that is, perpendicular to the onion's main axis. The largest parallel of latitude lies midway between the two poles and is designated the **equator**. The equator is a fundamental reference line for measuring the position of points on the globe.

Imagine now slicing the earth with a plane that passes through the axis of rotation, instead of across it. This is the way you might cut up a lemon to produce wedges. The cut outlines a circle on the globe passing through both poles. Half of this circular outline, connecting one pole to the other, is known as a meridian of longitude, or, more simply, a **meridian** (Figure 1.3*b*).

Meridians and parallels define compass directions. Meridians are north-south lines, so you are following a meridian if you walk north or south. Parallels are east-west lines, and so you are following a parallel if you walk east or west. There can be any number of parallels and meridians. Every point on the globe is associated with a unique combination of one parallel and one meridian.

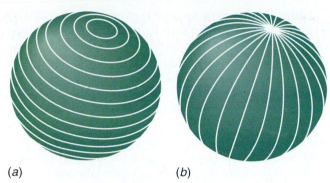

(a)   *(b)*

Figure 1.3 (*a*) Parallels of latitude divide the globe crosswise into rings. (*b*) Meridians of longitude divide the globe from pole to pole.

The position of the point is defined by their intersection. The total system of parallels and meridians forms a network of intersecting circles called the **geographic grid**.

> **The geographic grid consists of an orderly system of circles—meridians and parallels—that are used to locate position on the globe.**

Latitude and Longitude

We use a special system to label parallels and meridians—latitude and longitude. Parallels are identified by latitude, and meridians by longitude. Since parallels are circles and meridians are half-circles, we use *degrees of arc* as latitude and longitude measures to mark our place along particular parallels and meridians. What do we mean by degrees of arc? Figure 1.4 illustrates this concept. A full circle consists of a 360° angle. A lesser angle is associated with an arc, or portion of a circle. We can measure the arc by measuring the degrees of that angle.

Figure 1.4 A circle consists of 360°, measured as an angle from its center. An arc is a part of a circle. An angle is associated with an arc that measures the arc in degrees. Thus, we state that the arc angle is 60°.

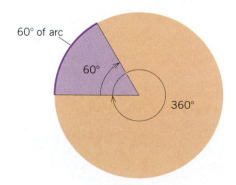

60° of arc

60°

360°

Latitude measures the position of a given point in terms of its angular distance from the equator (Figure 1.5). That is, latitude is an indicator of how far north or south of the equator a given point is situated. Latitude is measured in degrees of arc from the equator (0°) toward either pole, where the value reaches 90°. The equator divides the globe into two equal portions, or *hemispheres*. All points north of the equator—that is, in the *northern hemisphere*—are designated as north latitude, and all points south of the equator—in the *southern hemisphere*—are designated as south latitude.

Figure 1.5 shows how latitude is measured. The point *P* lies at the latitude of 50 degrees north, which we can abbreviate as lat. 50° N. Notice that latitude arc is actually measured along the meridian that passes through the point.

Longitude is a measure of the position of a point eastward or westward from a reference meridian, called the *prime meridian.* As Figure 1.5 shows, longitude is the arc angle, measured in degrees, between the meridian of a given point and the prime meridian. This arc is measured east to west along the parallel that passes through the point. In this example, the point *P* lies at longitude 60 degrees west (long. 60° W).

The prime meridian passes through the location of the old Royal Observatory at Greenwich, near London, England (Figure 1.6). For this reason it is also called the Greenwich meridian. It has the value long. 0°. The longitude of any given point on the globe is measured

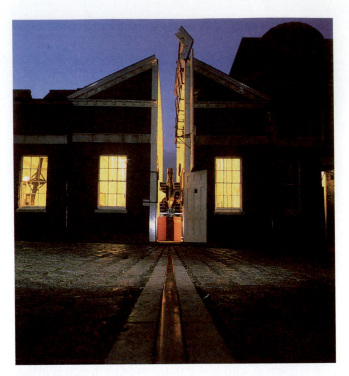

Figure 1.6 This photograph, taken at dusk at the old Royal Observatory at Greenwich, England, shows the prime meridian, which has been marked as a stripe on the forecourt paving. The dome covering the observatory is partly opened, and the telescope, located exactly on the meridian, is pointed toward the sky as if making an observation.

eastward or westward from the prime meridian, depending on which direction gives the shorter arc. Longitude thus ranges from 0° to 180°, east or west. When both the latitude and longitude of a place are known, it can be accurately and precisely located on the geographic grid.

When arcs of latitude or longitude are measured other than in full-degree increments, *minutes* and *seconds* can be used. A minute is 1/60 of a degree, and a second is 1/60 of a minute, or 1/3600 of a degree. Thus, the latitude 41°, 27 minutes ('), and 41 seconds (") north (lat. 41° 27' 41" N) means 41° north plus 27/60 of a degree plus 41/3600 of a degree. This cumbersome system has now largely been replaced by decimal notation. In this example, the latitude 41° 27' 41" N translates to 41.4614° N.

Statements of latitude and longitude do not describe distances in kilometers or miles directly. However, for latitude, you can estimate conversions from degrees into kilometers. One degree of latitude is approximately equivalent to 111 km (69 mi) of surface distance in the north-south direction. This value can be rounded off to 110 km (70 mi) for multiplying in your head. For example, if you live at lat. 40° N (on the 40th parallel north), you are located about $40 \times 110 = 4400$ km ($40 \times 70 = 2800$ mi) north of the equator.

East-west distances cannot be converted so easily from degrees of longitude into kilometers or miles because the meridians converge toward the poles

Figure 1.5 The point *P* has a latitude of 50°. This means that the arc between *P*'s parallel and the equator has an angle of 50°. We say that *P* is located on the 50th north parallel. *P* also has a longitude of 60°. Therefore, the arc between *P*'s meridian and the prime meridian, in Greenwich, England is 60°. We speak of *P* as being on the 60th west meridian.

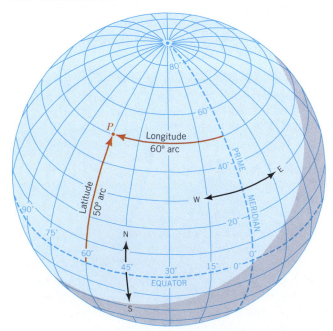

(Figure 1.3*b*). Only at the equator is a degree of longitude equivalent to 111 km (69 mi). At lat. 60° N or S, meridians are twice as close as at the equator. So, one degree of longitude is reduced to half its equatorial length, or about 55 km (about 35 mi).

Map Projections

With a working understanding of the geographic grid, we can now consider how to display the locations of continents, rivers, cities, islands, and other geographic features on maps. This will take us briefly into the realm of *cartography*, the art and science of making maps. Our discussion will focus on a few simple types of maps used in this text.

As we observed earlier, the earth's shape is nearly spherical, and so the earth's surface is curved. However, maps are flat. It is impossible to copy a curved surface onto a flat surface without cutting, stretching, or otherwise distorting the curved surface in some way. So, making a map means devising an orderly way of changing the globe's geographic grid of curved parallels and meridians into a grid that lies flat. We refer to a system for changing the geographic grid to a flat grid as a **map projection**.

A map projection is a system for displaying the curved surface of the geographic grid on a flat surface.

Associated with every map is a *scale fraction*—a ratio that relates distance on the map to distance on the earth's surface. For example, a scale fraction of 1:50,000 means that one unit of map distance equals 50,000 units of distance on the earth.

Because a curved surface cannot be projected onto a flat surface without some distortion, the scale fraction of a map holds only for one point or a single line on the map. At a location on the map away from that point or line, the scale fraction will be different. However, the variation in scale is only important for maps showing large regions, such as continents or hemispheres.

We will concentrate on the three most useful map projections. The first is the polar projection, which is essential today for scientific uses like weather maps of the polar regions. Second is the Mercator projection, a navigator's map invented in 1569 by Gerhardus Mercator. It is a classic that has never gone out of style. Third is the Goode projection, named for its designer, Dr. J. Paul Goode. It has special qualities not found in the other two projections.

Polar Projection

The *polar projection* (Figure 1.7) can be centered on either the north or the south pole. Meridians are straight lines radiating outward from the pole, and par-

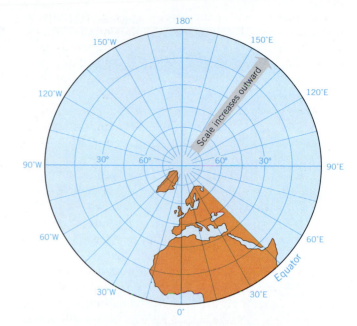

Figure 1.7 A polar projection. The map is centered on the north pole. All meridians are straight lines radiating from the center point, and all parallels are concentric circles. The scale fraction increases in an outward direction, making shapes toward the edges of the map appear larger.

allels are nested circles centered on the pole. Spacing of the parallels increases outward from the center. The map is usually cut off to show only one hemisphere so that the equator forms the outer edge of the map. Because the intersections of the parallels with the meridians always form true right angles, this projection shows the true shapes of all small areas. If you drew a small circle anywhere on the globe, it would reproduce as a perfectly shaped circle on this map. However, because the scale fraction increases in an outward direction, the circle would be larger toward the edge of the map than near the center.

Mercator Projection

The *Mercator projection* (Figure 1.8) is a rectangular grid of meridians as straight vertical lines and parallels as straight horizontal lines. Meridians are evenly spaced, but the spacing between parallels increases at higher latitude so that the spacing at 60° is double that at the equator. Closer to the poles, the spacing increases even more, and the map must be cut off at some arbitrary parallel, such as 80° N. This change of scale enlarges features when they near the pole, as can easily be seen in Figure 1.8. There Greenland appears larger than Australia and is nearly the size of Africa! In fact, it is very much smaller, as you can see on a globe.

The Mercator projection has several special properties. One is that a straight line drawn anywhere on the map is a line of constant compass direction. A navigator can therefore simply draw a line between any two points on the map and measure the *bearing*, or direc-

Figure 1.8 The Mercator projection. The diagram at the right shows how rapidly the map scale increases at higher latitudes. At lat. 60°, the scale is double the equatorial scale. At lat. 80°, the scale is six times greater than at the equator.

tion angle of the line, using a protractor. Once aimed in that compass direction, a ship or an airplane can be held to the same compass bearing to reach the final point or destination.

This line will not necessarily follow the shortest actual distance between two points. The shortest path on the globe follows the arc of a *great circle*. (A great circle connects two points on the globe and has the earth's center as the circle's center.) On a Mercator projection, a great circle line usually curves and can (falsely) seem to represent a much longer distance than a compass line.

Because the Mercator projection shows the true compass direction of any straight line on the map, it is used to show many types of straight-line features. Among these are flow lines of winds and ocean currents, directions of crustal features (such as chains of volcanoes), and lines of equal values, such as lines of equal air temperature or equal air pressure. This explains why the Mercator projection is chosen for maps of temperatures, winds, and pressures.

Goode Projection

The *Goode projection* (Figure 1.9) uses two sets of mathematical curves (sine curves and ellipses) to form its meridians. Between the 40th parallels, sine curves are used, and beyond the 40th parallel toward the poles, ellipses are used. Since the ellipses converge to meet at

the pole, the entire globe can be shown. The straight, horizontal parallels make it easy to scan across the map at any given level to compare regions most likely to be similar in climate.

The Goode projection has one very important property—it indicates the true sizes of areas of the earth's surface. That is, if we drew a small circle on a sheet of clear plastic and moved it over all parts of the Goode world map, all the areas enclosed by the circle would have the same value in square kilometers or square miles. Because of this property, we use the Goode map to show geographical features that occupy surface areas. Examples of useful Goode projections include maps of the world's climates, soils, and vegetation.

The Goode projection displays the relative areas of land masses correctly but distorts their shape—especially near the poles.

The Goode map suffers from a serious defect, however. The Goode projection distorts the shapes of areas, particularly in high latitudes and at the far right and left edges. To minimize this defect, Dr. Goode split his map apart into separate, smaller sectors, each centered on a different vertical meridian. These were then assembled at the equator. This type of split map is called an *inter-*

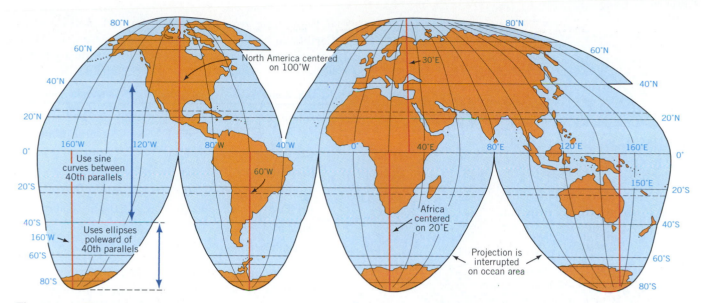

Figure 1.9   The Goode projection. The meridians in this projection follow sine curves between lat. 40° N and lat. 40° S, and ellipses between lat. 40° and the poles. Although the shapes of continents are distorted, their areas are properly shown. (Copyright © by the University of Chicago. Used by permission of the Committee on Geographical Studies, University of Chicago.)

rupted projection. Although the interrupted projection greatly reduces shape distortion, it separates parts of the earth's surface that actually lie close together, particularly in the high latitudes.

Maps and map projections are a practical application of the earth's geographic grid. Another practical application, which involves both the grid and the earth's rotation, is global time, a subject we turn to next.

GLOBAL TIME

Our planet requires 24 hours for a full rotation with respect to the sun. Put another way, humans long ago decided to divide the solar day into 24 units, called hours, and devised clocks to keep track of hours in groups of 12. Yet, different regions set their clocks differently. For example, when it is 10:03 A.M. in New York, it is 9:03 A.M. in Chicago, 8:03 A.M. in Denver, and 7:03 A.M. in Los Angeles. Note that these times differ by exactly one hour. How did this system come about? How does it work?

Our global time system is oriented to the sun. Think for a moment about how the sun appears to move across the sky. In the morning, the sun is low on the eastern horizon, and as the day progresses, it rises higher until at *solar noon* it reaches its highest point in the sky. If you check your watch at that moment, it will read a time somewhere near 12 o'clock (12:00 noon). After solar noon, the sun's elevation in the sky decreases. By late afternoon, the sun appears low in the sky, and at sunset it rests on the western horizon.

Imagine for a moment that you are in Chicago, the time is noon, and the sun is at or near its highest point in the sky. Further imagine that you call a friend in New York and ask about the time there and the position of the sun. You will receive a report that the time is 1:00 P.M. and that the sun is already past solar noon, its highest point. Calling a friend in Los Angeles, you hear that it is 10:00 A.M. there and that the sun is still working its way up to its highest point. However, a friend in Mobile, Alabama, will tell you that the time in Mobile is the same as in Chicago and that the sun is at about solar noon. How can we explain these different observations?

The difference in time between Chicago, New York, and Los Angeles makes sense because solar noon can occur simultaneously only at locations with the same longitude. In other words, only one meridian can be directly under the sun and experience solar noon at a given moment. Locations on meridians to the east of Chicago, like New York, already will have passed solar noon, and locations to the west of Chicago, like Los Angeles, will not yet have reached solar noon. Since Mobile and Chicago have nearly the same longitude, they experience solar noon at approximately the same time.

Figure 1.10 indicates how time varies with longitude. In this figure, the inner disk shows a polar projection of the world, centered on the north pole. Meridians are straight lines (radii) ranging out from the pole. The outer ring indicates the time in hours. The figure shows the moment in time when the prime meridian is directly under the sun—that is, the 0° meridian is directly on the 12:00 noon mark. This means that, at this instant, the sun is at the highest point of its path in the sky in Greenwich, England. The alignment of meridians with

Figure 1.10 The relation of longitude to time. The outer ring shows the time at locations identified by longitude meridians on the central map. The diagram is set to show noon conditions in Greenwich, England—that is, on the prime meridian. Clock time is earlier to the west of Greenwich and later to the east.

hour numbers tells us the time in other locations around the globe. For example, the time in New York, which lies roughly on the 75° W meridian, is about 7:00 A.M. In Los Angeles, which lies roughly on the 120° W meridian, the time is about 4:00 A.M.

Notice that 15° of longitude equates to an hour of time. Since the earth turns 360° in a 24-hour day, the rotation rate is 360 / 24 = 15° per hour.

Standard Time

We've just seen that locations with different longitudes experience solar noon at different times. But consider what would happen if each town or city set its clocks to read 12:00 at its own local solar noon. Because no two towns or cities lie exactly on the same meridian, all would have different local time systems. In these days of instantaneous global communication, chaos would soon result.

The use of standard time simplifies the global time-keeping problem. In the **standard time system**, the globe is divided into 24 **time zones**. All inhabitants within a zone keep time according to a *standard meridian* that passes through their zone. Since the standard meridians are usually 15 degrees apart, the difference in time between adjacent zones is normally one hour. In some states and nations, however, the difference is only one-half hour.

Eight time zones cover the United States and its Caribbean possessions. Five zones cover Canada. Their names and standard meridians of longitude are as follows:

U.S. Zones	Meridian	Canadian Zones
Atlantic	60°	
	$67\frac{1}{2}°$	Newfoundland
Eastern	75°	Eastern
Central	90°	Central
Mountain	105°	Mountain
Pacific	120°	Pacific–Yukon
Alaska	135°	
Hawaii, Bering	150°	

If carried out strictly, the standard time system would consist of belts exactly 15°, extending to meridians $7\frac{1}{2}°$ east and west of each standard meridian. However, this system could be inconvenient because the boundary meridians could divide a state, county, or city into two different time zones. As a result, time zone boundaries are often routed to follow agreed-upon natural or political boundaries.

In the standard time system, we keep global time according to nearby standard meridians that normally differ by one hour from each other.

Figure 1.11 presents a map of time zones for the contiguous United States and southern Canada. From this map, you can see that most time zone boundaries are conveniently located along an already existing and widely recognized line. For example, the eastern time–central time boundary line follows Lake Michigan down its center, and the mountain time–Pacific time boundary follows a ridge-crest line also used by the Idaho–Montana state boundary.

World Time Zones

Figure 1.12 shows the 24 principal standard time zones of the world. In the figure, 15° meridians are dashed lines, while the $7\frac{1}{2}°$ meridians, which form many of the boundaries between zones, are bold lines. The figure also shows the time of day in each zone when it is noon at the Greenwich meridian. The country spanning the greatest number of time zones from east to west is Russia, with 11 zones, but these are grouped into 8 standard time zones. China spans five time zones but runs on a single national time using the standard meridian of Beijing.

A few countries keep time by a meridian that is midway between standard meridians, so that their clocks depart from those of their neighbors by 30 or 90 minutes. India and Iran are examples. The Canadian

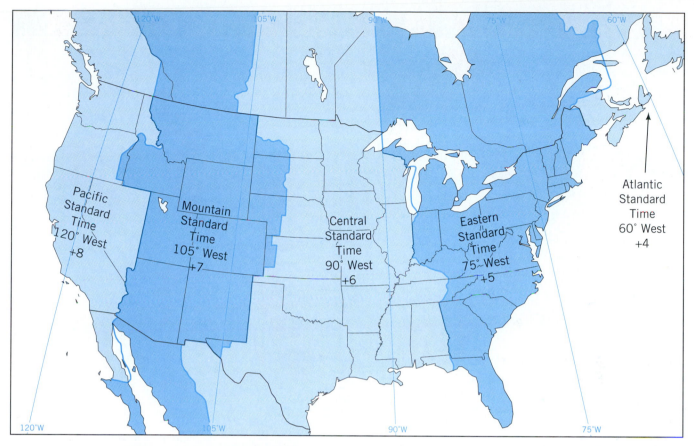

Figure 1.11 Time zones of the contiguous United States and southern Canada. The name, standard meridian, and number code are shown for each time zone. Note that time zone boundaries often follow preexisting natural or political boundaries.

province of Newfoundland and the interior Australian states of South Australia and Northern Territory are examples of regions within countries that keep time by $7\frac{1}{2}°$ meridians.

International Date Line

When we take a world map or globe with 15° meridians and count them in an eastward direction, starting with the Greenwich meridian as 0, we find that the 180th meridian is number 12 and that the time at this meridian is therefore 12 hours later than Greenwich time. Counting in a similar manner westward from the Greenwich meridian, we find that the 180th meridian is again number 12 but that the time is 12 hours earlier than Greenwich time. How can the same meridian be both 12 hours ahead of Greenwich time and 12 hours behind? This paradox is explained by the fact that on the 180th meridian at the exact instant of midnight, the same calendar day covers the entire globe. At any other time, the calendar day observed just west of the 180th meridian is one day ahead of that just to the east. For example, if the day on the Asian side of the dateline is Tuesday, June 8, then the day on the American side will be Monday, June 7. Over a century ago, an international

conference voted that the 180th meridian should be designated the *International Date Line*. To make this dividing line practical, it had to be modified by a zig-zag offset between Asia and North America, as well as an eastward offset in the South Pacific to keep clear of New Zealand and several island groups. (See Figure 1.12.)

Crossing the date line in either direction normally requires adjusting your calendar. On a cruise ship, you would either skip a whole day (going westward) or repeat a day (going eastward). Flying from North America to Asia may also require skipping a day. For example, flying from Los Angeles to Sydney, Australia, you may depart on a Tuesday evening and arrive on a Thursday morning after a flight that lasts only 14 hours. On a flight from Tokyo to San Francisco, you may actually arrive the day before you take off, taking the date change into account!

Daylight Saving Time

Especially in urban areas, many human activities begin well after sunrise and continue long after sunset. Therefore, we adjust our clocks during the part of the year that has a longer daylight period to correspond more closely with the modern pace of society. This

Figure 1.12 Time zones of the world. Dashed lines represent 15° meridians, and bold lines represent $7\frac{1}{2}°$ meridians. Alternate zones appear in color. (U.S. Navy Oceanographic Office.)

adjusted time system, called *daylight saving time*, is obtained by setting ahead all clocks by one hour. The effect of the time change is to transfer the early morning daylight period, theoretically wasted while schools, offices, and factories are closed, to the early evening, when most people are awake and busy. Daylight saving time also yields a considerable savings in power used for electric lights. In the United States, daylight saving time comes into effect on the first Sunday in April and is discontinued on the last Sunday of October.

THE EARTH'S REVOLUTION AROUND THE SUN

So far, we have discussed the importance of the earth's rotation on its axis. Another important motion of the earth is its **revolution**, or its movement in orbit around the sun.

The earth completes a revolution around the sun in about $365\frac{1}{4}$ days—one-fourth day more than the calendar year of 365 days. Every four years, the extra one-

fourth days add up to about one whole day. By inserting a 29th day in February in leap years, we largely correct the calendar for this effect. Further minor corrections are necessary to perfect the system.

The earth's orbit around the sun is shaped like an ellipse, or oval. This means that the distance between the earth and sun varies somewhat through the year. At *perihelion*, which occurs on or about January 3, the earth is nearest to the sun. At *aphelion*, on or about July 4, the earth is farthest away from the sun. However, the distance between sun and earth varies only by about 3 percent during one revolution, because the elliptical orbit is shaped very much like a circle. For most purposes we can regard the orbit as circular.

In which direction does the earth revolve? Imagine yourself in space, looking down on the north pole of the earth. From this viewpoint, the earth travels counterclockwise around the sun (Figure 1.13). This is the same direction as the earth's rotation. The moon rotates on its axis and revolves about the sun in the same direction as well. However, the moon's rate of rotation is such that one side of the moon is always directed toward the

Figure 1.13 Viewed from a point over the earth's north pole, the earth both rotates and revolves in a counterclockwise direction. From this point, the moon also rotates counterclockwise.

earth. Thus, astronomers had never seen the far side of the moon until a Soviet spacecraft spacecraft in lunar orbit transmitted photos back to earth in 1959.

Tilt of the Earth's Axis

The seasons we experience on earth are related to the orientation of the earth's axis of rotation and the position of the sun. We usually describe this situation by stating that "the earth's axis is tilted." What does this really mean? Let's refer to Figure 1.14. The figure shows the earth within a portion of the *plane of the ecliptic*—the plane containing the earth's orbit around the sun. Notice that the earth's orbit is shown lying within the plane.

Figure 1.14 The tilt of the earth's axis of rotation with respect to its orbital plane. As the earth moves in its orbit on the plane of the ecliptic around the sun, its rotational axis remains pointed toward Polaris, the north star, and makes an angle of $66\frac{1}{2}°$ with the ecliptic plane.

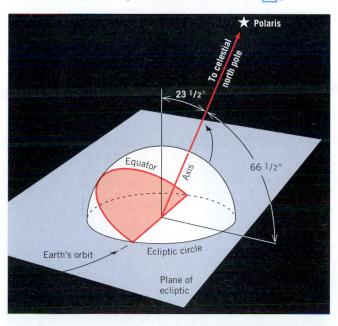

The axis of the earth's rotation is tilted by $23\frac{1}{2}°$ away from a right angle with the plane of the ecliptic.

Now consider the axis of the earth's rotation. Rather than being at a right angle to the plane of the ecliptic, the axis is tilted at an angle of $23\frac{1}{2}°$ away from a right angle. That is, the angle between the axis and the plane of the ecliptic is $66\frac{1}{2}°$, not 90°. In addition, the direction toward which the axis points is fixed in space—it aims toward Polaris, the north star. The direction of the axis does not change as the earth revolves around the sun. As a result, the north pole is tilted away from the sun during one part of the year (the situation shown in the diagram) and is tilted toward the sun during the remainder. Let's investigate this phenomenon in more detail.

Solstice and Equinox

Figure 1.15 diagrams the earth as it revolves in its orbit through the four seasons. Consider first the event on December 22, which is pictured on the far right. On this day, the earth is positioned so that the north polar end of its axis leans at the maximum angle away from the sun, $23\frac{1}{2}°$. This event is called the **winter solstice**. (While it is winter in the northern hemisphere, it is summer in the southern hemisphere, so you can use the term *December solstice* to avoid any confusion.) At this time, the southern hemisphere is tilted toward the sun and enjoys strong solar heating.

Six months later, on June 21, the earth is on the opposite side of its orbit in an equivalent position. At this event, known as the **summer solstice** (*June solstice*), the north polar end of the axis is tilted at its maximum angle of $23\frac{1}{2}°$ toward the sun. Thus, the north pole and northern hemisphere are tilted toward the sun, while the south pole and southern hemisphere are tilted away.

Midway between the solstice dates, the equinoxes occur. At an **equinox**, the earth's axis makes a right angle with a line drawn to the sun, and neither the north nor south pole is tilted toward the sun. The *vernal equinox* occurs on March 21, and the *autumnal equinox* occurs on September 23. Conditions are identical as far as earth–sun relationships are concerned on the two equinoxes. We should also note that the date of any solstice or equinox in a particular year may vary by a day or so, since the revolution period is not exactly 365 days.

Equinox Conditions

Let's look at equinoxes and solstices in more detail. The conditions at an equinox, shown in Figure 1.16, form the simplest case. The figure illustrates two important

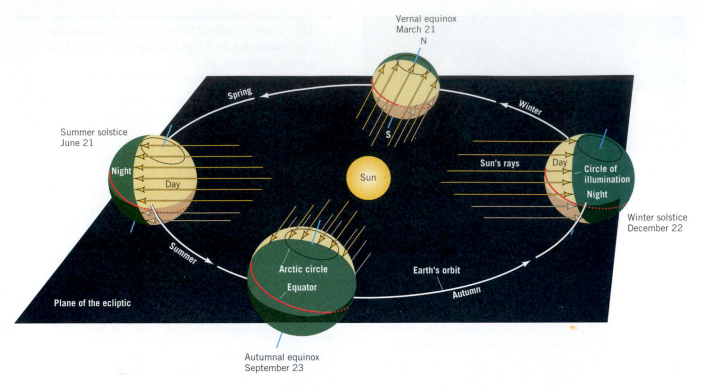

Figure 1.15 The four seasons occur because the earth's tilted axis keeps a constant orientation in space as the earth revolves about the sun. This tips the northern hemisphere toward the sun for the summer solstice, and away from the sun for the winter solstice. Both hemispheres are illuminated equally at the spring equinox and the fall equinox.

concepts of global illumination that we use for describing equinoxes and solstices. The first concept is the *circle of illumination.* Note that the earth is always divided into two hemispheres with respect to the sun's rays. One hemisphere (day) is lit by the sun, and the other (night) lies in the darkness of the earth's shadow. The circle of illumination is the circle that separates the day hemisphere from the night hemisphere. The second concept is the *subsolar point.* The sun is directly overhead at this single point on the earth's surface.

At equinox, the circle of illumination passes through the north and south poles, as we see in Figure 1.16. The sun's rays graze the surface at either pole, and the surface there receives little or no solar energy. The subsolar point falls on the equator. Here, the angle between the sun's rays and the earth's surface is 90°, and the solar illumination is received in full force. At an intermediate latitude, such as 40° N, the rays of the sun at noon strike the surface at a lesser angle. This *noon angle*—the elevation of the sun above the horizon at noon—can be easily determined. Some simple geometry shows that the noon angle is equal to 90° minus the latitude, or 50° in this example, for equinox conditions.

Imagine yourself at a point on the earth, say, at a latitude of 40° N. Visualize the earth rotating from left to right, so that you turn with the globe, completing a full circuit in 24 hours. At the equinox, you spend 12 hours

in darkness and 12 hours in sunlight. This is because the circle of illumination passes through the poles, dividing every parallel exactly in two. Thus, one important feature of the equinox is that day and night are of equal length everywhere on the globe (except at the poles, where the sun remains on the horizon for 24 hours).

Figure 1.16 Equinox conditions. At this time, the earth's axis of rotation is exactly at right angles to the direction of solar illumination. The subsolar point lies on the equator.

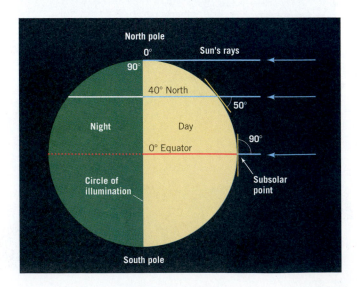

Solstice Conditions

Now examine the solstice conditions shown in Figure 1.17. Summer solstice is on the left. Consider yourself back at a point on the lat. 40° N parallel. The circle of illumination does not divide your parallel in equal halves because of the tilt of the northern hemisphere toward the sun. Instead, the larger part is in daylight. For you, the day is now considerably longer (about 15 hours) than the night (about 9 hours).

The farther north you go, the more the effect increases. In fact, the entire area of the globe north of lat. $66\frac{1}{2}°$ is on the daylight side of the circle of illumination. This parallel is known as the **arctic circle**. Even though the earth rotates through a full cycle during a 24-hour period, the area north of the arctic circle remains in continuous daylight. We also see that the subsolar point is at a latitude of $23\frac{1}{2}°$ N. This parallel is known as the **tropic of cancer**. Because the sun is directly over the tropic of cancer at this solstice, solar energy is most intense here.

> At the June solstice, regions poleward of the arctic circle experience continuous daylight.

At the winter solstice, conditions are exactly reversed from those of the summer solstice. If you imagine yourself back at lat. 40° N, you find that the night is now about 15 hours long while daylight lasts about 9 hours. All the area south of lat. $66\frac{1}{2}°$ S lies under the sun's rays, inundated with 24 hours of daylight. This parallel is known as the **antarctic circle**. The subsolar point has shifted to a point on the parallel at lat. $23\frac{1}{2}°$ S, known as the **tropic of capricorn**.

The solstices and equinoxes represent conditions at only four times of the year. Between these times, the subsolar point travels northward and southward in an annual cycle between the tropics of cancer and capricorn. We refer to the latitude of the subsolar point as the sun's *declination*. As the year progresses, the declination varies between $23\frac{1}{2}°$ S lat. at the December solstice and $23\frac{1}{2}°$ N lat. at the June solstice.

As the seasonal cycle progresses in polar regions, areas of 24-hour daylight or 24-hour night shrink and then grow. At other latitudes, the length of daylight varies a little from one day to the next. In this way, the earth experiences the rhythm of the seasons as it continues its revolution around the sun.

The earth's rotation on its axis and revolution around the sun are the most important determiners of changes in the amount of energy the earth receives from the sun on daily and annual time scales. But there are other earth–sun motions that are important over the longer term—thousands of years. Described in *Eye on the Environment • Earth–Sun Cycles and the Ice Ages*, these appear to have caused the ice ages that are a very important part of the earth's recent history.

A LOOK AHEAD

The flow of solar energy to the earth powers most of the natural processes that we experience every day, from changes in the weather to the work of streams in carving the landscape. Our next chapter examines in detail solar energy and its interaction with earth's atmosphere and surface. As you will see, the amount of incoming solar energy at a location depends on the length of the daylight portion of the day and on the height of the sun's path in the sky during the daylight period. Both of these factors depend on the latitude and season of the location. So, your understanding of the earth's rotation and revolution will come in quite handy in the chapter to come.

Figure 1.17 Solstice conditions. At the solstice, the north end of the earth's axis of rotation is fully tilted either toward or away from the sun. Because of the tilt, polar regions experience either 24-hour day or 24-hour night. The subsolar point lies on one of the tropics, at lat. $23\frac{1}{2}°$ N or S.

EYE ON THE ENVIRONMENT

Earth–Sun Cycles and Ice Ages

Stand back! The glaciers are coming! Well, perhaps not today or tomorrow, but during the past two to three million years, great sheets of ice—continental glaciers—have come and gone many times, covering large expanses of the northern hemisphere. A period when continental glaciers grow and cover the landscape is referred to as a *glaciation*. Right now, we are in an interglacial period—a period that started 12,000 years ago, when the last continental ice sheets melted away. Will the ice sheets return?

A glaciation sets in when global climates become colder and the high-latitude continents receive more snow. Each year, more snow accumulates than can be melted by warm summer weather, and eventually, an ice cap is formed. As the process continues over thousands of years, the ice cap can grow into a great continental ice sheet spreading over millions of square kilometers and covering an extremely large area of land.

What causes our climate to change so that glaciers form? Scientists have puzzled over this question since at least the middle of the nineteenth century, and many theories have been proposed. One theory that many scientists find reasonable states that glaciation is caused by slight changes in the amount of solar energy the earth receives. These slight changes are produced by minor changes in the earth's orbit and in the tilt of its axis.

The distance between the earth and sun at the time of summer solstice in the northern hemisphere varies slowly over time. This cycle of change lasts about 21,000 years. During the cycle, the average distance between earth and sun may vary from 1 to 5 percent greater to 1 to 5 percent less than the normal long-term average. When the distance is greater, less solar energy will reach earth and our earth will grow colder.

Another cycle slowly changes the tilt of the earth's axis. Just as a spinning top sometimes develops a wobble, so the earth's axis also has a slight wobble that changes the earth's tilt from 24° to 22° and back again. This cycle takes about 40,000 years. When the tilt is smaller (22°), it tends to make the polar regions slightly colder and the equatorial regions slightly warmer. When the tilt is greater (24°), we have the opposite effect—warmer polar regions and a cooler equatorial belt.

Since both cycles are going on simultaneously, they can either cancel or reinforce one another. When both cycles combine to make the polar regions colder and reduce summer snowmelt, ice cap formation is enhanced. When both cycles combine to warm the polar regions, the ice cap melts and shrinks away.

The figure at right shows the effect of the combined cycles. The vertical axis shows the change in solar energy reaching the earth at the top of the atmosphere. The amount of solar energy is calculated for lat. 65° N on the summer solstice, June 21. The horizontal axis shows the time, thousands of years before the present

date. The graph is known as the Milankovitch curve, named for Milutin Milankovitch, the astronomer who first calculated it in 1938.

As you can see, in the last 500,000 years about a dozen peaks of increased solar energy have alternated with dips to lower values. Some of these dips and peaks seem to correspond to the timings of glaciations and interglacial periods, as scientists attempt to date them using other evidence. The graph shows that our planet's solar energy is now decreasing from a peak occurring about 20,000 years ago.

One strongly supported theory is that these cycles in earth–sun relationships control the timing of our planet's ice ages. Other scientists however, have proposed that quite different mechanisms influence the onset of ice ages. These mechanisms include volcanic eruptions and cycles of change in the sun's brightness. Other factors, such as ocean current patterns, are also deemed important. We will touch on some of these in later chapters and *Eye on the Environment* features.

Questions

1. What two cycles are known to affect the amount of solar radiation received by polar regions of the earth?

2. What is the Milankovitch curve? What does it show about warm and cold periods during the last 500,000 years?

The Milankovitch curve. The vertical axis shows fluctuations in incoming solar energy at lat. 65° N on June 21, the summer solstice. These are calculated from mathematical models of the change in earth–sun distance and change in axial tilt with time. The zero value represents the present value. (Based on calculations by A. D. Vernekar, 1968. Copyright © A. N. Strahler.)

OUR PLACE IN THE SUN IN REVIEW

The rotation of the earth on its axis and the revolution of the earth in its orbit about the sun are fundamental topics in physical geography. The rotation provides the first great rhythm of our planet—the daily alternation of sunlight and darkness. The tides, and a sideward turning of ocean and air currents, are further effects of the earth's rotation.

The earth's axis of rotation provides a reference for the system of location on the earth's surface—the geographic grid, which consists of meridians and parallels. This system is indexed by our system of latitude and longitude, which uses the equator and the prime meridian as references. We require a map projection to display the earth's curved surface on a flat map. The polar projection is centered on either pole and pictures the globe as we might view it from the top or bottom. The Mercator projection converts the geographic grid into a flat, rectangular one and best displays directional features. The Goode projection distorts the shapes of continents and coastlines but preserves the areas of land masses in their correct proportion.

We monitor the earth's rotation by daily timekeeping. Each hour the earth rotates by 15°. In the standard time system, we keep time according to a nearby standard meridian. Since standard meridians are normally 15° apart, clocks around the globe usually differ by whole hours.

The seasons are the second great earthly rhythm. They arise from the revolution of the earth in its orbit around the sun, combined with the fact that the earth's rotational axis is tilted with respect to its orbital plane. The solstices and equinoxes mark the cycle of this revolution. At the summer (June) solstice, the northern hemisphere is tilted toward the sun. At the winter (December) solstice, it is the southern hemisphere that is tilted toward the sun. At the equinoxes, day and night are of equal length.

KEY TERMS

rotation	longitude	summer solstice
parallel	map projection	equinox
equator	standard time system	arctic circle
meridian	time zone	tropic of cancer
geographic grid	revolution	antarctic circle
latitude	winter solstice	tropic of capricorn

REVIEW QUESTIONS

1. What is the shape of the earth? How do you know?
2. What is meant by earth rotation? Describe its three physical effects.
3. Describe the geographic grid, including parallels and meridians.
4. How do latitude and longitude determine position on the globe? In what units are they measured?
5. Name three types of map projections and describe each briefly. Give reasons why you might choose different map projections to display different types of geographical information.
6. Explain the global timekeeping system. Define and use the terms *standard time*, *standard meridian*, and *time zone* in your answer.
7. What is meant by the "tilt of the earth's axis"? How does the tilt create the seasons?

VISUALIZING EXERCISES

1. Sketch a diagram of the earth at an equinox. Show the north and south poles, the equator, and the circle of illumination. Indicate the direction of the sun's incoming rays and shade the night portion of the globe.
2. Sketch a diagram of the earth at the summer (June) solstice, showing the same features. Also include the tropics of cancer and capricorn, and the arctic and antarctic circles.

A DEEPER VIEW OF EARTH–SUN CYCLES

Suppose that the earth's axis were tilted at 40° to the plane of the ecliptic instead of $23\frac{1}{2}°$. What would be the global effects of this change? How would the seasons change at your location?

Chapter 2

The Earth's Global Energy Balance

A beach on St. Croix, U. S. Virgin Islands.

The sun is deliciously warm on your back. The sound of the surf is subdued as you lie stretched full-length on your towel. The beach here at Cancun is everything you thought it would be—soft, fine sand and clear blue water, with hazy tropical skies overhead. What a place to spend spring break!

It's not long before the sun on your back seems a little too warm, and so you roll over, pulling the brim of your hat down over your eyes to screen out the bright sunlight. Soon you are hot all over, and the sound of the gently breaking waves seems very inviting. Shaking off a few clinging grains of sand, you head for the surf. A wave breaks over your feet, splashing cool drops of foam on your legs. A few quick steps, a shallow dive, and your skin is tingling with the shock of the Gulf waters. What a marvelous feeling!

Thoroughly refreshed, you emerge from the waves and walk back up to your place on the beach. As you towel the water from your skin, the tropical breeze takes on a cool edge. You lie down again, keeping low, and exposing your body to the direct rays of the sun overhead. Soon your skin dries, and you are warm once more.

Later, in the evening, you return to the beach. The stars are out now, and the white crescent is framed by palm trees behind you and the dark ocean in front. The gentle breakers splash phosphorescent foam up the beach, which quickly sinks into the fine sand. After a few minutes, you notice that your shoulders and arms are a bit cold. The sand, however, is still warm to your feet. You lie down, burrowing slightly into the soft surface to create a smooth support for your back and legs. Comfortable now, you relax and watch the sky overhead. Slowly, the stars grow brighter and the night behind them deepens to a velvet black. A shooting star flashes across the sky. This will be a spring break you'll always remember.

This description of a beach holiday probably seems far from the subject of this chapter—the earth's global energy balance. But, in fact, many of the principles involved in the global radiation balance are the same as those that warm or cool your skin on a Caribbean beach. Like your skin, the earth's surface absorbs sunlight and is warmed. At night, lacking sunlight, it loses heat and is cooled. The surface, as well as your skin, also loses heat when cooled by evaporation. And the surface can be cooled or warmed by direct contact with a cool or warm atmosphere—just as your skin can be cooled or warmed by direct contact with beach sand or ocean water.

No matter whether the surface is your skin or a layer of grass-covered soil, it is constantly emitting radiation. For example, our bodies emit radiation in the form of heat. Very hot objects, such as the sun or the filament of a light bulb, give off radiation that is nearly all in the form of light. The surface and atmosphere of our planet constantly emit heat energy. Of course, the earth also receives a flow of energy—radiation from the sun, largely in the form of light. Over the long run, the earth emits exactly as much energy as it absorbs, forming a global energy balance.

In this chapter, we examine the energy balance of the earth, including the energy balances of land and ocean surfaces and the atmosphere. These balances control the seasonal and daily changes in the earth's surface temperature. They also produce differences in energy flow rates from place to place that drive currents of air and ocean water. These, in turn, produce the changing weather and rich diversity of climates we experience on the earth's surface.

Chapter Outline

Electromagnetic Radiation
 Radiation and Temperature
 Solar Radiation
 Characteristics of Solar Energy
 Longwave Radiation from the Earth
 The Global Radiation Balance
Insolation over the Globe
 The Path of the Sun in the Sky
 Daily Insolation Through the Year
 Annual Insolation by Latitude
World Latitude Zones

Composition of the Atmosphere
 Ozone in the Upper Atmosphere
 Eye on the Environment • The Ozone Layer—Shield to Life
Sensible Heat and Latent Heat Transfer
The Global Energy System
 Insolation Losses in the Atmosphere
 Albedo
 Counterradiation and the Greenhouse Effect
 Global Energy Budgets of the Atmosphere and Surface
Net Radiation, Latitude, and the Energy Balance

Chapter 2

THE EARTH'S GLOBAL ENERGY BALANCE

This chapter explains how solar radiation is intercepted by our planet, flows through the earth's atmosphere, and interacts with the earth's land and ocean surfaces. Solar radiation, as we will see, is the driving power source for wind, waves, weather, rivers, and ocean currents. Most natural phenomena that take place at the earth's surface are directly or indirectly solar-powered—from the downhill flow of a river to the movement of a sand dune to the growth of a forest.

The sun's energy powers the processes of the life layer like the constant flow of gasoline powers a car cruising down the highway. But just as friction and air resistance constantly hold the car's speed in check, so the radiation of heat energy from our earth and atmosphere out to space constantly checks the buildup of energy absorbed from the sun. Thus, a balance is maintained between incoming solar radiation and outgoing terrestrial radiation.

A primary topic of this chapter is the *energy balance* of the earth, which includes land and ocean surfaces and the atmosphere. The energy balance of a surface describes the balance between the flows of energy reaching that surface and flows of energy leaving it. The earth's energy balance controls the seasonal and daily changes in the earth's surface temperature. Differences in energy flow rates from place to place, as measured by the surface energy balance, also drive currents of air and ocean water. These, in turn, produce the changing weather and rich diversity of climates we experience on the earth's surface.

Human activities now dominate many regions of the earth, and we have irreversibly modified our planet by farming much of its surface and adding carbon dioxide to its atmosphere. Have we shifted the balance of energy flows? Is our earth absorbing more solar energy and becoming warmer? Is it absorbing less and becoming cooler? We must examine the global energy balance in detail before we can understand human impact on the earth–atmosphere system.

ELECTROMAGNETIC RADIATION

Our study of the global energy system begins with the subject of radiation—that is, **electromagnetic radiation**. This form of radiation is emitted by all objects. Think of electromagnetic radiation as a collection, or spectrum, of waves of a wide range of wavelengths traveling quickly away from the surface of the object. *Wavelength* describes the distance separating one wave crest from the next wave crest (Figure 2.1). The unit we will use to measure wavelength is the *micrometer* (formerly called the micron). A micrometer is one millionth of a meter, or a ten-thousandth of a centimeter. This is such a small unit that the tip of your little finger is about 15,000 micrometers wide. In this text, we use the abbreviation µm for the micrometer. The first letter is the Greek letter µ, or mu. It is used in metric units to denote micro, meaning one-millionth.

Radiant energy can exist at any wavelength. Heat and light, two familiar examples of electromagnetic radiation, are identical except for their wavelengths. Light is radiation that is visible to our eyes. Heat radiation, though not visible, is easily felt when you hold your hand near a warm object, such as an oven or electric stove burner. In this chapter, we are concerned primarily with electromagnetic energy in the form of light and heat.

Radiation and Temperature

Two important physical principles concern the emission of electromagnetic radiation. The first is that there is an inverse relationship between the wavelength of the radiation that an object emits and the temperature of the object. For example, the sun, a very hot object, emits radiation with short wavelengths. In contrast, the earth, a much cooler object, emits radiation with longer wavelengths.

The second principle is that hot objects radiate more energy than cooler objects—much more. In fact, the

Figure 2.1 Electromagnetic radiation can be described as a collection of energy waves with different wavelengths. Wavelength *L* is the crest-to-crest distance between successive wave crests.

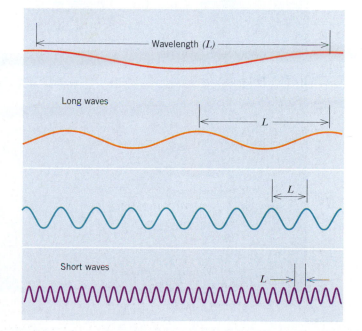

flow of radiant energy from the surface of an object is directly related to absolute temperature of the surface raised to the fourth power. (Absolute temperature is temperature measured on the Kelvin scale, with zero being the absence of all heat.) Thus, if the absolute temperature of an object is doubled, the flow of radiant energy from its surface will be 16 times larger. Because of this relationship, a small increase in temperature can mean a large increase in the rate at which radiation is given off by an object or surface. For example, water emits about one-third more radiation when warmed from the freezing point to room temperature.

Hotter objects radiate more energy, and at shorter wavelengths, than cooler objects.

Understanding these two principles—that hotter objects radiate more energy and that the energy is of shorter wavelengths—will help you understand the global energy balance.

Solar Radiation

Our sun is a ball of constantly churning gases that are heated by continuous nuclear reactions. It is about aver-

age in size compared to other stars, and it has a surface temperature of about 6000°C (about 11,000°F). Like all objects, it emits energy in the form of electromagnetic radiation. The energy travels outward in straight lines, or rays, from the sun at a speed of about 300,000 km (about 186,000 mi) per second. At that rate, it takes the energy about $8\frac{1}{3}$ minutes to travel the 150 million km (93 million mi) from the sun to the earth.

As solar radiation travels through space, none of it is lost. However, the rays spread apart as they move away from the sun. This means that a planet farther from the sun, like Mars, receives less radiation than one located nearer to the sun, like Venus. The earth intercepts only about one-half of one-billionth of the sun's total energy output.

The portion of the electromagnetic spectrum that contains radiation emitted by the sun and earth can be divided into four major portions, based on wavelength. These spectral regions are shown at the top of Figure 2.2. At the left lies the *ultraviolet* radiation portion of the spectrum. Its wavelengths range from about 0.2 to 0.4 μm. These short waves are high in energy. Solar ultraviolet radiation is nearly all absorbed by the gas molecules of the earth's atmosphere. However, when this radiation penetrates to the surface of the earth, it can damage living tissues, such as human skin.

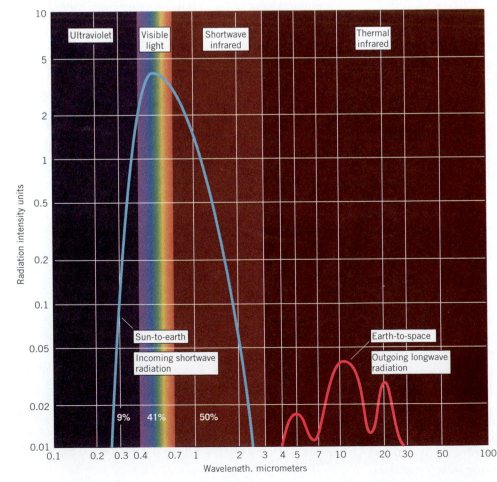

Figure 2.2 There are four important divisions within the electromagnetic spectrum—ultraviolet, visible light, shortwave infrared, and thermal infrared. The curve on the left shows the intensity of incoming solar shortwave radiation as it reaches the surface. The curve on the right shows the outgoing longwave radiation from earth and atmosphere to outer space.

Figure 2.3   A thermal infrared image of a suburban scene at night. Black and blue tones show lower temperatures, while yellow and red tones show higher temperatures. Ground and sky are coldest, while the heated windows of the homes are warmest.

Toward the center of the solar spectrum is the **visible light** portion of the spectrum, with wavelengths ranging from 0.4 to 0.7 μm. Our eyes sense this radiation as light energy. The color of the light is determined by its wavelength. The shorter wavelengths are perceived as violet, while the longer ones are perceived as red. In between is the rest of our familiar color spectrum, including blue, green, yellow, and orange. Only a small amount of visible light is absorbed by atmospheric gases or particles in passing through the atmosphere. Unless intercepted by clouds, nearly all of this radiation reaches the earth's surface.

Next is the *shortwave infrared* radiation region, ranging from 0.7 to 3 μm. This radiation acts much like visible light, although our eyes are not sensitive to it. Shortwave infrared radiation also penetrates the atmosphere easily. Taken together, we refer to ultraviolet, visible light, and shortwave infrared radiation as **shortwave radiation**.

At the right are the *thermal infrared* wavelengths, longer than 3 μm, which are emitted by cooler objects. We perceive this type of radiation as heat, such as you might feel sitting beside the glowing coals of a campfire. Thermal infrared radiation is also referred to as **longwave radiation**.

Thermal infrared radiation is no different in principle from visible light radiation. Although photographic film is not sensitive to thermal radiation, it is possible to acquire an image of thermal radiation using a special sensor. Figure 2.3 shows a thermal infrared image of a suburban scene obtained at night. Windows and pavement are warm and radiate more intensely. In contrast, the moist surfaces of lawns are cooler. Ground and sky are coldest.

Characteristics of Solar Energy

The sun does not emit all wavelengths of radiation equally. The tall curve at the left on the graph of Figure 2.2 shows the relative intensity of solar energy entering the outer atmosphere, plotted by wavelength. Energy intensity is shown on the graph on the vertical scale. Note that it is a logarithmic scale—that is, each whole unit marks an intensity 10 times greater than the one below.

The ultraviolet portion of the spectrum accounts for about 9 percent of the total incoming solar energy. Also included in the ultraviolet division are X rays and gamma rays, forms of radiation that have shorter wavelengths than the ultraviolet. They are not shown on the graph. The solar energy curve peaks in the visible light range, which accounts for about 41 percent of the total energy received by the earth. The shortwave infrared portion of the spectrum accounts for 50 percent of the total energy. Very little energy arrives in wavelengths longer than 2 μm.

The intensity of solar energy is strongest in visible wavelengths.

The sun's interior is the source generating solar energy. Here, hydrogen is converted to helium at very high temperatures and pressures. In this process of nuclear fusion, a vast quantity of energy is generated and finds its way to the sun's surface. Because the rate of production of energy is nearly constant, the output of solar radiation is also nearly constant. So, given the average distance of earth from the sun, the amount of solar energy received on a small, fixed area of surface held at right angles to the sun's rays is almost constant. This rate of incoming shortwave energy, known as the *solar constant*, is measured beyond the outer limits of the earth's atmosphere, before energy has been lost in passing through the earth's atmosphere. The solar constant has a value of about 1400 watts per square meter (W/m^2).

Figure 2.4 A diagram of the global energy balance. Shortwave radiation from the sun is transmitted through space, where it is intercepted by the earth. The absorbed radiation is then ultimately emitted as longwave radiation to outer space. (From A. N. Strahler, "The Life Layer," *Journal of Geography*, vol. 69, p. 72. Copyright 1970 © by The Journal of Geography. Used by permission.)

Let's take a moment to examine the units of energy flow for radiation, as used, for example, in the definition of the solar constant. The *watt* (W), which describes a rate of energy flow, is a familiar measure of power consumption. You've probably seen it applied to light bulbs or to home appliances, such as stereo amplifiers or microwave ovens. When we use the watt to measure the intensity of a flow of radiant energy, we must specify the unit of surface area that receives or emits the energy. This cross section is assigned the area of 1 square meter (1 m²). Thus, the measure of intensity of received (or emitted) radiation is given as watts per square meter (W/m²). Because there are no common equivalents for this energy flow rate in the English system, we will use only metric units.

Longwave Radiation from the Earth

Recall that both the wavelength and intensity of radiation emitted by an object depend on the object's temperature. Since the earth's surface and atmosphere are much colder than the sun's surface, we can deduce that our planet radiates less energy than the sun and that the energy emitted has longer wavelengths.

Figure 2.2 confirms these deductions. The curve on the right shows outgoing radiation flow from the earth as measured at the top of the atmosphere. Notice that the overall height is much lower than for the solar curve on the left, showing that the planet emits radiation at a lower rate. What about wavelength? The graph shows that the energy is emitted at wavelengths of about 3 to 30 μm, which is longer than the wavelengths of solar radiation and is consistent with the principle that a cooler surface emits radiation at longer wavelengths.

Note that the curve of outgoing longwave radiation shows three distinct peaks at wavelengths of about 5, 10, and 20 μm. Between these wavelengths, atmospheric gases, primarily water vapor and carbon dioxide, absorb much of the radiation leaving the surface.

The Global Radiation Balance

As we have seen, the earth constantly absorbs solar shortwave radiation and emits longwave radiation. Fig-

ure 2.4 presents a diagram of this energy flow process, which we refer to as the earth's **radiation balance**.

The sun provides a nearly constant flow of shortwave radiation toward earth. Part of this radiation, as we noted earlier, is scattered back into space without absorption. Clouds and dust particles in the atmosphere contribute to this scattering, as do land and ocean surfaces. The remaining shortwave energy from the sun is absorbed by either atmosphere, land, or ocean. Once absorbed, solar energy raises the temperature of the atmosphere, as well as the surfaces of the oceans and lands.

> **In the long run, the rate of solar energy absorption by the earth is balanced by the rate of longwave energy emission from the earth.**

The atmosphere, land, and ocean also emit energy in the form of longwave radiation. This radiation ultimately leaves the planet, headed for outer space. The longwave radiation outflow tends to lower the temperature of the atmosphere, ocean, and land, and thus cool the planet. In the long run, these flows balance—incoming energy absorbed and outgoing radiation emitted are equal. Since the temperature of a surface is determined by the amount of energy it absorbs and emits, the earth's overall temperature tends to remain constant.

INSOLATION OVER THE GLOBE

The flow of solar radiation to the earth as a whole remains constant, but it varies from place to place and from time to time. We turn now to this variation and its causes. In this discussion, we will refer to **insolation** (**in**coming **sol**ar radi**ation**)—the flow of solar energy intercepted by an exposed surface for the case of a uniformly spherical earth with no atmosphere. Insolation is a flow rate and has units of watts per square meter (W/m²). Daily insolation refers to the average of this flow rate over a 24-hour day. Annual insolation is the average flow rate over the entire year.

Figure 2.5   The angle of the sun's rays determines the intensity of insolation on the ground. The energy of vertical rays *A* is concentrated in square *a* by *c*, but the same energy in the slanting rays *B* is spread over a larger rectangle, *b* by *c*.

Insolation depends on the angle of the sun above the horizon. Figure 2.5 shows that insolation is greatest where the sun's rays strike vertically. When the sun's horizon angle diminishes, the same amount of solar energy spreads over a greater area of surface. So when the sun is low in the sky overhead, solar radiation is less intense. When the sun is high in the sky, solar radiation is greater.

Refer back to Chapter 1 and Figure 1.16, showing equinox conditions. Only at the subsolar point does the earth's spherical surface present itself at a right angle to the sun's rays. To a viewer on the earth at the subsolar point, the sun is directly overhead. But as we move away from the subsolar point toward either pole, the earth's curved surface becomes turned at an angle with respect to the sun's rays. To the earthbound viewer at a new latitude, the sun appears to be nearer the horizon. When the circle of illumination is reached, the sun's rays are parallel with the surface. That is, the sun is viewed just at the horizon. Thus, the angle of the sun in the sky at a particular location depends on the latitude of the location, the time of day, and the time of year.

The Path of the Sun in the Sky

How does the angle of the sun vary during the day? The angle depends on the sun's path in the sky. When the sun is high above the horizon—near noon, for example—the sun's angle is greater and so insolation will be greater. Figure 2.6 shows the sun's daily path in the sky and how this path changes from season to season. The diagram is drawn for latitude 40° N and is typical of conditions found in midlatitudes in the northern hemisphere—for example, at New York or Denver. The diagram shows a small area of the earth's surface bounded by a circular horizon. This is the way things appear to an observer standing on a wide plain. The earth's surface appears flat, and the sun seems to travel inside a vast dome in the sky.

At equinox, the sun rises directly to the east and sets directly to the west. The noon sun is positioned at an angle of 50° above the horizon in the southern sky. The sun is above the horizon for exactly 12 hours, as shown by the hour numbers on the sun's path. At summer solstice, the sun's path rises much higher in the sky—at

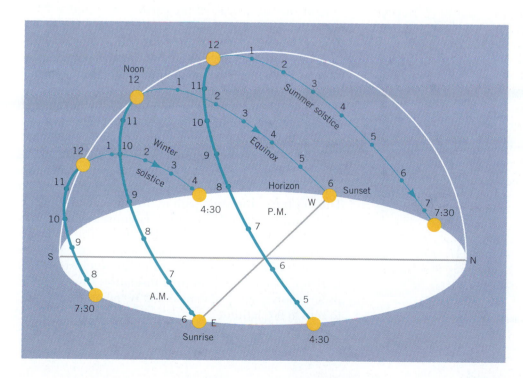

Figure 2.6   The sun's path in the sky changes greatly in position and height above horizon from summer to winter. This diagram is for a location at lat. 40° N (New York or Denver). At the winter solstice, the sun is low in the sky and the period of daylight is short. At the summer solstice, the sun is high and the daylight period is long.

noon it will be $73\frac{1}{2}°$ above the horizon. In addition, the sun is above the horizon for about 15 hours. At winter solstice, the sun's path is low in the sky, reaching only $26\frac{1}{2}°$ above the horizon, and the sun is visible for only about 9 hours. Clearly, the solar radiation reaching the surface is much greater during the day at summer solstice than at winter solstice.

Daily Insolation Through the Year

Daily insolation, which is the average insolation rate over a 24-hour day, depends on two factors: (1) the angle at which the sun's rays strike the earth, and (2) the length of time of exposure to the rays. Both of these factors depend on the sun's path in the sky, which is controlled by the latitude of the location and the time of year. Let's investigate daily insolation variations in more detail.

When daily insolation values are plotted for a full year, they form a wavelike curve on a graph, as shown in Figure 2.7. For example, the curve for latitude 40° N shows a much greater daily insolation rate at the summer solstice (June) and a much lower daily insolation rate at the winter solstice (December)—as we might expect from Figure 2.6.

Figure 2.7 Insolation curves at various latitudes in the northern hemisphere (assuming no atmosphere). Dashed lines mark the equinoxes and solstices. Latitudes between the equator (0°) and tropic of cancer ($23\frac{1}{2}°$ N) show two maximum values; others show only one. Poleward of the arctic circle ($66\frac{1}{2}°$ N), insolation is zero for at least some period of the year. (Copyright © A. N. Strahler.)

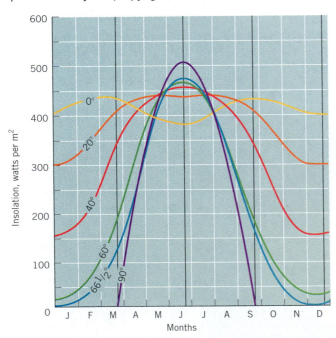

Not every latitude shows a simple progression from low to high to low daily insolation through the course of the year. Notice that the equator has two periods of maximum daily insolation. These periods occur near the equinoxes, when the sun is overhead at the equator. There are also two minimum periods near the solstices, when the subsolar point moves farthest north and south from the equator. All latitudes between the tropic of cancer ($23\frac{1}{2}°$ N) and the tropic of capricorn ($23\frac{1}{2}°$ S) have two maximum and minimum values. However, as either tropic is approached in a poleward direction, the two maximum periods get closer and closer in time, and then merge into a single maximum.

Poleward of the two tropics is a single daily insolation cycle with a maximum at one solstice and a minimum at the other. At the arctic circle ($66\frac{1}{2}°$ N), daily insolation is reduced to zero on the day of the winter solstice. At latitudes farther poleward, this seasonal period of no insolation increases from 1 day to 182 days.

From this discussion, we can see that the seasonal pattern of daily insolation is directly related to latitude. This pattern is important because it is the driving force for the annual cycle of climate. Nowhere on earth is insolation constant from day to day. Even at the equator, Figure 2.7 shows that there are significant seasonal differences in the solar energy available to warm the earth and atmosphere. Poleward of the tropics, the large differences produce the large variations in temperature from winter to summer.

Annual Insolation by Latitude

What is the effect of latitude on annual insolation—the rate of insolation averaged over an entire year? Figure 2.8 shows two curves of annual insolation by latitude—one for the actual case of the earth's axis tilted at $23\frac{1}{2}°$, the other for an earth with an untilted axis. Let's look first at the real case of a tilted axis. Annual insolation varies smoothly from the equator to the pole. As we might expect, annual insolation is greater at lower latitudes. But the high latitudes still receive a considerable amount of solar radiation—the annual insolation value at the pole is about 40 percent of the value at the equator.

Imagine for a moment that the earth's axis is untilted—that is, that the earth's axis is perpendicular to the plane of the ecliptic. In this case, there are no seasons. The sun traces the same path in the sky every day of the year at any location. Under these circumstances, annual insolation changes from a high value at the equator, where the sun is directly overhead at noon every day, to zero at the poles, where the sun is always just on the horizon. The curve of annual insolation thus shows a progression from a value near 450 W/m² at the equator to zero at 90° latitude.

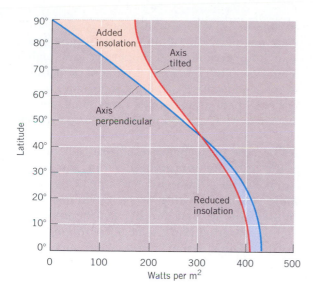

Figure 2.8 Total annual insolation from equator to pole for the earth at the top of the atmosphere. The solid line shows the actual curve. The dashed line shows what the annual insolation would be if the axis were perpendicular to the plane of the ecliptic.

The comparison of the two cases shows that the tilting of the earth's axis redistributes a significant portion of the earth's insolation from the equatorial regions toward the poles. Although the pole is in the dark for six months of the year, it still receives close to half the amount of solar radiation received at the equator. Without a tilted axis, our planet would be quite a different place!

WORLD LATITUDE ZONES

The seasonal pattern of insolation can be used as a basis for dividing the globe into broad latitude zones (Figure 2.9). The zone limits shown in the figure or specified below should not be taken as absolute and binding, however. Rather, this system of names is a convenient way to identify general world geographic belts throughout this book.

The *equatorial zone* encompasses the equator and covers the latitude belt roughly 10° north to 10° south. The sun provides intense insolation throughout the year, and days and nights are of roughly equal length within this zone. Spanning the tropics of cancer and capricorn are the *tropical zones*, ranging from latitudes 10° to 25° north and south. A marked seasonal cycle exists in these zones, combined with a large total annual insolation.

Moving toward the poles from each of the tropical zones are transitional regions called the *subtropical zones*. For convenience, we assign these zones the latitude belts 25° to 35° north and south. At times we may

extend the label "subtropical" a few degrees farther poleward or equatorward of these parallels.

The midlatitude zones lie between 35° and 55° latitudes in the northern and southern hemispheres.

The *midlatitude zones* are next, lying between 35° and 55° north and south latitude. In these belts, the sun's height in the sky shifts through a wide range annually. Differences in day length from winter to summer are also large. Thus, seasonal contrasts in insolation are quite strong. In turn, these regions can experience a large range in annual surface temperature.

Bordering the midlatitude zones on the poleward side are the *subarctic zone* and *subantarctic zone*, 55° to 60° north and south latitudes. Astride the arctic and antarctic circles from latitudes 60° to 75° N and S lie the *arctic* and *antarctic zones*. These zones have an extremely large yearly variation in day lengths, yielding enormous contrasts in insolation from solstice to solstice.

The *polar zones*, north and south, are circular areas between about 75° latitude and the poles. Here the polar regime of a six-month day and six-month night is predominant. These zones experience the greatest seasonal contrasts of insolation of any area on earth.

Figure 2.9 A geographer's system of latitude zones. These are based on the seasonal patterns of insolation observed over the globe.

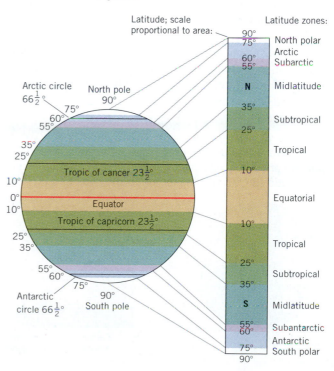

COMPOSITION OF THE ATMOSPHERE

A main function of this chapter is to explain the flows of energy within the atmosphere and between the atmosphere and the earth's surface. However, an understanding of this topic first requires some knowledge of basic facts about the atmosphere.

The earth's atmosphere consists of *air*—a mixture of various gases surrounding the earth to a height of many kilometers. This envelope of air is held to the earth by gravitational attraction. Almost all the atmosphere (97 percent) lies within 30 km (19 mi) of the earth's surface. The upper limit of the atmosphere is at a height of approximately 10,000 km (about 6000 mi) above the earth's surface, a distance approaching the diameter of the earth itself.

From the earth's surface upward to an altitude of about 80 km (50 mi), the chemical composition of air is highly uniform in terms of the proportions of its gases. Pure, dry air consists largely of nitrogen, about 78 percent by volume, and oxygen, about 21 percent (Figure 2.10). Other gases account for the remaining 1 percent.

Nitrogen in the atmosphere exists as a molecule consisting of two nitrogen atoms (N_2). Nitrogen gas does not enter easily into chemical union with other substances and can be thought of mainly as a neutral substance. Very small amounts of nitrogen are extracted by soil bacteria and made available for use by plants. Otherwise, it is largely a "filler," adding inert bulk to the atmosphere.

In contrast to nitrogen, *oxygen* gas (O_2) is highly active chemically. It combines readily with other elements in the process of oxidation. Combustion of fuels represents a rapid form of oxidation, whereas certain forms of rock decay (weathering) represent very slow forms of oxidation. Living tissues require oxygen to convert foods into energy.

Figure 2.10 Components of the lower atmosphere. Values show percentage by volume. Nitrogen and oxygen form 99 percent of our air, with other gases, principally argon and carbon dioxide, accounting for the final 1 percent.

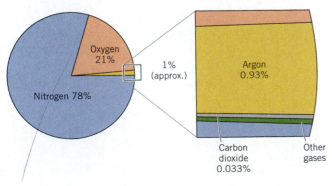

These two main component gases of the lower atmosphere are perfectly diffused so as to give pure, dry air a definite set of physical properties, as though it were a single gas. The remaining 1 percent of dry air is mostly argon, an inactive gas of little importance in natural processes. In addition, there is a very small amount of *carbon dioxide* (CO_2), amounting to about 0.033 percent. This gas is of great importance in atmospheric processes because of its ability to absorb radiant heat passing through the atmosphere from the earth. Carbon dioxide thus adds to the warming of the lower atmosphere. Carbon dioxide is also used by green plants in the photosynthesis process. During photosynthesis, CO_2 is converted into chemical compounds that build up the plant's tissues, organs, and supporting structures.

Water vapor is an important component of the atmosphere that varies in concentration from place to place and time to time.

Another important component of the atmosphere is *water vapor*, the gaseous form of water. Individual water molecules in the form of vapor are mixed freely throughout the atmosphere, just like the other atmospheric component gases. Unlike the other component gases, however, water vapor can vary highly in concentration. Usually, water vapor makes up less than 1 percent of the atmosphere. Under very warm, moist conditions, as much as 2 percent of the air can be water vapor. Since water vapor, like carbon dioxide, is a good absorber of heat radiation, it also plays a major role in warming the lower atmosphere. The atmosphere also contains dust and tiny floating particles that absorb and scatter radiation. These are discussed more fully in Chapter 4.

Ozone in the Upper Atmosphere

Another small, but important, constituent of the atmosphere is ozone. **Ozone** (O_3) is found mostly in the upper part of the atmosphere, in a layer termed the *stratosphere*. This layer of the atmosphere lies about 14–50 km (9–31 mi) above the surface and is described in more detail in the next chapter. Ozone in the stratosphere absorbs ultraviolet radiation from the sun as this radiation passes through the atmosphere, thus shielding the life layer from its harmful effects.

Ozone is most concentrated in a layer that begins at an altitude of about 15 km (9 mi) and extends to about 55 km (34 mi) above the earth. This *ozone layer* is produced by gaseous chemical reactions in the stratosphere. The details of these chemical reactions are complicated, but in summary they are quite simple. An oxygen molecule (O_2) absorbs ultraviolet light energy and splits into two oxygen atoms (O + O). A free oxygen atom (O) then combines with an O_2 molecule to form ozone, O_3. Once formed, ozone can also be destroyed. Like O_2, ozone absorbs ultraviolet radiation, and it splits to form O_2 + O. Furthermore, two oxygen atoms (O + O) can join to form O_2. The net effect is that ozone (O_3), molecular oxygen (O_2), and atomic oxygen (O) are constantly formed, destroyed, and reformed in the ozone layer, absorbing ultraviolet radiation with each transformation.

The absorption of ultraviolet radiation by the ozone layer protects the earth's surface from this damaging form of radiation. If the concentration of ozone is reduced, fewer transformations among O, O_2, and O_3 occur, and so ultraviolet absorption is reduced. If solar ultraviolet radiation were to reach the earth's surface at full intensity, most bacteria exposed on the earth's surface would be destroyed, and animal tissues would be severely damaged. The presence of the ozone layer is thus an essential protection in maintaining a viable environment for life on this planet.

Certain forms of air pollution are currently reducing ozone concentrations in some portions of the atmosphere at certain times of the year. This threat to the ozone layer is described in *Eye on the Environment • The Ozone Layer—Shield to Life.* **SCIENCE NEWS** 1-3

SENSIBLE HEAT AND LATENT HEAT TRANSFER

Thus far, we have discussed energy in the form of shortwave and longwave radiation. However, energy can be transported in other ways. Two of them—sensible heat and latent heat transfer—are extremely important in discussing the global energy balance.

The most familiar form of heat storage and transport is known as **sensible heat**—the quantity of heat held by an object that can be sensed by touching or feeling. This kind of heat is measured by a thermometer. When the temperature of an object or a gas increases, its store of sensible heat increases. When two objects of unlike temperature contact each other, heat energy moves by *conduction* from the warmer to the cooler. This type of heat flow is referred to as **sensible heat transfer**. Sensible heat transfer can also occur by *convection*. In this process, a fluid is heated by a warm surface, expands, and rises, creating an upward flow. This flow moves heat away from the surface and into the fluid. We will return to atmospheric convection in Chapter 4.

Unlike sensible heat, **latent heat**—or hidden heat—cannot be measured by a thermometer. It is heat that is taken up and stored when a substance changes state from a solid to a liquid, from a liquid to a gas, or from a solid directly to a gas. Water provides an important example. For liquid water to make the transition to water vapor, energy is required. As the water evaporates from liquid to gas, it draws up heat from the surroundings. This is the latent heat. It is carried in the form of the fast random motion of the free water vapor molecules as they enter the gaseous state during the process of evaporation. For example, consider the cooling you feel in a breeze when your skin is wet. As the water evaporates from liquid to gas, it draws up the heat required to make the change of state and carries it away from your skin, keeping you cool.

Although the latent heat of water vapor in air cannot be measured by a thermometer, energy is stored there just the same. When water vapor turns back to a liquid or solid, the latent heat is released. In the earth–atmosphere system, **latent heat transfer** occurs when water evaporates from a moist land surface or open water surface. This process transfers heat from the surface to the atmosphere. On a global scale, latent heat transfer by movement of moist air provides a very important mechanism for transporting large amounts of energy from one region of the earth to another.

THE GLOBAL ENERGY SYSTEM

The flow of energy from the sun to the earth and then back out into space is a complex system involving not only radiation, but also energy storage and transport. We will investigate this complex system in more detail to fully appreciate global weather and climate. Keep in mind that we are really studying the way the sun powers the earth's surface processes, ranging from wind systems to the cycle of precipitation.

Insolation Losses in the Atmosphere

Let's examine the flow of insolation through the atmosphere on its way toward the surface. As the shortwave radiation penetrates the atmosphere, its energy is absorbed or diverted in various ways. This is shown in Figure 2.11, which gives typical values for losses of incoming solar energy to the nearest 5 percent.

At an altitude of 150 km (93 mi), the solar radiation spectrum possesses almost 100 percent of its original energy. However, by the time rays penetrate to an altitude of 88 km (55 mi), absorption of X rays is nearly complete, and some of the ultraviolet radiation has been absorbed as well.

EYE ON THE ENVIRONMENT

The Ozone Layer • Shield to Life

When you hear about ozone on the nightly news, the news story is probably referring to damage to the atmosphere's ozone layer. A serious threat to the ozone layer is posed by the release of *chlorofluorocarbons*, or *CFCs*, from the earth's surface into the atmosphere. CFCs are synthetic industrial chemical compounds containing chlorine, fluorine, and carbon atoms. Before 1976, when the Environmental Protection Agency in the United States issued a ban on aerosol spray cans charged with CFCs, many households used such aerosols. CFCs are still in wide use as cooling fluids in refrigeration systems. When these appliances are disposed of, their CFCs are released to the air.

Molecules of CFCs in the atmosphere are very stable close to the earth. They move upward by diffusion without chemical change until they eventually reach the ozone layer. Here, they absorb ultraviolet radiation and are decomposed to chlorine oxide molecules. The chlorine oxide molecules in turn attack molecules of ozone, converting them in large numbers into ordinary oxygen molecules by a chain reaction. In this way, the ozone concentration within the ozone layer is reduced, and so there are fewer ozone molecules to absorb ultraviolet energy in the stratosphere.

Other gaseous molecules can act to reduce the concentrations of ozone in the stratosphere as well. Nitrogen oxides, bromine oxides, and hydrogen oxides are examples. As with chlorine oxides, the concentrations of these molecules in the stratosphere are influenced by human activity.

Not all threats to the ozone layer can be attributed to human activity. Aerosols inserted into the stratosphere by volcanic activity also can act to reduce ozone concentrations. The June 1991 eruption of Mount Pinatubo, in the Philippines, reduced global ozone in the stratosphere by 4 percent during the following year, with reductions over midlatitudes of up to 9 percent.

As studies of the ozone layer based on satellite data accumulated during the 1980s, a substantial decline in total global ozone was noted. By late in the decade, scientists had agreed that the decrease in the ozone layer was escalating far faster than had been predicted.

Compounding the problem of global ozone decrease was the discovery in the mid-1980s of a "hole" in the ozone layer over the continent of Antarctica. Here, seasonal thinning of the ozone layer occurs during the early spring of the southern hemisphere, and ozone reaches a minimum during the month of October. The thinning occurs after the formation of a polar vortex in the stratosphere during the winter period. This vast whirlpool of wind traps the air it contains and keeps this air out of the sun during the months of the long polar night. The air in the vortex eventually becomes very cold, and clouds of crystalline water (ice) or other water-containing compounds form within it. The crystals are important because they provide a surface on which chemical reactions can take place. One of these reactions allows chlorine to be converted from a stable form to chlorine oxide, ClO, which is highly reactive in the presence of sunlight. As the southern hemisphere spring approaches, the polar vortex slowly becomes illuminated by the sun, and the chlorine oxide reacts with ozone. Ozone concentrations are reduced, and the ozone hole is formed. The figure at right presents a satellite-derived map of ozone concentration over the south pole on October 1, 1998, clearly revealing the ozone hole.

In the northern hemisphere, a polar vortex forms as well, but is much weaker than in the southern hemisphere and is less stable. As a result, an early-spring ozone hole is not usually observed in the Arctic. However, in 1993, 1996, and 1997, notable arctic ozone holes occurred, and atmospheric computer models have projected even more severe losses in the period 2010–2019.

Although the most dramatic reductions in stratospheric ozone levels have been associated with arctic and antarctic ozone holes, reductions have been noted elsewhere as well—including the northern midlatitudes. In 1992, researchers reanalyzed 12 years of weather satellite data and conclud-

---

Molecules and particles in the air scatter radiation in all directions. When the rays are scattered back toward space, this process is referred to as diffuse reflection.

As radiation moves through deeper and denser atmospheric layers, gas molecules cause some visible light rays to be turned aside in all possible directions, a process known as **scattering**. Dust and other particles in the air also cause scattering. Scattered radiation is unchanged except for the direction of its path. Scattered radiation moving in all directions through the atmosphere is known as *diffuse radiation*. Because scattering occurs in all directions, some energy is sent back into outer space, and some flows down to the earth's surface. When the scattering turns radiation back to space, we call this process *diffuse reflection*. Under clear sky conditions, diffuse reflection sends about 5 percent of the incoming solar radiation back to space.

Another form of energy loss in the atmosphere is *absorption*. In this process, solar rays strike gas molecules and dust, and their energy is absorbed. This accounts for

Oct 1, 1998

Dobson Units

100   200   300   400   500

A satellite-derived map of ozone concentration for the south polar region from October 8, 1998. Ozone concentration is measured in Dobson units. Lowest values, over central Antarctica, are about 90 units. Prior to 1979, values were in the range of 250–300 units.

reduction of crop yields and the death of some forms of aquatic life.

Responding to the global threat of ozone depletion and its anticipated impact on the biosphere, 23 nations endorsed a United Nations plan in 1987 for cutting global CFC consumption by 50 percent by the year 1999. Since then, international agreements on CFC control have been strengthened, but some nations have not always fully complied with the terms of the treaty.

Nonetheless, the treaty has had an effect. In 1996, atmospheric scientists noted a reduction in atmospheric chlorine in the lower atmosphere beginning in 1994 and predicted that concentrations of chlorine oxide in the stratospheric ozone layer would begin to decrease in another four years. Although the ozone layer is not expected to be completely restored until the middle of the twenty-first century, this observation bodes well for the elimination of the threat of enhanced ultraviolet radiation to the earth's life layer.

Questions

ed that at various points in the northern hemisphere the ozone shield had thinned by as much as 6 percent. Damage to the ozone layer was noted as far south as Florida. In another study, ozone levels over North America dropped by about a half-percent per year between 1979 and 1990. By 1993, the total loss had reached 7.5 percent.

As the global ozone layer thins, the rate of incoming ultraviolet solar radiation reaching the earth's surface should increase. Recent studies of satellite data have documented this effect. Since 1978, the average annual exposure to ultraviolet radiation increased by 6.8 percent per decade along the 55° N parallel. At 55° S, the increase was 9.8 percent per decade. Over most of North America, the increase was about 4 percent per decade.

How might declining ozone concentrations affect the earth? An increase in the incidence of skin cancer in humans is one of the predicted effects. Other possible effects include

1. What are CFCs and how do they impact the ozone layer?

2. When and where have ozone reductions been reported? Have corresponding reductions in ultraviolet radiation been noted?

3. What is the outlook for the future health of the ozone layer?

about 15 percent of the incoming solar radiation. Thus, when skies are clear, diffuse reflection and absorption total about 20 percent, leaving as much as 80 percent of the solar radiation to reach the ground.

Both carbon dioxide and water vapor can directly absorb some wavelengths of solar radiation. Carbon dioxide is fairly uniform in the air, but, as we noted earlier, the water vapor content of the air can vary greatly. Thus, absorption can vary from one global environment to another. Absorption results in a rise in air temperature. In this way, incoming solar radiation causes some direct heating of the lower atmosphere.

Clouds can greatly increase the amount of incoming solar radiation reflected back to space. The bright white surfaces of thick, low clouds are extremely good reflectors of shortwave radiation. Cloud reflection can account for a direct turning back to space of 30 to 60 percent of incoming radiation (Figure 2.11). Clouds also absorb radiation, as much as 5 to 20 percent. Under conditions of a heavy cloud layer, as little as 10 percent of incoming solar radiation can reach the ground. The amount of global cloud cover, including cloud type and altitude, is an important factor in determining our global climate. In Chapter 6, we will discuss this topic further.

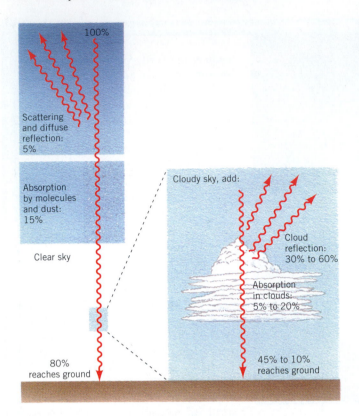

Figure 2.11 Losses of incoming solar energy are much lower with clear skies (*left*) than with cloud cover (*right*). (Copyright © A. N. Strahler.)

Albedo

The percentage of shortwave radiant energy scattered upward by a surface is termed its **albedo**. For example, a surface that reflects 40 percent of the shortwave radiation it receives has an albedo of 40 percent. Albedo is an important property of a surface, because it determines how fast the surface heats up when exposed to insolation. A surface with a high albedo, such as snow or ice (45 to 85 percent), reflects much or most of the solar radiation and so heats up slowly. A surface with a low albedo, such as black pavement (3 percent), absorbs nearly all the incoming insolation and so heats up quickly. The albedo of a water surface varies with the angle of incoming radiation. It is very low (2 percent) for nearly vertical rays. However, when the sun shines on a calm water surface at a low angle, much of the radiation is directly reflected as sun glint, producing a higher albedo. For fields, forests, and bare ground, albedos are of intermediate value, ranging from as low as 3 percent to as high as 25 percent.

The earth's overall albedo lies in the range of 29 to 34 percent.

Certain orbiting satellites are equipped with instruments to measure the energy levels of shortwave and longwave radiation above the top of the atmosphere. Both incoming radiation from the sun and outgoing radiation from the atmosphere and earth's surfaces below are measured. Data from these satellites have been used to estimate the earth's average albedo. This albedo includes reflection by the earth's atmosphere as well as its surfaces, so it includes diffuse reflection by clouds, dust, and atmospheric molecules. The albedo values obtained in this way vary between 29 and 34 percent. This means that the earth–atmosphere system directly returns back to space slightly less than one–third of the solar radiation it receives.

Counterradiation and the Greenhouse Effect

The ground can only radiate longwave energy upward, but the atmosphere radiates longwave energy in all directions—both upward to space, where it is lost, and downward to the ground. We have already seen that water vapor and carbon dioxide can directly absorb some wavelengths of shortwave radiation. In fact, water vapor and carbon dioxide in the atmosphere also absorb much of the longwave radiation emitted upward from the earth's surface. When they emit this radiation, a part is radiated back down toward the earth surface. This return of upwelling radiation is called **counterradiation**. Thus, the lower atmosphere, with its longwave-absorbing gases, acts like a blanket that returns heat to the earth. This mechanism helps to keep surface temperatures from dropping excessively during the night or in winter at middle and high latitudes. Cloud layers are even more important than carbon dioxide and water vapor in producing a blanketing effect to retain heat in the lower atmosphere, since clouds are excellent absorbers and emitters of longwave radiation.

The principle of counterradiation is shown in Figure 2.12. In part (*a*), no atmosphere is present. Solar energy is either absorbed or reflected upward by scattering, and longwave energy is emitted by the surface directly to space. In part (*b*), an atmospheric layer absorbs upwelling longwave radiation as well as a fraction of incoming shortwave radiation. The portion that is absorbed and reemitted downward is the counterradiation.

A similar principle is used in greenhouses and in homes with large windows to entrap solar heat. Here, the glass windows permit entry of shortwave energy but block the exit of longwave energy. This phenomenon is even better demonstrated by the intense heating of air in a closed automobile left parked in the sun. We use the expression **greenhouse effect** to describe this atmospheric heating principle.

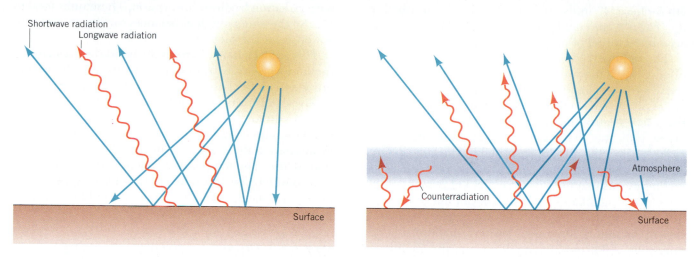

Figure 2.12 Effect of the atmosphere on the surface energy budget. Without an atmosphere (*a*), solar radiation is reflected by surface scattering back to space, and the surface emits longwave radiation to space. In (*b*), the atmosphere absorbs much upwelling longwave energy and emits some back to the surface as counterradiation. The atmosphere also scatters some shortwave radiation back to space.

Global Energy Budgets of the Atmosphere and Surface

The global energy budget for the earth's surface and atmosphere is diagrammed in Figure 2.13. We begin with a discussion of shortwave radiation, shown in the left part of the figure. Since shortwave radiation comes from the sun, we are concerned here only with the downward flow of solar radiation and its fate.

As shown on the left, reflection by molecules and dust, clouds, and the surface (including the oceans) totals 31 percentage units. This is the albedo of the earth–atmosphere system. The right side of this diagram shows values for absorption in the atmosphere. The combined losses through absorption by molecules, dust, and clouds average 20 percentage units. With 31 percentage units of the incoming solar energy flow reflected and 20 percentage units absorbed in the

Figure 2.13 Diagram of the global energy balance. Values are percentage units based on total insolation as 100. The left figure shows the fate of incoming solar radiation. The right figure shows longwave energy flows occurring between the surface and atmosphere and space. Also shown are the transfers of latent heat, sensible heat, and direct solar absorption that balance the budget for earth and atmosphere. Data of Kiehl and Trenberth, 1997.

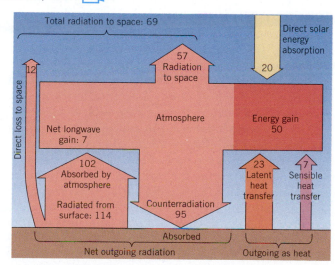

atmosphere, 49 units are left. This flow is absorbed by the earth's land and water surfaces. We have now accounted for all of the incoming flow of insolation.

The right part of Figure 2.13 shows the components of outgoing longwave radiation for the earth's surface, the atmosphere, and the planet as a whole. The large arrow on the left shows total longwave radiation leaving the earth's land and ocean surface. The long, thin arrow indicates that 12 percentage units are lost to space, while 102 units are absorbed by the atmosphere. The loss is therefore equivalent to 114 percentage units.

This means that the surface emits 114 percent of the total incoming solar radiation! How can this be correct? To answer this question, note that the surface receives a flow of 95 units of longwave counterradiation from the atmosphere. This will be in addition to the 49 units of shortwave radiation that are absorbed (upper figure). Thus, the surface receives 95 (longwave) + 49 (shortwave) = 144 units in all. Thus, a loss of 114 units by longwave radiation out of a gain of 144 units is certainly possible.

On the far right of the figure are two smaller arrows. These show the flow of energy away from the surface as latent heat and sensible heat. Evaporation of water from moist soil or ocean transfers latent heat from the surface to the atmosphere, as explained earlier. Sensible heat transfer occurs when heat is directly conducted from the surface to the adjacent air layer. When air is warmed by direct contact with a surface and rises, it takes its heat along with it. In this process, heat flows from the surface to the atmosphere.

Latent heat flows from a moist surface to the atmosphere when surface water evaporates.

These last two heat flows are not part of the radiation balance, since they are not in the form of radiation. However, they are a very important part of the total energy budget of the surface, which includes all forms of energy. Taken together, these two flows account for 30 units leaving the surface. With their contribution, the surface energy balance is complete. Total gains are 95 (longwave) + 49 (shortwave) = 144. Total losses are 114 (longwave) + 23 (latent heat) + 7 (sensible heat) = 144.

Like the earth's surface, the atmosphere also emits longwave radiation (Figure 2.12*b*). However, this radiation is divided into two parts. One part is directed into space (57 units), while the other part is counterradiation directed back to the earth's surface (95 units). The total of this emitted radiation is 57 + 95 = 152 percentage units. This figure seems absurdly large until we note that the atmosphere receives not only 102 units of longwave radiation, but also 23 units of latent heat, 7 units of sensible heat from the surface, and 20 units of short-

wave radiation by direct absorption. These units total to 152. So the energy budget balances for the atmosphere as well as for the surface.

This analysis helps us to understand the vital role of the atmosphere in trapping heat through the greenhouse effect. Because of counterradiation, the surface receives 144 units and radiates away 114 of them. In order to do so, the surface must be warm indeed. Without the counterradiation of the atmosphere, the surface would have only 49 units of absorbed shortwave radiation to lose. The temperature needed to radiate so few units is well below freezing. So without the greenhouse effect, our planet would be a cold, forbidding place.

NET RADIATION, LATITUDE, AND THE ENERGY BALANCE

Solar energy is intercepted by our planet, and because some of it is absorbed, the heat level of our planet tends to rise. At the same time, our planet radiates energy into outer space, a process that tends to reduce its level of heat energy. Over time, as we have seen, these flows must balance for the earth as a whole. However, incoming and outgoing radiation flows for any given surface do not have to be in balance. At night, for example, there is no incoming radiation, yet the earth's surface and atmosphere still emit outgoing radiation. In any one place and time, then, more radiant energy is being gained than lost, or vice versa.

Net radiation is the difference between all incoming radiation and all outgoing radiation. In places where radiant energy is flowing in faster than it is flowing out, net radiation is a positive quantity, providing a surplus. In other places, where radiant energy is flowing out faster than it is flowing in, net radiation is a negative quantity, yielding a radiant energy deficit. Our analysis of the earth's radiation balance has already shown that for the entire earth and atmosphere as a unit, the net radiation is zero on an annual basis.

In Figure 2.8, we saw that solar energy input varies strongly with latitude. What is the effect of this difference on net radiation? A study of the net radiation profile spanning the entire latitude range 90° N to 90° S will answer this question. In this analysis we will use yearly averages for each latitude, so that the effect of seasons is concealed.

The lower part of Figure 2.14 presents a global profile of net radiation from pole to pole. Between about lat. 40° N and lat. 40° S there is a net radiant energy gain labeled "energy surplus." In other words, incoming solar radiation exceeds outgoing longwave radiation throughout the year as long as there are other means to carry away the excess heat energy. Similarly, poleward of 40° N and 40° S, the net radiation is negative and is labeled

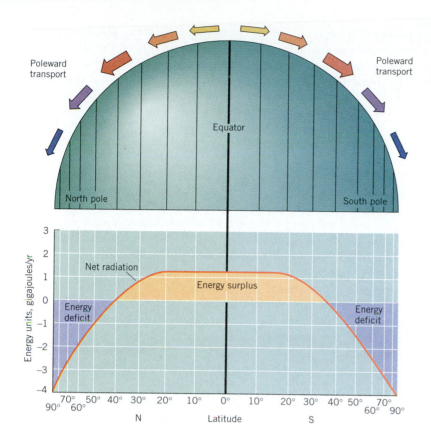

Figure 2.14 Annual surface net radiation from pole to pole. Where net radiation is positive, incoming solar radiation exceeds outgoing longwave radiation. There is an energy surplus, and so energy moves poleward as latent heat and sensible heat.

"deficit"—meaning that outgoing longwave radiation exceeds incoming shortwave radiation. This can happen only if energy is brought into these middle and high latitudes from the equatorial and tropical zones.

Net radiation at a surface measures the balance between incoming and outgoing radiation.

What mechanism brings energy from the lower latitudes poleward? Two processes are responsible. First, the circulation pattern of the oceans brings warm water poleward, moving heat energy across latitude boundaries in a sensible heat transfer. Second, the movement of warm, moist air poleward carries latent heat energy to higher latitudes in a latent heat transfer. These two energy transport processes taken together are called **poleward heat transport**, since the transport moves heat energy from lower to higher latitudes. The upper part of Figure 2.14 diagrams this flow.

If you carefully examine the lower part of Figure 2.14, you will find that the area on the graph labeled "surplus" is equal in size to the combined areas labeled "deficit." The matching of these areas confirms the fact that net radiation for the whole earth is zero and that global incoming shortwave radiation exactly balances global outgoing longwave radiation.

A LOOK AHEAD

Although our analysis of the earth's energy balance is not yet complete, it is clear that the balance is a sensitive one involving a number of factors that determine how energy is transmitted and absorbed. Has industrial activity already altered the components of the planetary radiation balance? An increase in carbon dioxide will increase the absorption of longwave radiation by the atmosphere, enhancing the greenhouse effect. Will this change cause a steady and permanent rise in average atmospheric temperature?

An increase in atmospheric particles at upper levels of the atmosphere will increase the scattering of incoming shortwave radiation and thus reduce the shortwave energy available to warm the surface. On the other hand, increased dust content at low levels will act to absorb more longwave radiation and so raise the surface temperature. Has either change occurred?

Human habitation, through cultivation and urbanization, has profoundly altered the earth's land surfaces. Have these changes, which affect surface albedo and the capacity of the ground to absorb and to emit longwave radiation, modified the global the energy balance? Answers to such questions require further study of the processes of heating and cooling of the earth's atmosphere, lands, and oceans. Our next chapter concerns air temperature and how and why it varies daily and annually depending on the surface energy balance.

THE EARTH'S GLOBAL ENERGY BALANCE IN REVIEW

Electromagnetic radiation is a form of energy emitted by all objects. The wavelength of the radiation determines its characteristics. The hotter an object, the shorter the wavelength of the radiation and the greater the amount of radiation that it emits. Shortwave radiation, emitted by the sun, includes ultraviolet, visible, and shortwave infrared radiation. Longwave radiation, emitted by earth surfaces, is familiar as heat. Radiation flows are measured in watts per square meter. The earth continuously absorbs solar (shortwave) radiation and emits longwave radiation. In the long run, the gain and loss of radiant energy remain in balance, and the earth's average temperature remains constant.

Insolation, the rate of solar radiation flow available at a location at a given time, is greater when the sun is higher in the sky. Daily insolation is also greater when the period of daylight is longer. Annual insolation is greatest at the equator and least at the poles. However, the poles still receive 40 percent of the radiation received at the equator. The pattern of annual insolation with latitude leads to a natural naming convention for latitude zones: equatorial, tropical, subtropical, midlatitude, subarctic (subantarctic), arctic (antarctic), and polar.

The earth's atmosphere is dominated by nitrogen and oxygen gases. Carbon dioxide and water vapor, though minor constituents by volume, are very important because they absorb longwave radiation and enhance the greenhouse effect. Ozone (O_3) is a tiny but important constituent of the atmosphere that is concentrated in a layer of the upper atmosphere. It is formed from oxygen (O_2) by chemical reactions that absorb ultraviolet radiation, thus sheltering the organisms of the earth's surface from the damaging effects of ultraviolet rays.

Latent heat and sensible heat are additional forms of energy. Latent heat is taken up or released when a change of state occurs. Sensible heat is contained within a substance. It can flow into a cooler substance that it contacts.

The global energy system includes a number of important pathways by which insolation is transferred and transformed. Part of the solar radiation passing through the atmosphere is absorbed or scattered by molecules, dust, and larger particles. Some of the scattered radiation returns to space as diffuse reflection. The land surfaces, ocean surfaces, and clouds also reflect some solar radiation back to space. The proportion of radiation that a surface absorbs is termed its albedo. The albedo of the earth and atmosphere as a whole planet is about 30 percent.

The atmosphere absorbs longwave energy emitted by the earth's surface, causing the atmosphere to counterradiate some of that longwave radiation back to earth, thereby creating the greenhouse effect. Because of this heat trapping, the earth's surface temperature is considerably warmer than we might expect for an earth without an atmosphere.

Although absorbed shortwave radiation balances longwave radiation emitted by the earth's surface and atmosphere for the earth as a whole, this net radiation does not have to balance at every surface location on the globe. At latitudes lower than 40 degrees, annual net radiation is positive, while it is negative at higher latitudes. This imbalance can be sustained only if heat is exported from low to high latitudes by poleward heat transport. This transport is achieved through sensible heat transfer in currents of warm ocean water, and as latent heat transfer in flows of warm, moist air.

KEY TERMS

electromagnetic radiation	insolation	scattering
visible light	ozone	albedo
shortwave radiation	sensible heat	counterradiation
longwave radiation	sensible heat transfer	greenhouse effect
radiation balance	latent heat	net radiation
	latent heat transfer	poleward heat transport

REVIEW QUESTIONS

1. What is electromagnetic radiation? How does the temperature of the hood of a car influence the nature of the electromagnetic radiation it emits?
2. Describe the types of radiation emitted by the sun and their characteristics.
3. Compare the terms *shortwave radiation* and *longwave radiation*. What are their sources?
4. What is the solar constant? What is its value? What are the units with which it is measured?
5. How does the sun's path in the sky influence the average daily insolation at a location?
6. What is the influence of latitude on the annual cycle of daily insolation? on annual insolation?
7. Why are carbon dioxide and water vapor important atmospheric constituents?
8. How does the ozone layer protect the life layer?
9. Describe latent heat transfer and sensible heat transfer.
10. What is the fate of incoming insolation? Define scattering and absorption, including the role of clouds.
11. Define albedo and give two examples.
12. Describe the counterradiation process and how it relates to the greenhouse effect.
13. Discuss the energy balance of the earth's surface. Identify the types and sources of energy that the surface receives and do the same for energy that it loses.
14. Discuss the energy balance of the atmosphere. Identify the types and sources of energy that the atmosphere receives and do the same for energy that it loses.
15. What is net radiation? How does it vary with latitude?

VISUALIZING EXERCISES

1. Place yourself in Figure 2.6. Imagine that you are standing in the center of the figure where the N–S and E–W lines intersect. Turn so that you face south. Using your arm to point at the sun's position, trace the path of the sun in the sky at the equinox. It will rise exactly to your left, swing upward to about a 50° angle, and then descend to the horizon exactly at your right. Repeat for the summer and winter solstices, using the figure as a guide.
2. Sketch a simple diagram of the sun above a layer of atmosphere above the earth's surface, somewhat like Figure 2.12*b*. Using Figure 2.13 as a guide, draw arrows indicating flows of energy among sun, atmosphere, and surface. Label each arrow using terms from Figure 2.13.
3. Sketch the world latitude zones on a circle representing the globe and give their approximate latitude ranges.

A DEEPER VIEW OF THE EARTH'S GLOBAL ENERGY BALANCE

1. Suppose the earth's axis of rotation were perpendicular to the orbital plane instead of tilted at $23\frac{1}{2}°$. How would global insolation be affected? How would insolation vary with latitude? How would the path of the sun in the sky change with the seasons?
2. Imagine that you are following a beam of either (a) shortwave solar radiation entering the earth's atmosphere heading toward the surface, or (b) a beam of longwave radiation emitted from the surface heading toward space. How will the atmosphere influence the beam?

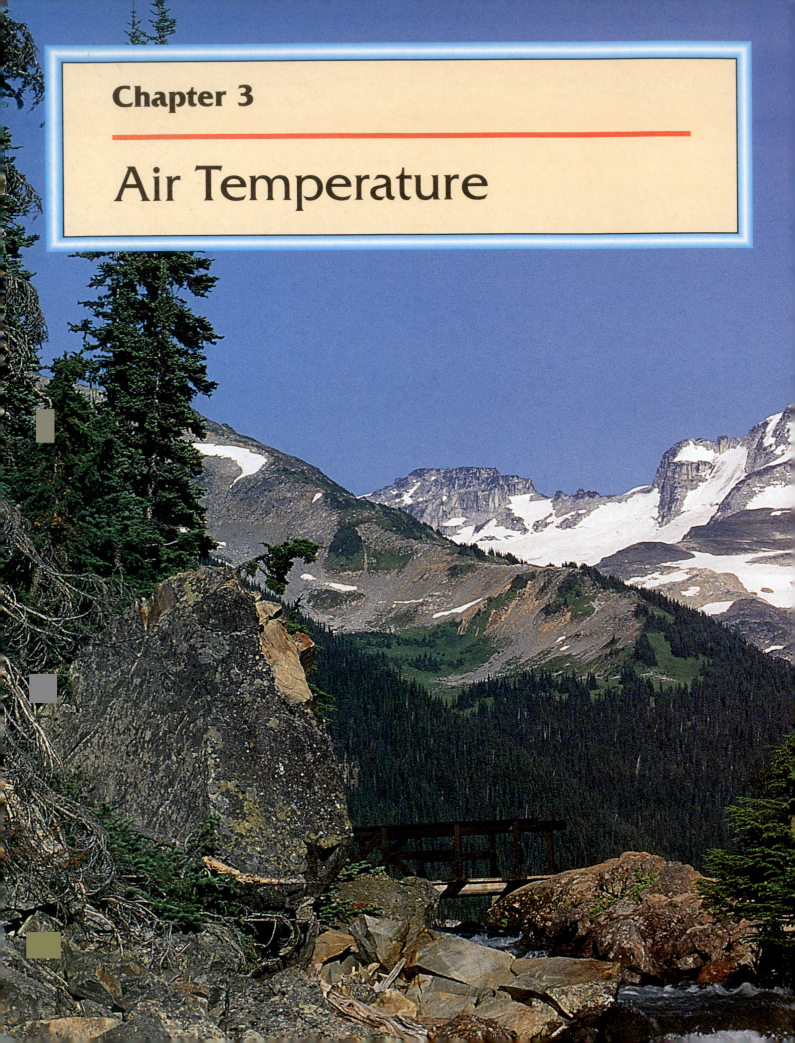

Chapter 3

Air Temperature

A hiker in Garibaldi Provincial Park, British Columbia, Canada.

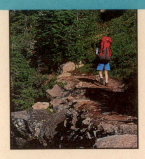

PUTTING YOU IN THE PICTURE

The stars are bright and clear tonight. From your vantage point here in the grassy meadow, it's easy to trace the outline of the Big Dipper, with the North Star high above its rim. In fact, you can even see the white blur of the Milky Way, arching across the sky and disappearing behind the peaks of the tall firs that surround the meadow. Brrr! It's getting chilly now. With regret, you decide to abandon your stargazing and return to the campsite, just a few minutes away in the midst of the conifer forest.

Entering the forest, you feel the chill leaving your body as the sheltering conifers shield you from the cold night sky. As you approach the camp, the fire is burning brightly. You step close to it, enjoying its warm glow. A friend offers you a mug of hot cocoa. "I'll bet we'll have frost in the meadow tonight," you think as you sip from the steaming cup. Sure enough, the next morning the grass in the meadow sparkles with white crystals that slowly turn to glistening beads as they absorb the sun's first rays.

A few hours later in the day, you are well along on your journey to the mountain summit. Here, the trail is steep and you are sweating. Although it was cool in the forest below, the low shrubs and scattered grasses nested in the cracks and crevices of the rocky, high terrain don't relieve the force of the sun high overhead

in the sky. A sudden breeze brings a pleasant cooling to your head and shoulders, as the moisture accumulating under your hat and shirt evaporates. You scramble over a low ledge, and the stark granite rocks are warm to your touch. Straightening up, you see the summit ahead, shimmering in the rising currents of hot air.

It is not long before you reach the summit, and the view is absolutely spectacular. In the thin, clear air, the surrounding peaks stand majestically against the deep blue sky, some with patches of white snow glistening in the intense sunlight. Below the peaks, the gray-brown tones of the barren summit slopes grade into the deep green mantle of conifer forests that line the valleys and lower slopes. It's been a long, hard climb, especially the last part up the crusted surface of the snowfield. The view, however, is worth every moment of exertion. You wouldn't have missed it for anything!

The subject of this chapter is air temperature, primarily the air temperature we experience every day within a few feet from the ground surface. What influences air temperature? Three factors are important. The first and most important one is the cyclic nature of solar radiation, produced by the two constant motions of the earth—its daily rotation on its axis and its annual revolution in its solar orbit. These cycles produce cycles of air temperature that follow daily and annual rhythms.

The second is the nature of the ground surface. In our example above, the forest is a thick layer covering the surface that acts to insulate the ground from temperature changes. Thus, the grassy meadow loses heat more rapidly at night than does the forest, and frost forms first on the meadow grass.

The third is location on the continent. Locations near the ocean experience a narrower range of temperatures than locations in continental interiors. The reason is that water heats and cools more slowly than soil, so air temperatures near water bodies tend to change more slowly as well. Understanding air temperatures will be easy if you keep these three factors in mind.

Chapter Outline

Measurement of Air Temperature
The Daily Cycle of Air Temperature
  Daily Insolation and Net Radiation
  Daily Temperature
  Temperatures Close to the Ground
  Urban and Rural Temperature Contrasts
  The Urban Heat Island
Temperature Structure of the Atmosphere
  Troposphere
  Stratosphere
  High-Mountain Environments
  Temperature Inversion and Frost

The Annual Cycle of Air Temperature
  Net Radiation and Temperature
  Land and Water Contrasts
World Patterns of Air Temperature
  Factors Controlling Air Temperature Patterns
  World Air Temperature Patterns for January and July
  The Annual Range of Air Temperatures
Global Warming and the Greenhouse Effect
  The Temperature Record
  Future Scenarios
Eye on the Environment • Carbon Dioxide—On the Increase

AIR TEMPERATURE

Temperature is a familiar concept. It is a measure of the level of sensible heat of matter, whether it is gaseous (air), liquid (water), or solid (rock or dry soil). We know from experience that when a substance receives a flow of heat energy, its temperature rises. As we saw in Chapter 2, the earth as a whole receives a constant flow of radiant shortwave energy from the sun. The earth also radiates longwave energy to space. During the day, the flow of shortwave radiation absorbed exceeds longwave energy emitted, and the surface temperature increases. At night, this net radiation balance reverses. No shortwave radiation strikes the darkened side of the earth, but longwave energy is still emitted from the surface. Therefore, surface temperatures decrease.

The temperature of a surface can be raised or lowered by processes other than absorption or emission of radiant energy. **Conduction** describes the flow of heat from a warmer substance to a colder one when the two are touching. In this process, the temperature of air in direct contact with the ground is raised when the ground receives strong sunlight. **Evaporation**, the process by which water changes from a liquid to a gas by absorbing heat, tends to lower the temperature of a wet surface. In the case of a patch of wet soil, the latent heat absorbed by the liquid water is taken away from the surface as the water vapor molecules leave the soil and move into the air layer above. Heat may also be transferred within a substance if the substance is a fluid— liquid or gas— and mixing occurs. This form of heat transfer, in which heat is distributed by mixing, is **convection**. (We'll return to convection in Chapters 4 and 5.)

The subject of this chapter is air temperature. The temperature of the layer of air close to the ground changes when heat flows into or out of the layer by the heat transfer mechanisms described above. Because these flows of heat change in natural cycles, air temperatures measured close to the ground surface follow natural cycles. These include both daily and seasonal rhythms of temperature rise and fall, which are produced by the daily and seasonal variations in insolation described in Chapter 2.

Because the daily and annual cycles of insolation vary systematically with latitude, we will see that some temperature cycles are closely related to latitude. For example, temperatures generally fall as we move poleward. Since the sun at higher latitudes does not rise as high in the sky, on average it provides less energy to warm earth surfaces there. Latitude also enhances seasonal contrasts. As much solar energy may be received at high latitudes in the summer as at the equator. In winter, however, much less energy is received. Because of this annual variation, high latitudes experience a much greater range in temperatures through the year.

Another important temperature difference occurs between oceanic and continental locations in mid- and high latitudes. It arises because water bodies heat and cool more slowly than continents. In the summer, we head for the beach when the weather gets hot because air temperatures there are cooler. Why? Because the water is cooler than the land, the air near the coast also tends to be cooler. In contrast, during the winter, coastal regions tend to have warmer temperatures, because the ocean water is often warmer than the land. However, keep in mind as you study this chapter that the main cause of air temperature change is still the change in the amount of the sun's energy that strikes the earth's surface from day to day, season to season, and latitude to latitude.

MEASUREMENT OF AIR TEMPERATURE

Air temperature is a piece of weather information that we encounter daily. In the United States, temperature is still widely measured and reported using the Fahrenheit scale. The freezing point of water on the Fahrenheit scale is 32°F, and the boiling point is 212°F. This represents a range of 180°F. In this book, we use the *Celsius* temperature scale, which is the international standard. On the Celsius scale, the freezing point of water is 0°C and the boiling point is 100°C. Thus, 100 Celsius degrees are the equivalent of 180 Fahrenheit degrees (1°C = 1.8°F; 1°F = 0.56°C). Conversion formulas between these two scales are given in Figure 3.1.

You are probably familiar with the *thermometer* as an instrument for measuring temperature. In this device, a liquid inside a glass tube expands and contracts with temperature, and the position of the liquid surface indicates the temperature. Since air temperature can vary with height, it is measured at a standard level—1.2 m (4 ft) above the ground. A *thermometer shelter* (Figure 3.2a) is a louvered box that holds thermometers or other weather instruments at proper height while sheltering them from the direct rays of the sun. Air circulates freely through the louvers, ensuring that temperatures inside the shelter are the same as the outside air.

Air temperatures are now automatically recorded by thermistors at a uniform height above the ground.

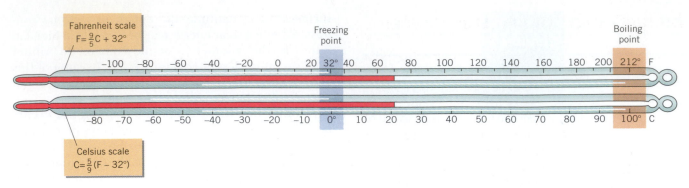

Figure 3.1 A comparison of the Celsius and Fahrenheit temperature scales. At sea level, the freezing point of water is at Celsius temperature (C) 0°, while it is 32° on the Fahrenheit (F) scale. Boiling occurs at 100°C, or 212°F.

Liquid-filled thermometers are being replaced by newer instruments for the routine measuring of temperatures. Perhaps you have used a digital fever thermometer, which reads out your body temperature directly as a number. It uses a device called a *thermistor*, which changes its electrical resistance with temperature. By measuring this resistance, the temperature may be obtained automatically. Many weather stations are now equipped with temperature measurement systems that use thermistors (Figure 3.2*b*).

Although some weather stations report temperatures hourly, most stations only report the highest and lowest temperatures recorded during a 24-hour period. These are most important values in observing long-term trends in temperature. This network comprises more than 5000 stations in the United States.

Daily, monthly, and yearly temperature statistics for a station are produced using the daily maximum and minimum temperature. The mean daily temperature is defined as the average of the maximum and minimum daily values. Records of mean daily temperatures averaged for monthly and yearly intervals of many years' duration are compiled for each observing station. These averages, along with other weather statistics, are used to describe the climate of the station and its surrounding area.

Figure 3.2 (*a*) A thermometer shelter. This white wooden louvered box houses maximum–minimum thermometers and other instruments. (*b*) Maximum–minimum temperature instrument system appears on the left side of this photo. To the right is a large recording rain gauge.

(*a*)

(*b*)

THE DAILY CYCLE OF AIR TEMPERATURE

Because the earth rotates on its axis, incoming solar energy at a location can vary widely throughout the 24-hour period. Insolation is greatest in the middle of the daylight period, when the sun is at its highest position in the sky, and falls to zero at night. Recall from Chapter 2 that net radiation is the balance between incoming energy absorbed by a surface and outgoing radiation that is emitted from that surface. Since received solar radiation varies greatly, so does net radiation. During the day, net radiation is positive, and the surface gains heat. At night, net radiation is negative, and the surface loses heat by radiating it to the sky and space. Since the air next to the surface is warmed or cooled as well, air temperatures follow the same cycle. This results in the daily cycle of rising and falling air temperatures.

Daily Insolation and Net Radiation

Let's look in more detail at how insolation, net radiation, and air temperature are linked in this daily cycle. The three graphs in Figure 3.3 show average curves of daily insolation, net radiation, and air temperature that we might expect for a typical observing station at lat. 40° to 45° N in the interior of North America. The time scale is set so that 12:00 noon occurs when the sun is at its highest elevation in the sky.

The daily cycle of temperature is controlled by the daily cycle of net radiation.

Graph (*a*) shows daily insolation. At the equinox (middle curve), insolation begins at about sunrise (6 A.M.), rises to a peak value at noon, and declines to zero at sunset (6 P.M.). At the June solstice, insolation begins about two hours earlier (4 A.M.) and ends about two hours later (8 P.M.). The June peak is much greater than at equinox, and the total insolation for the day is also much greater. At the December solstice, insolation begins about two hours later than the equinox curve (8 A.M.) and ends about two hours earlier (4 P.M.). Both the peak intensity and daily total insolation are greatly reduced in the winter solstice season.

Graph (*b*) shows net radiation for the surface. Recall that when net radiation is positive, the surface gains heat, and when negative, it loses heat. The curves for the solstices and equinox generally resemble those of insolation. Unlike insolation, net radiation begins the 24-hour day as a negative value—a deficit—at midnight. The deficit continues into the early morning hours. Net radiation shows a positive value—a surplus—shortly after sunrise and rises sharply to a peak at noon. In the afternoon, net radiation decreases as insolation decreases. A value of zero is reached shortly before sun-

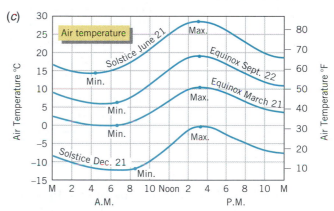

Figure 3.3 Idealized cycles of insolation, net radiation, and air temperature for a midlatitude station in the interior United States. Insolation (*a*) is a strong determiner of net radiation (*b*). Air temperatures (*c*) respond by generally increasing while net radiation is positive and decreasing when it is negative.

set. With no incoming insolation, net radiation then becomes negative, where it remains until midnight.

Although the three net radiation curves show the same general daily pattern—negative to positive to negative—they differ greatly in magnitude. For the June solstice, the positive values are quite large, and careful inspection of the figure will show that the area of surplus is much larger than the area of deficit. This means that for the day as a whole, net radiation is positive. At the December solstice, the surplus period is short and the surplus is small. The total deficit, which extends nearly 18 hours, outweighs the surplus. This means that the net radiation for the entire day is negative. As we will see, this pattern of positive daily net radiation in the summer and negative daily net radiation in the winter drives the annual cycle of temperatures.

Daily Temperature

Graph (*c*) shows the typical, or average, daily air temperature cycle. The minimum daily temperature usually occurs about a half hour after sunrise. Since net radiation has been negative during the night, heat has flowed from the ground surface, and the ground has cooled the surface air layer to its lowest temperature. As net radiation becomes positive, the surface warms quickly and transfers heat to the air above. Air temperature rises sharply in the morning hours and continues to rise long after the noon peak of net radiation.

We should expect the air temperature to rise as long as net radiation is positive. However, another process begins in the early afternoon. Mixing of the lower air by vertical currents distributes heat upward, offsetting the temperature rise. Therefore, the temperature peak usually occurs in the midafternoon. It is shown in the figure at about 3 P.M., but it usually occurs between 2 and 4 P.M., depending on local climatic conditions. By sunset, air temperature is falling rapidly. It continues to fall, but at a decreasing rate, throughout the night.

Temperatures are lowest just after sunrise and highest in midafternoon.

Note also that the general level of the temperature curves varies with the seasons. In the summer, the daily curve is high, showing warm temperatures. In winter, the curve is low, showing cold temperatures. In between are the equinoxes, with their intermediate temperatures. Since the temperatures lag behind the seasonal changes in net radiation, the September equinox shows temperatures considerably warmer than the March equinox. Even though net radiation is the same for the two, each curve lags behind, reflecting earlier seasonal conditions.

Temperatures Close to the Ground

The air temperatures we have discussed thus far are measured under standard conditions, at a short distance from the ground surface. As we get closer to the ground surface, the temperature trends get more extreme. Under the direct rays of the sun, a pavement surface heats to much higher temperatures than the air at standard height. At night, the pavement surface cools to temperatures that are lower than the air temperature at standard height. These effects are weaker under a forest cover where the ground is shaded and moist. On a dry desert floor or on the pavements of city streets and parking lots, however, the surface temperature extremes can be large.

Figure 3.4 presents a series of temperature profiles as they might be observed throughout the day close to the ground. At 8 A.M. (curve 1), the temperature of air and soil is uniform, producing a vertical line on the graph. By noon (curve 2), the surface is considerably warmer than the air at standard height, and the soil below the surface has been warmed as well. By 3 P.M. (curve 3), the soil surface is much warmer than the air at standard height. By 8 P.M. (curve 4), the surface is cooler than the air, and by 5 A.M. (curve 5), it is much colder. The daily cycle shows its greatest range near the soil surface and gradually weakens with depth.

Figure 3.4 A simplified diagram of the temperature profile close to the ground surface. The soil surface becomes very hot by midafternoon, but cools greatly at night.

Figure 3.5 Urban surfaces (*left*) are composed of asphalt, concrete, building stone, and similar materials. Sewers drain away rainwater, keeping urban surfaces dry. Rural surfaces (*right*) are composed of moist soil, largely covered by vegetation.

Urban and Rural Temperature Contrasts

Human activity has altered much of the earth's land surface. Cities are an obvious example. To build a city, vegetation is removed, and soils are covered with pavement or structures. Using the principles of the surface energy balance we have learned, we can understand the effects of these changes on the temperatures of urban areas as compared to rural regions.

Figure 3.5 illustrates some of the differences between urban and rural surfaces. In rural areas, the land surface is normally covered with a layer of vegetation. A natural activity of plants is **transpiration**. In this process, water is taken up by plant roots and moved to the leaves, where it evaporates. Since evaporation cools a surface by removing heat, we would expect the rural surface to be cooler.

The cooling effect of vegetation is even greater in a forest. Not only are transpiring leaves abundant, but also solar radiation is intercepted by a layer of leaves stretching from the tops of the trees to the forest floor. Thus, solar warming is not concentrated intensely at the ground surface, but rather serves to warm the whole forest layer.

Another feature of the rural environment is that the soil surfaces tend to be moist. During rainstorms, precipitation seeps into the soil. When sunlight reaches the soil surface, the water evaporates, thus removing heat and keeping the soil cool.

In contrast are the typical surfaces of the city. Rain runs easily off the impervious surfaces of roofs, sidewalks, and streets. The rainwater is channeled into sewer systems, where it flows directly into rivers, lakes, or oceans, instead of soaking into the soil beneath the city surfaces. Because the impervious surfaces are dry, the full energy of insolation warms the surfaces, with evaporation taking place only where street trees, lawns, and

bare soil patches occur. Thus, air temperatures measured above urban surfaces during the daytime are usually higher than those of rural surfaces.

> **Urban surfaces lack moisture and so are warmer than rural surfaces during the day. At night, urban materials conduct stored heat to the surface, also keeping temperatures warmer.**

Another factor contributing to warmer urban temperature is the ability of concrete, stone, and asphalt to conduct and hold heat better than soil, even when the soil is dry. Therefore, during the day, more heat can be stored by the city's building materials. At night, the heat is conducted back to the surface, keeping nighttime temperatures warmer. Heat absorption is enhanced by the many vertical surfaces in cities, which reflect radiation from one surface to another. Since some radiation is absorbed with each reflection, the network of vertical surfaces tends to trap heat more effectively than a single horizontal surface.

The Urban Heat Island

As a result of these effects, air temperatures in the central region of a city are typically several degrees warmer than those of the surrounding suburbs and countryside. Figure 3.6 shows city air temperatures on a typical summer evening. The lines of equal air temperature delineate a **heat island**. The heat island persists through the night because of the availability of a large quantity of heat stored in the ground during the daytime hours.

Another important factor in warming the city is fuel consumption. In summer, city temperatures are raised through the use of air conditioning. This equipment

Figure 3.6   A heat island over Washington, D.C., and surrounding areas. Air temperatures were taken at 10:00 P.M. on an evening in early August. (Data of H. E. Landsberg.)

pumps heat out of buildings, releasing the heat to the air. The power used to run the air conditioning systems is also released as heat. Although Figure 3.6 shows the urban heat island in summer, the heat island effect is usually more pronounced in winter. In that season, furnaces warm buildings. Interior heat is conducted to outside walls and roofs, which radiate heat into the urban environment.

Note that the heat island effect does not necessarily apply to cities in desert climates. In the desert, the evapotranspiration of the irrigated vegetation of the city may actually keep the city cooler than the surrounding barren region.

TEMPERATURE STRUCTURE OF THE ATMOSPHERE

Thus far, we have discussed temperatures in the range of the atmosphere that surrounds us—up to about 2 m (6 ft) above the ground surface. How different are air temperatures at increasing heights through the atmosphere? In general, temperatures are lower. The decrease in measured temperature is described by the **lapse rate**, which indicates the drop in temperature in degrees C per 1000 m (or degrees F per 1000 ft).

Why is air cooler at higher altitudes? Recall from Chapter 2 that most of the incoming solar radiation penetrates the atmosphere and is absorbed at the surface. Thus, the atmosphere is largely warmed from

below. In general, the farther the air is away from the earth's surface, the cooler the air will be.

In the lower atmosphere, air temperatures decrease with increasing altitude.

Figure 3.7 shows how temperature varies with altitude for a typical summer day in the midlatitudes. Altitude is plotted on the vertical axis and temperature on the horizontal axis. The curve is an average one—if we sent up a balloon that radioed the temperature of the air back to us every minute or two, and sent up many balloons over a long period of time, we would obtain an average profile of temperature very much like that shown in the figure. Temperature drops with altitude at an average rate of 6.4°C/1000 m (3.5°F/1000 ft). This average value is known as the **environmental temperature lapse rate**. For example, when the air temperature near the surface is a pleasant 21°C (70°F), the air at an altitude of 12 km (40,000 ft) will be a bone-chilling −55°C (−67°F). Keep in mind that the environmental temperature lapse rate is an average value and that on any given day the observed lapse rate might be quite different.

Figure 3.7 shows another important feature of the atmosphere. For the first 12 km (7 mi) or so, temperature falls with increasing elevation. However, at 12–15 km (7–9 mi) in height, the temperature stops decreasing. In fact, above that height, temperature slowly increases with elevation. This feature has led atmos-

Figure 3.7   A typical environmental temperature lapse-rate curve for a summer day in the midlatitudes. The rate of temperature decrease with elevation, or lapse rate, is shown at the average value of 6.4°C/1000 m (3.5°F/1000 ft). The tropopause separates the troposphere, where temperature decreases with increasing elevation, from the stratosphere, where temperature is constant or slightly increases with elevation.

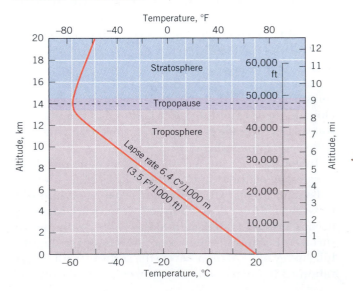

pheric scientists to distinguish two different layers in the lower atmosphere—the troposphere and the stratosphere.

Troposphere

The **troposphere** is the lowest atmospheric layer, in which temperature decreases with increasing elevation. Since almost all human activity occurs in this layer, it is of primary importance to us. Everyday weather phenomena, such as clouds or storms, occur mainly in this layer.

One important feature of the troposphere is that it contains significant amounts of water vapor. When the water vapor content is high, vapor can condense into water droplets (*condensation*), forming low clouds and fog, or be deposited as ice crystals (*deposition*), forming high clouds. When condensation or deposition is rapid, rain, snow, hail, or sleet—collectively termed *precipitation*—may be produced and fall to earth. Regions where water vapor content is high throughout the year will therefore have moist climates. In desert regions, where water vapor is present only in small amounts, precipitation is infrequent. In Chapter 2, we also described the important role of water vapor in absorbing and reradiating heat emitted by the earth's surface. In this way, water vapor helps to create the greenhouse effect, which warms the earth to habitable temperatures.

The troposphere contains countless tiny dust particles, so small and light that the slightest movements of the air keep them aloft. These are called *aerosols*. They are swept into the air from dry desert plains, lakebeds, and beaches, or they are released by exploding volcanoes. Oceans are also a source of aerosols. Strong winds blowing over the ocean lift droplets of spray into the air. These droplets of spray lose most of their moisture by evaporation, leaving tiny particles of salt as residues that are carried high into the air. Forest fires and brushfires are another important source of aerosols, contributing particles of soot as smoke. Meteors contribute dust particles as they vaporize from the heat of friction upon entering the upper layers of air. Industrial processes that burn coal or fuel oil incompletely release aerosols to the air as well.

The most important function of aerosols is that certain types serve as nuclei, or centers, around which water vapor condenses to form tiny droplets. When these droplets grow large and occur in high concentration, they are visible to the eye as clouds or fog.

Aerosols in the troposphere also scatter sunlight, thus brightening the whole sky while reducing slightly the intensity of the solar beam. This scattering behavior is strongest for the longer wavelengths in the visible part of the spectrum, and thus the red colors of sunrise and sunset are accentuated by dust.

The height at which the troposphere gives way to the stratosphere above is known as the *tropopause*. Here, temperatures stop decreasing with altitude and start to increase. The altitude of the tropopause varies somewhat with latitude and season. This means that the troposphere is not uniformly thick.

Stratosphere

Above the troposphere lies the **stratosphere** in which the air becomes slightly warmer as altitude increases. The stratosphere extends to a height of roughly 50 km (about 30 mi) above the earth's surface. It is the home of strong, persistent winds that blow from west to east. There is little mixing of air between the troposphere and stratosphere, and so the stratosphere normally holds very little water vapor or dust.

The troposphere and stratosphere are the two lowermost atmospheric layers.

One important feature of the stratosphere is that it contains the ozone layer. As we saw in Chapter 2, the ozone layer absorbs solar ultraviolet radiation and thus shields earthly life from this intense, harmful form of energy. In fact, the warming of the stratosphere with altitude is caused largely by the absorption of solar energy by ozone molecules.

High-Mountain Environments

Did you ever walk or drive to the top of a high mountain? Mount Whitney, in California, perhaps (Figure 3.8)? Or New Hampshire's Mount Washington? Or North Carolina's Mount Mitchell? If so, you have probably made the simple observation that temperatures generally drop as you go up in elevation. We've already noted this fact in the discussion of the environmental temperature lapse rate. But what else might you notice as you climb? You might notice that you sunburn a bit faster. Perhaps you also find yourself getting out of breath more quickly. And if you camp out, you may notice that the nighttime temperature gets lower than you might expect, even given that temperatures are generally cooler.

All these effects are related to the fact that air density decreases with elevation. At high elevations, the air is thin—that is, there are fewer gas molecules in a unit volume of air. Therefore, it makes sense that you can feel short of breath on a high mountain simply because the oxygen pressure in your lungs is lower than you are used to at lower elevations. With fewer air molecules and dust particles to scatter and absorb the sun's light, the sun's rays will feel stronger. And there is also less carbon dioxide and water vapor, and so the greenhouse effect is reduced. With the reduced warming effect, temperatures will tend to drop lower at night.

Figure 3.8 Mount Whitney (elevation 4418 m, 14,495 ft) and surrounding peaks of the Southern Sierra Nevada. The excellent visibility and dark blue sky are products of the thin air at this elevation.

At high elevations, air temperatures are generally cooler and show a greater day-to-night range.

What is the effect of elevation on the daily temperature cycle? Figure 3.9 shows temperature graphs for five stations ascending the Andes Mountain range in Peru. The July monthly mean for each station is shown as a horizontal line. Mean temperatures clearly decrease with elevation, from 16°C (61°F) at sea level to −1°C (30°F) at 4380 m (14,370 ft). The daily range clearly increases with elevation, except for Cuzco. This large city does not experience nighttime temperatures as low as you might expect because of its urban heat island and accompanying air pollution effects.

Temperature Inversion and Frost

So far, air temperatures seem to decrease with height above the surface. Is this always true? Think about what happens on a clear, calm night. Under clear conditions, the ground surface radiates longwave energy to the sky,

net radiation becomes negative, and the surface cools. This means that the air near the surface will also be cooled, so as you move away from the cool surface, the air becomes warmer.

This situation is illustrated in a graph of temperature with height (Figure 3.10). The straight, slanting line of the normal environmental lapse rate bends to the left in a "J" hook. In this example, the air temperature at the surface, point A, has dropped to −1°C (30°F). This value is the same as at point B, some 750 m (about 2500 ft) aloft. As we move up from ground level, temperatures become warmer up to about 300 m (about 1000 ft). Here the curve reverses itself, and the normal lapse rate takes over. The lower portion of the lapse rate curve is called a low-level **temperature inversion**. Here, the normal cooling trend is reversed and temperatures increase with height.

In a temperature inversion, air temperature increases with altitude.

In the case shown, the temperature of the lowermost air has fallen below the freezing point, 0°C (32°F). For

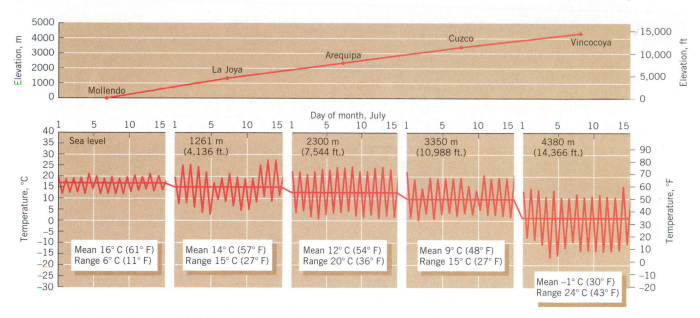

Figure 3.9 Daily maximum and minimum air temperatures for mountain stations in Peru, lat. 15° S. All data cover the same 15-day observation period in July. As elevation increases, the mean daily temperature decreases and the temperature range increases. (Data from Mark Jefferson.)

sensitive plants during the growing season, this temperature condition is called a killing frost (even though actual frost may not form). Perhaps you have seen news reports about killing frosts damaging fruit trees or crops in Florida or California. Growers commonly use several methods to reduce the inversion. Most commonly, large fans are used to mix the cool air at the surface with the warmer air above (Figure 3.11).

Where air pollution is not a problem, oil-burning heaters are sometimes used to warm the surface air layer and create air circulation.

Low-level temperature inversions often occur over snow-covered surfaces in winter. Inversions of this type are very intense and can extend thousands of meters into the air. They build up over many long nights in arctic and polar regions, where the solar heat of the short

Figure 3.10 A low-level temperature inversion with frost. Instead of temperature decreasing with elevation (dashed line) from 4°C (39°F), the surface temperature is at −1°C (30°F) and temperature increases with elevation (solid line) for several hundred meters (1000 ft or so) above the ground. Since the surface is below the freezing point, frost forms.

Figure 3.11 This large fan is used to mix the air at low levels in a navel orange orchard to protect blossoms and fruit from freezing.

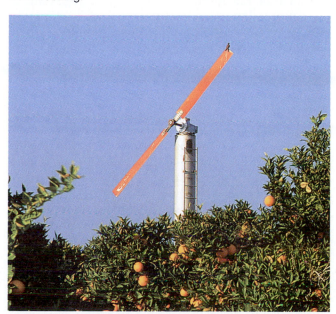

winter day cannot completely compensate for nighttime cooling. Inversions can also result when a warm air layer overlies a colder one. This type of inversion is often found along the west coasts of major continents, and we will discuss it in more detail in Chapter 4.

THE ANNUAL CYCLE OF AIR TEMPERATURE

As the earth revolves around the sun, the tilt of the earth's axis causes an annual cycle of variation in insolation. This cycle produces an annual cycle of net radiation, which, in turn, causes an annual cycle to occur in mean monthly air temperatures. Although the annual cycle of net radiation is the most important factor in determining the annual temperature cycle, another important consideration is location—maritime or continental. That is, places located well inland and far from oceans generally experience a stronger temperature contrast from winter to summer. We will begin our study of the annual temperature cycle, however, with the relationship between net radiation and temperature for four examples, ranging from the equator almost to the arctic circle.

> **The annual cycle of net radiation, which results from the variation of insolation with the seasons, drives the annual cycle of air temperatures.**

Net Radiation and Temperature

Graph (*a*) of Figure 3.12 shows the yearly cycle of the net radiation rate for four stations. The average value of the net radiation flow rate for the month is plotted in units of W/m². Graph (*b*) shows mean monthly air temperatures for these same stations. We will compare the net radiation graph with the air temperature graph for each station, beginning with Manaus, a city on the Amazon River in Brazil.

At Manaus, located nearly on the equator, the average net radiation rate is strongly positive in every month. However, there are two minor peaks. These coincide approximately with the equinoxes, when the sun is nearly straight overhead. A look at the temperature graph of Manaus shows uniform air temperatures, averaging about 27°C (81°F) for the year. The annual temperature range, or difference between the highest and lowest mean monthly temperature, is only 1.7°C (3°F). In other words, near the equator the temperature is similar each month. There are no temperature seasons.

We go next to Aswan, Egypt, a very dry desert location on the Nile River at lat. 24° N. The positive net radiation rate curve shows that a large radiation surplus

Place	Lat	Annual avg W/m²
Yakutsk	62°N	42
Hamburg	54°N	47
Aswan	24°N	87
Manaus	3°S	98

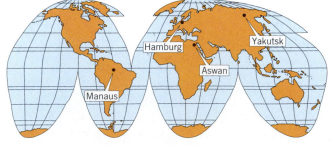

Figure 3.12 Net radiation (*a*) and temperature (*b*) cycles for four stations ranging from the equatorial zone to the arctic zone. Aswan (Egypt), Hamburg (Germany), and Yakutsk (Siberia) all show strong cycles with summer maxima and winter minima. In contrast is Manaus (Brazil), very near the equator, which shows two net radiation peaks and a nearly uniform temperature. (Data courtesy of David H. Miller.)

exists for every month. Furthermore, the net radiation rate curve has a much stronger annual cycle, with values for June and July that are almost double those of December and January. The temperature graph shows a corresponding annual cycle, with an annual range of about 17°C (30°F). June, July, and August are terribly hot, averaging over 32°C (90°F).

Moving farther north, we come to Hamburg, Germany, lat. 54° N. The net radiation rate cycle here is also strongly developed. The rate is positive for nine months, providing a radiation surplus. When the rate becomes negative for three winter months, a deficit is produced. The temperature cycle reflects the reduced total insolation at this latitude. Summer months reach a maximum of just over 16°C (60°F), while winter months reach a minimum of just about freezing (0°C or 32°F). The annual range is about 17°C (30°F), the same as at Aswan.

Finally, we travel to Yakutsk, Siberia, lat. 62° N. During the long, dark winters, the net radiation rate is negative, and there is a radiation deficit that lasts about six months. During this time, air temperatures drop to extremely low levels. For three of the winter months, monthly mean temperatures are between −35°C and −45°C (about −30°F and −50°F). This is actually one of the coldest places on earth. In summer, when daylight lasts most of a 24-hour day, the net radiation rate rises to a strong peak. In fact, this peak value is higher than those of the other three stations. As a result, air temperatures show a phenomenal spring rise to summer-month values of over 13°C (55°F). Because of Yakutsk's high latitude and continental interior location, its annual temperature range is enormous—over 60°C (108°F).

Land and Water Contrasts

Have you ever visited San Francisco? If so, you probably noticed that this magnificent city has quite a unique climate. Fog is frequent, and cool, damp weather prevails for most of the year (see Figure 4.13). In fact, Mark Twain is reputed to have said, "The coldest winter I ever spent in my life was summer in San Francisco!" The cool climate is due to its location—on the tip of a peninsula, with the Pacific Ocean on one side and San Francisco Bay on the other. Ocean and bay water temperatures are quite cool, since a southward-flowing current sweeps cold water from Alaska down along the northern California coast. Winds from the west move cool, moist ocean air, as well as clouds and fog, across the peninsula, keeping summer air temperatures low and winter temperatures above freezing. Figure 3.13 shows a typical record of temperatures for San Francisco for a week in the summer. Temperatures hover around 13°C (55°F) and change only a little from day to night.

In contrast is a location far from the water, like Yuma, Arizona, also shown in Figure 3.13. Located in the

Figure 3.13   A recording thermometer made these continuous records of the rise and fall of air temperature over a period of a week in summer at San Francisco, California, and at Yuma, Arizona. The daily cycle is strongly developed at Yuma, a station in the desert. In contrast, the graph for San Francisco, on the Pacific Ocean, shows a very weak daily cycle.

Sonoran Desert, air temperatures here are much warmer on the average—about 28°C (82°F). Clearly, no ocean cooling is felt in Yuma! The daily range is also much greater, nearly 20°C (36°F). Hot desert days become cool desert nights, with the clear, dry air allowing the ground to lose heat rapidly.

> **Land–water contrasts keep air temperatures at coastal locations more constant than at interior continental locations.**

It seems only natural that coastal locations should be cooler in summer and warmer in winter, and have a more limited day-to-night temperature range, than stations in the interior of continents. But did you ever wonder why? The important principle is this: the surface of any extensive, deep body of water heats more slowly and cools more slowly than the surface of a large body of land when both are subjected to the same intensity of insolation. Because of this principle, daily and annual air temperature cycles will be quite different at coastal locations than at interior locations.

Four important thermal differences between land and water surfaces account for the land–water contrast (see Figure 3.14). The most important difference is that much of the downwelling solar radiation penetrates water, distributing the absorbed heat throughout a substantial water layer. In contrast, solar radiation does not penetrate soil or rock, so its heating effect is concentrated at the surface. The radiation therefore warms a thick water surface layer only slightly, while a thin land surface layer is warmed more intensely.

Figure 3.14 Four differences that illustrate why a land surface heats more rapidly and more intensely than the surface of a deep water body. Locations near the ocean have more uniform air temperatures—cooler in summer and warmer in winter.

Oceans heat and cool more slowly than continents.

A second thermal factor is that water is slower to heat than dry soil or rock. The *specific heat* of a substance describes how the temperature of a substance changes with a given input of heat. Consider an experiment using a small volume of water and the same volume of rock. Let's suppose we warm each volume by the same amount—perhaps 1°C (1.8°F). If we measure the amount of heat required, we will find that it takes about five times as much heat to raise the temperature of the water one degree as it does to raise the temperature of the rock one degree. That is, the specific heat of water is about five times greater than that of rock. The same will be true for cooling—after losing the same amount of heat, the temperature of water falls less than the temperature of rock.

A third difference between land and water surfaces is related to mixing. That is, a warm water surface layer can mix with cooler water below, producing a more uniform temperature throughout. For the open ocean, the mixing is produced by wind-generated waves. Clearly, no such mixing occurs on land surfaces.

A fourth thermal difference is that an open water surface can be cooled easily by evaporation. Land surfaces can also be cooled by evaporation, but only if water is present at or near the soil surface. When the surface dries, evaporation stops. In contrast, a free water surface can always provide evaporation.

Daily Temperature Cycle

What effect does the contrast in thermal characteristics of land and water have on temperatures? Let's examine the daily cycle of air temperature first. Two sets of daily air temperature curves are shown in Figure 3.15—El Paso, Texas, and North Head, Washington. Four months are presented—January, April, July, and October.

The El Paso curves show the temperature environment of an interior desert in midlatitudes. Because soil moisture content is low and vegetation is sparse, evaporation and transpiration are not important cooling effects. Cloud cover is generally light. Under these circumstances, the ground surface heats intensely during the day and cools rapidly at night. Air temperatures show an average daily range of 11 to 14°C (20 to 25°F). This type of variation represents an interior temperature environment—typical of a station located in the interior of a continent, far from the ocean's influence.

North Head is located on the Washington coast. Here, prevailing westerly winds sweep cool, moist air off the adjacent Pacific Ocean. The average daily range at North Head is a mere 3°C (5°F) or less. Persistent fogs and cloud cover also contribute to the minimal daily range. Note further that the annual range is much

Figure 3.15 The average daily cycle of air temperature for four different months shows the effect of continental and maritime location. Daily and seasonal ranges are great at El Paso, a station in the continental interior, but only weakly developed at North Head, Washington, which is on the Pacific coast. The seasonal effect on overall temperatures is stronger at El Paso.

restricted, especially when compared to El Paso. North Head exemplifies a coastal temperature environment, typical of a station located in the path of oceanic air.

Annual Temperature Cycle

Let's now turn to the effects of land–water surface contrasts on the temperature cycle for the year. We have already noted for El Paso and North Head that the temperature cycle for the four months plotted shows a greater range for the interior station than for the coastal one. Let's look in more detail at the annual cycle for another pair of stations—Winnipeg, Manitoba, located in the interior of the North American continent, and the Scilly Islands, off the southwestern tip of England, which are surrounded by the waters of the Atlantic Ocean. This time, the two stations chosen are at the same latitude, 50° N. As a result, they have the same insolation cycle and receive the same potential amount of solar energy for surface warming. Figure 3.16 shows their annual cycles of temperature as well as the insolation curve common to both.

Monthly temperature maximums and minimums occur later at coastal stations than at interior stations.

The temperature graphs for the Scilly Islands and Winnipeg confirm the effects we have already noted for North Head and El Paso—that the annual range in temperature is much larger for the interior station (39°C, 70°F) than for the coastal station (8°C, 14°F). Note that the nearby ocean waters keep the air temperature at the Scilly Islands well above freezing in the winter, while January temperatures at Winnipeg fall to near −20°C (−4°F)!

Another important effect concerns the timing of maximum and minimum temperatures. Insolation reaches a maximum at summer solstice, but it is still strong for a long period afterward. This means that heat energy continues to flow into the ground well after the solstice. Therefore, the hottest month of the year for interior regions is July, the month following the solstice. Similarly, the coldest month of the year for large land areas is January, the month after the winter solstice. This is because the ground continues to lose heat even after insolation begins to increase.

Over the oceans and at coastal locations, maximum and minimum air temperatures are reached a month later than on land—in August and February, respectively. Because water bodies heat or cool more slowly than land areas, the air temperature changes more slowly. This effect is clearly shown in the Scilly Islands graph, where February is slightly colder than January.

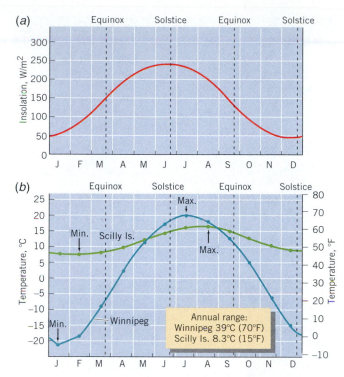

Figure 3.16 Annual cycles of insolation (*a*) and monthly mean air temperature (*b*) for two stations at lat. 50° N: Winnipeg, Canada, and Scilly Islands, England. Insolation is identical for the two stations. Winnipeg temperatures clearly show the large annual range and earlier maximum and minimum that are characteristic of its continental location. Scilly Islands temperatures show its maritime location in the small annual range and delayed maximum and minimum.

WORLD PATTERNS OF AIR TEMPERATURE

Thus far, we have learned some important principles about air temperatures in this chapter. Some of these principles are local—such as the effect of urban and rural surfaces on temperatures. Still others are global—such as the effect of latitude or elevation on temperature or coastal-interior location. We now turn to world temperature patterns, tying our discussion closely to these principles. First, however, we will need a quick explanation of air temperature maps and their meaning.

The distribution of air temperatures is often shown on a map by **isotherms**—lines drawn to connect locations having the same temperature. Figure 3.17 shows a map on which the observed air temperatures have been recorded and placed at their proper location on the map. These may be single readings, such as a daily maximum or minimum, or they may be averages of many years of records for a particular day or month of a year, depending on the purposes of the map. The isotherms are constructed by drawing smooth lines through and among the points in a way that best indicates a uniform temperature, given the observations at hand. Usually,

Figure 3.17 Isotherms are used to make temperature maps. Each line connects points having the same temperature. Where temperature changes along one direction, a temperature gradient exists. Where isotherms close in a tight circle, a center exists. This example shows a center of low temperature.

isotherms representing 5- or 10-degree differences are chosen, but they can be drawn at any convenient temperature interval.

Maps of isotherms show centers of high and low temperatures as well as temperature gradients.

Isothermal maps are valuable because they make the important features of the temperature pattern clearly visible. Centers of high or low temperatures are clearly outlined. Also visible are *temperature gradients*—directions along which temperature changes. Centers and gradients give the broad pattern of temperature, as shown in our discussion of world patterns below.

Factors Controlling Air Temperature Patterns

World patterns of isotherms are largely explained by three factors that we have already discussed. The first of these is latitude. As latitude increases, average annual insolation decreases, and so temperatures decrease as well, making the poles colder than the equator. The effect of latitude on seasonal variation is also important. For example, more solar energy is received at the poles at the summer solstice than at the equator. We must therefore consider the time of year as well as the latitude in understanding world temperature patterns.

Global air temperature patterns are controlled primarily by latitude, coastal-interior location, and elevation.

The second factor is that of coastal-interior contrasts. As we've noted, coastal stations that receive marine air from prevailing winds have more uniform temperatures—cooler in summer and warmer in winter. Interior stations, on the other hand, show a much larger annual variation in temperature. Ocean currents, discussed further in Chapter 5, can also have an effect. By keeping coastal waters somewhat warmer or cooler than expected, temperatures at maritime stations will be influenced in a similar manner.

Elevation is the third important factor. At higher elevations, temperatures will be cooler. Therefore, we expect world temperature maps to show the presence of mountain ranges, which will be cooler than surrounding regions.

World Air Temperature Patterns for January and July

With these factors in mind, let's look at some world temperature maps in more detail. Figure 3.18 consists of maps of world temperatures for two months—January and July. The Mercator projections show temperature trends from the equator to the midlatitude zones, and the polar projections give the best picture for high latitudes. From the maps, we can make six important points about the temperature patterns and the factors that produce them. Be sure to follow along by examining the maps carefully.

1. **Temperatures decrease from the equator to the poles.** Annual insolation decreases from the equator to the poles, thus causing temperatures to decrease. This temperature gradient is most clearly seen in the polar maps for the southern hemisphere in January and July. On these maps, the isotherms are nearly circular, decreasing to a center of low temperature on Antarctica near the south pole. The center is much colder in July, when most of the polar region is in perpetual night. We can also see this same general trend in the north polar maps, but the continents complicate the pattern. The general temperature gradient from the equator poleward is also evident on the Mercator maps.

2. **Large landmasses located in the subarctic and arctic zones develop centers of extremely low temperatures in winter.** The two large landmasses we have in mind are North America and Eurasia. The January north polar map shows these low-temperature centers very well. The cold center in Siberia, reaching −50°C (−58°F), is strong and well defined. The cold center over northern Canada is also quite cold (−35°C, −32°F) but is not as well defined. Both features are visible on

JANUARY

JULY

Figure 3.18 Mean monthly air temperatures (°C) for January and July, Mercator and polar projections. (Compiled by John E. Oliver.) (continued overleaf)

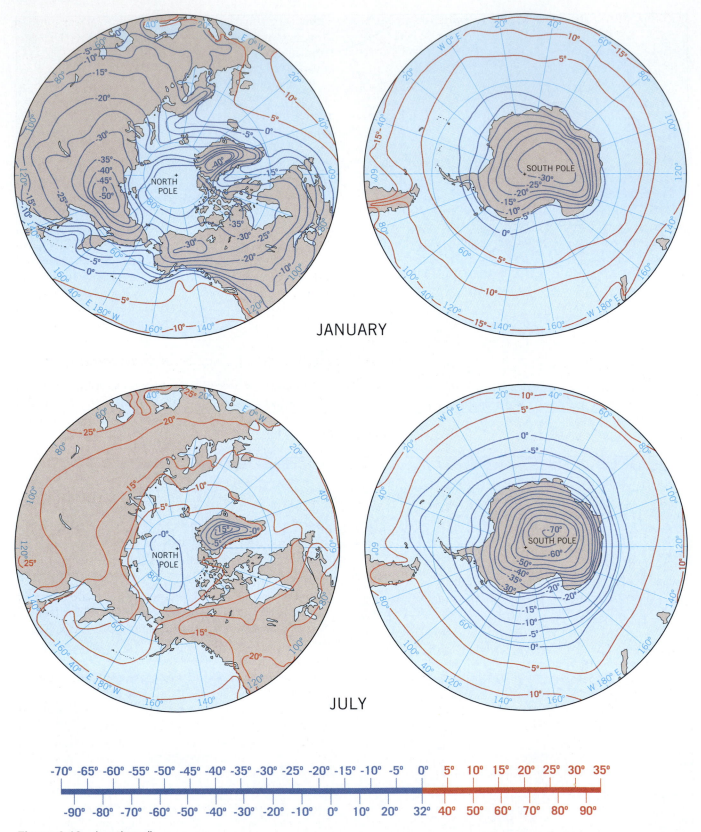

JANUARY

JULY

Figure 3.18   (continued)

the January Mercator map. Greenland shows a low-temperature center as well, but it has a high dome of glacial ice, as discussed in point 6. An important factor in keeping winter temperatures low in these regions is the

high albedo of snow cover, which reflects much of the winter insolation back to space.

3. **Temperatures in equatorial regions change little from January to July.** Note the broad space between

25°C (77°F) isotherms, which is evident on both January and July Mercator maps. In this region, the temperature is greater than 25°C (77°F) but less than 30°C (86°F). Although the two isotherms move a bit from winter to summer, the equator always falls between them. This demonstrates the uniformity of equatorial temperatures. (An exception is the northern end of the Andes Mountains in South America, where high elevations, and thus cooler temperatures, exist at the equator.) Equatorial temperatures are uniform primarily because insolation at the equator does not change greatly with the seasons.

4. **Isotherms make a large north-south shift from January to July over continents in the midlatitude and subarctic zones.** Figure 3.19 demonstrates this principle. In the winter, isotherms dip equatorward, while in the summer, they arch poleward. This effect is shown in North America and Eurasia in the January and July Mercator maps. In January the isotherms drop down over these continents, and in June they curve upward. For example, the 15°C (59°F) isotherm lies over central Florida in January. But by July this same isotherm has moved far north, cutting the southern shore of Hudson Bay and then looping far up into northwestern Canada. In contrast are the isotherms over oceans, which shift much less. This striking difference is due to the contrast between oceanic and continental surface properties, which cause continents to heat and cool more rapidly than oceans.

5. **Highlands are always colder than surrounding lowlands.** You can see this by looking at the pattern of isotherms around the Rocky Mountain chain, in western North America, on the Mercator maps. In winter, the –5°C (23°F) and –10°C (14°F) isotherms dip down around the mountains, indicating that the center of the range is colder. In summer, the 20°C (68°F) and 25°C (77°F) isotherms also dip down, showing the same

effect even though temperatures are much warmer. The Andes Mountains in South America show the effect even more strongly. The principle at work here is that temperatures decrease with an increase in elevation.

6. **Areas of perpetual ice and snow are always intensely cold.** Greenland and Antarctica contain our planet's two great ice sheets. Notice how they stand out on the polar maps as cold centers in both January and July. They are cold for two reasons. First, their surfaces are high in elevation, rising to over 3000 m (about 10,000 ft) in their centers. Second, the white snow surfaces reflect much of the insolation. Since little solar energy is absorbed, little is available to warm the snow surface and the air above it. The Arctic Ocean, bearing a cover of floating ice, also maintains its cold temperatures throughout the year. However, the cold is much less intense in January than on the Greenland Ice Sheet, since ocean water underneath the ice acts as a heat reservoir to keep the ice above from getting extremely cold.

The Annual Range of Air Temperatures

Figure 3.20 is a world Mercator map showing the annual range of air temperatures. The lines, resembling isotherms, show the difference between the January and July monthly means. We can explain the features of this map by using the same effects of latitude, interior-maritime location, and elevation.

> **The global pattern of annual temperature range shows the smallest range for tropical oceans and the largest for northern hemisphere continental interiors.**

1. **The annual range increases with latitude, especially over northern hemisphere continents.** This trend is most clearly shown for North America and Asia. This is due to the contrast between summer and winter insolation, which increases with latitude.

2. **The greatest ranges occur in the subarctic and arctic zones of Asia and North America.** The map clearly shows two very strong centers of large annual range—one in northeast Siberia and the other in northwest Canada–eastern Alaska. In these regions, summer insolation is nearly the same as at the equator, while winter insolation is very low.

3. **The annual range is moderately large on land areas in the tropical zone, near the tropics of cancer and capricorn.** These are regions of large deserts—North Africa (Sahara), southern Africa (Kalahari), and Australia (interior desert) are examples. Dry air and the absence of clouds and moisture allow these continental locations to cool strongly in winter and warm strongly in summer, even though insolation contrasts with the season are not as great as at higher latitudes.

Figure 3.19 The seasonal migration of isotherms is much greater over continents than over oceans. This difference occurs because oceans heat and cool much more slowly than continents.

Figure 3.20   Annual range of air temperature in Celsius degrees. Data show differences between January and July means. The inset box shows the conversion of Celsius degrees to Fahrenheit for each isotherm value. (Compiled by John E. Oliver.)

4. **The annual range over oceans is less than that over land at the same latitude.** This can be clearly seen by following a parallel of latitude—40° N, for example. Starting from the right, we see that the range is between 5 and 10°C (9 and 18°F) over the Atlantic but increases to about 30°C (54°F) in the interior of North America. In the Pacific, the range falls to 5°C (9°F) just off the California coast and increases to 15°C (27°F) near Japan. In Central Asia, the range is near 35°C (63°F). Again, these major differences are due to the contrast between land and water surfaces. Since water heats and cools much more slowly than land, a narrower range of temperatures is experienced.

5. **The annual range is very small over oceans in the tropical zone.** As shown on the map, the range is less than 3°C (5°F) since insolation varies little with the seasons near the equator and water heats and cools slowly.

GLOBAL WARMING AND THE GREENHOUSE EFFECT

Tune in to the evening news, and one of the recurring scientific subjects you will hear about is global warming. Is our climate changing? Many scientists have concluded that the temperature of the planet is warming signif-

icantly because of human activities. Other scientists argue that the warming of the last few years is part of a natural global cycle.

You may also hear that carbon dioxide (CO_2), produced by human activities, is a major cause of concern in climate warming. This gas is released to the atmosphere in large quantities by fossil fuel burning. Why is this a problem? Recall that the greenhouse effect is caused by atmospheric absorption of longwave radiation, largely by carbon dioxide and water vapor, that is emitted from the earth. Because some of this absorbed energy is counterradiated back from the atmosphere, the earth's surface is warmed by an extra amount.

Because human activities now continuously increase the amount of carbon dioxide in the air, many scientists are concerned that the greenhouse effect is escalating and that global temperatures are rising—or will rise soon—in response. (See, *Eye on the Environment • Carbon Dioxide—On the Increase*, in this chapter.) Also of concern are other gases that are normally present in very small concentrations—methane (CH_4), nitrous oxide (NO), ozone (O_3), and the chlorofluorocarbons. (See *Eye on the Environment • The Ozone Layer—Shield to Life*, in Chapter 2.) These further enhance the greenhouse effect, even though they are even less abundant than CO_2. Taken together with CO_2, they are referred to as **greenhouse gases**.

The Temperature Record

Has the buildup of greenhouse gases caused global temperatures to rise? If not, will this buildup warm the earth in the near future? There seems to be no doubt that an increased concentration of greenhouse gases can lead to global warming. The question seems to be whether or not warming will actually take place, given the cyclic nature of climate. If so, we also wish to know how soon it will occur and how great the warming will be.

Before addressing this important question, we need to review the record of the earth's surface temperature over the last few centuries. Figure 3.21 shows the earth's mean annual surface temperature from 1880 to 1996. The temperature is expressed as a difference from the average annual temperature for the period 1951–1980. Two curves are shown—yearly data and smoothed data, taken from a five-year moving average. Although temperature has increased since the last century, and especially toward the end, there have been wide swings in the mean annual surface temperature. **SCIENCE NEWS** 4, 5

Several theories have been proposed to explain the variations. For example, some measures of solar activity seem to suggest that varying solar output induces similarly varying temperatures. If this theory is correct, then the present warming may not be the result of greenhouse gas accumulation. Instead, solar cycles may be responsible.

Another factor to consider is volcanic activity. Dust and gases—especially sulfur dioxide, SO_2—are emitted in major eruptions and are typically carried upward into the stratosphere. SO_2 molecules rapidly take up water and form tiny sulfate particles. Strong stratospheric winds spread the volcanic particles quickly throughout the entire atmospheric layer. These particles have a cooling effect on the troposphere because they block incoming solar radiation.

For example, the eruption of Mount Pinatubo in the Philippines in the spring of 1991 lofted 15 to 20 million tons of sulfuric acid particles into the stratosphere. The layer of particles produced by the eruption reduced solar radiation reaching the earth's surface between 2 and 3 percent for the year or so following the blast. In response, global temperatures fell 0.5 to 0.7°C (0.9 to 1.3°F). Thus, volcanic activity may offset warming of the earth's surface from other causes.

Although the period of direct air temperature measurement does not extend past the middle of the last century, the record can be extended further back by using tree-ring analysis. The principle is simple. Each year, trees grow in diameter, and in climates where the seasons are distinct, this growth can appear as an annual ring. If growing conditions are good, the annual ring is wide. If poor, the annual ring is narrow. For trees along the timberline in North America, the width is related to temperature—the trees grow better when temperatures are warmer. Since only one ring is formed each year, the date of each ring is easy to determine by counting backward from the present. Because the trees are quite old, the temperature record may be extended backward several centuries. **SCIENCE NEWS** 6

Figure 3.22 shows a reconstruction of northern hemisphere temperatures from 1700 to about 1975 using tree-ring analysis. Like Figure 3.21, it is expressed as a difference from a recent average (1950–1965). From 1880 to present, the temperatures reconstructed from tree-ring analysis seem to fit the observed temperatures (Figure 3.21) quite well. Analysis of an earlier period shows us another cycle of temperature increases and declines. The low point, around 1840, marks a historic cold event during which European alpine glaciers became more active and advanced. Other evidence indicates that the two cycles in Figure 3.23 are part of a natural global cycle of temperature warming and cooling lasting about 150 to 200 years. These cycles have occurred regularly over the last thousand years. What causes these cycles? Again, a number of theories have been offered, but no consensus has yet been reached as to the actual cause or causes.

Future Scenarios

The year 1998 was the warmest year on record, with an average temperature of 14.6°C (58.2°F) compiled by NASA scientists at the Goddard Institute of Space Science. It was also the warmest year of the millennium, according to reconstructions of past temperatures using tree rings and glacial ice cores by University of Massachusetts scientists. In fact, 1990, 1995, 1997, and 1998 ranked as the warmest years since 1400.

Is this recent warming an effect of human activity? In 1995 the Intergovernmental Panel on Climate Change, a United Nations-sponsored group of more than 2000 scientists, concluded for the first time that human activ-

Figure 3.21   Mean annual surface temperature of the earth, 1856–1998. The vertical scale shows departures in degrees from a zero line of reference representing the average for the years 1951–1980. The yellow line shows the mean for each year. The red line shows a running five-year average. Note the effect of Mount Pinatubo in 1992–1993. (James Hansen/NASA Goddard Institute for Space Studies.)

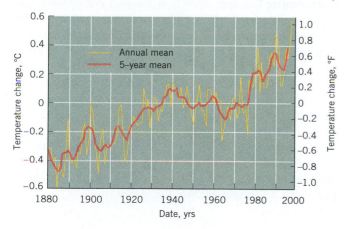

EYE ON THE ENVIRONMENT

Carbon Dioxide—On the Increase

In the centuries before global industrialization, carbon dioxide concentration in the atmosphere was at a level slightly less than 300 parts per million (ppm), or about 3/100ths of a percent by volume. During the last hundred years or so, that amount has been substantially increased. Why? The answer is fossil fuel burning. When fuels like coal, oil, or natural gas are burned, they yield water vapor and carbon dioxide. The release of water vapor does not present a problem, because a large amount of water vapor is normally present in the global atmosphere. But because the normal amount of CO_2 is so small, fossil fuel burning has raised the level to about 360 ppm. According to studies of bubbles of atmospheric gases trapped in galcial ice, this level is the highest attained in the last 420,000 years and nearly double the amount of CO_2 present during glaciations of the most recent Ice Age.

The figure at right shows how CO_2 has increased with time since 1860.

Until 1940 or so, the level remained nearly stable. But after 1940 CO_2 began a rapid rise and today, CO_2 levels are still rising. Even with future concerted global action to reduce CO_2 emissions, scientists estimate that levels will stabilize at a value not lower than about 550 ppm by late in the twenty-first century. This will nearly double preindustrial levels.

Predicting the buildup of CO_2 is difficult because not all the carbon dioxide emitted into the air by fossil fuel burning remains there. Instead, a complex cycle moves CO_2 throughout the life layer. Photosynthesis is one part of this *carbon cycle*. Plants use CO_2 in photosynthesis, the process by which they use light energy to build their tissues. So, plants take in carbon dioxide, removing it from the atmosphere. When plants die, their remains are digested by decomposing organisms, which release CO_2. This process returns CO_2 to the atmosphere. Normally, these processes are in balance.

Human activity has tilted this balance by clearing land and burning the vegetation cover as new areas of forest are opened for development. This practice increases the amount of CO_2 in the air. When agricultural land is allowed to return to its natural forest state, CO_2 is removed from the air by growing trees. At present, scientists calculate that forests are growing more rapidly than they are being destroyed in the midlatitude regions of the norther hemisphere. This plant growth helps to counteract the buildup of CO_2 produced by fossil fuel burning. But it may be outweighed by the tropical deforestation that is taking place in South America, Africa, and Asia. (See *Eye on the Environment • Exploitation of the Low-Latitude Rainforest Ecosystem*, Chapter 8.)

Laboratory and field experiments have shown that plants exposed to increased concentrations of CO_2 will grow faster and better. The faster they grow, the more CO_2 they can take in, which helps to reduce the amount in the atmosphere. However, scientists

ity has caused climatic warming. This judgment was based largely on computer simulations of global climate that account for the release of CO_2 and SO_2 from fossil fuel burning occurring since the turn of the century. The simulations agreed well with the patterns of warming observed over that period, leading to the conclusion that, in spite of natural variability, human influence has been felt in the climate record of the twentieth century. Since 1995, a number of new studies have increased confidence in this conclusion.

Given that CO_2 and other greenhouse gases are warming the planet, what will be the effect? Using various projections of the continuing release of greenhouse gases coupled with computer climate models, this same group of scientists projected that global temperatures will warm between 1°C (1.8°F) and 3.5°C (6.3°F) by the year 2100, with the most likely value of 2°C (3.6°F).

Why does this temperature rise cause concern? The problem is that many other changes may accompany a rise in temperature. One of these is a rise in sea level, as glaciers and sea ice melt in response to the warming. Current predictions call for a rise of 15 to 90 cm (5.9 to 35.4 in.) in sea level by the year 2100, with a best estimate of 48 cm (18.9 in.). This would place as many as 92 million people within the risk of annual flooding. Climate change could also promote the spread of insect-borne diseases such as malaria. Furthermore, climate boundaries may shift their positions, making some regions wetter while others become drier. Thus, agricultural patterns could shift, displacing large human populations as well as natural ecosystems.

A recently discovered effect of climatic warming is the enhancement of variability in climate. Events such as very high 24-hour precipitation—extreme snowstorms, rainstorms, sleet and ice storms, for example—appear to be occurring more frequently since 1980. More frequent and more intense spells of hot and cold weather may also be related to climatic warming. **SCIENCE NEWS** 7, 8

Increase in atmospheric carbon dioxide, observed to 1990 and predicted into the twenty-first century.

are unsure whether increased CO_2 will stimulate plant growth under natural conditions.

Another part of the cycle involves the oceans. The ocean's surface layer contains microscopic plant life that takes in carbon dioxide. The CO_2 in the ocean water initially comes from the atmosphere and is mixed into the ocean by surface waves. When these microscopic floating plants die, their bodies sink to the ocean bottom. There they decompose and release CO_2, enriching waters near the ocean floor.

This CO_2 eventually returns to the surface through a system of global ocean current flows in which CO_2-poor waters sink in the northernmost Atlantic and CO_2-rich bottom waters rise to the surface in the northernmost Pacific. In fact, the ocean acts like a slow conveyor belt, moving CO_2 from the surface to ocean depths and releasing it again in a cycle lasting about 1500 years.

At present, scientists estimate that ocean surface waters absorb more CO_2 than they release, owing to increased atmospheric levels of CO_2. Therefore, carbon dioxide may be accumulating in ocean depths. However, current studies of global climate computer simulations indicate that the oceans may not be as effective in removing excess CO_2 as scientists previously thought.

Although there is a great deal of uncertainty about the movements and buildup rate of excess CO_2 released to the atmosphere by fossil fuel burning, one thing is certain. Without conversion to solar, wind, and hydroelectric power, fuel consumption will continue to release carbon dioxide, and it is only a matter of time before the earth will feel its impact.

Questions

1. Why has the atmospheric concentration of CO_2 increased in recent years?

2. How does plant life affect the level of atmospheric CO_2?

3. What is the role of the ocean in influencing atmospheric levels of CO_2?

The world has become widely aware of the problem of the buildup in CO_2 and other greenhouse gases. At the Rio de Janeiro Earth Summit in 1992, nearly 150 nations signed a treaty limiting emissions of greenhouse gases. Many details were ironed out at a subsequent meeting in Kyoto, Japan, in 1997. The Kyoto Protocol calls on a group of 38 industrial nations to reduce their emissions of CO_2 and other greenhouse gases to about 5 percent below 1990 levels. Nations of the European Union, the United States, and Japan are to reduce average emissions during 2008 and 2012 to levels that are 6, 7, and 8 percent lower, respectively, than 1990 levels. Achieving such reductions could have economic consequences ranging from minor to severe, depending on the analysis.

Meeting again in Buenos Aires in 1998, the participating nations set timetables for the implementation of

Figure 3.22 A reconstruction of the departures of northern hemisphere temperatures from the 1950–1965 mean, based on analyses of tree rings sampled along the northern tree limit of North America. (Courtesy of Gordon C. Jacoby of the Tree-Ring Laboratory of the Lamont–Doherty Geological Observatory of Columbia University.)

plans for greenhouse gas reduction in 1999–2000. Important breakthroughs occurred when the conference adopted a plan to allow countries to trade emission rights, as well as a "clean development" mechanism that would permit industrial nations to invest in emissions-reducing enterprises in developing countries. For the first time, some developing countries pledged specific greenhouse gas reductions of their own.

Although these efforts will slow the buildup of greenhouse gases, the ultimate solution will certainly involve greater reliance on solar, wind, and geothermal energy sources, which produce power without releasing CO_2. Energy conservation and development of new methods of utilizing fossil fuels that produce reduced CO_2 emissions will also be important. **SCIENCE NEWS** 9

A LOOK AHEAD

In this chapter we have developed an understanding of both air temperatures and temperature cycles, along with the factors that influence them—latitude, continental-maritime location, and elevation. Air temperatures are a very important part of climate, and looking ahead, we will study the climates of the earth in Chapters 7–9.

The other key ingredient of climate is precipitation—the subject covered in the next chapter. Here temperature plays a very important role, since moist air must cool before condensation forms water droplets and, eventually, precipitation.

AIR TEMPERATURE IN REVIEW

The two major cycles of air temperature—daily and annual—are controlled by the cycles of insolation produced by the rotation and revolution of the earth. These cycles induce cycles of net radiation at the surface. When net radiation in the daily cycle is positive, air temperatures increase. When negative, air temperatures decrease. This principle applies for both daily and annual temperature cycles. Days are warm and nights are cool as net radiation goes from positive to negative. Summers are warm and winters are cold because average net radiation is high in the summer and low in the winter. Surface characteristics affect temperatures, too. Rural surfaces are generally moist and slow to heat, while urban surfaces are dry and absorb heat readily. This difference creates an urban heat island effect.

Air temperatures normally fall with altitude in the troposphere. At the tropopause, this decrease stops. In the stratosphere above, temperatures increase slightly with altitude. Air temperatures observed at mountain locations are lower with higher elevation, and day–night temperature differences increase with elevation.

Daily and annual air temperature cycles are influenced by maritime or continental location. Ocean temperatures vary less than land temperatures because water heats more slowly and can both mix and evaporate freely. Maritime locations that receive oceanic air therefore show smaller ranges of daily and annual temperature.

Global temperature patterns for January and July show the effects of latitude and maritime–continental location. Equatorial temperatures vary little from season to season. Poleward, temperatures decrease with latitude, and continental surfaces at high latitudes can become very cold in winter. Isotherms over continents swing widely north and south with the seasons, while isotherms over oceans move through a much smaller range of latitude.

Our planet's global temperature changes from year to year. Within the last few decades, global temperatures have been increasing. Most scientists agree that the human-induced buildup of greenhouse gases has begun to affect global climate. However, natural cycles, such as variations in the sun's output, still provide strong influences. If we continue to release large quantities of greenhouse gases at increasing rates, we can expect a significant rise in global temperatures that will be accompanied by shifts of climate zones and increasing sea level.

KEY TERMS

conduction	heat island	stratosphere
evaporation	lapse rate	temperature inversion
convection	environmental tempera-	isotherms
air temperature	ture lapse rate	greenhouse gases
transpiration	troposphere	

REVIEW QUESTIONS

1. How are mean daily temperature and mean monthly temperature determined?
2. How does the daily temperature cycle measured within a few centimeters or inches of the surface differ from the cycle at normal air temperature measurement height?
3. Compare the characteristics of urban and rural surfaces and describe how the differences affect urban and rural air temperatures. Include a discussion of the urban heat island.
4. What are the two layers of the lower atmosphere? How are they distinguished?
5. How and why are the temperature cycles of high mountain stations different from those of lower elevations?
6. Why do large water bodies heat and cool more slowly than land masses? What effect does this have on daily and annual temperature cycles for coastal and interior stations?
7. What three factors are most important in explaining the world pattern of isotherms? Explain how and why each factor is important, and what effect it has.
8. Turn to the January and July world temperature maps shown in Figure 3.18. Make six important observations about the patterns and explain why each occurs.
9. Turn to the world map of annual temperature range in Figure 3.20. What five important observations can you make about the annual temperature range patterns? Explain each.
10. Describe how global air temperatures have changed in the recent past. Identify some factors or processes that influence global air temperatures on this time scale.

VISUALIZING EXERCISES

1. Sketch graphs showing how insolation, net radiation, and temperature might vary from midnight to midnight during a 24-hour cycle at a midlatitude station such as Chicago.
2. Sketch a graph of air temperature with height showing a low-level temperature inversion. Where and when is such an inversion likely to occur?

A DEEPER VIEW OF AIR TEMPERATURE

1. Portland, Oregon, on the north Pacific coast, and Minneapolis, Minnesota, in the interior of the North American continent, are at about the same latitude. Sketch the annual temperature cycle you would expect for each location. How do they differ and why? Select one season, summer or winter, and sketch a daily temperature cycle for each location. Again, describe how they differ and why.
2. Many scientists have concluded that human activities are acting to raise global temperatures. What human processes are involved? How do they relate to natural processes? Are global temperatures increasing now? What other effects could be influencing global temperatures? What are the consequences of global warming?

Chapter 4

Atmospheric Moisture and Precipitation

Lightning heralds the approach of a summer thundershower.

Suddenly you notice that the thick, humid air is perfectly still. The songbirds and insects are strangely silent. Not a single leaf trembles in the huge, ancient oak shading your lawn chair. You look up from the book you've been reading and notice that clouds have quietly slipped in overhead, interrupting the flow of afternoon heat to the grassy field around you. Seen against the backdrop of graying clouds, the rambling old farmhouse where you are spending the holiday weekend with friends looks pale and somehow vacant of life.

A low, distant rumble shatters the stillness. A thunderstorm is approaching! Tucking your book under your arm, you fold up your lawn chair and head for the house. As you walk through the field, you begin to hear a rushing sound. Looking over your shoulder, you see the tops of the forest trees at the edge of the field shaking and swaying in a sudden wind. Then, the top of the old oak twists violently, its upper branches jerking chaotically as it is battered by strong gusts. A flood of cool air sweeps over you, and you move more quickly toward the farmhouse.

The first large, heavy drops begin to fall as you reach the wide front porch. The rushing sound deepens as the first wave of heavy rainfall approaches. The sky explodes with light, and a loud crack of thunder startles you. The rain now drives intensely toward the ground,

banging on the tin roof of the porch and swirling onto the porch in the strong gusts. The oak in the middle of the field is barely visible through the falling torrent.

A crackle pierces the din, and a blinding flash prints an image of the ancient oak onto your eyes. At the same moment, a deafening bang shatters the air. Lightning has struck the old oak! A huge branch crashes to the ground, the sound of its fall muffled by the wind and rain. That was too close for comfort! You retreat from the porch for the safety of the house to watch the storm from a window.

Later, you survey the damage. The top of the tree lies on the ground, its thick limbs shattered by the force of the lightning strike. The stout trunk is split almost to the base. Huge splinters are strewn on the surrounding ground. The smell of fresh-split green oak is pungent in the humid air. "Good thing I didn't decide to wait out the storm under the tree!" you think to yourself.

Not all precipitation events are this dramatic. Many involve nothing more

than a brief shower or a rainy day. But have you ever wondered why it rains? Or how a powerful thunderstorm is formed? These are the questions answered in this chapter.

Precipitation is the fall of liquid or solid water from the atmosphere to reach the earth's surface. It is formed when moist air is cooled, causing water vapor to form liquid droplets or solid ice particles.

What causes the cooling of moist air? The answer is upward motion. Whenever air moves upward in the atmosphere, it is cooled because of a simple physical principle stating that when air expands, it cools. And, because atmospheric pressure decreases with altitude, air moving upward expands and thus cools.

What causes air to move upward? There are three major causes of upward movement of air. One occurs when winds move air up and over a mountain barrier. The second takes place when unequal heating at the ground surface creates a bubble of warm air that is less dense and is buoyed upward. This chapter focuses on these two causes of uplift. The third occurs when a mass of cooler, denser air slides under a mass of warmer, lighter air, lifting the warmer air aloft. This type of upward movement occurs in weather systems, which we will cover in Chapter 6.

Chapter Outline

Water—The Global Perspective
The Hydrosphere and the Hydrologic Cycle
The Global Water Balance

Humidity
Specific Humidity
Relative Humidity

The Adiabatic Process
Dry Adiabatic Rate
Wet Adiabatic Rate

Clouds
Cloud Forms
Fog

Precipitation
Precipitation Processes
Orographic Precipitation
Convectional Precipitation
Thunderstorms

Eye on the Environment • Air Pollution
Smog and Haze
Fallout and Washout
Inversion and Smog
Climatic Effects of Urban Air Pollution
Acid Deposition
Air Pollution Control

ATMOSPHERIC MOISTURE AND PRECIPITATION

Water exists in the air in the form of humidity, clouds, fog, and precipitation. Humidity is the dampness in the air that can be so uncomfortable on hot, humid days. Clouds are visible as fleecy white or ominous gray masses in the sky overhead. Fog is a dense haze that surrounds you like a cold, wet blanket. And precipitation consists of water droplets or snowflakes that fall from the sky and reach the earth, providing our source of fresh water. How are humidity, clouds, fog, and precipitation related? How does water vapor, an invisible gas, become transformed into a torrent of rain that falls to earth? This subject is the main focus of this chapter.

Our chapter also includes a special feature: *Eye on the Environment • Air Pollution.* This environmental topic is often included in physical geography courses, as it provides a practical focus for much of your learning about the atmosphere. We present this special multipage feature at the close of our chapter.

Before we begin our study of atmospheric moisture and precipitation, we will briefly review the three states of water and the conversion of one state to another. As shown in Figure 4.1, water exists in three states—solid (ice), liquid (water), and gas (water vapor). A change of

Figure 4.1 A schematic diagram of the three states of water. Arrows show the ways that any one state can change into either of the other two states. Heat energy is absorbed or released, depending on the direction of change.

state from solid to liquid, liquid to gas, or solid to gas requires the input of heat energy. As we noted in Chapter 2, this energy is *latent heat*, which is drawn in from the surroundings. When the change goes the other way, from liquid to solid, gas to liquid, or gas to solid, this latent heat is released to the surroundings.

For each type of transition, there is a specific name, shown in the figure. Melting, freezing, evaporation, and condensation are all familiar terms. **Sublimation** describes the direct transition from solid to vapor. Perhaps you have noticed that old ice cubes in your refrigerator's freezer seem to shrink away from the sides of the ice cube tray and even get smaller over time. In this case, they are shrinking through sublimation, which is induced by the constant circulation of cold, dry air through the freezer. The ice cubes never melt, yet they lose bulk directly as vapor. **Deposition** is the reverse process, when water vapor crystallizes directly as ice.* Frost formed on a cold winter night is an example.

WATER—THE GLOBAL PERSPECTIVE

Water plays several key roles on our planet. First, the oceans cover more than two-thirds of the planet's surface. They act as a huge heat storage reservoir and redistribute heat from low to high latitudes by ocean currents. The oceans also act as a storage reservoir for dissolved compounds, ranging from salts to nutrients. Second, water falls on land in the form of rain or snow. As it runs off to the sea, water erodes rocks and soils and creates landscapes and landforms. This flow moves nutrients from one location to another, which also influences the distribution of plant and animal life. Third, water in the air moves huge quantities of heat from one place to another by absorbing surface heat in evaporation over warm oceans and releasing that latent heat in condensation or deposition over cooler regions. Like the flow of heat in ocean currents, this movement is generally poleward.

The Hydrosphere and the Hydrologic Cycle

Let's now turn to the nature of the hydrosphere and the flows of water among ocean, land, and atmosphere. Recall from Chapter 1 that the hydrosphere includes water on the earth in all its forms. About 97.2 percent of the hydrosphere consists of ocean saltwater, as shown in Figure 4.2. The remaining 2.8 percent is fresh water.

Deposition is a term used by chemists to describe the deposit of a solid directly from a vapor. Meteorologists generally use the term *sublimation* for both the change from vapor to solid and from solid to vapor, with the difference determined by the context. Since it is clearer to use a separate word for each process, we use the term *deposition* in this text.

Figure 4.2 Volumes of global water in each reservoir of the hydrosphere. Nearly all the earth's water is contained in the world ocean. Fresh surface and soil water make up only a small fraction of the total volume of global water.

The next largest reservoir of fresh water is stored as ice in the world's ice sheets and mountain glaciers. This water accounts for 2.15 percent of total global water.

Fresh liquid water is found both on top of and beneath the earth's land surfaces. Water occupying openings in soil and rock is called *subsurface water*. Most of it is held in deep storage as *ground water*, at a level where plant roots cannot access it. Ground water makes up 0.63 percent of the hydrosphere, leaving 0.02 percent of the water remaining.

Fresh liquid water on land is a very small part of the earth's total store of water.

The right-hand portion of Figure 4.2 shows how this small remaining proportion of the earth's water is distributed. This proportion is important to us because it includes the water available for plants, animals, and human use. *Soil water*, which is held in the soil within reach of plant roots, comprises 0.005 percent of the global total. Water held in streams, lakes, marshes, and swamps is called *surface water*. Most of this surface water is about evenly divided between freshwater lakes and saline (salty) lakes. An extremely small proportion is held in streams and rivers as they flow toward the sea or inland lakes.

Note that the quantity of water held as vapor and cloud water droplets in the atmosphere is also very small—0.001 percent of the hydrosphere. Though small, this reservoir of water is of enormous importance. It provides the supply of precipitation that replenishes all freshwater stocks on land. And, as we will see in the next chapter, the flow of water vapor from warm tropical oceans to cooler regions provides a global flow of heat, in latent form, from low to high latitudes.

The movements of water among the great global reservoirs constitute the **hydrologic cycle**. In this cycle, water moves from land and ocean to the atmosphere as water vapor and returns as precipitation. Because precipitation over land exceeds evaporation, water also runs off the land to the oceans. We will return to the hydrologic cycle in more detail in Chapter 15.

The hydrologic cycle describes the global flow of water among its reservoirs.

For now, we can summarize the main features of the hydrologic cycle in the global water balance, shown in Figure 4.3. The cycle begins with evaporation from water or land surfaces, in which water changes state from liquid to vapor and enters the atmosphere. In the case of land, this evaporation is enhanced by transpiration. Total evaporation is still about six times greater over oceans than land, however. This is because the oceans cover most of the planet and because land surfaces are not always wet enough to yield much evaporated water. Once in the atmosphere, water vapor can condense or deposit to form precipitation, which falls to earth as rain, snow, or hail. Precipitation over the

Figure 4.3 The global water balance. Figures give average annual water flows in and out of world land areas and world oceans. Values are given in thousands of cubic kilometers (cubic miles). Global precipitation equals global evaporation. (Based on data of John R. Mather).

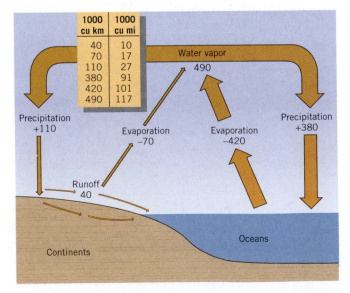

1000 cu km	1000 cu mi
40	10
70	17
110	27
380	91
420	101
490	117

oceans is nearly four times greater than precipitation over land.

Upon reaching the land surface, precipitation has three fates. First, it can evaporate and return to the atmosphere as water vapor. Second, it can sink into the soil and then into the surface rock layers below. As we will see in later chapters, this subsurface water emerges from below to feed rivers, lakes, and even ocean margins. Third, precipitation can run off the land, concentrating in streams and rivers that eventually carry it to the ocean or to a lake in a closed inland basin. This flow of water is known as *runoff.*

The Global Water Balance

Since our planet contains only a fixed amount of water, a global balance must be maintained among flows of water to and from the lands, oceans, and atmosphere. Let's examine this idea in more detail. For our analysis, we will assume that the volume of ocean waters and the overall volume of fresh water in surface and subsurface water remains constant from year to year. This is probably quite reasonable, unless climate is changing rapidly.

The global water balance is diagrammed in Figure 4.3. Let's consider the ocean first, shown on the right side of the diagram. There are three flows to and from the ocean: evaporation (out), precipitation (in), and runoff (in). For precipitation and runoff, water enters the ocean, while water leaves the ocean by evaporation. Since "in" and "out" must balance, we have

Precipitation + Runoff = Evaporation   (ocean)

Using the values in the figure, which are given as thousands of cubic kilometers, we find that 380 + 40 = 420.

Now let's turn to the continents, shown on the left side of the figure. Again, there are three flows—precipitation (in), evaporation (out), and runoff (out). This time, precipitation is entering (falling onto) the land, while evaporation and runoff are leaving. Thus,

Precipitation = Evaporation + Runoff   (land)

Using the values in the figure, we observe that 110 = 70 + 40.

If we rearrange this last equation to put evaporation first, we have

Evaporation = Precipitation – Runoff   (land)

compared to

Evaporation = Precipitation + Runoff   (ocean)

Taking the land and oceans together and considering only the flow of water from the planet's surface to the atmosphere and the reverse, we can conclude that the runoff terms within the last two equations cancel

out. This leaves total evaporation to equal total precipitation, as it must:

Total Evaporation	Total Precipitation
70 (land) + 420 (ocean) = cubic kilometers	110 (land) + 380 (ocean) cubic kilometers

The point of this simple analysis is not to write equations, but to show that we have constructed a budget that accounts for the flows of water between the reservoirs of the hydrologic cycle. Like a household budget, it must balance. If it does not, then there must be some flow that is not being considered. This is why scientists use budgets as tools to help understand natural processes, like the hydrologic cycle.

The global water balance describes the flows of water between ocean, atmosphere, and land.

For most of this chapter, we will be concerned with one aspect of the hydrologic cycle—the flow of water from the atmosphere to the surface in the form of precipitation. The examination of this aspect of the hydrologic cycle will involve descriptions of how the water vapor content of air is measured and how clouds and precipitation form.

HUMIDITY

The amount of water vapor present in the air varies widely from place to place and time to time. It ranges from almost nothing in the cold, dry air of arctic regions in winter to as much as 4 or 5 percent of a given volume of air in the warm, wet regions near the equator. The general term **humidity** refers to the amount of water vapor present in the air.

Understanding humidity and how the moisture content of air is measured involves an important principle—namely, that the maximum quantity of moisture that can be held at any time in the air is dependent on air temperature. Warm air can hold more water vapor than cold air—a lot more. Air at room temperature (20°C, 68°F) can hold about three times as much water vapor as air at freezing (0°C, 32°F).

Specific Humidity

The actual quantity of water vapor held by a parcel of air is its *specific humidity*. This quantity is stated as the mass of water vapor contained in a given mass of air and is expressed as grams of water vapor per kilogram of air (g/kg). Figure 4.4 shows the relationship between air temperature and the maximum amount of water vapor that air can hold. We see, for example, that

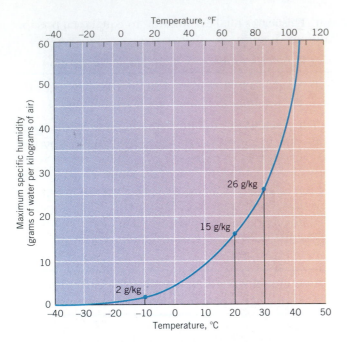

Figure 4.4 The maximum specific humidity of a mass of air increases sharply with rising temperature.

dard thermometer shelters the world over, and also on ships at sea.

Let's look first at specific humidity (*a*). The curve clearly shows the largest values for the equatorial zones, with values falling off rapidly toward both poles. This curve follows the pattern of insolation quite nicely (see Figure 2.7). More insolation is available at lower latitudes, on average, to evaporate water in oceans or on moist land surfaces. Therefore, specific humidity values are higher at low latitudes than at high latitudes.

The global profile of mean (average) surface air temperature, graph (*b*), shows a similar shape to the specific humidity profile. We would expect this to be the case, since air temperature and maximum specific humidity vary together, as shown in Figure 4.4.

Another way of describing the water vapor content of air is by its **dew-point temperature**. If air is slowly chilled, it eventually will reach *saturation*. At this temperature, the air holds the maximum amount of water vapor possible. If further cooling continues, condensation will begin and dew will form. The temperature at which saturation occurs is therefore known as the dew-point temperature. Moist air will have a higher dew-point temperature than drier air.

at 20°C (68°F), the maximum amount of water vapor that the air can hold—that is, the maximum specific humidity—is about 15 g/kg. At 30°C (86°F), it is almost doubled—about 26 g/kg. For cold air, the values are quite small. At −10°C (14°F), the maximum is only about 2 g/kg.

Specific humidity is often used to describe the moisture characteristics of a large mass of air. For example, extremely cold, dry air over arctic regions in winter may have a specific humidity as low as 0.2 g/kg. In comparison, the extremely warm, moist air of equatorial regions often holds as much as 18 g/kg. The total natural range on a worldwide basis is very wide. In fact, the largest values of specific humidity observed are from 100 to 200 times as great as the smallest.

Specific humidity is a geographer's yardstick for a basic natural resource—water—that can be applied from equatorial to polar regions. It is a measure of the quantity of water that can be extracted from the atmosphere as precipitation. Cold, moist air can supply only a small quantity of rain or snow, but warm, moist air is capable of supplying large amounts.

Warm air can hold much more water vapor than cold air.

Figure 4.5 is a set of global profiles showing how specific humidity varies with latitude and how it relates to mean surface air temperature. Both humidity and temperature are measured at the same locations in stan-

Figure 4.5 Pole-to-pole profiles of temperature (*above*) and specific humidity (*below*). The two are similar because the ability of air to hold water vapor (measured by specific humidity) is limited by temperature. (Data of J. von Hann, R. Süring, and J. Szava-Kovats as shown in Haurwitz and Austin, *Climatology.*)

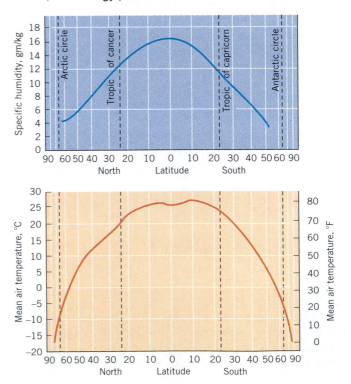

Relative Humidity

The measure of humidity that we encounter every day is *relative humidity*. This measure compares the amount of water vapor present to the maximum amount that the air can hold at that temperature, expressed as a percentage. For example, if the air currently holds half the moisture possible at the present temperature, then the relative humidity is 50 percent. When the humidity is 100 percent, the air holds the maximum amount possible, is saturated, and will be at the dew-point temperature.

Relative humidity compares the amount of water held by air to the maximum amount that can be held at that temperature.

A change in relative humidity of the atmosphere can happen in one of two ways. The first is by direct gain or loss of water vapor. For example, if an exposed water surface or wet soil is present, additional water vapor can enter the air, raising the specific humidity. This process is slow because the water vapor molecules must diffuse upward from the surface into the air layer above.

The second way is through a change of temperature. Even though no water vapor is added, a lowering of temperature results in a rise of relative humidity. Recall that the capacity of air to hold water vapor is dependent on temperature. When the air is cooled, this capacity is reduced. The existing amount of water vapor then represents a higher percentage of the total capacity.

An example may help to illustrate these principles (Figure 4.6). Beginning in the early morning hours, at 4 A.M. the temperature is 5°C (41°F), and the relative humidity of the air is 100 percent. That is, the air is saturated and cannot hold any additional water vapor. By

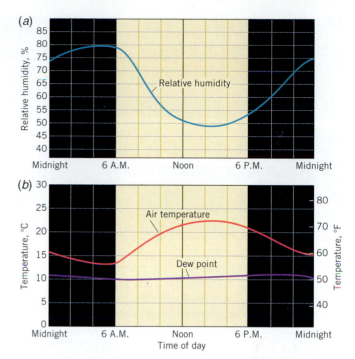

Figure 4.7 Relative humidity (*a*), and air temperature and dew-point temperature (*b*) throughout the day at Washington, D.C. The curves show average values for the month of May. (Data from National Weather Service.)

10 A.M., the temperature has risen to 16°C (61°F). The relative humidity has dropped to 50 percent, even though the amount of water vapor in the air remains the same. By 3 P.M., the air has been warmed by the sun to 32°C (90°F). The relative humidity has dropped to 20 percent, which is very dry air. The same amount of water vapor is present in the air, but the capacity of the air to hold water vapor has greatly increased. As air temperature falls in the evening, relative humidity again rises, and the cycle repeats itself.

The principle of relative humidity change caused by temperature change is further illustrated by a graph of these two properties throughout the day (Figure 4.7). Values shown are averages for May at Washington, D.C. As air temperature rises during the day, relative humidity falls, and vice versa. The dew-point temperature, which indicates the true moisture content, remains nearly constant.

How is humidity measured? A simple method uses two thermometers mounted together side by side in an instrument called a *sling psychrometer* (Figure 4.8). The wet-bulb thermometer bulb is covered with a sleeve of cotton fibers and is wetted. The dry-bulb thermometer remains dry. To operate the sling psychrometer, water is applied to the sleeve covering the wet bulb, and the two thermometers are then whirled in the open air. Evaporation cools the wet bulb. If the air is saturated, then evaporation cannot occur. In this

Figure 4.6 Relative humidity changes with temperature because the capacity of warm air to hold water vapor is greater than that of cold air. In this example, the amount of water vapor stays the same, and only the capacity changes.

Figure 4.8   This standard sling psychrometer uses paired thermometers. The cloth-covered wet-bulb thermometer projects beyond the dry-bulb thermometer. The handle is used to swing the thermometers in the free air. Evaporation lowers the wet-bulb temperature. Thus, the dryer the air, the greater the evaporation and the cooler the reading in relation to the dry thermometer. The sliding scale ruler, pictured below, enables rapid determination of relative humidity from wet and dry bulb readings.

case, no cooling results, and the temperature of the two thermometers is the same. If the air is not saturated, evaporation occurs and the wet-bulb thermometer is cooled. The drier the air, the greater is the cooling. To determine the relative humidity, a special sliding scale is used (Figure 4.8). Wet-bulb and dry-bulb temperatures are set on the scale, and relative humidity is read off directly.

THE ADIABATIC PROCESS

Given that ample water vapor is present in a mass of air, how is that related to precipitation? In other words, how is the water vapor turned into liquid or solid particles that can fall to earth? The answer is by natural cooling of the air. Since the ability of air to hold water vapor is dependent on temperature, the air must give up water vapor if it is cooled to the dew point and below. Think about a moist sponge. To extract the water, you have to squeeze the sponge—that is, reduce its ability to hold water. Chilling the air is like squeezing the sponge.

How is air chilled sufficiently to produce precipitation? One mechanism for chilling air is nighttime cooling. As we saw in Chapter 3, the ground surface can become quite cold on a clear night through loss of long-wave radiation. Thus, still air near the surface can be cooled below the condensation point, producing dew or frost. However, this mechanism is not sufficient to form precipitation. Precipitation is formed only when a substantial mass of air experiences a steady drop in temperature below the dew point. This happens when an air parcel is uplifted to a higher level in the atmosphere, as we will see in the following section.

Dry Adiabatic Rate

Have you ever pumped up a bicycle tire using a hand pump? If so, you may have noticed that as you pump vigorously, the metal pump gets hot. This occurs because air inside the pump is being compressed. An important principle of physics states that when a gas is compressed, its temperature increases. Conversely, when a gas is allowed to expand, its temperature drops. You may have noted this latter effect on an airplane, where a small nozzle overhead directs a stream of cool air toward you. In moving from higher pressure in the air hose that feeds the nozzle to lower pressure in the cabin, the air expands and cools. Physicists use the term **adiabatic process** to refer to a heating or cooling process that occurs solely as a result of pressure change.

> **The adiabatic principle states that compression warms a gas, while expansion cools it.**

Given that air cools or heats when the pressure on it changes, how does that relate to uplift and precipitation? The missing link is simply that atmospheric pressure decreases with an increase in altitude. So, if a parcel of air is uplifted, atmospheric pressure on the parcel will be lower, and it will expand and cool. Conversely, if a parcel of air descends, atmospheric pressure will be higher, and it will be compressed and warmed (Figure 4.9). The **dry adiabatic lapse rate** describes this behavior. This rate has a value of about 10°C per 1000 m (5.5°F per 1000 ft) of vertical rise. That is, if a mass of air is raised 1 km, its temperature will drop by 10°C. Or, in English units, if raised 1000 ft, its temperature will drop by 5.5°F. This is the "dry" rate because condensation does not occur.

Note that in the previous chapter we encountered the environmental temperature lapse rate. The temperature lapse rate is simply an expression of how the temperature of still air varies with altitude. This rate will vary from time to time and from one place to another, depending on the state of the atmosphere. This rate is quite different from the dry adiabatic lapse rate. The dry adiabatic rate is always constant and is determined by physical laws. It applies only to a mass of air in vertical motion. No motion of air is implied for the temperature lapse rate.

Wet Adiabatic Rate

With the adiabatic process firmly in mind, let's examine the fate of a parcel of moist air that is moved upward in the atmosphere (Figure 4.10). We will assume that the parcel starts with a temperature of 20°C (68°F). As the parcel is moved upward, its temperature drops at the dry adiabatic rate, 10°C/1000 m (5.5°F/1000 ft). At 500 m (1600 ft), the temperature has dropped by 5°C (9°F) to 15°C (59°F). At 1000 m (3300 ft), the temperature has fallen to 10°C (50°F).

If the rising process continues, the air will be cooled to saturation and condensation will start to occur. This is shown on the figure as the *lifting condensation level*. If cooling continues, water droplets will form, producing a cloud. If the parcel of saturated air continues to rise, however, a new principle comes into effect—latent heat release. That is, when condensation occurs, latent heat is released and warms the uplifted air. In other words, two effects are occurring at once. First, the uplifted air is being cooled by the reduction in atmospheric pres-

sure. Second, it is being warmed by the release of latent heat from condensation.

Rising air cools less rapidly when condensation is occurring, owing to the release of latent heat.

Which effect is stronger? As it turns out, the cooling effect is stronger, so the air will continue to cool as it is uplifted. But because of the release of latent heat, the cooling will occur at a lesser rate. This cooling rate is called the **wet adiabatic lapse rate**, and ranges between 4 and 9°C per 1000 m (2.2 and 4.9°F per 1000 ft). It is variable because it depends on the temperature and pressure of the air and its moisture content. For most situations, however, we can use a value of 5°C/1000 m (2.7°F/1000 ft). The higher rates apply only to cold, relatively dry air that contains little moisture and therefore little latent heat. In Figure 4.10, the wet adiabatic rate is shown as a slightly curving line to indicate that its value increases with altitude.

CLOUDS

A **cloud** is made up of water droplets or ice particles suspended in air. These particles have a diameter in the range of 20 to 50 μm (0.0008 to 0.002 in.). Recall from Chapter 2 that μm denotes the *micrometer*, or one-millionth of a meter. Each cloud particle is formed on a tiny center of solid matter, called a *condensation nucleus*.

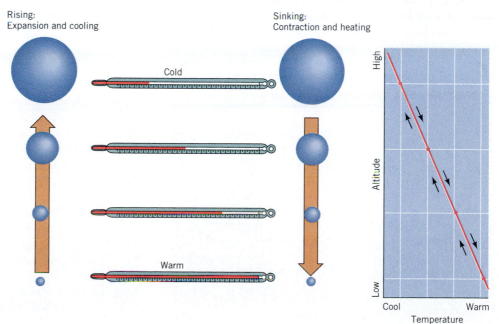

Figure 4.9 A schematic diagram of adiabatic cooling and heating that accompanies the rising and sinking of a mass of air. When air is forced to rise, it expands and its temperature decreases. When air is forced to descend, its temperature increases. (A. N. Strahler.)

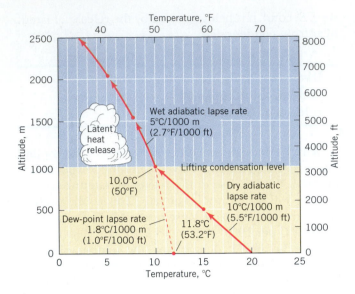

Figure 4.10   Adiabatic decrease of temperature in a rising parcel of air leads to condensation of water vapor into water droplets and the formation of a cloud. (Copyright © A. N. Strahler.)

This nucleus has a diameter in the range 0.1 to 1 μm (0.000004 to 0.00004 in.).

An important source of condensation nuclei is the surface of the sea. When winds create waves, droplets of spray from the crests of the waves are carried rapidly upward in turbulent air. Evaporation of sea water droplets leaves a tiny residue of crystalline salt suspended in the air. This aerosol strongly attracts water molecules. Even very clean and clear air contains enough condensation nuclei for the formation of clouds. However, the heavy load of dust carried by polluted air over cities substantially aids condensation and the formation of clouds and fog.

Cloud particles are tiny water droplets that form around condensation nuclei.

In our everyday life at the earth's surface, liquid water turns to ice when the surrounding temperature falls to the freezing point, 0°C (32°F), or below. However, the water in tiny cloud particles can remain in the liquid state at temperatures far below freezing. Such water is described as supercooled. Clouds consist entirely of water droplets at temperatures down to about −12°C (10°F). As cloud temperatures grow colder, a mix of water droplets and ice crystals occur. The coldest clouds, with temperatures below −40°C (−40°F), occur at altitudes of 6 to 12 km (20,000 to 40,000 ft) and are formed entirely of ice particles.

Cloud Forms

Clouds come in many shapes and sizes, from the small, white puffy clouds often seen in summer to the dark layers that produce a good, old-fashioned rainy day. Meteorologists classify clouds into four families, arranged by height: high, middle, and low clouds, and clouds with vertical development. These are shown in Figures 4.11 and 4.12. Some individual types with their names are also shown.

Clouds are grouped into two major classes on the basis of form—stratiform, or layered clouds, and cumuliform, or globular clouds. *Stratiform* clouds are blanket-like and cover large areas. A common type is *stratus*, which covers the entire sky. Stratus clouds are formed when large air layers are forced to rise gradually. This can happen when one air layer overrides another. As the rising layer is cooled, condensation occurs over a large area, and a blanket-like cloud forms. If the layer is quite moist and rising continues, dense, thick stratiform clouds result that can produce abundant rain or snow.

Stratiform clouds form horizontal layers, while cumuliform clouds are globular in shape.

Cumuliform clouds are globular masses of cloud that are associated with small to large parcels of rising air. The air parcels rise because they are warmer than the surrounding air. Like bubbles in a fluid, they move upward. Of course, as they are buoyed upward, they are cooled by the adiabatic process. When condensation occurs, a cloud is formed. The most common cloud of this type is the *cumulus* cloud (Figure 4.12). Sometimes the upward movement yields dense, tall clouds that produce thunderstorms. This form of cloud is the *cumulonimbus*. ("Nimbus" is the latin word for rain cloud, or storm.)

Fog

Fog is simply a cloud layer at or very close to the surface. In our industrialized world, fog is a major environmental hazard. Dense fog on high-speed highways can cause chain-reaction accidents, sometimes involving dozens of vehicles and causing injury or death to their occupants. At airports, landing delays and shutdowns because of fog bring economic losses to airlines and can inconvenience many thousands of travelers in a single day. For centuries, fog at sea has been a navigational hazard, increasing the danger of ship collisions and groundings. In addition, polluted fogs, like those of London in the early part of this century, can injure urban dwellers' lungs and take a heavy toll in lives.

One type of fog, known as *radiation fog*, is formed at

Classification of clouds according to height and form

Figure 4.11   Clouds are grouped into families on the basis of height. Individual cloud types are named according to their form.

Figure 4.12   (*a*) Rolls of altocumulus clouds, also known as a "mackerel sky." (*b*) High cirrus clouds lie above puffy cumulus clouds in this scene from Holland. (*c*) Fibrous cirrus clouds are drawn out in streaks by high-level winds to form "mare's tails." (*d*) Upper layer: lenticular clouds, a type of altostratus formed when moist air is uplifted as it crosses a range of hills. Lower layer: cumulus clouds, also triggered by the uplift.

night when temperature of the air layer at the ground level falls below the dew point. This kind of fog is associated with a low-level temperature inversion (Figure 3.10). Another fog type, *advection fog*, results when a warm, moist air layer moves over a colder surface. As the air layer loses heat to the surface, its temperature drops below the dew point, and condensation sets in. Advection fog commonly occurs over oceans where warm and cold currents occur side by side. When warm, moist air above the warm current moves over the cold current, condensation occurs. Fogs of the Grand Banks off Newfoundland are formed in this way, because here the cold Labrador current comes in contact with the warmer waters of the Gulf Stream.

Sea fog is frequently found along the California coast. It forms within a cool marine air layer in direct contact with cold water of the California current (Figure 4.13). Similar fogs are found on continental west coasts in the tropical latitude zones where cool, equatorward currents parallel the shoreline.

PRECIPITATION

With a firm understanding of the adiabatic process and how it produces cooling, condensation or deposition, and clouds, we can now turn to the final topic of our chapter—precipitation. Precipitation can form in two ways. In the first, cloud droplets collide and coalesce into larger and larger water droplets that fall as rain. In the second, ice crystals form and grow in a cloud that contains a mixture of both ice crystals and water droplets.

The first process occurs when saturated air rises rapidly, and cooling forces additional condensation. Cloud particles grow by added condensation and attain a diameter of 50 to 100 μm (0.002 to 0.004 in.). In collisions with one another, the particles grow into droplets of about 500 μm in diameter (about 0.02 in.). This is the size of water droplets in drizzle. Further collisions increase drop size and yield *rain*. Average raindrops have diameters of about 1000 to 2000 μm (about 0.04 to 0.1 in.), but they can reach a maximum diameter of about 7000 μm (about 0.25 in.). Above this value they become unstable and break into smaller drops while falling. This type of precipitation formation occurs in warm clouds typical of the equatorial and tropical zones.

Snow is produced by the second process in clouds that are a mixture of ice crystals and supercooled water droplets. When an ice crystal collides with a droplet of supercooled water, it induces freezing of the droplet. The ice crystals further coalesce to form snow particles, which can become heavy enough to fall from the cloud. Some ice crystals grow directly by deposition. Snowflakes formed entirely by deposition can have intricate crystal structures (Figure 4.14), but most particles of snow are simply tiny lumps of ice.

In warm clouds experiencing strong condensation, water droplets grow by condensation, then collide and coalesce to form raindrops.

When the underlying air layer is below freezing, snow reaches the ground as a solid form of precipitation. Otherwise, it melts and arrives as rain. A reverse process, the fall of raindrops through a cold air layer, results in the freezing of rain and produces pellets or grains of ice. These are commonly referred to in North

Figure 4.13 A layer of sea fog along the Big Sur coast of California, south of San Francisco. A patch of fog has burned off in the center embayment, revealing the water surface and a pocket beach.

Figure 4.14 These individual snow crystals, greatly magnified, were selected for their beauty. Most snowflakes are lumps of ice without such a delicate structure.

with a *rain gauge*. The simplest rain gauge is a straight-sided, flat-bottomed pan, which is set outside before a rainfall event. (An empty coffee can does nicely.) After the rain, the depth of water in the pan is measured.

A very small amount of rainfall, such as 2 mm (0.1 in.), makes too thin a layer to be accurately measured in a flat pan. To avoid this difficulty, a typical rain gauge is constructed from a narrow cylinder with a funnel at the top (Figure 4.15). The funnel gathers rain from a wider area than the mouth of the tube, so the cylinder fills more quickly. The water level gives the amount of precipitation, which is read on a graduated scale.

Precipitation is measured with a rain gauge.

Snowfall is measured by melting a sample column of snow and reducing it to an equivalent in rainfall. In this way rainfall and snowfall records may be combined in a single record of precipitation. Ordinarily, a 10 cm (or 10 in.) layer of snow is assumed to be equivalent to 1 cm (or 1 in.) of rainfall, but this ratio may range from 30 to 1 in very loose snow to 2 to 1 in old, partly melted snow.

Figure 4.15 This rain gauge consists of two clear plastic cylinders, here partly filled with red-tinted water. The inner cylinder receives rainwater from the funnel top. When it fills, the water overflows into the larger, outer cylinder.

America as *sleet*. (Among the British, sleet refers to a mixture of snow and rain.)

Perhaps you have experienced an *ice storm*. This occurs when the ground is frozen and the lowest air layer is also below freezing. Rain falling through the layer is chilled and freezes onto ground surfaces as a clear glaze. Ice storms cause great damage, especially to telephone and power lines and to tree limbs pulled down by the weight of the ice. In addition, roads and sidewalks are made extremely hazardous by the slippery glaze. Actually, the ice storm is more accurately named an "icing" storm, since it is not ice that is falling but supercooled rain.

Hail, another form of precipitation, consists of lumps of ice ranging from pea- to grapefruit-sized—that is, with a diameter of 5 mm (0.2 in.) or larger. Most hail particles are roughly spherical in shape. The formation of hail will be explained in our discussion of the thunderstorm.

Precipitation is measured in units of depth of fall per unit of time—for example, centimeters or inches per hour or per day. A centimeter (inch) of rainfall would cover the ground to a depth of 1 cm (1 in.), if the water did not run off or sink into the soil. Rainfall is measured

EYE ON THE ENVIRONMENT

Air Pollution

"Smog alert! Stay indoors and avoid exercising outside today!" Perhaps you've heard a warning like this on the radio, especially if you live in a metropolitan area with a smog problem. If so, you've had some first-hand experience with air pollution. In this *Eye on the Environment* feature, we focus on air pollution, including the sources of air pollutants, the natural processes that affect pollutant concentrations, and the environmental impacts of air pollutants.

In preceding chapters, we have discussed two kinds of substances in the air—aerosols and gases. Aerosols are small bits of matter in the air, so small that they float freely with normal air movements. Gases are molecular compounds that are mixed together to form the main body of air. To these two categories, we add a third category, *particulates*—larger, heavier particles that sooner or later fall back to earth.

An *air pollutant* is an unwanted substance injected into the atmosphere from the earth's surface by either natural or human activities. Air pollutants come as aerosols, gases, and particulates. Most pollutants generated by human activity arise in two ways. The first is through the day-to-day activities of large numbers of people, for example, in driving automobiles within urban centers. The second is through industrial activities, such as fossil fuel combustion or the smelting of mineral ores to produce metals. **SCIENCE NEWS** 10

Figure E4.1 shows the relative proportions of the most common air pollutants and their sources. The gases are carbon monoxide (CO), sulfur oxides, and nitrogen oxides. The nitrogen oxides can be NO, NO_2, or NO_3; this mixture is usually referred to as NO_x. Similarly, the sulfur oxides SO_2 and SO_3 are referred to as SO_x. Hydrocarbon compounds occur both as gases and as aerosols. The remaining pollutants are in the form of particulates. **SCIENCE NEWS** 11

Combustion of fossil fuel, in transportation or as stationary fuel combustion, is the most important source of all these pollutants. As shown in the right half of the figure, it accounts for 73 percent of the emissions. Exhausts from gasoline and diesel engines contribute most of the carbon monoxide, half the hydrocarbons, and about a third of the nitrogen oxides.

Stationary sources of fuel combustion include power plants that generate electricity and various industrial plants that also burn fossil fuels. They contribute most of the sulfur oxides because they burn coal or lower grade fuel oils. These fuels are richer in sulfur than gasoline and diesel fuel. Stationary sources also supply most of the particulate matter. Some of it is fly ash—coarse soot particles emitted from smokestacks of power plants. These particles settle out quite quickly within close range of the source. Smaller and finer carbon particles, however, can remain suspended almost indefinitely and become aerosols. Industrial processes, such as the smelting of ore, are also important sources of air pollutants (Figure E4.2). The burning of solid wastes makes a small contribution to air pollution.

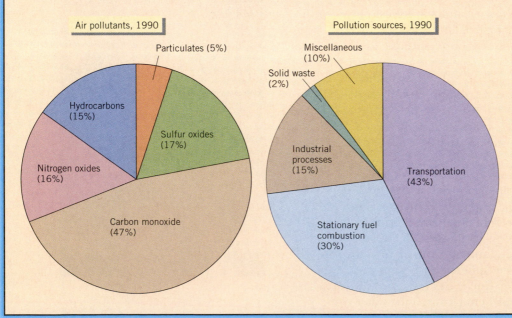

Air pollutants, 1990

Particulates (5%)
Hydrocarbons (15%)
Sulfur oxides (17%)
Nitrogen oxides (16%)
Carbon monoxide (47%)

Pollution sources, 1990

Miscellaneous (10%)
Solid waste (2%)
Industrial processes (15%)
Transportation (43%)
Stationary fuel combustion (30%)

Figure E4.1 Air pollution emissions in the United States in percentage by weight. Left: pollutants emitted. Right: sources of emission. (Data from National Air Pollution Control Administration, HEW.)

Figure E4.2 This satellite image shows the devastating effects of pollutants emitted by a smelter near Wawa, Ontario, Canada. Healthy vegetation is shown in red to pink tones. The blue streak on the right-hand side of the image is barren of vegetation.

Figure E4.3 Smog in the city. New York's World Trade Center projects skyward through a brownish-pink layer of smog.

Smog and Haze

When aerosols and gaseous pollutants are present in considerable density over an urban area, the resultant mixture is known as *smog* (Figure E4.3). This term was coined by combining the words "smoke" and "fog." Typically, smog allows hazy sunlight to reach the ground, but it may also hide aircraft flying overhead from view. Smog irritates the eyes and throat, and it can corrode structures over long periods of time.

Modern urban smog has three main toxic ingredients: nitrogen oxides, hydrocarbons, and ozone. Nitrogen oxides and hydrocarbons are largely automobile pollutants, although significant amounts can be released by industrial power plants and petroleum processing and storage facilities. Ozone is not normally produced directly by pollution sources in the city but forms through a photochemical reaction in the air. In this reaction, nitrogen oxides react with hydrocarbons in the presence of sunlight to form ozone. Ozone in urban smog can harm plant tissues and eventually kill sensitive plants. It is also harmful to human lung tissue, and it aggravates bronchitis, emphysema, and asthma. This is the same compound that absorbs ultraviolet radiation in the ozone layer within the stratosphere. So, ozone in the stratosphere is desirable, but if close to earth, it is undesirable, indeed. Photochemical reactions can also produce other toxic compounds in smog.

Haze is a condition of the atmosphere in which aerosols obscure distant objects. Haze builds up naturally in stationary air as a result of human and natural activity. When the air is humid and abundant water vapor is available, water films grow on suspended nuclei. This creates aerosol particles large enough to obscure and scatter light, reducing visibility. Natural haze aerosols include soil dust, salt crystals from the sea surface, hydrocarbon compounds from plants, plant pollen, and smoke from forest and grass fires. Some regions are noted for naturally occurring haze—the Great Smoky Mountains are an example.

Fallout and Washout

Pollutants generated by a combustion process are contained within hot exhaust air emerging from a factory smokestack or an auto tailpipe. Since hot air rises, the pollutants are at first carried aloft by convection. However, the larger particulates soon settle under gravity and return to the surface as *fallout*. Particles too small to settle out are later swept down to earth by precipitation in a process called *washout*. Through a combination of fallout and washout, the atmosphere tends to be cleaned of pollutants. Although a balance between input and output of pollutants is achieved in the long run, the quantities stored in the air at a given time fluctuate widely.

Pollutants are also eliminated from the air over their source areas by wind. Strong, through-flowing winds will disperse pollutants into large volumes of cleaner air in the downwind direction. Strong winds can quickly sweep away most pollutants from an urban area, but during periods when winds are light or absent, the concentrations can rise to high values.

Inversion and Smog

The concentration of pollutants over a source area rises to its highest levels when vertical mixing (convection) of the air is inhibited. This happens in an inversion—a condition in which the temperature of the air increases with altitude.

Why does the inversion inhibit mixing? Recall that a heated air parcel, perhaps emerging from a smokestack or chimney, will rise as long as it is warmer than the surrounding air. Recall, too, that as it rises, it is cooled according to the adiabatic principle. In an inversion, however, the surrounding air gets warmer, not colder, with altitude. So, the parcel will cool as it rises, while the surrounding air becomes warmer. Thus, the parcel will quickly arrive at the temperature of the surrounding air and uplift will stop. Under these conditions, heated air will move only a short distance upward. Pollutants in the air parcel will disperse at low levels, keeping concentrations high near the ground.

Two types of inversions are important in causing high air pollutant concentrations—low-level and high-level inversions. When a *low-level temperature inversion* develops over an urban area with many air pollution sources, pollutants are trapped under the "inversion lid." Heavy smog or highly toxic fog can develop. An example is the tragedy that occurred in Donora, Pennsylvania, in late October 1948, when a persistent low-level inversion developed. The city occupies a valley floor hemmed in by steeply rising valley walls. The walls prevented the free mixing of the lower air layer with that of the surrounding region (Figure E4.4) and helped keep the inversion intact. Industrial smoke and gases from factories poured into the inversion layer for five days, increasing the

pollution level. A poisonous fog formed and began to take its toll. Twenty persons died, and several thousand persons were stricken before a change in weather patterns dispersed the polluted layer.

Another type of inversion is responsible for the smog problem experienced in the Los Angeles basin and other California coastal regions, ranging north to San Francisco and south to San Diego. Here special climatic conditions produce prolonged inversions and smog accumulations (Figure E4.5). Off the California coast is a persistent fair-weather system, which is especially strong in the summer. This system produces a layer of hot, dry air at upper elevations. However, a cold current of upwelling ocean bottom water runs along the coast, just offshore. Moist ocean air at the surface moves across this cool current and is chilled, creating a cool, marine air layer.

Now, the Los Angeles basin is a low, sloping plain lying between the Pacific Ocean and a massive mountain barrier on the north and east sides. Weak winds from the south and southwest move the cool, marine air inland over the basin. Further landward movement is blocked by the mountain barrier. Since there is a warm layer above this cool marine air layer, the result is an inversion. Pollutants accumulate in the cool air layer and produce smog. The upper limit of the smog stands out sharply in contrast to the clear air above it, filling the basin like a lake and extend-

Figure E4.4 A low-level inversion held air pollutants close to the ground, inducing a poison fog accumulation at Donora, Pennsylvania, in October 1948.

Figure E4.5 A layer of hot, dry air from a persistent fair-weather system rides over a cool, moist marine air layer at the surface to create a persistent temperature inversion.

ing into valleys in the bordering mountains (Figure E4.6). Since this type of inversion persists to a higher level in the atmosphere, we can refer to it as a *high-level temperature inversion.*

An actual temperature inversion, in which temperature close to the surface increases with altitude, is not essential for building a high concen-tration of pollutants above a city. All that is required is light or calm winds and stable air. *Stable air* has a temper-ature profile that decreases with alti-tude, but at a slow rate. Some convec-tional mixing occurs in stable air, but convectional precipitation is inhibit-ed. At certain times of the year, slow-moving masses of dry, stable air occu-py the central and eastern portions of the North American continent. Under these conditions, a broad *pol-lution dome* can form over a city or region, and air quality will suffer (Fig-ure E4.7). When there is a regional wind, the pollution from a large city will be carried downwind to form a *pollution plume.*

Figure E4.6 A dense layer of smog and marine haze lying over the Los Angeles basin. The view is from a point over the San Gabriel Mountains, looking southwest.

Figure E4.7   If calm, stable air overlies a major city, a pollution dome can form. When a wind is present, pollutants are carried away as a pollution plume.

Climatic Effects of Urban Air Pollution

Urban air pollution reduces visibility and illumination. Specifically, a smog layer can cut illumination by 10 percent in summer and 20 percent in winter. Ultraviolet radiation is absorbed by ozone in smog. At times, ultraviolet radiation is completely prevented from reaching the ground. While this reduces the risk of human skin cancer, it may also permit increased viral and bacterial activity at ground level.

Winter fogs are much more frequent over cities than over the surrounding countryside. The fog is enhanced by the abundance of urban aerosols and particulates. Coastal airports, such as those of New York City, Newark, and Boston, suffer from an increased frequency of fog resulting from urban air pollution. Cities also show an increase in cloudiness and precipitation as compared to the surrounding countryside. This increase results from intensified convection generated when the lower air is heated by human activities.

Acid Deposition

Yet another air pollution problem is *acid deposition*. Perhaps you've heard about acid rain killing fish and poisoning trees. *Acid rain* is part of the phenomenon of acid deposition, which includes not only acid rain, but also dry acidic dust particles. Acid rain simply consists of raindrops that have been acidified by air pollutants. Dry acidic particles are dust particles that are acidic in nature. They fall to earth and coat the surface in a thin dust layer. When wetted by rain or fog, they acidify the water on leaves and soils. In winter, acid particles can mix with snow as it accumulates. In spring, when the snow melts, a surge of acid water is then released to soils and streams.

Acid deposition is produced by the release of sulfur dioxide (SO_2) and nitric oxide (NO_2) into the air. This occurs when fossil fuels are burned. Once in the air, SO_2 and NO_2 readily combine with oxygen and water in the presence of sunlight and dust particles to form sulfuric and nitric acid aerosols. These aerosols serve as con-

densation nuclei and acidify the tiny water droplets created. When the droplets coalesce in precipitation, acid raindrops or ice crystals result. Sulfuric and nitric acids can also be formed on dust particles, creating dry acid particles. These can be as damaging to plants, soils, and aquatic life as acid rain.

What are the effects of acid deposition? A primary effect is acidification of lakes and streams, injuring aquatic plants and animals and perhaps eventually sterilizing a lake, pond, or stream. Another problem is soil damage. Through a series of chemical effects, acid deposition can cause soils to lose nutrients, affecting the natural ecosystem within a region.

To measure acidity, scientists use the pH scale. Values on this scale range from 1 to 14. High values indicate alkalinity, and low values indicate acidity. Pure water has a pH of 7. Normally, rainwater is slightly acid, showing pH values of 5 to 6. In the 1960s, European water chemists observed that the pH of rain in northwestern Europe had dropped to values as low as 3 in some samples, similar to household vinegar. Because pH numbers are on a logarithmic scale, these values mean that rain in these samples was 100 to 1000 times more acid than normal.

In North America, scientists studying the chemical quality of rainwater have reported that since 1975 rainwater over a large area of the northeastern United States has had an average pH of about 4. Values less than pH 4 occur at times over many heavily industrialized U.S. cities, including Boston, New York, Philadelphia, Birmingham, Chicago, Los Angeles, and San Francisco. Values between pH 4 and 5 have been observed near smaller cities such as Tucson, Arizona; Helena, Montana; and Duluth, Minnesota—localities we do not usually associate with heavy air pollution.

Figure E4.8 is a map of the United States and Canada showing the acidity of rainwater averaged over a year. Compare this with Figure E4.9, showing emissions of sulfur dioxide and nitro-

Figure E4.8 Acidity of rainwater for the United States and Canada, averaged over a complete year. Values are pH units. The northeastern United States and southeastern Canada are most strongly affected.

The increased fish mortality observed in Canadian lakes has also been attributed to acidification. In 1980, Canada's Department of Environment reported that 140 Ontario lakes had no fish and that thousands of other lakes of that province were threatened with a similar fate. In 1990, American scientists estimated that 14 percent of Adirondack lakes were heavily acidic, along with 12 to 14 percent of the streams in the Mid-Atlantic states.

Forests, too, have been damaged by acid deposition (Figure E4.10). In western Germany, the impact has been especially severe in the Harz Mountains and the Black Forest. In 1983, it was reported that about one-third of West Germany's forests showed visible damage. In the eastern United States, pine and spruce trees have experienced damage in recent years. In the early 1990s, Europe continued to experience heavy losses in timber, resulting from fallout of sulfur

gen oxides by states. There is a strong relationship, especially between SO_2 and the average acidity of rainwater.

An important factor in the level of impact of acid deposition on the environment is the ability of the soil and surface water to absorb and neutralize acid. This factor ranges widely. In dry climates, acids can be readily neutralized because surface waters are normally somewhat alkaline. Areas where soil water is naturally somewhat acidic are most sensitive to acid deposition. Such areas are associated with moist climates generally and include the eastern United States, high-mountain regions of the western states, and the Pacific Northwest.

Acid deposition in Europe and North America has severely impacted some ecosystems. In Norway, acidification of stream water has virtually eliminated salmon runs by inhibiting salmon egg development. At present, the acidity of Norwegian streams has leveled off as air pollution levels have stabilized. However, salmon kills continue to occur, especially when rainstorms carry large amounts of sulfate into streams.

Figure E4.9 Emissions of sulfur dioxide and nitrogen oxides by states for the year 1993. Sulfur dioxide releases are especially heavy in the states that burn Appalachian coal. Releases of nitrogen oxides are high in Texas, due to oil processing, and in California, due to auto emissions. (Data from EPA, 1994.)

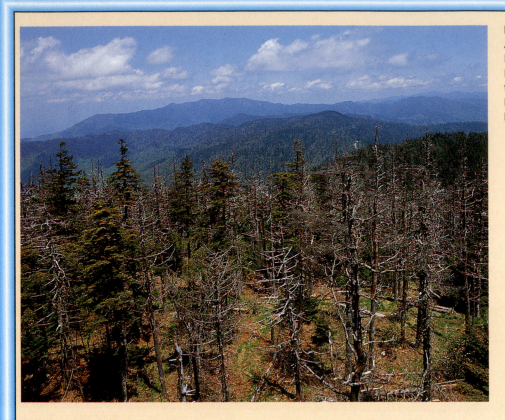

Figure E4.10 A view from the Appalachian Trail, near Klingman's Dome, in Great Smoky Mountains National Park, Tennessee. Death of the many leafless firs in the scene has been attributed to stress caused by acid deposition.

dioxide in combination with nitrogen oxide emissions from vehicles, industry, and farm wastes. Fallout of sulfur oxides was most heavily concentrated in Germany, the Czech Republic, Slovakia, Poland, and Hungary.

In 1990, the National Acid Precipitation Assessment Program (NAPAP) issued a report after its 10-year study of the causes and effects of acid precipitation in the United States. Emissions of sulfur dioxide and volatile organic compounds had declined 30 percent since 1970. Furthermore, nitrogen oxides had also declined over the period by 8 to 16 percent. These effects were due primarily to improved industrial emission controls. In 1990, electrical power plants continued to be the major source of sulfur oxides. However, new controls on sulfur dioxide emissions called for by the 1990 Clean Air Act should further reduce SO_2 releases and continue this improvement in air quality through the remainder of the decade.

Although the situation is improving, acid deposition is still an important problem in many parts of the world—especially Eastern Europe and the states of the former Soviet Union. There, air pollution controls have been virtually nonexistent for decades. Reducing pollution levels and cleaning up polluted areas will be a major task for these nations as they enter the next century.

Air Pollution Control

The United States and Canada have some of the strictest laws in the world limiting air pollution. As a result, many strategies have been developed to reduce emissions. Sometimes these strategies involve trapping and processing pollutants after they are generated. The "smog controls" on automobiles are an example. Sometimes the use of alternative, nonpolluting technology is in order—for example, substituting solar, wind, or geothermal power for coal burning to generate electricity.

In any event, both as individuals and as a society we can do much to preserve the quality of the air we breathe. Reducing fossil fuel consumption is an example. It will be a considerable challenge to our society and others to preserve and increase global air quality in the face of an expanding human population and increasing human resource consumption.

Questions

1. What are the most abundant air pollutants, and what are their sources?

2. What is smog? What important pollutant forms within smog, and how does this happen?

3. Distinguish between low-level and high-level inversions. How are they formed? What is their effect on air pollution? Give an example of a pollution situation of each type.

4. To what type of pollution does the term *acid deposition* refer? What are its causes? What are its effects?

Precipitation Processes

So far, we have seen how air that is moving upward will be chilled by the adiabatic process to the saturation point and then to condensation, and how eventually precipitation will form. However, one key piece of the precipitation puzzle has been missing—what causes air to move upward? Air can move upward in three ways. First, it can be forced upward as a through-flowing wind. Consider the case of a mass of air moving up and over a mountain range, for example. If the mountain range is high enough and the air is moist enough, then precipitation will occur. This is called **orographic precipitation**, and we'll describe it more fully in the following section.

The three types of precipitation are orographic, convectional, and cyclonic.

A second way for air to be forced upward is through convection. In this process, a parcel of air is heated, perhaps by a patch of warm ground, so that it is warmer and therefore less dense than the air around it. Like a bubble, it rises. If the air is quite moist and condensation sets in, then the release of latent heat of condensation will ensure that the bubble of air remains warmer than the surrounding air as it rises. This type of precipitation is **convectional precipitation**, and it will be described after orographic precipitation.

A third way for air to be forced upward is through the movement of air masses. As we will see in Chapter 6, air masses are large bodies of air, each with a set of relatively uniform temperature and moisture properties. Air masses move normally from west to east in midlatitudes, and one can overtake another. When this happens, one of them will be forced aloft. Since this overtaking usually occurs in a common type of storm called a cyclone, this type of precipitation is known as *cyclonic precipitation*. We will return to this subject in Chapter 6.

Orographic Precipitation

In orographic precipitation, through-flowing winds move moist air up and over a mountain barrier. (The term *orographic* means "related to mountains.") Figure 4.16 shows how this process works. Moist air arrives at the coast after passing over a large ocean surface (1). As the air rises on the windward side of the range, it is

Figure 4.16 The forced ascent of a warm, moist oceanic air mass over a mountain barrier produces precipitation and a rainshadow desert. As the air moves up the mountain barrier, it loses moisture through precipitation. As it descends the far slope, it is warmed.

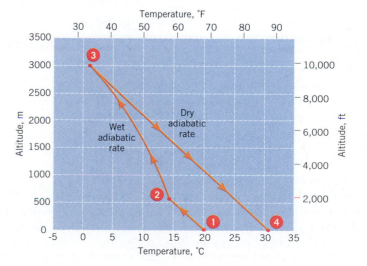

cooled at the dry adiabatic rate. When cooling is sufficient, the lifting condensation level is reached, condensation sets in, and clouds begin to form (2). Cooling now proceeds at the wet adiabatic rate. Eventually, precipitation begins. As the air continues up the slope, precipitation continues to fall.

After passing over the mountain summit, the air begins to descend down the leeward slopes of the range (3). However, as it descends, it is compressed. This means that it is warmed, according to the adiabatic principle. Since cooling is no longer occurring, cloud droplets and ice crystals evaporate or sublimate. The air clears rapidly. As it continues to descend, it warms further. At the base of the mountain on the far side (4), the air is now hot—and dry, since its moisture has been removed.

In orographic precipitation, moist air is forced up and over a mountain barrier, producing cooling, condensation, and precipitation.

The net result of the trip over the mountain is that the warm, moist air has become hot and dry. It was warmed by the latent heat released to the surrounding air as water vapor formed water droplets and ice crystals in the precipitation process. Its water was shed to the mountain slopes below, on the uphill journey. This effect creates a *rainshadow* on the far side of the mountain—a belt of dry climate that extends on down the leeward slope and beyond. Several of the earth's great deserts are of this type.

California's rainfall patterns provide an excellent example of orographic precipitation and the rainshadow effect. A map of California (Figure 4.17) shows mean annual precipitation. It uses lines of equal precipitation, called *isohyets*. Focus now on central California and follow the arrow on the map. Prevailing westerly winds bring moist air in from the Pacific Ocean, first over the Coast Ranges of central and northern California, and then, after rain is deposited on the Coast Ranges, the air descends into the broad Central Valley. Here, precipitation is low, averaging less than 25 cm (10 in.) per year.

Next the air continues up and over the great Sierra Nevada, whose summits rise to 4200 m (about 14,000 ft) above sea level. Heavy precipitation, largely in the form of winter snow, falls on the western slopes of these ranges and nourishes rich forests. Passing down the steep eastern face of the Sierra Nevada, the air descends quickly into the Owens Valley, at about 1000 m (3300 ft) elevation. The adiabatic heating warms and dries the air, creating a rainshadow desert. In the Owens Valley, annual precipitation is less than 10 cm (4 in.). In this way, the orographic effect on air moving across the

Figure 4.17   The effect of mountain ranges on precipitation is strong in the state of California, because of the prevailing flow of moist oceanic air from west to east. Centers of high precipitation coincide with the western slopes of mountain ranges, including the coast ranges and Sierra Nevada. To the east, in their rainshadows, lie desert regions.

mountains of California produces a part of America's great interior desert zone, extending from eastern California and across Nevada.

Convectional Precipitation

The second process of inducing uplift in a parcel of air is *convection*, the upward motion of a parcel of heated air. In this process, strong updrafts occur within *convection*

Figure 4.18 Rise of a bubble of heated air to form a cumulus cloud.

cells. Air rises in a convection cell because it is warmer, and therefore less dense, than the surrounding air.

The convection process starts when a surface is heated unequally. For example, consider an agricultural field surrounded by a forest. Since the field surface consists largely of bare soil and only a low layer of vegetation, it will be warmer under steady sunshine than the adjacent forest. This means that as the day progresses the air above the field will grow warmer than the air above the forest.

If a bubble of air is warmer than the surrounding air, it will be less dense and rise upward.

Now, the density of air depends on its temperature—warm air is less dense than cooler air. The hot-air baloon operates on this principle. The balloon is open at the bottom, and in the basket below a large gas burner forces heated air into the balloon. Because the heated air is less dense than the surrounding air, the balloon rises. The same principle will cause a bubble of air to form over the field, rise, and break free from the surface. Figure 4.18 diagrams this process.

As the bubble of air rises, it is cooled adiabatically and its temperature will decrease as it rises. However, we know that the temperature of the surrounding air will normally decrease with altitude as well. Nonetheless, as long as the bubble is still warmer than the surrounding air, it will be less dense and will therefore continue to rise.

If the bubble remains warmer than the surrounding air and uplift continues, adiabatic cooling chills the bubble below the dew point. Condensation occurs, and the rising air column becomes a puffy cumulus cloud. The flat base shows the level at which condensation begins. The bulging "cauliflower" top of the cloud is the top of the rising warm-air column, pushing into higher levels of the atmosphere. Normally, the small cumulus cloud will encounter winds aloft that mix it into the local air. After drifting some distance downwind, the cloud evaporates.

Unstable Air

Sometimes, however, convection continues strongly, and the cloud develops into a dense cumulonimbus mass, or thunderstorm, from which heavy rain will fall. Two conditions encourage the development of thunderstorms: (1) air that is very warm and moist, and (2) an environmental temperature lapse rate for which temperature decreases more rapidly with altitude than it does for either the dry or wet adiabatic lapse rates. Figure 4.19, which diagrams the convection process in unstable air, shows this type of lapse rate as having a flatter slope than the dry or wet adiabatic rates. Air with these characteristics is referred to as **unstable air**.

Unstable air—warm, moist, and heated at the surface—can produce abundant convectional precipitation.

For the first condition, recall that there are two adiabatic rates—dry and wet (Figure 4.10). The wet rate is about half the dry rate. While condensation is occurring, the lesser wet rates apply. Thus, air in which condensation is occurring cools less rapidly with uplift. Furthermore, the wet rate is smallest for warm, moist air. This means that the temperature decrease experienced by warm, moist, rising, condensing air is quite small. And, since the temperature decrease is small, the rising air is more likely to stay warmer than the surrounding air. Thus, uplift continues.

With regard to the second condition—an environmental temperature lapse rate value that exceeds the adiabatic rates—keep in mind that this rate describes the temperature of the surrounding, still air. Therefore, the temperature of the still air falls quite rapidly with altitude. This means that the condensing air in a rising parcel will stay warmer than the surrounding air. Again, uplift continues.

A simple example may make convection in unstable air clearer. Figure 4.19 shows how temperature changes with altitude for a parcel moving upward by convection in unstable air. The surrounding air is at a temperature of 26°C (78.8°F) at ground level and has a lapse rate of 12°C/1000 m (6.6°F/1000 ft). A parcel of air is heated by 1°C (1.8°F) to 27°C (80.6°F), and it begins to rise. At first, it cools at the dry adiabatic rate. At 500 m (1640 ft), the parcel is at 22°C (71.6°F), while the surrounding air is at 20°C (68°F). Since it is still warmer than the surrounding air, uplift continues. At 1000 m, the lifting condensation level is reached. The temperature of the parcel is 17°C (62.6°F).

As the parcel rises above the condensation level, the wet adiabatic rate applies, here using the value 5°C/1000 m (2.7°F/1000 ft). Now the parcel cools more slowly as it rises. At 1500 m (4920 ft), the parcel is 14.5°C (58.1°F), while the surrounding air is 8°C (46.4°F). Since the parcel is still warmer than the surrounding air, it continues to rise. Note that the difference in temperature between the rising parcel and the surrounding air now actually increases with altitude.

This means that the parcel will be buoyed upward ever more strongly, forcing even more condensation and precipitation.

The key to the convectional precipitation process is latent heat. When water vapor condenses into cloud droplets or ice particles, it releases latent heat to the rising air parcel. By keeping the parcel warmer than the surrounding air, this latent heat fuels the convection process, driving the parcel ever higher. When the parcel reaches a high altitude, most of its water will have condensed. As adiabatic cooling continues, less latent heat will be released. As a result, the convection cell will weaken. Eventually, uplift will stop, since the energy source, latent heat, is gone. The cell dies and dissipates into the surrounding air.

Latent heat release in condensation fuels the convectional precipitation process.

Unstable air is typical of summer air masses in the central and southeastern United States. As we will see later, summer weather patterns sweep warm, humid air from the Gulf of Mexico over the continent. The intense summer insolation, over a period of days, strongly heats the air layer near the ground, producing a steep lapse rate. Thus, both of the conditions that create unstable air are present. As a result, thunderstorms are very common in these regions during the summer.

Figure 4.19 Convection in unstable air. When the air is unstable, a parcel of air that is heated sufficiently to rise will continue to rise to great heights.

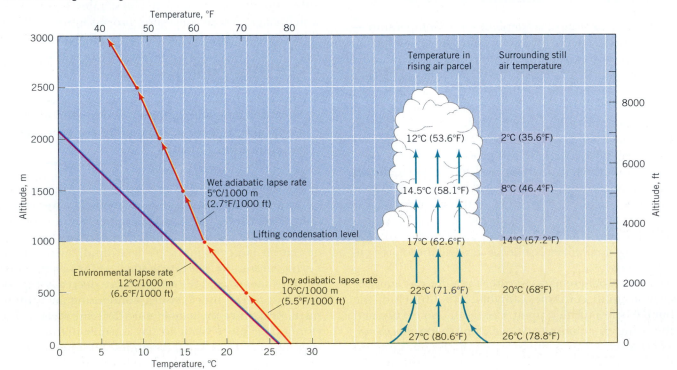

Unstable air is also likely to be found in the vast, warm and humid regions of the equatorial and tropical zones. In these regions, convective showers and thundershowers are common. At low latitudes, much of the orographic rainfall is in the form of heavy showers and thundershowers produced by convection. In this case, the forced ascent of unstable air up a mountain slope easily produces rapid condensation, which then triggers the convection mechanism.

Thunderstorms

The thunderstorm is an awesome event (Figure 4.20). Intense rain and violent winds, coupled with lightning and thunder, renew our respect for nature's power. What is the anatomy of a thunderstorm?

A **thunderstorm** is an intense local storm associated with a tall, dense cumulonimbus cloud in which there are very strong updrafts of air. A single thunderstorm consists of several individual convection cells. A single convection cell is diagrammed in Figure 4.21. Air rises

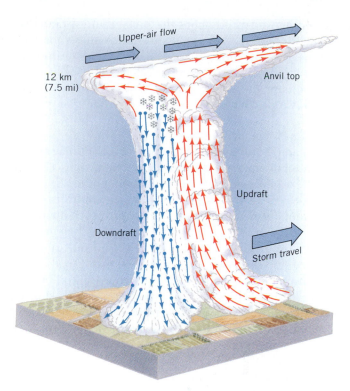

Figure 4.21 Anatomy of a thunderstorm cell. Successive bubbles of moist condensing air push upward in the cell. Their upward movement creates a corresponding downdraft, expelling rain, hail, and cool air from the storm as it moves forward.

Figure 4.20 This majestic cumulonimbus cloud marks a thunderstorm, moving from left to right. Lightning strikes lead the storm. A dark plume of falling rain follows.

within the cell as a succession of bubblelike air parcels. Intense adiabatic cooling within the bubbles produces precipitation. It can be in the form of water at the lower levels, mixed water and snow at intermediate levels, and snow at high levels where cloud temperatures are coldest.

As the rising air parcels reach high levels, which may be 6 to 12 km (about 20,000 to 40,000 ft) or even higher, the rising rate slows. Strong winds typically present at such high altitudes drag the cloud top downwind, giving the thunderstorm cloud its distinctive shape—resembling an old-fashioned blacksmith's anvil.

Thunderstorms are formed of convection cells in which rising bubbles of air are chilled to yield heavy precipitation.

Ice particles falling from the cloud top act as nuclei for freezing and deposition at lower levels. Large ice crystals form and begin to sink rapidly. Melting, they coalesce into large, falling droplets. The rapid fall of raindrops adjacent to the rising air bubbles pulls the air downward and feeds a downdraft. Emerging from the cloud base, the downdraft of cool air, laden with precipitation, approaches the surface and spreads out in all directions. Part of the downdraft moves forward, causing warm, moist surface air to rise and enter the updraft portion of the storm. This effect helps perpetuate the storm. The downdraft creates strong local winds. Wind gusts can sometimes be violent enough to topple trees and raise the roofs of weak buildings.

In addition to powerful wind gusts and heavy rains, thunderstorms can produce hail. Hailstones (Figure 4.22) are formed by the accumulation of ice layers on an ice pellet suspended in the strong updrafts of the thunderstorm. In this process, they can reach diameters of 3 to 5 cm (1.2 to 2.0 in.). When they become too heavy for the updraft to support, they fall to earth. Figure 4.23 shows a hailstorm in progress.

Annual losses from crop destruction caused by hailstorms amount to several hundred million dollars. Damage to wheat and corn crops is particularly severe in the Great Plains, running through Nebraska,

Figure 4.22   Close-up view of hailstones.

Figure 4.23   A hailstorm in progress, Santa Cruz County, Arizona. Although the hail is falling from a cloud directly overhead, the low afternoon sun still illuminates the scene at an angle. Marble-sized hailstones are accumulating on the ground.

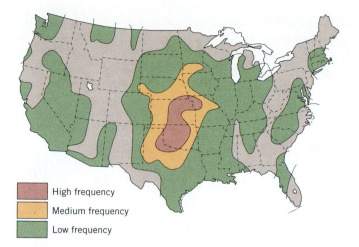

Figure 4.24 Map of frequency of severe hailstorms in the 48 contiguous states of the United States. A severe hailstorm is defined as a local convective storm producing hailstones equal to or greater than 1.9 cm (0.75 in.) in diameter. (From Richard H. Skaggs, *Proc. Assoc. American Geographers*, vol. 6, Figure 2. Used by permission.)

- High frequency
- Medium frequency
- Low frequency

Kansas, Missouri, Oklahoma, and northern Texas (Figure 4.24).

Another effect of convection cell activity is to generate lightning. This occurs when updrafts and downdrafts cause positive and negative static charges to accumulate within different regions of the cloud.

Lightning is a great electric arc—a series of gigantic sparks—passing between differently charged parts of the cloud mass or between cloud and ground (see Figure 4.20). During a lightning discharge, a current of as much as 60,000 to 100,000 amperes may develop. This current heats the air intensely, which makes it expand very rapidly—much like an explosion. This expansion sends out sound waves, which we recognize as a thunderclap. Most lightning discharges occur within the cloud, but a significant proportion strike land. In the United States, lightning causes a yearly average of about 150 human deaths and property damage of hundreds of millions of dollars, including loss by structural and forest fires set by lightning.

A LOOK AHEAD

Precipitation has been the major focus of this chapter. Keep in mind that no matter what the cause of precipitation, latent heat is always released. Because a significant amount of oceanic evaporation results in precipitation over land, there is a significant flow of latent heat from oceans to land. And, as we will see in the following chapter, global air circulation patterns move this latent heat poleward, helping to warm the continents and creating the distinctive pattern of climates that differentiates our planet.

ATMOSPHERIC MOISTURE AND PRECIPITATION IN REVIEW

Water moves freely between ocean, atmosphere, and land in the hydrologic cycle. The global water balance describes these flows. The fresh water in the atmosphere and on land in lakes, streams, rivers, and ground water is only a small portion of the total water in the hydrosphere.

Humidity describes the amount of water vapor present in air. The ability of air to hold water vapor depends on temperature. Warm air can hold much more water vapor than cold air. Specific humidity measures the mass of water vapor in a mass of air, in grams of water vapor per kilogram of air. Relative humidity measures water vapor in the air as the percentage of the maximum amount of water vapor that can be held at the given air temperature.

The adiabatic principle states that when a gas is compressed, it warms, and when a gas expands, it cools. When an air parcel moves upward in the atmosphere, it encounters a lower pressure and so expands and cools. The dry adiabatic rate describes the rate of cooling with altitude. If the air is cooled below the dew point, condensation or deposition occurs, and latent heat is released. This heat reduces the rate of cooling with altitude, which is described as the wet adiabatic rate if condensation or deposition is occurring.

Clouds are composed of droplets of water or crystals of ice that form on condensation nuclei. Clouds typically occur in layers, as stratiform clouds, or in globular masses, as cumuliform clouds. Fog occurs when a cloud forms at ground level. Precipitation from clouds occurs as rain, hail, snow, and sleet. When supercold rain falls on a surface below freezing, it produces an ice storm.

There are three types of precipitation—orographic, convectional, and cyclonic. In orographic precipitation, air moves up and over a mountain barrier. As it moves up, it is cooled adiabatically and rain forms. As it descends the far side of the mountain, it is warmed. In convectional precipitation, unequal heating of the surface causes an air parcel to become warmer and less dense than the surrounding air. Because it is less dense, it rises. As it moves upward, it cools, and condensation with precipitation may occur. Under conditions of unstable air, thunderstorms can form, yielding hail and lightning. Cyclonic precipitation is described in Chapter 6.

KEY TERMS

sublimation	adiabatic process	convectional precipitation
deposition	dry adiabatic lapse rate	unstable air
hydrologic cycle	wet adiabatic lapse rate	thunderstorm
humidity	cloud	
dew-point temperature	orographic precipitation	

REVIEW QUESTIONS

1. What is the hydrosphere? Where is water found on our planet? In what amounts? How does water move in the hydrologic cycle?
2. Define specific humidity. How is the moisture content of air influenced by air temperature?
3. Define relative humidity. How is relative humidity measured?
4. Use the terms *saturation*, *dew point*, and *condensation* in describing what happens when an air parcel of moist air is chilled.
5. What is the adiabatic process? Why is it important?
6. Distinguish between dry and wet adiabatic lapse rates. In a parcel of air moving upward in the atmosphere, when do they apply? Why is the wet adiabatic lapse rate ess than the dry adiabatic rate? Why is the wet adiabatic rate variable in amount?
7. How are clouds classified? Name four cloud families, two broad types of cloud forms, and three specific cloud types.
8. What is fog? Explain how radiation fog and advection fog form.
9. How is precipitation formed? Describe the process for warm and cold clouds.
10. Describe the orographic precipitation process. What is a rainshadow? Provide an example of the rainshadow effect.
11. What is unstable air? What are its characteristics?
12. Describe the convectional precipitation process. What is the energy source that powers this source of precipitation? Explain.

VISUALIZING EXERCISES

1. Lay out a simple diagram showing the main features of the hydrologic cycle. Include flows of water connecting land, ocean, and atmosphere. Label the flow paths.
2. Graph the temperature of a parcel of air as it moves up and over a mountain barrier, producing precipitation. Label each leg of the graph with the appropriate rate of temperature change.
3. Sketch the anatomy of a thunderstorm cell. Show rising bubbles of air, updraft, downdraft, precipitation, and other features.
4. Sketch a graph showing relative humidity and temperature through a 24-hour cycle.

A DEEPER VIEW OF ATMOSPHERIC MOISTURE AND PRECIPITATION

1. Water in the atmosphere is a very important topic for understanding weather and climate. Organize an essay or oral presentation on this topic, focusing on the following questions: What part of the global supply of water is atmospheric? Why is it important? What is its global role? How does the capacity of air to hold water vapor vary? How is the moisture content of air measured? Clouds and fog visibly demonstrate the presence of atmospheric water. What are they? How do they form?

2. Compare and contrast orographic and convectional precipitation. Begin with a discussion of the adiabatic process and the generation of precipitation within clouds. Then compare the two processes, paying special attention to the conditions that create uplift. Can convectional precipitation occur in an orographic situation? Under what conditions?

Chapter 5

Winds and Global Circulation

Sailboats on the Charles River, Boston.

Coming about! You push the tiller hard to the opposite side of the boat, and the small craft turns easily into the wind. The sail goes slack and begins to swing across the boat. You duck under the boom—the long pole that holds the bottom of the sail taut—and step neatly to the other side. The sail now catches the wind from the other side and begins to fill. You pull in the line attached to the sail and correct the rudder for your new course. The sail tightens, and the boat leans away from the wind, gathering speed. You quickly move to the boat's edge and, sitting on the rim, lean out over the water. Your weight is held in check by the strong pull of the sail on the line in your right hand. The boat swings upright, and you are skimming across the water again, thanks to the cool, southeast breeze.

What a marvelous setting for a sail! With the soaring skyscrapers of Boston on the south and the leafy Cambridge campus of MIT on the north, the mouth of the Charles River is the magnificent jewel in the center of the famed necklace of parks, green spaces, and beaches that frames the map of the city. But sailing on the Charles River is always something of a challenge—the wind can be a bit fickle, to say the least.

Unfortunately, you're not the only one out for a sail. There are perhaps a dozen other boats sharing the water with you, and since it's still early in the afternoon, the fleet is sure to increase. Watching out for them requires attention.

The river bank is fast approaching, and it's time to turn. You let out the sail and ease the tiller away from you. So far, so good. But wait, there's another boat about to sail into your bow! Turn, and fast! You kick the tiller over hard, and the boat pivots cleanly away from the oncoming craft. Oh, oh. The wind's shifting, just at the wrong moment. You're going to jibe! Duck! The sail comes flying across the boat, and the line attached to the boom jerks tight in your hand. You'd better move fast, now, you're on the wrong side of the boat.

Scrambling to the other side of the boat, you drop the tiller! A major mistake. Without the control of the rudder, the boat spins, tilting violently in the strong breeze. Water enters the boat as the rim tips below the waterline. Now the boom is in the water! You put all your weight on the opposite rim of the boat, trying to right the boat against the forces aligned to tip it over. Completing the spin, the boat comes to rest pointed into the wind, the unrestrained sail flapping noisily overhead. You are crouched in the bottom of the boat, breathless, under the wriggling boom. Ankle-deep water covers your feet. Humbled by the power of the wind, you think, "Welcome to the Charles!"

Wind is the subject of this chapter. Did you ever wonder why there is wind? Winds are caused by unequal heating of the earth's surface. As you know from Chapter 4, heating causes air to expand and become less dense. When this occurs, cooler, denser, surrounding air presses inward to buoy up the lighter, heated air, and a wind along the surface is created. This principle directly explains many types of local winds, such as the sea breeze you might experience on an afternoon at the beach. It is also the cause of global wind motions, when we consider that the equatorial and tropical regions are heated more intensely by the sun than are the mid- and higher latitudes. In response to this difference in heating, vast tongues of warm air move poleward, and huge pools of cool air shift equatorward, influencing our day-to-day weather strongly. For global wind motions, however, wind direction is strongly influenced by the earth's rotation—a factor that is also a subject of this chapter.

Chapter Outline

Atmospheric Pressure
  Measuring Atmospheric Pressure
  How Air Pressure Changes with Altitude
Winds and Pressure Gradients
  Sea and Land Breezes
  The Coriolis Effect and Winds
  Cyclones and Anticyclones
  Surface Winds on an Ideal Earth
Global Wind and Pressure Patterns
  Subtropical High-Pressure Belts
  The ITC and the Monsoon Circulation

Wind and Pressure Features of Higher Latitudes
Local Winds
Winds Aloft
  Global Circulation at Upper Levels
  Rossby Waves
  Jet Streams
Ocean Currents
  Surface Currents
Eye on the Environment • El Niño
  Deep Currents and Thermohaline Circulation
Poleward Transport of Heat and Moisture

Chapter 5

WINDS AND GLOBAL CIRCULATION

The air around us is always in motion. From a gentle breeze on a summer's day to a cold winter wind, we experience moving air much of the time. Why does the air move? What are the forces that cause winds to blow? Why do winds blow more often in some directions than others? What is the pattern of wind flow in the upper atmosphere, and how does it affect our weather? How does the global wind pattern produce ocean currents? These are some of the questions we will answer in this chapter. But first we need to examine atmospheric pressure and how wind develops in response to pressure differences. These are the subjects for the next few pages.

ATMOSPHERIC PRESSURE

Did you know that you are living at the bottom of a vast ocean? Yes, it's true! However, it's an ocean of air—the earth's atmosphere. Like the water in the ocean, the air in the atmosphere is constantly pressing on the solid or liquid surface beneath it. This pressure is exerted in all directions. For example, try taking a deep breath. What happens? As you expand your chest volume, air is forced into your lungs by atmospheric pressure. When you contract your chest volume and exhale, you push the air out against atmospheric pressure.

Pressure is a force per unit area. For the atmosphere, the force is produced by gravity acting on the mass of a column of air above a unit area. At sea level, this force is the weight of about 1 kilogram of mass of air above each square centimeter of surface (1 kg/cm²) (Figure 5.1). In the English system, the force is about 15 pounds for each square inch of surface (15 lb/in.²).

In the metric system, the unit of pressure is the *pascal* (Pa). At sea level, the average pressure of air is 101,320 Pa. Although the pascal is the proper metric unit for pressure, many atmospheric pressure measurements are based on the *bar* and the *millibar* (mb) (1 bar = 1000 mb), two former metric units that are still in wide use. In this text, we will use the millibar as the metric unit of atmospheric pressure, following this long tradition. One millibar is equal to 100 pa. Standard sea-level atmospheric pressure is 1013.2 mb.

Measuring Atmospheric Pressure

Perhaps you know that atmospheric pressure is measured by a **barometer**. But how does a barometer work? The principle is simple, and you have experienced it often. Visualize sipping a soda through a straw. By low-

Figure 5.1 Atmospheric pressure imagined as the weight of a column of air. *(a)* Metric system. The weight of a column of air 1 cm on a side is balanced by the weight of a mass of about 1 kg. *(b)* English system. The weight of a column of air 1 in. on a side is balanced by a weight of about 15 pounds.

ering your jaw and moving your tongue, you expand the volume of your mouth. This creates a partial vacuum, and air pressure forces the soda up through the straw and into your mouth.

A *mercury barometer* works on the same principle (Figure 5.2)—that atmospheric pressure will force a liquid up into a tube when a vacuum is present. To construct the barometer, a glass tube about 1 m (3 ft) long, sealed at one end, is completely filled with mercury. The open end is temporarily held closed. Then the tube is inverted, and the open end is immersed into a dish of mercury. When the opening of the tube is uncovered, the mercury in the tube falls some distance but then remains fixed at a level about 76 cm (30 in.) above the surface of the mercury in the dish.

What is happening here? At first, mercury flows out of the tube, under the force of gravity, and into the dish. Since there is no air in the top of the tube, a vacuum is formed. At the same time, atmospheric pressure pushes on the surface of the mercury in the dish, pressing the mercury upward into the tube. Since there is no air in the tube, there is no air pressure to push back against it. The pull of gravity on the mercury column then balances the pressure of the air, and the mercury level stays at a stable height. If air pressure changes, so will the

Figure 5.2 A simple diagram of a mercury barometer. Atmospheric pressure pushes the mercury upward into the tube, balancing the pressure exerted by the weight of the mercury column. As atmospheric pressure changes, the level of mercury in the tube changes. At sea level, the average height of the mercury level is about 76 cm (30 in.).

height of the mercury column. So, this device will measure air pressure directly—but as a height in centimeters or inches rather than a pressure. The mercury barometer has become the standard instrument for weather measurements because of its high accuracy.

The widespread use of the mercury barometer has led to the common measurement of air pressure in heights of mercury, as centimeters or inches. On this scale, standard sea-level pressure is 76 cm Hg (29.92 in. Hg). The letters Hg are the chemical symbol for mercury. In this text, we will use in. Hg as the English unit for atmospheric pressure.

In the mercury barometer, air pressure balances the pull of gravity of a column of mercury about 76 cm (30 in.) high.

Atmospheric pressures at a location vary from day to day by only a small proportion. On a cold, clear winter day, the barometric pressure at sea level might be as high as 1030 mb (30.4 in. Hg), while in the center of a storm system, pressure might drop to 980 mb (28.9 in. Hg), a difference of about 5 percent. Pressure changes are associated with traveling weather systems, a topic that we will discuss in Chapter 6.

How Air Pressure Changes with Altitude

Figure 5.3 shows how atmospheric pressure decreases with altitude. Moving upward from the earth's surface,

the decrease of pressure with altitude is initially quite rapid. At higher altitudes, the decrease is much slower. Because atmospheric pressure decreases rapidly with altitude near the surface, a small change in elevation will often produce a significant change in air pressure. For example, you may have noticed that your ears sometimes "pop" from a pressure change during an elevator ride in a tall building. You may also have noticed your ears popping on an airplane flight. Aircraft cabins are pressurized to about 800 mb (24 in. Hg), which corresponds to an elevation of about 1800 m (5900 ft).

The decrease of pressure with elevation also affects human physiology. Perhaps you've felt out of breath, or tired more easily, walking along a high mountain trail. With decreased air pressure, oxygen moves into lung tissues more slowly, producing shortness of breath and fatigue. These symptoms, along with headache, nosebleed, or nausea, are referred to as mountain sickness and are likely to occur at altitudes of 3000 m (about 10,000 ft) or higher. Most people adjust to the reduced air pressure of a high mountain environment after several days.

Figure 5.3 Atmospheric pressure decreases with increasing altitude above the earth's surface.

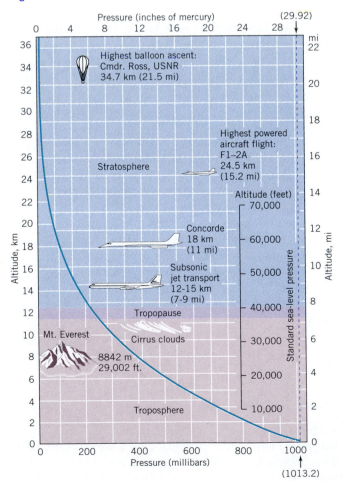

WINDS AND PRESSURE GRADIENTS

From a gentle, cooling breeze on a warm summer afternoon to the howling gusts of a major winter storm, wind is a familiar phenomenon. **Wind** is defined as air motion with respect to the earth's surface. Wind movement is dominantly horizontal. Air motions that are dominantly vertical are not called winds but are known by other terms, such as updrafts or downdrafts.

To describe winds fully, we must measure both wind speed and wind direction. Wind direction can be determined by a *wind vane*, the most common of the weather instruments. The design of the wind vane keeps it facing into the wind. Wind direction is given as the direction from which the wind is coming. So, an east wind comes from the east and moves to the west (Figure 5.4). Wind speed is measured by an *anemometer*. The most common type is the cup anemometer, which consists of three funnel-shaped cups mounted on the ends of spokes of a horizontal wheel. The cups rotate with a speed proportional to that of the wind. Units are either meters per second or miles per hour.

Wind is caused by differences in atmospheric pressure from place to place. Air tends to move from high to low pressure until the air pressures are equal. This follows the simple physical principle that any flowing fluid (such as air) subjected to gravity will move until the pressure at every level is uniform.

Figure 5.5 illustrates this principle. It is a simple map showing two locations, taken here as Wichita and Columbus. At Wichita, the pressure is high (H), and the barometer (adjusted to sea-level pressure) reads 1028 mb (30.4 in. Hg). At Columbus, the barometer is low (L) and reads 997 mb (29.4 in. Hg). Thus, there is a **pressure gradient** between Wichita and Columbus. The lines drawn on the map around Wichita and Columbus are lines of equal pressure, called **isobars**. As you move between Wichita and Columbus, you cross

Figure 5.5 Isobars and a pressure gradient. High pressure is centered at Wichita, Kansas, and low pressure is centered at Columbus, Ohio.

these isobars on the map, encountering changing pressures as a result of the drop from 1025 to 1000 mb (30.3 to 29.5 in. Hg).

Pressure gradients within the atmosphere cause winds.

Because atmospheric pressure is unequal at Wichita and Columbus, a *pressure gradient force* will push air from Wichita toward Columbus. The greater the pressure difference between the two locations, the greater this force will be. This pressure gradient will produce a wind. However, the wind will not necessarily move directly from Wichita to Columbus. It will be deflected by the Coriolis effect, as we will see later in this chapter.

Pressure gradients develop because of unequal heating of the earth's surface. This unequal heating produces air of different temperatures. Recall from Chapter 4 that air parcels of different temperatures will have different densities. That is, warm air is less dense than cool air. Thus, a column of warmer, less dense air over a location will weigh less, producing low pressure. Conversely, a column of colder, denser air will weigh more, producing high pressure.

Sea and Land Breezes

If you've ever vacationed at the beach in the summer, you may have noticed that in the afternoons a *sea breeze* often sets in. This wind brings cool air off the water, dropping temperatures and refreshing beachgoers and residents living close to the beach. Late at night, a *land breeze* may develop. This wind moves cooler air, which is chilled over land by nighttime radiant cooling, toward the water. Sea and land breezes are simple examples of how heating and cooling produce a pressure gradient that causes a wind. Figure 5.6 shows them in more detail.

On a clear day, the sun heats the land surface rapidly, and a shallow air layer near the ground is strongly warmed (Figure 5.6a). The warm air expands. Because

Figure 5.4 Winds are designated according to the compass point from which the wind comes. An east wind comes from the east, but the air is moving westward.

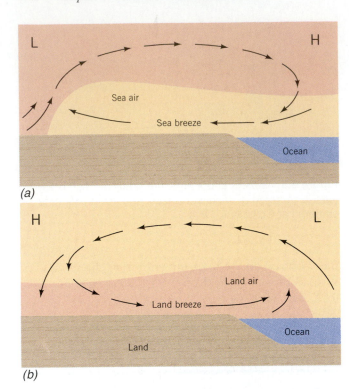

(a)

(b)

Figure 5.6 Sea breeze *(a)* and land breeze *(b)* alternate in direction from day to night. These are caused by heating and cooling of land, while the ocean remains at an even temperature. A convection loop is produced that circulates air in one direction at the surface and in the other direction aloft. (After R. G. Barry and R. J. Chorley, *Atmosphere, Weather and Climate,* 6th ed., Routledge. Used by permission.)

the warm air is less dense than the cool air offshore, it weighs less and exerts less pressure. Thus, low pressure is formed over the coastal belt. Pressure remains higher over the water. This means that there is a pressure gradient from ocean to land, and a wind is set in motion. The sea breeze can become very strong in a shallow air layer close to shore.

Because the air over the land is warmer and less dense, it will also tend to rise. Over the ocean, the air moving landward will be replaced by air moving downward from above, causing a sinking motion. The rising over land and the sinking over ocean will create a return flow aloft that is opposite in direction to the sea breeze. We refer to this as a *convection loop,* in which unequal heating causes a circuit of moving air.

At night, the land surface cools rapidly if skies are clear. A cooler, denser air layer of higher pressure develops (Figure 5.6*b*). Now the pressure gradient is from land to ocean, setting up a land breeze. Like the sea breeze, the convection loop is completed by a wind aloft moving in the opposite direction.

The Coriolis Effect and Winds

The pressure gradient force tends to move air from high pressure to low pressure. For sea and land

breezes, which are local in nature, this push produces a wind motion in about the same direction as the pressure gradient. For larger wind systems, however, the direction of air motion will be somewhat different. The difference is due to the earth's rotation, through the Coriolis effect. What is this effect, and how does it come about?

Through the **Coriolis effect**, an object in motion on the earth's surface always appears to be deflected away from its course. The object moves as though a force were pulling it sidewards. The apparent deflection is to the right in the northern hemisphere and to the left in the southern hemisphere. The effect is strongest near the poles and decreases to no apparent deflection at the equator. Figure 5.7 diagrams the effect. Note that the apparent force does not depend on the direction of motion—the deflection occurs whether the object is moving toward the north, south, east, or west.

The Coriolis effect causes the apparent motion of winds to be deflected away from the direction of the pressure gradient.

The Coriolis effect is a result of the earth's rotation. As our earth rotates, the compass directions (north,

Figure 5.7 The Coriolis effect acts to deflect the paths of winds or ocean currents to the right in the northern hemisphere and to the left in the southern hemisphere as viewed from the starting point.

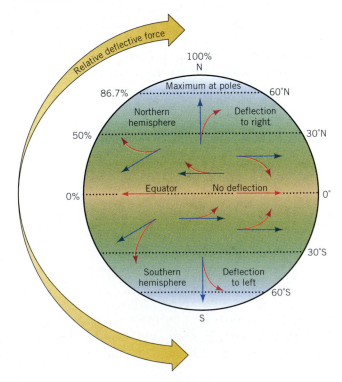

Figure 5.8 The Coriolis effect. The path of a rocket launched from the north pole toward New York will apparently be deflected to the right by the earth's rotation and might land near Chicago. If launched from New York toward the north pole, its path will apparently be deflected to the right as well.

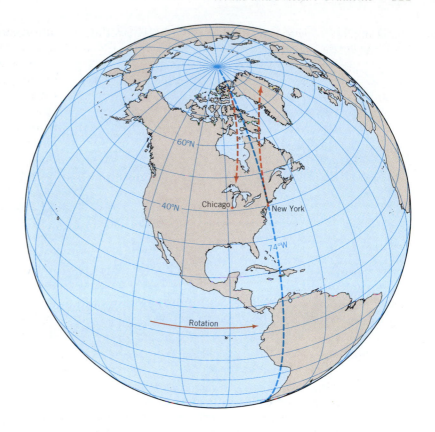

south, east, and west) at a point on the earth also rotate with respect to an observer in space. So motion that is independent of the earth's surface will not appear to follow a straight course to an observer on the ground. To visualize this effect, imagine that a rocket is launched from the north pole toward New York, following the 74° W longitude meridian (Figure 5.8). However, as it travels toward New York, the earth is rotating beneath its trajectory, moving from west to east. The rocket's path will curve to the right with respect to the earth underneath. Using reasonable values for calculation, we find that the rocket might well fall to earth near Chicago!

What would happen if the rocket were launched from New York heading along the 74° W meridian toward the pole? Because of the earth's rotation, a point on the earth's surface at the latitude of New York (about 40° N) moves eastward at about 1300 km/hr (about 800 mi/hr). Although the rocket is aimed properly along the meridian, its motion will have an eastward component of 1300 km/hr. However, as the rocket moves northward, it passes over land that is moving less swiftly eastward. At 60° N latitude, the eastward velocity of the earth's surface is only about 800 km/hr (about 500 mi/hr). So, the rocket will be moving more quickly eastward than the land underneath, and its path will appear to be deflected to the right.

Cyclones and Anticyclones

If the Coriolis effect deflects air moving from high to low pressure, how does that affect the wind pattern? A parcel of air in motion near the surface is subjected to three influences. First is the pressure gradient that propels the parcel toward low pressure. Second is the Coriolis effect that turns the parcel to the side. Third is friction with the ground surface. As a result of all these influences, air will move toward low pressure but at an angle to the pressure gradient.

Figure 5.9 shows wind patterns around surface high- and low-pressure centers in northern and southern hemispheres. When low pressure is at the center, the pressure gradient is straight inward (left side of figure). When high pressure is at the center, the gradient is straight outward (right side). But when air actually moves in response to these pressure gradients, it will move at an angle across the gradient.

In low-pressure centers (left side), air converges toward the center and moves upward. Winds are deflected into an inward-moving spiral pattern, or *inspiral*. In the northern hemisphere (upper portion), the inspiral will be counterclockwise, since the Coriolis deflection is to the right. In the southern hemisphere (lower portion), the inspiral will be clockwise, since the Coriolis deflection is to the left. Meteorologists use the

Figure 5.9 Surface winds spiral inward toward the center of a cyclone, but outward and away from the center of an anticyclone. Because the Coriolis effect deflects moving air to the right in the northern hemisphere and to the left in the southern hemisphere, the direction of inspiraling and outspiraling reverses from one hemisphere to the other.

term **cyclone** to refer to a center of low pressure with inspiraling air.

In high-pressure centers, the situation is reversed. Air descends from above and moves outward, forming an *outspiral*, clockwise in the northern hemisphere and counterclockwise in the southern hemisphere. This outspiral is referred to as an **anticyclone** (Figure 5.9).

In a cyclone, winds converge spiraling inward and upward. In an anticyclone, winds diverge, spiraling downward and outward.

Recall that in the case of the sea and land breezes, air moves upward at the site of low pressure and downward at the site of high pressure. The same is true of cyclones and anticyclones. In the cyclone, air spiraling inward converges, causing a rising motion at the center. Also recall from Chapter 4 that when air is forced aloft, it is cooled according to the adiabatic principle, and condensation and precipitation can occur. This is why cyclones, or low-pressure centers, are associated with cloudy or rainy weather.

In the anticyclone, air spiraling outward diverges, causing a sinking motion at the center. When air descends, we know from Chapter 4 that it is warmed according to the adiabatic principle, and thus conden-

sation cannot occur. So, anticyclones, or high-pressure centers, are associated with clear weather.

Cyclones and anticyclones are typically large features of the atmosphere—perhaps a thousand kilometers (about 600 mi) across, or more. For example, a fair weather system, which is an anticyclone, may stretch from the Rockies to the Appalachians. Cyclones and anticyclones can remain more or less in one location, or they can move, sometimes rapidly, to create weather disturbances.

Surface Winds on an Ideal Earth

Consider wind and pressure patterns on an ideal earth—one without a complicated pattern of land and water and no seasonal changes. Pressure and winds for such an ideal earth are shown in Figure 5.10. The surface wind patterns are shown on the globe's surface. On the right, a cross section through the atmosphere shows the wind patterns aloft.

In the Hadley cell convection loop, air rises at the intertropical convergence zone and descends in the subtropical high-pressure cells.

The most important features for understanding the pattern at low latitudes are shown on the cross section to the right, labeled "Hadley cells." Recall that when air is heated, low pressure forms and air rises. Since insolation is strongest when the sun is directly overhead, the equator will be heated more strongly than other places on this featureless earth. The result will be two convection loops, the **Hadley cells**, which form in the northern and southern hemispheres. In each Hadley cell, air rises over the equator, flows poleward, and descends at about a 30° latitude. (The Hadley cell is named for George Hadley, who first proposed its existence in 1735.)

Since air is heated and rises at the equator, a zone of low pressure will result. This zone is referred to as the *equatorial trough*. As shown by the wind arrows, air moves toward the equatorial trough, where it converges and moves aloft as part of the Hadley cell circulation. Convergence occurs in a narrow zone, named the **intertropical convergence zone (ITC)**. Because the air is generally rising, winds in the ITC are light and variable. In earlier centuries mariners on sailing vessels referred to this region as the *doldrums*, where they were sometimes becalmed for days at a time.

On the poleward side of the Hadley cell circulation, air descends and surface pressures are high. This produces two **subtropical high-pressure belts**, each centered at about 30° latitude. Within the belts, two, three, or four very large and stable anticyclones are formed. Because air descends at the centers of these anticyclones, winds in the subtropical high-pressure belts are weak.

Figure 5.10 This schematic diagram of global surface winds and pressures shows the features of an ideal earth, without the disrupting effect of oceans and continents and the variation of the seasons. Surface winds are shown on the disk of the earth, while the cross section at the right shows winds aloft.

Calm prevails as much as one-quarter of the time. Because of the high frequency of calms, mariners named this belt the *horse latitudes.* The name is said to have originated in colonial times from the experiences of New England traders carrying cargoes of horses to the West Indies. When their ships were stranded for long periods of time, the freshwater supplies ran low and the horses were thrown overboard.

Winds around the subtropical high-pressure centers are outspiraling and move toward equatorial as well as middle latitudes. The winds moving equatorward are the strong and dependable *trade winds.* North of the equator, these are from the northeast and are referred to as the *northeast trades.* To the south of the equator, they are from the southeast and are the *southeast trades.* Poleward of the subtropical anticyclones, air spiraling outward produces southwesterly winds in the northern hemisphere and northwesterly winds in the southern hemisphere.

Between about 30° and 60° latitude, the pressure and wind pattern becomes more complex. This latitudinal belt is a zone of conflict between air bodies with different characteristics. Masses of cool, dry air move into the region, heading eastward and equatorward. These form *polar outbreaks.* The border of the polar outbreak is known as the *polar front.* (We'll have much more to say about air masses and the polar front in Chapter 6.) As a

result of this activity, pressures and winds can be quite variable in the midlatitudes from day to day and week to week. On average, however, winds are more often from the west, so the region is said to have *prevailing westerlies.*

At the poles, the air is intensely cold. As a result, high pressure occurs. Outspiraling of winds around a polar anticyclone should create surface winds from a generally easterly direction, known specifically as *polar easterlies.* As we will see in discussing actual pressure and wind patterns, this situation exists only in the south polar region. In the north polar region, winds tend to have an eastward component, but there is too much variation in direction to consider polar easterlies as dominant winds there.

GLOBAL WIND AND PRESSURE PATTERNS

So far, we've discussed the wind pattern for a seasonless, featureless earth. Let's turn now to actual global wind and pressure patterns. Here, we will use the global maps of wind and pressure for January and July, shown in Figure 5.11. The pressures and winds shown are averages over many years for all daily observations in either January or July. They are corrected for the elevation of the recording station, so that pressures are shown for sea level. Average barometric pressure is 1013 mb (29.2

JANUARY

JULY

Figure 5.11 Mean monthly atmospheric pressure and prevailing surface winds for January and July. Pressure units are millibars reduced to sea level. Many of the wind arrows are inferred from isobars. (Data compiled by John E. Oliver.)

Figure 5.11 (continued)

in. Hg). Values greater than this value are "high" (red lines), while those lower are "low" (green lines). The maps make use of both Mercator and polar projections. As we discuss the various features of global pressure and winds, be sure to examine both sets of maps.

Subtropical High-Pressure Belts

The largest and most prominent features of the maps are the subtropical high-pressure belts, created by the Hadley cell circulation. The high-pressure belt in the southern hemisphere conforms well to the pattern of Figure 5.10. It has three large high-pressure cells, each developed over oceans, that persist year round. A fourth, weaker high-pressure cell forms over Australia in July, as the continent cools during the southern hemisphere winter.

In the northern hemisphere, the situation is somewhat different. The subtropical high-pressure belt shows two large anticyclones centered over oceans—the Hawaiian High in the Pacific and the Azores High in the Atlantic. From January to July, these intensify and move northward. From their positions off the east and west coasts of the United States, they have a dominant influence on summer weather in North America.

Figure 5.12 is a schematic map of two large high-pressure cells bounded by continents, showing the features of circulation around the cells. As we noted earlier, the outspiraling circulation produces the trade winds in the tropical and equatorial zones, and the westerlies in the subtropical zones and poleward. In the high-pressure cells, air on the east side subsides more intensely. This

Figure 5.12 Over the oceans, surface winds spiral outward from the subtropical high-pressure cells, feeding the trades and the westerlies. On the eastern side of the cells, air subsides more strongly, producing dry winds. On the western sides, subsidence is not as strong, and a long passage over oceans brings warm, moist air to the continents.

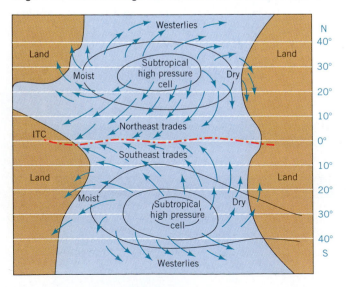

means that winds spiraling outward on the eastern side are drier. On the west side, subsidence is less strong. In addition, these winds travel long distances across warm, tropical ocean surfaces before reaching land, and so are warm and moist.

When the Hawaiian and Azores Highs intensify and move northward in the summer, our east and west coasts feel their effects. On the west coast, dry subsiding air from the Hawaiian High dominates, so fair weather and rainless conditions prevail. On the east coast, warm, moist air from the Azores High flows across the continent from the southeast, producing generally hot, humid weather for the central and eastern United States. In winter, these two anticyclones weaken and move to the south—leaving our weather to the mercy of colder winds and air masses from the north and west.

The ITC and the Monsoon Circulation

Recall from Chapter 2 that insolation is most intense when the sun is directly overhead. Remember, too, that the latitude at which the sun is directly overhead changes with the seasons, migrating between the tropics of cancer and capricorn. Since the Hadley cell circulation is driven by this heating, we can expect that the elements of the Hadley cell circulation—ITC and subtropical high-pressure belts—will also shift with the seasons. We have already noted this shift for the Hawaiian and Azores Highs, which intensify and move northward as July approaches.

In general, the ITC also shifts with the seasons, as we might expect. The shift is moderate in the western hemisphere. The ITC moves a few degrees north from January to July over the oceans. In South America, the ITC lies across the Amazon in January and swings northward by about 20° to the northern part of the continent. But what happens in Africa and Asia? By comparing the two Mercator maps of Figure 5.11, you can see that there is a huge shift indeed! In January, the ITC runs south across eastern Africa and crosses the Indian Ocean to northern Australia at a latitude of about 15° S. In July, it swings north across Africa along the southern side of the Sahara, and then it rises further north to lie along the south rim of the Himalayas, in India, at a latitude of about 25° N. This is a shift of about 40 degrees of latitude!

Why does such a large shift occur in Asia? To answer this question, look at the pressure and wind pattern over Asia. In January, an intense high-pressure system, the Siberian High, is found there. Recall from Chapter 3 that, in winter, temperatures in northern Asia are very cold, so this high-pressure is to be expected. In July, this high-pressure center is absent, replaced instead by a low centered over the Middle Eastern desert region. The Asiatic low is produced by the intense summer heating of the landscape.

The Asian monsoon wind pattern consists of a cool, dry air flow from the northeast during the low-sun season, and a warm, moist airflow from the southwest during the high-sun season.

The movement of the ITC and the change in the pressure pattern with the seasons create a reversing wind pattern in Asia known as the **monsoon** (Figure 5.13). In the winter, there is a strong outflow of dry, continental air from the north across China, Southeast Asia, India, and the Middle East. During this *winter monsoon*, dry conditions prevail. In the summer, warm, humid air from the Indian Ocean and the southwestern Pacific moves northward and northwestward into Asia, passing over India, Indochina, and China. This airflow is known as the *summer monsoon* and is accompanied by heavy rainfall in southeastern Asia.

North America does not have the remarkable extremes of monsoon winds experienced in Asia. Even so, in summer there is a prevailing tendency for warm,

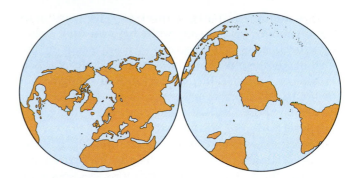

Figure 5.14 Northern and southern hemispheres—a study in contrasts in the distribution of lands and oceans. One pole bears a deep ocean and the other a great continental land mass.

moist air originating in the Gulf of Mexico to move northward across the central and eastern part of the United States. This Gulf air sometimes moves westward across the southwestern deserts of New Mexico and Arizona, occasionally reaching the California coast. This airflow is referred to locally as a "monsoon." In winter, the airflow across North America generally reverses, and dry, continental air from Canada moves south and eastward.

Wind and Pressure Features of Higher Latitudes

The northern and southern hemispheres are quite different in their geography. As you can easily see from Figure 5.14, the northern hemisphere has two large continental masses, separated by oceans. An ocean is also at the pole. In the southern hemisphere, we find a large ocean with a cold, glacier-covered continent at the center. These differing land–water patterns strongly influence the development of high- and low-pressure centers with the seasons.

Let's examine the northern hemisphere first (Figure 5.11). Recall from Chapter 3 that continents will be cold in winter and warm in summer, as compared to oceans at the same latitude. But we know from our analysis earlier in the chapter that cold air will be associated with high pressure and warm air with low pressure. Thus, continents will show high pressure in winter and low pressure in summer.

This pattern is very clear in the northern hemisphere polar maps of Figure 5.11. In winter, the strong Siberian High is found in Asia, and a weaker cousin, the Canadian High, is found in North America. From these high-pressure centers, air spirals outward, bringing cold air to the south. Over the oceans, two large centers of low pressure are striking—the Icelandic Low and the Aleutian Low. These two low-pressure centers are not actually large stable features that we would expect to find on every daily world weather map in January. Rather, they are regions of average low pressure where winter storm systems are spawned.

Figure 5.13 The Asiatic monsoon winds alternate in direction from January to July, responding to reversals of barometric pressure over the large continent.

JANUARY

JULY

In summer, the pattern reverses. The continents show generally low pressure, while high pressure builds over the oceans. This pattern is easily seen on the northern hemisphere July polar map. The Asiatic Low is strong and intense. Inspiraling winds, forming part of the monsoon cycle, bring warm, moist Indian Ocean air over India and Southeast Asia. A lesser low forms over the deserts of the southwestern United States and northwestern Mexico. The two subtropical highs, the Hawaiian Highs and the Azores Highs, strengthen and dominate the Atlantic and Pacific Ocean regions. Outspiraling winds, following the pattern of Figure 5.12, keep the west coasts of North America and Europe warm and dry, and the east coasts of North America and Asia warm and moist (see Figure 5.11 also).

The higher latitudes of the southern hemisphere present a polar continent surrounded by a large ocean. Since Antarctica is covered by a glacial ice sheet and is cold at all times, a permanent anticyclone, the South Polar High, is centered there. Easterly winds spiral outward from the high-pressure center. Surrounding the high is a band of deep low pressure, with strong, inward-spiraling westerly winds. As early mariners sailed southward, they encountered this band, in which wind strength intensifies toward the pole. Because of the strong prevailing westerlies, they named these southern latitudes the "roaring forties," "flying fifties," and "screaming sixties."

LOCAL WINDS

Not every wind is produced by global pressure gradients in large weather systems. In some locations, *local winds* are important. These are often influenced by local terrain. We have already discussed one class of local winds—sea and land breezes—earlier in this chapter.

Mountain winds and *valley winds* are local winds that alternate in direction in a manner similar to the land and sea breezes. During the day, mountain hill slopes are heated intensely by the sun, causing air to rise. To take the place of the rising air, a current moves up valleys from the plains below—upward over rising mountain slopes, toward the summits. At night, the hill slopes are chilled by radiation. The cooler, denser hill slope air then moves valleyward, down the hill slopes, to the plain below. These winds are responding to local pressure gradients, set up by heating or cooling of the local air.

Another group of local winds are known as *drainage winds*, in which cold, dense air flows under the influence of gravity from higher to lower regions. In a typical situation, cold, dense air accumulates in winter over a high plateau or high interior valley. Under favorable conditions, some of this cold air spills over low divides or through passes, flowing out on adjacent lowlands as a strong, cold wind.

Figure 5.15 Strong downslope winds of hot, dry air fan this brushfire in Oakland, California.

Local winds include land and sea breezes, mountain and valley winds, drainage winds, and Santa Ana winds.

Drainage winds occur in many mountainous regions of the world and go by various local names. The *mistral* of the Rhône Valley in southern France is a well-known example—it is a cold, dry local wind. On the ice sheets of Greenland and Antarctica, powerful drainage winds move down the gradient of the ice surface and are funneled through coastal valleys. Picking up loose snow, these winds produce blizzardlike conditions that last for days at a time.

Another type of local wind occurs when the outward flow of dry air from an anticyclone is combined with the local effects of mountainous terrain. An example is the Santa Ana—a hot, dry easterly wind that sometimes blows from the interior desert region of southern California across coastal mountain ranges to reach the Pacific coast. This wind is funneled through local mountain gaps or canyon floors where it gains great force. At times the Santa Ana wind carries large amounts of dust. Because this wind is dry, hot, and strong, it can easily fan wildfires in brush or forest out of control (Figure 5.15).

Still another type of local wind, sometimes bearing the name *chinook*, results when strong regional winds pass over a mountain range and descend on the lee side. (See Figure 4.16.) As we know from studying orographic precipitation in Chapter 4, the descending air is heated and dried. A chinook wind can sublimate snow or dry out soils very rapidly.

WINDS ALOFT

The surface wind systems we have examined describe airflows at or near the surface. How does air move at the higher levels of the troposphere? Just as we saw for air near the surface, air aloft will move in response to pressure gradients and will be influenced by the Coriolis effect.

Pressure gradients aloft are generated by large, slow-moving, high- and low-pressure systems. Figure 5.16 shows a typical pattern of high and low pressure and winds that might be found at an altitude of 5 to 7 km (about 16,000 to 23,000 ft). Notice that the wind arrows are shown as following the isobars, not cutting across them as they do at the surface. The difference is that a parcel of air at high altitude moves freely, without the drag of surface friction. As a result, air at upper levels in the atmosphere flows smoothly around centers of high and low pressure, following a curving path. This type of wind is called the *geostrophic wind.*

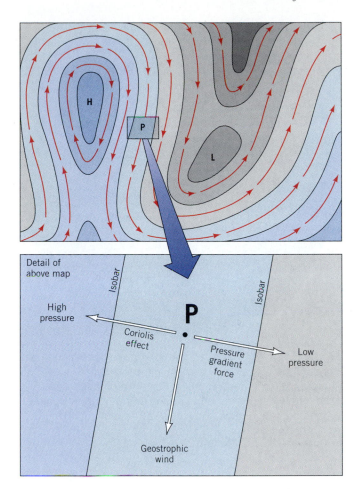

Figure 5.16 At high levels above the earth's surface, the wind blows parallel to the isobars. The northern hemisphere is shown here.

Global Circulation at Upper Levels

Figure 5.17 sketches the general pattern of airflows at higher levels in the troposphere. The pattern has four major features—equatorial easterlies, tropical high-pressure belts, upper-air westerlies, and a polar low.

The *upper-air westerlies* blow in a complete circuit about the earth, from about 25° latitude almost to the poles. They comprise the first major tropospheric circulation system. At high latitudes these westerlies form a huge circumpolar spiral, circling a great polar low-pressure center. Toward lower latitudes, atmospheric pressure rises steadily, forming a *tropical high-pressure belt* at 15° to 20° N and S latitude. This is the high-altitude part of the surface subtropical high-pressure belt, but it is shifted somewhat equatorward. Between the high-pressure ridges is a zone of weak low pressure in which the winds are easterly. These winds comprise the second major circulation system of the troposphere. They are called the *equatorial easterlies.*

Figure 5.17 A generalized plan of global winds high in the troposphere. Easterly winds flow in the band in between the subtropical high-pressure belts. Poleward of these belts, winds are westerly, but often sweep to the north or south around centers of high and low pressure aloft. (Copyright © A. N. Strahler.)

The general pattern of winds aloft is a band of equatorial easterly winds, flanked by tropical high-pressure belts, and a spiral of westerly winds to poleward.

Thus, the overall picture of upper-air wind patterns is really quite simple—a band of easterly winds in the equatorial zone, belts of high pressure near the tropics of cancer and capricorn, and westerly winds, with some variation in direction, spiraling around polar lows.

Rossby Waves

The smooth westward flow of the upper-air westerlies frequently forms undulations, called **Rossby waves**. Figure 5.18 shows how these waves develop and grow in the northern hemisphere. The waves arise in a zone of contact between cold polar air and warm tropical air.

For a period of several days or weeks, the flow may be fairly smooth. Then, an undulation develops, and warm air pushes poleward while a tongue of cold air is brought to the south. Eventually, the tongue is pinched off, leaving a pool of cold air at a latitude far south of its normal location. This cold pool may persist for some days or weeks, slowly warming with time.

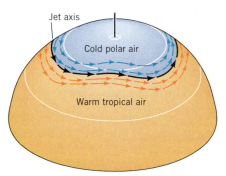

The jet stream begins to undulate.

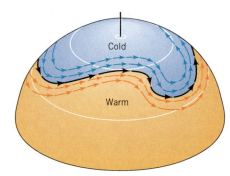

Rossby waves begin to form.

Figure 5.18 Development of upper-air Rossby waves in the westerlies of the northern hemisphere. (Copyright © A. N. Strahler.)

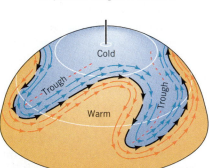

Waves are strongly developed. The cold air occupies troughs of low pressure.

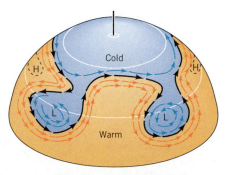

When the waves are pinched off, they form cyclones of cold air.

Jet Streams

Jet streams are important features of upper-air circulation. They are narrow zones at a high altitude in which wind streams reach great speeds. They occur where atmospheric temperature gradients are strong. Along a jet stream, pulselike movements of air follow broadly curving tracks (Figure 5.19). The greatest wind speeds occur in the center of the jet stream, with velocities decreasing away from the center. There are three kinds of jet streams. Two are westerly wind streams. The third is a weaker jet with easterly winds that develops in Asia as part of the summer monsoon circulation. These are shown in Figure 5.20.

The most poleward type of jet stream is located along the polar front—the fluctuating boundary between cold polar air and warmer tropical air (Figure 5.10). It is designated the *polar-front jet stream* (or simply, the "polar jet"). Located generally between 35° and 65° latitude, it is present in both hemispheres. The polar jet normally follows the edges of Rossby waves, since these mark a strong temperature boundary between cold polar air and warm subtropical air. The polar jet is typically found at altitudes of 10 to 12 km (about 30,000 to 40,000 ft), and wind speeds in the jet range from 350 to 450 km/hr (about 200 to 250 mi/hr). In the northern hemisphere, traveling aircraft often use the polar jet to increase ground speed when flying eastward. In the westward direction, flight paths are chosen to avoid the strong winds of the polar jet.

A second type of jet stream forms in the subtropical latitude zone—the *subtropical jet stream*. It occupies a position at the tropopause just above the Hadley cell in northern and southern hemispheres (Figure 5.21). Here, westerly wind speeds reach maximum values of 345 to 385 km/hr (215 to 240 mph). This jet is associat-

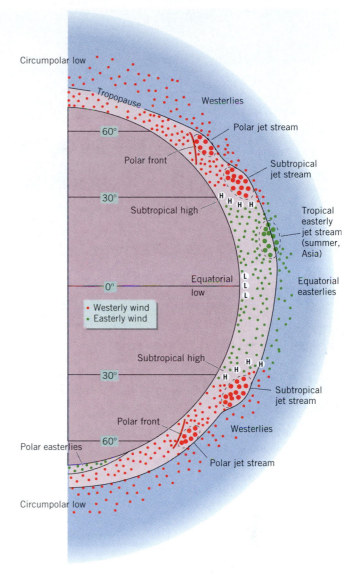

Figure 5.20   A schematic diagram of wind directions and jet streams along an average meridian from pole to pole. The four polar and subtropical jets are westerly in direction, in contrast to the single tropical easterly jet. (Copyright © A. N. Strahler.)

ed with a high-altitude temperature contrast at the poleward edge of the Hadley cell. Cloud bands of the subtropical jet stream are pictured in Figure 5.21.

A third type of jet stream is found at even lower latitudes. Known as the *tropical easterly jet stream*, it runs from east to west—opposite in direction to that of the polar-front and subtropical jet streams. The tropical easterly jet occurs only in the summer season and is limited to a northern hemisphere location over Southeast Asia, India, and Africa.

Figure 5.19   The polar jet stream is shown on this map by lines of equal wind speed. (National Weather Service.)

Figure 5.21   A strong subtropical jet stream is marked in this space photo by a narrow band of cirrus clouds, generated by turbulence in the air stream. This jet stream is moving from west to east at an altitude of about 12 km (40,000 ft). The cloud band lies at about 25° N lat. In this view, astronauts aimed their camera toward the southeast, taking in the Nile River Valley and the Red Sea. At the left is the tip of the Sinai Peninsula.

OCEAN CURRENTS

Just as there is a circulation pattern to the atmosphere, so there is a circulation pattern to the oceans. An *ocean current* is any persistent, dominantly horizontal flow of ocean water. Current systems act to exchange heat between low and high latitudes and are essential in sustaining the global energy balance. We can recognize both surface currents and deep currents. Surface currents are driven by prevailing winds. Deep currents are powered by changes in temperature and density occurring in surface waters that cause them to sink.

Surface Currents

The patterns of surface ocean currents are strongly related to prevailing surface winds. Energy is transferred from wind to water by the friction of the air blowing over the water surface. Because of the Coriolis effect, the actual direction of water drift is deflected about 45° from the direction of the driving wind.

Because ocean currents move warm waters poleward and cold waters toward the equator, they are important regulators of air temperatures. Warm surface currents keep winter temperatures in the British Isles from falling much below freezing in winter. Cold surface currents keep weather on the California coast cool, even in the height of summer.

Figure 5.22 presents a world map that shows surface ocean currents for the month of January. The primary features are large circular movements, called **gyres**, that are centered at latitudes of 20° to 30°. These motions track the movements of air around the subtropical high-pressure cells.

In the equatorial region, ocean currents flow westward, pushed by northeast and southeast trade winds. As these equatorial currents approach land, they are turned poleward along the west sides of the oceans, forming warm currents paralleling the coast. Examples are the Gulf Stream of eastern North America and the Kuroshio Current of Japan.

In the zone of westerly winds, a slow eastward motion of water is named the *west-wind drift*. As west-wind drift

Figure 5.22 Surface drifts and currents of the oceans in January. (Based on data from U.S. Navy Oceanographic Office. Redrawn and revised by A. N. Strahler.)

waters approach the western sides of the continents, they are deflected equatorward along the coast. The equatorward flows are cool currents. They are often accompanied by *upwelling* along continental margins. In this process, colder water from greater depths rises to the surface. Examples of cool currents with upwelling are the Humboldt (or Peru) Current, off the coast of Chile and Peru, the Benguela Current, off the coast of southern Africa, and the California Current.

Global surface currents are dominated by huge circular gyres centered near the subtropical high-pressure cells.

West-wind drift water also moves poleward to join arctic and antarctic circulations. In the northeastern Atlantic Ocean, the west-wind drift forms a relatively warm current. This is the North Atlantic Drift, which spreads around the British Isles, into the North Sea, and along the Norwegian coast. The Russian port of Murmansk, on the arctic circle, remains ice-free year round because of this warm drift current.

In the northern hemisphere, where the polar sea is largely landlocked, cold currents flow equatorward along the east sides of continents. Two examples are the Kamchatka Current, which flows southward along the Asian coast across from Alaska, and the Labrador Current, which flows between Labrador and Greenland to reach the coasts of Newfoundland, Nova Scotia, and New England.

The strong west winds around Antarctica produce an Antarctic circumpolar current of cold water. Some of this flow branches equatorward along the west coasts of South America and Africa, adding to the Humboldt and Benguela Currents.

Figure 5.23 shows a satellite image of ocean temperature along the east coast of North America for a week in April. The Gulf Stream stands out as a tongue of red and yellow (warm) color, moving along the coast and heading off from North Carolina in a northeasterly direction. Cooler water from the Labrador Current, in green and blue tones, hugs the northern Atlantic coast. This current heads south and then turns to follow the Gulf Stream to the northeast. Instead of mixing, the two flows remain quite distinct. The boundary between them

EYE ON THE ENVIRONMENT

El Niño

At intervals of about three to eight years, a remarkable disturbance of ocean and atmosphere occurs. It begins in the eastern Pacific Ocean and spreads its effects widely over the globe. This disturbance lasts for more than a year, bringing extreme and sometimes severe weather to many global locations. This phenomenon is called *El Niño*. The expression comes from Peruvian fishermen, who refer to the *Corriente del Niño*, or the "Current of the Christ Child," in describing an invasion of warm surface water that occurs once every few years around Christmas time and greatly depletes their catch of fish. El Niño occurs at irregular intervals and with varying degrees of intensity. Notable El Niño events occurred in 1891, 1925, 1940–1941, 1965, 1972–1973, 1982–1983, and 1989–1990, 1991–1992, 1994–1995, and 1997–1998.

Normally, the cool Humboldt (Peru) Current flows northward off the South American coast, and then at about the equator it turns westward across the Pacific as the south equatorial current (see Figure 5.22). The Humboldt Current is fed by upwelling of cold, deep water, bringing with it nutrients that serve as food for marine life. With the onset of El Niño, upwelling ceases, the cool water is replaced by warm, sterile water from the west, and the abundant marine life disappears.

In an El Niño year, a major change in barometric pressures occurs across the entire stretch of the equatorial zone as far west as southeastern Asia. Normally, low pressure prevails over northern Australia, the East Indies, and New Guinea, where the largest and warmest body of ocean water can be found (left figure, facing page). Abundant rainfall occurs in this area during December, which is the high-sun period in the southern hemisphere. During an El Niño event, the low-pressure system is replaced by weak high pressure and drought ensues. Pressures drop in the equatorial zone of the eastern Pacific, strengthening the equatorial trough. Rainfall is abundant in this new low-pressure region (figure below, right). The shift in barometric pressure patterns is known as the *Southern Oscillation*.

Surface winds and currents also change with this change in pressure. During normal conditions, the strong, prevailing trade winds blow westward, causing very warm ocean water to move to the western Pacific and to "pile up" near the western equatorial low. This westward motion causes the normal upwelling along the South American coast, as bottom water is carried up to replace the water dragged to the west. During an El Niño event, the easterly trade winds die with the change in atmospheric pressure. A weak westerly wind flow sometimes occurs, completely reversing the normal wind direction. Without the pressure of the trade winds to hold them back, warm

waters surge eastward. Sea-surface temperatures and actual sea levels rise off the tropical western coasts of the Americas.

The major change in sea-surface temperatures that accompanies an El Niño can also shift weather patterns across large regions throughout the globe. The winter of 1997–1998 experienced one of the strongest El Niños of the century. Torrential rains drenched Peruvian and Ecuadorean coast ranges, producing mudflows, debris avalanches, and extnesive river flooding. Large protions of Australia and the East Indies went rainless for months, and forest fires burned out of control in Sumatra, Borneo, and Malysia. In east Afraica, Kenya experienced rainfall 100 cm (80 in.) above normal. In North America, a series of powerful winter storms lashed the Pacific coast, doing extensive damage in California. Monstrous tronadoes ripped through Florida, killing over 40 people and destroying more than 800 homes. Meanwhile, mild winter conditions east of the Rockies saved vast amounts of fossil fuel while generating an ice storm that put 4 million people without power in Quebec and the northeastern United States. All told, the deranged weather of the El Niño of 1997–1998 did property damage estimated at $33 billion and killed an estimated 2100 people.

Somewhat rarer, but also capable of altering global weather patterns, is La Niña (the girl child), a condition

shows a wavelike flow, much like Rossby waves in the atmosphere. Warm and cold bodies of water are cut off to float freely, forming warm-core and cold-core rings.

Another phenomenon of ocean surface currents is *El Niño*. During an El Niño event, Pacific surface currents shift into an unusual pattern. Pacific upwelling along the Peruvian coast ceases, trade winds weaken, and a weak equatorial eastward current develops. Global patterns of precipitation also change during El Niño events, bringing floods to some regions and droughts to others. In contrast to El Niño is *La Niña*, in which normal Peruvian coastal upwelling is enhanced,

Pressure, winds, and currents for normal and El Niño conditions in the Pacific.

roughly opposite to El Niño. During a La Niña period, sea-surface temperatures in the central and western Pacific Ocean fall to lower than average levels. This happens because the South Pacific subtropical high becomes very strongly developed during the high-sun season. The result is abnormally strong southeast trade winds. The force of these winds drags a more-than-normal amount of warm surface water westward, which enhances upwelling along western continental coasts. Figure 5.24 provides a pair of satellite images of sea-surface temperature that contrast El Niño and La Niña conditions.

The El Niño of 97–98 was rapidly followed by the La Niña of 98–99. The result was heavier monsoon rains in India and more rain in Australia. In North America, winter conditions were colder than normal in the Northwest and upper Midwest. The eastern Atlantic region endured drought through spring and early summer. The hurricane season of 1998, spawning the monster storm Hurricane Mitch, was the deadliest in the past two centuries.

El Niño and its alter ego La Niña show how dynamic our planet really is. As a grand-scale, global phenomenon, El Niño–La Niña shows how the circulation patterns of the ocean and atmosphere are linked and interact to provide teleconnections capable of producing extreme events affecting millions of people throughout the world.

Questions

1. Compare the normal pattern of wind, pressure, and ocean currents in the equatorial Pacific with the pattern during an El Niño event.

2. What are some of the weather changes reported for El Niño events?

3. What is La Niña, and how does it compare with the normal pattern?

trade winds strengthen, and cool water is carried far westward in an equatorial plume.

Figure 5.24 shows two satellite images of sea-surface temperature observed during El Niño and La Niña years. During La Niña conditions (left), the cool water of the Humboldt Current, moving northward along the

coast of Chile and Peru, is carried westward into the Pacific in a long plume. Much cold, upwelling water (dark green) is brought to the surface along the Peruvian coast. During an El Niño year (right), the eastward motion of warm water holds the Humboldt Current in check. Some slight upwelling occurs (yellow color), but

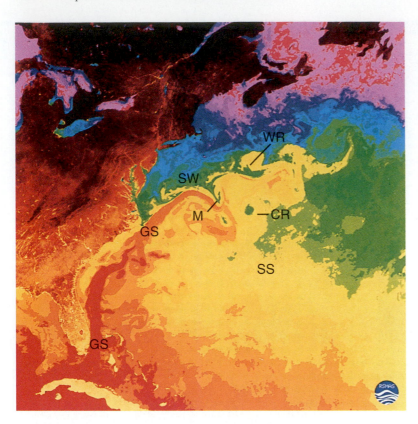

Figure 5.23 A satellite image showing sea-surface temperatures for a week in April from data acquired by the NOAA-7 orbiting satellite. Cold water appears in green and blue tones, warm water in red and yellow tones. The image shows the Gulf Stream (GS) and its interactions with cold water of the Continental Slope (SW), brought down by the Labrador Current, and the warm water of the Sargasso Sea (SS). Other features include a meander (M), a warm-core ring (WR), and a cold-core ring (CR).

Figure 5.24 Sea-surface temperatures in the eastern tropical Pacific during *(a)* La Niña and *(b)* El Niño years as measured by the NOAA-7 satellite. Green tones indicate cooler temperatures, while red tones indicate warmer temperatures.

the amount is greatly reduced compared to normal or La Niña years. *Eye on the Environment • El Niño* provides a fuller description of this phenomenon. **SCIENCE NEWS 12, 13**

Deep Currents and Thermohaline Circulation

Deep currents move ocean waters in a slow circuit across the floors of the world's oceans. They are generated when surface waters become more dense and slowly sink downward. Coupled with these deep currents are very broad and slow surface currents on which the more rapid surface currents, described above, are superimposed. Figure 5.25 diagrams this slow flow pattern, which links all of the world's oceans. It is referred to as *thermohaline circulation*, since it depends on the temperature and salinity of North Atlantic Ocean waters. ("Haline" means "salty.") These factors act through changes to the density of sea water. Since warm water is slightly less dense than cold water, surface waters do not normally mix with cold bottom waters. Follow the figure carefully as we explain the process.

Beginning at A in the figure, warm Atlantic surface water slowly moves northward through the equatorial and tropical zones. As this surface layer is warmed, evaporation occurs and it becomes more salty. This increases the density of the layer slightly. As the water reaches the northern boundaries of the Atlantic, it loses heat to the atmosphere, and so becomes colder and still more dense. Eventually, the surface layer becomes dense enough to sink (B). This sinking occurs all along the northern boundary of the North Atlantic.

Carried along the bottom, cold, dense water eventually reaches the Southern Ocean (C). Here, upwelling and mixing occur and the deep waters are carried into the surface waters of the Indian and southern Pacific oceans (D, E). A coupled circulation loop moves surface water from the Pacific through the Indonesian sea (F) and into the Indian Ocean (G). Most of this flow connects with a westward flow link along the Southern Ocean, poleward of the African and Australian continents. However, some of the flow escapes to the west around the southern tip of Africa to enter the South Atlantic (H), completing the entire circuit.

Figure 5.25   Deep ocean currents, generated by the sinking of cold, salty water in the northern Atlantic, circulate sea water in slowly moving coupled loops involving the Atlantic, Pacific, Indian, and Southern oceans. Actual flows are much broader than the narrow flow paths shown here. (After A. J. Gordon, *Nature* 382:399–400, August 1, 1996. Used by permission.)

Thermohaline circulation plays an important role in the carbon cycle by moving CO_2-rich surface waters into the ocean depths. As noted in Chapter 3 in *Eye on the Environment • Carbon Dioxide—On the Increase*, deep ocean circulation provides a conveyor belt for storage and release of CO_2 in a cycle of about 1500 years' duration. This allows the ocean to moderate rapid changes in atmospheric CO_2 concentration, such as those produced by human activity through fossil fuel burning.

Some scientists have observed that thermohaline circulation could be slowed or stopped by inputs of fresh water into the North Atlantic. Such freshwater inputs could come from the sudden drainage of large lakes formed by melting ice at the close of the last Ice Age. The fresh water would decrease the density of the ocean water, keeping the water from becoming dense enough to sink. Without sinking, circulation would stop. In turn, this would interrupt a major flow pathway for transfer of heat from equatorial regions to the northern midlatitudes. This mechanism could result in relatively rapid climatic change. In fact, it is one explanation for periodic cycles of warm and cold temperatures experienced since the melting of continental ice sheets about 12,000 years ago.

POLEWARD TRANSPORT OF HEAT AND MOISTURE

Recall from Chapter 2 that a vast global heat flow extends from the equatorial and tropical regions, where the sun's energy is concentrated and net radiation is positive, to the middle and upper latitudes, where net radiation is negative. This heat flow was referred to as *poleward heat transport*. With our knowledge of the global circulation patterns of the atmosphere and oceans, and our knowledge of precipitation processes, we can now examine this flow of heat in more detail.

> **Heat and moisture are moved poleward from equatorial and tropical regions by global winds and ocean currents.**

Figure 5.26 is a schematic diagram of global heat and moisture flow within the atmosphere. In the tropical and equatorial zones, the Hadley cell circulation acts like a heat pump to contain and concentrate heat between the two subtropical high-pressure belts. Tracing the flow of air in the Hadley cell loop, we start at the surface near the subtropical high-pressure belt. Air moves toward the equator and absorbs moisture from

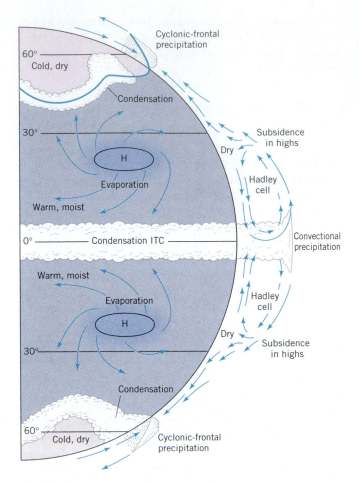

Figure 5.26 Global atmospheric transport of heat and moisture.

the ocean surface, which adds to the latent heat content of the air. As the air converges and rises in the intertropical convergence zone (ITC), the latent heat is released. This heat adds to the sensible heat content of the air. Much of the added heat content is retained as the air completes the cycle and subsides in the subtropical high-pressure belts. There are two net effects. The first is to move water from the tropical regions, where evaporation takes place, to the equatorial zone, where it falls as precipitation. The second is to move sensible heat at high levels from the equatorial zone to the tropical zones.

The Hadley cell pump leaks rather badly, however, and some of its sensible heat also flows poleward. This occurs when some diverging air from the subtropical high-pressure belts heads poleward into the midlatitudes. This air also acquires latent heat by picking up moisture as it moves over warm oceans. Through the Rossby wave mechanism, occasional lobes of cold, dry polar air plunge toward the equator, while tongues of warm, moist air flow toward the poles. These exchanges provide a heat flow that warms the mid- and higher latitudes well beyond the capabilities of the sun. Water is transported in this part of the cycle, too, from warm subtropical oceans where it is evaporated, to midlatitude, arctic, antarctic, and even polar zones where it condenses or deposits and falls as precipitation.

Thermohaline circulation plays a very important role in moving heat from one region of the globe to another. The poleward movement of warm water to the northern edges of the Atlantic provides a flow of heat to northern latitudes. This flow is absorbed by the atmosphere and then carried eastward on prevailing westerly winds to the European continent. The amount of heat conveyed to Europe is quite large. In fact, a recent calculation shows that it is about 35 percent of the total annual insolation received by the Atlantic Ocean north of 40° latitude! This warming is sufficient to shift several climate zones in Europe northward by as much as 10° latitude over their counterparts in North America. This type of circulation does not occur in the Pacific or Indian Ocean.

A LOOK AHEAD

The global circulation of winds and currents paves the way for our next subject—weather systems. Recall from Chapter 4 that when warm, moist air rises, precipitation can occur. This happens in the centers of cyclones, where air converges. Although cyclones and anticyclones are generally large features, they move from day to day and are steered by the global pattern of winds. Thus, your knowledge of global winds and pressures will help you to understand how weather systems and storms develop and migrate. Your knowledge will also be very useful for the study of climate, which we take up in Chapter 7.

WINDS AND GLOBAL CIRCULATION IN REVIEW

Atmospheric pressure is a measurement of the weight of air pressing on a unit of surface area. Atmospheric pressure decreases rapidly as altitude increases. Wind occurs when air moves with respect to the earth's surface. Air motion is produced by pressure gradients that form between one region and another. A pressure gradient is formed when air in one location is heated to a temperature that is warmer than another. Because warm air is less dense than cold air, low pressure is associated with warm air and high pressure with cold air. The sea breeze is an example of a wind caused by a pressure gradient that is produced by a difference in surface heating. A convection loop is formed as warm air rises over land and sinks over the nearby water surface.

At a global scale, atmospheric circulation is strongly influenced by the earth's rotation. The Coriolis effect deflects winds and ocean currents, producing circular or spiraling flow paths around cyclones (centers of low pressure) and anticyclones (centers of high pressure).

Because the equatorial and tropical regions are heated more intensely than the higher latitudes, two convection loops develop—the Hadley cells. These loops drive the northeast and southeast trade winds, the convergence and lifting of air at the intertropical convergence zone (ITC), and the sinking and divergence of air in the subtropical high-pressure cells. The monsoon circulation of Asia responds to a reversal of atmospheric pressure over the continent with the seasons. A winter monsoon flow of cool, dry air from the northeast alternates with a summer monsoon flow of warm, moist air from the southwest.

In the midlatitudes and poleward, westerly winds prevail. In winter, continents develop high pressure, and intense oceanic low-pressure centers are found off the Aleutian Islands and near Iceland in the northern hemisphere. In the summer, the continents develop low pressure as oceanic subtropical high-pressure cells intensify and move poleward.

Local winds are generated by local pressure gradients. The sea and land breezes, as well as mountain and valley winds, are examples caused by local surface heating. Other local winds include drainage winds, Santa Ana, and chinook winds.

Winds aloft follow a simple basic pattern—easterlies between two belts of tropical high pressure, and westerlies to the north and south of these belts. Rossby waves develop in the westerlies, bringing cold, polar air equatorward and warmer air poleward. The polar and subtropical jet streams are concentrated westerly wind streams with high wind speeds.

Ocean surface currents are dominated by huge gyres that are driven by the global surface wind pattern. Equatorial currents move warm water westward and then poleward along the east coasts of continents. Return flows bring cold water equatorward along the west coasts of continents. El Niño events occur when an unusual flow of warm water in the equatorial Pacific moves eastward to the coasts of Central and South America, suppressing the normal northward flow of the Humboldt Current. Upwelling along the Peruvian coast is greatly reduced. El Niño events normally occur on a three- to eight-year cycle and affect global patterns of precipitation in many regions. The causes of El Niño are not well understood.

Slow, deep ocean currents are driven by the sinking of cold, salty water in the northern Atlantic. This circulation pattern involves nearly all the earth's ocean basins, and also acts to moderate the buildup of atmospheric CO_2 by moving CO_2-rich surface waters to ocean depths.

Global air and ocean circulation provides the mechanism for poleward heat transport by which excess heat moves from the equatorial and tropical regions toward the poles. These heat flows help make northern and southern climates warmer than we might expect based on solar heating alone.

KEY TERMS

barometer
wind
pressure gradient
isobars
Coriolis effect
cyclone

anticyclone
Hadley cells
intertropical convergence zone (ITC)
subtropical high-pressure belts

monsoon
Rossby waves
jet streams
gyres

REVIEW QUESTIONS

1. Explain atmospheric pressure. Why does it occur? How is atmospheric pressure measured, and in what units? What is the normal value of atmospheric pressure at sea level? How does atmospheric pressure change with altitude?
2. Describe land and sea breezes. How do they illustrate the concepts of pressure gradient and convection loop?
3. What is the Coriolis effect, and why is it important? What produces it? How does it influence the motion of wind and ocean currents in the northern hemisphere? in the southern hemisphere?
4. Define cyclone and anticyclone. How does air move within each? What is the direction of circulation of each in the northern and southern hemispheres? What type of weather is associated with each and why?
5. What is the Asian monsoon? Describe the features of this circulation in summer and winter. How is the ITC involved? How is the monsoon circulation related to the high- and low-pressure centers that develop seasonally in Asia?
6. Compare the winter and summer patterns of high and low pressure that develop in the northern hemisphere with those that develop in the southern hemisphere.
7. List four types of local winds and describe three of them.
8. Describe the basic pattern of global atmospheric circulation at upper levels.
9. What are Rossby waves? How are they formed?
10. Identify five jet streams. Where do they occur? In which direction do they flow?
11. What is the general pattern of ocean surface current circulation? How is it related to global wind patterns? to air patterns?
12. How does thermohaline circulation induce deep ocean currents?
13. How does the global circulation of the atmosphere and oceans provide poleward heat transport?

VISUALIZING EXERCISES

1. Sketch an ideal earth (without seasons or ocean–continent features) and its global wind system. Label the following on your sketch: doldrums, equatorial trough, Hadley cell, ITC, northeast trades, polar easterlies, polar front, polar outbreak, southeast trades, subtropical high-pressure belts, and westerlies.
2. Draw four spiral patterns showing outward and inward flow in clockwise and counterclockwise directions. Label each as appropriate to cyclonic or anticyclonic circulation in the northern or southern hemisphere.

A DEEPER VIEW OF WINDS AND GLOBAL CIRCULATION

1. An airline pilot is planning a nonstop flight from Los Angeles to Sydney, Australia. What general wind conditions can the pilot expect to find in the upper atmosphere as the airplane travels? What jet streams will be encountered? Will they slow or speed the aircraft on its way?
2. You are planning to take a round-the-world cruise, leaving New York in October. Your vessel's route will take you through the Mediterranean Sea to Cairo, Egypt, in early December. Then you will pass through the Suez Canal and Red Sea to the Indian Ocean, calling at Bombay, India, in January. From Bombay, you will sail to Djakarta, Indonesia, and then go directly to Perth, Australia, arriving in March. Rounding the southern coast of Australia, your next port of call is Auckland, New Zealand, which you will reach in April. From Auckland, you head directly to San Francisco, your final destination, arriving in June. Describe the general wind and weather conditions you will experience on each leg of your journey.

Chapter 6

Weather Systems

Hurricane Andrew approaching the Louisiana coast on August 25, 1992, as depicted in a computer-enhanced satellite image.

PUTTING YOU IN THE PICTURE

Our day-to-day weather may occasionally be stormy, bringing the inconvenience of rain or snow to our daily activities. But rare storms, with high winds, high coastal ocean waters, and torrential rains or blizzards of snow can cause great destruction and loss of life. What is it like to experience a severe storm?

On August 23, 1992, Hurricane Andrew approached the coast of south Florida, near Miami. Following the path of the storm was veteran storm chaser and photographer, Warren Faidley, who decided to ride out the storm with two friends, Steve and Mike. For their shelter, they selected a sturdy parking garage, built of concrete and steel, in a coastal community south of Miami. Here is part of Faidley's account of the terrifying night of the storm.

At sunset, my friends and I gathered at the garage. Throughout the evening we monitored radio and television broadcasts. With ominous accuracy, Robert Sheets, director of the National Hurricane Center, gave hourly updates on Andrew, now a Category 4…. Like sentries awaiting an advancing army, we took turns scanning the Miami skyline from our borrowed fortress, searching for early signs of Andrew. Bright flashes from downed power lines on the horizon marked Andrew's imminent arrival.

Steve and I held an anemometer through the opening on the fourth floor. "Fifty-five miles per hour!" I yelled over the howling. "A gust to 65!" The meter almost flew out of my hand. The electrical explosions across the city grew more intense, yet the fury of the wind swallowed up all sound save the car alarms whining inside the garage. Eventually even the alarms were overwhelmed by the whistling wind.

Around 4 A.M. all hell broke loose. I have experienced severe storms with winds in excess of 75 m.p.h., but these gusts were blowing at well over 140 m.p.h. The anemometer and later damage surveys indicated even stronger gusts. By now only an occasional thud— some building collapsing or losing a roof—would punctuate the noise. Windows from the surrounding buildings imploded, scattering glass everywhere. I looked down at my arm and saw blood, not knowing when I'd been cut. Putting on my reading glasses to protect my eyes, I made a mental note: next time, bring goggles.

My senses were on red alert, a self-preservation mode I adopt only in the most extreme danger. Everything I see, hear, feel—even taste—goes by so quickly that my mind simply slows it all down to preserve rational thought. The alternative is sheer panic.

The infamous "hurricane wail" people talk about is in many ways unreal— "the scream of the devil," Steve and I agreed. The piercing sound alone is wicked enough, but mixed with breaking glass and crashing debris, it rattled me inside. I will never forget that sound….

In the garage, not all was well. The elevator shafts had begun to flood; water poured out between the doors and cascaded down the emergency stairwells. The lights in the garage flickered on and off and the fire alarm blared, yet it was hardly audible in the din…. I trusted the structural integrity of our fortress, but under Andrew's full force, the garage began to oscillate. After two hours of this swaying, I began to worry one of the floors might collapse….

For more than five hours Andrew pounded the garage and the rest of South Florida with everything it had. As the building shook and the storm raged all around, I felt as though I was inside a tornado and riding an earthquake at the same time.*

Hurricane Andrew was the most costly natural disaster to strike the United States, with some damage estimates as high as $30 billion. *Eye on the Environment • Hurricane Andrew—Killer Cyclone* provides more information on this devastating storm. Hurricanes, or tropical cyclones, as they are termed by meteorologists, are but one of a number of types of weather systems that are the subject of this chapter.

* From *Weatherwise*, vol. 45, no. 6, Copyright © Warren Faidley. Used by permission.

Chapter Outline

Traveling Cyclones and Anticyclones
Air Masses
North American Air Masses
Cold, Warm, and Occluded Fronts
Wave Cyclones
Weather Changes Within a Wave Cyclone
Cyclone Tracks and Cyclone Families
The Tornado

Tropical and Equatorial Weather Systems
Easterly Waves and Weak Equatorial Lows
Tropical Cyclones
Impacts of Tropical Cyclones
Eye on the Environment • Hurricane Andrew—Killer Cyclone
Cloud Cover, Precipitation and Global Warming

WEATHER SYSTEMS

What are cold fronts and warm fronts? You may have heard these terms while watching the weather on television or listening to weather reports on the radio. You probably know that cold fronts generally bring cold weather and that warm fronts generally bring warm weather. But did you know that cold and warm fronts are often organized together as part of the same weather system? Cold fronts and warm fronts are found together in *wave cyclones*—low-pressure centers in the midlatitudes that move eastward and provide much of North America's precipitation.

Wave cyclones are just one type of **weather system**—an organized state of the atmosphere associated with a characteristic weather pattern. Other weather systems we will encounter in this chapter include traveling anticyclones, tornadoes, tropical cyclones (including hurricanes), easterly waves, and weak equatorial lows.

TRAVELING CYCLONES AND ANTICYCLONES

What's the difference between a cyclone and an anticyclone? If the sun is shining and skies are blue, it's an anticyclone. If the day is overcast and gray, it's a cyclone. Most types of cyclones and anticyclones move slowly across the earth's surface, bringing changes in the weather as they move. These are referred to as *traveling cyclones* and *traveling anticyclones*.

In a cyclone, low pressure at the center draws air inward and carries it aloft. And when moist air rises, condensation or deposition can occur. Thus, cyclones normally have dense, stratiform clouds, often releasing rain or snow. This is **cyclonic precipitation**. Many cyclones are weak and pass overhead with little more than a period of cloud cover and light precipitation. However, when pressure gradients are steep and the inspiraling motion is strong, intense winds and heavy rain or snow can accompany the cyclone. In this case, the disturbance is called a **cyclonic storm**.

> **In middle and higher latitudes, traveling cyclones and anticyclones bring changing weather systems.**

Traveling cyclones fall into three types. First is the wave cyclone of midlatitude, arctic, and antarctic zones. This type of cyclone ranges in intensity from a weak disturbance to a powerful storm. Second is the tropical cyclone of tropical and subtropical zones. This type of cyclone ranges in intensity from a mild disturbance to the highly destructive hurricane or typhoon. A third type is the tornado, a small, intense cyclone of enormously powerful winds. The tornado is much, much smaller in size than other cyclones, and it is related to strong convectional activity.

In an anticyclone, high pressure at the center pushes surface air outward, causing air to descend from above. Since descending air warms adiabatically, condensation does not occur. Skies are fair, except for occasional puffy cumulus clouds that can develop in a moist surface air layer. Because of these characteristics, anticyclones are sometimes termed fair-weather systems. Toward the center of an anticyclone, the pressure gradient is weak, and winds are light and variable. Traveling anticyclones are found in the midlatitudes. They are typically associated with ridges or domes of clear, dry air that move eastward and equatorward.

Air Masses

Cyclonic and anticyclonic weather systems are often associated with the motion of air masses. An **air mass** is a large body of air with fairly uniform temperature and moisture characteristics. It can be several thousand kilometers or miles across, and extend upward to the top of the troposphere. A given air mass is characterized by a distinctive combination of surface temperature, environmental temperature lapse rate, and surface-specific humidity. Air masses range widely in temperature—from searing hot to icy cold—as well as in moisture content.

Air masses acquire their characteristics in *source regions*. In a source region, air moves slowly or stagnates, which allows the air to acquire temperature and moisture characteristics from the surface. For example, an air mass with warm temperatures and a high water vapor content develops over a warm equatorial ocean. Over a large tropical desert, slowly subsiding air forms a hot air mass with low relative humidities. A very cold air mass with a low water vapor content is generated over cold, snow-covered land surfaces in the arctic zone in winter.

Air masses move from one region to another under the influence of barometric pressure gradients and are sometimes pushed or blocked by high-level jet stream winds. When an air mass moves to a new area, its properties will begin to change because it is influenced by the new surface environment. For example, the air mass may lose heat or take up water vapor.

Air masses are classified on the basis of the latitudinal position and the nature of the underlying surface of their source regions. Latitudinal position primarily determines surface temperature and environmental temperature lapse rate of the air mass, while the nature of the underlying surface—continent or ocean—usually determines the moisture content. For latitudinal position, five types of air masses are distinguished as shown in the following table.

Air Mass	Symbol	Source Region
Arctic	A	Arctic ocean and fringing lands
Antarctic	AA	Antarctica
Polar	P	Continents and oceans, lat. 50–60° N and S
Tropical	T	Continents and oceans, lat. 20–35° N and S
Equatorial	E	Oceans close to equator

For the type of underlying surface, two subdivisions are used:

Air Mass	Symbol	Source Region
Maritime	m	Oceans
Continental	c	Continents

Combining these two types of labels produces a list of six important types of air masses, shown in Table 6.1. The table also gives some typical values of surface temperature and specific humidity, although these can vary widely, depending on season. Air mass temperatures can range from −46°C (−51°F) for continental arctic (cA) air masses to 27°C (81°F) for maritime equatorial air masses (mE). Specific humidities show a very high range—from 0.1 g/kg for the cA air mass to as much as 19 g/kg for the mE air mass. In other words, maritime equatorial air can hold about 200 times as much moisture as continental arctic air.

Figure 6.1 shows schematically the global distribution of source regions of these air masses. An idealized continent is shown in the center of the figure surrounded by ocean. Note from the figure that the polar air masses (mP, cP) originate in the subarctic latitude zone, not in

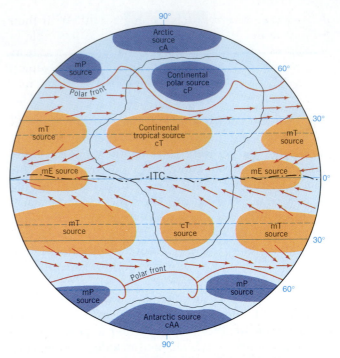

Figure 6.1 A schematic diagram of global air masses and source regions. An idealized continent, producing continental (c) air masses, is shown at the center. It is surrounded by oceans, producing maritime air masses (m). Tropical (T) and equatorial (E) source regions provide warm or hot air masses, while polar (P), arctic (A), and antarctic (AA) source regions provide colder air masses with lower specific humidities.

the polar latitude zone. Recall that in Chapter 2 we defined "polar" as a region containing one of the poles. However, meteorologists use the word "polar" to describe air masses from the subarctic and subantarctic zones, and we will follow their usage when referring to air masses.

The maritime tropical air mass (mT) and maritime equatorial air mass (mE) originate over warm oceans in the tropical and equatorial zones. They are quite similar

Table 6.1 Properties of Typical Air Masses

Air Mass	Symbol	Source Region	Properties	Temperature °C	(°F)	Specific Humidity (gm/kg)
Maritime equatorial	mE	Warm oceans in the equatorial zone	Warm, very moist	27°	(81°)	19
Maritime tropical	mT	Warm oceans in the tropical zone	Warm, moist	24°	(75°)	17
Continental tropical	cT	Subtropical deserts	Warm, dry	24°	(75°)	17
Maritime polar	mP	Midlatitude oceans	Cool, moist (winter)	4°	(39°)	4.4
Continental polar	cP	Northern continental interiors	Cold, dry (winter)	−11°	(12°)	1.4
Continental arctic (and continental antarctic)	cA (cAA)	Regions near north and south poles	Very cold, very dry (winter)	−46°	(−51°)	0.1

in temperature and water vapor content. With their high values of specific humidity, both are capable of very heavy yields of precipitation. The continental tropical air mass (cT) has its source region over subtropical deserts of the continents. Although this air mass may have a substantial water vapor content, it tends to be stable and has low relative humidity when highly heated during the daytime.

The maritime polar air mass (mP) originates over midlatitude oceans. Since the quantity of water vapor it holds is not as large as maritime tropical air masses, the mP air mass yields only moderate precipitation. Much of this precipitation is orographic and occurs over mountain ranges on the western coasts of continents. The continental polar air mass (cP) originates over North America and Eurasia in the subarctic zone. It has low specific humidity and is very cold in winter. Last is the continental arctic (and continental antarctic) air mass type (cA, cAA), which is extremely cold and holds almost no water vapor.

North American Air Masses

The air masses that form in and near North America have a strong influence on North American weather. The source regions of these air masses are shown in Figure 6.2. The continental polar (cP) air mass of North America originates over north-central Canada. This air mass forms tongues of cold, dry air that periodically

Figure 6.2 North American air mass source regions and trajectories. (Data from U.S. Department of Commerce.)

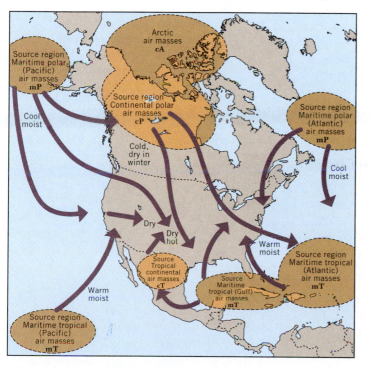

extend south and east from the source region to produce anticyclones accompanied by cool or cold temperatures and clear skies. The arctic air mass (cA) that develops over the Arctic Ocean and its bordering lands of the arctic zone in winter is extremely cold and stable. When this air mass moves southward, it produces a severe cold wave.

The maritime polar air mass (mP) originates over the North Pacific and Bering Strait, in the region of the persistent Aleutian low-pressure center. This air mass is characteristically cool and moist, with a tendency in winter to become unstable, giving heavy precipitation over coastal ranges. Another maritime polar air mass of the North American region originates over the North Atlantic Ocean. It, too, is cool and moist, and is felt especially in Canada's maritime provinces in winter.

Of the tropical air masses, the most common visitor to the central and eastern United States is the maritime tropical air mass (mT) from the Gulf of Mexico. It moves northward, bringing warm, moist, unstable air over the eastern part of the country. In the summer, particularly, this air mass brings hot, sultry weather to the central and eastern United States. It also produces many thunderstorms. Closely related is a maritime tropical air mass from the Atlantic Ocean east of Florida, over the Bahamas.

Over the Pacific Ocean, a source region of another maritime tropical air mass (mT) lies in the cell of high pressure located to the southwest of lower California. Occasionally, in summer, this moist unstable air mass penetrates the southwestern desert region, bringing severe thunderstorms to southern California and southern Arizona. In winter, a tongue of mT air frequently reaches the California coast, bringing heavy rainfall that is intensified when forced to rise over coastal mountain ranges.

A hot, dry continental tropical air mass (cT) originates over northern Mexico, western Texas, New Mexico, and Arizona during the summer. This air mass does not travel widely, but governs weather conditions over the source region.

Cold, Warm, and Occluded Fronts

A given air mass usually has a sharply defined boundary between itself and a neighboring air mass. This boundary is termed a **front**. We saw an example of a front in the contact between polar and tropical air masses, shown in Figures 5.10, 5.18, and 5.20. This feature is the *polar front*, and it is located below the axis of the jet stream in the upper-air waves. If one air mass is advancing on another, the surface of contact between the two will be at a relatively low angle. If the two air masses are stationary with respect to each other, the frontal contact will be nearly vertical.

Figure 6.3 A cold front, along which a cold air mass lifts a warm air mass aloft. The upward motion sets off a line of thunderstorms. The frontal boundary is actually much less steep than shown in this schematic drawing. (Drawn by A. N. Strahler.)

In an occluded front, a cold front overtakes a warm front, lifting a pool of warm, moist air upward.

Figure 6.3 shows the structure of a front along which a cold air mass invades a zone occupied by a warm air mass. A front of this type is called a **cold front**. Because the colder air mass is denser, it remains in contact with the ground. As it moves forward, it forces the warmer air mass to rise above it. If the warm air is unstable, severe thunderstorms may develop. Thunderstorms near a cold front often form a long line of massive clouds stretching for tens of kilometers.

Figure 6.4 diagrams a **warm front** in which warm air moves into a region of colder air. Here, again, the cold

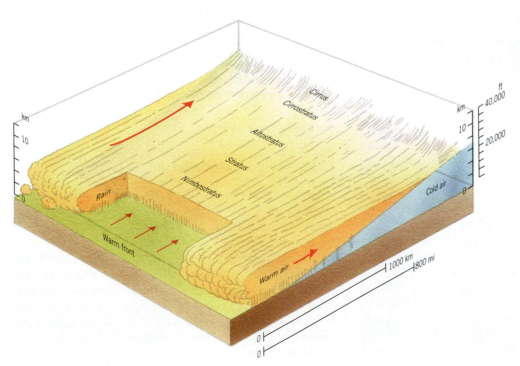

Figure 6.4 A warm front. Warm air advances toward cold air and rides up and over the cold air. A notch of cloud is cut away to show rain falling from the dense stratus cloud layer. (Drawn by A. N. Strahler.)

Figure 6.5 An occluded front. A warm front is overtaken by a cold front. The warm air is pushed aloft, and it no longer contacts the ground. Abrupt lifting by the denser cold air produces precipitation. (Drawn by A. N. Strahler.)

air mass remains in contact with the ground because it is denser. The warm air mass is forced to rise on a long ramp over the cold air below. The rising motion causes stratus clouds to form, and precipitation often follows. If the warm air is stable, the precipitation will be steady (Figure 6.4). If the warm air is unstable, convection cells can develop, producing cumulonimbus clouds with heavy showers or thunderstorms (not shown in the figure).

Cold fronts normally move along the ground at a faster rate than warm fronts. Thus, when both types are in the same neighborhood, a cold front can overtake a warm front. The result is an **occluded front**, diagrammed in Figure 6.5. The colder air of the fast-moving cold front remains next to the ground, forcing both the warm air and the less cold air to rise over it. The warm air mass is lifted completely free of the ground.

Wave Cyclones

In middle and high latitudes, the dominant form of weather system is the **wave cyclone**. The wave cyclone is a large inspiral of air that repeatedly forms, intensifies, and dissolves along the polar front. Figure 6.6 shows a situation favorable to the formation of a wave cyclone. Two large anticyclones are in contact on the polar front. One contains a cold, dry polar air mass, and the other a warm, moist maritime air mass. Airflow converges from opposite directions on the two sides of the front, setting up an unstable situation. The wave cyclone will begin to

Figure 6.6 Conditions for formation of a wave cyclone. Warm subtropical air and cold polar air are in contact on the polar front. The shaded area is shown in Block A of the following figure as the early stage of development of a wave cyclone.

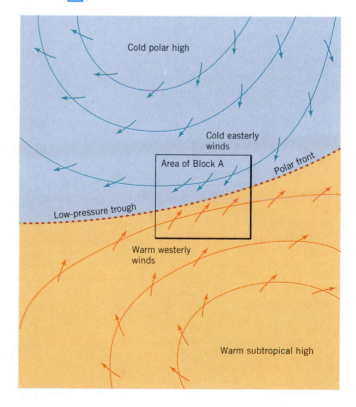

form between the two high-pressure cells in a zone of lower pressure referred to as a *low-pressure trough.*

How does a wave cyclone form, grow, and eventually dissolve? Figure 6.7 shows the life history of a wave cyclone. In Block A (early stage), the polar-front region shows a wave beginning to form. Cold air is turned in a southerly direction and warm air in a northerly direction, so that each advances on the other. As these frontal motions develop, precipitation will form.

In Block B (open stage), the wave disturbance along the polar front has deepened and intensified. Cold air actively pushes southward along a cold front, and warm air actively moves northeastward along a warm front. The zones of precipitation along the two fronts are now strongly developed but are wider along the warm front than along the cold front.

In Block C (occluded stage), the cold front has overtaken the warm front, producing an occluded front. The warm air mass at the center of the inspiral is forced off the ground, intensifying precipitation. Eventually, the polar front is reestablished (Block D, dissolving stage), but a pool of warm, moist air remains aloft. As the moisture content of the pool is reduced, precipitation dies out, and the clouds gradually dissolve.

Keep in mind that a wave cyclone is quite a large feature—1000 km (about 600 mi) or more across. Also, the cyclone normally moves eastward as it develops, propelled by prevailing westerlies aloft. Therefore, Blocks A–D are like three-dimensional snapshots taken at intervals along an eastbound track.

Cyclonic precipitation is associated with air mass movements, fronts, and cyclonic storms.

You may recall from Chapter 4 that we recognized three types of precipitation—orographic, convectional, and cyclonic. With your knowledge of how precipitation occurs in cold fronts, warm fronts, and occluded fronts, you now understand cyclonic precipitation as the precipitation associated with air mass movements, fronts, and cyclonic storms.

Weather Changes Within a Wave Cyclone

How does weather change as a wave cyclone passes through a region? Figure 6.8 shows two simplified weather maps of the eastern United States depicting

(a) Early stage

(b) Open stage

(c) Occluded stage

(d) Dissolving stage

Figure 6.7 Development of a wave cyclone. In (*a*), a cyclonic motion begins at a point along the polar front. Cold and warm fronts intensify in (*b*). In (*c*), the cold front overtakes the warm front, producing an occluded front in the center of the cyclone. Later, the polar front is reestablished with a mass of warm air isolated aloft (*d*).

Figure 6.8 Simplified surface weather maps and cross sections through a wave cyclone. In the open stage (*left*), cold and warm fronts pivot around the center of the cyclone. In the occluded stage (*right*), the cold front has overtaken the warm front, and a large pool of warm, moist air has been forced aloft.

conditions on successive days. The structure of the storm is defined by the isobars, labeled in millibars. The three kinds of fronts are shown by special line symbols. Areas of precipitation are shown in gray.

The map on the left shows the cyclone in an open stage, similar to Figure 6.7, Block B. The isobars show that the cyclone is a low-pressure center with inspiraling winds. The cold front is pushing south and east, supported by a flow of cold, dry continental polar air from the northwest filling in behind it. Note that the wind direction changes abruptly as the cold front passes. There is also a sharp drop in temperature behind the cold front as cP air fills in. The warm front is moving north and somewhat east, with warm, moist maritime tropical air following. The precipitation pattern includes a broad zone near the warm front and the central area of the cyclone. A thin band of precipitation extends down the length of the cold front. Cloudiness generally prevails over much of the cyclone.

A cross section along the line A–A' shows how the fronts and clouds are related. Along the warm front is a broad layer of stratus clouds. These take the form of a wedge with a thin leading edge of cirrus. (See Figures 4.11 and 4.12 for cloud types.) Westward, this wedge thickens to altostratus, then to stratus, and finally to nimbostratus with steady rain. Within the sector of warm air, the sky may partially clear with scattered cumulus. Along the cold front are cumulonimbus clouds associated with thunderstorms. These yield heavy rains but only along a narrow belt.

The weather map on the right shows conditions 24 hours later. The cyclone has moved rapidly northeastward, its track shown by the red line. The center has moved about 1600 km (1000 mi) in 24 hours—a speed of just over 65 km (40 mi) per hour. The cold front has overtaken the warm front, forming an occluded front in the central part of the disturbance. A high-pressure area, or tongue of cold polar air, has moved in to the area west and south of the cyclone, and the cold front has pushed far south and east. Within the cold air tongue, the skies are clear. A cross section below the map shows conditions along the line B–B', cutting through the occluded part of the storm. Notice that the warm air mass is lifted well off the ground and yields heavy precipitation.

Cyclone Tracks and Cyclone Families

Wave cyclones tend to form in certain areas and travel common paths until they dissolve. Figure 6.9 is a world map showing common paths of wave cyclones and tropical cyclones. (We will discuss tropical cyclones in detail later in this chapter.) The western coast of North America commonly receives wave cyclones arising in the North Pacific Ocean. Wave cyclones also originate over land, shown by the tracks starting in Alaska, the Pacific Northwest, the south-central United States, and along the Gulf coast. Most of these tracks converge toward the northeast and pass out into the North

Atlantic, where they tend to concentrate in the region of the Icelandic Low.

In the northern hemisphere, wave cyclones are heavily concentrated in the neighborhood of the Aleutian and Icelandic Lows. These cyclones commonly form in a succession, traveling as a chain across the North Atlantic and North Pacific oceans. Figure 6.10, a world weather map, shows several such cyclone families. Each cyclone moves northeastward, deepening in low pressure and occluding to form an upper-air low. For this reason, intense cyclones arriving at the western coasts of North America and Europe are usually occluded.

In the southern hemisphere, storm tracks are more nearly along a single lane, following the parallels of latitude. Three such cyclones are shown in Figure 6.10. This track is more uniform because of the uniform pattern of ocean surface circling the globe at these latitudes. Only the southern tip of South America projects southward to break the monotonous expanse of ocean.

Figure 6.9 Typical paths of tropical cyclones (red lines) and wave cyclones of middle latitudes (blue lines). (Based on data of S. Pettersen, B. Haurwitz and N. M. Austin, J. Namias, M. J. Rubin, and J-H. Chang.)

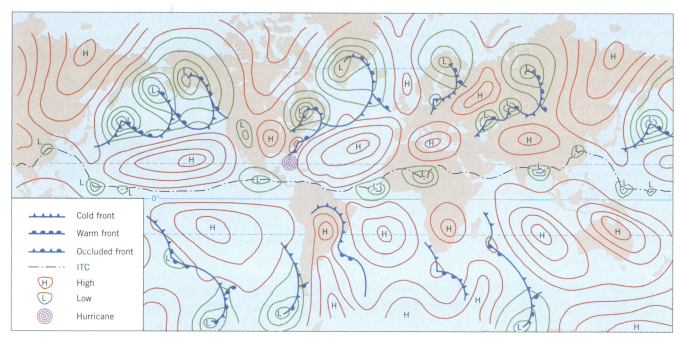

Figure 6.10 A daily weather map of the world for a given day during July or August might look like this map, which is a composite of typical weather conditions. (After M. A. Garbell.)

The Tornado

A **tornado** is a small but intense cyclonic vortex in which air spirals at tremendous speed. It is associated with thunderstorms spawned by fronts in midlatitudes of North America. Tornadoes can also occur inside tropical cyclones (hurricanes).

The tornado appears as a dark funnel cloud hanging from the base of a dense cumulonimbus cloud (Figure 6.11). At its lower end, the funnel may be 100 to 450 m (about 300 to 1500 ft) in diameter. The base of the funnel appears dark because of the density of condensing moisture, dust, and debris swept up by the wind. Wind speeds in a tornado exceed speeds known in any other

Figure 6.11 This tornado touched down near Clearwater, Kansas, on May 16, 1991. Surrounding the funnel is a cloud of dust and debris carried into the air by the violent winds.

storm. Estimates of wind speed run as high as 400 km (about 250 mi) per hour. As the tornado moves across the countryside, the funnel writhes and twists. Where it touches the ground, it can cause the complete destruction of almost anything in its path. **SCIENCE NEWS** 14

Tornado winds are the highest of all storm winds and can cause great local damage.

Tornadoes occur as parts of cumulonimbus clouds traveling in advance of a cold front. They seem to originate where turbulence is greatest. They are most common in the spring and summer, but can occur in any month. Where a cold front of maritime polar air lifts warm, moist maritime tropical air, conditions are most favorable for tornadoes. As shown in Figure 6.12, they occur in greatest numbers in the central and southeastern states and are rare over mountainous and forested regions. They are almost unknown west of the Rocky Mountains and are relatively less frequent on the east-

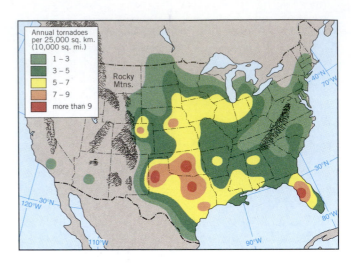

Figure 6.12 Frequency of occurrence of observed tornadoes in the 48 contiguous United States. The data span a 30-year record, 1960–1989. (Courtesy of Edward W. Ferguson, National Severe Storms Forecast Center, National Weather Service.)

Figure 6.13 A swath of destruction left by a tornado that swept through Kissimmee, Florida, in February of 1998.

ern seaboard. Tornadoes are a typically American phenomenon, since they are most frequent and violent in
the United States. They also occur in Australia in substantial numbers and are occasionally reported from
other midlatitude locations. Tornadoes are most likely
to occur from early spring to late summer. April, May
and June are the three months in which tornadoes are
most common.

Devastation from a tornado is often complete within
the narrow limits of its path. Only the strongest buildings constructed of concrete and steel can resist major
structural damage from the extremely violent winds.
The National Weather Service maintains a tornado forecasting and warning system. Whenever weather conditions favor tornado development, the danger area is
alerted, and systems for observing and reporting a tornado are set in readiness.

Tornado frequency varies from year to year. In the
United States, the three years of highest tornado frequency are 1992, 1995, and 1998, with counts of
1200–1300 tornadoes each year. Perhaps not so coinci-
dentally, these years immediately followed strong El Niño
events. By altering weather patterns, an El Niño may
increase the frequency of conditions favoring the development of tornadoes. Figure 6.13 shows damage from
the tornado outbreak of February 22–23, 1998, in central
Florida—the most devastating in the state's history. The
seven tornadoes spawned that night killed 41 persons,
injured 260, and damaged nearly 300 structures.

TROPICAL AND EQUATORIAL WEATHER SYSTEMS

So far, the weather systems we have discussed are those
that are important in the midlatitudes and poleward.
Weather systems of the tropical and equatorial zones
show some basic differences from those of the midlatitudes. The Coriolis effect is weak close to the equator,
and there is a lack of strong contrast between air masses. So, clearly defined fronts and large, intense wave
cyclones are missing. On the other hand, there is
intense convectional activity because of the high moisture content of low-latitude maritime air masses. Slight
convergence and uplifting can be sufficient to trigger
convectional precipitation in these very moist air
masses.

Easterly Waves and Weak Equatorial Lows

One of the simplest forms of tropical weather systems is
an *easterly wave,* a slowly moving trough of low pressure
within the belt of tropical easterlies (trades). These
waves occur in latitudes 5° to 30° N and S over oceans,
but not over the equator itself. Figure 6.14 is a simplified
upper-air map of an easterly wave showing wind patterns, the axis of the wave, isobars, and the zone of

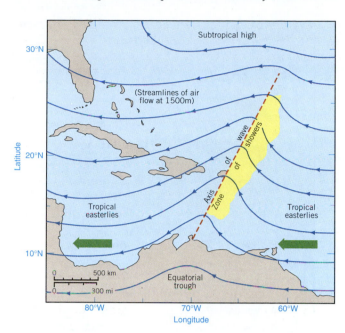

Figure 6.14   An easterly wave passing over the West
Indies. (Data from H. Riehl, *Tropical Meteorology,* McGraw-
Hill, New York.)

showers. At the surface, a zone of weak low pressure
underlies the axis of the wave. The wave travels westward
at a rate of 300 to 500 km (about 200 to 300 mi) per day.
Surface airflow converges on the eastern, or rear, side of
the wave axis. This convergence causes the moist air to
be lifted, producing scattered showers and thunderstorms. The rainy period may last a day or two as the
wave passes.

Another related weather system is the *weak equatorial
low,* a disturbance that forms near the center of the
equatorial trough. Moist equatorial air masses converge
on the center of the low, causing rainfall from many
individual convectional storms. Several such weak lows
are shown on the world weather map (Figure 6.10),
lying along the ITC. Because the map is for a day in July
or August, the ITC is shifted well north of the equator.
At this season, the rainy monsoon is in progress in
Southeast Asia. It is marked by a weak equatorial low in
northern India.

Tropical Cyclones

The most powerful and destructive type of cyclonic
storm is the **tropical cyclone,** which is known as the *hurricane* in the Atlantic and eastern Pacific Oceans or the
typhoon in the western Pacific and Indian oceans. This
type of storm develops over oceans in 8° to 15° N and S
latitudes, but not closer to the equator. It originates as
an easterly wave or weak low, which then intensifies and
grows into a deep, circular low. High sea-surface temperatures, over 27°C (81°F), are required for tropical
cyclones to form. Once formed, the storm moves west-

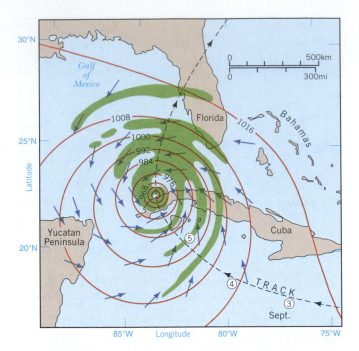

Figure 6.15 A simplified weather map of a hurricane passing over the western tip of Cuba. Daily locations, beginning on September 3, are shown as circled numerals. The recurving path will take the storm over Florida. Shaded areas show dense rain clouds as seen in a satellite image.

Table 6.2 Simpson–Saffir Scale of Tropical Cyclone Intensity

Category	Central Pressure mb (in. Hg)	Storm Surge m (ft)	Mean Wind m/s (mph)
1 Weak	>980 (>29.0)	1.2–1.7 (4–5)	33–42 (74–95)
2 Moderate	965–979 (28.5–29.0)	1.8–2.6 (6–8)	43–49 (96–110)
3 Strong	945-964 (27.9–28.5)	2.7–3.8 (9–12)	50–58 (111–130)
4 Very Strong	920–944 (27.2–27.9)	3.9–5.6 (13–18)	59–69 (131–155)
5 Devastating	<920 (<27.2)	>5.6 (>18)	>69 (>155)

storm center commonly falls to 950 mb (28.1 in. Hg) or lower.

A characteristic feature of the tropical cyclone is its central eye, in which clear skies and calm winds prevail (Figure 6.16 and chapter opener photo). The eye is a cloud-free vortex produced by the intense spiraling of the storm. In the eye, air descends from high altitudes and is adiabatically warmed. As the eye passes over a site, calm prevails, and the sky clears. Passage of the eye may take about half an hour, after which the storm strikes with renewed ferocity, but with winds in the opposite direction.

The intensity of tropical cyclones is rated on the Simpson–Saffir scale, shown in Table 6.2. This scale ranks storms from weak to devastating based on the central pressure of the storm, mean wind speed, and height of accompanying storm surge. Category 1 storms are weak, while category 5 storms are devastating.

Tropical cyclones (hurricanes or typhoons) are the most powerful and destructive of cyclonic storms.

---

ward through the trade-wind belt, often intensifying as it travels. It can then curve northwest, north, and northeast, steered by westerly winds aloft. Tropical cyclones can penetrate well into the midlatitudes, as many residents of the southern and eastern coasts of the United States who have experienced hurricanes can attest.

An intense tropical cyclone is an almost circular storm center of extremely low pressure. Winds spiral inward at high speed, accompanied by very heavy rainfall (Figure 6.15). The storm gains its energy through the release of latent heat as the intense precipitation forms. The storm's diameter may be 150 to 500 km (about 100 to 300 mi). Wind speeds can range from 120 to 200 km (75 to 125 mi) per hour (34 to 56 m/s) and sometimes much higher. Barometric pressure in the

Figure 6.16 Schematic diagram of a hurricane. Cumulonimbus (Cb) clouds in concentric rings rise through dense stratiform clouds. Cirrus clouds (Ci) fringe out ahead of the storm. Width of diagram represents about 1000 km (about 600 mi). (Redrawn from NOAA, National Weather Service.)

Refer to Figure 6.9, presented earlier, which shows typical paths of wave cyclones and tropical cyclones. As you can see, tropical cyclones always form over oceans. In the western hemisphere, hurricanes originate in the Atlantic off the west coast of Africa, in the Caribbean Sea, or off the west coast of Mexico. Curiously, tropical cyclones do not form in the South Atlantic or southeast Pacific regions, so South America is never threatened by these severe storms. In the Indian Ocean, typhoons originate both north and south of the equator, moving north and east to strike India, Pakistan, and Bangladesh, as well as south and west to strike the eastern coasts of Africa and Madagascar. Typhoons of the western Pacific also form both north and south of the equator, moving into northern Australia, Southeast Asia, China, and Japan.

Tracks of tropical cyclones of the North Atlantic are shown in detail in Figure 6.17. Most of the storms originate at 10° to 20° N latitude, travel westward and northwestward through the trades, and then turn northeast at about 30° to 35° N latitude into the zone of the westerlies. Here their intensity lessens, especially if they move over land. In the trade-wind belt, the cyclones travel 10 to 20 km (6 to 12 mi) per hour. In the zone of the westerlies, their speed is more variable.

Tropical cyclones occur only during certain seasons. For hurricanes of the North Atlantic, the season runs from May through November, with maximum frequency in late summer or early autumn. In the southern hemisphere, the season is roughly the opposite. These periods follow the annual migrations of the ITC to the north and south with the seasons, and correspond to periods when ocean temperatures are warmest.

For convenience, tropical cyclones are given names as they are tracked by weather forecasters. Male and female names are alternated in an alphabetical sequence renewed each season. Two sets of names are used—one for hurricanes of the Atlantic, and another for typhoons of the Pacific. Names are reused, but the names of storms that cause significant damage or destruction are retired from further use.

Impacts of Tropical Cyclones

Tropical cyclones can be tremendously destructive storms. Islands and coasts feel the full force of the high winds and flooding as tropical cyclones move onshore. For example, Hurricane Andrew, striking the Florida coast near Miami, was the most devastating storm ever to occur in the United States, claiming as much as $25 billion in property damage and 43 lives. *Eye on the Environment • Hurricane Andrew—Killer Cyclone* provides more information about this costly natural disaster.

The most serious effect of tropical cyclones is usually coastal destruction by storm waves and very high tides. Since the atmospheric pressure at the center of the cyclone is so low, sea level rises toward the center of the storm. High winds create a damaging surf and push water toward the coast, raising sea level even higher. Waves attack the shore at points far inland of the normal tidal range. Low pressure, winds, and the underwater shape of a bay floor can combine to produce a sudden rise of water level, known as a **storm surge**. During surges, ships are lifted by the high waters and can be stranded far inland.

The strong onshore winds of a cyclonic storm, coupled with high tides and a favorable offshore sea bottom configuration, can create a devastating storm surge.

If high tide accompanies the storm, the limits reached by inundation are even higher. This flooding can create enormous death tolls. At Galveston, Texas, in 1900, a sudden storm surge generated by a severe hurricane flooded the low coastal city and drowned about 6000 persons. Low-lying coral atolls of the western Pacific may be entirely swept over by wind-driven sea water, washing away palm trees and houses and drowning the inhabitants.

Also important is the large amount of rainfall produced by tropical cyclones. For some coastal regions, these storms provide much of the summer rainfall. Although this rainfall is a valuable water resource, it can also produce fresh water flooding, raising rivers and streams out of their banks. On steep slopes, soil saturation and high winds can topple trees and produce disastrous mudslides and landslides.

Hurricane Mitch, one of the deadliest Atlantic tropi-

Figure 6.17 Tracks of some typical hurricanes occurring during August. The storms arise in warm tropical waters and move northwest. On entering the region of prevailing westerlies, the storms change direction and move toward the northeast.

EYE ON THE ENVIRONMENT

Hurricane Andrew—Killer Cyclone

Tropical cyclones, or hurricanes, are among the most destructive of natural phenomena. With sustained high winds and rain, they can wreak a pattern of destruction across vast areas of land. Coastal areas are particularly vulnerable. Since hurricanes develop and reach their maximum intensity over ocean waters, coastal regions experience the full brunt of the storm. Storm surges caused by strong onshore winds can raise sea levels and bring storm-ravaged surf inland to batter coastal structures well back from the beach. Low coastal islands can be completely submerged, drowning their inhabitants.

Hurricane Andrew, striking the south Florida peninsula on August 23–24, 1992, and moving on to ravage the Louisiana coast on August 26, was the most damaging natural disaster ever to hit the United States. In the path of Andrew in south Dade County, Florida, was a population of 355,000 people. By the time the storm was over, 80,000 dwellings were demolished or destroyed. Another 55,000 dwellings were seriously damaged. Estimates of property loss ranged between $20 and $30 billion. The death toll was small, however—only 43 victims were claimed by the storm. Evacuations and emergency warnings reduced the toll significantly. In Louisiana, Andrew struck a rural area of agricultural fields, marshlands, and small towns, so damage was less intense. Another 15 people died, and additional damage was estimated at about $2 billion.

Hurricane Andrew, like many other tropical cyclones, began as a tropical easterly wave of showers off the African coast near the Cape Verde Islands on August 14. Moving westward it became a tropical depression with winds of 40 miles per hour (18

m/s) on August 17 and was named Andrew—the first tropical cyclone of the 1992 season. By August 20, the storm had moved westward but gained little strength. Opposing the westward surface motion of the storm was an upper air flow from west to east, creating a high-level wind shear that kept the storm from developing a strong vertical structure. However, by August 21, an unusual easterly upper air flow took hold, eliminating the wind shear. This allowed Andrew to gain strength. By 11 P.M. on the 21st, Andrew's winds had increased to 60 m.p.h (27 m/s).

Andrew continued to intensify. On Saturday morning, August 22, the storm became a category 1 hurricane on the Simpson–Saffir scale, and later in the day, it attained category 2 status. The storm continued its rapid development, reaching category 3, then category 4, on Sunday, August 23. Bearing winds of 160 m.p.h. (72 m/s) and a central low pressure of

27.33 in. Hg (924 mb), the hurricane swept through the southern Bahamas that evening, killing four and leaving 1700 homeless. Between 5:00 and 6:00 A.M. Monday morning, Andrew's eye crossed the coast of Florida just south of Miami. Sustained winds were in the range of 145 m.p.h. (65 m/s). A few miles to the north of the eye, the anemometer atop the National Hurricane Center measured a peak of 164 m.p.h. (73 m/s) before breaking at the height of the storm. Other anemometers measured gusts of about 175 m.p.h. (78 m/s). The storm surge brought sea level up to 16.9 ft (5.2 m) above normal.

Crossing Florida at a speed of 16 to 18 m.p.h. (7 to 8 m/s) and entering the Gulf of Mexico, Andrew headed west and slightly north. Intensifying over the warm Gulf water, the hurricane reattained category 4 status and struck the Louisiana coast on Wednesday, August 26. Low barrier

Hurricane Andrew blew the roofs from these homes in the suburbs of Miami, Florida, exposing the contents to wind and water damage.

The entrance to Pinelands, in the Everglades National Park, is blocked by fallen pine trees brought down by the high winds of Hurricane Andrew.

islands were overtopped, eroded, and cut with new channels. Marshes and shrimping grounds were filled with sand and debris. Sugar cane crops were flattened. Many dwellings were severely damaged. Moving inland and weakening, Andrew headed northeast, finally disintegrating into a region of showers in the Ohio River valley on August 28.

Although Andrew's winds in south Florida were measured in the 135 to 165 m.p.h. (60 to 74 m/s) range, an analysis of the damage patterns afterward indicated that winds were still higher in some locations. These winds may have resulted from a particularly intense convectional cell that developed on the north side of the storm as it crossed the coastline.

Another possibility, according to Professor Ted Fujita of the University of Chicago, is a "suction vortex"—a swirl of faster winds that occurs within the storm's eye wall. Although the vortex normally achieves wind speeds of only 20 to 30 m.p.h. (9 to 13 m/s), the vortex can be sucked upward by a strong local updraft, causing it to nar-row and spin more intensely. In fact, the vortex can attain wind speeds of up to 80 m.p.h. (36 m/s). Where the swirling motion reinforces the overall storm wind of perhaps 120 m.p.h. (54 m/s), total wind speed can approach 200 m.p.h. (89 m/s). No normal structure can survive such winds for more than a minute or two. Fortunately, these winds are restricted to a narrow, tornadolike swath of up to 300 ft (about 100 m) in width.

Not all the damage was sustained by human settlements. Hurricane Andrew delivered a solid blow to the Everglades as well. Most "hammocks"—tree-covered islands in the watery grassland—were stripped bare of leaves, their trees broken and splintered. Mangrove swamps on the western edge of the Everglades were also defoliated.

Although the Everglades ecosystem has endured the periodic assaults of hurricanes for centuries, and is well adjusted to surviving them, ecologists are concerned about the regrowth process. A number of imported species threaten to take over, crowd-ing out the native ones. It is not certain how the presence of these exotic plants will influence the recovery of the hammocks and other areas opened up by Andrew's high winds and waters.

Andrew is an example of the brutal power of the tropical cyclone to change the face of the earth. Only a massive earthquake in a major urban area is capable of doing more damage. While little can be done to stop such disasters from occurring, much can be done to reduce the damage they cause. Warnings and evacuations have reduced death rates to very low numbers. What is needed is better attention to building structures that are more wind-resistant and have such simple protections as storm shutters. Preserving shallow bays and mangrove swamps in natural states can also help by brunting the force of storm surges. Keeping beachfront development in check can reduce the number and size of structures exposed to the full force of hurricane winds and waves. Although Andrew was the first major hurricane to hit south Florida in nearly 30 years, the next hurricane will probably hit much sooner. Let's hope that this region will fare better next time, thanks to the lessons learned from Andrew.

Questions

1. Where did Hurricane Andrew first form, and how did it develop and move?

2. What damage was sustained by south Florida and coastal Louisiana from Hurricane Andrew?

3. What were Andrew's maximum wind speeds? How were they achieved?

4. How did Hurricane Andrew affect the Everglades?

cal cyclones in history, killed most of its 9086 victims with its torrential rains and their aftermath. The monster category 5 storm attained a central low pressure of 905 mb with winds of 77.2 m/s (172 mi/hr) in the western Caribbean on October 26, 1998. This central pressure is the fourth lowest ever measured in an Atlantic hurricane and ties Hurricane Mitch with Hurricane Camille of 1969. Moving at an average speed of only 2.2 m/s (5 mi/hr), Mitch hovered over the seven-nation Central American isthmus for nearly a week, dumping about 60 cm (24 in.) of rain at most locations. As much as 91 cm (36 in.) was reported at one mountain weather station, and it is likely that higher rainfalls went unrecorded. Countless mudflows and debris avalanches coursed down the rugged slopes of the western ranges of Honduras and Nicaragua, burying entire villages and villagers alike. Major rivers quickly flooded, sweeping floodplain settlements away on roiling waters. Damage was truly devastating. Honduras and Nicaragua suffered losses of about half their annual gross national product. Guatemala and El Salvador were also hard hit.

CLOUD COVER, PRECIPITATION, AND GLOBAL WARMING

Thus far in this text we have examined climatic change from several different perspectives. Let's now focus on how global climate might be influenced by an increase in clouds and precipitation. Recall that global temperatures have been rising over the last 20 years. As part of this global warming, satellite data have detected a rise in temperature of the global ocean surface of about 1°C (1.8°F) over the past decade. Any rise in sea-surface temperature increases the rate of evaporation, which in turn raises the water content of the atmosphere. What effect will this have on climate?

Water has several roles in global climate. First, water vapor is one of the most important greenhouse gases, enhancing the warming effect of the atmosphere above the earth's surface. Thus, one result of an increase in global water content of the atmosphere should be to enhance warming.

Second, water vapor can form clouds. Clouds reflect a large proportion of incoming shortwave radiation back to space, thus acting to cool global temperatures. Present estimates of this cooling show an energy loss rate for the earth as a whole of about 50 W/m². But cloud droplets and ice particles also enhance the greenhouse effect, in fact more strongly than water vapor. This greenhouse warming by clouds creates an energy gain of about 30 W/m². The net energy flow is therefore about 20 W/m² of cooling, and so present-day clouds act to cool the earth. Will clouds continue to have a cooling effect as climate warms? Yes, but the effect will not be as strong, according to computer models of global climate. This in turn will lead to temperatures that are even warmer.

What about precipitation? With more water vapor and more clouds in the air, more precipitation should result. If precipitation increases in arctic and subarctic zones, more of the earth's surface could be covered by snow and ice. Since snow is a good reflector of solar energy, this would tend to reduce global temperatures and thus moderate warming.

At this time, scientists are unsure how the global climate system will respond to global warming induced by the CO_2 increases predicted for the twenty-first century. Perhaps an increased amount of water in the atmosphere will enhance the warming trend—or perhaps it will slow it. As time goes by, however, our understanding of global climate and our ability to predict its changes are certain to increase.

A LOOK AHEAD

With a knowledge of the global circulation patterns of the atmosphere and oceans, as well as an understanding of weather systems and how they produce precipitation, the stage is set for global climate—which is the topic of Chapter 7. As we will see, the annual cycles of temperature and precipitation that most regions experience are quite predictable, given the changes in wind patterns, air mass flows, and weather systems that occur with the seasons. The result will be a description of the world's climates that grows easily and naturally from the principles you have mastered in your study of physical geography thus far.

WEATHER SYSTEMS IN REVIEW

A weather system is an organized state of the atmosphere associated with a characteristic weather pattern. Weather systems include wave cyclones, traveling anticyclones, tornadoes, easterly waves, weak equatorial lows, and tropical cyclones.

Air masses are distinguished by the latitudinal location and type of surface of their source regions. The boundaries between air masses are termed fronts. Wave cyclones typically form in the midlatitudes at the boundary between cool, dry air masses and warm, moist air masses. In the wave cyclone, a vast inspiraling motion produces cold and warm fronts, and eventually an occluded front. Precipitation normally occurs with each type of front. Tornadoes are very small, intense cyclones that occur as a part of thunderstorm activity. Their high winds can be very destructive.

Tropical weather systems include easterly waves and weak equatorial lows. Easterly waves occur when a weak low-pressure trough develops in the easterly wind circulation of the tropical zones, producing convergence, uplift, and shower activity. Weak equatorial lows occur near the intertropical convergence zone. They are convergence zones where convectional precipitation is frequent.

Tropical cyclones can be the most powerful of all storms. They develop over very warm tropical oceans and can intensify to become vast inspiraling systems of very high winds with very low central pressures. As they move onto land, they can bring heavy surf and storm surges of very high waters. Tropical cyclones have caused great death and destruction in coastal regions.

KEY TERMS

weather system	front	wave cyclone
cyclonic precipitation	cold front	tornado
cyclonic storm	warm front	tropical cyclone
air mass	occluded front	storm surge

REVIEW QUESTIONS

1. Define air mass. What two features are used to classify air masses?
2. Compare the characteristics and source regions for mP and cT air mass types.
3. Describe a tornado. Where and under what conditions do tornadoes typically occur?
4. Identify three weather systems that bring rain in equatorial and tropical regions. Describe each system briefly.
5. Describe the structure of a tropical cyclone. What conditions are necessary for the development of a tropical cyclone? Give a typical path for the movement of a tropical cyclone in the northern hemisphere.
6. Why are tropical cyclones so dangerous?

VISUALIZING EXERCISES

1. Identify three types of fronts and draw a cross section through each. Show the air masses involved, the contacts between them, and the direction of air mass motion.
2. Sketch two weather maps, showing a wave cyclone in open and occluded stages. Include isobars on your sketch. Identify the center of the cyclone as a low. Lightly shade areas where precipitation is likely to occur.

A DEEPER VIEW OF WEATHER SYSTEMS

1. Compare and contrast midlatitude and tropical weather systems. Be sure to include the following terms or concepts in your discussion: air mass, convectional precipitation, cyclonic precipitation, easterly wave, polar front, stable air, traveling anticyclone, tropical cyclone, unstable air, wave cyclone, and weak equatorial low.
2. Prepare a description of the annual weather patterns that are experienced through the year at your location. Refer to the general temperature and precipitation pattern as well as the types of weather systems that occur in each season.

Chapter 7

Global Climates

These residents of Bombay dress in light, bright clothing to keep cool and use umbrellas to ward off the strong tropical sun.

PUTTING YOU IN THE PICTURE

London, Rome, Cairo, Delhi, Singapore, Tokyo, Honolulu, and Anchorage! An around-the-world tour! It was hard to believe that you were the winner of the sweepstakes drawing sponsored by the credit card company, but now that you've picked up your first-class air ticket, it's finally hitting home. Although you'll only spend a few days in each of these famous cities, it will be a real thrill to experience them first-hand.

But what kind of weather should you expect? Will it be hot or cold, sunny or rainy? Should you pack your overcoat and sweaters? Or will your lightweight clothes be more appropriate? Will you need an umbrella in any of these cities? You'll be leaving in January, so you know what to expect for your trip to the airport, but what happens after that? Fortunately, you still have your college physical geography textbook to refer to. That evening, you brush up on the climates you will encounter.

"Let's see, London. Oh yes, that's a marine west-coast climate type. Lots of occluded wave cyclones in January. Hmm. I'd better have my raincoat and umbrella. Wool would be good. Perhaps that lightweight gray sweater will fit in my carry-on bag. Then on to Rome. That's the Mediterranean climate. Well, winter is the wet season for the Mediterranean climate, but the rainfall is so much lighter than in London that I'm likely to get away with just an umbrella. Still, I'll need some kind of light coat, since it won't be very warm. The rain-coat will probably do fine.

"Cairo—now, that's the dry subtropical climate. Not much chance of rain in that climate, and the hottest temperatures aren't till later in the year, when the sun moves to the tropic of cancer and is nearly overhead. Better have something light and cool. And I'll need my swimsuit for lounging by the hotel pool. Then there's Delhi. It's dry tropical there as well. Hmm. With the winter monsoon bringing cold, continental air down the slope of the Himalayas, it should be pretty clear, with warm days and cool nights.

"Then on from Delhi to Singapore. Oh oh, Singapore is right on the equator, so that would be the wet equatorial climate. I'd better make sure my umbrella isn't buried too deeply in my suitcase. Those convectional showers don't last long, but they can be pretty intense. It'll certainly be hot and steamy. Now, Tokyo is in the moist continental climate, at about 35° N latitude. That's the same as New York. In January? Looks like I'll need those wool clothes again. And I'd better keep that umbrella handy.

"Now, Honolulu! That's the wet equatorial climate, the same as Singapore. But Hawaii's at about 20° N, so the ocean will be cooler and the daily temperature should be just perfect! Last of all is Anchorage. That's the boreal forest climate—and in January! I'll certainly need something warm. But I hate to lug my heavy winter coat around the world just for the last few days. Here's an idea! I've been putting off buying that new ski suit for a couple of years, but how about buying it in Anchorage? And skis, too. What a way to end the trip! Swimming in the warm Pacific and skiing in Alaska!"

Who said learning physical geography wouldn't come in handy some day? After finishing this chapter, you'll be set to pack your clothes for any destination, any time!

As we examine the earth's climates, you will find that your earlier studies of air temperatures, precipitation, global winds and circulation patterns, air masses, and weather systems will all combine to help you understand the nature of global climates as they are located over the globe. Thus, this chapter will help you to bring together much of the knowledge of physical geography that you have already acquired in a focus on the breadth and diversity of the world's climatic regions.

Chapter Outline

Global Precipitation

Climate Classification

Low-Latitude Climates

The Wet Equatorial Climate ①

The Monsoon and Trade-Wind Coastal Climate ②

Special Supplement • The Köppen Climate System

The Wet-Dry Tropical Climate ③

Eye on the Environment • Drought and Land Degradation in the African Sahel

The Dry Tropical Climate ④

Midlatitude Climates

The Dry Subtropical Climate ⑤

The Moist Subtropical Climate ⑥

The Mediterranean Climate ⑦

The Marine West-Coast Climate ⑧

The Dry Midlatitude Climate ⑨

The Moist Continental Climate ⑩

High-Latitude Climates

The Boreal Forest Climate ⑪

The Tundra Climate ⑫

The Ice Sheet Climate ⑬

GLOBAL CLIMATES

What do we mean by climate? In its most general sense, **climate** is the average weather of a region. To describe the average weather of a region, we could use many of the measures describing the state of the atmosphere that we have already encountered. These might include daily net radiation, barometric pressure, wind speed and direction, cloud cover and type, presence of fog, precipitation type and intensity, incidence of cyclones and anticyclones, frequency of frontal passages, and other such items of weather information. However, observations as detailed as these are not made regularly at most weather stations around the world. To study climate on a worldwide basis, we must turn to the two simple measurements that are made daily at every weather station—temperature and precipitation. We will be concerned with both their average values and their variation within a period of time.

Annual cycles of mean monthly temperature and precipitation are used to determine the climate type.

Note that temperature and precipitation are factors that strongly influence the natural vegetation of a region—for example, forests generally occur in moist regions and grasslands in dry regions. The natural vegetation cover is often a distinctive feature of a climatic region and typically influences the human use of the area. Temperature and precipitation are also important factors in cultivation of crop plants—a necessary process for human survival. And the development of soils, as well as the types of processes that shape landforms, is partly dependent on temperature and precipitation. For these reasons, we will find that climates defined on the basis of temperature and precipitation also help set apart many features of the environment, not just climate alone. This is why the study of global climates is such an important part of physical geography.

A few simple principles discussed in earlier chapters are very helpful in understanding the global scope of climate. First, recall that two major factors influence the annual cycle of air temperature experienced at a station: latitude and coastal versus continental location.

- *Latitude.* The annual cycle of temperature at a station depends on its latitude. Near the equator, temperatures are warmer and the annual range is low. Toward the poles, temperatures are colder and the annual range is greater. These effects are produced by the annual cycle of insolation, which varies with the seasons.

- *Coastal-continental location.* Coastal stations show a smaller annual variation in temperature, while the variation is larger for stations in continental interiors. This effect occurs because ocean surface temperatures vary less with the seasons than do temperatures of land surfaces.

Air temperature also has an important effect on precipitation. Recall this principle from Chapter 4:

- *Warm air can hold more moisture than cold air.* This means that colder regions generally have lower precipitation than warmer regions. In addition, precipitation will tend to be greater during the warmer months of the temperature cycle.

Keeping these three key ideas in mind as you read this chapter will help make the climates easy to understand and explain.

GLOBAL PRECIPITATION

Figure 7.1 is a map of mean annual precipitation over the world. This map uses **isohyets**—lines drawn through all points having the same annual precipitation. For regions where all or most of the precipitation is rain, we use the word "rainfall." For regions where snow is a significant part of the annual total, we use the word "precipitation." Let's look at moist and dry regions in more detail.

Orographic precipitation occurs on mountainous coasts exposed to moist trade winds or frequent cyclonic storms.

Rainfall is generally high in a broad belt around the equator, including large regions of South America, Africa, and Southeast Asia. This is largely convectional precipitation, which occurs in the warm, moist, unstable air that converges in the intertropical zone. Farther from the equator, easterly waves and tropical cyclones also generate precipitation. Rainfall is especially high in Southeast Asia. Recall that this is produced by the monsoon system, which brings warm, moist air from the Indian Ocean across the subcontinent during the rainy season.

Orographic precipitation occurs along mountainous coasts in the belt of the trade winds. In these regions,

Figure 7.1 World precipitation.
Isohyets are labeled with cm (in.).

MEAN ANNUAL PRECIPITATION
OF THE WORLD

Simplified and modified from Plate 3, World Climatology,
Volume I, The Times Atlas, Editor John Bartholomew,
The Times Publishing Company, Ltd., London, 1958

Centimeters 0   10   30   50   100   200   500 above 500

Inches 0   4   12   20   40   80   200 above 200

Isohyets labeled in centimeters, inches (approximate) in parentheses.

Kilometers
0   1000   2000   3000

Miles
0   1000   2000

(True distances on mid-meridians and parallels 0 to 40)
Interrupted homolosine projection
Based on Goode Base Map

southeast or northeast trades bring warm, moist air
ashore, where it rises and releases precipitation as it
crosses the coastal mountains. The east coast of Central
America, the northeast coast of South America, the
Philippines, Malaysia, and the eastern coast of
Indochina are examples.

Beyond the equatorial and tropical regions, the
world map shows high precipitation in a few narrow
bands in the midlatitude region. These are all located
along the western edges of continents. The Pacific
Northwest coast, including parts of British Columbia
and the Alaska panhandle, provides the best example.
Bands of high coastal precipitation are also found in
southern South America and in northwestern Europe.

The bands occur because large numbers of cyclonic
storms develop over oceans to the west and pass across
these coasts, headed eastward. In North and South
America, mountain belts cause the rainfall to be intensified by the orographic effect.

Let's now look at the dry regions of the globe. Dry
regions occur along the central coast of western South
America (Chile), along the western coast of southern
Africa, and in the American Southwest. These locations
correspond well with the eastern sides of subtropical
high-pressure cells, where subsiding air is generally
clear, dry, and stable (see Figure 5.12).

The largest dry region consists of a broad band
ranging from the Sahara Desert of North Africa,

through the Middle East, and well into Central Asia. The western part, including the Sahara and other Middle Eastern deserts, owes its dryness to descending air in the subtropical high-pressure belt. Farther to the east, in Central Asia, conditions are dry largely because the region is far from oceans. Moist oceanic air loses its water vapor in precipitation as it passes over land before reaching this region. This inland dryness effect also occurs in the high plains of western North America. Here moist Pacific air is dried out as it moves eastward over the mountain ranges and western high country, producing a dry interior region.

The high-latitude regions are also areas of low precipitation. Examples are the Arctic islands of Canada, northern Greenland, and eastern Siberia, which are cold and show low precipitation nearly year round.

CLIMATE CLASSIFICATION

Mean monthly values of air temperature and precipitation can describe the climate of a weather station and its nearby region quite accurately. To study climates from a global viewpoint, climatologists classify these values into a set of climate types. This requires developing a set of rules to use in examining monthly temperature and precipitation values. By applying the rules, the climatologist can use each station's data to determine the climate to which it belongs. This textbook recognizes 13 distinctive climates based on principles of the soil water balance as derived from temperature and precipitation. The climate types are designed to be understood and explained by air mass movements and frontal zones. The types follow quite naturally from the understanding of global temperature and precipitation processes that you have acquired in prior chapters.

An alternative climate classification to that provided in this textbook is that of Vladimir Köppen, devised in 1918, and modified by R. Geiger and W. Pohl in 1953. It uses a system of letters to label climates. It is presented in a special supplement in this chapter.

Recall from Chapter 6 that air masses are classified according to the general latitude of their source regions and their surface type—land or ocean—within that region (see Table 6.1) and that air masses have typical temperatures and moisture content values based on their source regions. Thus, air mass characteristics control the two most important climate variables—temperature and precipitation—experienced at a location.

We also know from Chapter 6 that frontal zones are regions in which air masses are in contact. When unlike air masses are in contact, cyclonic precipitation is likely to develop. However, the position of frontal zones changes with the seasons. For example, the polar-front zone lies generally across the middle latitudes of the United States in winter, but it retreats northward to Canada during the summer.

Figure 7.2 shows our schematic diagram of air mass source regions. We have subdivided this diagram into global bands that contain three broad groups of climates: low-latitude (Group I), midlatitude (Group II), and high-latitude (Group III). They are described briefly as follows.

- *Group I: Low-Latitude Climates.* The region of low-latitude climates (Group I) is dominated by the source regions of continental tropical (cT), maritime tropical (mT), and maritime equatorial (mE) air masses. These source regions are related to the three most obvious atmospheric features that occur within their latitude band—the two subtropical high-pressure belts and the equatorial trough at the intertropical convergence zone (ITC). Air of polar origin occasionally invades regions of low-latitude climates. Easterly waves and tropical cyclones are important in this climate group.

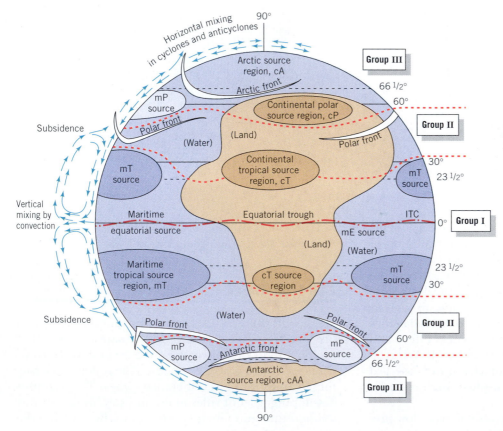

Figure 7.2 Using the map of air mass source regions, we can identify five global bands associated with three major climate groups. Within each group is a set of distinctive climates with unique characteristics that are explained by the movements of air masses and frontal zones.

Special Supplement • The Köppen Climate System

Air temperature and precipitation data have formed the basis for several climate classifications. One of the most important of these is the Köppen climate system, devised in 1918 by Dr. Vladimir Köppen of the University of Graz in Austria. For several decades, this system, with various later revisions, was the most widely used climate classification among geographers. Köppen was both a climatologist and plant geographer, so that his main interest lay in finding climate boundaries that coincided approximately with boundaries between major vegetation types.

Under the Köppen system, each climate is defined according to assigned values of temperature and precipitation, computed in terms of annual or monthly values. Any given station can be assigned to its particular climate group and subgroup solely on the basis of the records of temperature and precipitation at that place.

The Köppen system features a shorthand code of letters designating major climate groups, subgroups within the major groups, and further subdivisions to distinguish particular seasonal characteristics of temperature and precipitation. Five major climate groups are designated by capital letters as follows.

A Tropical rainy climates

Average temperature of every month is above 18°C (64.4°F). These climates have no winter season. Annual rainfall is large and exceeds annual evaporation.

B Dry climates

Evaporation exceeds precipitation on the average throughout the year. There is no water surplus; hence, no permanent streams originate in B climate zones.

C Mild, humid (mesothermal) climates

The coldest month has an average temperature of under 18°C (64.4°F) but above −3°C (26.6°F); at least one month has an average temperature above 10°C (50°F). The C climates thus have both a summer and a winter.

D Snowy-forest (microthermal) climates

The coldest month has an average temperature of under −3°C (26.6°F). The average temperature of the warmest month is above 10°C (50°F). (Forest is not generally found where the warmest month is colder than 10°C [50°F].)

E Polar climates

The average temperature of the warmest month is below 10°C (50°F). These climates have no true summer.

Note that four of these five groups (A, C, D, and E) are defined by temperature averages, whereas one (B) is defined by precipitation-to-evaporation ratios. Groups A, C, and D have sufficient heat and precipitation for the growth of forest and woodland vegetation. Figure S7.1 shows the boundaries of the five major climate

Figure S7.1 Highly generalized world map of major climate regions according to the Köppen classification. Highland areas are in black. (Based on Goode Base Map.)

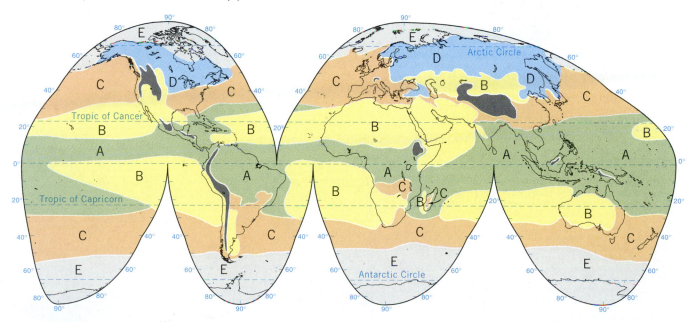

groups, and Figure S7.2 is a world map of Köppen climates.

Subgroups within the five major groups are designated by a second letter according to the following code.

S *Semiarid (steppe)*

W *Arid (desert)*

(The capital letters *S* and *W* are applied only to the dry *B* climates.)

f *Moist, adequate precipitation in all months, no dry season. This modifier is applied to A, C, and D groups.*

w *Dry season in the winter of the respective hemisphere (low-sun season).*

s *Dry season in the summer of the respective hemisphere (high-sun season).*

m *Rainforest climate, despite short, dry season in monsoon type of precipitation cycle. Applies only to A climates.*

From combinations of the two letter groups, 12 distinct climates emerge:

Af *Tropical rainforest climate*

The rainfall of the driest month is 6 cm (2.4 in.) or more.

Figure S7.2 World map of climates according to the Köppen-Geiger-Pohl system.

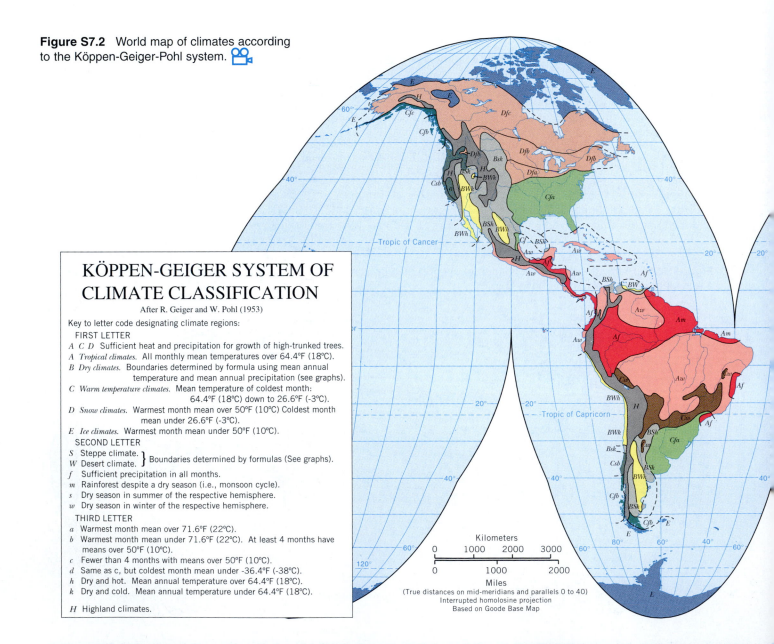

KÖPPEN-GEIGER SYSTEM OF CLIMATE CLASSIFICATION
After R. Geiger and W. Pohl (1953)

Key to letter code designating climate regions:

FIRST LETTER
A C D Sufficient heat and precipitation for growth of high-trunked trees.
A Tropical climates. All monthly mean temperatures over 64.4°F (18°C).
B Dry climates. Boundaries determined by formula using mean annual temperature and mean annual precipitation (see graphs).
C Warm temperature climates. Mean temperature of coldest month: 64.4°F (18°C) down to 26.6°F (-3°C).
D Snow climates. Warmest month mean over 50°F (10°C) Coldest month mean under 26.6°F (-3°C).
E Ice climates. Warmest month mean under 50°F (10°C).

SECOND LETTER
S Steppe climate. } Boundaries determined by formulas (See graphs).
W Desert climate.
f Sufficient precipitation in all months.
m Rainforest despite a dry season (i.e., monsoon cycle).
s Dry season in summer of the respective hemisphere.
w Dry season in winter of the respective hemisphere.

THIRD LETTER
a Warmest month mean over 71.6°F (22°C).
b Warmest month mean under 71.6°F (22°C). At least 4 months have means over 50°F (10°C).
c Fewer than 4 months with means over 50°F (10°C).
d Same as c, but coldest month mean under -36.4°F (-38°C).
h Dry and hot. Mean annual temperature over 64.4°F (18°C).
k Dry and cold. Mean annual temperature under 64.4°F (18°C).

H Highland climates.

Kilometers
0   1000   2000   3000

0   1000   2000
Miles
(True distances on mid-meridians and parallels 0 to 40)
Interrupted homolosine projection
Based on Goode Base Map

Am *Monsoon variety of Af*

The rainfall of the driest month is less than 6 cm (2.4 in.). The dry season is strongly developed.

Aw *Tropical savanna climate*

At least one month has rainfall less than 6 cm (2.4 in.). The dry season is strongly developed.

Figure S7.3 shows the boundaries between *Af*, *Am*, and *Aw* climates as

Figure S7.3
Boundaries of the *A* climates.

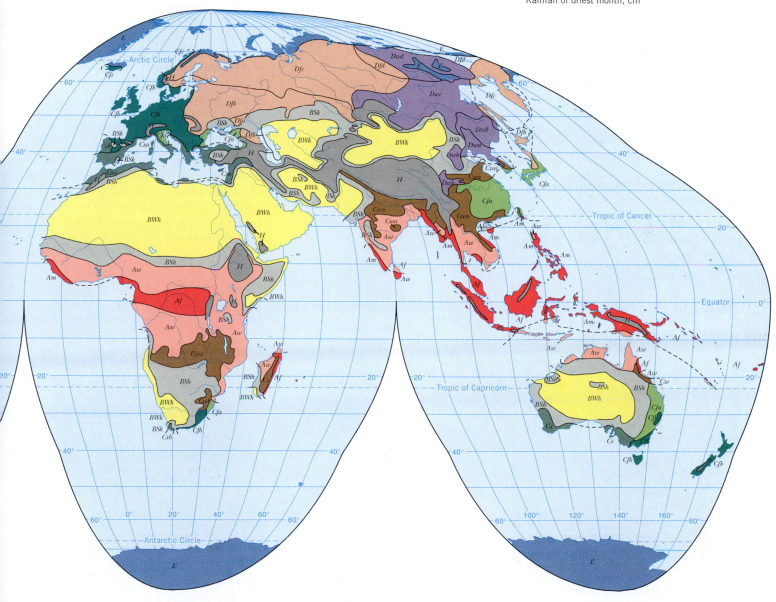

determined by both annual rainfall and rainfall of the driest month.

BS   *Steppe climate*

A semiarid climate characterized by grasslands, it occupies an intermediate position between the desert climate (*BW*) and the more humid climates of the *A, C,* and *D* groups. Boundaries are determined by formulas given in Figure S7.4.

BW   *Desert climate*

Desert has an arid climate with annual precipitation of usually less than 40 cm (15 in.). The boundary with the adjacent steppe climate (*BS*) is determined by formulas given in Figure S7.4.

Cf   *Mild humid climate with no dry season*

Precipitation of the driest month averages more than 3 cm (1.2 in.).

Cw   **Mild humid climate with a dry winter**

The wettest month of summer has at least 10 times the precipitation of the driest month of winter. (Alternative definition: 70 percent or more of the mean annual precipitation falls in the warmer six months.)

Cs   *Mild humid climate with a dry summer*

Precipitation of the driest month of summer is less than 3 cm (1.2 in.). Precipitation is at least three times as much as the driest month of summer. (Alternative definition: 70 percent or more of the mean annual precipitation falls in the six months of winter.)

Df   *Snowy-forest climate with a moist winter*

No dry season.

Dw   *Snowy-forest climate with a dry winter*

ET   *Tundra climate*

The mean temperature of the warmest month is above 0°C (32°F) but below 10°C (50°F).

EF   *Perpetual frost climate*

In this ice sheet climate, the mean monthly temperatures of all months are below 0°C (32°F).

To denote further variations in climate, Köppen added a third letter to the code group. The meanings are as follows:

a With hot summer; warmest month is over 22°C (71.6°F); *C* and *D* climates.

b With warm summer; warmest month is below 22°C (71.6°F); *C* and *D* climates.

c With cool, short summer; less than four months are over 10°C (50°F); *C* and *D* climates.

d With very cold winter; coldest month is below –38°C (–36.4°F); *D* climates only.

h Dry-hot; mean annual temperature is over 18°C (64.4°F); *B* climates only.

k Dry-cold; mean annual temperature is under 18°C (64.4°F); *B* climates only.

As an example of a complete Köppen climate code, *BWk* refers to a cool desert climate, and *Dfc* refers to a cold, snowy-forest climate with cool, short summer.

Figure S7.4 Boundaries of the *B* climates. Upper labels: metric system. Lower labels: English system.

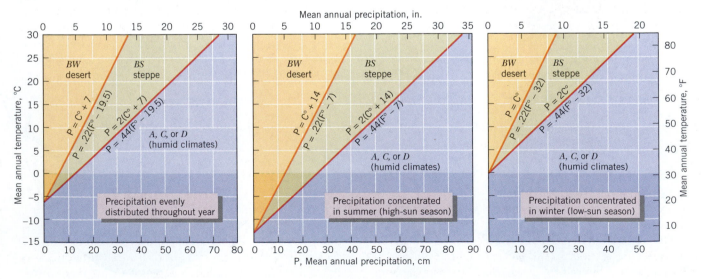

- *Group II: Midlatitude Climates.* The region of midlatitude climates (Group II) lies in the polar-front zone—a zone of intense interaction between unlike air masses. Here tropical air masses moving poleward and polar air masses moving equatorward are in conflict. Wave cyclones are normal features of the polar front, and this zone may contain as many as a dozen wave cyclones around the globe in a single day.

- *Group III: High-Latitude Climates.* The region of high-latitude climates (Group III) is dominated by polar and arctic (including antarctic) air masses. In the arctic belt of the 60th to 70th parallels, continental polar air masses meet arctic air masses along an *arctic-front zone*, creating a series of eastward-moving wave cyclones. In the southern hemisphere, there are no source regions in the subantarctic belt for continental polar air—just a great single oceanic source region for maritime polar (mP) air masses. The pole-centered continent of Antarctica provides a single great source of the extremely cold, dry antarctic air mass (cAA). These two air masses interact along the *antarctic-front zone.*

Within each of these three climate groups are a number of climate types (or simply, climates)—four low-latitude climates (Group I), six midlatitude climates (Group II), and three high-latitude climates (Group III)—for a total of 13 climate types. In this textbook, the climates are numbered for ease in identification on maps and diagrams. We will refer to each climate by name, because the names describe the general nature of the climate and also suggest its global location. For convenience, we include the climate number next to its name in the text.

The 13 climate types can be understood and explained by air mass movements and frontal zones.

In presenting the climates, we will make use of a pictorial device called a **climograph**. It shows the annual cycles of monthly mean air temperature and monthly mean precipitation for a location, along with some other useful information. Figure 7.3 is an example of a climograph. The data plotted are for Kayes, a station in the African country of Mali. At the top of the climograph, the mean monthly temperature is plotted as a line graph. At the bottom, the mean monthly precipitation is shown as a bar graph. The annual range in temperature and the total annual precipitation are stated on every climograph as well. Most climographs also display dominant weather features, which are shown using picture symbols. For Kayes, the two dominant features

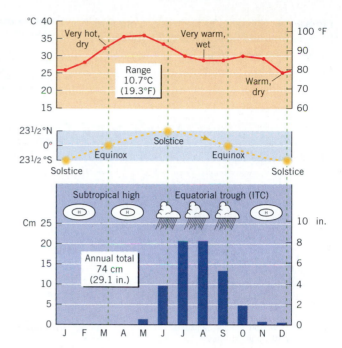

Figure 7.3 Climograph for Kayes, Mali. This station is located in western Africa, just south of the Sahara Desert. Here, the climate is nearly rainless for seven months of the year. During these months, dry subsiding air from subtropical high-pressure cells dominates. Rainfall occurs when the intertropical convergence zone moves northward and reaches the vicinity of Kayes in June or July. Temperatures are hottest in April and May, before the rainy season begins.

are the subtropical high and equatorial trough (ITC). Many climographs also include a small graph of the sun's declination in order to help show when solstices and equinoxes occur.

The world map of climates, Figure 7.4, shows the actual distribution of climate types on the continents. This map, based on data collected at a large number of observing stations, is simplified because the climate boundaries are uncertain in many areas where observing stations are thinly distributed.

LOW-LATITUDE CLIMATES

The low-latitude climates lie for the most part between the tropics of cancer and capricorn. In terms of world latitude zones, the low-latitude climates occupy all of the equatorial zone (10° N–10° S), most of the tropical zone (10–15° N and S), and part of the subtropical zone (15–25° N and S). In terms of prevailing pressure and wind systems, the region of low-latitude climates includes the equatorial trough of the intertropical convergence zone (ITC), the belt of tropical easterlies

Figure 7.4 Climates of the world. Compiled from station data by A. N. Strahler. (Based on Goode Base Map.)

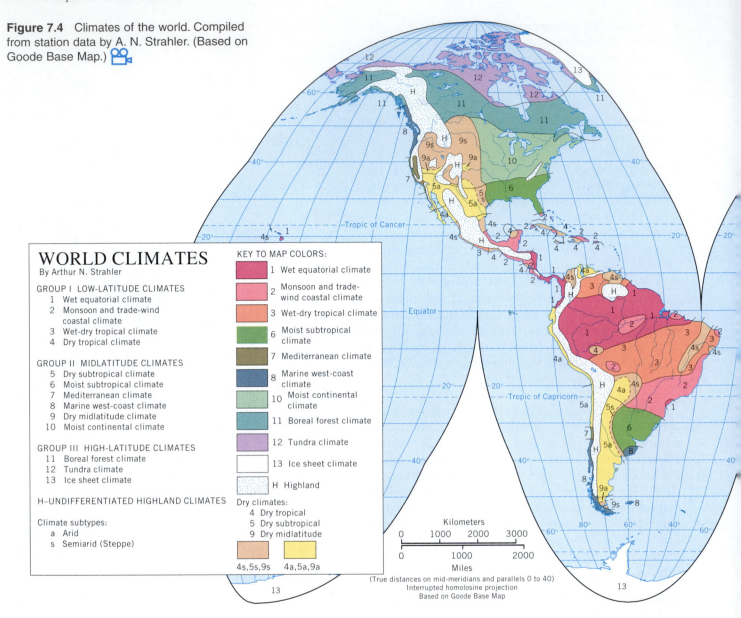

WORLD CLIMATES
By Arthur N. Strahler

GROUP I LOW-LATITUDE CLIMATES
 1 Wet equatorial climate
 2 Monsoon and trade-wind coastal climate
 3 Wet-dry tropical climate
 4 Dry tropical climate

GROUP II MIDLATITUDE CLIMATES
 5 Dry subtropical climate
 6 Moist subtropical climate
 7 Mediterranean climate
 8 Marine west-coast climate
 9 Dry midlatitude climate
 10 Moist continental climate

GROUP III HIGH-LATITUDE CLIMATES
 11 Boreal forest climate
 12 Tundra climate
 13 Ice sheet climate

H—UNDIFFERENTIATED HIGHLAND CLIMATES

Climate subtypes:
 a Arid
 s Semiarid (Steppe)

KEY TO MAP COLORS:
 1 Wet equatorial climate
 2 Monsoon and trade-wind coastal climate
 3 Wet-dry tropical climate
 6 Moist subtropical climate
 7 Mediterranean climate
 8 Marine west-coast climate
 10 Moist continental climate
 11 Boreal forest climate
 12 Tundra climate
 13 Ice sheet climate
 H Highland
Dry climates:
 4 Dry tropical
 5 Dry subtropical
 9 Dry midlatitude

 4s,5s,9s   4a,5a,9a

Kilometers
0   1000   2000   3000

Miles
0   1000   2000

(True distances on mid-meridians and parallels 0 to 40)
Interrupted homolosine projection
Based on Goode Base Map

(northeast and southeast trades), and large portions of the oceanic subtropical high-pressure belt (Figure 7.2).

Figure 7.5 shows climographs for the four low-latitude climates. These climates range from extremely moist—the wet equatorial climate ① —to extremely dry—the dry tropical climate ④. They also vary strongly in the seasonality of their rainfall. In the wet equatorial climate ①, rainfall is abundant all year round. But in the wet-dry tropical climate ③, rainfall is abundant for only part of the year. During the remainder of the year, little or no rain falls. The seasonal temperature cycle also varies among these climates. In the wet equatorial climate ①, temperatures are nearly uniform throughout the year. In the dry tropical climate ④, there is a strong annual temperature cycle. Table 7.1 summarizes some of the characteristics of these climates.

The Wet Equatorial Climate ① (Köppen: *Af*)

The **wet equatorial climate**① is a climate of the intertropical convergence zone (ITC), which is nearby for most of the year. The climate is dominated by warm, moist maritime equatorial (mE) and maritime tropical (mT) air masses that yield heavy convectional rainfall. Rainfall is plentiful in all months, and the annual total often exceeds 250 cm (about 100 in.). However, there is usually a seasonal pattern to the rainfall, so that rainfall is greater during some part of the

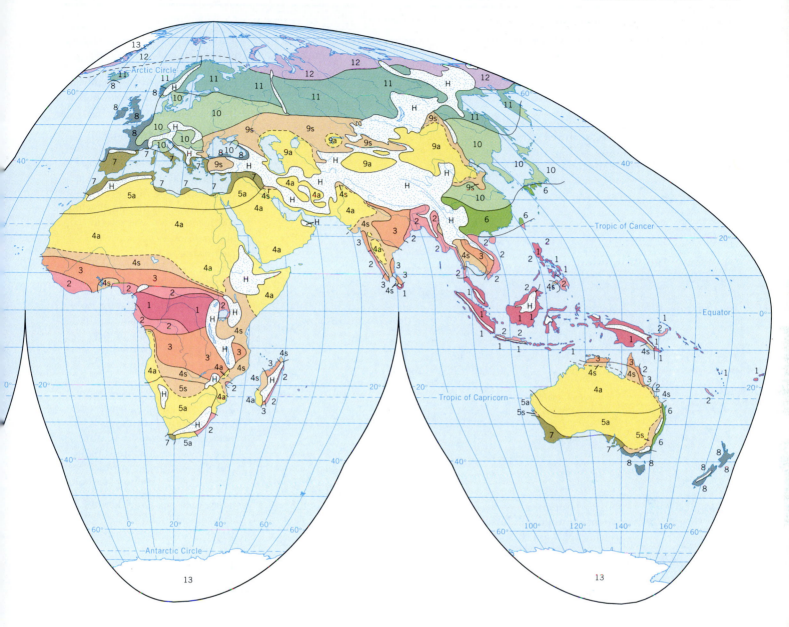

Figure 7.5 Climographs for the four low-latitude climates.

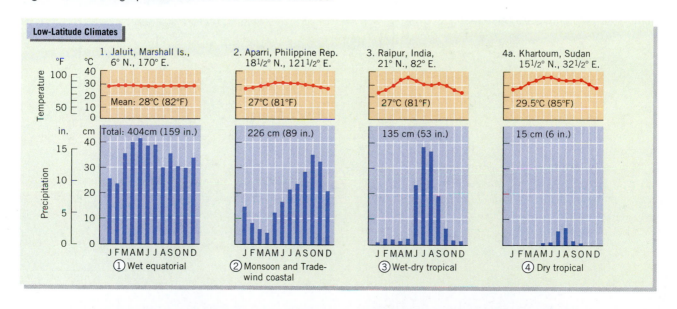

Table 7.1 Low-Latitude Climates

Climate	Temperature	Precipitation	Explanation
Wet equatorial ①	Uniform temperatures, mean near 27° C (81°F).	Abundant rainfall, all months, from mT and mE air masses. Annual total may exceed 250 cm (100 in.).	The ITC dominates this climate, with abundant convectional precipitation generated by convergence in weak equatorial lows. Rainfall is heaviest when the ITC is nearby.
Monsoon and trade-wind coastal ②	Temperatures show an annual cycle, with warmest temperatures in the high-sun season.	Abundant rainfall but with a strong seasonal pattern.	Trade-wind coastal: Rainfall from mE and mT air masses is heavy when the ITC is nearby, lighter when the ITC moves to the opposite hemisphere. Asian monsoon coasts: dry air flowing southwest in low-sun season alternates with moist oceanic air flowing northeast, producing a seasonal rainfall pattern on west coasts.
Wet-dry tropical ③	Marked temperature cycle, with hottest temperatures before the rainy season.	Wet high-sun season alternates with dry low-sun season.	Subtropical high pressure moves into this climate in the low sun-season, bringing very dry conditions. In the high-sun season, the ITC approaches and rainfall occurs. Asian monsoon climate: alternation of dry continental air in low-sun season, with moist oceanic air in low-sun season with moist oceanic air in high-sun season brings a strong pattern of dry and wet seasons.
Dry tropical ④	Strong temperature cycle, with intense hot temperatures during high-sun season.	Low precipitation. Sometimes rainfall occurs when the ITC is near.	This climate is dominated by subtropical high pressure, which provides clear, stable air for much or all of year. Insolation is intense during high-sun period.

Figure 7.6 World map of wet equatorial ① and monsoon and trade-wind coastal climate ②. (Based on Goode Base Map.)

year. This period of heavier rainfall occurs when the ITC migrates into the region. Remarkably uniform temperatures prevail throughout the year. Both mean monthly and mean annual temperatures are typically close to 27°C (81°F).

Figure 7.6 shows the world distribution of the wet equatorial climate ①. This climate is found in the latitude range 10° N to 10° S. Its major regions of occurrence include the Amazon lowland of South America, the Congo basin of equatorial Africa, and the East Indies, from Sumatra to New Guinea.

Figure 7.7 is a climograph for Iquitos, Peru (located in Figure 7.6), a typical wet equatorial station situated close to the equator in the broad, low basin of the upper Amazon River. Notice the very small annual range in temperature and the very large annual rainfall total. Monthly air temperatures are extremely uniform in the wet equatorial climate①. Typically, mean monthly air temperature will range between 26° and 29°C (79° and 84°F) for stations at low elevation in the equatorial zone.

The Monsoon and Trade-Wind Coastal Climate ② (Köppen: *Af, Am*)

Like the wet equatorial climate ①, the **monsoon and trade-wind coastal climate** ② has abundant rainfall. But unlike the wet equatorial climate ①, the rainfall of the monsoon and trade-wind coastal climate ② always shows a strong seasonal pattern. This seasonal pattern is due to the migration of the intertropical convergence zone (ITC). In the high-sun season ("summer," depending on the hemisphere), the ITC is near and monthly rainfall is greater. In the low-sun season, when the ITC has

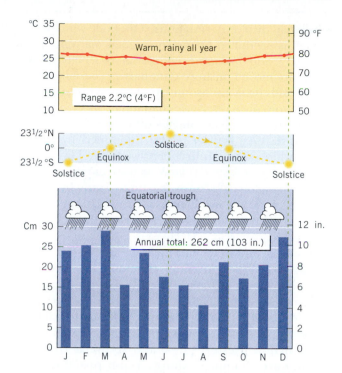

Figure 7.7 Wet equatorial climate ①. Iquitos, Peru, is located in the upper Amazon lowland, close to the equator. Temperatures differ very little from month to month, and there is abundant rainfall throughout the year.

migrated to the other hemisphere, subtropical high pressure dominates and monthly rainfall is less. Figure 7.6 shows the global distribution of the monsoon and trade-wind coastal climate ②. The climate occurs over latitudes from 5° to 25° N and S.

As the name of the monsoon and trade-wind coastal climate ② suggests, this climate type is produced by two

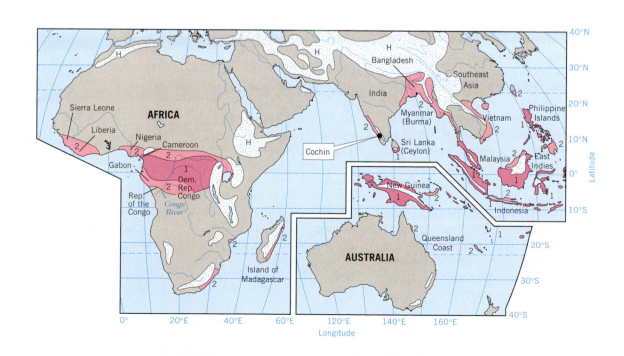

somewhat different situations. On trade-wind coasts, rainfall is produced by moisture-laden maritime tropical (mT) and maritime equatorial (mE) air masses. These are moved onshore onto narrow coastal zones by trade winds or by monsoon circulation patterns. As the warm, moist air passes over coastal hills and mountains, the orographic effect touches off convectional shower activity. Shower activity is also intensified by easterly waves, which are more frequent when the ITC is nearby. The east coasts of land masses experience this trade-wind effect, because the trade winds blow from east to west. Trade-wind coasts are found along the east sides of Central and South America, the Caribbean Islands, Madagascar (Malagasy), Southeast Asia, the Philippines, and northeast Australia (see Figure 7.6).

The coastal precipitation effect also applies to the summer monsoon of Asia, when the monsoon circulation brings mT air onshore. However, the onshore monsoon winds blow from southwest to northeast, so it is the western coasts of land masses that are exposed to this moist airflow. Western India and Myanmar (formerly Burma) are examples. Moist air also penetrates well inland in Bangladesh, providing the very heavy monsoon rains for which the region is well known.

In central and western Africa, as well as southern Brazil, the monsoon pattern shifts the intertropical convergence zone over 20° of latitude, or more (see Figure 5.13). Here, heavy rainfall occurs in the high-sun season, when the ITC is nearby. Drier conditions prevail in the low-sun season, when the ITC is far away.

Temperatures in the monsoon and trade-wind coastal climate ②, though warm throughout the year, also show an annual cycle. Warmest temperatures occur in the high-sun season, just before arrival of the ITC brings clouds and rain. Minimum temperatures occur at the time of low sun.

Figure 7.8 is a climograph for the city of Belize in the Central American country of Belize (Figure 7.6). This east coast city, located at latitude 17° N, is exposed to the tropical easterly trade winds. Rainfall is abundant from June through November, when the ITC is nearby. Easterly waves are common in this season, and on occasion a tropical cyclone will bring torrential rainfall. Following the December solstice, rainfall is greatly reduced, with minimum values in March and April. At this time, the ITC lies farthest away, and the climate is dominated by subtropical high pressure. Air temperatures show an annual range of 5° C (9° F), with a maximum in the high-sun months.

The Asiatic monsoon shows a similar pattern, but there is an extreme peak of rainfall during the high-sun period and a well-developed dry season with two or three months of only small rainfall amounts. The climograph for Cochin, India, provides an example (Figure 7.9). Located at lat. 10° N on the west coast of

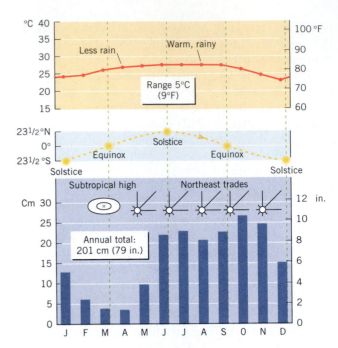

Figure 7.8 Trade-wind coastal climate ②. This climograph for Belize, a Central American east coast city at lat. 17° N, shows a marked season of low rainfall following the period of low sun. For the remainder of the year, precipitation is high, produced by warm, moist northeast trade winds.

lower peninsular India (see Figure 7.6), Cochin receives the warm, moist southwest winds of the summer monsoon. In this season, monthly rainfall is extreme in both June and July. A strongly pronounced season of low rainfall occurs at time of low sun—December through March. Air temperatures show only a very weak annual cycle, cooling a bit during the rains, so the annual range is small at this low latitude.

The Low-Latitude Rainforest Environment

Our first two climates—wet equatorial ① and monsoon and trade-wind coastal ②—are quite uniform in temperature and have a very high annual rainfall. These factors create a special environment—the *low-latitude rainforest environment*. In the rainforest, streams flow abundantly throughout most of the year, and river channels are lined with dense forest vegetation (Figure 7.10). The channels serve as the main arteries of trade, with water craft ranging from dugout canoes to large, shallow-draft river boats.

Low-latitude rainforests, unlike higher latitude forests, possess a great diversity of plant and animal species. The low-latitude rainforest can contain as many as 3000 different tree species in an area of only a few square kilometers, whereas midlatitude forests possess fewer than one-tenth that number. The number of types of animals found in the rainforest is also very large. A 16

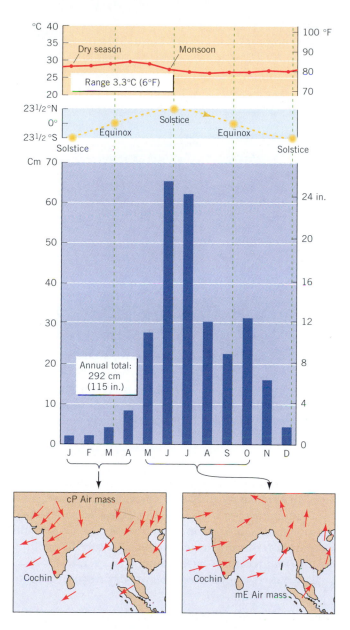

sawood, is an important export. Quinine, cocaine, and other drugs come from the bark and leaves of tropical plants. Cocoa is derived from the seed kernel of the cacao plant. Natural rubber is made from the sap of the rubber tree, which originated in South America.

The Wet-Dry Tropical Climate ③ (Köppen: *Aw, Cwa*)

In the monsoon and trade-wind coastal climate ②, we noted that the movements of the intertropical convergence zone into and away from the climate region produce a seasonal cycle of rainfall and temperature. As we move farther poleward, this cycle becomes stronger, and the monsoon and trade-wind coastal climate ② grades into the **wet-dry tropical climate** ③.

The wet-dry tropical climate ③ is distinguished by a very dry season at low sun that alternates with a very wet season at high sun. During the low-sun season, when the

Figure 7.10   Rainforest of the western Amazon lowland, near Manaus, Brazil. The river is a tributary of the Amazon.

Figure 7.9   Monsoon coastal climate ②. Cochin, India, on a windward coast at lat. 10° N, shows an extreme peak of rainfall during the rainy monsoon, contrasting with a short dry season at time of low sun.

km² (6.17 mi²) area in Panama near the Canal, for example, contains about 20,000 species of insects, whereas all of France has only a few hundred.

Taken together, the wet equatorial ① and monsoon and trade-wind coastal ② climates comprise the low-latitude rainforest environment.

Many products of the rainforest are of economic value. Rainforest lumber, such as mahogany, ebony, or balequatorial trough is far away, dry continental tropical (cT) air masses prevail. In the high-sun season, when the ITC is nearby, moist maritime tropical (mT) and maritime equatorial (mE) air masses dominate. Cooler temperatures accompany the dry season but give way to a very hot period before the rains begin.

Figure 7.11 shows the global distribution of the wet-dry tropical climate ③. It is found at latitudes of 5° to 20° N and S in Africa and the Americas, and at 10° to 30° N in Asia. In Africa and South America, the climate occupies broad bands poleward of the wet equatorial ① and monsoon and trade-wind coastal climates ②. Because these regions are farther away from the ITC, less rainfall is triggered by the ITC during the rainy season, and subtropical high pressure can dominate more strongly during the low-sun season. In central India and Indochina, the regions of wet-dry tropical climate are somewhat protected by mountain barriers from the warm, moist mE and mT airflows provided by trade and monsoon winds. These barriers create a rainshadow effect, so that even less rainfall occurs during the rainy season and the dry season is drier still.

Figure 7.12 is a climograph for Timbo, Guinea, at lat. 10° N in West Africa (Figure 7.11). Here the rainy season begins just after the March equinox and reaches a peak in August, about two months following the June solstice. At this time, the ITC has migrated to its most northerly position, and moist mE air masses flow into the region from the ocean lying to the south. Monthly rainfall then decreases as the low-sun season arrives and the ITC moves to the south. Three months—December through February—are practically rainless. During this season, subtropical high pressure dominates the climate, and stable, subsiding continental tropical (cT) air pervades the region. The temperature cycle is closely

Figure 7.12 Wet-dry tropical climate ③. Timbo, Guinea, at lat. 10° N, is in West Africa. A long wet season at time of high sun alternates with an almost rainless dry season at time of low sun.

linked to both the solar cycle and the precipitation pattern. In February and March, insolation increases, and air temperature rises sharply. A brief hot season occurs. As soon as the rains set in, the effect of cloud cover and evaporation of rain causes the temperature to decline. By July, temperatures have resumed an even level.

The native vegetation of the wet-dry tropical climate

Figure 7.11 World map of the wet-dry tropical climate ③. (Based on Goode Base Map.)

Figure 7.13   Savanna woodland of the Serengeti Plains, Tanzania, East Africa. Acacia trees with flattened crowns remain green, although the coarse grasses have turned to straw in the dry season.

③ must survive alternating seasons of very dry and very wet weather. Most plants enter a dormant phase during the dry period, then burst forth into leaf and bloom with the coming of the rains. For this reason, the native plant cover can be described as *rain-green vegetation*. A dominant type of rain-green vegetation is the *savanna*—a grassy plain with scattered trees (Figure 7.13).

In some regions of wet-dry tropical climate ③, the annual variation in rainfall is quite large. The African Sahel is such a region. *Eye on the Environment • Drought*

and Land Degradation in the African Sahel documents how human usage interacting with climate variability can produce a permanent deterioration of the environment.

The Dry Tropical Climate ④ (Köppen: *BWh, BSh*)

The **dry tropical climate** ④ is found in the center and east sides of subtropical high-pressure cells. Here, air strongly subsides, warming adiabatically and inhibiting

EYE ON THE ENVIRONMENT

Drought and Land Degradation in the African Sahel

The wet-dry tropical climate ③ is subject to years of devastating drought as well as to years of abnormally high rainfall that can result in severe floods. Climate records show that two or three successive years of abnormally low rainfall (a drought) typically alternate with several successive years of average or higher than average rainfall. This variability is a permanent feature of the wet-dry tropical climate ③, and the plants and animals inhabiting this region have adjusted to the natural variability in rainfall, with one exception: the human species.

The wet-dry tropical climate ③ of West Africa, including the adjacent semiarid southern belt of the dry tropical climate ④ to the north, provides a lesson on the human impact on a delicate ecological system. Countries of this perilous belt, called the *Sahel*, or *Sahelian zone*, are shown in the figure at right. From 1968 through 1974, all these countries were struck by a severe drought. Both nomadic cattle herders and grain farmers share this zone. During the drought, grain crops failed and foraging cattle could find no food to eat. In the worst stages of the Sahel drought, nomads were forced to sell their remaining cattle. Because the cattle were their sole means of subsistence, the nomads soon starved. Some 5 million cattle perished, and it has been estimated that 100,000 people died of starvation and disease in 1973 alone.

The Sahelian drought of 1968–1974 was associated with a special phenomenon, which at that time was called *desertification*—the permanent transformation of the land surface by human activities to resemble a desert, largely through the destruction of grasses, shrubs, and trees by grazing animals and fuelwood harvesting. That term has now been aban-

The Sahel, or Sahelian zone, shown in color, lies south of the great Sahara Desert of North Africa.

doned in favor of *land degradation*. Also visible are the effects of accelerated soil erosion, such as gullying of slopes and accumulations of sediment in stream channels. The removal of soil by wind is also intensified.

Periodic droughts throughout past decades are well documented in the Sahel, as they are in other world regions of wet-dry tropical climate ③. To the right is a graph showing the percentage of departures from the long-term mean of each year's rainfall

in the western Sahel from 1901 through 1995. Note the wide year-to-year variation. Since about 1950, the duration of periods of continuous departures both above and below the mean seems to have increased substantially. The period of sustained high-rainfall years in the 1950s contrasts sharply with a series of severe drought episodes starting in 1971. For an earlier record, scientists have examined fluctuations in the level of Lake Chad. In times of drought, the

At the height of the Sahelian drought, vast numbers of cattle had perished and even the goats were hard pressed to survive. Trampling of the dry ground prepared the region for devastating soil erosion in the rains that eventually ended the drought.

lake's shoreline retreats, while in times of abundant rainfall, it expands. The changes document periods of rainfall deficiency and excess—1820–1840, below normal; 1870–1895, above normal; 1895–1920, below normal.

The sustained drought periods of the past show that droughts and wet periods are a normal phenomenon in the Sahel. In prior eras, the landscape was able to recover from the droughts during periods of abundant rainfall. Today, the pressure of increased pop-

ulations of humans and cattle keeps the land degraded. As long as these populations remain high, the land degradation will be permanent.

The effect of an accelerated global warming in the coming decades on regions of wet-dry tropical climate ③ is difficult to predict. Some regions may experience greater swings between drought and surplus precipitation. At the same time, these regions may migrate poleward as a result of intensification of the Hadley cell circulation. Thus, the Sahel region may move northward, into the desert zone of the Sahara. However, such changes are highly speculative. At this time, there is no scientific consensus on how climatic change will affect the Sahel or other regions of wet-dry tropical climate ③.

Questions

1. What is meant by the term land degradation? Provide an example of a region in which land degradation has occurred.

2. Examine the Sahel rainfall graph carefully. Compare the pattern of rainfall fluctuations for the periods 1901–1950 and 1951–present. How do they differ?

Rainfall fluctuations for stations in the western Sahel, 1901–1995, expressed as a percent departure from the long-term mean. (Courtesy of Sharon E. Nicholson, Department of Meteorology, Florida State University, Tallahassee.)

condensation. Rainfall is very rare and occurs only when unusual weather conditions move moist air into the region. Since skies are clear most of the time, the sun heats the surface intensely, keeping air temperatures high. During the high-sun period, heat is extreme. During the low-sun period, temperatures are cooler. Given the dry air and lack of cloud cover, the daily temperature range is very large.

The driest areas of the dry tropical climate ④ are near the tropics of cancer and capricorn. As we travel from the tropics toward the equator, we find that somewhat more rain falls. Continuing in this direction, we encounter a short rainy season at the time of year when the ITC is near, and the climate grades into the wet-dry tropical ③ type.

Figure 7.14 shows the global distribution of the dry tropical climate ④. Nearly all of the dry tropical climate ④ areas lie in the latitude range 15° to 25° N and S. The largest region is the Sahara–Saudi Arabia–Iran–Thar desert belt of North Africa and southern Asia. This vast desert expanse includes some of the driest regions on earth. Another large region of dry tropical climate ④ is the desert of central Australia. The west coast of South America, including portions of Ecuador, Peru, and Chile, also exhibits the dry tropical climate ④. However, temperatures are moderated by the cool marine air layer that blankets the coast.

Figure 7.15 is a climograph for a dry tropical station in the heart of the North African desert. Wadi Halfa, Sudan (Figure 7.14), lies at lat. 22° N, almost on the tropic of cancer. The temperature record shows a strong annual cycle with a very hot period at the time of high

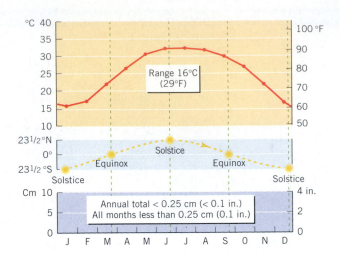

Figure 7.15 Dry tropical climate ④, dry desert. Wadi Halfa is a city on the Nile River in Sudan at lat. 22° N, close to the Egyptian border. Too little rain falls to be shown on the graph. Air temperatures are very high during the high-sun months.

sun, when three consecutive months average 32°C (90°F). Daytime maximum air temperatures are frequently between 43° and 48°C (about 110° to 120°F) in the warmer months. There is a comparatively cool season at the time of low sun, but the coolest month averages a mild 16°C (61°F) and freezing temperatures are rarely recorded. No rainfall bars are shown on the climograph because precipitation averages less than 0.25 cm (0.1 in.) in all months. Over a 39-year period, the maximum rainfall recorded in a 24-hour period at Wadi Halfa was only 0.75 cm (0.3 in.).

Figure 7.14 World map of the dry tropical ④, dry subtropical ⑤, and dry midlatitude ⑨ climates. (Based on Goode Base Map.)

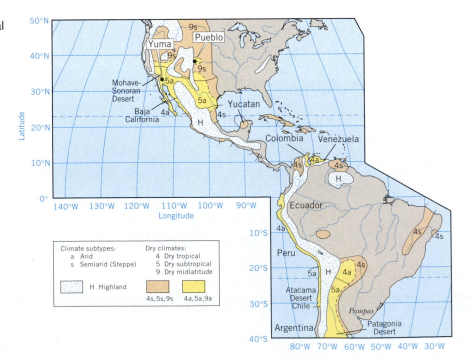

Figure 7.16 This strange-looking baobab tree is a common inhabitant of the thorntree semi-desert of Botswana, in the Kalahari region of southern Africa.

The world's dry climates consist largely of extremely arid deserts. In addition, there are broad zones at the margins of the desert that are best described as *semiarid*. These zones have a short wet season that supports the growth of grasses on which animals (both wild and domestic) graze. Geographers also call these semiarid regions **steppes**. Nomadic tribes and their herds of animals visit these areas during and after the brief moist period. In Figures 7.4 (world climate map) and 7.14 (world map of dry climates), the two subdivisions of dry climates are distinguished with the letters *a* (arid) and *s* (semiarid).

An example of the semiarid dry tropical climate ④ is that of Kayes, Mali, which we presented earlier as a sample climograph (Figure 7.3). Located in the Sahel region of Africa, this station has a distinct rainy season that occurs when the intertropical convergence moves north in the high-sun season. This precipitation pattern shows the semiarid subtype as a transition between the arid dry tropical climate ④ and the wet-dry tropical climate ③.

Much of the arid desert landscape consists of barren areas of drifting sand or sterile salt flats. However, in semiarid regions, thorny trees and shrubs are often abundant, since the climate includes a small amount of regular rainfall. Figure 7.16 presents a scene from the semiarid part of the Kalahari Desert in southwest Africa (Figure 7.14) that shows this type of vegetation in the dry season.

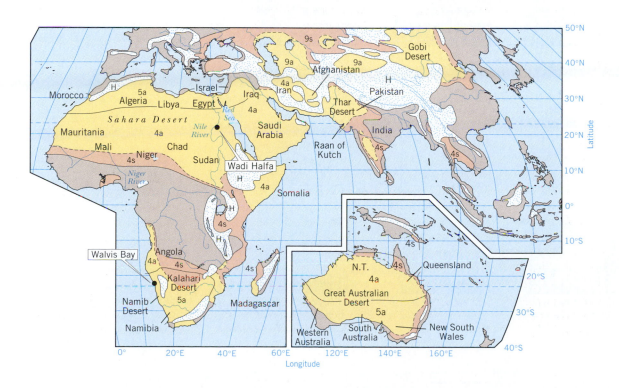

MIDLATITUDE CLIMATES

The *midlatitude climates* almost fully occupy the land areas of the midlatitude zone and a large proportion of the subtropical latitude zone. Along the western fringe of Europe, they extend into the subarctic latitude zone as well, reaching to the 60th parallel. In the southern hemisphere, the land area poleward of the 40th parallel is so small that the climates are dominated by a great southern ocean and do not develop the continental characteristics of their counterparts in the northern hemisphere.

The midlatitude climates of the northern hemisphere lie in a broad zone of intense interaction between two groups of very unlike air masses (Figure 7.2). From the subtropical zone, tongues of maritime tropical (mT) air masses enter the midlatitude zone. There, they meet and conflict with tongues of maritime polar (mP) and continental polar (cP) air masses along the discontinuous and shifting polar-front zone.

In terms of prevailing pressure and wind systems, the midlatitude climates include the poleward halves of the great subtropical high-pressure systems and much of the belt of prevailing westerly winds. As a result, weather systems, such as traveling cyclones and their fronts, characteristically move from west to east. This dominant global eastward airflow influences the distribution of climates from west to east across the North American and Eurasian continents.

> **The midlatitude climates lie in a zone of frontal interaction between tropical air masses and polar air masses.**

The six midlatitude climate types range from two that are very dry to three that are extremely moist (see Figure 7.17). The midlatitude climates span the range from those with strong wet and dry seasons—the Mediterranean climate ⑦—to those with precipitation

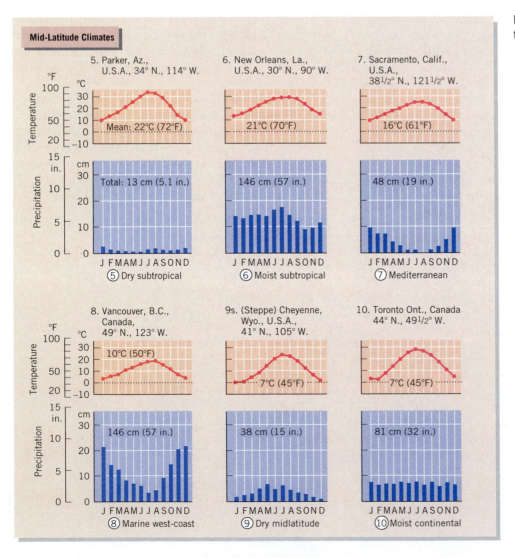

Figure 7.17 Climographs for the six midlatitude climates.

that is more or less uniformly distributed through the year. Temperature cycles are also quite varied. Low annual ranges are seen along the windward west coasts. In contrast, annual ranges are large in the continental interiors. Table 7.2 summarizes the important features of these climates.

The Dry Subtropical Climate ⑤ (Köppen: *BWh, BWk, BSh, BSk*)

The **dry subtropical climate** ⑤ is simply a poleward extension of the dry tropical climate ④, caused by somewhat similar air mass patterns. A point of differ-

Table 7.2   Midlatitude Climates

Climate	Temperature	Precipitation	Explanation
Dry subtropical ⑤	Distinct cool or cold season at low-sun period.	Precipitation is low in nearly all months.	This climate lies poleward of the subtropical high-pressure cells and is dominated by dry cT air most of the year. Rainfall occurs when moist mT air reaches the region, either in summer monsoon flows or in winter frontal movements.
Moist subtropical ⑥	Temperatures show strong annual cycle, but with no winter month below freezing.	Abundant rainfall, cyclonic in winter and convectional in summer. Humidity generally high.	The flow of mT air from the west sides of subtropical high-pressure cells provides moist air most of the year. cP air may reach this region during the winter.
Mediterranean ⑦	Temperature range is moderate, with warm to hot summers and mild winters.	Unusual pattern of wet winter and dry summer. Overall, drier when nearer to subtropical high pressure.	The poleward migration of subtropical high-pressure cells moves clear, stable air cT into this region in the summer. In winter, cyclonic storms and polar frontal precipitation reach the area.
Marine west-coast ⑧	Temperature cycle is moderated by marine influence.	Abundant precipitation but with a winter maximum.	Moist mP air, moving inland from the ocean to the west, dominates this climate most of the year. In the summer, subtropical high pressure reaches these regions, reducing precipitation.
Dry midlatitude ⑨	Strong temperature cycle with large annual range. Summers warm to hot, winters cold to very cold.	Precipitation is low in all months but usually shows a summer maximum.	This climate is dry because of its interior location, far from mP source regions. In winter, cP dominates. In summer, a local dry continental air mass develops.
Moist continental ⑩	Summers warm, winters cold with three months below freezing. Very large annual temperature range.	Ample precipitation, with a summer maximum.	This climate lies in the polar-front zone. In winter, cP air dominates, while mT invades frequently in summer. Precipitation is abundant, cyclonic in winter, and convectional in summer.

ence is in the annual temperature range, which is greater for the dry subtropical climate ⑤. The lower latitude portions have a distinct cool season, and the higher latitude portions, a cold season. The cold season, which occurs at a time of low sun, is due in part to invasions of cold continental polar (cP) air masses from higher latitudes. Precipitation occurring in the low-sun season is produced by midlatitude cyclones that occasionally move into the subtropical zone. As in the dry tropical climate ④, both arid and semiarid subtypes are recognized.

Figure 7.14 shows the global distribution of the dry subtropical climate ⑤. A broad band of this climate type is found in North Africa, connecting with the Near East. Southern Africa and southern Australia also contain regions of dry subtropical climate ⑤ poleward of the dry tropical climate ④. In South America, a band of dry subtropical climate ⑤ occupies Patagonia, a region east of the Andes in Argentina. In North America, the Mojave and Sonoran deserts of the American Southwest and northwest Mexico are of the dry subtropical type.

Figure 7.18 is a climograph for Yuma, Arizona, a city within the arid subtype of the dry subtropical climate ⑤, close to the Mexican border at lat. 33° N. The pattern of monthly temperatures shows a strong seasonal cycle with a dry, hot summer. A cold season brings monthly means as low as 13°C (55°F). Freezing temperatures—0°C (32°F) and below—can be expected at

night in December and January. The annual range is 20°C (36°F). Precipitation, which totals about 8 cm (3 in.), is small in all months but has peaks in late winter and late summer. The August maximum is caused by the invasion of maritime tropical (mT) air masses, which bring thunderstorms to the region. Higher rainfalls from December through March are produced by midlatitude wave cyclones following a southerly path. Two months, May and June, are nearly rainless.

The Subtropical Desert Environment

The environment of the dry subtropical climate ⑤ is similar to that of the dry tropical climate ④ in that both are very dry. The boundary between these two climate types is gradational. But if we were to travel northward in the subtropical climate zone of North America, arriving at about 34° N in the interior Mojave Desert of southeastern California, we would encounter environmental features significantly different from those of the low-latitude deserts of tropical Africa, Arabia, and northern Australia. Although the great summer heat of the low-elevation regions of the Mojave Desert is comparable to that experienced in the Sahara Desert, the low sun brings a winter season that is not found in the tropical deserts. Here, cyclonic precipitation can occur in most months, including the cool low-sun months.

In the Mojave Desert and adjacent Sonoran Desert, plants are often large and numerous, in some places giving the appearance of an open woodland. One example is the occurrence of forestlike stands of the tall, cylindrical saguaro cactus. Another is the woodland of Joshua trees found in higher parts of the Mojave Desert (Figure 7.19). Other distinctive plants include the prickly pear cactus, the ocotillo plant, and the creosote bush.

The deserts of the dry subtropical climate ⑤ are marked by a distinctive cover of desert plants and an assemblage of animals adapted to drought.

Both plants and animals of deserts are adapted to the dry environment. As in the dry tropical environment, many annual plants remain dormant as seeds during long dry periods and then spring to life, flower, and bloom very quickly when rain falls. Certain invertebrate animals adopt the same life pattern. For example, the tiny brine shrimp of the North American desert may wait many years in dormancy until normally dry lakebeds fill with water, an event that occurs perhaps three or four times per century. The shrimp then emerge and complete their life cycles before the lake evaporates.

Figure 7.18 Dry subtropical climate ⑤. Yuma, Arizona, lat. 33° N, has a strong seasonal temperature cycle. Compare with Wadi Halfa (Figure 7.15).

Figure 7.19 A Mojave Desert landscape. The Joshua tree, shown in the foreground, is abundant at higher elevations in the Mojave Desert.

The Moist Subtropical Climate ⑥ (Köppen: *Cfa*)

Recall that circulation around the subtropical high-pressure cells provides a flow of warm, moist air onto the eastern side of continents. This flow of maritime tropical (MT) air dominates the **moist subtropical climate** ⑥. Summer in this climate sees abundant rainfall, much of it convectional. Occasional tropical cyclones further enhance summer precipitation. In Southeast Asia, this climate is characterized by a strong monsoon effect, with summer rainfall much increased above winter rainfall. Summer temperatures are warm, with persistent high humidity.

Winter precipitation in the moist subtropical climate ⑥ is also plentiful, produced in midlatitude cyclones. Invasions of continental polar (cP) air masses are fre-

quent in winter, bringing spells of subfreezing weather. No winter month has a mean temperature below 0°C (32°F).

Figure 7.20 presents a global map of the moist subtropical climate ⑥. It is found on the eastern sides of continents in the latitude range 20° to 35° N and S. In South America, it includes parts of Uruguay, Brazil, and Argentina. In Australia, it consists of a narrow band between the eastern coastline and the eastern interior ranges. Southern China, Taiwan, and southernmost Japan are regions of the moist subtropical climate ⑥ in Asia. In the United States, the moist subtropical climate ⑥ covers most of the Southeast, from the Carolinas to east Texas.

Figure 7.21 is a climograph for Charleston, South Carolina (Figure 7.21), located on the eastern seaboard

Figure 7.20 World map of the moist sub-tropical climate ⑥. (Based on Goode Base Map).

at lat. 33° N. In this region, a marked summer maximum of precipitation is typical. Total annual rainfall is abundant—120 cm (47 in.)—and ample precipitation falls in every month. The annual temperature cycle is

Figure 7.21 Moist subtropical climate ⑥. Charleston, South Carolina, lat. 33° N, has a mild winter and a warm summer. There is ample precipitation in all months, but a definite summer maximum.

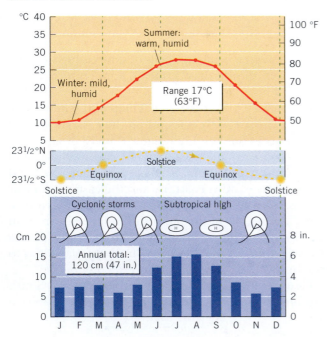

strongly developed, with a large annual range of 17°C (31°F). Winters are mild, with the January mean temperature well above the freezing mark.

With its abundant rainfall in all seasons, the dominant vegetation type of the moist subtropical climate ⑥ is forest. However, large areas of forest within this climate region have been replaced with agricultural croplands. The warm temperatures favor crop growth, but the heavy rainfall can wash nutrients from soils, requiring fertilization for crop growth. In Asia, staples such as rice and tea are commonly planted.

The Mediterranean Climate ⑦ (Köppen: *Csa*, *Csb*)

The **Mediterranean climate** ⑦ is unique among the climate types because its annual precipitation cycle has a wet winter and a very dry summer. The reason for this precipitation cycle lies in the poleward movement of the subtropical high-pressure cells during the summer season. The Mediterranean climate ⑦ is located along the west coasts of continents, just poleward of the dry, eastern side of the subtropical high-pressure cells (see Figures 5.11, 5.12). When the subtropical high-pressure cells move poleward in summer, they enter the region of this climate. Dry continental tropical (cT) air then dominates, producing the dry summer season. In winter, the moist mP air mass invades with cyclonic storms and generates ample rainfall.

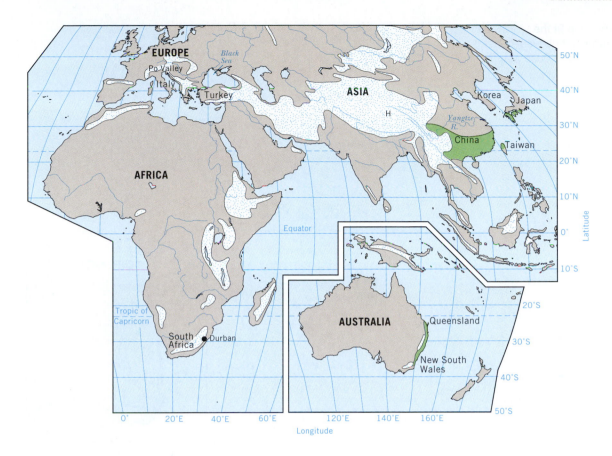

In terms of total annual rainfall, the Mediterranean climate ⑦ spans a wide range from arid to humid, depending on location. Generally, the closer an area is to the tropics, the stronger the influence of subtropical high pressure will be, and thus the drier the climate. The temperature range is moderate, with warm to hot summers and mild winters. Coastal zones between lat. 30° and 35° N and S, such as southern California, show a smaller annual range, with very mild winters.

Figure 7.22 is a climograph for Monterey, California, a Pacific coastal city at lat. 36° N. The annual temperature cycle is very weak. The small annual range and cool summer reflect the strong control of the cold California current and its cool, marine air layer. Fogs are frequent. This temperature regime is found only in a narrow coastal zone. Rainfall drops to nearly zero for four consecutive summer months but rises to substantial amounts in the rainy winter season. In California's Central Valley, the dry summers persist, but the annual temperature range is greater, so that daily high temperatures match those of the adjacent dry climates.

A global map of the Mediterranean ⑦ climate is shown in Figure 7.23. It is found in the latitude range 30° to 45° N and S. In the southern hemisphere, it occurs along the coast of Chile, in the Cape Town region of South Africa, and along the southern and western coasts of Australia. In North America, it is found in central and southern California. In Europe, this cli-

Figure 7.22 Mediterranean climate ⑦. Monterey, California, lat. 36° N, has a very weak annual temperature cycle because of its closeness to the Pacific Ocean. The summer is very dry.

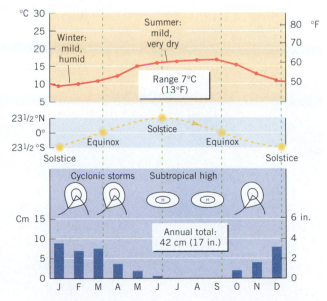

Figure 7.23 World map of the Mediterranean ⑦ and marine west-coast ⑧ climates. (Based on Goode Base Map.)

mate type surrounds the Mediterranean Sea, which gives the climate its distinctive name.

The vegetation of the Mediterranean environment is adapted to the pattern of wet winters and dry summers. Many plant species rely on thick leaves and bark to keep from losing scarce water in the dry season. Coastal California's chaparral, a type of tough, sometimes-spiny shrub cover, is an example (Figure 7.24). The hot, dry summers can make the brush extremely flammable, potentially leading to destructive fires.

Figure 7.24 California coastal chaparral. Wiry shrubs cover the steep slopes of a coast range near Acton, California.

The Marine West-Coast Climate ⑧
(Köppen: *Cfb, Cfc*)

The **marine west-coast climate** ⑧ occupies midlatitude west coasts. These locations receive the prevailing westerlies from over a large ocean and experience frequent cyclonic storms involving cool, moist mP air masses. Where the coast is mountainous, the orographic effect causes a very large annual precipitation. In this moist climate, precipitation is plentiful in all months, but there is often a distinct winter maximum. In summer, subtropical high pressure extends poleward into the region, reducing rainfall. The annual temperature range is comparatively small for midlatitudes. The marine influence keeps winter temperatures mild, as compared with inland locations at equivalent latitudes.

The global map of the marine west-coast climate ⑧ (Figure 7.23) shows the areas in which this climate occurs. In North America, the climate occupies the western coast from Oregon to northern British Columbia. In Western Europe, the British Isles, Portugal, and much of France fall into the marine west-coast climate.

New Zealand and the southern tip of Australia, as well as the island of Tasmania, are marine west-coast climate ⑧ regions found in the southern hemisphere, as is the Chilean coast south of 35° S. The general latitude range of this climate is 35° to 60° N and S.

Figure 7.25 is a climograph for Vancouver, British Columbia, just north of the U.S.-Canadian border. The annual precipitation is very great, and most of it falls during the winter months. Notice the greatly reduced rainfall in the summer months. The temperature cycle shows a remarkably small range for this latitude. Even the winter months have averages above the freezing mark.

With abundant rainfall and mild temperatures, the marine west-coast climate ⑧ in North America is associated with dense, evergreen forests of fir, cedar, hemlock, and spruce (Figure 7.26). These forests dominate the mountainous terrain of the region. In Western Europe, extensive forests disappeared many centuries ago, yielding to cultivation for crops and dairy farming. The remaining small plots of forest remain under intense management, providing lumber for construction and wood for fuel.

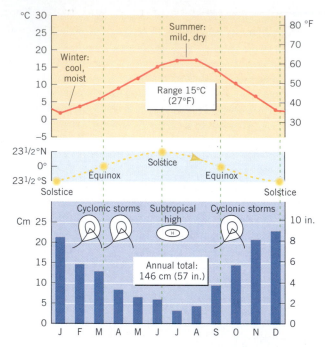

Figure 7.25 Marine west-coast climate ⑧. Vancouver, British Columbia, lat. 49° N, has a large annual total precipitation, but with greatly reduced amounts in the summer. The annual temperature range is small, and winters are very mild for this latitude.

The Dry Midlatitude Climate ⑨ (Köppen: *BWk, BSk*)

The **dry midlatitude climate** ⑨ is limited almost exclusively to interior regions of North America and Eurasia, where it lies within the rainshadow of mountain ranges on the west or south. Maritime air masses are effectively blocked out much of the time, so that the continental polar (cP) air mass dominates the climate in winter. In summer, a dry continental air mass of local origin is dominant. Summer rainfall is mostly convectional and is caused by occasional invasions of maritime air masses. The annual temperature cycle is strongly developed, with a large annual range. Summers are warm to hot, but winters are cold to very cold.

The largest expanse of the dry midlatitude climate ⑨ is in Eurasia, stretching from the southern republics of the former Soviet Union to the Gobi Desert and northern China (Figure 7.14). In the central portions of this region lie true deserts of the arid climate subtype, with very low precipitation. Extensive areas of highlands occur here as well. In North America, the dry western interior regions, including the Great Basin, Columbia Plateau, and the Great Plains, are of the semiarid subtype. A small area of dry midlatitude climate ⑨ is found in southern Patagonia, near the tip of South America. The latitude range of this climate is 35° to 55° N.

Figure 7.26 Needleleaf forest of Sitka spruce and hemlock in the Hoh rainforest, Olympic National Park, Washington. Ferns and mosses provide a lush ground cover in this very wet environment.

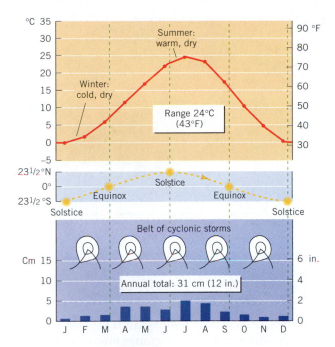

Figure 7.27 Dry midlatitude ⑨ climate. Pueblo, Colorado, lat. 38° N, shows a marked summer maximum of rainfall in the summer months. Note also the cold, dry winter season. Figure 7.14 shows the location of this station.

Figure 7.27 is a climograph for Pueblo, Colorado, a semiarid station located at lat. 38° N, just east of the Rocky Mountains. Total annual precipitation is 31 cm (12 in.). Most of this precipitation is in the form of convectional summer rainfall, which occurs when moist maritime tropical (mT) air masses invade from the south and produce thunderstorms. In winter, snowfall is light and yields only small monthly precipitation averages. The temperature cycle has a large annual range, with warm summers and cold winters. January, the coldest winter month, has a mean temperature just below freezing.

The low precipitation and cold winters of this semiarid climate produce a steppe landscape dominated by hardy perennial short grasses. In North America, this cover is termed *short-grass prairie*, rather than steppe. Wheat is a major crop of the semiarid, dry midlatitude steppelands, but wheat harvests are at the mercy of rainfall variations from year to year. With good spring rains, there is a good crop, but if spring rains fail, so does the wheat crop.

The Moist Continental Climate ⑩ (Köppen: *Dfa, Dfb, Dwa, Dwb*)

The **moist continental climate** ⑩ is located in central and eastern parts of North America and Eurasia in the midlatitudes. This climate lies in the polar-front

zone—the battleground of polar and tropical air masses. Seasonal temperature contrasts are strong, and day-to-day weather is highly variable. Ample precipitation throughout the year is increased in summer by invading maritime tropical (mT) air masses. Cold winters are dominated by continental polar (cP) and continental arctic (cA) air masses from subarctic source regions.

In eastern Asia—China, Korea, and Japan—the seasonal precipitation pattern shows more summer rainfall and a drier winter than in North America. This is an effect of the monsoon circulation, which moves moist maritime tropical (mT) air across the eastern side of the continent in summer and dry continental polar southward through the region in winter. In Europe, the moist continental climate ⑩ lies in a higher latitude belt (45° to 60° N) and receives precipitation from mP air masses coming from the North Atlantic.

Madison, Wisconsin, lat. 43° N, in the American Midwest (Figure 7.28), provides an example of the moist continental climate ⑩. The annual temperature range is very large. Summers are warm, but winters are cold, with three consecutive monthly means well below freezing. Precipitation is ample in all months, and the annual total is large. There is a summer maximum of precipitation when the maritime tropical (mT) air mass invades, and thunderstorms are formed along moving cold fronts and squall lines. Much of the winter precipitation is in the form of snow, which remains on the ground for long periods.

Figure 7.29 shows the global locations of the moist continental climate ⑩. It is restricted to the northern hemisphere, occurring in latitudes 30° to 55° N in

Figure 7.28 Moist continental climate ⑩. Madison, Wisconsin, lat. 43° N, has cold winters and warm summers, making the annual temperature range very large.

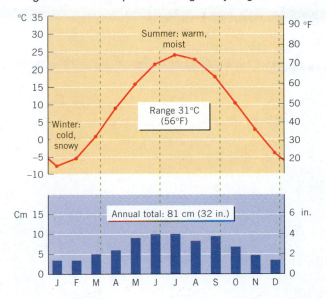

Figure 7.29 World map of the moist continental climate ⑩. (Based on Goode Base Map.)

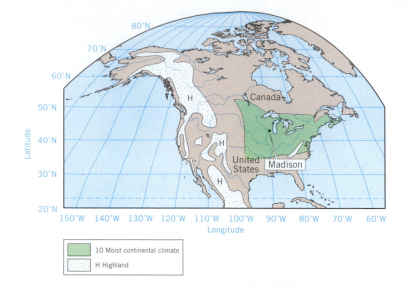

North America and Asia, and in latitudes 45° to 60° N in Europe. In Asia, it is found in northern China, Korea, and Japan. Most of Central and Eastern Europe has a moist continental climate ⑩, as does most of the eastern half of the United States from Tennessee to the north, as well as the southernmost strip of eastern Canada.

Throughout most of the moist continental climate ⑩, forests are the dominant natural vegetation cover. However, where the climate grades into drier climates, such as the dry midlatitude climate ⑨, tall, dense grasses may be the natural cover. The *tall-grass prairie* of Illinois, Iowa, and Nebraska is an example. These lands are now the heart of the American corn belt. In Europe, lands of the moist continental climate ⑩ have been in

Figure 7.30 The Mosel River in its winding, entrenched meandering gorge through the Rhineland–Pfalz province of western Germany. On the steep, undercut valley wall at the right are vineyards and forested land.

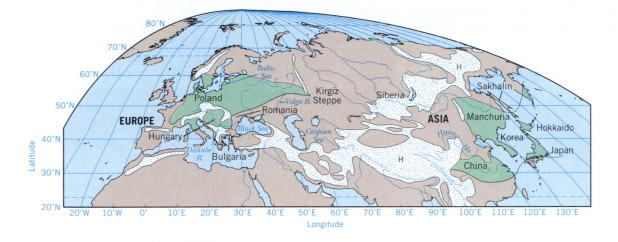

cultivation for many centuries, providing a multitude of crops. Forests are planted and managed for timber yield. The result is a landscape that shows human influence everywhere (Figure 7.30).

HIGH-LATITUDE CLIMATES

By and large, the *high-latitude climates* are climates of the northern hemisphere (Figure 7.31). They occupy the northern subarctic and arctic latitude zones, but also extend southward into the midlatitude zone as far south as about the 47th parallel in eastern North America and eastern Asia. One of these, the ice sheet climate ⑬, is present in both hemispheres in the polar zones. Table 7.3 provides an overview of the three high-latitude climates.

The high-latitude climates coincide closely with the belt of prevailing westerly winds that circles each pole. In the northern hemisphere, this circulation sweeps maritime polar (mP) air masses, formed over the northern oceans, into conflict with continental polar (cP) and continental arctic (cA) air masses on the continents. Rossby waves form in the westerly flow, bringing lobes of warmer, moister air poleward into the region in exchange for colder, drier air moved equatorward. The result of these processes is frequent wave cyclones, produced along a discontinuous and constantly fluctuating arctic-front zone. In summer, tongues of maritime tropical air masses (mT) reach the subarctic latitudes to interact with polar air masses and yield important amounts of precipitation.

The Boreal Forest Climate ⑪
(Köppen: *Dfc, Dfd, Dwc, Dwd*)

The **boreal forest climate** ⑪ is a continental climate with long, bitterly cold winters and short, cool summers. It occupies the source region for cP air masses, which are cold, dry, and stable in the winter. Invasions of the very cold cA air mass are common. The annual range of temperature is greater than that of any other climate and is

Figure 7.31 Climographs of the three high-latitude climates.

Table 7.3 High-Latitude Climates

Climate	Temperature	Precipitation	Explanation
Boreal forest ⑪	Short, cool summers and long, bitterly cold winters. Greatest annual range of all climates.	Annual precipitation small, falling mostly in summer months.	This climate falls in the source region for cold, dry stable cP air masses. Traveling cyclones, more frequent in summer, bring precipitation from mP air.
Tundra ⑫	No true summer but a short mild season. Otherwise, cold temperatures.	Annual precipitation small, falling mostly during mild season.	The coastal arctic fringes occupied by this climate are dominated by cP, mP, and cA air masses. The maritime influence keeps winter temperatures from falling to the extreme lows of interiors.
Ice sheet ⑬	All months below freezing, with lowest global temperatures on earth experienced during Antarctic winter.	Very low precipitation, but snow accumulates since temperatures are always below freezing.	Ice sheets are the source regions of cA and cAA air masses. Temperatures are intensely cold.

greatest in Siberia. Precipitation increases substantially in summer, when maritime air masses penetrate the continent with traveling cyclones, but the total annual precipitation is small. Although much of the boreal forest climate ⑪ is moist, large areas in western Canada and Siberia have low annual precipitation and are therefore cold and dry.

The global extent of the boreal forest climate ⑪ is presented in Figure 7.32. In North America, it stretches from central and western Alaska, across the Yukon and Northwest Territories to Labrador on the Atlantic coast. In Europe and Asia, it reaches from the Scandinavian Peninsula eastward across all of Siberia to the Pacific. In latitude, this climate type ranges from 50° to 70° N.

Figure 7.33 is a climograph for Fort Vermilion, Alberta, at lat. 58° N. The very great annual temperature range shown here is typical for North America. Monthly mean air temperatures are below freezing for

seven consecutive months. The summers are short and cool. Precipitation shows a marked annual cycle with a summer maximum, but the total annual precipitation is small. Although precipitation in winter is small, a snow cover remains over solidly frozen ground through the entire winter. On the same climograph, temperature data are shown for Yakutsk, a Siberian city at lat. 62° N. in the Russian Republic. The enormous annual range is evident, as well as the extremely low means in winter months. January reaches a mean of about −42°C (−45°F). Except for the ice sheet interiors of Antarctica and Greenland, this region is the coldest on earth. Precipitation is not shown for Yakutsk, but the annual total is very small.

The Boreal Forest Environment

Land surface features of much of the region of boreal forest climate ⑪ were shaped beneath the great ice

Figure 7.32 World map of the boreal forest climate ⑪. (Based on Goode Base Map.)

Figure 7.33 Boreal forest climate ⑪. Extreme winter cold and a very great annual range in temperature characterizes the climate of Fort Vermilion, Alberta. The temperature range of Yakutsk, Siberia, is even greater.

mate region is boreal forest, consisting of needleleaf trees.

Although the growing season in the boreal forest climate ⑪ is short, crop farming is still possible. It is largely limited to lands surrounding the Baltic Sea, bordering Finland and Sweden. Crops grown in this area include barley, oats, rye, and wheat. Along with dairying, these crops primarily supply food for subsistence. The needleleaf trees of the boreal forest provide pulpwood and construction lumber.

The Tundra Climate ⑫ (Köppen: *ET*)

The **tundra climate** ⑫ occupies arctic coastal fringes and is dominated by polar (cP, mP) and arctic (cA) air masses. Winters are long and severe. A moderating influence of the nearby ocean water prevents winter temperatures from falling to the extreme lows found in the continental interior. There is a very short mild season, which many climatologists do not recognize as a true summer.

The world map of the tundra climate ⑫ (Figure 7.34) shows the tundra ringing the Arctic Ocean and extending across the island region of northern Canada. It includes the Alaskan north slope, the Hudson Bay region, and the Greenland coast in North America. In Eurasia, this climate type occupies the northernmost fringe of the Scandinavian Peninsula and Siberian coast. The Antarctic Peninsula (not shown in Figure 7.34) belongs to the tundra climate ⑫. The latitude range for this climate is 60° to 75° N and S., except for the northern coast of Greenland, where tundra occurs at latitudes greater than 80° N.

Figure 7.35 is a climograph for Upernavik, located on the west coast of Greenland at lat. 73° N. A short mild period, with above-freezing temperatures, is equivalent to a summer season in lower latitudes. The long winter is very cold, but the annual temperature range is not as large as that for the boreal forest climate ⑪ to the south. Total annual precipitation is small. Increased precipita-

sheets of the Ice Age. Severe erosion by the moving ice exposed hard bedrock over vast areas and created numerous shallow rock basins. Bouldery rock debris mantles the rock surface in many places. Many of the shallower rock basins have been filled by organic bog materials. The dominant upland vegetation of this cli-

Figure 7.34 World map of the tundra climate ⑫.

tion beginning in July is explained by the melting of the sea-ice cover and a warming of ocean water temperatures. This increases the moisture content of the local air mass, allowing more precipitation.

Figure 7.35 Tundra climate ⑫. Upernavik, Greenland, lat. 73° N, shows a smaller annual range than Fort Vermilion (Figure 7.33).

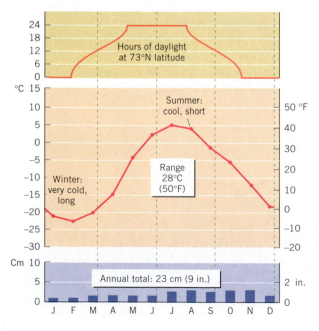

The Arctic Tundra Environment

The term **tundra** describes both an environmental region and a major class of vegetation (Figure 7.36). (An equivalent climatic environment—called *alpine tundra*—prevails in many global locations in high mountains above the timberline.) Trees exist in the tundra only as small, shrublike plants. They are stunted because of the seasonal damage to roots by freeze and thaw of the soil layer and to branches exposed to the abrading action of wind-driven snow. Vegetation of the treeless tundra consists of grasses, sedges, and lichens, along with shrubs of willow. Peat bogs are numerous. Because soil water is solidly and permanently frozen not far below the surface, the summer thaw brings a condition of water saturation to the soil.

At high latitudes, perennially frozen ground, or permafrost, occurs. Human disturbance can cause permafrost to thaw.

Because of the cold temperatures experienced in the tundra and northern boreal forest climate zones, the ground is typically frozen to great depth. This perennially frozen ground, or **permafrost**, prevails over the tundra region and a wide bordering area of boreal forest climate ⑪. Normally, a top layer of the ground will thaw each year during the mild season. This active

Figure 7.36 Caribou migration across the arctic tundra of northern Alaska.

layer of seasonal thaw is from 0.6 to 4 m (2 to 13 ft) thick, depending on latitude and the nature of the ground.

The permafrost environment is in a stable, delicate equilibrium that can be easily disturbed. When human activities remove the cover of moss or peat and living plants that insulates the permafrost, the summer thaw is extended to a greater depth. Ice wedges or ice masses beneath the surface melt, causing the soil surface to sink. Meltwater mixes with the fine soil to form mud, which is then eroded and transported by water in streams. This activity is called thermal erosion and is reversed only with great difficulty.

The Ice Sheet Climate ⑬
(Köppen: *EF*)

The **ice sheet climate** ⑬ coincides with the source regions of arctic (A) and antarctic (AA) air masses, situated on the vast, high ice sheets of Greenland and Antarctica and over polar sea ice of the Arctic Ocean. Mean annual temperature is much lower than that of any other climate, with no monthly mean above freezing. Strong temperature inversions develop over the ice sheets. In Antarctica and Greenland, the high surface altitude of the ice sheets intensifies the cold. Strong cyclones with blizzard winds are frequent. Precipitation, almost all occurring as snow, is very low, but accumulates because of the continuous cold. The latitude range for this climate is 65° to 90° N and S.

Figure 7.37 shows temperature graphs for several representative ice sheet stations. The graph for Eis-

mitte, a research station on the Greenland ice cap, shows the northern hemisphere temperature cycle, whereas the other four examples are all from Antarctica. Temperatures in the interior of Antarctica have

Figure 7.37 Ice sheet climate ⑬. Temperature graphs for five ice sheet stations.

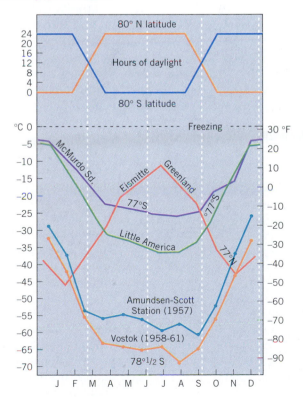

proved to be far lower than those at any other place on earth. A Russian meteorological station at Vostok, located about 1300 km (about 800 mi) from the south pole at an altitude of about 3500 m (11,500 ft), may be the world's coldest spot. Here a low of −88.3°C (−127°F) was observed in 1958. At the pole itself (Amundsen–Scott Station), July, August, and September of 1957 had averages of about −60°C (−76°F). Temperatures are considerably higher, month for month, at Little America in Antarctica because it is located close to the Ross Sea and is at a low altitude.

A LOOK AHEAD

The climates of the earth are remarkably diverse, ranging from the hot, humid wet equatorial climate ① at the equator to the bitterly cold and dry ice sheet climate ⑬ at the poles. Between these extremes are the other climates, each with distinctive features—such as dry summers or dry winters, or uniform or widely varying temperatures. Yet, these distinctive features can be explained using the principles controlling air temperatures (Chapter 3), atmospheric moisture and precipitation (Chapter 4), global circulation (Chapter 5), and weather systems (Chapter 6). In this way, the study of climate in this chapter acts to focus our existing knowledge on the problem of how the world's regions are differentiated by the annual patterns of weather they experience.

Climate exerts strong controls on vegetation and soils, especially at the global level. As we will see in Chapter 8, some types of plants are better adapted to certain environments than others. Only a few hardy species are found in extreme climates, while in climates with abundant heat, moisture, and nutrients, many plant species can flourish. To understand global patterns of vegetation, we will need to investigate species–environment relationships briefly—which is how we begin our next chapter.

GLOBAL CLIMATES IN REVIEW

Climate classification makes use of the annual cycles of temperature and precipitation as they are observed at many locations around the world. The world precipitation map shows rainfall concentrated in a broad belt around the equator. Mountain regions also show high precipitation, especially where they are swept by moist winds off nearby oceans. The driest regions occur where air continually subsides in subtropical high-pressure cells. Continental interiors are also dry. Precipitation generally decreases toward the poles.

Low-latitude climates are controlled by the characteristics and annual movements of the ITC and the subtropical highs. In the wet equatorial climate ①, the ITC is always nearby, so rainfall is abundant throughout the year. The monsoon and trade-wind coastal climate ② shows a dry period when the ITC has migrated toward the opposite tropic. In the wet-dry tropical climate ③, this dry period is more pronounced. The dry tropical climate ④ is dominated by subtropical high pressure and is often nearly rainless. The annual cycle of temperature is nearly uniform in the wet equatorial climate ①, but becomes more pronounced as dry conditions come to dominate the climate type.

Midlatitude climates are more varied. The dry subtropical ⑤ and dry midlatitude ⑨ climates extend the dry tropical ④ climate poleward into continental interiors, where they are far from oceanic sources of moist air. Both the Mediterranean ⑦ and marine west-coast ⑧ climates show a summer precipitation minimum, produced by the poleward movement of subtropical high pressure at that time of year. The moist subtropical ⑥ and moist continental ⑩ climates fall on the east sides of continents. They can receive moist tropical air or dry polar air at any time of year, but the moist subtropical climate ⑥ receives warm moist air more frequently and is located further equatorward. Midlatitude climate temperature cycles show strong seasonal effects. Temperature range increases in a poleward direction and as the climate becomes drier. The west-coast climates have a smaller range because of their maritime location.

High-latitude climates have low precipitation since air temperatures are low. The precipitation generally occurs during the short warm period. The boreal forest climate ⑪ occupies the cP source region in continental interiors and experiences a great annual range of temperatures. The tundra climate ⑫ is strongly influenced by nearby ocean and so experiences a lesser range. The ice sheet climate ⑬ is the coldest on earth.

KEY TERMS

climate
isohyet
climograph
wet equatorial climate ①
monsoon and trade-wind
   coastal climate ②
wet-dry tropical
   climate ③
dry tropical climate ④

steppes
dry subtropical climate ⑤
moist subtropical
   climate ⑥
Mediterranean climate ⑦
marine west-coast
   climate ⑧
dry midlatitude
   climate ⑨

moist continental
   climate ⑩
boreal forest climate ⑪
tundra climate ⑫
tundra
permafrost
ice sheet climate ⑬

REVIEW QUESTIONS

1. Why are annual cycles of temperature and precipitation useful measures of climate?
2. Identify two moist regions and two dry regions of the world, and explain why each region is either moist or dry.
3. Why is the annual temperature cycle of the wet equatorial climate ① so uniform?
4. The wet-dry tropical climate ③ has two distinct seasons. What factors produce the dry season? The wet season?
5. Why are the dry tropical ④, dry subtropical ⑤, and dry midlatitude ⑨ climates dry? How do they differ in temperature and precipitation cycles?
6. Both the moist subtropical ⑥ and moist continental ⑩ climates are found on eastern sides of continents in the midlatitudes. What are the major factors that determine their temperature and precipitation cycles? How do these two climates differ?
7. Both the Mediterranean ⑦ and marine west-coast ⑧ climates are found on the west coasts of continents. Why do they experience more precipitation in winter than in summer? How do the two climates differ?
8. The subtropical high-pressure cells influence several climate types in the low latitudes and midlatitudes. Identify the climates and describe the effects of the subtropical high-pressure cells on them.
9. Both the boreal forest ⑪ and tundra ⑫ climate are climates of the northern regions, but the tundra is found fringing the Arctic Ocean and the boreal forest is located further inland. Compare these two climates from the viewpoint of coastal-continental effects.
10. What is the coldest climate on earth? How is the annual temperature cycle of this climate related to the cycle of insolation?
11. Using Figure 7.4 or S7.2, identify the climate region you live in. What are the major temperature and precipitation cycles of this climate?

VISUALIZING EXERCISES

1. Sketch the temperature and rainfall cycles for a typical station in the monsoon and trade-wind coastal climate ②. What factors contribute to the seasonality of the two cycles?
2. Sketch the temperature and rainfall cycles for the climate type in which you live. Explain the causes for the main features of this climate

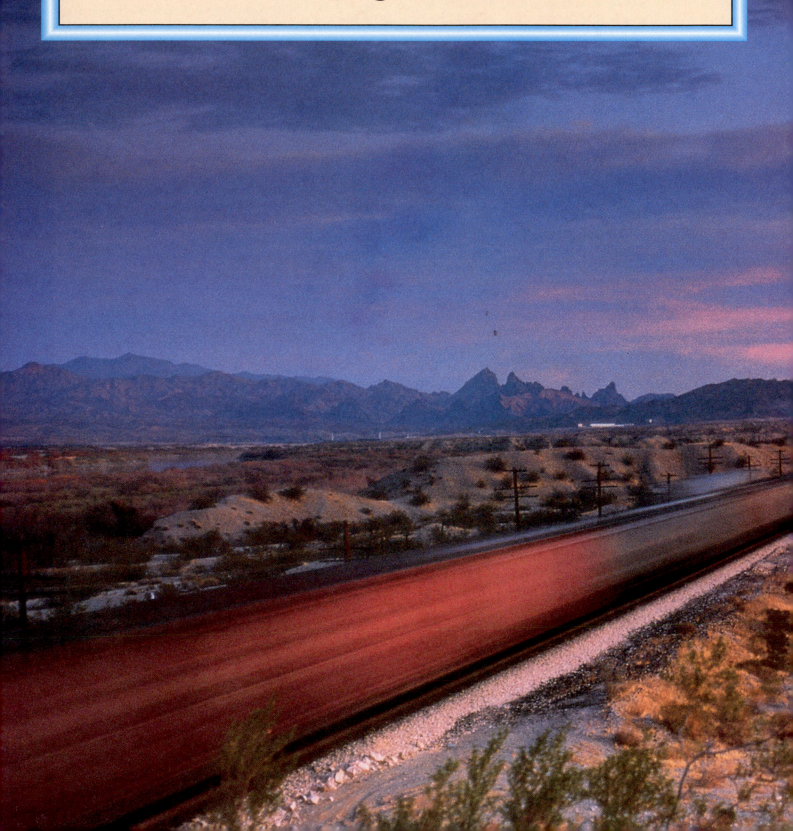

Chapter 8

Global Ecosystems

A train streaks across the Mojave Desert.

PUTTING YOU IN THE PICTURE

All aboard! The conductor's cry resounds through Union Station, Los Angeles, and at precisely 11:50 A.M., your train begins to move. At last your journey is underway! You're headed for Chicago on a three-day train ride that will take you through some of the most breathtaking scenery in America.

An hour later, you are relaxing in the comfortable armchair of your roomette, gazing out the large picture window. The Mojave Desert flashes by. The strange forms of Joshua trees contrast against the gray and tan colors of the desert earth. Overhead a bird with a large wingspan glides effortlessly. You wonder if it could be a California condor, far from its coastal habitat. The coach rushes across a dry lakebed of white salt sparkling in the sun. Sharp peaks loom in the far distance, their angular foothills terminating the broad plains of grayish creosote bush that sweep toward the mountain fronts.

Early that evening, the train pulls out of Las Vegas, headed up toward the high desert of the Colorado Plateau. As the sun slowly sets, the reds and browns of the rocks and soils intensify. Soon the desert sky darkens to a blue-black. Sagebrush carpets the landscape, looking soft and inviting in the moonlight. Reluctantly, you draw the shade and turn in for the night.

"More coffee?" The steward expertly refills your breakfast coffee cup, swaying with the motion of the dining car. During the night, the train has slowly gained elevation to enter the rugged upper reaches of the Colorado basin. Twisted crowns of piñon pines and junipers dot the hills. A frightened squirrel gathering pine nuts leaps away quickly. Gray-green sagebrush blankets the lowlands, with green, willowy saltbush lining the dry creek beds.

By early afternoon, you are well into your crossing of the Rocky Mountains. The track follows the narrow canyon of the upper Colorado. Steep, conifer-clad slopes enclose the rushing train. Soon you are in the high country. Ragged, windswept peaks jut skyward, spread with a green and white quilt of snowfields, conifer forests, and aspen groves. You glimpse a pair of mule deer, half hidden in a small stand of firs. Suddenly the train plunges into the Moffat Tunnel for the long passage under the continental divide. Emerging on the east side, you descend steeply toward the high plains and the skyscrapers of Denver.

Later that evening, you sit alone in your darkened compartment, gazing out the window. The short-grass prairie sweeps by, its moonlit surface like a table spread with faded lace. A prairie dog town flashes past, with burrows like blisters on the brown soil. Soon, the rocking motion of the train lulls you to sleep.

Noises in the station at Omaha awaken you early in the morning. After breakfast, you watch Iowa roll by. Maturing corn and soybean crops form a dense, green blanket on the earth. Cattle graze stubbly green fields where hay was harvested earlier in the season. As the train heads eastward, the density of settlement increases. By late afternoon, you reach the Chicago area, and precisely at 4:15 P.M., your train pulls into the station. Regretfully, you hoist your suitcase and step off the train. Your trip is over, but you will long remember the magnificent landscapes and scenic vistas of the American West.

This narrative records some of the variations in plant and animal life that you might encounter on a train trip across the American West. Crossing through so many different and distinctive biological regions, you might be tempted to wonder how they relate to physical geography. In fact, they are *part* of the physical geography. Through time, each region develops a distinct assemblage of plants and animals, depending on climate and landform. This chapter explores the relationship between organisms and their environment and maps the major biological regions of the globe.

Chapter Outline

Energy Flow in Ecosystems
  The Food Web
  Biomass, Net Primary Production, and Climate
Organisms and Environment
  Habitat
  Organisms and Water Need
  Organisms and Temperature
  Other Climatic Factors
  Ecological Succession
Natural Vegetation
  Structure and Life-Form of Plants

Terrestrial Ecosystems—The Biomes
  Forest Biome
Eye on the Environment • Exploitation of the Low-Latitude Rainforest Ecosystem
  Savanna Biome
  Grassland Biome
  Desert Biome
  Tundra Biome
  Altitude Zones of Vegetation
  Climatic Gradients and Vegetation Types

GLOBAL ECOSYSTEMS

The life layer that we discussed in the Prologue contains the greater part of the *biosphere*—the living organisms of the earth, including both plants and animals. The earth's life-forms come in a bewildering array. Higher plants are largely fixed in position and utilize solar energy as their power source. Animals are largely mobile and utilize plants or other organisms as their power source. Many organisms are very specialized, both in form and in the way in which they interact with and exploit their environment.

The study of the interactions between life-forms and their environment is the science of **ecology**. The total assemblage of components—living and nonliving—entering into the interactions of a group of organisms is known as an ecological system, or more simply, an **ecosystem**. An ecosystem tends to achieve a set of balances of the various processes and activities that occur within it. Although many of these balances are robust and self-regulating, some are quite sensitive and can be easily upset or destroyed.

Biogeography is the study of the variation of ecosystems from place to place. It also includes understanding the nature and functions of ecosystems and how they are related to such variables as climate and soil.

From the viewpoint of human use, ecosystems are natural resource systems. Food, fiber, fuel, and structural material are products of ecosystems and are manufactured by organisms using energy derived from the sun. The products and productivity of ecosystems depend to a large degree on climate. Where temperature and rainfall cycles permit, ecosystems provide a rich bounty for human use. Where temperature or rainfall cycles restrict ecosystems, human activities can also be limited. Of course, humans, too, are part of the ecosystem. A persistent theme of human geography is the study of how human societies function within ecosystems, utilizing ecosystem resources and modifying ecosystems for human benefit.

Ecosystems fall into two major groups—aquatic and terrestrial. *Aquatic ecosystems* include marine environments and the freshwater environments of the lands. Marine ecosystems include the open ocean, coastal estuaries, and coral reefs. Freshwater ecosystems include lakes, ponds, streams, marshes, and bogs. Our survey of physical geography will not include these aquatic environments. Instead, we will focus on the **terrestrial ecosystems**, which are dominated by land plants spread widely over the upland surfaces of the continents. The terrestrial ecosystems are directly influenced by climate and interact with the soil. In this way, they are closely woven into the fabric of physical geography.

Within terrestrial ecosystems, the largest recognizable subdivision is the **biome**. Although the biome includes the total assemblage of plant and animal life interacting within the life layer, the green plants dominate the biome physically because of their enormous mass, as compared with that of other organisms. Thus, biomes are typically distinguished by the characteristics of their vegetation. This chapter is devoted to an overview of ecosystems and in particular a global study of biomes. As we will see, biomes are often strongly related to climate, so the knowledge of climate you obtained in Chapter 7 will be very useful as you study this chapter.

ENERGY FLOW IN ECOSYSTEMS

The biosphere can be regarded as a layer on the earth's surface that traps light energy from the sun and uses that energy to sustain itself. Thus, the great diversity of life-forms that we encounter on the earth's surface are powered by sunlight, directly or indirectly. We also note from experience that temperature and precipitation are important determiners of the kinds of plants and animals found at a location. To understand how climate influences biogeography, we must first understand how the biosphere uses its power source—sunlight—to provide energy to all organisms and their ecosystems.

The Food Web

A salt marsh provides a good example of an ecosystem (Figure 8.1). A variety of organisms are present—algae and aquatic plants, microorganisms, insects, snails, and crayfish, as well as larger organisms such as fishes, birds, shrews, mice, and rats. Inorganic components will also be found—water, air, clay particles and organic sediment, inorganic nutrients, trace elements, and light energy. Energy transformations in the ecosystem occur by means of a series of steps or levels, referred to as a **food chain** or **food web**.

Photosynthesizing plants are the primary producers of the food web.

The plants and algae in the food web are the **primary producers**. They use light energy to convert carbon dioxide and water into carbohydrates (long chains of sugar molecules) and eventually into other biochemical molecules needed for the support of life. This process of

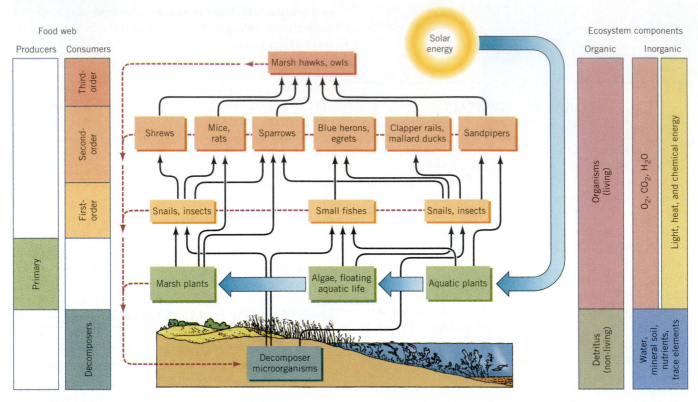

Figure 8.1 Flow diagram of a salt-marsh ecosystem in winter. The arrows show how energy flows from the sun to producers, consumers, and decomposers. (Food chain after R. L. Smith, *Ecology and Field Biology*, Harper and Row, New York.)

energy conversion is called *photosynthesis.* Organisms engaged in photosynthesis form the base of the food web.

The primary producers support the **consumers—** organisms that ingest other organisms as their food source. At the lowest level of consumers are the *primary consumers* (the snails, insects, and fishes). At the next level are the *secondary consumers* (the mammals, birds, and larger fishes), which feed on the primary consumers. Still higher levels of feeding occur in the salt marsh ecosystem as marsh hawks and owls consume the smaller animals below them in the food web. The **decomposers** feed on *detritus,* or decaying organic matter, derived from all levels. They are largely microscopic organisms (microorganisms) and bacteria. Among consumers, it is useful to distinguish *herbivores* (plant eaters), *carnivores* (animal eaters), and *omnivores,* which consume both plants and animals. Carnivores feed higher in the food web than herbivores, while omnivores obtain their nourishment from multiple levels.

The food web is really an energy flow system, tracing the path of solar energy through the ecosystem. Solar energy is absorbed by the primary producers and stored in the chemical products of photosynthesis. As these organisms are eaten and digested by consumers, chemical energy is released. This chemical energy is used to power new biochemical reactions, which again produce stored chemical energy in the bodies of the consumers.

At each level of energy flow in the food web, energy is lost to *respiration.* Respiration can be thought of as the burning of fuel to keep the organism operating. Energy expended in respiration is ultimately lost as waste heat and cannot be stored for use by other organisms higher up in the food chain. This means that, generally, both the numbers of organisms and their total amount of living tissue must decrease greatly up the food chain. In general, only 10 to 50 percent of the energy stored in organic matter at one level can be passed up the chain to the next level. Normally, there are about four levels of consumers.

Figure 8.2 is a bar graph showing the percentage of energy passed up the chain when exactly 10 percent moves from one level to the next. The horizontal scale is in powers of 10. In ecosystems of the lands, the mass of organic matter and the number of individuals of the consuming animals decrease with each upward step. In the food chain shown in Figure 8.1, there are only a few marsh hawks and owls in the third level of consumers, while countless individuals are found in the primary level.

For individual species, the number of individuals of a species present in an ecosystem ultimately depends on

Figure 8.2 Percentage of energy passed up the steps of the food chain, assuming 90 percent is lost energy at each step.

the level of resources available to support the species population. If these resources provide a steady supply of energy, the population size will normally attain a steady level. In some cases, however, resources vary with time, for example, in an annual cycle. The population size of a species depending on these resources may then also fluctuate in a corresponding cycle. Population sizes and the change in size of species populations with time have been studied extensively by ecologists.

Biomass, Net Primary Production, and Climate

The earth's ecosystems are very diverse and can differ greatly in their characteristics. One characteristic used by ecologists to describe an ecosystem is its **biomass**—the dry weight of the organic matter it contains. Biomass could be stated for a single plant or animal, but it is normally expressed as the biomass per unit of surface area within the ecosystem—that is, kilograms of biomass per square meter or (metric) tons of biomass per hectare (1 hectare = 10^4 m^2). Of all ecosystems, forests have the greatest biomass because of the large amount of wood that the trees accumulate through time. The biomass of grasslands and croplands is much smaller in comparison. For freshwater bodies and the oceans, the biomass is even smaller—on the order of one-hundredth that of the grasslands and croplands.

> **Day length, air and soil temperature, and water availability are the most important climatic factors that control net primary productivity.**

Although the amount of biomass present per unit area is an important indicator of the amount of photosynthetic activity, it can be misleading. In some ecosystems, biomass is broken down very quickly by consumers and decomposers, so the amount maintained is less. From the viewpoint of ecosystem productivity, what is important is the annual yield of useful energy produced by the plants of an ecosystem, or the *net primary production.*

What climatic factors control net primary productivity? Light intensity and duration, as well as temperature, are the most important factors influencing net photosynthesis. Since light energy is used in photosynthesis, light at sufficient levels must be present. The photosynthetic process is also temperature dependent. When temperatures are too hot or too cold, photosynthesis slows and stops.

Another important factor is the availability of water. Water is essential to the process of photosynthesis. Transpiration cools the leaf in the presence of strong sunlight so that photosynthesis can proceed at an optimal temperature. When leaf water becomes scarce, transpiration shuts down. Leaf pores close to conserve water, and carbon dioxide can no longer enter the leaf. So photosynthesis must shut down as well. From the viewpoint of global climate, precipitation is the single factor most directly related to soil water availability—the greater is the precipitation, the less likely it is that water will limit photosynthesis. However, very high rainfall environments are often less productive. Apparently, large volumes of precipitation can remove soil nutrients and in this way limit productivity.

Combining the effects of light intensity, temperature, and precipitation, we can assign rough values of productivity to each of the climates as follows (units are kilograms of carbon per square meter per year):

Highest (over 800)	Wet equatorial ①
Very high (600–800)	Monsoon and trade-wind coastal ②
	Wet-dry tropical ③
High (400–600)	Wet-dry tropical ③ (Southeast Asia)
	Moist subtropical ⑥
	Marine west coast ⑧
Moderate (200–400)	Mediterranean ⑦
	Moist continental ⑩
Low (100–200)	Dry tropical, semiarid ④s
	Dry midlatitude, semiarid ⑨s
	Boreal forest ⑩
Very low (0–100)	Dry tropical, desert ④d
	Dry midlatitude, desert ⑨d
	Boreal forest ⑩
	Tundra ⑫

Remote sensing can be used to monitor net primary productivity of the earth's land surface. Figure 8.3 provides an image of global productivity for both oceans and land as estimated using measurements from satellite instruments.

Figure 8.3 Global productivity as inferred using remote sensing. *Lands*: Displayed is a vegetation greenness index for the period April 1982 to March 1985, obtained from the Advanced Very High Resolution Radiometer (AVHRR) instrument aboard the NOAA-7 spacecraft. The most productive regions are shown in dark green tones, ranging through medium and light green tones to yellow-brown tones (least productive). *Oceans*: Displayed is phytoplankton pigment concentration averaged over the period 1978 to 1986, derived by processing data of the Coastal Zone Color Scanner instrument on the *Nimbus-7* satellite. Pigment concentrations are indicated by colors: red (highest) through yellow, green, and blue, to purple (lowest).

Net primary production represents a source of renewable energy derived from the sun that can be exploited to fill human energy needs. The use of biomass as an energy source involves releasing solar energy that has been fixed in plant tissues through photosynthesis. This process can take place in a number of ways. The simplest is direct burning of plant matter as fuel, as in a campfire or a wood-burning stove. Other approaches involve the generation of intermediate fuels from plant matter—methane gas, charcoal, and alcohol, for example. Biomass energy conversion is not highly energy efficient. Typical values of net annual primary production of plant communities range from 1 to 3 percent of available solar energy. However, the abundance of terrestrial biomass is so great that biomass utilization could provide the energy equivalent to 3 million barrels of oil per day for the United States with proper development.

ORGANISMS AND ENVIRONMENT

With an examination of the "big picture," relating energy flow in ecosystems to climate, we can now turn to a closer examination of the relationship between organisms and their environment. This will be helpful in understanding how vegetation is related to climate and soils, as well as how animals are related to the patterns of vegetation types that provide their food and shelter.

Habitat

The *habitat* of an organism is the physical environment in which it is most likely to be found. Thus, salt marsh is the habitat of cord grass, and boreal forest and tundra are the habitat of the caribou. For vegetation, habitat is largely determined by landform and soil. (As we will see in Chapter 10, landform refers to the configuration of the land surface, including features such as hills, valleys, ridges, or cliffs.) Vegetation in an upland habitat—relatively high ground with thick soil through which water drains easily—is often quite different from that in an adjacent valley floor habitat, where soils are wet much of the time. Vegetation is also strikingly different in habitats of rocky ridges and steep cliffs, where water drains away rapidly and soil is thin or largely absent. Figure 8.4 presents an example of habitats within a Canadian forest. Here, there are six distinctive habitats: upland, bog, bottomland, ridge, cliff, and active sand dune. Each habitat supports a different type of vegetation cover.

Of course, animals have habitats, too. The California condor is at home on the steep slopes and cliffs of the coastal ranges of California (Figure 8.5), while the

Figure 8.4 Habitats within the Canadian forest. (After P. Dansereau.)

Deciduous forest

Needleleaf forest

Moving dune

Bottomland

Ridge

Bog

Upland

Cliff

Canadian forest habitats

shortgrass prairie is the habitat of the prairie dog (Figure 8.6).

Organisms and Water Need

Since every living cell requires water as the medium for its metabolic activities, water is a vital quantity for both plants and animals. As you know from the study of climate in Chapter 7, many regions of the world have climates that are dry or have a significant dry season. Many species of plants have developed adaptations to help survive dry conditions. Plants that are adapted to drought conditions are termed **xerophytes**. The word "xerophyte" comes from the Greek roots *xero-*, meaning "dry," and *phyton*, meaning "plant." Because they are highly tolerant of drought, xerophytes can survive in habitats that dry quickly following rapid drainage of precipitation (for example, sand dunes, beaches, and bare rock surfaces). The adaptations of desert plants to dry conditions make them xerophytes as well.

Xerophytes and xeric animals are plants and animals that are adapted to a dry and sometimes hot environment.

In some xerophytes, water loss is reduced by a thick layer of wax or waxlike material on leaves and stems.

Figure 8.5 A California condor on a perch. The natural habitat of this large carnivore is the coastal ranges of Southern California.

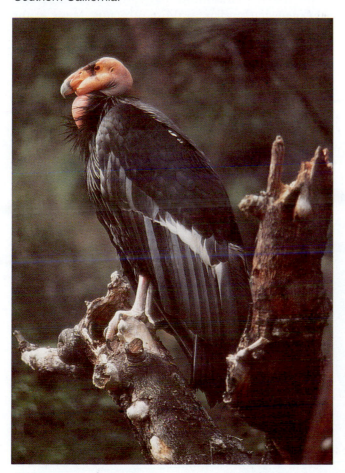

Figure 8.6 A prairie dog guards its burrow in the semiarid steppelands of Utah. This particular species, the Utah white-tailed prairie dog, is threatened with extinction.

The wax helps to seal water vapor inside the leaf or stem. Still other xerophytes adapt to a desert environment by greatly reducing their leaf area or by bearing no leaves at all. Needlelike leaves, or spines in place of leaves, are also adaptations of plants to conserve water. In cactus plants, the foliage leaf is not present, and transpiration is limited to thickened, water-filled stems that store water for use during long, dry periods (Figure 8.7).

Adaptations of plants to water-scarce environments also include improved abilities to obtain and store water. Roots may extend deeply to reach soil moisture far from the surface. In cases where the roots reach to the ground water zone, a steady supply of water is assured. Plants drawing from ground water may be found along dry stream channels and valley floors in desert regions. In these environments, ground water is usually near the surface. Other desert plants produce a widespread, but shallow, root system. This enables them to absorb water from short desert downpours that saturate only the uppermost soil layer.

Another adaptation to extreme aridity is a very short life cycle. Many small desert plants will germinate from seed, then leaf out, bear flowers, and produce seed in the few weeks immediately following a heavy rain shower. In this way, they complete their life cycle when soil moisture is available, and then survive the dry period as seeds that require no moisture.

Certain climates, such as the wet-dry tropical climate ③ and the moist continental climate ⑩, have a yearly cycle with one season in which water is unavailable to plants because of lack of precipitation or because the soil water is frozen. This season alternates with one in which there is abundant water. Many plants respond to this pattern by dropping their leaves at the close of the moist season and becoming dormant during the dry or cold season. When water is again available, they leaf out and grow at a rapid rate. Trees and shrubs that shed their leaves seasonally are termed *deciduous*. In contrast are *evergreen* plants, which retain most of their leaves in a green state through one or more years.

The Mediterranean climate ⑦ also has a strong seasonal wet-dry alternation, with dry summers and wet winters. Plants in this climate are often xerophytic and characteristically have hard, thick, leathery leaves. An example is the live oak, which holds most of its leaves through the dry season (Figure 8.8). Such hard-leaved evergreen trees and woody shrubs are called *sclerophylls*. (The prefix *scler-* is from the Greek root for "hard" and is combined with the Greek word for leaf, *phyllon*.) Plants that hold their leaves through a dry or cold season have the advantage of being able to resume photosynthesis quickly when growing conditions become favorable, whereas the deciduous plants must grow a new set of leaves.

To cope with shortage of water, *xeric animals*, those adapted to dry conditions, have evolved methods that are somewhat similar to those of plants. Many of the invertebrates exhibit the same pattern as desert annuals—evading the dry period in dormant stages. When rain falls, they emerge to take advantage of the new veg-

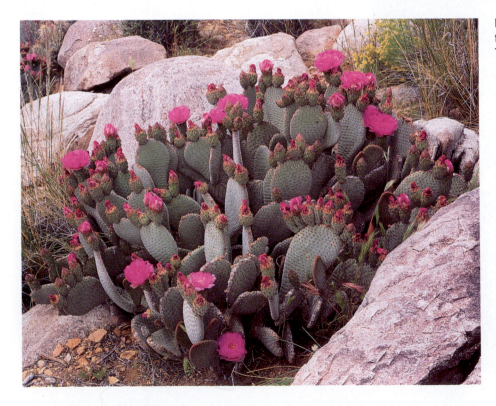

Figure 8.7 Beavertail cactus in flower in the Mojave Desert. Joshua Tree National Monument, California.

Figure 8.8 This California live oak is an example of a sclerophyll—a plant with thick, leathery leaves that is adapted to an environment with a very dry season.

etation. For example, the tiny brine shrimp of the Great Basin may wait many years in dormancy until normally dry lakebeds fill with water, an event that occurs perhaps three or four times a century. The shrimp then emerge, feed, and complete their life cycles before the lake evaporates. Spadefoot toads exhibit similar adaptations, emerging from burrows during a rainy period and breeding in the nearest puddle. The tadpoles quickly mature, and by the time the puddle dries they are capable of digging their own burrows to pass the next long dry period. Many species of birds regulate their behavior to nest only when rains occur, the time of most abundant food for their offspring.

The mammals are by nature poorly adapted to desert environments. Yet many survive there, employing a variety of mechanisms to avoid water loss. Just as plants reduce transpiration to conserve water, so many desert mammals do not sweat through skin glands. Instead, they rely on other methods of cooling. For example, the huge ears of the jackrabbit serve as efficient radiators of heat to the sky. Calculations have shown that a jackrabbit sitting in a shaded depression would dispose of about one-third of its metabolic heat load through long-wave radiation from its uninsulated ears to the clear desert sky above. Many of the desert mammals conserve water by excreting highly concentrated urine and relatively dry feces. In addition, many are able to use to their advantage metabolic water, formed as a byproduct when food is processed to produce energy. The desert mammals also evade the heat by avoiding daytime activity. In this respect, they are joined by most of the rest of the desert fauna, spending their days in cool burrows in the soil and their nights foraging for food.

Organisms and Temperature

Temperature is the second important climatic factor that influences organisms. It acts directly by influencing the rates at which physiological processes take place. For plants, each species has an optimum temperature associated with each of its functions, such as photosynthesis, flowering, fruiting, or seed germination. There are also limiting lower and upper temperatures for these individual functions and for the total survival of the plant itself.

Temperature also acts as an indirect factor. For example, higher air temperatures increase the water vapor holding capacity of the air. This means that liquid water will evaporate more readily if air temperature is increased. Therefore, plants lose more water through increased transpiration at higher temperatures. Soils also dry out more readily, reducing later availability to plants utilizing the soil.

Only a relatively small number of plants are adapted to survival in subfreezing temperatures.

In general, the colder the climate, the fewer the species that are capable of surviving. Among the plants, a large number of tropical species cannot survive below-freezing temperatures for more than a few hours. Freezing causes ice crystals to form inside plant cells, disrupting the internal structure of the cell and damaging the fine cell structure. In the severely cold arctic and alpine environments of high latitudes and high altitudes, only a few plant species are found. This principle explains why a forest in the equatorial zone has many species of trees, whereas a forest of the subarctic zone will be dominated by only a few.

The effects of temperature variations on animals are moderated by their physiology and by their ability to seek sheltered environments. Most animals lack a physiological mechanism for internal temperature regulations. These animals, including reptiles, invertebrates, fish, and amphibians, are *cold-blooded* animals. Their body temperatures passively follow the environment, although they can influence their temperatures by such behaviors as lying in the sun. With a few exceptions (notably fish and some social insects), these animals are active only during the warmer parts of the year. They survive the cold weather of the midlatitude zone winter by becoming dormant.

Some vertebrates enter a dormant state termed hibernation, in which metabolic processes virtually halt and body temperatures closely parallel those of the surroundings. Most hibernators seek out burrows, nests, or other environments where winter temperatures do not reach extremes or fluctuate rapidly. Because the annual range of soil temperatures is greatly reduced below the uppermost layers, soil burrows are particularly suited to hibernation.

Other animals maintain their tissues at a constant temperature by internal metabolism. This group includes the birds and mammals. These *warm-blooded* animals possess a veriety of adaptations to maintain a constant internal temperature. Fur, hair, and feathers act as insulation by trapping dead air spaces next to the skin surface, reducing heat loss to the surrounding air or water. A thick layer of fat will also provide excellent insulation (Figure 8.9). Other adaptations are for cool-

ing: for example, sweating or panting uses the high latent heat of vaporization of water to remove heat. Heat loss is also facilitated by exposing blood-circulating tissues to the cooler surroundings. The seal's flippers and bird's feet serve this function.

Other Climatic Factors

Light is also an important factor regulating plants and animals. The amount of light available to a plant will depend in large part on the plant's position. While tree crowns in the upper layer of a forest receive maximum light, they reduce the amount of light available to lower layers. In some cases, the trees cut off light so effectively that the forest floor is almost free of shrubs and smaller plants. In many midlatitude deciduous forests, smaller plants go through a rapid life cycle in early spring, when the leaves have not yet emerged and light intensities are high at ground level.

Length of the daylight portion of the day, or *photoperiod*, depends on the latitude and season of the year. Photoperiod provides useful cues to plants and animals. Many plant development stages, such as budding, flowering, fruiting, and leaf shedding, are triggered by changes in photoperiod. In animals, behaviors are often conditioned by photoperiod. As autumn days grow shorter and shorter, squirrels and other rodents hoard food for the coming winter season. Later, increasing photoperiod will signal such activities as mating and reproduction in the spring.

The day–night cycle controls the activity patterns of many animals. Birds, for example, are generally active during the day, while small foraging mammals, such as weasels, skunks, and chipmunks, are more active at night.

Figure 8.9 A harbor seal lounges on an ice flow in southeastern Alaska, with the Le Conte Glacier in the background. This warm-blooded mammal's thick layer of body fat provides insulation against heat loss to the cold waters of its habitat.

Wind is also an important environmental factor for vegetation in highly exposed positions. Close to the timberline in high mountains and along the northern limits of tree growth in the arctic zone, trees are deformed by wind in such a way that the branches project from the sheltered side of the trunk only, attaining a flag shape (see Figure 8.37, later in this chapter). Some trees may show trunks and branches that are flattened and facing away from the prevailing wind direction. The effect of wind is to cause excessive drying, damaging the exposed side of the plant.

Ecological Succession

Although climatic influences are important in determining the general type of ecosystem within a region, local ecosystem patterns often depend on the developmental stage of the vegetation cover. For example, a country drive in the southeastern United States reveals patches of vegetation in many stages of development—from open fallow fields to grassy shrublands to forests. Each stage will have its characteristic assemblage of animals—field mice and voles in fallow fields, deer and turkey in forest. As we will see shortly, these are stages in a process by which abandoned fields become mature forests. On a longer time scale, clear lakes gradually fill in with sediment from the rivers that drain into them and become marshes. The marshes may eventually dry out as drainage improves and be replaced by a tall, mature forest.

The development process in which plant and animal communities succeed one another on the way to a stable endpoint is termed an *ecological succession*. Succession may begin on a newly constructed mineral deposit, such as a sand dune or river bar. It may also occur on a previously vegetated area that has been recently disturbed by agents such as fire, flood, and windstorms, or by human activity. In general, succession leads to formation of the most complex community of organisms possible in an area, given its physical controlling factors of climate, soil, and water. The stable community that is the endpoint of succession is the *climax*.

An important case of succession occurs on farmlands that are intensively cultivated for decades and then abandoned to the natural sequence of vegetation development. An example of this "old-field" succession comes from the southeastern United States, where the moist subtropical climate ⑥ provides ample soil water much of the year and the growing season is long. Figure 8.10 shows the succession, starting at the left with the bare field.

The first plants to take hold in the hostile environment are called *pioneers*. They are annual herbs that most persons would call weeds. When they die, these plants supply organic matter to the soil. They also begin to shade the soil and reduce the extremes of soil temperature. Next, grasses and shrubs move in, and these occupy the old field during the first two decades or so. Pine seedlings now enter the habitat and, as these grow, a shade cover develops. This greatly changes the climate near the ground.

In the process of ecological succession, an ecosystem proceeds through regular stages of change to reach a climax.

In the next half-century, as the pine forest matures, broad-leaved deciduous trees begin to displace the pines. By the middle of the second century, a mature forest of oak and hickory has dominated the habitat. This climax forest has a lower layer of shrubs and small trees, a basal herb layer, and a substantial accumulation of organic matter on the ground in various stages of decomposition. In this environment, soil water tends to be conserved, and the thermal environment is protected from extremes of heat or cold.

Figure 8.10   Old-field succession in the southeastern United States, following abandonment of cultivated fields. This is a pictorial graph of continuously changing plant composition spanning about 150 years.

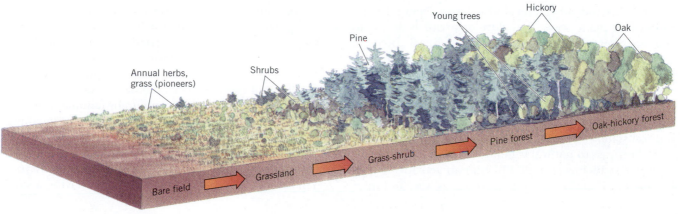

As the composition of the plant cover changes with time, so does the composition of associated animal communities. At first, the insects of cultivated fields appear in large numbers. The old-field mouse, harvest mouse, and least shrew then appear and become abundant, accompanied by the bobwhite and eastern meadowlark. Cottontail rabbits and cotton rats soon follow. As pines develop dominance, these rodents become less frequent. A different group of birds appears. Forest species, such as the white-tailed deer and the gray fox, begin to visit the area. As the pines mature and develop, their nuts support squirrel populations. In the oak–hickory phase, nuts and acorns provide sustenance for wild turkeys.

Although the oak–hickory climax forest is generally stable in composition and structure, serious natural upsets to the successional sequence may occur through fires and severe storms. Insect hordes and disease epidemics may also radically alter the climax forest. Such disturbances will be followed by an appropriate succession, with stages leading to the return of the oak–hickory forest.

Disturbances may also result from human activity. Cutting and clearing of forest trees to reclaim the land for agriculture removes the climax forest. Or a disease introduced from a foreign continent may cause the extinction of a particular plant species, changing the climax. An example is the chestnut blight, introduced from Europe, which eliminated the American chestnut from the forests of the northeastern United States. Imported insects may also wipe out most of the mature individuals of a plant species if no native predator is available to combat the invasion. These are just a few of the ways in which humans influence natural vegetation and the succession process.

NATURAL VEGETATION

Over the last few thousand years, human societies have come to dominate much of the land area of our planet. In many regions, humans have changed the natural vegetation—sometimes drastically, other times subtly. What do we mean by natural vegetation? **Natural vegetation** is a plant cover that develops with little or no human interference. It is subject to natural forces of modification and destruction, such as storms or fires. Natural vegetation can still be seen over vast areas of the wet equatorial climate ①, although the rainforests there are being rapidly cleared. Much of the arctic tundra and the boreal forest of the subarctic zones is in a natural state.

In contrast to natural vegetation is *human-influenced vegetation*, which is modified by human activities. Much of the land surface in midlatitudes is totally under human control, through intensive agriculture, grazing, or urbanization. Some areas of natural vegetation appear to be untouched but are actually dominated by human activity in a subtle manner. For example, most American national parks and national forests have been protected from fire for many decades. However, periodic burning is part of the natural cycle in many regions. One vital function of fire is to release nutrients that are stored in plant tissues. When the vegetation burns, the ashes containing the nutrients remain. These enrich the soil for the next cycle of vegetation cover. In recent years, managers of some parks and forests have stopped suppressing wildfires, allowing the return of more natural periodic burning.

Most areas of the earth's land surface are influenced subtly or strongly by human activity.

Our species has influenced vegetation in yet another way—by moving plant species from their original habitats to foreign lands and foreign environments. Such species are called *exotics*. The Australian eucalyptus tree is a striking example—species of eucalyptus have been transplanted and naturalized in such far-off lands as California, North Africa, and India. Sometimes exotic plants thrive like weeds, forcing out natural species and becoming a major nuisance. The kudzu vine, which invades roadbanks and forest edges, is a familiar example from the southeastern United States. Few of the grasses that clothe the coastal ranges of California are native species, yet a casual observer might think that these represent native vegetation.

Even so, all plants have limited tolerance to the environmental conditions of soil water, heat and cold, and soil nutrients. Consequently, the structure and outward appearance of the plant cover conforms to basic environmental controls, and each vegetation type is associated with a characteristic geographical region—whether forest, grassland, or desert.

Structure and Life-Form of Plants

Plants come in many types, shapes, and sizes. Botanists recognize and classify plants by species. However, the biogeographer is less concerned with individual species and more concerned with the plant cover as a whole. In describing the plant cover, biogeographers refer to the **life-form** of the plant—its physical structure, size, and shape. Although the life-forms go by familiar names, we will review them to establish a uniform set of meanings.

Both trees and shrubs are erect, woody plants. They are *perennial*, meaning that their woody tissues endure from year to year. Most have life spans of many years. *Trees* are large, woody perennial plants having a single

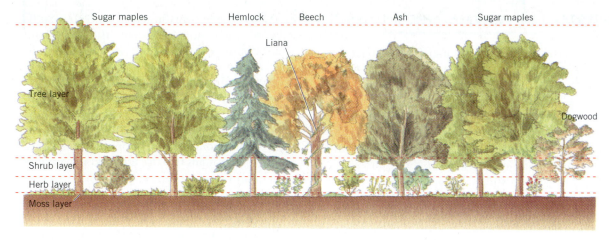

Sugar maples   Hemlock   Beech   Ash   Sugar maples

Liana

Tree layer

Dogwood

Shrub layer

Herb layer

Moss layer

Figure 8.11   This schematic diagram shows the layers of a beech–maple–hemlock forest. The vertical dimensions of the lower layers are greatly exaggerated. (After P. Dansereau.)

upright main trunk, often with few branches in the lower part, but branching in the upper part to form a crown. *Shrubs* are woody perennial plants that have several stems branching from a base near the soil surface, so as to place the mass of foliage close to ground level.

Lianas are also woody plants, but they take the form of vines supported on trees and shrubs. Lianas include not only the tall, heavy vines of the wet equatorial and tropical rainforests (see Figure 8.16), but also some woody vines of midlatitude forests. Poison ivy and the tree-climbing form of poison oak are familiar American examples of lianas.

Herbs comprise a major class of plant life-forms. They lack woody stems and so are usually small, tender plants. They occur in a wide range of shapes and leaf types. Some are *annuals,* living only for a single season, while others are perennials. Some herbs are broad-leaved, and others are narrow-leaved, such as grasses. Herbs as a class share few characteristics in common except that they usually form a low layer as compared with shrubs and trees.

Forest is a vegetation structure in which trees grow closely together. Crowns are in contact, so that the foliage largely shades the ground. Many forests in moist climates show at least three layers of life-forms (Figure 8.11). Tree crowns form the uppermost layer, shrubs an intermediate layer, and herbs a lower layer. There is sometimes a fourth, lowermost layer that consists of mosses and related very small plants. In **woodland**, crowns of trees are mostly separated by open areas that often have a low herb or shrub layer.

Lichens are another life-form seen in a layer close to the ground. They are plant forms in which algae and fungi live together to form a single plant structure. In some alpine and arctic environments, lichens grow in profusion and dominate the vegetation (Figure 8.12).

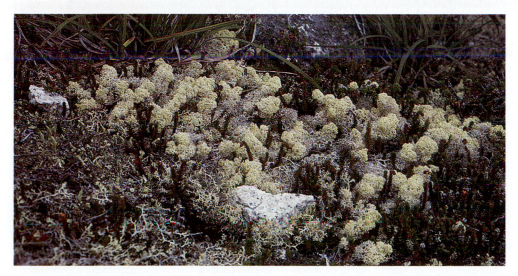

Figure 8.12   Reindeer moss, a white variety of lichen, seen here on rocky tundra of Alaska. This interesting plant is a xerophyte with no true root system. It obtains its moisture directly from the air.

TERRESTRIAL ECOSYSTEMS—THE BIOMES

Among terrestrial ecosystems, ecologists and biogeographers recognize five principal types, known as *biomes*, that are distinguished primarily based on their vegetation cover. The **forest biome** is dominated by trees, which form a closed or moderately open canopy. Forest requires an abundance of soil water, so forests are found in the moist climates. Temperatures must also be suitable, requiring at least a warm season, if not warm temperatures the year round. The **savanna biome** is transitional between forest and grassland. It exhibits an open cover of trees with grasses and herbs underneath. The **grassland biome** develops in regions with moderate shortages of soil water. The semiarid regions of the dry tropical ④, dry subtropical ⑤, and dry midlatitude ⑨ climates are the home of the grassland biome. Temperatures must also provide adequate warmth during the growing season. The **desert biome** includes organisms that can survive a moderate to severe water shortage for most, if not all, of the year. Temperatures can range from very hot to cool. Plants are often xerophytes, showing adaptations to the dry environment. The **tundra biome** is limited by cold temperatures. Only plants and animals that are adapted to a very short warm season and a very long, very cold one can survive here.

Ecologists recognize five primary biome types: forest, grassland, savanna, desert, and tundra.

Figure 8.13 Natural vegetation of the world. (Based on maps of S. R. Eyre, 1968.)

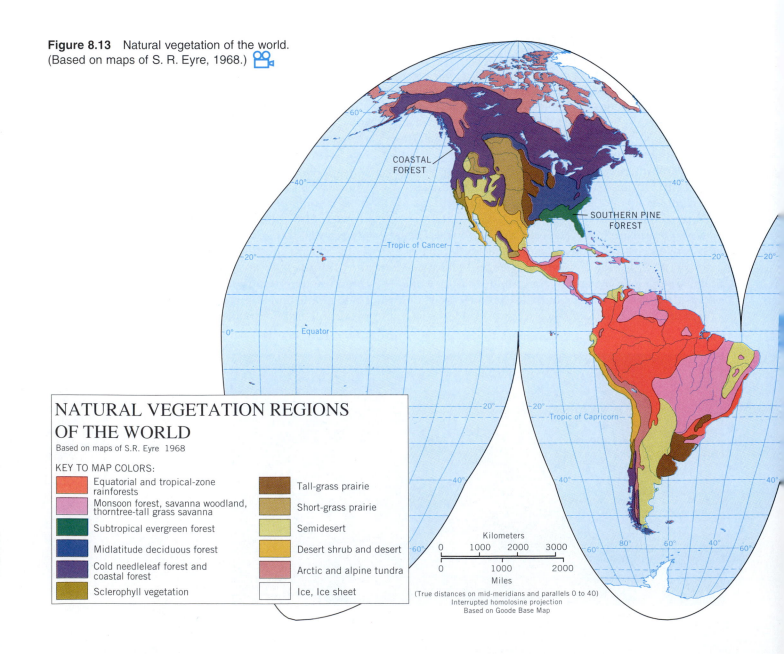

NATURAL VEGETATION REGIONS OF THE WORLD
Based on maps of S.R. Eyre 1968

KEY TO MAP COLORS:

- Equatorial and tropical-zone rainforests
- Monsoon forest, savanna woodland, thorntree-tall grass savanna
- Subtropical evergreen forest
- Midlatitude deciduous forest
- Cold needleleaf forest and coastal forest
- Sclerophyll vegetation
- Tall-grass prairie
- Short-grass prairie
- Semidesert
- Desert shrub and desert
- Arctic and alpine tundra
- Ice, Ice sheet

COASTAL FOREST

SOUTHERN PINE FOREST

Kilometers
0   1000   2000   3000

0   1000   2000
Miles

(True distances on mid-meridians and parallels 0 to 40)
Interrupted homolosine projection
Based on Goode Base Map

Biogeographers break the biomes down further into smaller units, called *formation classes*. For example, at least four and perhaps as many as six kinds of forests are easily distinguished within the forest biome. At least three kinds of grasslands are easily recognizable. Deserts, too, span a wide range in terms of the abundance and life-form of plants. Although the formation classes are primarily distinguished by their vegetation types, each also has a somewhat different group of associated animal species. The formation classes described in the remaining portion of this chapter are major, widespread types that are clearly associated with specific climate types. Figure 8.13 is a generalized world map of the formation classes. It shows large uniform regions in which a given formation class might be expected to occur. Table 8.1 lists the biomes and formation classes we discuss in this chapter.

Forest Biome

Within the forest biome, we can recognize six major formations: low-latitude rainforest, monsoon forest, subtropical evergreen forest, midlatitude deciduous forest, needleleaf forest, and sclerophyll forest.

Low-Latitude Rainforest

Low-latitude rainforest, found in the equatorial and tropical latitude zones, consists of tall, closely set trees. Crowns form a continuous canopy of foliage and provide dense shade for the ground and lower layers

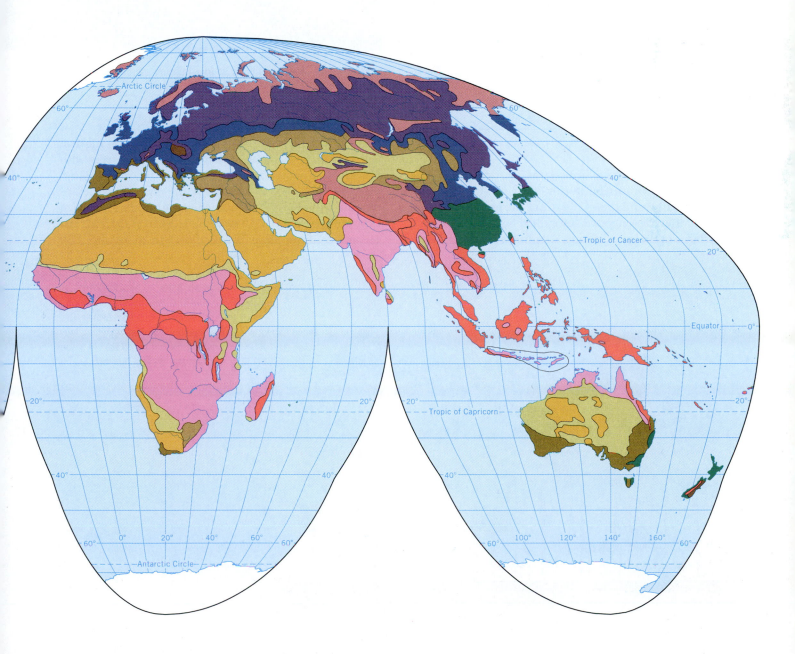

Table 8.1 Global Biomes and Major Formation Classes

Biome	Formation
Forest Biome	Low-Latitude Rainforest
	Monsoon Forest
	Subtropical Evergreen Forest
	Subtropical Evergreen Broadleaf Forest (Laurel Forest)
	Subtropical Evergreen Needleleaf Forest
	Midlatitude Deciduous Forest
	Needleleaf Forest
	Boreal Forest
	Coastal Forest
	Sclerophyll Forest
	Sclerophyll Woodland
	Sclerophyll Scrub (Chaparral)
Savanna Biome	Savanna Woodland
	Thorntree–Tall-Grass Savanna
Grassland Biome	Tall-Grass Prairie
	Steppe (Short-Grass Prairie)
Desert Biome	Semidesert
	Thorntree Semidesert
	Dry Desert
Tundra Biome	Arctic Tundra
	Alpine Tundra

Figure 8.14 An aerial view of low-latitude rainforest. Middle Mazaruni River near Bartica, Guyana.

(Figure 8.14). The trees are characteristically smooth-barked and unbranched in the lower two-thirds. Tree leaves are large and evergreen; thus, the equatorial rainforest is often described as "broadleaf evergreen forest."

Crowns of the trees of the low-latitude rainforest tend to form two or three layers (Figure 8.15). The highest layer consists of scattered *emergents*. These are trees with crowns that protrude from the closed canopy below, often rising to 40 m (130 ft). Some emergent species develop wide buttress roots, which aid in their physical support (Figure 8.16). Below the layer of emergents is a second, continuous layer, which is 15 to 30 m (about 50

Figure 8.15 This diagram shows the typical structure of equatorial rainforest. (After J. S. Beard, *The Natural Vegetation of Trinidad*, Clarendon Press, Oxford.)

to 100 ft) high. A third, lower layer consists of small, slender trees 5 to 15 m (about 15 to 50 ft) high with narrow crowns.

Typical of the low-latitude rainforest are thick, woody lianas supported by the trunks and branches of trees. Some are slender, like ropes, while others reach thicknesses of 20 cm (8 in.). They climb high into the trees to the upper canopy, where light is available, and they develop numerous branches of their own. *Epiphytes* ("air plants") are also common in low-latitude rainforest. These plants attach themselves to the trunk, branches, or foliage of trees and lianas. Their "host" is used solely as a means of physical support. Epiphytes include plants of many different types—ferns, orchids, mosses, and lichens (Figure 8.17).

A particularly important characteristic of the low-latitude rainforest is the large number of species of trees that coexist. In equatorial regions of rainforest, as many as 3000 species may be found in a few square kilometers. Individuals of a given species are often widely separated. Many species of plants and animals in this diverse ecosystem still have not been identified or named by biologists. **SCIENCE NEWS** 15

Equatorial and tropical rainforests are not "jungles" of impenetrable plant thickets. Rather, the floor of the low-latitude rainforest is usually so densely shaded that plant foliage is sparse close to the ground. This gives the forest an open aspect, making it easy to travel within its interior. The ground surface is covered only by a thin litter of leaves. Dead plant matter rapidly decomposes because the warm temperatures and abundant moisture promote its breakdown by invertebrates and bacteria. Nutrients released by decay are quickly absorbed by roots. As a result, the soil is low in organic matter.

Animal life of the tropical rainforest is most abundant in the upper layers of vegetation. Above the canopy, birds and bats are important carnivores, feeding on the insects above and within the topmost canopy. Below this level are found a wide variety of birds, mammals, reptiles, and invertebrates; they feed on the leaves, fruit, and nectar abundantly available in the main part of the canopy. Ranging between the canopy and the ground are the climbing mammals, which forage in both layers. At the surface are the larger ground mammals, including herbivores, which graze the low leaves and fallen fruits, and carnivores, which prey upon the abundant vertebrates and invertebrates found at the surface.

Large herbivores are uncommon in the low-latitude rainforest. They include the African okapi and the tapir of South America and Asia. Most herbivores are climbers, and they include many primates—monkeys and apes. Toucans, parrots, and tinamous join fruit-eating bats as flying grazers of the forest. Tree sloths spend their lifetimes hanging upside down as they browse the forest canopy. There are few large predators. Notable are the leopards of African and Asian forests and the jaguars and ocelots of the South American forests.

Figure 8.16 Buttress roots at the base of a large rainforest tree. Note also the thick, heavy vines (lianas) climbing the trunk. From Daintree National Park, Queensland, Australia.

Figure 8.17 In this photo from the El Yunque rainforest, Caribbean National Forest, Puerto Rico, red-flowering epiphytes adorn the trunks of sierra palms.

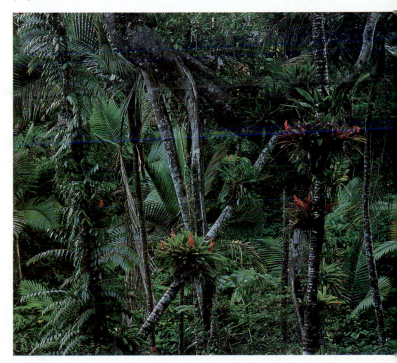

Low-latitude rainforest develops in a climate that is continuously warm, frost-free, and has abundant precipitation in all months of the year (or, at most, has only one or two dry months). These conditions occur in the wet equatorial climate ① and the monsoon and trade-wind coastal climate ②. In the absence of a cold or dry season, plant growth goes on throughout the entire year. Some plant species grow new leaves continuously, shedding old ones continuously as well. Still other plant species shed their leaves according to their own seasons, responding to the slight changes in photoperiod that occur with the seasons.

Low-latitude rainforests are very diverse, containing large numbers of plant and animal species.

World distribution of the low-latitude rainforest is shown in Figure 8.18. A large area of rainforest lies astride the equator. In South America, this equatorial rainforest includes the Amazon lowland. In Africa, the *equatorial rainforest* is found in the Congo lowland and in a coastal zone extending westward from Nigeria to Guinea. In Indonesia, the island of Sumatra, on the west, bounds a region of equatorial rainforest that stretches eastward to the islands of the western Pacific.

The low-latitude rainforest extends poleward through the tropical zone (lat. 10° to 25° N and S) along monsoon and trade-wind coasts. The monsoon and trade-wind coastal climate ② in which this *tropical-zone rainforest* thrives has a short dry season. However, the dry season is not intense enough to deplete soil water.

In the northern hemisphere, trade-wind coasts that have rainforests are found in the Philippine Islands as well as along the eastern coasts of Central America and the West Indies. These highlands receive abundant orographic rainfall in the belt of the trade winds. A good example of tropical-zone rainforest is the rainforest of the eastern mountains of Puerto Rico. In Southeast Asia, tropical-zone rainforest is extensive in monsoon coastal zones and highlands that have heavy rainfall and a very short dry season. These occur in Vietnam and Laos, southeastern China, and on the western coasts of India and Myanmar. In the southern hemisphere, belts of tropical-zone rainforest extend down the eastern Brazilian coast, the Madagascar coast, and the coast of northeastern Australia.

Within the regions of low-latitude rainforest are many islandlike highland regions where climate is cooler and rainfall is increased by the orographic effect. Here, rainforest extends upward on the rising mountain slopes. Between 1000 and 2000 m (about 3300 and 6500 ft), the rainforest gradually changes in structure and becomes *montane forest* (Figure 8.18). The canopy of montane forest is more open, and tree heights are lower, than in the rainforest. With increasing elevation, the forest canopy height becomes even lower. Tree ferns and bamboos are numerous, and epiphytes are particularly abundant. Mist and fog eventually become persistent, giving high-elevation montane forest the name *cloud forest*.

Low-latitude rainforest is under increasing human pressure. Slowly but surely, the rainforest is being conquered by logging, clearcutting, and conversion to grazing and farming. *Eye on the Environment • Exploitation of the Low-Latitude Rainforest Ecosystem* describes this process in more detail. **SCIENCE NEWS** 16, 17

Figure 8.18 World map of low-latitude rainforest, showing equatorial and tropical rainforest types. (Data source same as Figure 8.13. Based on Goode Base Map.)

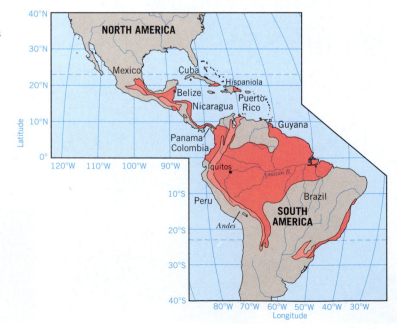

Figure 8.19 Monsoon woodland in the Bandipur Wild Animal Sanctuary in the Nilgiri Hills of southern India. The scene is taken in the rainy season, with trees in full leaf.

Monsoon Forest

Monsoon forest of the tropical latitude zone differs from tropical rainforest because it is deciduous, with most of the trees of the monsoon forest shedding their leaves during the dry season. Shedding of leaves results from the stress of a long dry season that occurs at the time of low sun and cooler temperatures. In the dry season, the forest resembles the deciduous forests of the midlatitudes during their leafless winter season.

Monsoon forest is typically open. It grades into woodland, with open areas occupied by shrubs and grasses (Figure 8.19). Because of its open nature, light easily reaches the lower layers of the monsoon forest.

EYE ON THE ENVIRONMENT

Exploitation of the Low-Latitude Rainforest Ecosystem

Many of the world's equatorial and tropical regions are home to the rainforest ecosystem. This ecosystem is perhaps the most diverse on earth. That is, it possesses more species of plants and animals than any other. Very large tracts of rainforest still exist in South America, south Asia, and some parts of Africa. Ecologists regard this ecosystem as a genetic reservoir of many species of plants and animals. But as human populations expand and the quest for agricultural land continues, low-latitude rainforests are being threatened with clearing, logging, and cultivation for cash crops and animal grazing.

In the past, low-latitude rainforests were farmed by native peoples using the *slash-and-burn* method—cutting down all the vegetation in a small area, then burning it (see photo below). In a rainforest ecosystem, most of the nutrients are held within living plants rather than in the soil. Burning the vegetation on the site releases the trapped nutrients, returning a portion of them to the soil. Here, the nutrients are available to growing crops. The supply of nutrients derived from the original vegetation cover is small, however, and the

harvesting of crops rapidly depletes the nutrients. After a few seasons of cultivation, a new field is cleared, and the old field is abandoned. Rainforest plants reestablish their hold on the abandoned area. Eventually, the rainforest returns to its original state. This cycle shows that primitive, slash-and-burn agriculture is fully compatible with maintenance of the rainforest ecosystem.

On the other hand, modern intensive agriculture uses large areas of land and is not compatible with the rainforest ecosystem. When large areas are abandoned, seed sources are so far away that the original forest species cannot take hold. Instead, secondary species dominate, often accompanied by species from other vegetation types. These species are good invaders, and once they enter an area, they tend to stay. The dominance of these secondary species is permanent, at least on the human time scale. Thus, we can regard the rainforest ecosystem as a resource that, once cleared, will never return. The loss of low-latitude rainforest will result in the disappearance of thousands of species of organisms from the rainforest environment—a loss of mil-

lions of years of evolution, together with the destruction of the most complex ecosystem on earth.

In Amazonia, transformation of large areas of rainforest into agricultural land uses heavy machinery to carve out major highways, such as the Trans-Amazon Highway in Brazil, and innumerable secondary roads and trails. Large fields for cattle pasture or commercial crops are created by cutting, bulldozing, clearing, and burning the vegetation. In some regions, the great broadleaved rainforest trees are removed for commercial lumber.

What are the effects of large-scale clearing? A study using a computer model, completed in 1990 by scientists of the University of Maryland and the Brazilian Space Research Institute, indicated that when the Amazon rainforest is entirely removed and replaced with pasture, surface and soil temperatures will be increased 1 to 3°C (2 to 5°F). Precipitation in the region will decline by 26 percent and evaporation by 30 percent. The deforestation will change weather and wind patterns so that less water vapor enters the Amazon basin from outside sources, making the basin even drier.

As a result, these lower layers are better developed than in the rainforest. Tree heights are also lower. Typically, many tree species are present—as many as 30 to 40 species in a small tract—although the rainforest has many more. Tree trunks are massive, often with thick, rough bark. Branching starts at a comparatively low level and produces large, round crowns.

Figure 8.20 is a world map of the monsoon forest and closely related formation classes. Monsoon forest develops in the wet-dry tropical climate ③ in which a long rainy season alternates with a dry, rather cool season. These conditions, though most strongly developed in the Asiatic monsoon climate, are not limited to that area. The typical regions of monsoon forest are in

Myanmar, Thailand, and Cambodia. In the monsoon forest of southern Asia, the teakwood tree was once abundant and was widely exported to the Western world to make furniture, paneling, and decking. Now this great tree is logged out, and the Indian elephant, once trained for this purpose, is unemployed. Large areas of monsoon forest also occur in south-central Africa and in Central and South America, bordering the equatorial and tropical rainforests.

Subtropical Evergreen Forest

Subtropical evergreen forest is generally found in regions of moist subtropical climate ⑥, where winters are mild and

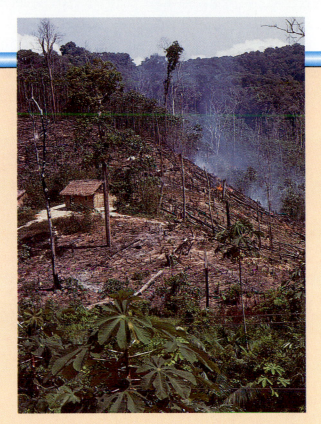

This rainforest in Maranhão, Brazil, has been felled and burned in preparation for cultivation.

hectares (about 7300 mi²) are converted every year. Africa's loss of rainforest was estimated at about 470,000 hectares per year (1800 mi²). Among individual countries, Brazil and Indonesia are the loss leaders, accounting for nearly half of the rainforest area converted to other uses. Note that these values do not include even larger losses of moist deciduous forests in these regions. Deforestation in low-latitude dry deciduous forests and hill and montane forests is also very serious.

Although deforestation rates are very rapid in some regions, many nations are now working to reduce the rate of loss of rainforest environment. However, because the rainforest can provide agricultural land, minerals, and timber, the pressure to allow deforestation continues.

In areas where a marked dry season occurs, that season will be lengthened. Although such models contain simplifications and are subject to error, the results confirm the pessimistic conclusion that once large-scale deforestation has occurred, artificial restoration of a rainforest comparable to the original one may be impossible to achieve.

According to a report issued by the United Nations Food and Agriculture Organization in 1993, about 0.6 percent of the world's rainforest is lost annually by conversion to other uses. More rainforest land, 2.2 million hectares (about 8500 mi²), is lost annually in Asia than in Latin America and the Caribbean, where 1.9 million

Questions

1. How do traditional agricultural practices in the low-latitude rainforest compare to present-day practices? What are the implications for the rainforest environment?

2. What are the effects of large-scale clearing on the rainforest environment?

there is ample rainfall throughout the year. This forest occurs in two forms: broadleaf and needleleaf. The *subtropical broadleaf evergreen forest* differs from the low-latitude rainforests, which are also broadleaf evergreen types, in having relatively few species of trees. Trees are not as tall as in the low-latitude rainforests. Their leaves tend to be smaller and more leathery, and the leaf canopy less dense. The subtropical broadleaf evergreen forest often has a well-developed lower layer of vegetation. Depending on the location, this layer may include tree ferns, small palms, bamboos, shrubs, and herbaceous plants. Lianas and epiphytes are abundant.

Figure 8.21 is a map of the subtropical evergreen forests of the northern hemisphere. Here subtropical evergreen forest consists of broad-leaved trees such as evergreen oaks and trees of the laurel and magnolia families. The name "laurel forest" is applied to these forests, which are associated with the moist subtropical climate ⑥ in the southeastern United States, southern China, and southern Japan. However, these regions are under intense crop cultivation because of their favorable climate. In these areas, the land has been cleared of natural vegetation for centuries, and little natural laurel forest remains.

Most of the subtropical evergreen broadleaf forest has been lost to cultivation.

Figure 8.20 World map of monsoon forest and related types—savanna woodland and thorn–tree-tall grass savanna. (Data source same as Figure 8.13. Based on Goode Base Map.)

The *subtropical needleleaf evergreen forest* occurs only in the southeastern United States (Figure 8.21). Here it is referred to as the southern pine forest, since it is dominated by species of pine. It is found on the wide belt of sandy soils that fringes the Atlantic and Gulf coasts. Because the soils are sandy, water drains away quickly, leaving the soils quite dry. In infrequent drought years, these forests may burn. Since pines are well adapted to droughts and fires, they form a stable vegetation cover for large areas of the region. Timber companies have taken advantage of this natural preference for pines, creating many plantations yielding valuable lumber and pulp (Figure 8.22).

Midlatitude Deciduous Forest

Midlatitude deciduous forest is the native forest type of eastern North America and Western Europe. It is dominated by tall, broadleaf trees that provide a continuous and dense canopy in summer but shed their leaves completely in the winter. Lower layers of small trees and shrubs are weakly developed. In the spring, a lush layer of lowermost herbs quickly develops but soon fades after the trees have reached full foliage and shaded the ground.

Figure 8.23 is a map of midlatitude deciduous forests, which are found almost entirely in the northern hemi-

Figure 8.21 Northern hemisphere map of subtropical evergreen forests, including the southern pine forest. (Data source same as Figure 8.13. Based on Goode Base Map.)

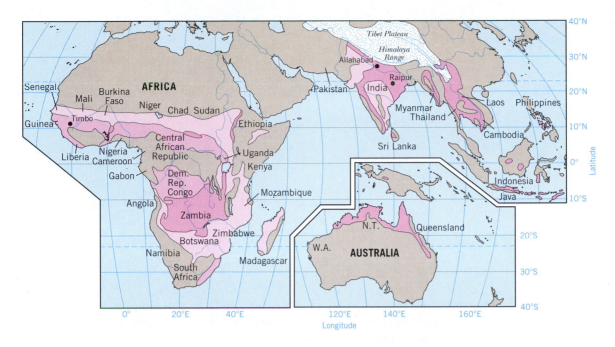

Figure 8.22 This plantation of longleaf pine grows on the sandy soil of the southeastern coastal plain. Near Waycross, Georgia.

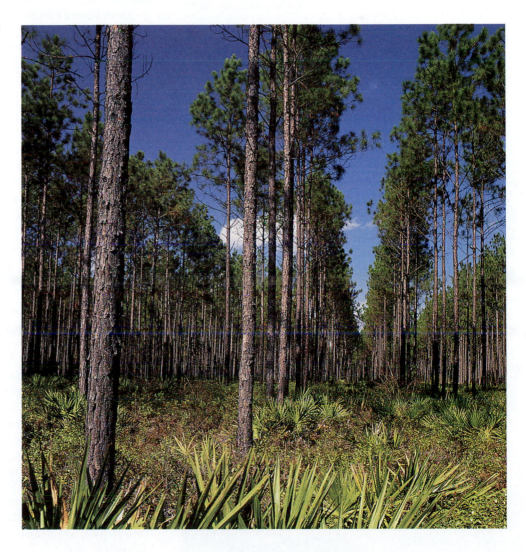

Figure 8.23 Northern hemisphere map of midlatitude deciduous forests. (Data source same as Figure 8.13. Based on Goode Base Map.)

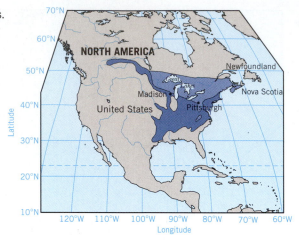

sphere. Throughout much of its range, this forest type is associated with the moist continental climate ⑩. Recall from Chapter 7 that this climate receives adequate precipitation in all months, normally with a summer maximum. There is a strong annual temperature cycle with a cold winter season and a warm summer.

Figure 8.24 This deciduous forest is located in the Catskill mountain region of New York. Needleleaf evergreen species of pine and hemlock are also found in the deciduous forest of this region.

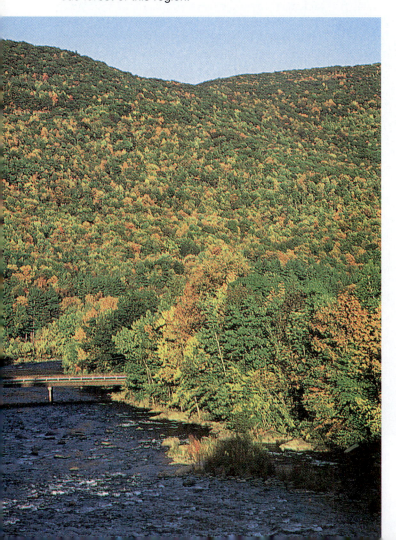

Deciduous forest often includes a few evergreen species.

Common trees of the deciduous forest of eastern North America, southeastern Europe, and eastern Asia are oak, beech, birch, hickory, walnut, maple, elm, and ash. A few needleleaf trees are often present as well—hemlock, for example (Figure 8.24). Where the deciduous forests have been cleared by lumbering, pines readily develop as second-growth forest.

In Western Europe, the midlatitude deciduous forest is associated with the marine west-coast climate ⑧. Here the dominant trees are mostly oak and ash, with beech in cooler and moister areas. In Asia, the midlatitude deciduous forest occurs as a belt between the boreal forest to the north and steppelands to the south. A small area of deciduous forest is found in Patagonia, near the southern tip of South America.

The deciduous forest includes a great variety of animal life, much of it stratified according to canopy layers. As many as five layers can be distinguished: the upper canopy, lower canopy, understory, shrub layer, and ground layer. Because the ground layer presents a more uniform environment in terms of humidity and temperature, it contains the largest concentration of organisms and the greatest diversity of species. Many small mammals burrow in the soil for shelter or food in the form of soil invertebrates. Among this burrowing group are ground squirrels, mice, and shrews, as well as some larger animals—foxes, woodchucks, and rabbits. Most of the larger mammals feed on ground and shrub layer vegetation, except for some, such as the brown bear, which are omnivorous and prey upon the small animals as well.

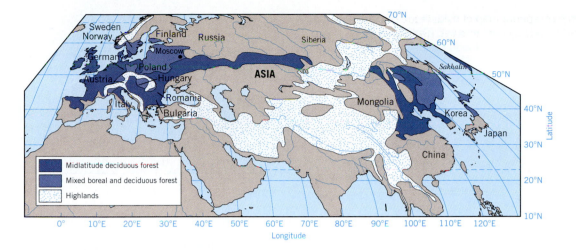

Even though birds possess the ability to move through the layers at will, most actually restrict themselves to one or more layers. For example, the wood peewee is found in the lower canopy, and the red-eyed vireo is found in the understory. Above them are the scarlet tanagers and blackburnian warblers, which are upper canopy dwellers. Below are the ground dwellers, such as grouse, warblers, and ovenbirds. Flying insects often show similar stratification patterns.

Among the large herbivores that graze in the deciduous forest are the red deer and roe deer of Eurasia and the white-tailed deer of North America. Smaller herbivores include voles, mice, and squirrels. Predators include bears, wildcats, lynx, wolves, foxes, weasels, and, among birds, owl species.

Needleleaf Forest

Needleleaf forest refers to a forest composed largely of straight-trunked, cone-shaped trees with relatively short branches and small, narrow, needlelike leaves. These trees are termed *conifers*, since their seeds are contained within cones borne on the branches. Most conifers are evergreen, retaining their needles for several years before shedding them. When the needleleaf forest is dense, it provides continuous and deep shade to the ground. Lower layers of vegetation are sparse or absent, except for a thick carpet of mosses that may occur. Species are few; in fact, large tracts of needleleaf forest consist almost entirely of only one or two species.

Boreal forest is the cold-climate needleleaf forest of high latitudes. It occurs in two great continental belts, one in North America and one in Eurasia (Figure 8.25). These belts span their landmasses from west to east in latitudes 45° N to 75° N, and they closely correspond to the region of boreal forest climate ⑪. The boreal forest of North America, Europe, and western Siberia is composed of such evergreen conifers as spruce and fir

(Figure 8.26). The boreal forest of north-central and eastern Siberia is dominated by larch. The larch tree sheds its needles in winter and is thus a deciduous needleleaf tree.

Broadleaf deciduous trees, such as aspen, balsam poplar, willow, and birch, tend to take over rapidly in areas of needleleaf forest that have been burned over. These species can also be found bordering streams and in open places. Between the boreal forest and the midlatitude deciduous forest lies a broad transition zone of mixed boreal and deciduous forest.

In its northernmost range, the boreal needleleaf forest grades into cold woodland. This form of vegetation is limited to the northern portions of the boreal forest climate ⑪ and the southern portions of the tundra climate ⑫. Trees are low in height and spaced well apart. A shrub layer may be well developed. The ground cover of lichens and mosses is distinctive. Cold woodland is often referred to as *taiga*. It is transitional into the treeless tundra at its northern fringe.

Mammals of the boreal needleleaf forest in North America include deer, moose, elk, black bear, marten, mink, wolf, wolverine, and fisher. Common birds include jays, ravens, chickadees, nuthatches, and a number of warblers. The caribou, lemming, and snowshoe rabbit inhabit both needleleaf forest and the adjacent tundra biome. The boreal forest often experiences large fluctuations in animal species populations, a result of the low diversity and highly variable environment.

Needleleaf forests span the high latitudes of North America and Eurasia and extend southward on high mountains and plateaus.

Needleleaf evergreen forest extends into lower latitudes wherever mountain ranges and high plateaus

Figure 8.25 Northern hemisphere map of cold-climate needleleaf forests, including coastal forest. (Data source same as Figure 8.13. Based on Goode Base Map.)

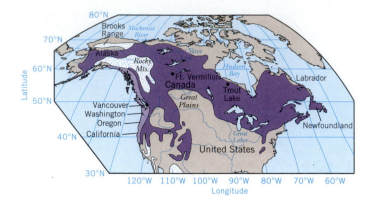

exist. For example, in western North America this formation class extends southward into the United States on the Sierra Nevada and Rocky Mountain ranges and over parts of the higher plateaus of the southwestern states (Figure 8.27, *left*). In Europe, needleleaf evergreen forests flourish on all the higher mountain ranges.

Coastal forest is a distinctive needleleaf evergreen forest of the Pacific Northwest coastal belt, ranging in latitude from northern California to southern Alaska (Figure 8.25). Here, in a band of heavy orographic precipitation, mild temperatures, and high humidity, are perhaps the densest of all conifer forests, with the world's largest trees (Figure 8.27, *right*). Individual redwood trees attain heights of over 100 m (about 325 ft) and girths of over 20 m (about 65 ft).

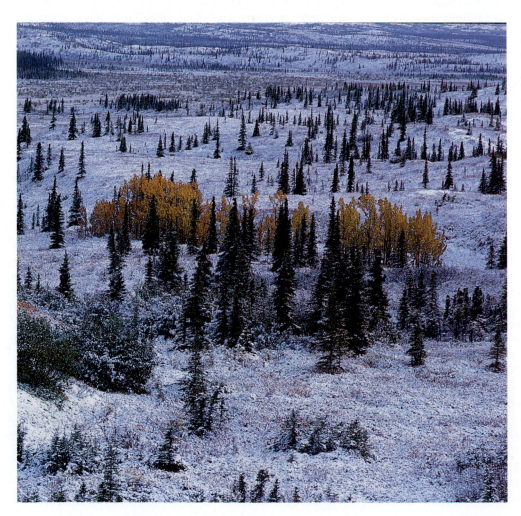

Figure 8.26 A view of the boreal forest, Denali National Park, Alaska, pictured here just after the first snowfall of the season. At this location near the northern limits of the boreal forest, the tree cover is sparse. The golden leaves of aspen mark the presence of this deciduous species.

Figure 8.27   Two kinds of needleleaf forest of the western United States. (*left*) Open forest of western yellow pine (ponderosa pine), in the Kaibab National Forest, Arizona. (*right*) A grove of great redwood trees in Humboldt State Park, California.

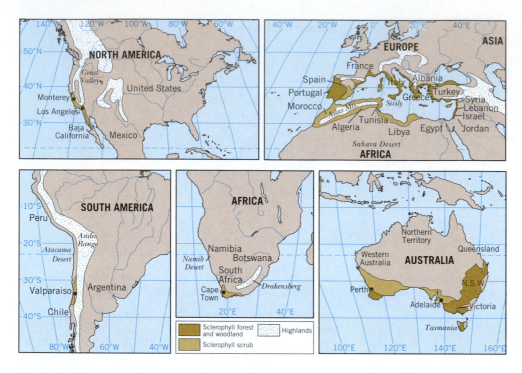

Figure 8.28 World map of sclerophyll vegetation. (Data source same as Figure 8.13. Based on Goode Base Map.)

Sclerophyll Forest

The native vegetation of the Mediterranean climate ⑦ is adapted to survival through the long summer drought. Shrubs and trees that can survive such drought are characteristically equipped with small, hard, or thick leaves that resist water loss through transpiration. As we noted earlier in the chapter, these plants are called sclerophylls.

Sclerophyll forest consists of trees with small, hard, leathery leaves. The trees are often low-branched and gnarled, with thick bark. The formation class includes *sclerophyll woodland*, an open forest in which only 25 to 60 percent of the ground is covered by trees. Also included are extensive areas of *scrub*, a plant formation type consisting of shrubs covering somewhat less than half of the ground area. The trees and shrubs are evergreen, retaining their thickened leaves despite a severe annual drought.

Figure 8.28, our map of sclerophyll vegetation, includes forest, woodland, and scrub types. Sclerophyll forest is closely associated with the Mediterranean

Figure 8.29 Evergreen oak woodland with grassland, Santa Ynez Valley, Santa Barbara County, California. At the end of the long, dry summer (*left*), the grasses are dormant, but the oak trees are green. At the end of the cool, wet winter (*right*), grasses are a lush green, whereas a few deciduous oaks in the foreground are nearly leafless.

climate ⑦ and is narrowly limited to west coasts between 30° and 40° or 45° N and S latitude. In the Mediterranean lands, the sclerophyll forest forms a narrow, coastal belt ringing the Mediterranean Sea. Here, the Mediterranean forest consists of such trees as cork oak, live oak, Aleppo pine, stone pine, and olive. Over the centuries, human activity has reduced the sclerophyll forest to woodland or destroyed it entirely. Today, large areas of this former forest consist of dense scrub.

The other northern hemisphere region of sclerophyll vegetation is the California coast ranges. Here, the sclerophyll forest or woodland is typically dominated by live oak and white oak. Grassland occupies the open ground between the scattered oaks (Figure 8.29). Much of the remaining vegetation is sclerophyll scrub or "dwarf forest," known as *chaparral* (Figure 8.30). It varies in composition with elevation and exposure. Chaparral may contain wild lilac, manzanita, mountain mahogany, poison "oak," and live oak.

In central Chile and in the Cape region of South Africa, sclerophyll vegetation has a similar appearance, but the dominant species are quite different. Important areas of sclerophyll forest, woodland, and scrub are also found in southeast, south-central, and southwest Australia, including many species of eucalyptus and acacia.

Figure 8.30 Chaparral, a vegetation type dominated by woody shrubs. Santa Lucia Mountains Wilderness, California.

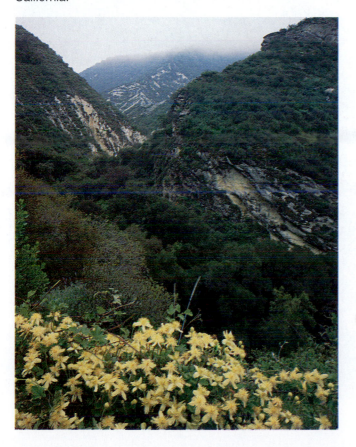

Savanna Biome

The savanna biome is usually associated with the tropical wet-dry climate ③ of Africa and South America. It includes vegetation formation classes ranging from woodland to grassland. In *savanna woodland*, the trees are spaced rather widely apart, because soil moisture during the dry season is not sufficient to support a full tree cover. The open spacing permits development of a dense lower layer, which usually consists of grasses. The woodland has an open, parklike appearance. Savanna woodland usually lies in a broad belt adjacent to equatorial rainforest.

Savanna vegetation is adapted to a strong wet-dry annual cycle. Grazing and periodic burning help maintain the openness of the savanna by suppressing tree seedlings.

In the tropical savanna woodland of Africa, the trees are of medium height. Tree crowns are flattened or umbrella-shaped, and the trunks have thick, rough bark (see Figure 7.13). Some species of trees are xerophytic forms with small leaves and thorns. Others are broadleaved deciduous species that shed their leaves in the dry season. In this respect, savanna woodland resembles monsoon forest.

Fire is a frequent occurrence in the savanna woodland during the dry season, but the tree species of the savanna are particularly resistant to fire. Periodic burning of the savanna grasses is thought to maintain the grassland against the invasion of forest. Fire does not kill the underground parts of grass plants, but it limits tree growth to individuals of fire-resistant species. Many rainforest tree species that might otherwise grow in the wet-dry climate regime are prevented by fires from invading. The browsing of animals, which kills many young trees, is also a factor in maintaining grassland at the expense of forest.

The regions of savanna woodland are shown in Figure 8.20, along with monsoon forest (discussed earlier). In Africa, the savanna woodland grades into a belt of *thorntree–tall-grass savanna*, a formation class transitional to the desert biome. The trees are largely of thorny species. Trees are more widely scattered, and the open grassland is more extensive than in the savanna woodland. One characteristic tree is the flat-topped acacia, seen in Figure 7.13. Elephant grass is a common species. It can grow to a height of 5 m (16 ft) to form an impenetrable thicket.

The thorntree–tall-grass savanna is closely identified with the semiarid subtype of the dry tropical and subtropical climates (④s, ⑤s). In the semiarid climate, soil water storage is adequate for the needs of plants only during the brief rainy season. The onset of the rains is quickly followed by the greening of the trees and grasses. For this reason, vegetation of the savanna biome is described as rain-green, an adjective that also applies to the monsoon forest.

The African savanna is widely known for the diversity of its large grazing mammals, which include more than a dozen antelopes. Careful studies have shown that each species has a particular preference for different parts of the grasses they consume—blade, sheath, and stem. Grazing stimulates the grasses to continue to grow, and so the ecosystem is more productive when grazed than when left alone. With these grazers comes a large variety of predators—lions, leopards, cheetahs, hyenas, and jackals. Elephants are the largest animals of the savanna and adjacent woodland regions.

Grassland Biome

The grassland biome includes two major formation classes that we will discuss here—tall-grass prairie and steppe. *Tall-grass prairie* consists largely of tall grasses. *Forbs*, which are broad-leaved herbs, are also present. Trees and shrubs are absent from the prairie but may occur in the same region as narrow patches of forest in stream valleys. The grasses are deeply rooted and form a thick and continuous turf (Figure 8.31).

Prairie grasslands are best developed in regions of the midlatitude and subtropical zones with well-developed winter and summer seasons. The grasses flower in spring and early summer, and the forbs flower in late summer. Tall-grass prairies are closely associated with the drier areas of moist continental climate ⑩. Here, soil water is in short supply during the summer months.

When European settlers first arrived in North America, the tall-grass prairies were found in a belt extending from the Texas Gulf coast northward to southern Saskatchewan. A broad peninsula of tall-grass prairie extended eastward into Illinois, where conditions are somewhat more moist. Since the time of settle-

Figure 8.31 Tall-grass prairie, Iowa. In addition to grasses, tall-grass prairie vegetation includes many forbs, such as the flowering species shown in this photo.

Figure 8.32 Bison grazing on short-grass prairie in the northern Great Plains.

ment, these prairies have been converted entirely to agricultural land. Another major area of tall-grass prairie is the Pampas region of South America, a zone of prairies (pampas) that occupies parts of Uruguay and eastern Argentina. The Pampas region falls into the moist subtropical climate ⑥, with mild winters and abundant precipitation.

Tall-grass prairie provides rich agricultural land suited to cultivation and cropping. Short-grass prairie, or steppe, occupies vast regions of semidesert and is suited to grazing.

Recall from Chapter 7 that **steppe**, also called *short-grass prairie*, is a vegetation type consisting of short grasses occurring in sparse clumps or bunches (Figure 8.32). Scattered shrubs and low trees may also be found in the steppe. The plant cover is poor, and much bare soil is exposed. Many species of grasses and forbs occur. A typical grass of the American steppe is buffalo grass. Other typical plants are the sunflower and loco weed.

Steppe grades into semidesert in dry environments and into prairie where rainfall is higher.

Our map of the grassland biome (Figure 8.33) shows that steppe grassland is concentrated largely in midlatitude areas of North America and Eurasia. The only southern hemisphere occurrence shown on the map is the "veldt" region of South Africa, a highland steppe in Orange Free State and Transvaal.

Steppe grasslands correspond well with the semiarid subtype of the dry continental climate ⑨. Spring rains nourish the grasses, which grow rapidly until early summer. By midsummer, the grasses are usually dormant. Occasional summer rainstorms cause periods of revived growth.

The animals of the grassland are distinctive. Before the exploitation of the North American grassland for cattle ranching, large grazing mammals were abundant. These included the pronghorn antelope, the elk, and the buffalo, which ranged widely from tall-grass prairie to steppe. Now nearly extinct, the buffalo (*Bison*) once numbered 60 million and roamed the grasslands from the Rockies to the Shenandoah Valley of Virginia. By

Figure 8.33 World map of the grassland biome in subtropical and midlatitude zones. (Data source same as Figure 8.13. Based on Goode Base Map.)

1889, however, this herd had been reduced to 800 individuals, or approximately the same number that are alive today. Most of the present buffalo population is confined to Yellowstone National Park. Today, rodents, and rabbits join cattle as the major grazers in the grasslands ecosystem.

The grassland ecosystem supports some rather unique adaptations to life. A common adaptive mechanism is jumping or leaping locomotion, assuring an unimpeded view of the surroundings. Jackrabbits and jumping mice are examples of jumping rodents. The pronghorn combines the leap with great speed, which

Figure 8.34 World map of the desert biome, including desert and semidesert formation classes. (Data source same as Figure 8.13. Based on Goode Base Map.)

allows it to avoid predators and fire. Burrowing is also another common life habit, for the soil provides the only shelter in the exposed grasslands. Examples are burrowing rodents, including prairie dogs, gophers, and field mice. Rabbits exploit old burrows, using them for nesting or shelter. Invertebrates also seek shelter in the soil, and many are adapted to living with the burrows of rodents, where extremes of moisture and temperature are substantially moderated.

Desert Biome

The desert biome includes several formation classes that are transitional from grassland and savanna biomes into vegetation of the arid desert. Here we recognize two basic formation classes: semidesert and dry desert. Figure 8.34 is a world map of the desert biome.

Semidesert is a transitional formation class found in a wide latitude range—from the tropical zone to the mid-

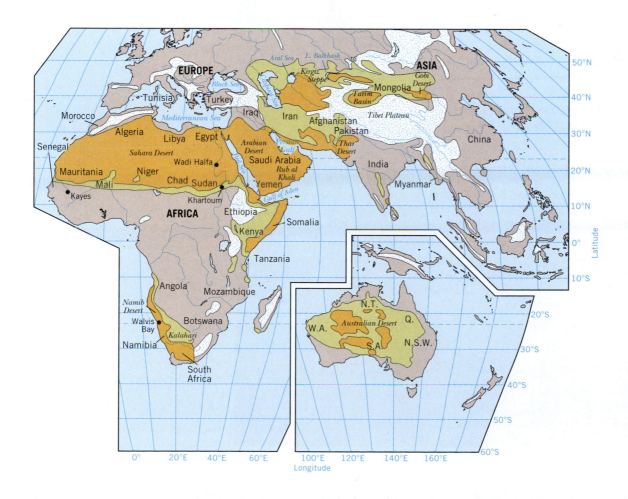

latitude zone. It is identified primarily with the arid subtypes of all three dry climates. Semidesert consists of sparse xerophytic shrubs. One example is the sagebrush vegetation of the middle and southern Rocky Mountain region and Colorado Plateau (Figure 8.35). As a result of overgrazing and trampling by livestock, semidesert shrub vegetation seems to have recently expanded into wide areas of the western United States that were formerly steppe grasslands.

Thorntree semidesert of the tropical zone consists of xerophytic trees and shrubs that are adapted to a climate with a very long, hot dry season and only a very brief, but intense, rainy season. These conditions are found in the semiarid and arid subtypes of the dry tropical ④ and dry subtropical ⑤ climates. The thorny trees and shrubs are known locally as thorn forest, thornbush, or thornwoods (see Figure 7.16). Many of these are deciduous plants that shed their leaves in the dry season. The shrubs may be closely intergrown to form dense thickets. Cactus plants are present in some localities.

Dry desert is a formation class of xerophytic plants that are widely dispersed over only a small proportion of the ground. The visible vegetation of dry desert consists of small, hard-leaved, or spiny shrubs, succulent plants (such as cactus), or hard grasses. Many species of small annual plants may be present but appear only after a rare, but heavy, desert downpour. Much of the world map area assigned to desert vegetation has no plant cover at all, because the surface consists of shifting dune sands or sterile salt flats.

Desert plants differ greatly in appearance from one part of the world to another. In the Mojave and Sonoran deserts of the southwestern United States, plants are often large, giving the appearance of a woodland (Figure 8.36). Examples are the treelike saguaro cactus, the prickly pear cactus, the ocotillo, creosote bush, and smoke tree.

Desert animals, like the plants, are typically adapted to the dry conditions of the desert. We have already mentioned some adaptions of xeric animals to the water scarcity of the desert. Important herbivores in American deserts include kangaroo rats, jackrabbits, and grasshopper mice. Insects are abundant, as are insect-eating bats and birds such as the cactus wren. Reptiles, especially lizards, are also common.

Figure 8.35 Sagebrush semidesert in Monument Valley, Utah.

Figure 8.36 A desert scene near Phoenix, Arizona. The tall, columnar plant is saguaro cactus; the delicate wandlike plant is ocotillo. Small clumps of prickly pear cactus are seen between groups of hard-leaved shrubs.

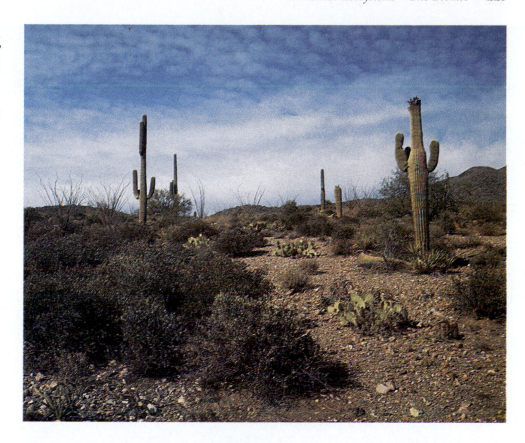

Tundra Biome

Arctic tundra is a formation class of the tundra climate ⑫. (See Figure 7.34 for a polar map of the arctic tundra.) In this climate, plants grow during the brief summer of long days and short (or absent) nights. At this time, air temperatures rise above freezing, and a shallow surface layer of ground ice thaws. The permafrost beneath, however, remains frozen, keeping the meltwater at the surface. These conditions create a marshy environment for at least a short time over wide areas. Because plant remains decay very slowly within the cold meltwater, layers of organic matter can build up in the marshy ground. Frost action in the soil fractures and breaks large roots, keeping tundra plants small. In winter, wind-driven snow and extreme cold also injure plant parts that project above the snow.

> **Tundra vegetation consists of low plants that are adapted to survival through a harshly cold winter. They grow, bloom, and set seed during a short summer thaw.**

Plants of the arctic tundra are mostly low herbs, although dwarf willow, a small woody plant, occurs in places. Sedges, grasses, mosses, and lichens dominate the tundra in a low layer (Figure 7.36). Typical plant species are ridge sedge, arctic meadow grass, cotton grasses, and snow lichen. There are also many species of forbs that flower brightly in the warmer season. Tundra composition varies greatly as soils range from wet to well-drained. One form of tundra consists of sturdy hummocks of plants with low, water-covered ground between. Some areas of arctic scrub vegetation composed of willows and birches are also found in tundra.

Tundra vegetation is also found at high elevations. This *alpine tundra* develops above the limit of tree growth and below the vegetation-free zone of bare rock and perpetual snow (Figure 8.37). Alpine tundra resembles arctic tundra in many physical respects.

As is most often true in particularly dynamic environments, species diversity in the tundra is low, but the abundance of individuals is high. Among the animals, vast herds of caribou in North America and reindeer (their Eurasian relatives) roam the tundra, lightly grazing the lichens and plants and moving constantly. A smaller number of musk-oxen are also primary consumers of the tundra vegetation. Wolves and wolverines, as well as arctic foxes and polar bears, are predators. Among the smaller mammals, snowshoe rabbits and lemmings are important herbivores. Invertebrates are scarce in the tundra, except for a small number of insect species. Black flies, deerflies, mosquitoes, and "no-see-ums" (tiny biting midges) are all abundant and can

Figure 8.37 Alpine tundra near the summit of the Snowy Range, Wyoming. Flag-shaped spruce trees (left side of photo), shaped by prevailing winds, mark the upper limit of tree growth.

make July on the tundra most uncomfortable. Reptiles and amphibians are also rare. The boggy tundra, however, presents an ideal summer environment for many migratory birds such as waterfowl, sandpipers, and plovers.

The food web of the tundra ecosystem is simple and direct. The important producer is reindeer moss, the lichen *Cladonia rangifera*. In addition to the caribou and reindeer, lemmings, ptarmigan (arctic grouse), and snowshoe rabbits are important lichen grazers. The important predators are the fox, wolf, and lynx, although all these animals may feed directly on plants as well. During the summer, the abundant insects help support the migratory waterfowl populations.

Altitude Zones of Vegetation

In earlier chapters, we described the effects of increasing elevation on climatic factors, particularly air temperature and precipitation. We noted that, with elevation, temperatures decrease and precipitation generally increases. These changes produce systematic changes in the vegetation cover as well, yielding a sequence of vegetation zones related to altitude.

The vegetation zones of the Colorado Plateau region in northern Arizona and adjacent states provide a striking example of this altitude zonation. Figure 8.38 is a diagram showing a cross section of the land surface in this region. Elevations range from about 700 m (about

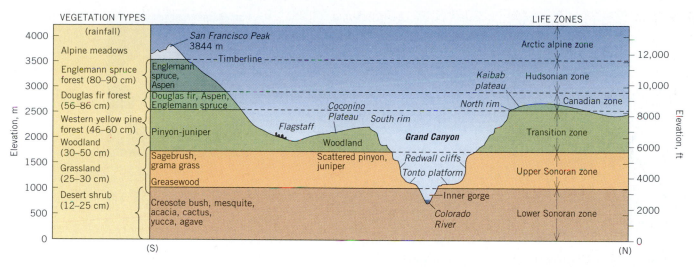

Figure 8.38 Altitude zone of vegetation in the arid southwestern United States. The profile shows the Grand Canyon–San Francisco Mountain district of northern Arizona. (Based on data of G. A. Pearson, C. H. Merriam, and A. N. Strahler.)

2300 ft) at the bottom of the Grand Canyon to 3844 m (12,608 ft) at the top of San Francisco Peak. The vegetation cover and rainfall range are shown on the left. The vegetation zonation includes desert shrub, grassland, woodland, pine forest, Douglas fir forest, Engelmann spruce forest, and alpine meadow. Annual rainfall ranges from 12 to 25 cm (about 5 to 10 in.) in the desert shrub vegetation type to 80 to 90 cm (about 30 to 35 in.) in the Engelmann spruce forest.

Ecologists have used the vegetation zonation in this region to set up a series of *life zones*, which are also shown in the figure. These range from the lower Sonoran life zone, which includes the desert shrubs typical of the Sonoran Desert, to the arctic-alpine life zone, which includes the alpine meadows at the top of San Francisco Peak. Other life zones take their names from typical regions of vegetation cover. For example, the Hudsonian zone, 2900 to 3500 m (about 9500 to 11,500 ft), bears a needleleaf forest that is similar in form to the needleleaf boreal forest of the subarctic zone near Hudson's Bay, Canada.

Climatic Gradients and Vegetation Types

In discussing the major formation classes of vegetation, we have emphasized the importance of climate. As climate changes with latitude or longitude, vegetation also changes. Figure 8.39 shows three transects across portions of continents that illustrate this principle. (For these transects, we will ignore the effects of mountains or highland regions on climate and vegetation.)

The upper transect stretches from the equator to the tropic of cancer in Africa. Across this region, climate ranges through all four low-latitude climates: wet equatorial ①, monsoon and trade-wind coastal ②, wet-dry tropical ③, and dry tropical ④. Vegetation grades from equatorial rainforest, savanna woodland, and savanna grassland to tropical scrub and tropical desert.

The middle transect is a composite from the tropic of cancer to the arctic circle in Africa and Eurasia. Climates include many of the mid- and high-latitude types: dry subtropical ⑤, Mediterranean ⑦, moist continental ⑩, boreal forest ⑪, and tundra ⑫. The vegetation cover grades from tropical desert through subtropical steppe to sclerophyll forest in the Mediterranean. Further north is the midlatitude deciduous forest in the region of moist continental climate ⑩, which grades into boreal needleleaf forest, subarctic woodland, and finally tundra.

The lower transect ranges across the United States, from Nevada to Ohio. On this transect, the climate begins as dry midlatitude ⑨. Precipitation gradually increases eastward, reaching moist continental ⑩ near the Mississippi River. The vegetation changes from midlatitude desert and steppe to short-grass prairie, tallgrass prairie, and midlatitude deciduous forest.

The changes on these transects are largely gradational rather than abrupt. Yet, the global maps of both biome types and climates show distinct boundaries from one region to the next. Which is correct? The true situation is gradational rather than abrupt. Maps must necessarily have boundaries to communicate information. But climate and vegetation know no specific boundaries. Instead, they are classified into specific types for convenience in studying their spatial patterns. When studying any map of natural features, keep in mind that boundaries are almost always approximate and gradational.

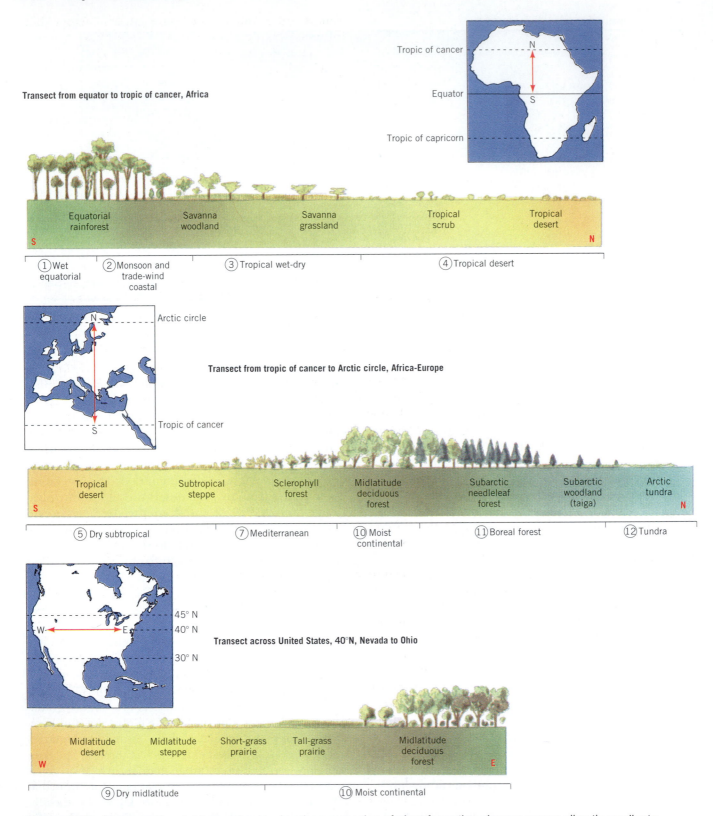

Figure 8.39 Three continental transects showing the succession of plant formation classes across climatic gradients.

A LOOK AHEAD

In this chapter, we have seen that the global pattern of ecosystems depends largely on climate. However, the biosphere is not the only component of the physical landscape that is influenced by climate. Soils also respond to climate, as we will see in our next chapter.

Temperature and moisture are critical factors that influence the development of soils. Also important is the nature of the plant cover and animal activities in which a soil develops. Thus, global soils will complete our suite of three chapters on climate and climate-related phenomena.

GLOBAL ECOSYSTEMS IN REVIEW

Ecology is the science of interactions among organisms and their environment. Its focus is the ecosystem, which by interaction among components provides pathways for flows of energy and cycles of matter. Biogeography is the study of how and why ecosystems vary from place to place. The biome is the largest unit of terrestrial ecosystems.

The food web of an ecosystem details how food energy flows from primary producers through consumers and on to decomposers. Because energy is lost at each level, only a relatively few top-level consumers are normally present. The biomass of an ecosystem measures its average dry weight per unit surface area. Net photosynthesis increases with increasing light and temperature, up to a point. Forests and estuaries are ecosystems with high rates of net primary production, while grasslands and agricultural lands are generally lower. Oceans are most productive in coastal and upwelling zones near continents. Among climate types, those with abundant rainfall and warm temperatures are most productive.

Plants and animals need water to survive. Plants lose water through transpiration. Xerophytes and xeric animals are plants and animals, respectively, that are adapted to dry environments. Temperature acts both directly and indirectly to limit the spread of plants and animals. Both plants and animals can possess adaptations to excessively cold or hot environments. Light, especially as photoperiod, and wind also are important climatic variables that influence ecosystems. In the process of ecological succession on cleared land or new ground, one ecosystem type succeeds another until the stable climax type is reached.

Natural vegetation is a plant cover that develops with little or no human interference. Although much vegetation appears to be in a natural state, humans influence the vegetation cover by fire suppression and introduction of new species. The life-form of a plant refers to its physical structure, size, and shape. Life-forms include trees, shrubs, lianas, herbs, and lichens.

The largest unit of terrestrial ecosystems is the biome: forest, grassland, savanna, desert, and tundra. The forest biome includes a number of important forest formation classes. The low-latitude rainforest exhibits a dense canopy with a very large number of species and an open floor. Most of its animals dwell and forage in the layers of the canopy. Subtropical evergreen forest occurs in broadleaf and needleleaf forms in the moist subtropical climate ⑥. Monsoon forest is largely deciduous, with most species shedding their leaves after the wet season. Midlatitude deciduous forest is associated with the moist continental climate ⑩. Its species shed their leaves before the cold season. Animals frequent both the canopy layer and the understory layer below. Needleleaf forest consists largely of evergreen conifers. It includes the coastal forest of the Pacific Northwest, the boreal forest of high latitudes, and needle-leaved mountain forests. Deer, moose, elk, and black bear are typical large mammals of the boreal forest. Sclerophyll forest is comprised of trees with small, hard, leathery leaves, and is found in the Mediterranean climate ⑦ region.

The savanna biome consists of widely spaced trees with an understory, often of grasses. Dry-season fire is frequent in the savanna biome, limiting the number of trees and encouraging the growth of grasses. African savanna is widely known for the diversity of its large grazing mammals. The grassland biome of midlatitude regions includes tall-grass prairie, in moister environments, and short-grass prairie, or steppe, in semiarid areas. North American grasslands were once dominated by the buffalo, pronghorn antelope, and elk.

Vegetation of the desert biome ranges from thorny shrubs and small trees to dry desert vegetation comprised of drought-adapted species. Small herbivores and reptiles are commonly desert dwellers that are well adapted to the dry environment. Tundra biome vegetation is limited largely to lichens and low herbs that are adapted to the severe drying cold experienced on the fringes of the Arctic Ocean. Tundra grazers include caribou in North America and reindeer in Eurasia. Since climate changes with altitude, vegetation typically occurs in altitudinal zones. Climate also changes gradually with latitude, and so biome changes are typically gradual, without abrupt boundaries.

KEY TERMS

ecology	consumers	forest biome
ecosystem	decomposers	savanna biome
biogeography	biomass	grassland biome
terrestrial ecosystems	xerophytes	desert biome
biome	natural vegetation	tundra biome
food chain	life-form	steppe
food web	forest	
primary producers	woodland	

REVIEW QUESTIONS

1. Define the terms *biosphere, ecology, ecosystem, biogeography,* and *biome.*
2. What is a food web or food chain? What are its essential components? How does energy flow through the food web of an ecosystem?
3. Contrast photosynthesis and respiration. How are these processes related to position in the food web?
4. How is net primary production related to biomass? What parts of the food chain have the highest biomass?
5. How is net primary production related to climate? Which climates are most productive? least productive?
6. The climates of the globe provide great variations in annual moisture and temperature cycles. Discuss the various adaptations that plants and animals have developed for life in stressful climates.
7. Define and illustrate succession using the example of abandoned fields in the southeasern United States.
8. What is natural vegetation? How do humans influence vegetation?
9. Plant geographers describe vegetation by its overall structure and by the life-forms of individual plants. Define and differentiate the following terms: forest, woodland, tree, shrub, herb, liana, perennial, deciduous, evergreen, broadleaf, needleleaf.

10. What are the five main biome types that ecologists and biogeographers recognize? Describe each briefly.
11. Low-latitude rainforests occupy a large region of the earth's land surface. What are the characteristics of these forests? Include forest structure, types of plants and animals, diversity, and climate in your answer.
12. Monsoon forest and midlatitude deciduous forest are both deciduous, but for different reasons. Compare the characteristics of these two formation classes and their climates.
13. Describe the layering of the deciduous forest. How do different types of animals exploit the various layers?
14. Subtropical broadleaf evergreen forest and tall-grass prairie are two vegetation formation classes that have been greatly altered by human activities. How was this done and why?
15. Distinguish among the types of needleleaf forest. What characteristics do they share? How are they different? How do their climates compare?
16. Which type of forest, with related woodland and scrub types, is associated with the Mediterranean climate ⑦? What are the features of these vegetation types? How are they adapted to the Mediterranean climate ⑦?
17. Describe the formation classes of the savanna biome and their characteristics. Where is this biome found and in what climate types? What role does fire play in the savanna biome?
18. Compare the two formation classes of the grassland biome. How do their climates differ? What are the characteristic animals of North American grasslands? How are they adapted to conditions of the grassland biome?
19. Describe the vegetation types of the desert biome.
20. What are the features of arctic and alpine tundra? How does the cold tundra climate ⑫ influence the vegetation cover? Identify some of the common mammals of the tundra.

VISUALIZING EXERCISES

1. Using a sketch of the vegetation sequence, describe succession on old fields in the southeastern United States, including how and when it occurs.
2. How does elevation influence vegetation? Sketch a hypothetical mountain peak in the southwestern U.S. desert that rises from a plain at about 500 m (about 1600 ft) elevation to a summit at about 4000 m (about 13,000 ft) and label the vegetation zones you might expect to find on its flanks.

A DEEPER VIEW OF GLOBAL VEGETATION

1. Figure 8.39 presents a vegetation transect from Nevada to Ohio. Expand the transect on the west so that it begins in Los Angeles. On the east, extend the transect northeast from Ohio through Pennsylvania, New York, western Massachusetts, and New Hampshire, to end in Maine. Sketch the biome formation types in your additions and label them, as in the diagram. Below your vegetation transect, draw a long bar subdivided to show the climate types.
2. Construct a similar transect of climate and vegetation from Miami to St. Louis, Minneapolis, and Winnipeg.
3. In what biome and formation class would you place your own area? Identify some of the common plants and animals of this type occurring in your region. How are the local vegetation and animal life influenced by human activity?

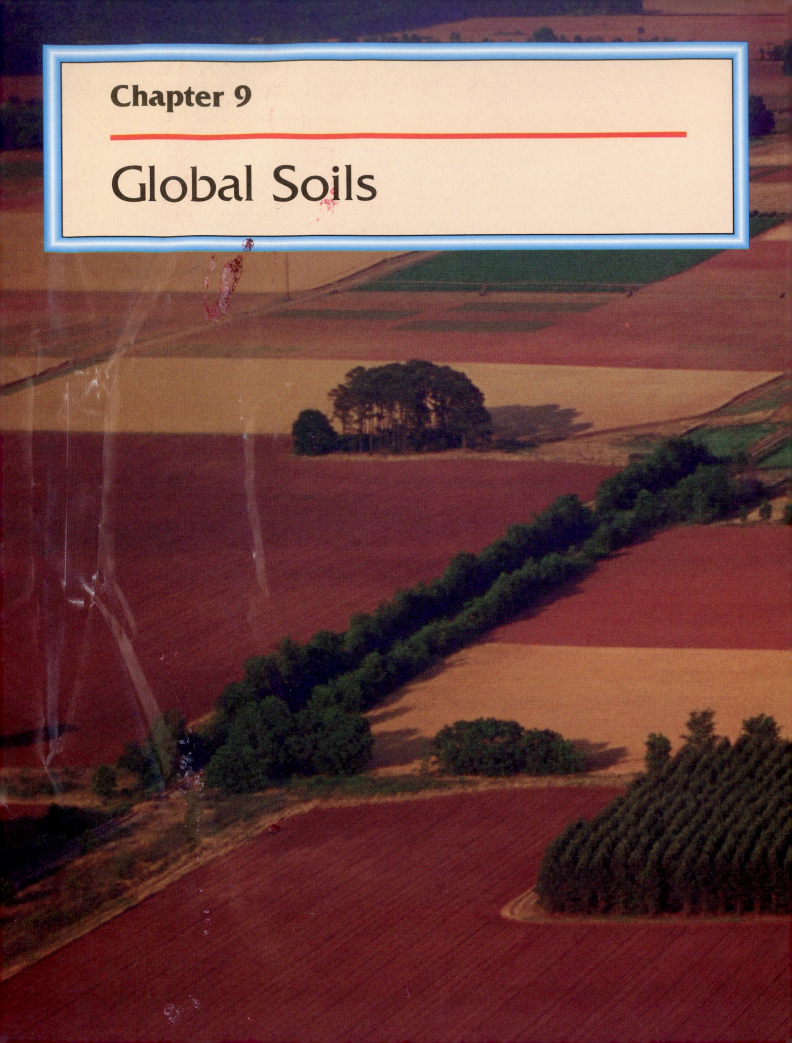

Chapter 9

Global Soils

Piedmont farmland in southern Georgia. The ancient soils of this region are highly productive with proper treatment.

This chapter is about soil—the loose surface layer of earth that supports the growth of vegetation. North American soils provide the base for the most efficient and productive agriculture in the world. But this has not always been the case.

At the close of the eighteenth century, the fertility of the agricultural lands of the Southeast was declining rapidly. Romantically, we picture George Washington, happily retired on his Virginia estate Mount Vernon, reaping a richly deserved bounty from expansive farmlands under his personal supervision. This vision is far from the truth. In fact, agricultural yields at Mount Vernon, and elsewhere in the Southeast, were dropping. In 1834, 35 years after Washington's death, a visitor to Mount Vernon declared that "a more widespread and perfect agricultural ruin could not be imagined."

The mystery of failing agricultural fertility was solved by Edmund Ruffin (1794–1865), who owned lands on the coastal plain of Virginia that were rapidly declining in the early 1800s. Quite by chance, Ruffin obtained a copy of Sir Humphry Davy's *Agricultural Chemistry*, published in 1813. Despite Ruffin's lack of formal education in science, he was quick to grasp the significance of one statement: "any acid matter… may be ameliorated by application of quicklime."

So it came about that on a February morning in 1818, Ruffin directed his field-hands to haul marl from pits in low areas of his lands. Marl is a soft, lime-rich mud that occurs widely on the eastern coastal plain. The workers spread 200 bushels of marl over several acres of newly cleared ridge land of poor quality. In the spring, Ruffin planted this area in corn to test the effect of the marl. In the words of historian Avery Craven, this is what happened: "Eagerly he waited. As the season advanced, he found reason for joy. From the very start the plants on marled ground showed marked superiority, and at harvest time they yielded an advantage of fully forty per cent. The carts went back to the pits. Fields took on fresh life. A new era in agricultural history of the region had dawned."*

Why did the application of lime enhance soil fertility so dramatically? As we will see in this chapter, nutrients are held in soils on the surfaces of fine particles. In an acid environment, these surfaces do not hold nutrients well, and so downward-moving water from precipitation washes nutrients out of the reach of plant roots. Lime reduces acidity and restores the ability of soil particles to hold nutrients, so that they remain available to plants.

Why didn't American farmers understand the importance of liming their fields? The colonists who settled this region were largely from England, where soils are formed on freshly ground mineral matter left by the great ice sheets. The breakdown products of this fresh mineral matter act like lime to reduce soil acidity, so that liming is not needed. In contrast, upland soils of the eastern seaboard from Virginia to Georgia have been continuously exposed to a mild, moist climate for tens of thousands of years. Fresh mineral matter in these soils has long been broken down, and the residual mineral matter that remains is highly acid. Thus, the colonists had no experience with these fundamentally different southeastern soils.

Soils are influenced by the rock material from which they develop, by the climate of the region (especially precipitation and temperature), by the vegetation cover that develops on them, and by the length of time that the soil has been in place. Your new knowledge of global climate and vegetation, acquired in the last two chapters, will be quite helpful in understanding soils and the processes that form them.

*Avery Craven, *Edmund Ruffin, Southerner* (New York; Appleton, 1932), p. 55.

Chapter Outline

The Nature of the Soil
  Soil Color and Texture
  Soil Colloids
  Soil Acidity and Alkalinity
  Soil Structure
  Minerals of the Soil
  Soil Moisture
Soil Development

  Soil Horizons
  Soil-Forming Processes
  Soil Temperature
The Global Scope of Soils
  Soil Orders
  Desert and Tundra Soils
Eye on the Environment • Death of a Civilization

GLOBAL SOILS

Soil is the uppermost layer of the land surface that plants use and depend on for nutrients, water, and physical support. Soils can vary greatly from continent to continent, region to region, and even from field to field. This is because they are influenced by factors and processes that can vary widely from place to place. For example, a field near a large river that floods regularly may acquire a layer of nutrient-rich silt, making its soil very productive. A nearby field at a higher elevation, without the benefit of silt enrichment, may be sandy or stony and require the addition of fertilizers to grow crops productively.

Vegetation is an important factor in determining soil qualities. For example, some of America's richest soils developed in the Middle West under a cover of thick grass sod. The deep roots of the grass, in a cycle of growth and decay, deposited nutrients and organic matter throughout the thick soil layer. In the Northeast, conifer forests provided a surface layer of decaying needles that kept the soil quite acid. This acidity allowed nutrients to be washed below root depth, out of the reach of plants. When farmed today, these soils need applications of lime to reduce their acidity and enhance their fertility.

Climate, measured by precipitation and temperature, is also an important determiner of soil properties. Precipitation controls the downward movement of nutrients and other chemical compounds in soils. If precipitation is abundant, water tends to wash soluble compounds, including nutrients, deeper into the soil and out of reach of plant roots.

Temperature acts to control the rate of decay of organic matter that falls to the soil from the plant cover or that is provided to the soil by the death of roots. When conditions are warm and moist, decay organisms work efficiently, consuming organic matter readily. Thus, organic matter and nutrients in soils of the tropical and equatorial zones are generally low. Where conditions are cooler, decay proceeds more slowly, and organic matter is more abundant in the soil. Of course, if the climate is very dry or desertlike, then vegetation growth is slow or absent. No matter what desert temperatures are like, organic matter will be low.

Time is also an important factor. The characteristics and properties of soils require time for development. For example, a fresh deposit of mineral matter, like the clean, sorted sand of a dune, may require hundreds to thousands of years to acquire the structure and properties of a sandy soil.

Geographers are keenly interested in the differences in soils from place to place over the globe. The ability of the soils and climate of a region to produce food largely determines the size of the population it will support. In spite of the growth of cities, most of the world's inhabitants still live close to the soil that furnishes their food. And many of those same inhabitants die prematurely when the soil does not furnish enough food for all.

THE NATURE OF THE SOIL

Soil, as the term is used in soil science, is a natural surface layer that contains living matter and can support plants. The soil consists of matter in all three states—solid, liquid, and gas. It includes both *mineral matter* and *organic matter*. Mineral matter is largely derived from rock material, whereas organic matter is of biological origin and may be living or dead. Living matter in the soil consists not only of plant roots, but also of many kinds of organisms, including microorganisms.

Soil scientists use the term *humus* to describe finely divided, partially decomposed organic matter in soils. Some humus rests on the soil surface, and some is mixed through the soil. Humus particles of the finest size are gradually carried downward to lower soil layers by rainfall that sinks in and moves through the soil. When abundant, humus particles can give the soil a brown or black coloration.

Both air and water are found in soil. Water may tend to contain high levels of dissolved substances, such as nutrients. Air in soils may have high levels of such gases as carbon dioxide or methane, and low levels of oxygen.

The solid, liquid, and gaseous matter of the soil are constantly changing and interacting through chemical and physical processes. This makes the soil a very dynamic layer. Because of these processes, soil science, often called *pedology*, is a highly complex body of knowledge.

Although we may think of soil as occurring everywhere, large expanses of continents possess a surface layer that cannot be called soil. For example, dunes of moving sand, bare rock surfaces of deserts and high mountains, and surfaces of fresh lava near active volcanoes do not have a soil layer.

The characteristics of soils are developed over a long period of time through a combination of many processes acting together. Physical processes act to break down rock fragments of regolith into smaller and smaller pieces. An example is the growth of fine roots, or crystals of ice or salts, in rock fissures to wedge rock fragments apart. Chemical processes alter the mineral composition of the original rock, producing new minerals. Taken together, these physical and chemical processes are referred to as *weathering*. Weathering occurs in soils and is part of the process by which soils develop their

Figure 9.1 In this cross section through the land surface, vegetation and forest litter lie atop the soil. Below is regolith, produced by the breakup of the underlying bedrock.

properties and characteristics. (We will return to the topic of weathering in Chapters 10 and 13.)

In most soils, the inorganic material of the soil consists of fine particles of mineral matter. The term **parent material** describes all forms of mineral matter that are suitable for transformation into soil. Parent material may be derived from the underlying *bedrock*, which is solid rock below the soil layer (Figure 9.1). Over time, weathering processes soften, disintegrate, and break bedrock apart, forming a layer of *regolith*, or residual mineral matter. Regolith is one of the common forms of parent material. Other kinds of regolith consist of mineral particles transported to a place of rest by the action of streams, glaciers, waves, water currents, or winds. For example, dunes formed of sand transported by wind are a type of regolith on which soil may be formed.

Soil Color and Texture

The most obvious feature of a soil or soil layer is probably its color. Some color relationships are quite simple. For example, the soils of the Midwest prairies have a black or dark brown color because they contain abundant particles of humus. Red or yellow colors often mark the soils of the Southeast. These colors are created by the presence of iron-containing oxides.

In some areas, soil color may be inherited from the mineral parent material, but, more generally, soil color is generated by soil-forming processes. For example, a white surface layer in soils of dry climates often indicates the presence of mineral salts, brought upward by evapo-

ration. A pale, ash-gray layer near the top of soils of the boreal forest climate results when organic matter and various colored minerals are washed downward, leaving only pure, light-colored mineral matter behind. As we explain soil-forming processes and describe the various classes of soils, soil color will take on more meaning.

The major factors influencing soil and soil development are parent material, climate, vegetation, and time.

The mineral matter of the soil consists of individual mineral particles that vary widely in size. The term **soil texture** refers to the proportion of particles that fall into each of three size grades—*sand, silt,* and *clay.* The diameter range of each of these grades is shown in Figure 9.2. Millimeters are the standard units. Each unit on the scale represents a power of ten, so that clay particles of 0.000,001 millimeter diameter are one-millionth the size of sand grains 1 mm in diameter. The finest of all soil particles are termed *colloids.* In measuring soil texture, gravel and larger particles are eliminated, since these play no important role in soil processes.

Soil texture is described by a series of names that emphasize the dominant particle size, whether sand, silt, or clay (including colloids). Figure 9.3 gives examples of five soil textures with typical percentage compositions. A *loam* is a mixture containing a substantial proportion of each of the three grades. Loams are classified as sandy, silty, or clay-rich when one of these grades is dominant.

Why is soil texture important? Texture largely determines the ability of the soil to retain water. Coarse-textured (sandy) soils have many small passages between touching mineral grains that quickly conduct water

Figure 9.2 Size grades, which are names like sand, silt, and clay, refer to mineral particles within a specific size range. They are defined using the metric system. English equivalents are also shown.

Figure 9.3 These diagrams show the proportions of sand, silt, and clay in five different soil texture classes.

Figure 9.4 Seen here enlarged about 20,000 times are tiny flakes of clay minerals of colloidal dimensions. These particles have settled from suspension in San Francisco Bay.

through to deeper layers. If the soil consists of fine particles, passages and spaces are much smaller. Thus, water will penetrate more slowly and also tend to be retained. We will return to the important topic of the water-holding ability of soils in a later section.

Soil Colloids

Soil colloids consist of particles smaller than one hundred-thousandth of a millimeter (0.000,01 mm). Like other soil particles, some colloids are mineral, while others are organic. Mineral colloids are usually very fine particles of clay minerals. If you examine mineral colloids under a microscope, you will find that they consist of thin, platelike bodies (Figure 9.4). When well mixed in water, particles this small remain suspended indefinitely, giving the water a murky appearance. Organic colloids are tiny bits of organic matter that are resistant to decay.

Soil colloids are important because their surfaces attract soil nutrients, which are in the form of ions dissolved in soil water. An *ion* is an electrically charged atom or group of atoms. Among the many ions found in soil water, one important group consists of **bases**, which are ions of four elements: calcium (Ca^{++}), magnesium (Mg^{++}), potassium (K^+), and sodium (Na^+). Because plants require these elements, they are among the *plant nutrients*—ions or chemical compounds that are needed for plant growth. Colloids hold these ions but also give them up to plants when in close contact with root membranes. Without this ion-holding ability of soil colloids, most of the vital nutrients would be carried out of the soil by percolating water and would be taken out of the region in streams, eventually reaching the sea. This leaching process goes on continually in moist climates, but loss is greatly retarded by the ion-holding capacity of soil colloids.

Soil Acidity and Alkalinity

The soil solution also contains hydrogen (H^+) and aluminum (Al^{+++}) ions. But unlike the bases, they are not considered to be plant nutrients. They tend to make the soil solution acid in chemical balance. An important principle of soil chemistry is that these acid ions have the power to replace the nutrient bases clinging to the surfaces of the soil colloids. As acid ions accumulate, the bases are released to the soil solution. They are gradually washed downward below rooting level, reducing soil fertility. When this happens, the soil acidity is increased.

The degree of acidity or alkalinity of a solution is designated by the pH value. The lower the pH value, the greater the degree of acidity. A pH value of 7 represents a neutral state—for example, pure water has a pH of 7. Lower values are in the acid range, while higher values are in the alkaline range.

Colloids are the finest particles in the soil. The surfaces of colloids attract base ions, which plants use as nutrients.

Table 9.1 shows the natural range of acidity and alkalinity found in soils. High soil acidity is typical of cold, humid climates. In arid climates, soils are typically alkaline. Acidity can be corrected by the application of lime, a compound of calcium, carbon, and oxygen ($CaCO_3$), which removes acid ions and replaces them with the base calcium.

Table 9.1 Soil Acidity and Alkalinity

pH	4.0	4.5	5.0	5.5	6.0	6.5	6.7	7.0	8.0	9.0	10.0	11.0
Acidity	Very strongly acid	Strongly acid	Moderately acid		Slightly acid		Neutral	Weakly alkaline	Alkaline	Strongly alkaline		Excessively alkaline
Lime Requirements	Lime needed except for crops requiring acid soil	Lime needed for all but acid-tolerant crops			Lime generally not required		No lime needed					
Occurrence	Rare	Frequent	Very common in cultivated soils of humid climates					Common in subhumid and arid climates		Limited areas in deserts		

Source: C. E. Millar, L. M. Turk, and H. D. Foth, *Fundamentals of Soil Science*, 3rd edition, (New York: John Wiley & Sons). See Chart 4.

Soil Structure

Soil structure refers to the way in which soil grains are grouped together into larger masses, called *peds*. Peds range in size from small grains to large blocks. They are bound together by soil colloids. Small peds, roughly shaped like spheres, give the soil a granular structure or crumb structure (see Figure 9.5). Larger peds provide an angular, blocky structure. Peds form when colloid-rich clays shrink in volume as they dry out. Shrinkage results in the formation of soil cracks, which define the surfaces of the peds.

Soils with a well-developed granular or blocky structure are easy to cultivate. This is an important agricultural factor in lands where primitive plows, drawn by animals, are still widely used. Soils with a high clay content can lack peds. These soils are sticky and heavy when wet and are difficult to cultivate. When dry, they become too hard to be worked.

Minerals of the Soil

Soil scientists recognize two classes of minerals abundant in soils: primary minerals and secondary minerals. **Primary minerals** are compounds present in unaltered rock. They are mostly silicate minerals—compounds of silicon and oxygen, with varying proportions of aluminum, calcium, sodium, iron, and magnesium. (The silicate minerals are described more fully in Chapter 10.) Primary minerals form a large fraction of the solid matter of many kinds of soils, but they play no important role in sustaining plant or animal life.

When primary minerals are exposed to air and water at or near the earth's surface, they are slowly altered in chemical composition. This process is part of *mineral alteration*, a chemical weathering process that is explained in more detail in Chapter 10. The primary minerals are altered into **secondary minerals**, which are essential to soil development and to soil fertility.

Figure 9.5 This soil shows a granular texture. The grains are referred to as peds.

In terms of the properties of soils, the most important secondary minerals are the *clay minerals*. They form the majority of fine mineral particles in soils. From the viewpoint of soil fertility, the ability of a clay mineral to hold base ions is its most important property. This ability varies with the particular type of clay mineral—some hold bases tightly, and others loosely.

The nature of the clay minerals in a soil determines its *base status*. If the clay minerals can hold abundant base ions, the soil is of *high base status* and generally will be highly fertile. If the clay minerals hold a smaller supply of bases, the soil is of *low base status* and is generally less fertile. Humus colloids have a high capacity to hold bases, so the presence of humus is usually associated with potentially high soil fertility.

Mineral oxides are secondary minerals of importance in soils. They occur in many kinds of soils, particularly those that remain in place in areas of warm, moist climates over very long periods of time (hundreds of thousands of years). Under these conditions, minerals are ultimately broken down chemically into simple oxides, compounds in which a single element is combined with oxygen.

Oxides of aluminum and iron are the most common oxides in soils. Aluminum oxide sometimes forms the mineral *bauxite*, which is a combination of aluminum oxide and water molecules bound together. It occurs as hard, rocklike lumps and layers below the soil surface. Where bauxite layers are thick and uniform, they are sometimes strip-mined as aluminum ore.

Iron oxide (Fe_2O_3), again held in combination with water molecules, is *limonite*, a yellowish to reddish mineral that supplies the typical reddish to chocolate-brown colors of soils and rocks. Some shallow accumulations of limonite were formerly mined as a source of iron. Limonite and bauxite occur in close association in soils of warm, moist climates in low latitudes.

Soil Moisture

Besides providing nutrients for plant growth, the soil layer serves as a reservoir for the moisture that plants require. Soil moisture is a key factor in determining how the soils of a region support vegetation and crops.

The soil receives water from rain and from melting snow. Where does this water go? First, some of the water can run off the soil surface and not sink in. Instead, it flows into brooks, streams, and rivers, eventually reaching the sea. What of the water that sinks into the soil? Some of this water is returned to the atmosphere as water vapor. This happens when soil water evaporates and when transpiration by plants lifts soil water from roots to leaves, where it evaporates. Taken together, we can term these last two losses *evapotranspiration*. Some water can also flow completely through the soil layer to recharge supplies of ground water at depths below the reach of plant roots.

When precipitation infiltrates the soil, the water wets the soil layer. This process is called *soil water recharge*. Eventually, the soil layer holds the maximum possible quantity of water, even though the larger pores may remain filled with air. Water movement then continues downward.

Suppose now that no further water enters the soil for a time. Excess soil water continues to drain downward, but some water will cling to the soil particles. This water resists the pull of gravity because of the force of *capillary tension*. To understand this force, think about a droplet of condensation that has formed on the cold surface of a glass of ice water. The water droplet seems to be enclosed in a "skin" of surface molecules, drawing the droplet together into a rounded shape. The "skin" is produced by capillary tension. This force keeps the drop clinging to the side of the glass indefinitely, defying the force of gravity. Similarly, tiny films of water adhere to soil grains, particularly at the points of grain contacts. They remain until they evaporate or are absorbed by plant rootlets.

When a soil has first been saturated by water and then allowed to drain under gravity until no more water moves downward, the soil is said to be holding its *storage capacity* of water. For most soils drainage takes no more than two or three days. Most excess water is drained out within one day.

The storage capacity of a soil measures the largest amount of water that the soil can hold.

Storage capacity is measured in units of depth, usually centimeters or inches, as with precipitation. It depends largely on the texture of the soil, as shown in Figure 9.6. Finer textures hold more water than coarser textures. This occurs because fine particles have a much larger surface area in a unit of volume than coarse particles. Thus, a sandy soil has a small storage capacity, while a clay soil has a large storage capacity.

The figure also shows the *wilting point*, which is the water storage level below which plants will wilt. The wilting point depends on soil texture. Because fine particles hold water more tightly, it is more difficult for plants to extract moisture from fine soils. Thus, plants wilt in fine-textured soils at higher water storage levels than in coarse-textured soils. The difference between the storage capacity of a soil and its wilting point is the *available water capacity*—that is, the maximum amount of water available to plants when the soil is at storage capacity. The available water capacity is greatest in loamy soils.

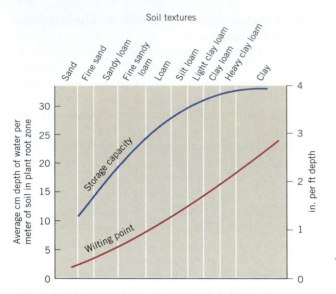

Figure 9.6 Storage capacity and wilting point vary according to soil texture. Finer textured soils hold more water. They also hold water more tightly, so that plants wilt more quickly.

SOIL DEVELOPMENT

How do soils develop their distinctive characteristics? Let's turn to the processes that act to form soils and soil layers. We begin with soil horizons.

Soil Horizons

Most soils possess **soil horizons**—distinctive horizontal layers that differ in physical composition, chemical composition, organic content, or structure (Figure 9.7). Soil horizons are developed by the interactions through time of climate, living organisms, and the configuration of the land surface. Horizons usually develop by either selective removal or accumulation of certain ions, colloids, and chemical compounds. The removal or accumulation is normally produced by water seeping down through the soil profile from the surface to deeper layers. Horizons are often distinguished by their color. The display of horizons on a cross section through the soil is termed a **soil profile**.

Let's review briefly the main types of horizons and their characteristics. For now, this discussion will apply to the types of horizons and processes that are found in moist forest climates. These are shown in Figure 9.8. Soil horizons are of two types: organic and mineral. Organic horizons, designated by the capital letter O, overlie the mineral horizons and are formed from accumulations of organic matter derived from plants and animals. Soil scientists recognize two possible layers. The upper O_i *horizon* contains decomposing organic

matter that is recognizable as leaves or twigs. The lower O_a *horizon* contains material that is broken down beyond recognition by eye. This material is humus, which we described earlier.

Mineral horizons lie below the organic horizons. Four main horizons are important—A, E, B, and C. Plant roots readily penetrate A, E, and B horizons and influence soil development within them. Soil scientists limit the term *soil* to refer to the A, E, and B horizons. The A horizon is the uppermost mineral horizon. It is rich in organic matter, consisting of numerous plant roots and downwashed humus from the organic horizons above. Next is the E horizon. Clay particles and oxides of aluminum and iron are removed from the E *horizon* by downward-seeping water, leaving behind pure grains of sand or coarse silt.

The B horizon receives the clay particles, aluminum and iron oxides, as well as organic matter washed down from the A and E horizons. It is made dense and tough by the filling of natural spaces with clays and oxides. Transitional horizons are present at some locations.

Beneath the B horizon is the C horizon, which is not considered part of the soil. It consists of the parent mineral matter of the soil and has been described earlier in this chapter as regolith (see Figure 9.1). Below the regolith lies bedrock or sediments of much older age than the soil.

Figure 9.7 A column of soil will normally show a series of horizons—horizontal layers with different properties.

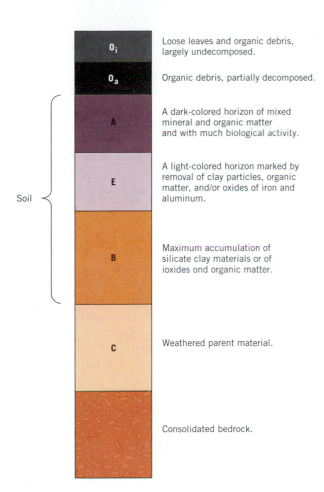

Loose leaves and organic debris, largely undecomposed. (O_i)

Organic debris, partially decomposed. (O_a)

A dark-colored horizon of mixed mineral and organic matter and with much biological activity. (A)

A light-colored horizon marked by removal of clay particles, organic matter, and/or oxides of iron and aluminum. (E)

Maximum accumulation of silicate clay materials or of ioxides ond organic matter. (B)

Weathered parent material. (C)

Consolidated bedrock.

Figure 9.8 A sequence of horizons that might appear in a forest soil developed under a cool, moist climate. (Soil Conservation Service, U.S. Department of Agriculture.)

Soil-Forming Processes

There are four classes of soil-forming processes. The first includes processes of *soil enrichment*, which add material to the soil body. For example, inorganic enrichment occurs when sediment is brought from higher to lower areas by overland flow. Stream flooding also deposits fine mineral particles on low-lying soil surfaces. Wind is another source of fine material that can accumulate on the soil surface. Organic enrichment occurs when humus accumulating in *O* horizons is carried downward to enrich the *A* horizon.

The second class includes processes that remove material from the soil body. This *removal* occurs when surface erosion carries sediment away from the uppermost layer of soil. Another important process of loss is *leaching*, in which seeping water dissolves soil materials and moves them to deep levels or to ground water.

The third class of soil-forming processes involves *translocation*, in which materials are moved within the soil body, usually from one horizon to another. Two processes of translocation that operate simultaneously

are eluviation and illuviation. **Eluviation** consists of the downward transport of fine particles, particularly the clays and colloids, from the uppermost part of the soil. Eluviation leaves behind grains of sand or coarse silt, forming the *E* horizon. **Illuviation** is the accumulation of materials that are brought down downward, normally from the *E* horizon to the *B* horizon. The materials that accumulate may be clay particles, humus, or iron and aluminum oxides.

Four classes of soil-forming processes include enrichment, removal, translocation, and transformation.

Figure 9.9 is a soil profile developed under a cool, humid forest climate. It shows the effects of both soil enrichment and translocation processes. The topmost layer of the soil (most visible in the upper left part of the photo) is a thin deposit of wind-blown silt and dune sand that has enriched the soil profile. The next layer is the *A* horizon, darkened by the presence of organic matter. Below it is the whitened *E* horizon, which shows the effect of eluviation, a translocation process. The colloids and oxides lost by the *E* horizon have been carried downward, and illuviation has added them to the *B* horizon, which displays the orange-red colors of iron oxide.

Figure 9.9 A forest soil profile on outer Cape Cod. The pale grayish *E* horizon overlies a reddish *B* horizon. A thin layer of wind-deposited silt and dune sand (pale brown layer) has been deposited on top.

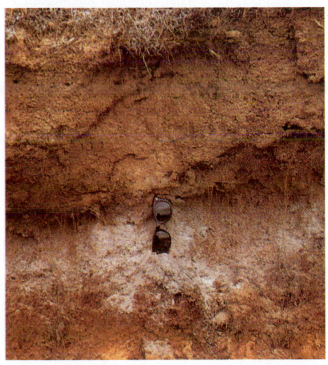

The translocation of calcium carbonate is another important process. In pure form, this secondary mineral is calcite ($CaCO_3$). In many areas, the parent material of the soil contains a substantial proportion of calcium carbonate derived from the disintegration of limestone, a common variety of bedrock. Carbonic acid, which forms when carbon dioxide gas dissolves in rainwater or soil water, readily reacts with calcium carbonate. The products of this reaction remain dissolved in solution as ions.

In moist climates, a large amount of surplus soil water moves downward to the ground water zone. This water movement leaches calcium carbonate from the entire soil in a process called *decalcification*. Soils that have lost most of their calcium are also usually acid in chemical balance and so are low in bases. Addition of lime or pulverized limestone not only corrects the acid condition, but also restores the calcium, which is used as a plant nutrient.

In dry climates, calcium carbonate is dissolved in upper layers of the soil during periods of rain or snowmelt when soil water recharge is taking place. The dissolved carbonate matter is carried down to the *B* horizon, where water penetration reaches its limits. Here, the carbonate matter is precipitated (deposited in crystalline form) in the *B* horizon, a process called *calcification*. Calcium carbonate deposition takes the form of white or pale-colored grains, nodules, or plates in the *B* or *C* horizons.

A last process of translocation occurs in desert climates. In some low areas, a layer of ground water lies close to the surface, producing a flat, poorly drained area. Evaporation of water at or near the soil surface draws up a continual flow of ground water by capillary tension, much like a cotton wick draws oil upward in an oil lamp. Moreover, the ground water is often rich in dissolved salts. When evaporation occurs, the salts precipitate and accumulate as a distinctive *salic horizon*. This process is called *salinization*. Most of the salts are compounds of sodium, of which ordinary table salt (sodium chloride, or halite, NaCl) is a familiar example. Sodium in large amounts is associated with highly alkaline conditions and is toxic to many kinds of plants. When salinization occurs in irrigated lands in a desert climate, the soil can be ruined for further agricultural use. (See *Eye on the Environment • Death of a Civilization*, in this chapter.)

The last class of soil-forming processes involves the *transformation* of material within the soil body. An example is the conversion of minerals from primary to secondary types, which we have already described. Another example is decomposition of organic matter to produce humus, a process termed *humification*. In warm, moist climates, humification can decompose organic matter completely to yield carbon dioxide and water, leaving virtually no organic matter in the soil.

Soil Temperature

Soil temperature is another important factor in determining the chemical development of soils and the formation of horizons. Temperature acts as a control over biologic activity and also influences the intensity of chemical processes affecting soil minerals. Below 10°C (50°F), biological activity is slowed, and at or below the freezing point (0°C, 32°F), biological activity stops. Chemical processes affecting minerals are inactive. The root growth of most plants and germination of their seeds require soil temperatures above 5°C (41°F). For plants of the warm, wet low-latitude climates, germination of seeds may require a soil temperature of at least 24°C (75°F).

The temperature of the uppermost soil layer and the soil surface strongly affects the rate at which organic matter is decomposed by microorganisms. Thus, in cold climates, where decomposition is slow, organic matter in the form of fallen leaves and stems tends to accumulate to form a thick *O* horizon. As we have already described, this material becomes humus, which is carried downward to enrich the *A* horizon.

In warm, moist climates of low latitudes, the rate of decomposition of plant material is rapid, so that nearly all the fallen leaves and stems are disposed of by bacterial activity. Under these conditions, the *O* horizon may be missing, and the entire soil profile will contain very little organic matter.

Soil and Surface Configuration

The configuration, or shape, of the ground surface is an important factor in soil formation. Configuration includes the steepness of the ground surface, or *slope*, as well as its compass orientation, or *aspect*. Generally speaking, soil horizons are thick on gentle slopes but thin on steep slopes. This is because the soil is more rapidly removed by erosion processes on the steeper slopes.

Aspect acts to influence soil temperatures and the soil water regime. Slopes facing away from the sun are sheltered from direct insolation and tend to have cooler, moister soils. Slopes facing toward the sun are exposed to direct solar rays, raising soil temperatures and increasing evapotranspiration.

Biological Processes in Soil Formation

The presence and activities of living plants and animals, as well as their nonliving organic products, have an important influence on soil. We have already noted the role that organic matter as humus plays in soil fertility. The colloidal structure of humus holds bases, which are needed for plant growth. In this way, humus helps keep nutrients cycling through plants and soils, and ensures that nutrients will not be lost by leaching to ground water. Humus also helps bind the soil into crumbs and clumps. This structure allows water and air to penetrate the soil freely, providing a healthy environment for plant roots.

Animals living in the soil include many species and come in many individual sizes—from bacteria to burrowing mammals. The total role of animals in soil formation is extremely important in soils that are warm and moist enough to support large animal populations. For example, earthworms continually rework the soil not only by burrowing, but also by passing soil through their intestinal tracts. They ingest large amounts of decaying leaf matter, carry it down from the surface, and incorporate it into the mineral soil horizons. Many forms of insect larvae perform a similar function. Tubelike openings are made by larger animals—moles, gophers, rabbits, badgers, prairie dogs, and other species.

Human activity also influences the physical and chemical nature of the soil. Large areas of agricultural soils have been cultivated for centuries. As a result, both the structure and composition of these agricultural soils have undergone great changes. These altered soils are often recognized as distinct soil classes that are just as important as natural soils.

THE GLOBAL SCOPE OF SOILS

An important aspect of soil science for physical geography is the classification of soils into major types and subtypes that are recognized in terms of their distribution over the earth's land surfaces. Geographers are particularly interested in the linkage of climate, parent material, time, biologic process, and landform with the distribution of types of soils. Geographers are also interested in the kinds of natural vegetation associated with each of the major soil classes. The geography of soils is thus essential in determining the quality of environments of the globe. It is important because soil fertility, along with availability of fresh water, is a basic measure of the ability of an environmental region to produce food for human consumption.

The soils of the world have been classified according to a system developed by scientists of the U.S. Soil Conservation Service, in cooperation with soil scientists of many other nations. Here we are concerned only with the two highest levels of this classification system. The top level contains 11 **soil orders** summarized in Table 9.2.

Table 9.2   Soil Order

Group I

Soils with well-developed horizons or with fully weathered minerals, resulting from long-continued adjustment to prevailing soil temperature and soil water conditions.

Oxisols	Very old, highly weathered soils of low latitudes, with a subsurface horizon of accumulation of mineral oxides and very low base status.
Ultisols	Soils of equatorial, tropical, and subtropical latitude zones, with a subsurface horizon of clay accumulation and low base status.
Vertisols	Soils of subtropical and tropical zones with high clay content and high base status. Vertisols develop deep, wide cracks when dry, and the soil blocks formed by cracking move with respect to each other.
Alfisols	Soils of humid and subhumid climates with a subsurface horizon of clay accumulation and high base status. Alfisols range from equatorial to subartic latitude zones.
Spodosols	Soils of cold, moist climates, with a well-developed *B* horizon of illuviation and low base status.
Mollisols	Soils of semiarid and subhumid midlatitude grasslands, with a dark, humus-rich epipedon and very high base status.
Aridisols	Soils of dry climates, low in organic matter, and often having subsurface horizons of accumulation of carbonate minerals or soluble salts.

Group II

Soils with a large proportion of organic matter.

Histosols	Soils with a thick upper layer very rich in organic matter.

Group III

Soils with poorly developed horizons or no horizons, and capable of further mineral alteration.

Entisols	Soils lacking horizons, usually because their parent material has accumulated only recently.
Inceptisols	Soils with weakly developed horizons, having minerals capable of further alteration by weathering processes.
Andisols	Soils with weakly developed horizons, having a high proportion of glassy volcanic parent material produced by erupting volcanoes.

The second level consists of *suborders*, of which we need to mention only a few.

Soil orders and suborders are often distinguished by the presence of a *diagnostic horizon*. Each diagnostic horizon has some unique combination of physical properties (color, structure, texture) or chemical properties (minerals present or absent). The two basic kinds of diagnostic horizons are (1) a horizon formed at the surface and called an *epipedon* (from Greek *epi*, meaning "over" or "upon") and (2) a subsurface horizon formed by the removal or accumulation of matter. In our descriptions of soil orders, we will refer to a number of diagnostic horizons.

The presence of diagnostic horizons distinguishes most soil orders and suborders.

We can recognize three groups of soil orders (Table 9.2). The largest group includes seven orders with well-developed horizons or fully weathered minerals. A second group includes a single soil order that is very rich in organic matter. The last group includes three soil orders with poorly developed horizons or no horizons. Although each order has several suborders, we will refer to only a few.

The world soils map presented in Figure 9.10 shows

Figure 9.10 Soils of the world.

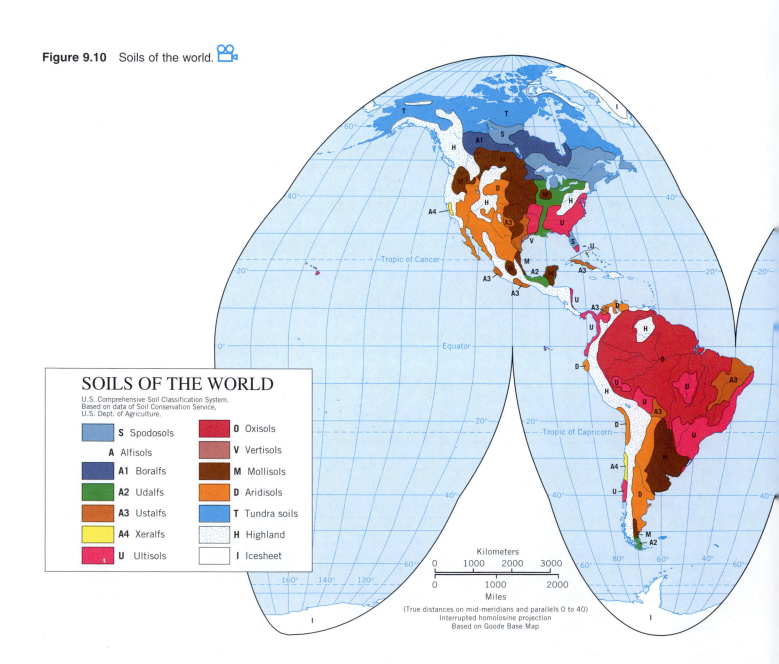

SOILS OF THE WORLD
U.S. Comprehensive Soil Classification System.
Based on data of Soil Conservation Service,
U.S. Dept. of Agriculture.

S Spodosols
A Alfisols
A1 Boralfs
A2 Udalfs
A3 Ustalfs
A4 Xeralfs
U Ultisols
O Oxisols
V Vertisols
M Mollisols
D Aridisols
T Tundra soils
H Highland
I Icesheet

Kilometers
0   1000   2000   3000
0   1000   2000
Miles
(True distances on mid-meridians and parallels 0 to 40)
Interrupted homolosine projection
Based on Goode Base Map

The Global Scope of Soils • 249

the major areas of occurrence of the soil orders. The Alfisols (Table 9.2) have been subdivided into four important suborders that correspond well to four basic climate zones. The map is quite general, indicating those areas where a given soil order is likely to be found. The map does not show many important areas of Entisols, Inceptisols, Histosols, and Andisols (Table 9.2). These orders are largely of local occurrence, since they are found on recent deposits such as floodplains, glacial landforms, sand dunes, marshlands, bogs, or volcanic ash deposits. The map also shows areas of highlands. In these regions, the soil patterns are too complex to show at a global scale.

Soil Orders

Three soil orders dominate the vast land areas of low latitudes: Oxisols, Ultisols, and Vertisols. Soils of these orders have developed over long time spans in an environment of warm soil temperatures and soil water that is abundant in a wet season or lasts throughout the year. We will discuss these orders first.

Oxisols

Oxisols have developed in equatorial, tropical, and subtropical zones on land surfaces that have been stable over long periods of time. During soil development, the

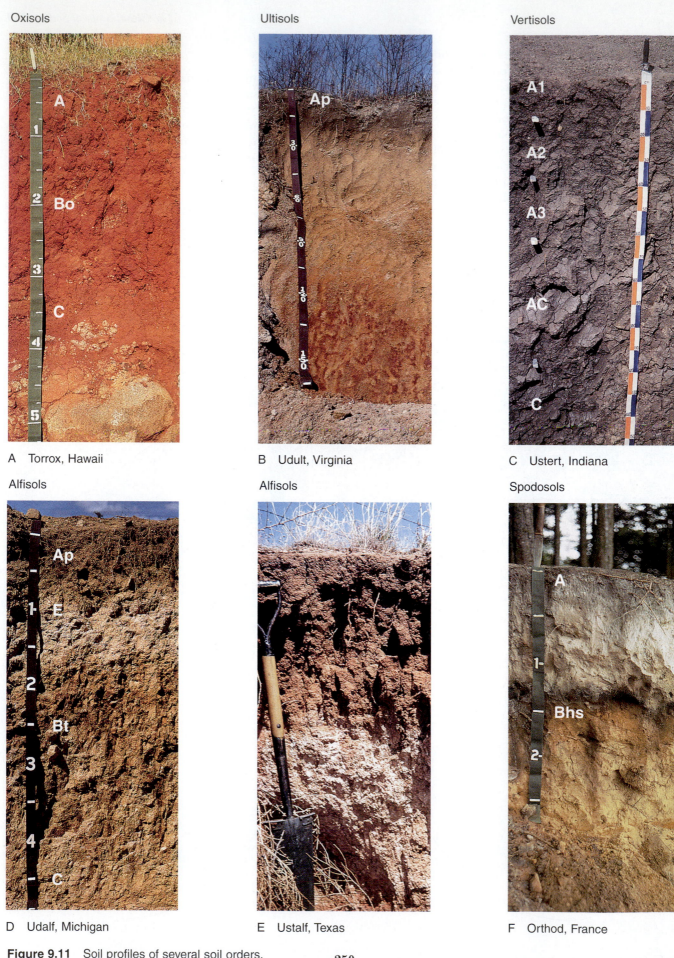

Oxisols

A   Torrox, Hawaii

Ultisols

B   Udult, Virginia

Vertisols

C   Ustert, Indiana

Alfisols

D   Udalf, Michigan

Alfisols

E   Ustalf, Texas

Spodosols

F   Orthod, France

Figure 9.11   Soil profiles of several soil orders.

Mollisols

G Boroll, USSR

Mollisols

H Udoll, Argentina

Mollisols

I Ustoll, Colorado

Mollisols

J Rendoll, Argentina

Aridisols

K Argid, Colorado

Histosols

L Fibrist, Minnesota

Figure 9.12   An Oxisol in Hawaii. Sugarcane is being cultivated here.

Figure 9.13   Ultisol profile in North Carolina. The thin, pale layer at the top is an *E* horizon, showing the effects of removal of materials by eluviation. Below the base of the thick, reddish *B* horizon is a blotchy-colored zone of plinthite.

climate has been moist, with a large water surplus. Oxisols have developed over vast areas of South America and Africa in the wet equatorial climate ①. Here the native vegetation is rainforest. The wet-dry tropical climate ③ with its large seasonal water surplus is also associated with Oxisols in South America and Africa.

Oxisols usually lack distinct horizons, except for darkened surface layers. Soil minerals are weathered to an extreme degree and are dominated by stable aluminum and iron oxides. Red, yellow, and yellowish-brown colors are normal (Figures 9.11*a* and 9.12). The base status of the Oxisols is very low, since nearly all the bases required by plants have been removed from the soil profile. A small store of nutrient bases occurs very close to the soil surface. The soil is quite easily broken apart and allows easy penetration by rainwater and plant roots.

Ultisols

Closely related to the Oxisols in outward appearance and environment of origin are the **Ultisols**. They are reddish to yellowish in color (Figures 9.11*b* and 9.13) and have a subsurface horizon of clay accumulation, called an *argillic horizon*, which is not found in the Oxisols. It is a *B* horizon and has developed through accumulation of clay in the process of illuviation. Although forest is the characteristic native vegetation, the base status of the Ultisols is low. As in the Oxisols, most of the bases are found in a shallow surface layer where they are released by the decay of plant matter. They are quickly taken up and recycled by the shallow roots of trees and shrubs.

In a few areas, the Ultisol profile contains a subsurface horizon of aluminum and iron oxides. This horizon is capable of hardening to a rocklike material if it becomes exposed at the surface and is subjected to repeated wetting and drying. This material is referred to as *plinthite* (from the Greek word *plinthos*, meaning "brick"). In the hardened state, plinthite is referred to as *laterite* (from the Latin *later*, or brick). In Southeast Asia, plinthite is quarried and cut into building blocks (Figure 9.14). These blocks harden into laterite blocks when they are exposed to the air.

Ultisols are widespread throughout Southeast Asia and the East Indies. Other important areas are in eastern Australia, Central America, South America, and the southeastern United States. Ultisols extend into the lower midlatitude zone in the United States, where they correspond quite closely in extent with the area of moist subtropical climate ⑥. In lower latitudes, Ultisols are identified with the wet-dry tropical climate ③ and the monsoon and trade-wind coastal climate ②. Note that all these climates have a dry season, even though it may be short.

Oxisols and Ultisols are soils of low-latitude regions. Ultisols have an argillic horizon, which Oxisols lack.

Both Oxisols and Ultisols of low latitudes were used for centuries under shifting agriculture prior to the advent of modern agricultural technology. This primitive agricultural method, known as slash-and-burn, is still widely practiced (Chapter 8). Without fertilizers, these soils can sustain crops on freshly cleared areas for only two or three years, at most, before the nutrient bases are exhausted and the garden plot must be abandoned. Substantial use of lime, fertilizers, and other industrial inputs is necessary for high, sustained crop yields. Another drawback to agriculture on these soils is that the exposed soil surface is vulnerable to devastating soil erosion, particularly on steep hill slopes.

Figure 9.14 Laterite, formed by hardening of plinthite, is being quarried for building stone in this scene from India.

Vertisols

A unique set of properties distinguishes the **Vertisols** from the Oxisols and Ultisols. Vertisols are black in color and have a high clay content (Figures 9.11*c* and 9.15). Much of the clay consists of a particular mineral that shrinks and swells greatly with seasonal changes in soil water content. Wide, deep vertical cracks develop in the soil during the dry season. As the dry soil blocks are wetted and softened by rain, some fragments of surface soil drop into the cracks before they close, so that the soil "swallows itself" and is constantly being mixed.

Vertisols typically form under grass and savanna vegetation (Chapter 8) in subtropical and tropical climates with a pronounced dry season. These climates include the semiarid subtype of the dry tropical steppe climate ④ and the wet-dry tropical climate ③. Because Vertisols require a particular clay mineral as a parent material, the major areas of occurrence are scattered and show no distinctive pattern on the world map. An important region of Vertisols is the Deccan Plateau of western

Figure 9.15 A Vertisol in Texas. The clay minerals that are abundant in Vertisols shrink when they dry out, producing deep cracks in the soil surface.

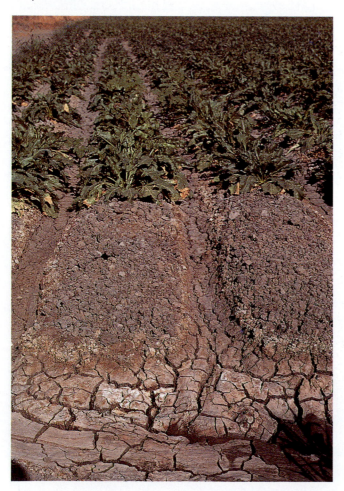

India, where basalt, a dark variety of igneous rock, supplies the silicate minerals that are altered into the necessary clay minerals.

Vertisols are high in base status and are particularly rich in such nutrient bases as calcium and magnesium. The soil solution is nearly neutral in pH, and a moderate content of organic matter is distributed through the soil. The soil retains large amounts of water because of its fine texture, but much of this water is held tightly by the clay particles and is not available to plants. Where soil cultivation depends on human or animal power, as it does in most of the developing nations where the soil occurs, agricultural yields are low. This is because the moist soil becomes highly plastic and is difficult to till with primitive tools. For this reason, many areas of Vertisols have been left in grass or shrub cover, providing grazing for cattle. Soil scientists think that the use of modern technology, including heavy farm machinery, could result in substantial production of food and fiber from Vertisols that are not now in production.

Alfisols

The **Alfisols** are soils characterized by an argillic horizon, produced by illuviation. Unlike the argillic horizon of the Ultisols, this *B* horizon is enriched by silicate clay minerals that have an adequate capacity to hold bases such as calcium and magnesium. The base status of the Alfisols is therefore generally quite high.

Above the *B* horizon of clay accumulation is a horizon of pale color, the *E* horizon, that has lost some of the original bases, clay minerals, and iron and aluminum oxides by the process of eluviation. These materials have become concentrated by illuviation in the *B* horizon. Alfisols also have a gray, brownish, or reddish surface horizon.

The world distribution of Alfisols is extremely wide in latitude (see Figure 9.10). Alfisols range from latitudes as high as 60° N in North America and Eurasia to the equatorial zone in South America and Africa. Obviously, the Alfisols span an enormous range in climate types. For this reason, we need to recognize four of the important suborders of Alfisols, each with its own climate affiliation.

Boralfs are Alfisols of cold (boreal) forest lands of North America and Eurasia. They have a gray surface horizon and a brownish subsoil. *Udalfs* are brownish Alfisols of the midlatitude zone. They are closely associated with the moist continental climate ⑩ in North America, Europe, and eastern Asia (Figure 9.11*d*).

Ustalfs are brownish to reddish Alfisols of the warmer climates (Figure 9.11*e*). They range from the subtropical zone to the equator and are associated with the wet-dry tropical climate ③ in Southeast Asia, Africa, Australia, and South America. *Xeralfs* are Alfisols of the

Mediterranean climate ⑦, with its cool moist winter and dry summer. The Xeralfs are typically brownish or reddish in color.

Spodosols

Poleward of the Alfisols in North America and Eurasia lies a great belt of soils of the order **Spodosols**, formed in the cold boreal forest climate ⑪ beneath a needleleaf forest. Spodosols have a unique property—a *B* horizon of accumulation of reddish mineral matter with a low capacity to hold base cations (Figure 9.11*f*). This horizon is called the spodic horizon (Figure 9.16). It is made up of a dense mixture of organic matter and compounds of aluminum and iron, all brought downward by eluviation from an overlying *E* horizon. Because of the intensive removal of matter from the *E* horizon, it has a bleached, pale gray to white appearance (see Figure 9.9). This conspicuous feature led to the naming of the soil as *podzol* (ash-soil) by Russian peasants. In modern terminology, this pale layer is an *albic horizon*. The *O* horizon, a thin, very dark layer of organic matter, overlies the *A* horizon.

Spodosols are strongly acid and are low in plant nutrients such as calcium and magnesium. They are also low in humus. Although the base status of the Spodosols is low, forests of pine and spruce are supported through the process of recycling of the bases.

Spodosols are closely associated with regions recently covered by the great ice sheets of the Ice Age (see Chapter 18). These soils are therefore very young. Typically, the parent material is coarse sand consisting largely of the mineral quartz (silicon dioxide, SiO_2). This mineral cannot weather to form clay minerals.

Spodosols are naturally poor soils in terms of agricultural productivity. Because they are acid, application of lime is essential. Heavy applications of fertilizers are also required. With proper management and the input of the required industrial products, Spodosols can be highly productive, if the soil texture is favorable. High yields of potatoes from Spodosols in Maine and New Brunswick are examples. Another factor unfavorable to agriculture is the shortness of the growing season in the more northerly parts of the Spodosol belt.

Mollisols

The **Mollisols** are soils of grasslands that occupy vast areas of semiarid and subhumid climates in midlatitudes. Mollisols are unique in having a very thick, dark brown to black surface horizon called the *mollic epipedon* (Figures 9.11*g* to *j* and 9.17). This layer lies within the *A* horizon and is always more than 25 cm (10 in.) thick. The soil has a loose, granular structure or a soft consistency when dry. Other important qualities of the Mollisols are the dominance of calcium among the bases of the *A* and *B* horizons and the very high base status of the soil.

Most areas of Mollisols are closely associated with the semiarid subtype of the dry midlatitude climate ⑨ and the adjacent portion of the moist continental climate ⑩. In North America, Mollisols dominate the Great Plains region, the Columbia Plateau, and the northern Great basin. In South America, a large area of Mollisols covers the Pampa region of Argentina and Uruguay. In Eurasia, a great belt of Mollisols stretches from Rumania eastward across the steppes of Russia, Siberia, and Mongolia. Russians refer to the Mollisols as *chernozems*, a

Figure 9.16 Diagram of a Spodosol profile.

Figure 9.17 Diagram of a Mollisol profile.

term that has gained widespread use by Western soil scientists as well.

Because of their loose texture and very high base status, Mollisols are among the most naturally fertile soils in the world. They now produce most of the world's commercial grain crop. Most of these soils have been used for crop production only in the last century. Prior to that time, they were used mainly for grazing by nomadic herds. The Mollisols have favorable properties for growing cereals in large-scale mechanized farming and are relatively easy to manage. Production of grain varies considerably from one year to the next because seasonal rainfall is highly variable.

A brief mention of four suborders of the Mollisols as they occur in the United States and Canada will help you to understand important regional soil differences related to climate. *Borolls*, the cold-climate suborder of the Mollisols, are found in a large area extending on both sides of the US-Canadian border east of the Rocky Mountains (Figure 9.11*g*). *Udolls* are Mollisols of a relatively moist climate, as compared with the other suborders (Figure 9.11*h*). Formerly, the Udolls supported tall-grass prairie, but today they are closely identified with the corn belt in the American Midwest.

Ustolls are Mollisols of the semiarid subtype of the dry midlatitude climate ⑨, with a substantial soil water shortage in the summer months. The Ustolls underlie much of the short-grass prairie region east of the Rockies (Figure 9.18). Xerolls are Mollisols of the Mediterranean climate ⑦, with its tendency to cool, moist winters and rainless summers.

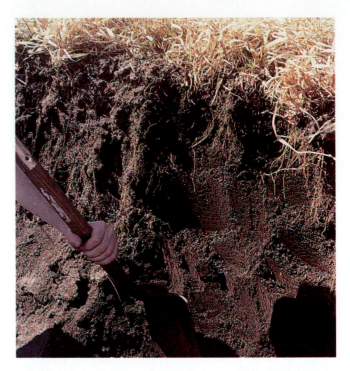

Figure 9.18 A Mollisol developed on wind-blown silt. Dry prairie grasses are seen on the surface.

Histosols

Throughout the northern regions of Spodosols are countless patches of **Histosols**. This unique soil order has a very high content of organic matter in a thick, dark upper layer (Figures 9.11*l* and 9.19). Most

Figure 9.19 Profile of a Histosol seen in the wall of a pit cut into a bog near Belle Glade, Florida. Water is being pumped from the floor of the pit.

Histosols go by such common names as peats or mucks. They have formed in shallow lakes and ponds by accumulation of partially decayed plant matter. In time, the water is replaced by a layer of organic matter, or *peat*, and becomes a bog (Figure 9.20). Peat bogs are used extensively for cultivation of cranberries (cranberry bogs). Sphagnum peat from bogs is dried and baled for sale as a mulch for use on suburban lawns and shrubbery beds. For centuries, Europe has used dried peat from bogs of glacial origin as a low-grade fuel.

Histosols are soils composed almost completely of organic matter. Peat is an example of a histosol.

Some Histosols are mucks—organic soils composed of fine black materials of sticky consistency. These are agriculturally valuable in midlatitudes, where they occur as beds of former lakes in glaciated regions. After appropriate drainage and application of lime and fertilizers, these mucks are remarkably productive for truck garden vegetables (Figure 9.21). Histosols are also found in low latitudes, where conditions of poor drainage have favored thick accumulations of plant matter.

Entisols

Two key characteristics that **Entisols** have in common are a mineral soil and the absence of distinct horizons. Entisols are soils in the sense that they support plants, but they may be found in any climate and under any vegetation. Entisols lack distinct horizons for two reasons. It may be the result of a parent material, such as quartz sand, in which horizons do not readily form. Or it may be the result of lack of time for horizons to form in recent deposits of alluvium or on actively eroding slopes.

Entisols occur from equatorial to arctic latitude zones. From the standpoint of agricultural productivity, Entisols of the subarctic zone and tropical deserts (along with arctic areas of Inceptisols) are the poorest of all soils. In contrast, Entisols and Inceptisols of floodplains and delta plains in warm and moist climates are among the most highly productive agricultural soils in

Figure 9.20 This peat bog in Galway, Ireland, has been trenched to reveal a Histosol profile. Peat blocks, seen to the sides of the trench, are drying for use as fuel.

EYE ON THE ENVIRONMENT

Death of a Civilization

What causes a civilization to collapse after flourishing for thousands of years? Wars and conquests come to mind, but sometimes the cause is less obvious. A case in point is the decline of the civilization of Sumer, and its successor, Babylon. Perhaps you have already learned something of these ancient Middle Eastern cultures, which were nourished by the Tigris and Euphrates rivers thousands of years ago. These two famous rivers drain the Zagros and Taurus Mountains of Turkey and Iran, and flow out through a broad valley across the deserts of Iraq to the Persian Gulf.

The Sumerian civilization evolved in the lower part of the Tigris–Euphrates Valley, beginning with a village culture dating to about five millennia B.C. By 3000 B.C., the Sumerian civilization was well established. The agriculture that supported Sumer depended on a system of irrigation canals utilizing the waters of the Tigris and Euphrates. Supported by this highly productive agricultural base, urban culture progressed to a high level, including the development of a written language. Skilled craftsmen produced pottery, jewelry, and ornate weapons of copper, silver, and gold (see figure at right).

Trouble set in for the Sumerians in about 2400 B.C. It came in the form of a deterioration of their croplands caused by an accumulation of salt in the soil. In this process, today called salinization, a small increment of salt is added to the soil each year by the evaporation of irrigation water containing a low concentration of salts. Over time, the salt content of the soil increases until it reaches levels high enough to affect crop plants. Wheat, which was a staple crop of Sumer, is quite sensitive to salt, and its yield declines sharply as the salt concentration rises in the soil. Barley, another staple crop of that era, is somewhat less sensitive to salt.

Archaeologists and historians have inferred the onset of salinization in Sumerian agricultural lands by a shift in the proportions of wheat and barley in the national agricultural output. It seems that around 3500 B.C. these two grains were grown in about equal amounts. By 2500 B.C., however, wheat accounted for only about one-sixth of the total grain production. Evidently, the more salt-tolerant barley had replaced wheat over much of the area.

At this point in time, the Sumerian grain yield also began to decline seriously. Ancient records show that in 2400 B.C., around the Sumerian city of Girsu, fields were yielding an average of about 300 kg of grain per hectare (270 lb/acre)—a high rate, even by modern standards. By 2100 B.C. the grain yield had declined to 180 kg per hectare (160 lb/acre), and by 1700 B.C., to a mere 100 kg/ha (90 lb/acre). Cities so declined in population that they finally dwindled into villages. As Sumerian civilization withered in the south, Babylonia, in the northern part of the valley, became more important. Soon, political control of the region passed from Sumer to Babylon.

Historians chronicle an event that also seems to have played a part in the decline of the Sumerian civilization. A group of Sumerian cities along the Euphrates River needed more irrigation water than they had. After a long and fruitless series of conflicts with upstream cities to gain additional water, the problem was solved by running a canal across the valley from the Tigris River. Now there was plenty of water. However, it was used so abundantly that a rise of the ground water table set in. As the water table came closer to the ground surface, capillary tension drew the ground water to the surface, where it evaporated rapidly in the hot sun. The result was an even faster accumulation of salt. When the salty water table came up still further, it saturated the roots of the barley plants, devastating the crops. Unknowingly, the Sumerians destroyed their own land in their quest for water.

The rising civilization of Babylon

---

the world because of their favorable texture, ample nutrient content, and large soil water storage.

Inceptisols

Where the soil is quite young, we often find **Inceptisols**—soils with horizons that are weakly developed. These areas occur within some of the regions shown on the world map as Ultisols and Oxisols. Especially important are the Inceptisols of river floodplains and delta plains in Southeast Asia that support dense populations of rice farmers.

In these regions, annual river floods cover low-lying plains and deposit layers of fine silt. This sediment is rich in primary minerals that yield bases as they weather chemically over time. The constant enrichment of the soil explains the high soil fertility in a region where uplands develop only Ultisols of low fertility. Inceptisols of these floodplain and delta lands are of a suborder called *Aquepts*—Inceptisols of wet places. Much closer to home is another prime example of Aquepts within the domain of the Ultisols—the lower Mississippi River floodplain and delta plain.

Andisols

Andisols are soils in which more than half of the parent mineral matter is volcanic ash, spewed high into the air

An ancient enamelled panel commorating the victory in war of a king in the first dynasty of Ur, Mesopotamia. Note the rich agricultural spoils depicted in the second and third rows—cattle, sheep, fishes, sacks of grain.

was doomed to fall in turn. Along the Tigris River, east of the modern city of Baghdad, an extremely elaborate system of irrigation canals was built. This system was begun in a pre-Babylonian period, 3000 to 2400 B.C. After a long history of abandonments and reconstructions, it was superseded between A.D. 200 and 500 by a final irrigation system based on a central canal, the Nahrwan Canal.

A problem soon developed, however. The irrigation system featured long, branching channels that proved to be traps for silt, which settled out of muddy river water. Frequent cleaning of silt from the channels was required, and silt was piled up in great embank-ments and mounds beside the channels. Gradually, silt from the mounds was washed by rains into nearby fields, adding layer upon layer to the land surface. According to one estimate, about 1 m (3 ft) of silt accumulated over the fields in a 500-year period. This rise in level of the farmland by silting made its irrigation all the more difficult. Until the strong central government authority collapsed, things went along well enough. Afterward, however, the entire system gradually broke down, and by the twelfth century it was abandoned completely. Not long after, Mongol hordes invaded the valley. By then, little remained of the once prosperous civilization that had dominated the region for over 4000 years.

Salinization, waterlogging, and silt accumulation have affected nearly every major irrigation system developed in a desert climate. Today, Pakistan struggles with salinization and waterlogging in a vast irrigation system in the Indus Valley. The lower Colorado River lands, including areas in Mexico, have suffered from salinization. Israel faces the threat of salinization of vital agricultural lands that have only recently been placed in irrigation. Some new methods may help. For example, soaking the ground through tubes, while keeping the soil covered under a plastic sheet, can reduce the buildup of salt.

Small wonder, then, that new proposals for desert irrigation systems arouse little enthusiasm. Perhaps a record of nearly total failure, spread over the entire span of civilization, is beginning to get its message across.

Questions

1. Describe the processes of salinization, waterlogging, and siltation and their impact on agriculture.

2. A technological "fix" for siltation is to build a dam upstream of an irrigation canal complex, so that silt will be trapped behind the dam. What other effects might building such a dam produce, both desirable and undesirable?

from the craters of active volcanoes and coming to rest in layers over the surrounding landscape. The fine ash particles are glasslike shards. A high proportion of carbon, formed by the decay of plant matter, is also typical, so that the soil usually appears very dark in color. Andisols form over a wide range of latitudes and climates. They are for the most part fertile soils, and in moist climates they support a dense natural vegetation cover.

Andisols do not appear on our world map because they are found in small patches associated with individual volcanoes that are located mostly in the "Ring of Fire"—the chain of volcanic mountains and islands that surrounds the great Pacific Ocean. Andisols are also found on the island of Hawaii, where volcanoes are presently active.

Desert and Tundra Soils

Desert and tundra soils are soils of extreme environments. Aridisols characterize the desert climate. As might be expected, they are low in organic matter and high in salts. Tundra soils are poorly developed because they are formed on very recent parent material, left in place by glacial activity during the Ice Age. Cold temperatures have also restricted soil development in tundra regions.

Figure 9.21   Garden crops cultivated on a Histosol of a former glacial lakebed, southern Ontario, Canada.

Aridisols

The **Aridisols** are soils of desert climates and are dry for long periods of time. Because the climate supports only a very sparse vegetation, humus is lacking and the soil color ranges from pale gray to pale red (Figure 9.11*k*). Soil horizons are weakly developed, but there may be important subsurface horizons of accumulated calcium carbonate (petrocalcic horizon) or soluble salts (salic horizon) (Figure 9.22). The salts, which are high in sodium, give the soil a very high degree of alkalinity. The Aridisols are closely correlated with the arid subtype of the dry tropical ④, dry subtropical ⑤, and dry midlatitude ⑨ climates.

Aridisols are desert soils. They are low in organic matter and may contain horizons of calcium carbonate or salt accumulation.

Most Aridisols are used for nomadic grazing, as they have been through the ages. This use is dictated by the limited rainfall, which is inadequate for crops without irrigation. Locally, where water supplies from mountain streams or ground water permit, Aridisols can be highly productive for a wide variety of crops under irrigation (Figure 9.23). Great irrigation systems, such as those of the Imperial Valley of the United States, the Nile Valley of Egypt, and the Indus Valley of Pakistan, have made Aridisols highly productive, but not without problems of salt buildup and waterlogging. *Eye on the Environment • Death of a Civilization* shows how the ancient cultures of Babylon and Sumer declined as their irrigated aridisols deteriorated from salinization and waterlogging.

Figure 9.22   A salic horizon, appearing as a white layer, lies close to the surface in this Aridisol profile in the Nevada desert. The scale is marked in feet (1 ft = 30 cm).

Figure 9.23 This gray desert soil, an Aridisol, has proved highly productive when cultivated and irrigated. The locality is near Palm Springs, California, in the Coachella Valley.

Tundra Soils

Soils of the arctic tundra fall largely into the order of Inceptisols, soils with weakly developed horizons that are usually associated with a moist climate. Inceptisols of the tundra climate belong to the suborder of Aquepts, which as we noted earlier are Inceptisols of wet places. More specifically, the tundra soils can be assigned to the *Cryaquepts*, a subdivision within the Aquepts. The prefix cry- is derived from the Greek word *kryos*, meaning "icy cold." We may refer to these soils simply as *tundra soils*.

Tundra soils are formed largely of primary minerals ranging in size from silt to clay that are broken down by frost action and glacial grinding. Layers of peat are often present between mineral layers (Figure 9.24). Beneath the tundra soil lies perennially frozen ground (permafrost), described in Chapter 7. Because the annual summer thaw affects only a shallow surface layer, soil water cannot easily drain away. Thus, the soil is saturated with water over large areas. Repeated freezing and thawing of this shallow surface layer disrupts plant roots, so that only small, shallow-rooted plants can maintain a hold.

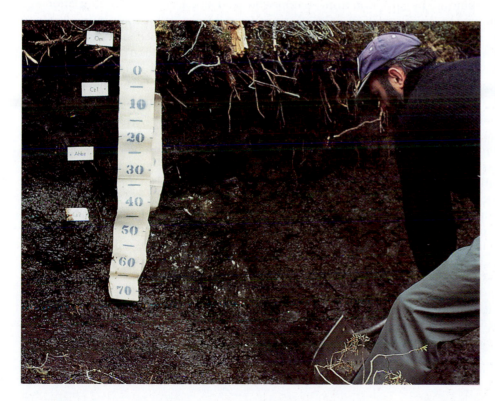

Figure 9.24 Profile of a tundra soil (Cryaquept) in southern Yukon Territory, Canada. Permafrost (perennially frozen soil water) appears at a depth of between 40 and 60 cm.

A LOOK AHEAD

This chapter completes a sequence of chapters describing three interrelated topics—climate, vegetation, and soils. The climate of a region is determined by the characteristics of air masses and weather systems that it experiences through the year. The nature of the vegetation cover depends largely on the climate, varying from desert to tundra biomes. Soils depend on both climate and vegetation to develop their unique characteristics. But soils can also influence vegetation development. Similarly, the climate at a location can be influenced by its vegetation cover. Thus, there is a strong interaction between climate, soils, and vegetation.

The next three chapters also comprise a set that are strongly related. They deal with the lithosphere—the realm of the solid earth, as we described in the Prologue. Our survey of the lithosphere will begin with the nature of earth materials, then move to a discussion of how continents and ocean basins are formed and how they are continually changing, even today. Finally, we will discuss landforms occurring within continents that are produced by such lithospheric processes as volcanic activity and earthquake faulting.

GLOBAL SOILS IN REVIEW

The soil layer is a complex mixture of solid, liquid, and gaseous components. It is derived from parent material, or regolith, that is produced from rock by weathering. The major factors influencing soil and soil development are parent material, climate, vegetation, and time. Soil texture refers to the proportions of sand, silt, and clay that are present. Colloids are the finest particles in soils and are important because they help retain nutrients, or bases, that are used by plants. Soils show a wide range of pH values, from acid to alkaline. Soils with granular or blocky structures are most easily cultivated.

In soils, primary minerals are chemically altered to secondary minerals, which include oxides and clay minerals. The nature of the clay minerals determines the soil's base status. If base status is high, the soil retains nutrients. If low, the soil can lack fertility. When a soil is fully wetted by heavy rainfall or snowmelt and allowed to drain, it reaches its storage capacity. Evaporation from the surface and transpiration from plants draws down soil water storage until precipitation occurs again.

Soils possess distinctive horizontal layers called horizons. These layers are developed by processes of enrichment, removal, translocation, and transformation. In downward translocation, materials such as humus, clay particles, and mineral oxides are removed by eluviation from an upper horizon and accumulate by illuviation in a lower one. In salinization, salts are translocated upward by evaporating water to form a salic horizon. In humification, a transformation process, organic matter is broken down by bacterial decay. Where soil temperatures are warm, this process can be highly effective, leaving a soil low in organic content. Animals, such as earthworms, can be very important in soil formation where they are abundant.

Global soils are classified into 11 soil orders, often by the presence of one or more diagnostic horizons. Oxisols are old, highly weathered soils of low latitudes. They have a horizon of mineral oxide accumulation and a low base status. Ultisols are also found in low latitudes. They have a horizon of clay accumulation and are also of low base status. Vertisols are rich in a type of clay mineral that expands and contracts with wetting and drying, and has a high base status. Alfisols have a horizon of clay accumulation, but they are of high base status. They are found in moist climates from equatorial to subarctic zones. Spodosols, found in cold, moist climates, exhibit a horizon of illuviation and low base status. Mollisols have a thick upper layer rich in humus. They are soils of midlatitude grasslands. Aridisols are soils of arid regions, marked by horizons of accumulation of carbonate minerals or salts. Histosols have a thick upper layer formed almost entirely of organic matter.

Three soil orders have poorly developed horizons or no horizons—Entisols, Inceptisols, and Andisols. Entisols are composed of fresh parent material and have no horizons. The horizons of Inceptisols are only weakly developed. Andisols are weakly developed soils occurring on young volcanic deposits.

KEY TERMS

soil	soil profile	Spodosols
parent material	eluviation	Mollisols
soil texture	illuviation	Histosols
soil colloids	soil orders	Entisols
bases	Oxisols	Inceptisols
primary minerals	Ultisols	Andisols
secondary minerals	Vertisols	Aridisols
soil horizons	Alfisols	

REVIEW QUESTIONS

1. Which important factors condition the nature and development of the soil?
2. Soil color, soil texture, and soil structure are used to describe soils and soil horizons. Identify each of these three terms, showing how they are applied.
3. Explain the concepts of acidity and alkalinity as they apply to soils.
4. Identify two important classes of secondary minerals in soils and provide examples of each class.
5. How does the ability of soils to hold water vary, and how does this ability relate to soil texture?
6. What is a soil horizon? How are soil horizons named? Provide two examples.
7. Identify four classes of soil-forming processes and describe each.
8. What are translocation processes? Identify and describe four translocation processes.
9. How many soil orders are there? Try to name them all.
10. Name three soil orders that are especially associated with low latitudes. For each order, provide at least one distinguishing characteristic and explain it.
11. Compare Alfisols and Spodosols. What features do they share? What features differentiate them? Where are they found?
12. Where are Mollisols found? How are the properties of Mollisols related to climate and vegetation cover? Name four suborders within the Mollisols.
13. Desert and tundra are extreme environments. Which soil order is characteristic of each environment? Briefly describe desert and tundra soils.

VISUALIZING EXERCISES

1. Sketch the profile of a spodosol, labeling *O*, *A*, *E*, *B*, and *C* horizons. Diagram the movement of materials from the zone of eluviation to the zone of illuviation.
2. Examine the world soils map (Figure 9.10) and identify three soil types that are found near your location. Develop a short list of characteristics that would help you tell them apart.

A DEEPER VIEW OF GLOBAL SOILS

1. Document the important role of clay particles and clay mineral colloids in soils. What is meant by the term *clay*? What are colloids? What are their properties? How does the type of clay mineral influence soil fertility? How does the amount of clay influence the water-holding capacity of the soil? What is the role of clay minerals in horizon development?
2. Using the world maps of global soils, global vegetation, and global climate, compare the pattern of soils on a transect along the 20° E longitude meridian with the patterns of vegetation and climate encountered along the same meridian. What conclusions can you draw about the relationship between soils, vegetation, and climate? Be specific.

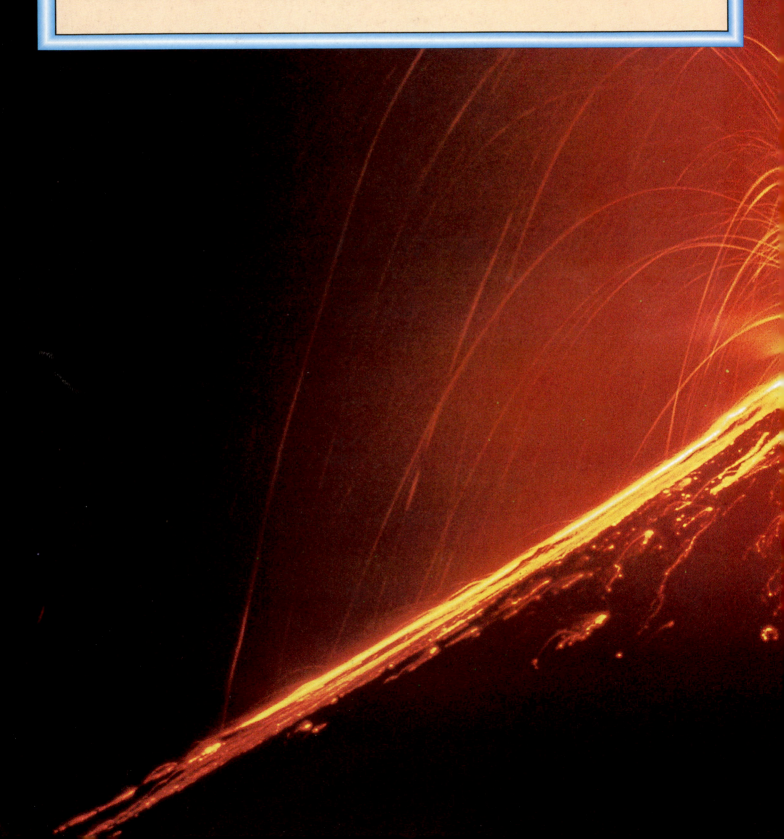

Chapter 10

Earth Materials

Arenal volcano, Costa Rica, erupting in July 1991.

Perhaps from collecting pebbles along a beach or stones from the bed of a stream, you may be aware that there are many different kinds of rocks. The primary objective of this chapter is to describe the more important types of rocks and minerals and how they are formed. Although most processes of rock formation are slow and steady, some are rapid and even a bit dangerous to witness. Imagine yourself on a journey to a pit crater on the Hawaiian volcano of Kilauea to watch the formation of new rock from molten lava.

The "Visitor Parking" sign emerges quickly from the fog, and you nose the car to the left, pulling into a vacant space in the parking lot. You expected better weather, but the whole eastern half of the "Big Island"—Hawaii—is covered by trade-wind clouds. At this elevation, about 1200 m (4000 ft), you are in the middle of the wet, gray, cloud layer. You open the trunk to get the sweatshirt from your backpack and then change into your sturdy hiking boots. Your destination is Mauna Ulu, a nearly circular pit crater that contains a pool of molten lava.

At first, the trail winds through a scrubby forest of low trees and shrubs. The tough rock underfoot is broken into angular blocks, separated by cracks and fissures. The footing is difficult. Each step requires attention to keep your boot from becoming wedged into a crevice.

Soon you emerge from the woodland onto a surface of fresh lava. The trail is marked by small flags every 20 feet or so. The fog thickens a bit. As you walk carefully from one flag to the next, you seem to be inside a dim, gray bubble on a strange planet. There are no landmarks—nothing familiar is within sight—as you pick your way across the rough terrain.

You gradually become aware of a noise, like a distant waterfall. As you continue along the trail, it gets louder and you begin to feel the ground shaking beneath your feet. The fog takes on a reddish glow. The acrid scent of sulfur catches in your nose and tightens your throat. Dimly, something emerges ahead. You are nearing the rim of the crater. You make out a wooden platform on the edge, with perhaps a half-dozen people clustered toward the back. As you reach the platform, the noise builds to a great, low-pitched roar. You mount the trembling platform and approach the edge. The heat from the lava strikes your face with an unexpected intensity.

The sides of the crater drop away steeply. At the bottom of the great pit is a black surface crust of congealed lava, covering the hot magma. Long cracks gash the crust, glowing red from deep within. Suddenly, a huge rock mass falls from the side of the crater into the pit, fracturing the black surface into jagged pieces that are tossed outward by the impact. Molten rock splashes slowly upward, spattering the rock wall with red and white spray. The heat of the fresh molten rock is unbearable, and you turn away.

After watching the incredible spectacle for as long as you can endure the heat, you leave the platform and head back. As you walk, the roaring fades, and the fog loses its red glow. The lava landscape seems mysterious rather than ominous and threatening. The scrubby woods seems somehow welcoming, its twisted trees beckoning as outposts of earthly life on your return journey to familiar ground. You feel small and insignificant, humbled by the immense physical power of the earth.

The cooling of molten lava that is released at the earth's surface is one way in which new rock is formed. This chapter explores other ways that are perhaps not so dramatic, but equally important. Our objective will be to lay the groundwork necessary for understanding how the earth's crustal processes control landforms and landscapes, as well as provide the raw materials for the land-forming agents of water, wind, and glacial ice that we discuss in later chapters.

Chapter Outline

The Crust and Its Composition
Rocks and Minerals
Igneous Rocks
Common Igneous Rocks
Intrusive and Extrusive Igneous Rocks
Eye on the Environment • Battling Iceland's Heimaey Volcano
Chemical Alteration of Igneous Rocks

Sediments and Sedimentary Rocks
Clastic Sedimentary Rocks
Chemically Precipitated Sedimentary Rocks
Hydrocarbon Compounds in Sedimentary Rocks
Metamorphic Rocks
The Geologic Time Scale
The Cycle of Rock Change

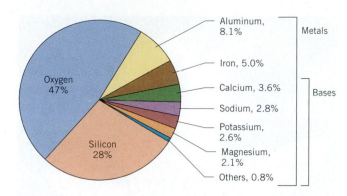

Figure 10.1 The eight most abundant elements in the earth's crust, measured by percentage of weight. Oxygen and silicon dominate, with aluminum and iron following.

EARTH MATERIALS

This chapter is the first in a sequence of three chapters that focus on the solid mineral realm of our planet. The first chapter (Chapter 10, "Earth Materials") deals with the materials that compose the earth's outer rock layer—minerals and rocks that are formed in various ways both deep within the earth and near or at its surface. In the second chapter (Chapter 11, "The Lithosphere and Plate Tectonics"), we will see that the outermost layer of the solid earth is a hardened crust of rock that is fractured, broken into large plates, and moved by forces deep within the earth. In the third chapter (Chapter 12, "Volcanic and Tectonic Landforms"), we will show how surface features reveal the history of the crust and the forces that have shaped it. The topics in these three chapters are drawn from *geology*, the science of the solid earth and its history.

Why are physical geographers concerned with the solid earth? The earth's outer layer serves as a platform, or base, for life on the lands. It provides the continental surfaces that are carved into landforms by moving water, wind, and glacial ice. Landforms, in turn, influence the distribution of ecosystems and exert strong controls over human occupation of the lands. Landforms made by water, wind, and ice are the subject of the closing chapters of this book.

We begin our study of the solid earth with an examination of earth materials—the minerals and rocks that are found at or near the earth's surface. We will be concerned only with a few of the great number of minerals and rocks, focusing on those that are most important for a broad understanding of the continents, ocean basins, and their varied features.

THE CRUST AND ITS COMPOSITION

The thin, outermost layer of our planet is the *earth's crust*. This mineral skin ranges from about 8 to 40 km (about 5 to 25 mi) thick and contains the continents and ocean basins. It is the source of soil on the lands, salts of the sea, gases of the atmosphere, and the water of the oceans, atmosphere, and lands.

Figure 10.1 displays the eight most abundant elements of the earth's crust in terms of percentage by weight. Oxygen, the predominant element, accounts for a little less than half the total weight. Second is silicon, which accounts for a little more than a quarter. Together they account for 75 percent of the crust, by weight.

Aluminum accounts for approximately 8 percent and iron for about 5 percent of the earth's crust. These metals are very important to our industrial civilization, and, fortunately, they are relatively abundant. Four metallic elements that we recognized as bases (Chapter 9)—calcium, sodium, potassium, and magnesium—make up the remaining 12 percent. All four occur at about the same order of abundance (2 to 4 percent). These elements are essential for plant and animal life and are described in Chapter 9 as plant nutrients.

Of the remaining dozens of chemical elements, a few are very important from the viewpoint of powering the cycles of the solid earth. These are the radioactive forms of certain elements that, through slow radioactive decay, provide a nearly infinite source of heat that seeps slowly outward from the earth's interior. We will return to the topic of radioactive decay and radiogenic heating in Chapter 11.

Rocks and Minerals

The elements of the earth's crust occur in chemical compounds that we recognize as minerals. A **mineral** is a naturally occurring, inorganic substance that usually possesses a definite chemical composition and characteristic atomic structure. Most minerals have a crystalline structure. Gemstones such as diamonds or rubies are examples, although their crystalline shape is enhanced by the stonecutter. Quartz is an example of a very common crystalline mineral. It usually occurs as a clear, six-sided prism (Figure 10.2).

Minerals are combined into **rock**, which we can broadly define as an assemblage of minerals in the solid state. Rock comes in a very wide range of compositions, physical characteristics, and ages. A given variety of rock is usually composed of two or more minerals, and often many different minerals are present. However, a few rock varieties consist almost entirely of one mineral. Most rock of the earth's crust is extremely old by human

Figure 10.2 These large crystals of quartz form six-sided, translucent columns.

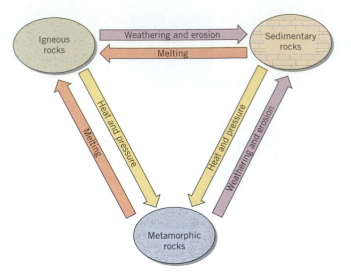

Figure 10.3 The cycle of rock change. In this simplified version of the cycle of rock change, the three classes of rock are transformed into one another by weathering and erosion, melting, and exposure to heat and pressure.

standards, with the age of formation often dating back many millions of years. But rock is also being formed at this very hour as active volcanoes emit lava that solidifies on contact with the atmosphere.

The three major categories of rocks are igneous, sedimentary, and metamorphic.

Rocks of the earth's crust fall into three major classes. (1) **Igneous rocks** are solidified from mineral matter in a high-temperature molten state. (2) **Sedimentary rocks** are layered accumulations of mineral particles derived mostly by weathering and erosion of preexisting rocks. (3) **Metamorphic rocks** are formed from igneous or sedimentary rocks that have been physically or chemically changed, usually by application of heat and pressure during crustal mountain-making.

The three classes of rocks are constantly forming from one another in a continuous circuit—the **cycle of rock change**—through which the crustal minerals have been recycled during many millions of years of geologic time (Figure 10.3). In the process of melting, preexisting rock of any class is melted and then later cools to form igneous rock. In weathering and erosion, preexist-ing rock is broken down and accumulated in layers that become sedimentary rock. Heat and pressure convert igneous and sedimentary rocks to metamorphic rock. At the close of this chapter, after we have taken a more detailed look at rocks, minerals, and their formation processes, we will return to a more complete version of the cycle of rock change.

IGNEOUS ROCKS

Igneous rocks are formed when molten material moves from deep within the earth to a position within or atop the crust. There the molten material cools, forming rocks composed of mineral crystals. Table 10.1 presents the more important igneous rock types that we will refer to in this chapter.

Most igneous rock consists of *silicate minerals*, chemical compounds that contain silicon and oxygen atoms. Most of the silicate minerals also have one, two, or more of the metallic elements listed in Figure 10.1—that is, aluminum, iron, calcium, sodium, potassium, and magnesium. Although there are many silicate minerals, we will only be concerned with seven of them, which are shown in Figure 10.4.

Among the most common minerals of all rock classes is **quartz**, which is silicon dioxide (SiO_2) (Figure 10.2). It is quite hard and resists chemical breakdown. Two silicate-aluminum minerals called *feldspars* follow. One type, *potash feldspar*, contains potassium as the dominant metal besides aluminum. A second type, *plagioclase feldspar*, is rich in sodium, calcium, or both. Quartz and

Table 10.1 Some Common Igneous Rock Types

Subclass	Rock Type	Composition
Intrusive	Granite	Felsic minerals, typically quartz, feldspars, and mica
(Cooling at depth, producing coarse crystal texture)	Diorite	Felsic minerals without quartz, usually including plagioclase feldspar and amphibole
	Gabbro	Mafic minerals, typically plagioclase feldspar, pyroxene and olivine
	Peridotite	An ultramafic rock of pyroxene and olivine
Extrusive	Rhyolite	Same as granite
(Cooling at the surface, producing fine crystal texture)	Andesite	Same as diorite
	Basalt	Same as gabbro

Figure 10.4 Silicate minerals and igneous rocks. Only the most important silicate mineral groups are listed, along with four common igneous rock types. The patterns shown for mineral grains indicate their general appearance through a microscope.

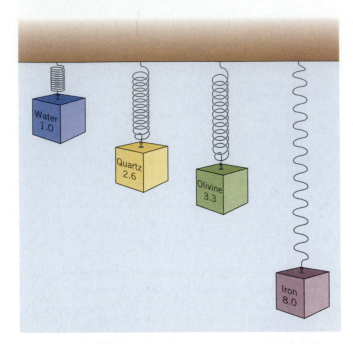

Figure 10.5 The concept of density is illustrated by several cubes of the same size, but of different materials, hung from coil springs under the influence of gravity. The stretching of the coil spring is proportional to the density of each material, shown in thousands of kilograms per cubic meter. Quartz and olivine are common minerals. (Copyright © A. N. Strahler.)

feldspar form a silicate mineral group described as **felsic** ("fel" for feldspar; "si" for silicate).

Quartz and the feldspars are light in color (white, pink, or grayish) and lower in density than the other silicate minerals. By density, we mean the quantity of matter contained in a unit volume. Figure 10.5 shows the density of four earth materials: water, quartz, the mineral olivine (discussed later in this chapter), and pure iron.

The next three silicate minerals are actually mineral groups, with a number of mineral varieties in each group. They are the *mica, amphibole,* and *pyroxene* groups. All three are silicates containing aluminum, magnesium, iron, and potassium or calcium. The seventh mineral, *olivine,* is a silicate of only magnesium and iron that lacks aluminum. Altogether, these minerals are described as **mafic** ("ma" for magnesium; "f" from the chemical symbol for iron, Fe). The mafic minerals are dark in color (usually black) and are denser than the felsic minerals.

Common Igneous Rocks

Igneous rocks solidify from rock in a hot, molten state, known as **magma.** From pockets a few kilometers below the earth's surface, magma makes its way upward

through older solid rock and eventually solidifies as igneous rock. No single igneous rock is made up of all seven silicate minerals listed in Figure 10.4. Instead, a given rock variety contains three or four of those minerals as the major ingredients.

The column in the center of Figure 10.4 shows four common igneous rocks. Each rock is connected by arrows to the principal minerals it contains. These four rock types are carefully selected to be used in our later explanation of features of the earth's crust.

Granite and diorite are examples of felsic igneous rocks.

The first igneous rock is **granite**. The bulk of granite consists of quartz (27 percent), potash feldspar (40 percent), and plagioclase feldspar (15 percent). The remainder is mostly biotite and amphibole. Because most of the volume of granite is of felsic minerals, we classify granite as a *felsic igneous rock.* Granite is a mixture of white, grayish or pinkish, and black grains, but the overall appearance is a light gray or pink color (Figure 10.6).

Diorite, the second igneous rock on the list, lacks quartz. It consists largely of plagioclase feldspar (60 percent) and secondary amounts of amphibole and pyroxene. Diorite is a light-colored felsic rock that is only slightly denser than granite. When its grains are coarse, it may have a speckled appearance (Figure 10.7).

The third igneous rock is *gabbro,* in which the major mineral is pyroxene (60 percent). A substantial amount of plagioclase feldspar (20 to 40 percent) is present, and, in addition, there may be some olivine (0 to 20 per-

Figure 10.6 Coarse-grained granite. Dark grains are amphibole and biotite; light grains are feldspars; clear grains are quartz.

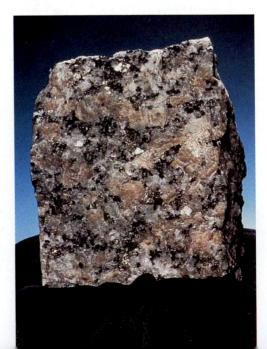

cent). Since the mafic minerals pyroxene and olivine are dominant, gabbro is classed as a *mafic igneous rock*. It is dark in color and denser than the felsic rocks.

The fourth igneous rock, *peridotite*, is dominated by olivine (60 percent). The rest is mostly pyroxene (40 percent). Peridotite is classed as an *ultramafic igneous rock*, denser even than the mafic types. The mineral grains in igneous rocks are very tightly interlocked, and the rock is normally very strong.

These four common varieties of igneous rock show an increasing range of density, from felsic through mafic to ultramafic types. This arrangement is duplicated on a grand scale in the principal rock layers that comprise the solid earth, with the least dense layer (mostly felsic rocks) near the surface and the densest layer (ultramafic rocks) deep in the earth's interior. We will stress this layered arrangement again in describing the earth's crust and the deeper interior zones in Chapter 11.

Intrusive and Extrusive Igneous Rocks

Magma that solidifies below the earth's surface and remains surrounded by older, preexisting rock is called **intrusive igneous rock**. The process itself is *intrusion*. Where magma reaches the surface, it emerges as **lava**, which solidifies to form **extrusive igneous rock** (Figure 10.8). The process is one of *extrusion*.

Although both intrusive rock and extrusive rock can solidify from the same original body of magma, their outward appearances are quite different when you compare freshly broken samples of each. Intrusive igneous rocks cool very slowly and, as a result, develop large mineral crystals—that is, they are *coarse-textured*. A good example is the granite pictured in Figure 10.6.

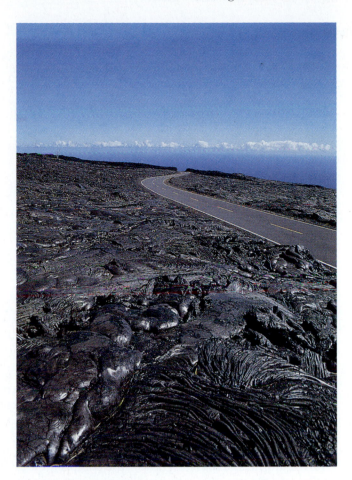

Figure 10.8   This fresh black lava shows a glassy surface and flow structures. Chain of Craters Road, Hawaii Volcanoes National Park, Hawaii.

Figure 10.7   Coarse-grained diorite. Feldspar grains are light; amphibole and pyroxene grains are dark. Compare with granite—clear quartz grains are lacking.

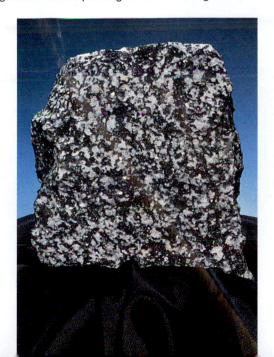

In an intrusive igneous rock, you can see individual mineral crystals with the unaided eye or with the help of a simple magnifying lens. In an extrusive rock, which cools very rapidly, the individual crystals are very small. They can only be seen through a microscope. Rocks with such small crystals are termed *fine-textured*. If the lava contains dissolved gases, it can cool to form a rock with a frothy, bubble-filled texture, called *scoria* (Figure 10.9, *right*). Sometimes a lava cools to form a shiny natural volcanic glass (Figure 10.9, *left*). Most lava solidifies simply as a dense, uniform rock with a dark, dull surface.

Since the outward appearance of intrusive and extrusive rocks formed from the same magma are so different, they are named differently. The igneous rock types we have discussed so far—granite, diorite, gabbro, and peridotite—are intrusive rocks. Except for peridotite, all have counterparts as extrusive rocks. They are named in the right column of Figure 10.4. Each one has the same mineral composition as the intrusive rock named at the left. *Rhyolite* is the name for lava of the same composi-

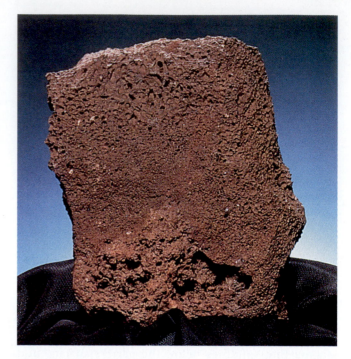

Figure 10.9 (*left*) Obsidian, or volcanic glass. The smooth, glassy appearance is acquired when a gas-free lava cools very rapidly. This sample shows red and black streaks, caused by minor variations in composition. (*right*) A specimen of scoria, a form of lava containing many small holes and cavities produced by gas bubbles.

tion as granite; **andesite** is lava with the mineral composition of diorite; and **basalt** is lava of the composition of gabbro. Andesite and basalt are the two most common types of lavas. Rhyolite and andesite are pale grayish or pink in color, whereas basalt is black. Lava flows, along with particles of solidified lava blown explosively from narrow vents, often accumulate as isolated hills or mountains that we recognize as volcanoes. They are described in Chapter 12.

A body of intrusive igneous rock is called a **pluton**. Granite typically accumulates in enormous plutons, called *batholiths.* Figure 10.10 shows the relationship of a batholith to the overlying rock. As the hot fluid magma rises, it melts and incorporates the older rock lying above it. A single batholith extends down several kilometers and may occupy an area of several thousand square kilometers.

Sills and dikes are examples of plutons— bodies of intrusive igneous rocks that cool at depth.

Figure 10.10 shows two other common forms of plutons. One is a *sill*, a platelike layer formed when magma forces its way between two preexisting rock layers. In the example shown, the sill has lifted the overlying rock layers to make room. A second kind of pluton is the *dike*, a wall-like body formed when a vertical rock frac-

ture is forced open by magma. Commonly, these vertical fractures conduct magma to the land surface in the process of extrusion. Figure 10.11 shows a dike of mafic rock cutting across layers of older rock. The dike rock is fine-textured because of its rapid cooling. Magma entering small, irregular, branching fractures in the surrounding rock solidifies in a branching network of thin *veins*.

Figure 10.10 This block diagram illustrates various forms of intrusive igneous rock plutons as well as an extrusive lava flow.

Figure 10.11   A dike of mafic igneous rock with nearly vertical parallel sides, cutting across flat-lying sedimentary rock layers. Arrows mark the contact between igneous rock and sedimentary rock. Spanish Peaks region, Colorado.

The formation of extrusive igneous rock can be witnessed today where volcanic processes are active. *Eye on the Environment • Battling Iceland's Heimaey Volcano* describes the struggle of courageous Icelanders to save their volcanic island village and harbor from the onslaught of a volcanic eruption.

Chemical Alteration of Igneous Rocks

The minerals in igneous rocks are formed at high temperatures, and often at high pressures, as magma cools. When igneous rocks are exposed at or near the earth's surface, the conditions are quite different. Temperatures and pressures are low. Also, the rocks are exposed to soil water and ground water solutions that contain dissolved oxygen and carbon dioxide. In this new environment, the minerals within an igneous rock may no longer be stable. Instead, most of these minerals undergo a slow chemical change that weakens their structure. Chemical change in response to this alien environment is called **mineral alteration**. It is a process of weathering, which we defined in Chapter 9 as the physical and chemical processes that break down rocks.

Weathering also includes the physical forces of disintegration that break up igneous rock into small fragments and separate the component minerals, grain by grain. This breakup, or fragmentation, is essential for the chemical reactions of mineral alteration. The reason is that fragmentation results in a great increase in mineral surface area exposed to chemically active solutions. (We will take up the processes of physical disintegration of rocks in Chapter 13.)

Oxygen dissolved in soil or ground water can oxidize minerals. *Oxidation* occurs when oxygen is added in a chemical reaction. Oxidation is the normal fate of most silicate minerals exposed at the surface. With oxidation, the silicate minerals are converted to *oxides*, in which silicon and the metallic elements—such as calcium, magnesium, and iron—each bond completely with oxygen. Oxides are very stable. Quartz, with the composition silicon dioxide (SiO_2), is a mineral oxide that also occurs naturally in granite. Because it is stable, quartz is very long-lasting. As we will see shortly, it is a major constituent of sedimentary rocks.

Water combines with some silicate minerals in a reaction known as *hydrolysis*. This process is not merely a soaking or wetting of the mineral, but a true chemical change that produces a different mineral compound. The products of hydrolysis are stable and long-lasting, as are the products of oxidation.

Some of the alteration products of silicate minerals are **clay minerals**, described in Chapter 9. A clay mineral is one that has plastic properties when moist, because it consists of thin flakes of colloidal size that become lubricated by layers of water molecules. Clay minerals formed by mineral alteration are abundant in common types of sedimentary rocks.

When carbon dioxide dissolves in water, a weak acid—*carbonic acid*—is formed. Carbonic acid can dissolve certain minerals, especially calcium carbonate. In addition, where decaying vegetation is present, soil water contains many complex organic acids that are capable of reacting with minerals. Certain common minerals, such as rock salt (sodium chloride), dissolve directly in water, but simple solution is not particularly effective for the silicate minerals.

SEDIMENTS AND SEDIMENTARY ROCKS

We can now turn to the second great rock class, the sedimentary rocks. The mineral particles in sedimentary rocks can be derived from preexisting rock of any of the three rock classes as well as from newly formed organic matter. However, igneous rock is the most important original source of the inorganic mineral matter that makes up sedimentary rock. For example, a granite can weather to yield grains of quartz and particles of clay minerals derived from feldspars, thus contributing sand and clay to a sedimentary rock. Sedimentary rocks include rock types with a wide range of physical and chemical properties. We will only touch on a few of the most important kinds of sedimentary rocks, which are shown in Table 10.2.

In the process of mineral alteration, solid rock is weakened, softened, and fragmented, yielding particles of many sizes and mineral compositions. When trans-

EYE ON THE ENVIRONMENT

Battling Iceland's Heimaey Volcano

Normally, the processes of rock formation are very slow. But when a volcanic eruption occurs, large volumes of fresh rock can be created very quickly. In 1973, inhabitants of Heimaey, a small island very near the southern coast of Iceland, found themselves witnessing the birth of new rock from a vantage point that was a little too close.

It was around 2:00 A.M., early on a January morning, that the eruption began. First, a fissure split the eastern side of the island from one coast to the other, sending up a curtain of fire in a pyrotechnical display nearly 2 km (1.2 mi) long. Soon, however, the spraying fountains of volcanic debris became restricted to a small area not far from Helgafell, an older volcanic cone. The lava and lava fragments of ash and cinders, called *tephra*, poured out at a rate of 100 m³/sec (about 3500 ft³/sec), building a cone that soon reached a height of about 100 m (about 300 ft) above sea level. It was dubbed Kirkjufell after a farmstead, Kirkjubaer, which lay beneath the debris.

It wasn't long before strong easterly winds set in, and Iceland's main fishing port, Vestmannaeyjar, located within a kilometer (about a half-mile) from the eruption, began to receive a "snowfall" of fine, black cinders. Houses on the east side of the town were buried under the tephra. Many collapsed under the weight of the hot ash. Still others were set afire. Lava flows soon reached the village, bury-

During the eruption of Heimaey's volcano, many buildings were covered with black volcanic ash.

This map of Heimaey Island, Iceland, shows the extent of lava flows from the eruptions of the Kirkjufell crater. (Modified from U.S. Geological Survey.)

ing and burning still other buildings (see photo).

Over the passing weeks, the emissions of tephra ceased, but lava continued to flow from the cone, which soon reached 200 m (about 600 ft) in height. Unfortunately, the lava began to flow northward, narrowing the harbor and threatening the future livelihood of the town and its evacuated inhabitants. Fishing is a major industry in Iceland, accounting for nearly 80 percent of Iceland's foreign exchange. Vestmannaeyjar normally lands and processes 20 percent of the nation's fish catch.

Because of the importance of the harbor and fishing port to the nation's economy, Icelanders embarked on a bold plan to save the harbor by altering the course of the lava, diverting the flow eastward. This was to be accomplished by cooling the flow on its north edge with water streams, creating a natural wall to channel the flow alongside the harbor instead of across it. Within a few weeks, the first pumping of water onto the flowing lava was begun. By early March, a pump ship came into operation in the harbor, providing a steady flow of sea water to cool the slowly moving flows. It was joined in April by as many as 47 high-capacity pumps floating on barges, delivering in total as much as 1 cu m of sea water per second (35 ft³/sec).

The most effective technique for slowing the lava at a particular location began with cooling the edge and nearby surface of the flow with water from hoses. This allowed bulldozers to build a crude road up and over the slowly moving flow, using nearby tephra as the road material. Large plastic pipes were then laid along the road and across the flow, with small holes spraying water on hot spots. As long as sea water was flowing in the plastic pipes, they remained cool enough not to melt. After a day or so, the flow typically began to slow. Pumping usually continued for about two weeks, until the lava stopped steaming at spray points. The result of the water applications was to build a broad wall of cool, rubbly lava with thickening lava behind it.

The huge undertaking successfully stabilized the northern front of the lava flow, keeping the harbor from becoming closed (see map). Although the lava had indeed reached the harbor, it had merely narrowed its entrance. In fact, the new flow improved the harbor's ability to shelter the boats within it (see map). Within about five months, the eruption was over, and the digging out began in earnest. Within a year, life was back to normal on Heimaey.

The diversion of the lava and reconstruction of the village of Vestmannaeyjar were extremely costly. To pay for the cost, Iceland passed a special tax increase, requiring the average Icelandic family to pay about 10 percent of its annual income for one year. Generous foreign aid also helped cover the enormous expenses borne by the tiny nation.

Questions

1. What were the initial effects of the eruption of Heimaey's new volcano?

2. What important industry was threatened by the volcanic activity and why?

3. How was sea water used to keep lava from blocking the harbor?

Table 10.2   Some Common Sedimentary Rock Types

Subclass	Rock Type	Composition
Clastic (Composed of rock and/or mineral fragments	Sandstone	Cemented sand grains
	Siltstone	Cemented silt particles
	Conglomerate	Sandstone containing pebbles of hard rock
	Mudstone	Silt and clay, with some sand
	Claystone	Clay
	Shale	Clay, breaking easily into flat flakes and plates
Chemically precipitated (Formed by chemical precipitation from sea water or salty inland lakes)	Limestone	Calcium carbonate, formed by precipitation on sea or lake floors
	Dolomite	Magnesium and calcium carbonates, similar to limestone
	Chert	Silica, a microcrystalline form of quartz
	Evaporites	Minerals formed by evaporation of salty solutions in shallow inland lakes or coastal lagoons
Organic (Formed from organic material)	Coal	Rock formed from peat or other organic deposits; may be burned as a mineral fuel
	Petroleum (mineral fuel)	Liquid hydrocarbon found in sedimentary deposits; not a true rock but a mineral fuel
	Natural gas (mineral fuel)	Gaseous hydrocarbon found in sedimentary deposits; not a true rock but a mineral fuel

ported by a fluid medium—air, water, or ice—these particles are known collectively as **sediment**. Used in its broadest sense, sediment includes both inorganic and organic matter. Dissolved mineral matter in the form of ions in solution must also be included.

Streams and rivers carry sediment to lower land levels, where sediment can accumulate. The most favorable sites of sediment accumulation are shallow sea floors bordering continents. But sediments also accumulate in inland valleys, lakes, and marshes. Thick accumulations of sediment may become deeply buried under newer (younger) sediments. Wind and glacial ice also transport sediment, but not necessarily to lower elevations or to places suitable for accumulation. Over long spans of time, the sediments can undergo physical or chemical changes, becoming compacted and hardened to form sedimentary rock.

The three major classes of sediment are clastic, chemically precipitated, and organic.

There are three major classes of sediment. First is **clastic sediment**, which consists of inorganic rock and mineral fragments. Examples are the materials in a sand bar of a river bed, or on a sandy ocean beach. Second is **chemically precipitated sediment**, which consists of inorganic mineral compounds precipitated from a saltwater solution or as hard parts of organisms. In the process of chemical precipitation, ions in solution combine to form solid mineral matter separate from the solution. A

layer of rock salt, such as that found in dry lakebeds in arid regions, is an example. A third class is **organic sediment**, which consists of the tissues of plants and animals as accumulated and preserved after the death of the organism. An example is a layer of peat in a bog or marsh.

Sediment accumulates in more-or-less horizontal layers, called **strata**, or simply "beds." Individual strata are separated from those below and above by surfaces called stratification planes or bedding planes. These separation surfaces allow one layer to be easily removed from the next. Strata of widely different compositions can occur alternately, one above the next.

Clastic Sedimentary Rocks

Clastic sediments are derived from one or more of the rock groups—igneous, sedimentary, metamorphic—and thus include a very wide range of minerals for sedimentary rock formation. Silicate minerals are the most important, both in original form and as altered by oxidation and hydrolysis. Quartz and feldspar usually dominate. Because quartz is hard and is immune to alteration, it is the most important single component of the clastic sediments. Second in abundance are fragments of unaltered fine-grained parent rocks, such as tiny pieces of lava rock. Feldspar and mica are also commonly present. Clay minerals are major constituents of the finest clastic sediments.

The range of particle sizes in a clastic sediment determines how easily and how far the particles are transported by water currents. The finer the particles, the more easily they are held suspended in the fluid. Coarser particles tend to settle to the bottom of the fluid layer. In this way, a separation of size grades, called *sorting*, occurs. Sorting determines the texture of the sediment deposit and of the sedimentary rock derived from that sediment. Colloidal clay particles do not settle out unless they are made to clot together into larger clumps. This clotting process normally occurs when river water carrying clay mixes with the saltwater of the ocean.

When sediments accumulate in thick sequences, the lower strata are exposed to the pressure produced by the weight of the sediments above them. This pressure compacts the sediments, squeezing out excess water. Cementation occurs as dissolved minerals recrystallize where grains touch and in the spaces between mineral particles. Silicon dioxide (quartz, SiO_2) is very slightly soluble in water, and so the cement is often a form of quartz, called *silica*, which lacks a true crystalline form. Calcium carbonate ($CaCO_3$) is another common material that cements clastic sedimentary rocks. Compaction and cementation produce sedimentary rock.

The important varieties of clastic sedimentary rock are distinguished by the size of their particles. They include sandstone, conglomerate, siltstone, mudstone, claystone, and shale. **Sandstone** is formed from fine to coarse sand (Figure 10.12). The cement may be silica or calcium carbonate. The sand grains are commonly of quartz. (Refer to Figure 9.2 for the names and diameters of the various grades of sediment particles.) Sandstone containing numerous rounded pebbles of

Figure 10.12   An eroded sandstone formation on the Colorado Plateau. These rocks were formed from thin layers of wind-blown sand. Iron oxides have stained some of the layers red, while others remain white.

Figure 10.13 A piece of quartzitic conglomerate, cut through and polished, reveals rounded pebbles of quartz (clear and milky colors) and chert (grayish). It is about 12 cm (4.7 in.) in diameter.

hard rock is called *conglomerate* (Figure 10.13). Sandstone and conglomerate are both typically very resistant to erosion, and often form cliffs or ridges (Figure 10.14). Similar to sandstone is **siltstone**, which is a clastic sedimentary rock formed from silt particles.

A mixture of water with particles of silt and clay, along with some sand grains, is called *mud*. The sedimentary rock hardened from such a mixture is called *mudstone*. Compacted and hardened clay layers become *claystone*. Sedimentary rocks of mud composition are commonly layered in such a way that they easily break apart into small flakes and plates. The rock is then described as being *fissile* and is given the name **shale**. Shale, the most abundant of all sedimentary rocks, is formed largely of clay minerals. The compaction of the mud to form mudstone and shale involves a considerable loss of volume as water is driven out of the clay. Often shales are weak rocks and erode relatively quickly. Figure 10.15 shows a sequence of colored shales being actively eroded in the Utah desert.

Chemically Precipitated Sedimentary Rocks

Under favorable conditions, mineral compounds are deposited from the salt solutions of sea water and of salty inland lakes in desert climates. One of the most common sedimentary rocks formed by chemical precipitation is **limestone**, composed largely of the mineral *calcite*. Calcite is calcium carbonate ($CaCO_3$). Marine limestones—limestone strata formed on the sea floor—accumulated in thick layers in many ancient seaways in past geologic eras (Figure 10.16). A closely related rock is *dolomite*, composed of calcium-magnesium carbonate. Limestone and dolomite are grouped together as the *carbonate rocks*. They are dense rocks, ranging in color from white to pale gray or even black.

Carbonate rocks—limestone and dolomite—are formed by precipitation from sea water.

Sea water also yields sedimentary layers of silica in a hard, microcrystalline form called *chert*. Chert is a variety of sedimentary rock, but it also commonly occurs combined with limestone (cherty limestone).

Figure 10.14 Cliffs and ledges of sedimentary rock formations in the North Rim of the Grand Canyon, Arizona. The great sheer cliff is the Coconino Formation, an ancient dune deposit. (A. N. Strahler.)

Figure 10.15 These flat-lying strata consist of layers of different colored shales that are easily eroded. Near Glen Canyon City in southern Utah. (A. N. Strahler.)

Shallow water bodies acquire a very high level of salinity where evaporation is sustained and intense. One type of shallow water body is a bay or estuary in a coastal desert region. Another type is the salty lake of inland desert basins (see Figure 14.16). Sedimentary minerals and rocks deposited from such concentrated solutions are called **evaporites**. Ordinary rock salt, the mineral halite (sodium chloride), has accumulated in this way in thick sedimentary rock layers. These layers are mined as major commercial sources of halite.

Figure 10.16 The Etretât cliffs of Normandy, France, are composed of soft limestone, known as chalk.

Figure 10.17 A coal seam near Sheridan, Wyoming, being strip mined by heavy equipment.

Hydrocarbon Compounds in Sedimentary Rocks

Hydrocarbon compounds (compounds of carbon, hydrogen, and oxygen) form a most important type of organic sediment—one on which human society increasingly depends for fuel. These substances occur both as solids (peat and coal) and as liquids and gases (petroleum and natural gas). Only coal qualifies physically as a rock. *Peat* is a soft, fibrous substance of brown to black color. It accumulates in a bog environment where the continual presence of water inhibits the decay of plant remains. We have already mentioned peat as an example of a Histosol in Chapter 9. (See Figure 9.20.)

At various times and places in the geologic past, plant remains accumulated on a large scale, accompanied by sinking of the area and burial of the compacted organic matter under thick layers of inorganic clastic sediments. *Coal* is the end result of this process (Figure 10.17). Individual coal seams are interbedded with shale, sandstone, and limestone strata. *Petroleum* (or *crude oil*, as the liquid form is often called) includes many hydrocarbon compounds. *Natural gas*, which is found in close association with accumulations of liquid petroleum, is a mixture of gases. The principal gas is methane (marsh gas, CH_4). Natural gas and petroleum commonly occupy open interconnected pores in a thick sedimentary rock layer—a porous sandstone, for example. Geologists generally agree that petroleum and natural gas are of organic origin, but the nature of the process is not fully understood. They are not classed as minerals, but rather are separately classed as mineral fuels.

Hydrocarbon compounds in sedimentary rocks are important because they provide an energy resource on which modern human civilization depends. These **fossil fuels**, as they are called collectively, have required millions of years to accumulate. However, they are being consumed at a very rapid rate by our industrial society. These fuels are nonrenewable resources. Once they are gone, there will be no more, because the quantity produced by geologic processes even in a thousand years is scarcely measurable in comparison to the quantity stored through geologic time.

METAMORPHIC ROCKS

Any type of igneous or sedimentary rock may be altered by the tremendous pressures and high temperatures that accompany the mountain-building processes of the earth's crust. The result is a rock so changed in texture and structure as to be reclassified as **metamorphic rock**.

Table 10.3   Some Common Metamorphic Rock Types

Rock Type	Description
Slate	Shale exposed to heat and pressure that splits into hard flat plates
Schist	Shale exposed to intense heat and pressure that shows evidence of shearing
Quartzite	Sandstone that is "welded" by a silica cement into a very hard rock of solid quartz
Marble	Limestone exposed to heat and pressure, resulting in larger, more uniform crystals
Gneiss	Rock resulting from the exposure of clastic sedimentary or intrusive igneous rocks to heat and pressure

Mineral components of the parent rock are, in many cases, reconstituted into different mineral varieties. Recrystallization of the original minerals can also occur. Our discussion of metamorphic rocks will mention only five common types—slate, schist, quartzite, marble, and gneiss (Table 10.3).

Slate is formed from shale that is heated and compressed by mountain-making forces. This fine-textured rock splits neatly into thin plates, which are familiar as roofing shingles and as patio flagstones. With application of increased heat and pressure, slate changes into **schist**, representing the most advanced stage of metamorphism. Schist has a structure called foliation, consisting of thin but rough, irregularly curved planes of parting in the rock (Figure 10.18). These are evidence of *shearing*—a compressional stress that pushes the layers sideways, much as a deck of cards can be pushed into a fan with the sweep of a palm. Schist is set apart from slate by the coarse texture of the mineral grains, the abundance of mica, and occasionally the presence of scattered large crystals of newly formed minerals, such as garnet.

The metamorphic equivalent of conglomerate, sandstone, and siltstone is **quartzite**, which is formed by the addition of silica to fill completely the open spaces between grains. This process is carried out by the slow movement of underground waters carrying silicate into the rock, where it is deposited.

Limestone, after undergoing metamorphism, becomes *marble*, a rock of sugary texture when freshly broken. During the process of internal shearing, calcite in the limestone reforms into larger, more uniform crystals than before. Bedding planes are obscured, and

Figure 10.18 A specimen of foliated schist.

masses of mineral impurities are drawn out into swirling bands.

Finally, the important metamorphic rock **gneiss** may be formed either from intrusive igneous rocks or from clastic sedimentary rocks that have been in close contact with intrusive magmas. A single description will not fit all gneisses because they vary considerably in appearance, mineral composition, and structure. One variety of gneiss is strongly banded into light and dark layers or lenses (Figure 10.19), which may be bent into wavy folds. These bands have differing mineral compositions. They are thought to be relics of sedimentary strata, such as shale and sandstone, to which new mineral matter has been added from nearby intrusive magmas.

Figure 10.19 A surf-washed rock surface exposing banded gneiss. Pemaquid Point, Maine.

THE GEOLOGIC TIME SCALE

The earth has a long and complex history. Through the slow processes of the cycle of rock change, rocks are constantly being formed and reformed, but different formation processes are active at different places and at different times. Because of this, rocks now exposed at the surface are of varying age. Most are millions of years old, sometimes tens or hundreds of millions years old.

To describe the age of various rocks, geologists have developed a scale of geologic time. Table 10.4 lists the major geologic time divisions. All time older than 570 million years (m.y.) before the present is *Precambrian time.* Three eras of time follow: *Paleozoic, Mesozoic,* and *Cenozoic.* These eras saw the evolution of life-forms in the oceans and on the lands. The geologic eras are subdivided into *periods.* Their names, ages, durations, and associated life forms are also given in Table 10.4.

The Cenozoic Era is particularly important in terms of the continental surfaces, because nearly all landscape features seen today have been produced in the 65 million years since that era began. The Cenozoic Era is comparatively short in duration, scarcely more than the average duration of a single period in older eras. It is subdivided directly into seven lesser time units called epochs. (Note that the terms "Tertiary Period" and "Quaternary Period" are sometimes applied to the Cenozoic epochs Paleocene through Pliocene, and to the Pleistocene Epoch, respectively.) Details of the Pleistocene and Holocene epochs are given in Chapter 18.

The human genus *Homo* evolved during the late Pliocene Epoch and throughout the Pleistocene Epoch. As you can see, the period of human occupation of the earth's surface—a few million years at best—is but a fleeting moment in the vast duration of our planet's history.

THE CYCLE OF ROCK CHANGE

The processes that form rocks, when taken together, constitute a single system that cycles and recycles earth materials over geologic time from one form to another. The *cycle of rock change,* a concept that we introduced toward the beginning of this chapter, describes this system. Figure 10.20 shows a more complete version of this cycle. There are two environments—a surface environment of low pressures and temperatures and a deep

Table 10.4 Table of Geologic Time

Era	Period	Epoch	Duration (millions of years)	Age (millions of years)	Evolution of Life forms
Cenozoic	Neogene	Holocene	(10,000 yr)		
		Pleistocene	2	2	Human genus
		Pliocene	3	5	
		Miocene	19	24	Hominoids
	Paleogene	Oligocene	13	37	Whales
		Eocene	21	58	Bats
		Paleocene	8	66	Mammals
Mesozoic	Cretaceous		78	144	Flowering plants
	Jurassic		64	208	Dinosaurs (extinct)
	Triassic		37	245	Turtles
Paleozoic	Permian		41	286	Frogs
	Carboniferous		74	360	Conifers; higher fishes
	Devonian		48	408	Vascular plants; primitive fishes
	Silurian		30	438	
	Ordovician		67	505	
	Cambrian		65	570	Invertebrates

Precambrian Time (Extends to oldest known rocks, about 4 billion years)
Age of earth as a planet: 4.6 to 4.7 billion years.
Age of universe: 17 to 18 billion years.

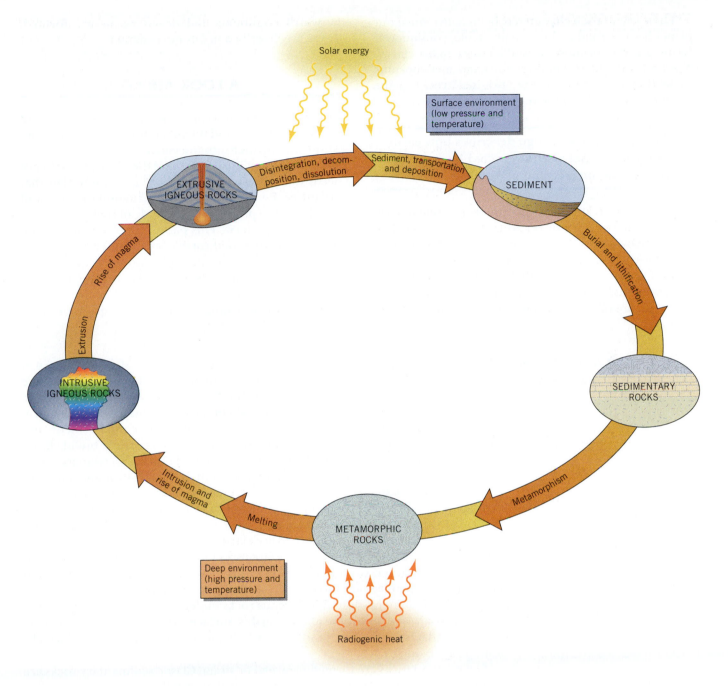

Figure 10.20 The cycle of rock change. It is powered by solar energy at the surface and by radiogenic heat, arising from spontaneous decay of radioactive atoms, deep within the earth's crust.

environment of high pressures and temperatures. The surface environment is the site of rock alteration and sediment deposition. In this environment, igneous, sedimentary, and metamorphic rocks are uplifted and exposed to air and water. Their minerals are altered chemically and broken free from the parent rock, yielding sediment. The sediment accumulates in basins, where deeply buried sediment layers are compressed and cemented into sedimentary rock.

Sedimentary rock, entering the deep environment, is heated and confined by high pressure. Here it is transformed into metamorphic rock. Pockets of magma are formed in the deep environment and move upward, melting and incorporating surrounding rock as they rise. Upon reaching a higher level, magma cools and solidifies, becoming intrusive igneous rock. Or it may emerge at the surface to form extrusive igneous rock. Either way, the cycle is completed.

The cycle of rock change has been active since our planet became solid and internally stable, continuously forming and reforming rocks of all three major classes. Not even the oldest dated igneous and metamorphic rocks thus far found are the "original" rocks of the earth's crust, for they were recycled eons ago.

The cycle of rock change describes how earth materials are cycled and recycled by processes acting at the surface and at depth.

There are two key questions about the cycle of rock change that we have not yet addressed, however. The first is, "Why is the earth's interior hot—hot enough to melt rock?" The second is, "How do sediments and rocks descend into the earth?" To answer these questions, we must look in more detail at the composition and structure of the earth's interior and at the process by which continents and ocean basins are formed. These are described in our next chapter.

A LOOK AHEAD

This chapter has described the minerals and rocks of the earth's surface and the processes of their formation. As we look further into the earth's outermost layers in the next chapter, we will see that these processes do not occur everywhere. Instead, there is a grand plan that organizes the formation and destruction of rocks and distributes the processes of the cycle of rock change in a geographic pattern. This pattern is controlled by the way in which the solid earth's brittle outer layer is fractured into great plates that split and separate and also converge and collide. The grand plan is plate tectonics, the scheme for understanding the dynamics of the earth's crust over millions of years of geologic time.

EARTH MATERIALS IN REVIEW

The elements oxygen and silicon dominate the earth's crust. Metallic elements, which include aluminum, iron, and the base elements, account for nearly all the remainder. These elements are found in minerals—naturally occurring, inorganic substances. Each has an individual chemical composition and atomic structure.

Silicate minerals make up the bulk of igneous rocks. They contain silicon and oxygen together with some of the metallic elements. There are three broad classes of igneous rocks, depending on their mineral content. Felsic rocks contain mostly felsic minerals and are least dense. Mafic rocks, containing mostly mafic minerals, are denser. Ultramafic rocks are most dense. Because felsic rocks are least dense, they are generally found in the upper layers of the earth's crust. Mafic and ultramafic rocks are more abundant in the layers below. If magma erupts on the surface to cool rapidly as lava, the rocks formed are extrusive and have a fine crystal texture. If the magma cools slowly below the surface as a pluton, the rocks are intrusive and the crystals are larger. Granite (felsic, intrusive), andesite (felsic, extrusive), and basalt (mafic, extrusive) are three very common igneous rock types.

Most silicate minerals found in igneous rocks undergo mineral alteration when exposed to air and moisture at the earth's surface. Mineral alteration occurs through oxidation, hydrolysis, or solution. Clay minerals are commonly produced by mineral alteration.

Sedimentary rocks are formed in layers, or strata. Clastic sedimentary rocks are composed of fragments of rocks and minerals that usually accumulate on ocean floors. As the layers are buried more and more deeply, water is pressed out and particles are cemented together. Sandstone and shale are common examples. Chemical precipitation also produces sedimentary rocks, such as limestone. Coal, petroleum, and natural gas are hydrocarbon compounds occurring in sedimentary rocks that are used as mineral fuels.

Metamorphic rocks are formed when igneous or sedimentary rocks are exposed to heat and pressure. Shale is altered to schist, sandstones and siltstones become quartzite, and intrusive igneous rocks or clastic sediments are metamorphosed into gneiss.

Geologists trace the history of the earth through the geologic time scale. Precambrian time includes the earth's earliest history. It is followed by three major divisions—the Paleozoic, Mesozoic, and Cenozoic eras. Most landforms and life-forms seen today have their roots in the Cenozoic era.

In the cycle of rock change, rocks are exposed at the earth's surface, and their minerals are broken free and altered to form sediment. The sediment accumulates in basins, where the layers are compressed and cemented into sedimentary rock. Deep within the earth, rock melts into magma, which can move upward into the crust to form igneous rocks that cool at or below the surface. Rocks deep in the crust are exposed to heat and pressure, forming metamorphic rock. Mountain-building forces move deep igneous, sedimentary, and metamorphic rocks upward to the surface, providing new material for surface alteration and breakup and completing the cycle.

KEY TERMS

mineral	lava	strata
rock	extrusive igneous rock	sandstone
igneous rocks	andesite	siltstone
sedimentary rocks	basalt	shale
metamorphic rocks	pluton	limestone
cycle of rock change	mineral alteration	evaporites
quartz	clay mineral	fossil fuels
felsic	sediment	metamorphic rock
mafic	clastic sediment	schist
magma	chemically precipitat-	quartzite
granite	ed sediment	gneiss
intrusive igneous rock	organic sediment	

REVIEW QUESTIONS

1. What is the earth's crust? What elements are most abundant in the crust?
2. Define the terms *mineral* and *rock*. Name the three major classes of rocks.
3. What are silicate minerals? Describe two classes of silicate minerals.
4. Name four types of igneous rocks and arrange them in order of density.
5. How do igneous rocks differ when magma cools (*a*) at depth, and (*b*) at the surface?
6. How are igneous rocks chemically altered? Identify and describe three processes of chemical alteration.
7. What is sediment? Define and describe three types of sediments.
8. Describe two processes that produce sedimentary rocks, and identify at least three important varieties of clastic sedimentary rocks.
9. How are sedimentary rocks formed by chemical precipitation?
10. What types of sedimentary deposits consist of hydrocarbon compounds? How are they formed?
11. What are metamorphic rocks? Describe at least three types of metamorphic rocks and how they are formed.
12. More recent geologic time is divided into three eras. Name them in order from oldest to youngest. How do geologists use the terms *period* and *epoch*? What age is applied to time before the earliest era?

VISUALIZING EXERCISES

1. Sketch a cross section of the earth showing the following features: batholith, sill, dike, veins, lava, and volcano.
2. Sketch the cycle of rock change and describe the processes that act within it to form igneous, sedimentary, and metamorphic rocks.

A DEEPER VIEW OF EARTH MATERIALS

A granite is exposed at the earth's surface, high in the Sierra Nevada mountain range. Describe how mineral grains from this granite might be released, altered, and eventually become incorporated in a sedimentary rock. Trace the route and processes that would incorporate the same grains in a metamorphic rock.

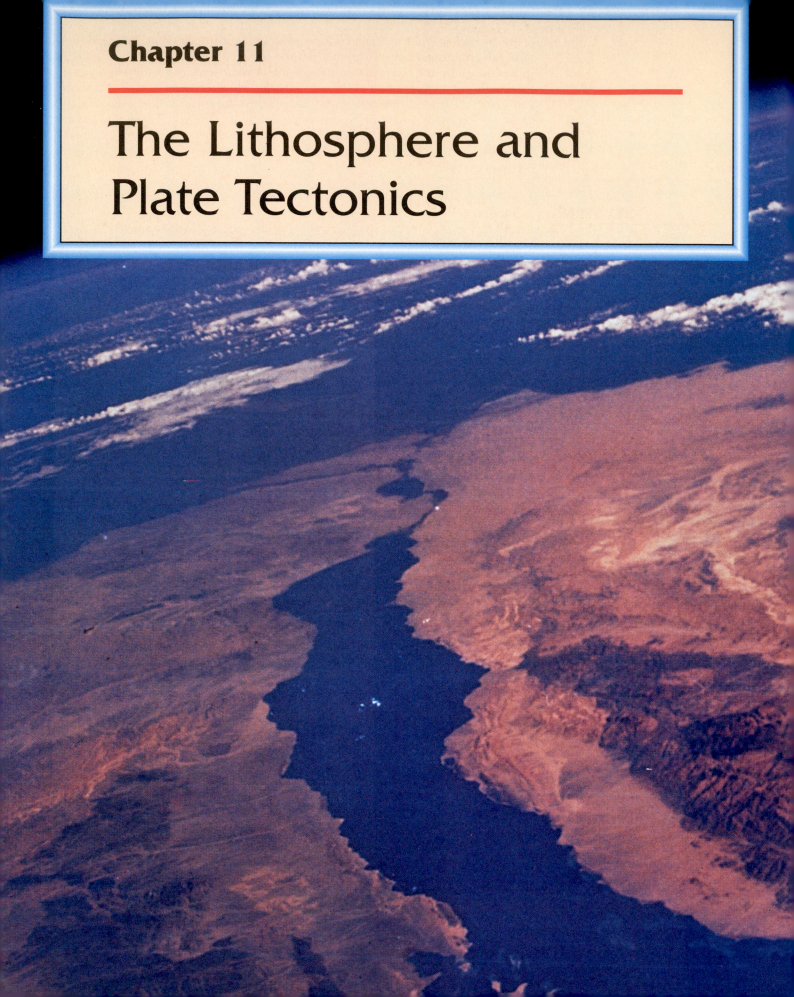

Chapter 11

The Lithosphere and Plate Tectonics

Sinai Peninsula, seen from space. Red Sea, with Gulfs of Suez and Aqaba to left and right.

PUTTING YOU IN THE PICTURE

Perhaps it was in that eighth grade geography class when you began to stare at the wall map of the world that hung in the front of the room, and you noticed that some parts of the map looked like a puzzle coming apart. The upper right corner of Africa was particularly suspect. Twisting the continent toward the northeast would close the Red Sea and Persian Gulf, making the Arabian Peninsula join Africa and the Middle East in a continuous broad land bridge.

Then you noticed the Mediterranean. With a little squeezing, it could shut quite nicely, uniting Europe and Africa. Greece would have to tuck in under Turkey, which would move northward to fill in the Black Sea. Italy and its islands would have to move south, filling in the obvious gap in the African coast of Libya and Tunisia. And then there was that funny bulge on the east coast of South America that would lie rather nicely along the west coast of Africa, with the Guianas butting against the west African republics from Liberia to Nigeria.

North America was a lot tougher, with those messy Caribbean islands and Central America to worry about. Could North and South America have been closer? If the Yucatan and Honduran-Nicaraguan peninsulas curled north and west, they would tuck in along the Gulf Coast. Then Florida would nestle up against Panama, with Cuba, Jamaica, and Hispaniola lying along the Venezuelan coast. Now, with North America closer to South America, the east coast would fit against the bulge of western Africa rather nicely.

Perhaps, you wondered, all these lands actually did once fit together in a single world continent. Fun to think about, but utterly fantastic.

Or was it? The idea that the continents were once united and then broken apart is an old one. In fact, in 1668, a Frenchman interpreted the matching coastlines of eastern South America and western Africa as proof that the two continents were separated during the biblical flood!

Within the last 30 years, the slow movement of continents over long spans of geologic time has become widely accepted by geologists and physical geographers. Evidence for the fracturing of the earth's outer rock layer and movement of the resulting rock plates has come from many sources—earthquake analysis, studies of ancient collections of plants and animals as they may have existed on united continents, and the matching of geologic structures, such as mountain chains and rock sequences, from continent to continent. Most recently, the rates of movement of continents have actually been measured using laser beams bounced from continents to orbiting satellites and back. Thus, many lines of evidence now converge on *plate tectonics*—the body of knowledge describing the movements and changes through time of lithospheric plates carrying continents and ocean basins.

Plate tectonics provides the plan of the earth's major relief features, from the locations of mountain ranges and basins to the positions of oceans and inland seas. An understanding of plate tectonics is thus important to the understanding of the global patterns of the earth's landscapes—including climate, soils, vegetation, landforms, and, ultimately, human activity. In other words, the processes of physical geography are superimposed on a global geography given by plate tectonics. For this reason, the motions and histories of continental plates—splitting and separating to form ocean basins, closing and colliding to form mountain ranges—can be taken as a grand plan to organize a vast body of geographic and geologic knowledge. This is why plate tectonic theory ranks with the theory of evolution as one of the great milestones in scientific study of the earth.

Chapter Outline

The Structure of the Earth
  The Earth's Interior
  The Lithosphere and Asthenosphere
Major Relief Features of the Earth's Surface
  Relief Features of the Continents
  Relief Features of the Ocean Basins
Plate Tectonics
  Tectonic Processes

Plate Motions and Interactions
The Global System of Lithospheric Plates
Subduction Tectonics
Orogens and Collisions
Continental Rupture and New Ocean Basins
The Power Source for Plate Movements
Continents of the Past
Eye on the Environment • Alfred Wegener and Continental Drift

THE LITHOSPHERE AND PLATE TECTONICS

On the globes and maps we've seen since early childhood, the outline of each continent is so unique that we would never mistake one continent for another. But why are no two continents even closely alike? As we will see in this chapter, the continents have had a long history. In fact, in each of the continents some regions of metamorphic rocks date back more than 2 billion years. As part of that history, the continents have been fractured and split apart, as well as pushed together and joined.

Perhaps you've visited an old New England farmhouse that was constructed over hundreds of years—first the small two-room house, to which the kitchen shed was added at the back, and then the parlor wing at the side. Later the roof was raised for a second story, the tool house and barn were joined to the main house, the carriage house was built, and so it went. The earth's continents have that kind of history. They are composed of huge masses of continental crust that have been assembled at different times in each continent's history. The theory describing the movements and evolution of the continents and ocean basins, and the processes that fracture and fuse them, is called *plate tectonics*.

In this chapter we will survey the major geologic features of our planet, starting with the layered structure of its deep interior. We will then examine the outermost layer, or crust, and compare the crustal features of the continents with those of the ocean basins. Lastly, we will turn to plate tectonics and describe how plate movements have created broad regions of igneous, sedimentary, and metamorphic rocks, as well as the uparching of mountains and the downdropping of basins. We will find that plate tectonics furnishes us with a complete scenario of earth history on a grand scale in both time and spatial dimensions—a scenario that sets the stage for our study of landform-shaping processes to come in future chapters.

THE STRUCTURE OF THE EARTH

What lies deep within the earth? From studies of earthquake waves, reflected from deep earth layers, scientists have learned that our earth is far from uniform from its outer crust to its center. Instead, it consists of a central core with several concentric surrounding layers. The densest matter is at the center, and each layer above it is increasingly less dense. This structure is inherited from the earth's earliest history as it was formed by accretion from a mass of gas and dust orbiting the sun.

The Earth's Interior

Figure 11.1 is a cutaway diagram of the earth showing its interior. The earth as a whole is an almost spherical body approximately 6400 km (about 4000 mi) in radius. The center is occupied by the **core**, which is about 3500

Figure 11.1 This cutaway diagram of the earth shows the inner core of iron, which is solid, surrounded by a liquid outer core. The mantle, which surrounds the core, is a thick layer of ultramafic rock. The crust is too thin to show to correct scale. Thicknesses of layers are rounded to the nearest 100 units.

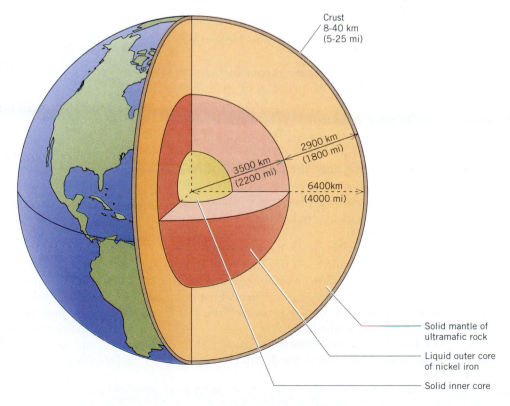

Crust
8-40 km
(5-25 mi)

2900 km
(1800 mi)

3500 km
(2200 mi)

6400km
(4000 mi)

Solid mantle of ultramafic rock

Liquid outer core of nickel iron

Solid inner core

km (about 2200 mi) in radius. Because of the sudden change in behavior of earthquake waves upon reaching the core, scientists have concluded that the outer core has the properties of a liquid. However, the innermost part of the core is in the solid state. Based on earthquake waves (and other kinds of data), it has long been inferred that the core consists mostly of iron, with some nickel. The core is very hot—its temperature ranges from about 2800°C to 3100°C (about 5100°F to 5600°F).

Enclosing the metallic core is the **mantle**, a rock shell about 2900 km (about 1800 mi) thick. Earthquake waves indicate that mantle rock is composed of mafic minerals similar to olivine (a silicate of magnesium and iron) and may resemble the ultramafic igneous rock peridotite, which is found exposed here and there on the continental surface (see Figure 10.4 and Table 10.1). Temperatures in the mantle range from about 2800°C (about 5100°F) near the core to about 1800°C (about 3300°F) near the crust.

The layers of the earth's interior include the crust, mantle, liquid core, and solid core.

The outermost and thinnest of the earth shells is the **crust**, a layer normally about 8 to 40 km (about 5 to 25 mi) thick (Figure 11.2). It is formed largely of igneous rock, but it also contains substantial proportions of metamorphic rock and a comparatively thin upper layer of sedimentary rock. The base of the crust is sharply defined where it contacts the mantle. This contact is detected by the way in which earthquake waves abruptly change velocity at that level. The boundary surface between crust and mantle is called the *Moho*, a simplification of the name of the scientist, Andrija Mohorovičić, who discovered it in 1909.

The continental crust is quite different from the crust beneath the oceans. From studies of earthquake waves, geologists have concluded that the **continental crust** consists of two continuous zones—a lower, continuous rock zone of mafic composition, which is more dense, and an upper zone of felsic rock, which is less dense (Figure 11.2). Because the felsic portion has a chemical composition similar to that of granite, it is commonly described as being *granitic rock*. Much of the granitic rock is metamorphic rock. There is no sharply defined separation between the felsic and mafic zones.

While continental crust typically has two layers of different composition, **oceanic crust** has only one. It consists almost entirely of mafic rocks with the composition of basalt and gabbro. Basalt, as lava, forms an upper zone, whereas gabbro, an intrusive rock of the same composition, lies beneath the basalt.

Another key distinction between continental and oceanic crust is that the crust is much thicker beneath

(a)

(b)

Figure 11.2 (a) Idealized cross section of the earth's crust and upper mantle. (b) Details of the crust and mantle at the edge of a continent, including the types of rocks found there. Also shown are the lithosphere and asthenosphere. (Copyright © A. N. Strahler.)

the continents than beneath the ocean floors (Figure 11.2). While 35 km (22 mi) is a good average figure for crustal thickness beneath the continents, 7 km (4 mi) is a typical figure for thickness of the basalt and gabbro crust beneath the deep ocean floors. Later in this chapter we will relate this difference in thickness to the processes that form the two types of crust.

The Lithosphere and Asthenosphere

Geologists use the term **lithosphere** to mean an outer earth zone, or shell, of rigid, brittle rock. This usage of the term is different from that found in the Prologue, which identified the lithosphere as one of the four great realms of earth—atmosphere, hydrosphere, lithosphere, and biosphere. The word does not carry any special meaning in terms of chemical makeup of the rock.

The lithosphere is a much thicker zone than the crust, as shown in Figure 11.2. It includes not only the crust, but also the cooler, upper part of the mantle that is composed of brittle rock. The lithosphere ranges in thickness from 60 to 150 km (40 to 95 mi). It is thickest under the continents and thinnest under the ocean basins.

> **The lithosphere is the solid, brittle outer portion of the earth. It includes the crust and the uppermost part of the mantle.**

Some tens of kilometers deep in the earth, the brittle condition of the lithospheric rock gives way gradually to a plastic, or "soft," layer named the **asthenosphere** (Figure 11.2). (This word is derived from the Greek root *asthenes*, meaning "weak.") However, at still greater depth in the mantle, the strength of the rock material again increases. Thus, the asthenosphere is a soft layer sandwiched between the "hard" lithosphere above and a "strong" mantle rock layer below. In terms of states of matter, the asthenosphere is not a liquid, even though its temperature reaches 1400°C (about 2600°F).

The lithosphere resting atop the asthenosphere can be thought of as a hard, brittle shell resting on a soft, plastic underlayer. Because the asthenosphere is soft and plastic, the rigid lithosphere can easily move over it. The lithospheric shell consists of large pieces called **lithospheric plates**. A single plate can be as large as a continent and can move independently of the plates that surround it. Like great slabs of floating ice on the polar sea, lithospheric plates can separate from one another at one location, while elsewhere they may collide in crushing impacts that raise great ridges. Along these collision ridges, one plate can be found diving down beneath the edge of its neighbor. These varied sorts of plate movements will be described shortly.

MAJOR RELIEF FEATURES OF THE EARTH'S SURFACE

Before we begin our discussion of plate tectonics, it is important to examine the major relief features of the earth—its continents and ocean basins—since they are produced by plate tectonic processes. Detailed global maps show that about 29 percent of the earth's surface is land and 71 percent oceans (Figure 11.3). If the seas were to drain away, however, we would see that broad areas lying close to the continental shores are actually covered by shallow water, less than 150 m (500 ft) deep. From these relatively shallow continental shelves, the ocean floor drops rapidly to depths of thousands of meters. In other words, the ocean basins seem to be

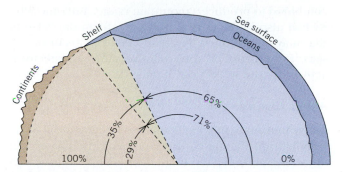

Figure 11.3 Actual global percentages of land and ocean areas compared with percentages if sea level were to drop 180 m (about 600 ft), exposing the continental shelves. (Copyright © A. N. Strahler.)

more than "brim full" of water—the oceans have spread over the margins of ground that would otherwise be assigned to the continents. If the ocean level were to drop by 150 m (about 500 ft), the shelves would be exposed, adding about 6 percent to the area of the continents. The surface area of continents would then be increased to 35 percent, and the ocean basin area decreased to 65 percent. These revised figures represent the true relative proportions of continents and oceans.

Figure 11.4 is a map showing some of the major relief features of continents and oceans, including mountain arcs, island arcs, ocean trenches, and the midoceanic ridge. We will refer to this map as we discuss the relief features of continents and oceans in turn.

Relief Features of the Continents

Broadly viewed, the continents consist of two basic subdivisions: active belts of mountain-making, and inactive regions of old, stable rock. The mountain ranges in the active belts grow through one of two very different geologic processes. First is *volcanism*, the formation of massive accumulations of volcanic rock by extrusion of magma. Many lofty mountain ranges consist of chains of volcanoes built of extrusive igneous rocks.

> **The two basic subdivisions of continental masses are active belts of mountain-making and inactive regions of old, stable rock.**

The second mountain-building process is *tectonic activity*, the breaking and bending of the earth's crust under internal earth forces. This tectonic activity usually occurs when great lithospheric plates come together in titanic collisions. (We will return to this topic later in this chapter.) Crustal masses that are raised by tectonic activity form mountains and plateaus. In some instances, both volcanism and tectonic activity have combined to produce a mountain range. Tectonic activity can also lower crustal masses to form depressions that may be occupied by ocean embayments or inland seas. Landforms produced by volcanic and tectonic activity are the subject of Chapter 12.

Alpine Chains

Active mountain-making belts are narrow zones that are usually found along continental margins. These belts are sometimes referred to as *alpine chains*, because they are characterized by high, rugged mountains, such as the Alps of Central Europe. These mountain belts were formed in the Cenozoic Era by volcanism, or tectonic activity, or a combination of both. Alpine mountain-building continues even today in many places.

The alpine chains are characterized by broadly curved lines on the world map (Figure 11.4). Each curved section of an alpine chain is referred to as a **mountain arc**. The arcs are linked in sequence to form two principal mountain belts. One is the *circum-Pacific belt* (shown in green), which rings the Pacific Ocean basin. In North and South America, this belt is largely on the continents and includes the Andes and coastal ranges of North America. In the western part of the Pacific basin, the mountain arcs lie well offshore from the continents and take the form of **island arcs**. Partly submerged, they join the Aleutians, Kurils, Japan, the Philippines, and other smaller islands. These island arcs are the result of volcanic activity. Between the larger islands, the arcs are represented by volcanoes rising above the sea as small, isolated islands.

The second chain of major mountain arcs forms the *Eurasian-Indonesian belt*, shown in blue in Figure 11.4. It starts in the west at the Atlas Mountains of North Africa and continues through the European Alps and the ranges of the Near East and Iran to join the Himalayas. The belt then continues through Southeast Asia into Indonesia, where it abruptly meets the circum-Pacific belt. Later we will return to these active belts of mountain-making and explain them in terms of lithospheric plate motions.

Continental Shields

Belts of recent and active mountain-making account for only a small portion of the continental crust. The remainder consists of comparatively inactive regions of much older rock. Within these stable regions we can recognize two types of crustal structure—continental shields and mountain roots. **Continental shields** are low-lying continental surfaces beneath which lie igneous

Figure 11.4 Principal mountain arcs, island arcs, and trenches of the world and the midoceanic ridge. (Midoceanic ridge map copyright © A. N. Strahler.)

Figure 11.5 Generalized world map of continental shields, exposed and covered. Continental centers of early Precambrian age (Archean) lie within the areas encircled by a broken red line. Heavy brown lines show mountain roots of Caledonian and Hercynian orogenies. (Based in part on data of R. E. Murphy, P. M. Hurley, and others. Copyright © A. N. Strahler.)

and metamorphic rocks in a complex arrangement. Figure 11.5 is a generalized map showing the shield areas of the continents. Two classes of shields are shown: exposed shields and covered shields.

Exposed shields include very old rocks, mostly from Precambrian time, and have had a very complex geologic history. An example of an exposed shield is the Canadian Shield of North America. Exposed shields are also extensive in Scandinavia, South America, Africa, Asia, peninsular India, and Australia. The exposed shields are largely regions of low hills and low plateaus, although there are some exceptions where large crustal blocks have been recently uplifted. Many thousands of meters of rock have been eroded from these shields throughout the past half-billion or more years.

Covered shields are areas of the continental shields that are covered by younger sedimentary layers, ranging in age from Paleozoic through Cenozoic eras. These strata accumulated at times when the shields were inundated by shallow seas. Marine sediments were laid down on the ancient shield rocks in thicknesses ranging from hundreds to thousands of meters. These shield areas were then broadly arched upward to become land surfaces again. Erosion has since removed large sections of their sedimentary cover, but it still remains intact over vast areas. The covered shields are shown in Figure 11.5.

Some core areas of the shields are composed of rock as old as early Precambrian time, dating back to a time period called the Archean Eon, 2.5 to 3.5 billion years ago. On our map, these ancient areas are shown encircled by bold red lines. The ancient cores are exposed in some areas but covered in others.

Ancient Mountain Roots

Remains of older mountain belts lie within the shields in many places. These *mountain roots* are mostly formed of Paleozoic and early Mesozoic sedimentary rocks that have been intensely bent and folded, and in some locations changed into metamorphic rocks—slate, schist, and quartzite, for example. Thousands of meters of overlying rocks have been removed from these old tectonic belts, so that only lowermost structures remain. Roots appear as chains of long, narrow ridges, rarely rising over a thousand meters above sea level. (Landforms of mountain roots are described in Chapter 16.)

Mountain roots, long eroded to low highlands and ridges, mark ancient plate collisions.

One important system of mountain roots was formed in the Paleozoic Era, during a great collision between two enormous lithospheric plates that took place about 400 million years ago. What were then high alpine mountain chains have since been worn down to belts of subdued mountains and hills. Today these roots, called Caledonides, form a highland belt across the northern British Isles and Scandinavia. They are also present in the Maritime Provinces of eastern Canada and in New England. A second, but younger, root system was formed during another great collision of plates near the close of the Paleozoic Era, about 250 million years ago. In North America, this highland system is represented by the Appalachian Mountains. The Caledonides and Appalachians are shown as mountain roots in Figure 11.5.

Relief Features of the Ocean Basins

The relief features of ocean basins are quite different from those of the continents. Crustal rock of the ocean floors consists almost entirely of basalt, which is covered over large areas by a comparatively thin accumulation of sediments. Age determinations of the basalt and its sediment cover show that the oceanic crust is quite young, geologically speaking. Much of the oceanic crust is less than 60 million years old, although some large areas have ages of about 65 to 135 million years. When we consider that the great bulk of the continental crust is of Precambrian age—mostly over 1 billion years old—the young age of the oceanic crust is quite remarkable. However, we will soon see how plate tectonic theory explains this young age.

The Midoceanic Ridge and Ocean Basin Floor

Figure 11.6 shows the important relief features of ocean basins. The ocean basins are characterized by a central ridge structure that divides the basin in about half. The *midoceanic ridge* consists of submarine hills that rise gradually to a rugged central zone. Precisely in the center of the ridge, at its highest point, is the *axial rift*, which is a narrow, trenchlike feature. The location and form of this rift suggest that the crust is being pulled apart along the line of the rift.

The midoceanic ridge and its principal branches can be traced through the ocean basins for a total distance of about 60,000 km (about 37,000 mi). Figure 11.4 shows the extent of the ridge. Beginning in the Arctic Ocean, it then divides the Atlantic Ocean basin from Iceland to the South Atlantic. Turning east, it enters the Indian Ocean, where one branch penetrates Africa. The other branch continues east between Australia and Antarctica, and then swings across the South Pacific. Nearing South America, it turns north and reaches North America at the head of the Gulf of California.

On either side of the midoceanic ridge are broad, deep plains and hill belts that belong to the *ocean basin floor* (Figure 11.6). Their average depth below sea level is about 5000 m (about 16,000 ft). The flat surfaces are called *abyssal plains*. They are extremely smooth because they have been built up of fine sediment that has settled slowly and evenly from ocean water above.

Continental Margins

The *continental margin*, shown on the left and right sides of Figure 11.6, is the narrow zone in which oceanic lithosphere is in contact with continental lithosphere (see Figure 11.2*b*). As the continental margin is approached from deep ocean, the ocean floor begins to slope gradually upward, forming the *continental rise* (Figure 11.6). The floor then steepens greatly on the *continental slope*. At the top of this slope we arrive at the

Figure 11.6   This schematic block diagram shows the main features of ocean basins. It applies particularly well to the North and South Atlantic oceans.

edge of the *continental shelf,* a gently sloping platform that is about 120 to 160 km (about 75 to 100 mi) wide along the eastern margin of North America. Water depth is about 150 m (about 500 ft) at the outer edge of the shelf.

The symmetrical model illustrated in Figure 11.6 is nicely shown in the North Atlantic and South Atlantic Ocean basins. It also applies rather well to the Indian Ocean and Arctic Ocean basins. The margins of these symmetrical basins are described as *passive continental margins,* which means they have not been subjected to strong tectonic and volcanic activity during the last 50 million years. Both continental and oceanic lithosphere at a passive continental margin are part of the same lithospheric plate and move together, away from the axial rift.

Over time, passive continental margins accumulate great quantities of continental sediments.

Many passive continental margins have accumulated great thicknesses of continental sediments. Near the continent, the strata of the shelf form a wedge-shaped deposit, thickening oceanward. It is formed by river-borne sediments that are spread by currents over the shallow floor of the shelf. A block diagram, Figure 11.7, shows details of this deposit. Beneath the continental rise and its adjacent abyssal plain is another thick sediment deposit. It is formed from sediments carried down the continental slope by swift muddy currents. The weight of these sediments causes them to sink, depressing the oceanic crust at its margin with the continental crust.

Deltas built by rivers contribute a great deal of the shelf sediment. Tonguelike turbid currents thickened with fine silt and clay from these deltas carve submarine canyons into the continental shelf. Where these currents emerge from the canyons, they deposit their sediment and build *deep sea cones* (Figure 11.8).

The margins of the Pacific Ocean basin are very different from those of the Atlantic. Although the Pacific has a midoceanic ridge with ocean basin floors on either side, its margins are characterized by mountain arcs or island arcs with deep offshore *oceanic trenches.* Geologists refer to these trenched ocean-basin edges as *active continental margins.*

The locations of the major trenches are shown in Figure 11.4. Trench floors can reach depths of nearly 11,000 m (about 36,000 ft), although most range from 7,000 to 10,000 m (about 23,000 to 33,000 ft). Many lines of scientific evidence show that the oceanic crust is bent down sharply to form these trenches and that they mark the boundary between two lithospheric plates that are being brought together.

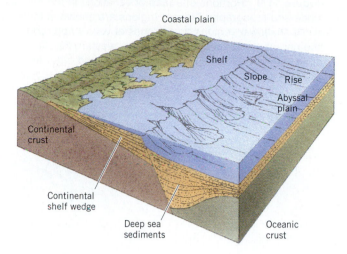

Figure 11.7 This block diagram shows an inner wedge of sediments beneath the continental shelf and an outer wedge of deep-sea sediments beneath the continental rise and abyssal plain.

Many details of the ocean basins and their submarine landforms are shown in Figure 11.9. This image of the ocean floor was constructed from data of a U.S. Navy satellite that measured the surface height of the ocean very precisely using radar. Because small variations in gravity caused by undersea ridges and valleys produce corresponding small variations in the ocean height, it is possible to infer the undersea topography from height measurement. The right page of the figure shows the North Atlantic basin. The prominent central feature is the Mid-Atlantic Ridge. The left page shows the deep trenches of the western Pacific that mark the positions of subduction arcs—features of plate motion that we will discuss shortly.

Figure 11.8 A deep-sea cone and its relationship to the continental shelf (Copyright © A. N. Strahler.)

Figure 11.9 Portions of a map of variations in the pull of gravity, which indicate undersea topography. Deeper regions are shown in tones of purple, blue, and green, while shallower regions are in tones of yellow and reddish brown. On the left page, the ring of subduction trenches of the western Pacific basin is a prominent feature. On the right page, the Mid-Atlantic Ridge and other features of sea-floor spreading are visible. Data were acquired by the U.S. Navy Geosat satellite altimeter. (Copyright © 1995, David T. Sandwell. Used by permission.)

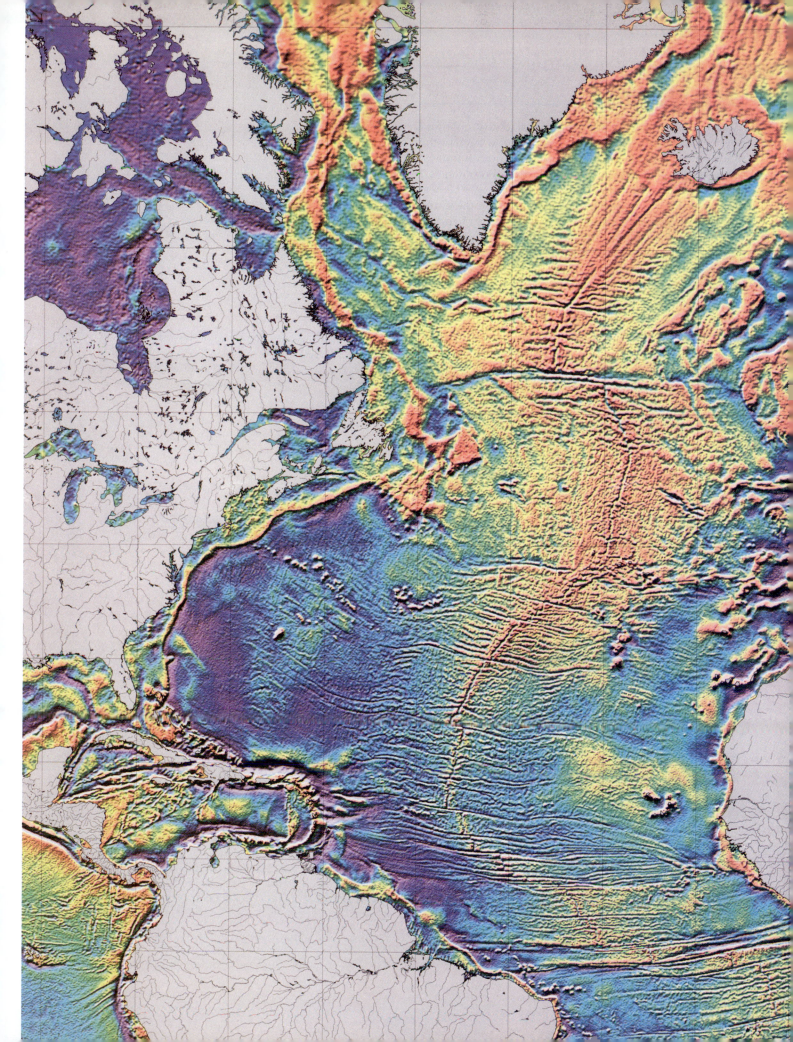

PLATE TECTONICS

With the structures and features of continents and ocean basins as a background, we can now turn to the motion of lithospheric plates and their interactions at boundaries—that is, to **plate tectonics**. **Tectonics** is a noun meaning "the study of tectonic activity." Here it refers to all forms of breaking and bending of the entire lithosphere, including the crust. Breaking and bending occur on a very large scale when lithospheric plates collide. Sedimentary strata that lay flat and undisturbed for tens of millions of years on ocean floors and continental margins are crumpled and sheared into fragments by such collisions. The collision process itself may take millions of years to complete, but in the perspective of geologic time, it's only a brief episode.

Tectonic Processes

Generally speaking, prominent mountain masses and mountain chains (other than volcanic mountains) are elevated by one of two basic tectonic processes: *compression* and *extension* (Figure 11.10). Extensional tectonic activity—"pulling apart"—occurs where oceanic plates are separating or where a continental plate is undergoing breakup into fragments. As the crust thins, it is fractured and pushed upward, producing block mountains. We will return to a more detailed description of this process in a later section on continental rupture.

Figure 11.10 Two basic forms of tectonic activity. Flat-lying rock layers may be compressed to form folds or pulled apart to form faults by rifting.

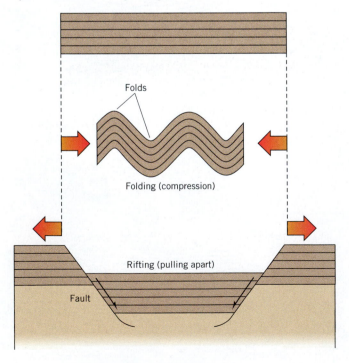

Folds

Folding (compression)

Rifting (pulling apart)

Fault

Overturned fold

Recumbent fold

Overthrusting

Thrust sheet

Overthrust fault

Figure 11.11 Schematic diagrams to show the development of a recumbent fold, broken by a low-angle overthrust fault to produce a thrust sheet, or nappe, in alpine structure. (Based on diagrams by A. Heim, 1922, *Geologie der Schweiz*, vol. II–1, Tauschnitz, Leipzig.)

Compressional tectonic activity—"squeezing together" or "crushing"—acts at converging plate boundaries. The result is often an alpine mountain chain consisting of intensely deformed strata of marine origin. The strata are tightly compressed into wavelike structures called *folds*.

A simple experiment you can do will demonstrate how folding occurs in mountain-building. Take an ordinary towel of thick, limp cloth and lay it out on a smooth table top. Using both hands, palms down, bring the ends of the cloth slowly together. First, a simple upfold will develop and grow, until it overturns to one side or the other. Then more folds will form, grow, and overturn, giving you a simple model of a new mountain range.

Orogens are huge deformed rock masses formed by continent–continent collisions.

Typically, alpine folds can be traced through a history. First, they are overturned, and then they become *recumbent* as they are further overturned upon themselves (Figure 11.11). Eventually, the rock layers are broken and separated, forming a *fault*. (We will return to faults in Chapter 12 in more detail.) In this type of faulting, slices of rock move over the underlying rock on fault surfaces with gentle inclination angles. These are *overthrust faults*. Individual rock slices, called *thrust sheets*, are carried many tens of kilometers over the underlying rock. In the European Alps, thrust sheets of this kind

were named *nappes* (from the French word meaning "cover sheet" or "tablecloth"). Nappes may be thrust one over the other to form a great pile. The entire deformed rock mass produced by such compressional mountain-making is called an *orogen*, and the event that produced it is an *orogeny*.

Plate Motions and Interactions

Figure 11.12 shows the major features of plate interactions. The vertical dimension of the block diagram (*a*) is greatly exaggerated, as are the landforms. A true-scale cross section (*b*) shows the correct relationships between crust and lithosphere, but surface relief features are too small to be shown.

Note that the figure shows two different kinds of lithospheric plates (see Figure 11.12*b*). Plates that lie beneath the ocean basins consist of *oceanic lithosphere*. This form of lithosphere is comparatively thin (about 50 km or 30 mi thick). The diagrams show two plates, X and Y, both made up of oceanic lithosphere. Plate Z bears continental crust and is made up of *continental lithosphere*, which is much thicker (about 150 km or 95 mi).

An important scientific principle applies to the relative surface heights of the two kinds of lithosphere. The lithosphere can be thought of as "floating" on the soft asthenosphere. Consider two blocks of wood, one thicker than the other, floating in a pan of water. The surface of the thick block will ride higher above the water surface than that of the thin block. This princi-

Figure 11.12 Schematic cross sections showing some of the important elements of plate tectonics. Part *a* is greatly exaggerated in vertical scale, and emphasizes surface and crustal features. Only the uppermost 30 km (20 mi) is shown. Part *b* is drawn to true scale and shows conditions to a depth of 250 km (155 mi). Here the actual relationships between lithospheric plates can be examined, but surface features are too small to be shown. (Copyright © A. N. Strahler.)

ple explains why the continental surfaces rise high above the ocean floors.

As shown in Figure 11.12, plates X and Y are pulling apart along their common boundary, which lies along the axis of a midoceanic ridge. This pulling apart tends to create a gaping crack in the crust, but magma continually rises from the mantle beneath to fill it. The magma appears as basaltic lava in the floor of the rift and quickly congeals. At greater depth under the rift, magma solidifies into gabbro, an intrusive rock of the same composition as basalt. Together, the basalt and gabbro continually form new oceanic crust. This type of boundary between plates is termed a *spreading boundary*.

At the right, the oceanic lithosphere of plate Y is moving toward the thick mass of continental lithosphere that comprises plate Z. Where these two plates collide, they form a *converging boundary*. Because the oceanic plate is comparatively thin and dense, in contrast to the thick, buoyant continental plate, the oceanic lithosphere bends down and plunges into the soft layer, or asthenosphere. The process in which one plate is carried beneath another is called **subduction**.

SCIENCE NEWS 18

The leading edge of the descending plate is cooler and therefore more dense than the surrounding hot, soft asthenosphere. As a result, the slab "sinks under its own weight," once subduction has begun. However, the slab is gradually heated by the surrounding hot rock and thus eventually softens. The underportion, which is mantle rock in composition, simply reverts to mantle rock as it softens.

Sea-floor spreading builds lithospheric plates by accretion. Plate material is lost to consumption when it sinks into plastic mantle rock and melts.

However, the descending plate is covered by a thin upper layer of less dense mineral matter derived from oceanic and continental sediments. This material can melt and become magma. The magma tends to rise because it is less dense than the surrounding material. Figure 11.12 shows some magma pockets formed from the upper edge of the slab. They are pictured as rising like hot-air balloons through the overlying continental lithosphere. When they reach the earth's surface, they form a chain of volcanoes lying about parallel with the deep oceanic trench that marks the line of descent of the oceanic plate.

Viewing plate Y as a unit in Figure 11.12, we see that this single lithospheric plate is simultaneously undergoing both *accretion* (growth by addition) and *consumption* (loss by softening and melting in subduction zones). If

Figure 11.13 A transform fault involves the horizontal motion of two adjacent lithospheric plates, one sliding past the other. (Copyright © by A. N. Strahler.)

rates of accretion and consumption are equal, the plate will maintain its overall size. If consumption is slower, the plate will expand. If accretion is slower, the plate will shrink.

We have yet to consider a third type of lithospheric plate boundary. Two lithospheric plates may be in contact along a common boundary on which one plate merely slides past the other with no motion that would cause the plates either to separate or to converge (Figure 11.13). This is a *transform boundary*. The plane along which motion occurs is a nearly vertical fracture extending down through the entire lithosphere, and it is called a *transform fault*. A *fault* is a rock plane along which there is motion of the rock mass on one side with respect to that on the other. (More about faults will appear in Chapter 12.) Transform boundaries are often associated with midoceanic ridges and are shown in Figures 11.7 and 11.12.

In summary, there are three major kinds of active plate boundaries:

- *Spreading boundaries.* New lithosphere is being formed by accretion. Example: Sea-floor spreading along the axial rift.

- *Converging boundaries.* Subduction is in progress, and lithosphere is being consumed. Example: Active continental margin.

- *Transform boundaries.* Plates are gliding past one another on a transform fault. Example: Transform boundary associated with midoceanic ridge.

The Global System of Lithospheric Plates

The global system of lithospheric plates consists of six great plates. These are listed in Table 11.1 and are shown on a world map in Figure 11.14. Several lesser plates and subplates are also recognized. They range in size from intermediate to comparatively small. Plate

Table 11.1 The Lithospheric Plates

Great plates	Lesser Plates
Pacific	Nazca
American (North, South)	Cocos
Eurasian	Philippine
Persian subplate	Caribbean
African	Arabian
Somalian subplate	Juan de Fuca
Austral-Indian	Caroline
Antarctic	Bismark
	Scotia

boundaries are shown by symbols, explained in the key accompanying the map.

The great Pacific plate occupies much of the Pacific Ocean basin and consists almost entirely of oceanic lithosphere. Its relative motion is northwesterly, so that it has a subduction boundary along most of the western and northern edge. The eastern and southern edge is mostly a spreading boundary. A sliver of continental lithosphere is included and makes up the coastal portion of California and all of Baja California. The California portion is bounded by an active transform fault (the San Andreas fault).

The earth's lithospheric shell is divided into six great lithospheric plates and at least nine lesser plates.

The American plate includes most of the continental lithosphere of North and South America as well as the entire oceanic lithosphere lying west of the midoceanic ridge that divides the Atlantic Ocean basin down the middle. For the most part, the western edge of the American plate is a subduction boundary, with oceanic lithosphere diving beneath the continental lithosphere. The eastern edge is a spreading boundary. (Some classifications recognize separate North American and South American plates.)

The Eurasian plate is mostly continental lithosphere, but it is fringed on the west and north by a belt of ocean-

Figure 11.14a World maps of lithospheric plates.

Figure 11.14 World maps of lithospheric plates (continued).

(b)

(c)

ic lithosphere. The African plate has a central core of continental lithosphere nearly surrounded by oceanic lithosphere.

The Austral-Indian plate takes the form of a long rectangle. It is mostly oceanic lithosphere but contains two cores of continental lithosphere—Australia and peninsular India. Recent evidence shows that these two continental masses are moving independently and may actually be considered to be parts of separate plates. The Antarctic plate has an elliptical shape and is almost completely enclosed by a spreading plate boundary. This means that the other plates are moving away from the pole. The continent of Antarctica forms a central core of continental lithosphere completely surrounded by oceanic lithosphere.

Figure 11.15 is a schematic circular cross section of the lithosphere along a great circle in low latitudes. It shows several of the great plates and their boundaries. The great circle is tilted by about 30° with respect to the equator. The section crosses the African plate, heads northeast across the Eurasian plate through the Himalayas to Japan and Korea, then dips southeast across the Pacific plate, cutting across the South American plate, and finally returns to Africa. Three spreading boundaries at midoceanic ridges are encountered, with two subduction zones (the Japan and Peru trenches) and a continent-to-continent collision, where the Austral-Indian plate dives under the Eurasian plate.

Subduction Tectonics

Converging plate boundaries, with subduction in progress, are zones of intense tectonic and volcanic

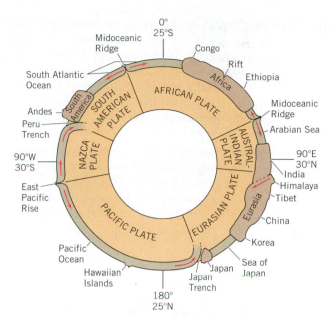

Figure 11.15   Schematic circular cross section of the major plates on a great circle tilted about 30 degrees with respect to the equator. (A. N. Strahler.)

activity. The narrow zone of a continent that lies above a plate undergoing subduction is therefore an active continental margin. (See plates Y and Z in Figure 11.12.) Let's look at this boundary in more detail (Figure 11.16).

The oceanic trench receives sediment coming from two sources. Carried along on the moving oceanic plate is deep-ocean sediment—fine clay and ooze—that has settled to the ocean floor. From the continent comes

Figure 11.16   Some typical features of an active subduction zone. The diagram uses a great vertical exaggeration to show surface and crustal details. Sediments scraped off the moving plate form tilted wedges that accumulate in a rising tectonic mass. Near the mainland is a shallow trough in which sediment brought from the land is accumulating. Metamorphic rock is forming above the descending plate. Magma rising from the top of the descending plate reaches the surface to build a chain of volcanoes. (Copyright © A. N. Strahler.)

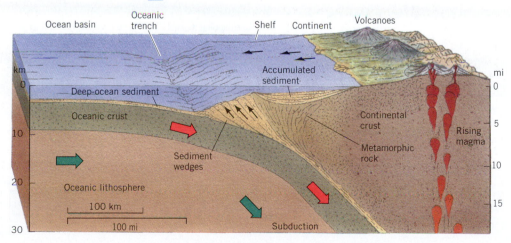

terrestrial sediment in the form of sand and mud brought by streams to the shore and then swept into deep water by currents. In the bottom of the trench, both types of sediment are intensely deformed and are dragged down with the moving plate. The deformed sediment is then scraped off the plate and shaped into wedges that ride up, one over the other, on steep fault planes. The sediment wedges accumulate into a large mass in which metamorphism takes place. In this way, the continental margin is built outward as new continental crust of metamorphic rock is formed.

Intense heating of the upper surface of the descending plate melts the oceanic crust, forming magma. Some of this magma rises and reaches the surface to form volcanoes.

Orogens and Collisions

Visualize, now, a situation in which two continental lithospheric plates converge along a subduction boundary. Ultimately, the two masses must collide because the impacting masses are too thick and too buoyant to allow either plate to slip under the other. The result is an orogeny in which various kinds of crustal rocks are crumpled into folds and sliced into nappes. This process has been called "telescoping" after the behavior of a folding telescope that collapses from a long tube into a short cylinder. Collision per-

manently unites the two plates, terminating further tectonic activity along that collision zone. Appropriately, the collision zone is named a **continental suture**. Our world map of plates, Figure 11.14, uses a special symbol for these sutures, shown in northern Africa and near the Aral Sea.

Continent–continent collisions occurred in the Cenozoic Era along a great tectonic line that marks the southern boundary of the Eurasian plate. (See Figure 11.14, map *b*.) The line begins with the Atlas Mountains of North Africa, and it runs across the Aegean Sea region into western Turkey. Beyond a major gap in Turkey, the line takes up again in the Zagros Mountains of Iran. Jumping another gap in southeastern Iran and Pakistan, the collision line sets in again in the great Himalayan Range, where it is still active.

Each segment of this collision zone represents the collision of a different north-moving plate against the single and relatively immobile Eurasian plate. A European segment containing the Alps was formed when the African plate collided with the Eurasian plate in the Mediterranean region. A Persian segment resulted from the collision of the Arabian plate with the Eurasian plate. A Himalayan segment represents the collision of the Indian continental portion of the Austral-Indian plate with the Eurasian plate.

Figure 11.17 is a series of cross sections in which the tectonic events of a typical continent–continent colli-

Figure 11.17 Schematic cross sections showing continent–continent collision and the formation of a suture zone with nappes. (Copyright © A. N. Strahler.)

sion are reconstructed. Diagram *a* shows a passive margin at the left and an active subduction margin at the right. As the ocean between the converging continents is eliminated, a succession of overlapping thrust faults cuts through the oceanic crust in diagram *b*. The thrust slices ride up, one over the other, telescoping the oceanic crust and the sediments above it. As the slices become more and more tightly squeezed, they are forced upward. The upper part of each thrust sheet assumes a horizontal attitude to form a nappe in diagram *c*, which then glides forward under gravity on a low downgrade. A mass of metamorphic rock is formed between the joined continental plates, welding them together. This new rock mass is the continental suture. It is a distinctive type of orogen.

Continent–continent collisions have occurred many times since the late Precambrian time. Several ancient sutures have been identified in the continental shields. The Ural Mountains, which divide Europe from Asia, are one such suture, formed near the end of the Paleozoic Era.

Continental Rupture and New Ocean Basins

We have already noted that the continental margins bordering the Atlantic Ocean basin on both its eastern and western sides are very different from the active margin of a subduction zone. At present, the Atlantic margins have no important tectonic activity and are passive continental margins. Even so, they represent the contact between continental lithosphere and oceanic lithosphere, with continental crust meeting oceanic crust, as shown in Figure 11.8. Passive margins are formed when a single plate of continental lithosphere is rifted apart. This process is called *continental rupture.*

Figure 11.18 uses three schematic block diagrams to show how continental rupture takes place and leads to the development of passive continental margins. At first, the crust is both lifted and stretched apart as the lithospheric plate is arched upward. The crust fractures and moves along faults—upthrown blocks form mountains, while downdropped blocks form basins.

Eventually a long narrow valley, called a *rift valley*, appears (block *a*). The widening crack in its center is continually filled in with magma rising from the mantle below. The magma solidifies to form new crust in the floor of the rift valley. Crustal blocks on either side slip down along a succession of steep faults, creating a mountainous landscape. The rift valley system of East Africa—described in more detail in Chapter 12—is a notable example of this stage of continental rupture. As separation continues, a narrow ocean appears, with a spreading plate boundary running down its center (block *b*). Plate accretion takes place in the central rift to produce new oceanic crust and lithosphere. The

union of continental lithosphere and oceanic lithosphere on either side of the spreading zone creates young passive margins. These margins move away from the spreading zone as spreading continues.

The Red Sea is a good example of a recent continental rupture.

The Red Sea is a narrow ocean formed by a continental rupture. Its straight coasts mark the edges of the rupture. The widening of such an ocean basin can continue until a large ocean has formed and the continents are widely separated (Figure 11.18, block *c*). Each continent now shows a well-developed passive margin where continental crust and lithospheric crust meet and sediment accumulates.

During the process of opening of an ocean basin, the spreading boundary develops a series of offsets, one of

Figure 11.18 Schematic block diagrams showing stages in continental rupture and the opening up of a new ocean basin. The vertical scale is greatly exaggerated to emphasize surface features. (*a*) The crust is uplifted and stretched apart, causing it to break into blocks that become tilted on faults. (*b*) A narrow ocean is formed, floored by new oceanic crust. (*c*) The ocean basin widens, while the passive continental margins subside and receive sediments from the continents. (Copyright © A. N. Strahler.)

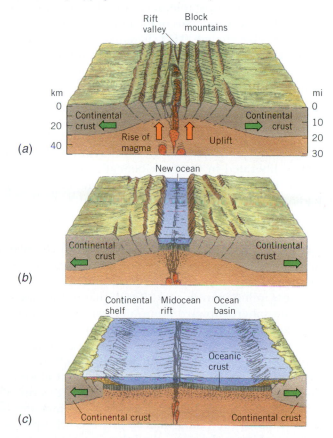

which is shown in the upper left-hand part of diagram *a* of Figure 11.12. The offset ends of the axial rift are connected by an active transform fault. As spreading continues, a scarlike feature is formed on the ocean floor as an extension of the transform fault. These *transform scars* take the form of narrow ridges or clifflike features and may extend for hundreds of kilometers across the ocean floors. Before the true nature of these scars was understood, they were called "fracture zones." That name still persists and can be seen on maps of the ocean floor. The scars are not, for the most part, associated with active faults.

The Power Source for Plate Movements

The system of lithospheric plates in motion represents a huge flow system of dense mineral matter. Its operation requires enormous power. It is generally agreed that the source of this power lies in the heat released by radioactivity. You may recall from basic chemistry that some elements are found in forms that differ with respect to their atomic mass. These forms are known as *isotopes*. For example, the most common isotope of uranium (U) is uranium-238, a form with an atomic mass of 238. Another form is uranium-235, which has an atomic mass number of 235. Chemists distinguish these two forms by writing them as ^{238}U and ^{235}U.

A key to the understanding of radioactivity is that certain isotopes are unstable, meaning that the composition of the atomic nucleus of the isotope can experience an irreversible change. This change process is known as *radioactive decay*. When a nucleus decays, it emits matter and energy. The energy is absorbed by the surrounding matter and thus ultimately takes the form of sensible heat. The flow of heat from radioactive decay is termed *radiogenic heat*. Nearly all of this heat is generated by the decay of isotopes of uranium (^{238}U, ^{235}U), thorium (^{232}Th), and potassium (^{40}K). Most of the radiogenic heat is liberated in the rock beneath the continents, within the uppermost 100 km or so (about 60 mi). Radiogenic heat is sufficient to keep earth layers below the crust close to the melting point, and thus provides the power source for the formation of magma.

How does radiogenic heating produce plate motions? The answer is not completely known, but one theory is that plate motions are produced by convection currents in hot mantle rock. Recall that in the case of the atmosphere, upward motion of warmer, less dense air takes place by convection. It is thought that, in a somewhat similar type of convection, unequal heating produces streams of upwelling mantle rock that rise steadily beneath spreading plate boundaries. Geologists do not yet have a full understanding of how this rise of heated rock causes plates to move. One hypothesis states that, as the rising mantle lifts the lithospheric plate to a higher elevation, the lithospheric plate tends

to move horizontally away from the spreading axis under the influence of gravity.

At the opposite edge of the plate, subduction occurs because the oceanic plate is colder and denser than the asthenosphere through which it is sinking. The motion of the oceanic plate exerts a drag on the underlying asthenosphere, setting in motion flow currents in the upper mantle. Thus, slow convection currents probably exist in the asthenosphere beneath the moving plates, but their pathways and depths of operation are not well understood.

CONTINENTS OF THE PAST

The continents are moving today. Data from orbiting satellites have shown that rates of separation, or of convergence, between plate pairs are on the order of 5 to 10 centimeters (2 to 4 in.) per year, or 50 to 100 km per million years. At that rate, global geography must have been very different in past geologic eras than it is today. Many continental riftings and many plate collisions have taken place over the past 2 billion years. Single continents have fragmented into smaller ones, while at other times, small continents have merged to form large ones.

A brief look at the changing continental arrangements and locations over the past 250 million years (m.y.) is well worthwhile in our study of physical geography because those changes brought with them changes in climate, soils, and vegetation on each continent as the continents moved across the parallels of latitude.

Modern reconstructions of the global arrangements of past continents have been available since the mid-1960s. Global maps drawn for each geologic period have been repeatedly revised in the light of new evidence, but differences of interpretation affect only the minor details.

> **A single great continent, Pangea, surrounded by a single great ocean, Panthalassa, broke apart during millions of years to form the present-day continents.**

Figure 11.19 shows stages in the separations and travels of our continents, starting in the Mesozoic Era. Geologists now agree that in the Permian Period, ending about 250 million years ago, there existed a single great supercontinent named *Pangea*. Thus, there was only one worldwide superocean, named *Panthalassa*. In map *a* Pangea lies astride the equator and extends nearly to the two poles. Regions that are now North America and western Eurasia lie in the northern hemisphere. Jointly they are called *Laurasia*. Regions that are now South America, Africa, Antarctica, Australia, New Zealand, Madagascar, and peninsular India lie south of

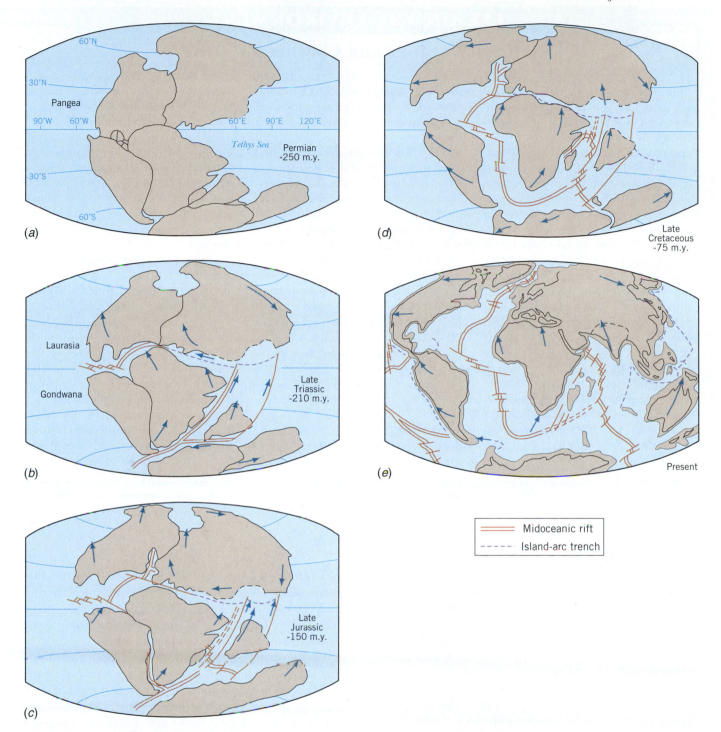

Figure 11.19   The breakup of Pangea is shown in five stages. Inferred motion of lithospheric plates is indicated by arrows. (Redrawn and simplified from maps by R. S. Dietz and J. C. Holden, *Journal of Geophysical Research*, vol. 75, pp. 4943–4951, Figures 2 to 6. Copyrighted by the American Geophysical Union. Used by permission.)

the equator. Jointly, they go by the name *Gondwana.* Subsequent maps (*b* through *d*) show the breaking apart and dispersal of the Laurasia and Gondwana plates to yield their modern components and locations, as seen in map *e.*

Note in particular that North America traveled from a low-latitude location into high latitudes, finally closing off the Arctic Ocean as a largely landlocked sea. This change may have been a major factor in bringing on the Ice Age in late Cenozoic time. (See Chapter 18.) Another traveler was the Indian Peninsula, which started out from a near-subarctic southerly location in Permian time and streaked northeast across the Thethys Sea to collide with Asia in the northern tropical savanna

EYE ON THE ENVIRONMENT

Alfred Wegener and Continental Drift

Although modern plate tectonic theory is only a few decades old, the concept of a breakup of an early supercontinent into fragments that drifted apart dates back to the nineteenth century and beyond. Almost as soon as good navigational charts became available to show the continental outlines, geographers became intrigued with the close correspondence in outline between the eastern coastline of South America and the western coastline of Africa. In 1858 Antonio Snider-Pelligrini produced a map to show that the American continents nested closely against Africa and Europe. He went beyond the purely geometrical fitting to suggest that the reconstructed single continent explains the close similarity of fossil plant types in coal-bearing rocks in both Europe and North America.

Moving ahead to the early twentieth century, we come to the ideas of two Americans, Frank B. Taylor and Howard B. Baker, whose published articles presented evidence favoring the hypothesis that the New World and Old World continents had drifted apart. Credit for a full-scale hypothesis of the breakup of a single supercontinent and the drifting apart of individual continents belongs to Alfred Wegener, the German meteorologist and geophysicist who offered geologic evidence as early as 1915 that the continents had once been united. He reconstructed a supercontinent named *Pangea*, which existed intact as early as about 300 million years ago in the Carboniferous Period. One of his maps is reproduced here (see figure at right). It is strikingly similar to the one we have shown as Figure 11.19*a*.

A storm of controversy followed, and many American geologists denounced Wegener's "continental drift" hypothesis. However, he had some loyal supporters in Europe, South Africa, and Australia. Several lines of hard scientific evidence presented by Wegener and his followers strongly favored the existence of Pangea. Those arguments remain strong today. But the actual physical process of separation of the continents that Wegener proposed was strongly criticized on valid physical grounds.

Wegener proposed that a continental layer of less dense rock had moved like a great floating "raft" through a "sea" of denser oceanic crustal rock. Note that both the "raft" and the "sea" are composed of strong, brittle rock. Geologists could show, by use of established principles of physics, that this proposed mechanism was impossible because two strong, rigid crustal rock layers, one embedded in the other, could not behave in such a fashion. In 1930, Wegener perished of cold and

zone. India's bedrock still bears the grooves and scratches of a great glaciation in Permian time, when it was nearer to the south pole. Looking ahead, we may safely predict that the Atlantic Ocean basins will become much wider, and the westward motion of the Americas will cause a reduction in the width of the Pacific basin.

Alfred Wegener (1880–1930), a German meteorologist and geophysicist, is generally credited with developing the first complete hypothesis describing the formation of present-day continents from a single large continent. His ideas are described in more detail in *Eye on the Environment • Alfred Wegener and Continental Drift*.

A LOOK AHEAD

Over the long spans of geologic time, our planet's surface is shaped and reshaped endlessly. Oceans open and close, mountains rise and fall. Rocks are folded, faulted, and fractured. Arcs of volcanoes spew lava to build chains of lofty peaks. Deep trenches consume sediments that are carried beneath adjacent continents. These activities are byproducts of the motions of lithospheric plates and are described by plate tectonics.

For geographers, the great value of plate tectonics is that it provides a grand scheme for understanding the nature and distribution of the largest and most obvious features of our planet's surface—its continents and ocean basins and their major relief features. However, large features, such as mountain ranges, are made up of many smaller features—individual peaks, for example. In the next chapter, we will look at the continental surface in more detail, examining the volcanic and tectonic landforms that result when volcanoes erupt and rock layers are folded and faulted. As we will see, the formation of these landforms is often marked by powerful earthquakes—phenomena that can produce major impacts indeed on modern technological society.

Alfred Wegener's 1915 map fitting together the continents that today border the Atlantic Ocean basin. The sets of dashed lines show the fit of Paleozoic tectonic structures between Europe and North America and between southernmost Africa and South America. (From A. Wegener, 1915, *Die Entstehung der Kontinente und Ozeane*, F. Vieweg, Braunschweig.)

exhaustion while on an expedition into the Greenland Ice Sheet. During the following three decades, only a small minority of scientists promoted his hypothesis.

In the 1960s, however, geophysicists showed beyond doubt that thick lithospheric plates are in motion, both along the midoceanic ridge and beneath the deep offshore trenches of the continents. Other scientists used magnetic data "frozen" into crustal rock to conclude that the continents had moved great distances apart. Within only a few years, Wegener's scenario was validated, but only by applying a mechanism never dreamed of in his time.

Questions

1. Provide a brief history of the idea of "drifting continents."

2. What was Wegener's theory about "continental drift?" Why was it opposed at the time?

THE LITHOSPHERE AND PLATE TECTONICS IN REVIEW

At the center of the earth lies the core—a dense mass of liquid iron and nickel that is solid at the very center. Enclosing the metallic core is the mantle, composed of ultramafic rock. The outermost layer is the crust. Continental crust consists of two zones—a lighter zone of felsic rocks atop a denser zone of mafic rocks. Oceanic crust consists only of denser, mafic rocks. The lithosphere, the outermost shell of rigid, brittle rock, includes the crust and an upper layer of the mantle. Below the lithosphere is the asthenosphere, a region of the mantle in which mantle rock is soft, or plastic.

Continental masses consist of active belts of mountain-making and inactive regions of old, stable rock. Mountain-building occurs by volcanism and tectonic activity. Alpine chains occur in two principal mountain belts—the circum-Pacific and Eurasian-Indonesian belts. Continental shields are regions of low-lying igneous and metamorphic rocks. They may be exposed or covered by layers of sedimentary rocks. Ancient mountain roots lie within some shield regions.

The ocean basins are marked by a midoceanic ridge with its central axial rift. This ridge occurs at the site of crustal spreading. Most of the ocean floor is abyssal plain, covered by fine sediment. As passive continental margins are approached, the continental rise, slope, and shelf are encountered. Thick wedges of sediment underlie the shelf, slope, and rise. At active continental margins, deep oceanic trenches lie offshore.

The two basic tectonic processes are compression and extension. Both processes can lead to the formation of mountains. Under the compression resulting from the collision of lithospheric plates, rock layers can be shaped into folds, then break and move atop one another along thrust faults.

Continental lithosphere includes the thicker, lighter continental crust and a rigid layer of mantle rock beneath. Oceanic lithosphere is comprised of the thinner, denser oceanic crust and rigid mantle below. The lithosphere is fractured and broken into a set of plates, large and small, that move with respect to each other. Where plates move apart, a spreading boundary occurs. At converging boundaries, plates collide. At transform boundaries, plates move past one another on a transform fault. There are six major lithospheric plates.

When oceanic lithosphere and continental lithosphere collide, the denser oceanic lithosphere plunges beneath the continental lithospheric plate, a process called subduction. A trench marks the site of downplunging. Some subducted oceanic crust melts and rises to the surface, producing volcanoes. Under the severe compression that occurs with continent–continent collision, the two continental plates are welded together in a zone of metamorphic rock named a continental suture.

In continental rupture, extensional tectonic forces move a continental plate in opposite directions, creating a rift valley. Eventually the rift valley widens and opens to the ocean, and new oceanic crust forms as spreading continues.

Plate movements are thought to be powered by radiogenic heat, released by decay of naturally occurring radioactive isotopes found in the upper layers of the crust and mantle. The heat is thought to generate convection currents in the plastic mantle rock of the asthenosphere. Upwelling mantle rock lifts lithospheric plates, causing them to fracture and slide away from the rift. During the Permian Period, the continents were joined in a single, large supercontinent—Pangea—that broke apart, leading eventually to the present arrangement of continents and ocean basins.

KEY TERMS

core	lithosphere	continental shields
mantle	asthenosphere	plate tectonics
crust	lithospheric plates	tectonics
continental crust	mountain arc	subduction
oceanic crust	island arcs	continental suture

REVIEW QUESTIONS

1. Describe the earth's inner structure, from the center outward. What types of crust are present? How are they different?
2. How do geologists use the term *lithosphere*? What layer underlies the lithosphere, and what are its properties? Define the term *lithospheric plate*.
3. What proportion of the earth's surface is in ocean? in land? Do these proportions reflect the true proportions of continents and oceans? If not, why not?
4. What are the two basic subdivisions of continental masses?
5. What term is attached to belts of active mountain-making? What are the two basic processes by which mountain belts are constructed? Provide examples of mountain arcs and island arcs.
6. What is a continental shield? How old are continental shields? What two types of shields are recognized?
7. Describe how compressional mountain-building produces folds, faults, overthrust faults, and thrust sheets (nappes).

8. Name the six great lithospheric plates. Identify an example of a spreading boundary by general geographic location and the plates involved. Do the same for a converging boundary.
9. Describe the process of subduction as it occurs at a converging boundary of continental and oceanic lithospheric plates. How is the continental margin extended? How is subduction related to volcanic activity?
10. How does continental rupture produce passive continental margins? Describe the process of rupturing and its various stages.
11. What are transform faults? Where do they occur?
12. What forces are thought to power plate tectonics? Explain.
13. Briefly summarize the history of our continents that geologists have reconstructed beginning with the Permian Period.

VISUALIZING EXERCISES

1. Sketch a cross section of an ocean basin with passive continental margins. Label the following features: midoceanic ridge, axial rift, abyssal plain, continental rise, continental slope, and continental shelf.
2. Identify and describe two types of lithospheric plates. Sketch a cross section showing a collision between the two types. Label the following features: oceanic crust, continental crust, mantle, oceanic trench, and rising magma. Indicate where subduction is occurring.
3. Sketch a continent–continent collision and describe the formation process of a continental suture. Provide a present-day example where a continental suture is being formed, and give an example of an ancient continental suture.

A DEEPER VIEW OF THE LITHOSPHERE AND PLATE TECTONICS

1. Figure 11.14 is a Mercator map of lithospheric plates, whereas Figure 11.15 presents a cross section of plates on a great circle. Construct a similar cross section of plates on the 30° S parallel of latitude. As in Figure 11.15, label the plates and major geographic features. Refer to Figure 11.4 for arcs, trenches, and midocean ridges.
2. Suppose astronomers discover a new planet that, like earth, has continents and oceans. They dispatch a reconnaissance satellite to photograph the new planet. What features would you look for, and why, to detect past and present plate tectonic activity on the new planet?

Diamond Head with Waikiki Beach, Honolulu, Hawaii, in the distance. Diamond Head is a peak on the rim of an extinct, eroded volcanic crater.

Some of the most devastating natural phenomena are eruptions of volcanoes and great earthquakes, both of which are subjects of this chapter. You've probably seen videos or television news coverage of these spectacular events. But did you ever think what it would be like to experience a tsunami—a seismic sea wave generated by an earthquake or explosive volcanic eruption?

Imagine yourself on the balcony of an ocean-view room at an exclusive beach hotel at Waikiki, taking in a view that is postcard-perfect. The arc of hotels stretches to your right and left, fringed with palms and pools. Beyond is the beach, lined with sun worshippers. Sailboats glide lazily back and forth in the lagoon, while outrigger canoes move more deliberately under paddle power.

As you gaze out beyond the surfriders and sailors, you notice a white foaming of the surf out at the edge of the reef. Slowly the foaming intensifies, and within a minute or so, the reef is a white crescent of churning, seething water. Something very wrong is happening. The water level in the lagoon is rising, and very rapidly, as ocean water pours in over the reef. Accompanying the water are ocean waves of full force, no longer damped by the effects of the reef and lagoon. Waves are breaking higher and higher on the beach, and within a few moments, tongues, then torrents, of sea water begin to pour toward the hotels.

Bathers running landward are overtaken and tumbled among the tables and chairs swept from beachside terraces. Palm trees bend and fall under the onslaught.

As the first waves hit your hotel, you feel a shaking motion, like an earthquake. The beach has completely disappeared, and the ocean waves are crashing full-force against the hotel, shattering glass windows and doors and breaking interior partitions. You can feel the thumping and tearing of steel and concrete as the building shudders with the blows.

Just as you are beginning to think about what to do, the water begins to retreat, first slowly, then as a great outgoing flood. Splintered woodwork, uprooted plants, and even floating cars flash past as the ebbing waters carry them seaward. The beach is visible once again, now gullied by the ocean-bound water. The surf withdraws far out in the lagoon, leaving a vast sandy plain littered with debris and shallow water pools, glistening in the sun. Awestruck by the devastation spread in front of you, you thank your lucky stars that you were not a casualty of the great tsunami. But then you glance seaward, and stare with horror. The reef is awash with foam once more. Another great sea wave is headed for the beach!

Seem far-fetched? Not a bit. On April 1, 1946, a major earthquake close to the Aleutian Islands generated a tsunami that struck the Hawaiian Islands five hours later, killing 170 persons. An observer on the north shore of Oahu documented the wave as a rise and fall of sea level, repeated several times at 12-minute intervals. The first wave inundated the coast to a height of 3.9 m (13 ft) above tide level. The water then retreated, but it was followed minutes later by an even higher landward surge, this time reaching 5.1 m (17 ft) above tide level. A third, even higher surge, followed the second. On the island of Hawaii, the city of Hilo was ravaged by breaking waves at a height of 9 m (30 ft) above sea level.

A tsunami is a very dramatic example of earth forces at work—in this case, a secondary effect of a huge, submarine earthquake. This chapter focuses on how earth forces act to shape the landscape, from the building of steep-sided volcanoes by volcanic eruptions to the lifting of huge volumes of rock by motions along earthquake faults. As we will see, the power of the earth in building these landforms can be awesome indeed.

Chapter Outline

Landforms
Volcanic Activity
  Stratovolcanoes
  Shield Volcanoes
  Volcanic Activity over the Globe
Landforms of Tectonic Activity
  Fold Belts

Faults and Fault Landforms
The Rift Valley System of East Africa
Earthquakes
  Earthquakes and Plate Tectonics
  Seismic Sea Waves
  Earthquakes along the San Andreas Fault
Eye on the Environment • The Northridge Earthquake of 1994

Chapter 12

VOLCANIC AND TECTONIC LANDFORMS

Have you ever seen an active volcano? If you live on the west coast, perhaps you've visited Mount St. Helens to view the devastation of its huge explosion in 1980. If you've been to the island of Hawaii, you may have seen flowing lava at first-hand. Even if you've never visited a volcano, you've probably seen pictures or films of erupting volcanoes. These lofty mountain forms are unmistakable and are one example of the landforms we will study in this chapter.

Even if you've never seen a real volcano, you've probably visited mountains. Some mountains are simply masses of rock that are resistant to erosion, and so they stand out against a surrounding area of weaker rocks. Others, however, are the direct result of tectonic forces that have recently uplifted them. The mountains of the western United States contain many ranges that are examples of uplifted mountain blocks bounded by faults—planes of slippage along which rock masses move in opposite directions. Tectonic activity can also produce mountains by bending rock layers into long, upfolded ridges, with downfolded valleys in between.

By looking in detail at landforms created by volcanic and tectonic activity, we are moving down from the global perspective of enormous lithospheric plates that collide to form vast mountain arcs or separate to form yawning ocean basins. We are now "zooming in" to examine landscape features that are actually small enough to view from a single vantage point. As we will see, the global perspective of plate tectonics helps explain how these landscape features are created and where they occur.

LANDFORMS

Landforms are the surface features of the land—for example, mountain peaks, cliffs, canyons, plains, beaches, and sand dunes. Landforms are created by many processes, and much of the remainder of this book describes landforms and how they are produced. **Geomorphology** is the scientific study of the processes that shape landforms. In this chapter, we will examine landforms produced directly by volcanic and tectonic processes.

The shapes of continental surfaces reflect the "balance of power," so to speak, between two sets of forces. Internal earth forces act to move crustal materials upward through volcanic and tectonic processes, thus creating new rock at the earth's surface. In opposition are processes and forces that act to lower continental surfaces by removing and transporting mineral matter through the action of running water, waves and currents, glacial ice, and wind. We can refer to these processes of land sculpture collectively as **denudation.**

Seen in this perspective, landforms in general fall into two basic groups—initial landforms and sequential landforms. *Initial landforms* are produced directly by volcanic and tectonic activity. They include volcanoes and lava flows, as well as downdropped rift valleys and elevated mountain blocks in zones of recent crustal deformation. Figure 12.1*a* shows a mountain block uplifted by crustal activity. The energy for lifting molten rock and rigid crustal masses to produce the initial landforms has an internal heat source. As we explained in Chapter 11, this heat energy is generated largely by natural radioactivity in rock of the earth's crust and mantle. It is the fundamental energy source for the motions of lithospheric plates.

Landforms are produced by the endless struggle between internal earth forces that lift crustal masses upward and external forces of denudation that wear them down.

Landforms shaped by processes and agents of denudation belong to the group of *sequential landforms.*

Figure 12.1   Initial and sequential landforms. (Drawn by A. N. Strahler.)

Mountain block created:
an initial landform

(a)

Earth's crust fractured, dislocated by internal earth forces

Mountain block carved into sequential landforms

(a) Erosional:
Canyon
Divide

(b)

Earth forces dormant

(b) Depositional (fan)

The word "sequential" means that they follow in sequence after the initial landforms have been created, and a crustal mass has been raised to an elevated position. Figure 12.1*b* shows an uplifted crustal block (an initial landform) that has been attacked by agents of denudation and carved up into a large number of sequential landforms.

You can think of any landscape as representing the existing stage in a great contest of opposing forces. As lithospheric plates collide or pull apart, internal earth forces periodically elevate parts of the crust to create initial landforms. The external agents of denudation—gravity, running water, waves, wind, and glacial ice—persistently wear these masses down and carve them into vast numbers of smaller sequential landforms.

Today we can observe the many stages of this endless struggle between internal and external forces by traveling to different parts of the globe. Where we find high alpine mountains and volcanic chains, internal earth forces have recently dominated the contest. In the rolling low plains of the continental interiors, the agents of denudation have won a temporary victory. At other locations, we can find many intermediate stages. Because the internal earth forces act repeatedly and violently, new initial landforms keep coming into existence as old ones are subdued.

This chapter focuses on initial landforms that are created directly by volcanic activity and tectonic activity. After we have described the most important agents of denudation in Chapters 13 and 15, we will return in Chapter 16 to the sequential landforms that develop on volcanic and tectonic landforms.

VOLCANIC ACTIVITY

Volcanism, or volcanic activity, is one of the two principal processes of mountain-building. As lava is expelled at the surface to form volcanoes and thick lava flows, volcanic cones, ridges, and plateaus can combine to construct imposing mountain ranges.

A **volcano** is a conical or dome-shaped initial landform built by the emission of lava and its contained gases from a constricted vent in the earth's surface (Figure 12.2). The magma rises in a narrow, pipelike conduit from a magma reservoir lying beneath. Upon reaching the surface, igneous material may pour out in tongue-like lava flows. Magma may also be ejected in the form of solid fragments driven skyward under pressure of confined gases. Ejected fragments, ranging in size from boulders to fine dust, are collectively called *tephra*. Forms and dimensions of a volcano are quite varied, depending on the type of lava and the presence or absence of tephra.

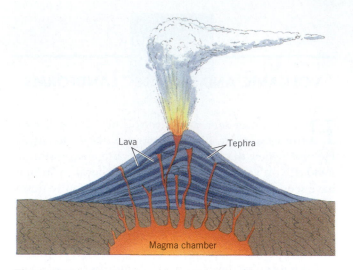

Figure 12.2 Idealized cross section of a stratovolcano with feeders from magma chamber beneath. The steep-sided cone is built up from layers of lava and tephra. (Copyright © A. N. Strahler.)

Stratovolcanoes

The nature of volcanic eruption, whether explosive or quiet, depends on the type of magma. Recall from Chapter 10 that there are two main types of igneous rocks: felsic and mafic. The felsic lavas (rhyolite and andesite) have a high degree of viscosity—that is, they are thick and gummy, and resist flow. So, volcanoes of felsic composition typically have steep slopes, and lava usually does not flow long distances from the volcano's vent. When the volcano erupts, tephra falls on the area surrounding the crater and contributes to the structure of the cone. *Volcanic bombs* are also included in the tephra. These solidified masses of lava range up to the size of large boulders and fall close to the crater.

The interlayering of sluggish streams of felsic lava and falls of tephra produces a **stratovolcano.** These tall, steep-sided volcanic cones usually steepen toward the summit, where a bowl-shaped depression—the *crater*—is located. Felsic lavas usually hold large amounts of gas under high pressure. As a result, these lavas can produce explosive eruptions. The eruption of Mount St. Helens in 1980 is an example of an explosive eruption of felsic lavas (Figure 12.3). Very fine volcanic dust from such eruptions can rise high into the troposphere and stratosphere, traveling hundreds or thousands of kilometers before settling to the earth's surface.

Lofty, conical stratovolcanoes are well known for their scenic beauty. Fine examples are Mount Hood and Mount St. Helens in the Cascade Range (Figure 12.3), Fujiyama in Japan, Mount Mayon in the Philippines (Figure 12.4), and Mount Shishaldin in the Aleutian Islands. Stratovolcanoes are sometimes referred to as *composite volcanoes*, or composite cones, since they are formed as composites of ash and lava.

Figure 12.3   Mount St. Helens, a stratovolcano of the Cascade Range in southwestern Washington, violently erupted on the morning of May 18, 1980, emitting a great cloud of condensed steam, heated gases, and ash from the summit crater. Within a few minutes, the plume had risen to a height of 20 km (12 mi), and its contents were being carried eastward by stratospheric winds. The eruption was initiated by explosive demolition of the northern portion of the cone, which is concealed from this viewpoint.

Figure 12.4   Mount Mayon, in southeastern Luzon, the Philippines, is often considered the world's most nearly perfect composite volcanic cone. It is an active volcano, and its summit rises to an altitude of nearly 2400 m (about 8000 ft). A series of eruptions began here in February 1993.

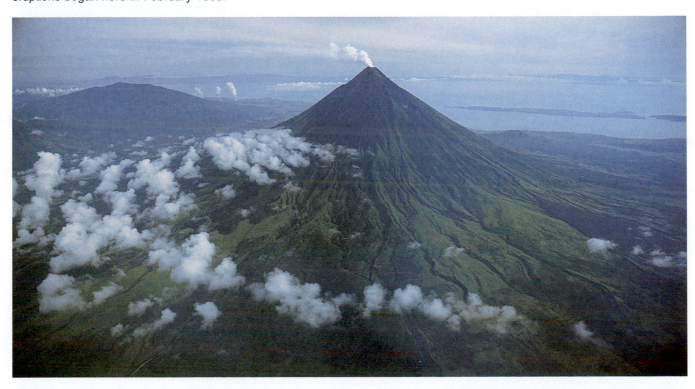

Another important form of emission from explosive stratovolcanoes is a cloud of white-hot gases and fine ash. This intensely hot cloud, or "glowing avalanche," travels rapidly down the flank of the volcanic cone, searing everything in its path. On the Caribbean island of Martinique, in 1902, a glowing cloud issued without warning from Mount Pelée. It swept down on the city of St. Pierre, destroying the city and killing all but two of its 30,000 inhabitants. More recently, glowing avalanches spawned by a 1997 eruption of the Soufrière Hills Volcano on Montserrat, another island of the Lesser Antilles, killed about 20 people in small villages. Plymouth, the island's capital, was flooded with hot ash and debris, causing extensive fires that devastated the evacuated city. The southern two-thirds of the small island was left uninhabitable. **SCIENCE NEWS** 19

Calderas

One of the most catastrophic of natural phenomena is a volcanic explosion so violent that it destroys the entire central portion of the volcano. Vast quantities of ash and dust are emitted and fill the atmosphere for many hundreds of square kilometers around the volcano. Only a great central depression, named a *caldera*, remains after the explosion. Although some of the upper part of the

volcano is blown outward in fragments, most of it settles back into the cavity formed beneath the volcano by the explosion.

Krakatoa, a volcanic island in Indonesia, exploded in 1883, leaving a huge caldera. Great seismic sea waves generated by the explosion killed many thousands of persons living in low coastal areas of Sumatra and Java. About 75 km³ (18 mi³) of rock was blown out of the crater during the explosion. Vast quantities of gas and fine particles of dust were carried into the stratosphere, where they contributed to a rosy glow of sunrises and sunsets that was seen around the world for several years afterward.

A classic example of a caldera produced in prehistoric times is Crater Lake, Oregon (Figure 12.5). The former volcano, named Mount Mazama, is estimated to have risen 1200 m (about 4000 ft) higher than the present caldera rim. The great explosion and collapse occurred about 6600 years ago.

Stratovolcanoes and Subduction Arcs

Most of the world's active stratovolcanoes lie within the circum-Pacific belt. Here, subduction of the Pacific, Nazca, Cocos, and Juan de Fuca plates is active. In Chapter 11, we explained how magmas rise beneath vol-

Figure 12.5 Crater Lake, Oregon, is surrounded by the high, steep wall of a great caldera. Wizard Island (center foreground) is an almost perfect basaltic cinder cone with basalt lava flows. It was built on the floor of the caldera after the major explosive activity had ceased.

Figure 12.6 Basaltic shield volcanoes of Hawaii. At lower left is the now-cold Halemaumau fire pit, formed in the floor of the central depression of Kilauea volcano. On the distant skyline is the snow-capped summit of Mauna Kea volcano, its elevation over 4000 m (about 13,000 ft).

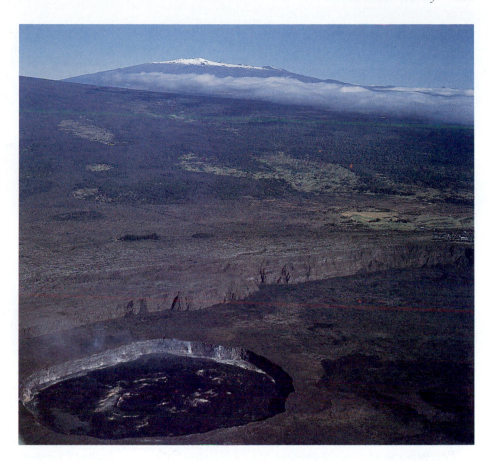

canic arcs of active continental margins (see Figure 11.14). One good example is the volcanic arc of Sumatra and Java, lying over the subduction zone between the Australian plate and the Eurasian plate. Another is the Aleutian volcanic arc, located where the Pacific plate dives beneath the North American plate. The Cascade Mountains of northern California, Oregon, and Washington form a similar chain. Important segments of the Andes Mountains in South America consist of stratovolcanoes.

Shield Volcanoes

In contrast to thick, gassy felsic lava, mafic lava (basalt) is often highly fluid. It typically has a low viscosity and holds little gas. As a result, eruptions of basaltic lava are usually quiet, and the lava can travel long distances to spread out in thin layers. Typically, then, large basaltic volcanoes are broadly rounded domes with gentle slopes. They are referred to as **shield volcanoes**. Hawaiian volcanoes are of this type.

The shield volcanoes of the Hawaiian Islands are characterized by gently rising, smooth slopes that flatten near the top, producing a broad-topped volcano (Figure 12.6). Domes on the island of Hawaii rise to summit elevations of about 4000 m (about 13,000 ft) above sea level. Including the basal portion lying below sea level, they are more than twice that high. In width

they range from 16 to 80 km (10 to 50 mi) at sea level and up to 160 km (about 100 mi) at the submerged base. The basalt lava of the Hawaiian volcanoes is highly fluid and travels far down the gentle slopes. Most of the lava flows issue from fissures (long, gaping cracks) on the flanks of the volcano.

Hawaiian lava domes have a wide, steep-sided central depression that may be 3 km (2 mi) or more wide and several hundred meters deep. These large depressions are a type of collapse caldera. Molten basalt is sometimes seen in the floors of deep pit craters that occur on the floor of the central depression or elsewhere over the surface of the lava dome.

Hotspots, Sea-Floor Spreading, and Shield Volcanoes

The chain of Hawaiian volcanoes were created by the motion of the Pacific plate over a *hotspot*—a plume of upwelling basaltic magma deep within the mantle, arising far down in the asthenosphere. Figure 12.7 diagrams this process. As the hot mantle rock rises, magma forms in bodies that melt their way through the lithosphere and reach the sea floor. Each major pulse of the plume sets off a cycle of volcano formation. However, the motion of the oceanic lithosphere eventually carries the volcano away from the location of the deep plume, and so it becomes extinct. Erosion processes wear the volcano away, and ultimately it becomes a low island.

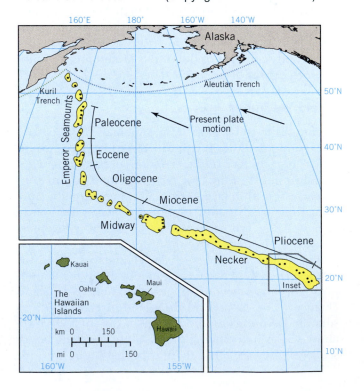

Figure 12.7 A chain of volcanic islands is formed as oceanic crust moves across a hotspot. (Copyright © A. N. Strahler.)

Continued attack by waves and slow settling of the island reduce it to a coral-covered platform. Eventually only a sunken island, or *guyot*, exists.

The process of crustal motion over a hotspot thus produces a long trail of islands and guyots. The Hawaiian trail, shown in Figure 12.8, trends northwestward. It is 2400 km (about 1500 mi) long and includes a sharp bend to the north caused by a sudden change of direction of the Pacific plate. This distant leg consists of the Emperor Seamounts. Several other long trails of volcanic seamounts cross the Pacific Ocean basin. They, too, follow parallel paths that reveal the plate motion.

A few basaltic volcanoes also occur along the midoceanic ridge, where sea-floor spreading is in progress. Perhaps the outstanding example is Iceland, in the North Atlantic Ocean. Iceland is constructed entirely of basalt. Basaltic flows are superimposed on older basaltic rocks as dikes and sills formed by magma emerging from deep within the spreading rift. Mount Hekla, an active volcano on Iceland, is a shield volcano somewhat similar to those of Hawaii. In Chapter 10, *Eye on the Environment • Battling Iceland's Heimaey Volcano* documents how valiant Icelanders coped with a recent eruption. Other islands consisting of basaltic volcanoes located along or close to the axis of the Mid-Atlantic Ridge are the Azores, Ascension, and Tristan da Cunha.

Where a mantle plume lies beneath a continental lithospheric plate, the hotspot may generate enormous volumes of basaltic lava that emerge from numerous vents and fissures and accumulate layer upon layer. The basalt may ultimately attain a thickness of thousands of meters and cover thousands of square kilometers. These accumulations are called *flood basalts*.

An important American example is found in the Columbia Plateau region of southeastern Washington, northeastern Oregon, and westernmost Idaho. Here, basalts of Cenozoic age cover an area of about 130,000

Figure 12.8 Sketch map of the Hawaiian seamount chain in the northwest Pacific Ocean basin. Dots are summits; the enclosing colored area marks the base of the volcano at the ocean floor. (Copyright © A. N. Strahler.)

Figure 12.9 Basalt lava flows exposed in cliffs bordering the Columbia River in Washington. Each set of cliffs is a major lava flow. In cooling, the lava acquires a vertical structure that makes it crack into tall columns.

km² (about 50,000 mi²)—nearly the same area as the state of New York. Individual basalt flows, exposed in the walls of river gorges, are expressed as cliffs, in which vertical joint columns are conspicuous (Figure 12.9).

Associated with flood basalts, shield volcanoes, and scattered occurrences of basaltic lava flows is a small volcano known as a *cinder cone* (Figure 12.10). Cinder cones form when frothy basalt magma is ejected under high

Figure 12.10 This young cinder cone is surrounded by rough-surfaced basalt lava flows. Lava Beds National Monument, northern California.

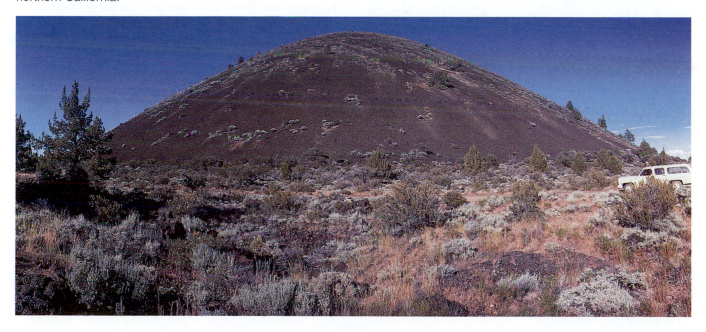

pressure from a narrow vent, producing tephra. The rain of tephra accumulates around the vent to form a roughly circular hill with a central crater. Cinder cones rarely grow to heights of more than a few hundred meters. An exceptionally fine example of a cinder cone is Wizard Island, built on the floor of Crater Lake long after the caldera was formed (Figure 12.5).

Hot Springs and Geysers

Where hot rock material is near the earth's surface, it can heat nearby ground water to high temperatures. When the ground water reaches the surface, it provides *hot springs* at temperatures not far below the boiling point of water (Figure 12.11). At some places, jetlike emissions of steam and hot water occur at intervals from small vents—producing *geysers* (Figure 12.12).

Figure 12.11 Mammoth Hot Springs, an example of geothermal activity in Yellowstone National Park, Wyoming. Terraces ringed by mineral deposits hold steaming pools of hot water.

Figure 12.12 Old Faithful Geyser in Yellowstone National Park, Wyoming.

Since the water that emerges from hot springs and geysers is largely ground water that has been heated in contact with hot rock, this water is recycled surface water. Little, if any, is water that was originally held in rising bodies of magma. **SCIENCE NEWS** 20

Volcanic Activity over the Globe

Figure 12.13 shows the locations of volcanoes that have been active within the last 12,000 years. By comparing this map with that of Figure 11.4, it is easy to see that many volcanoes are located along subduction boundaries. In fact, the "ring of fire" around the Pacific Rim is the most obvious feature on the map. Other volcanoes are located on or near oceanic spreading centers. Iceland is an example. In continental regions, spreading in the East Africa rift zone has also produced volcanoes. (We'll return to the East African rift later in this chapter.) Hotspot activity, as in the location of the Hawaiian Islands, is also present.

Figure 12.13   Volcanic activity of the earth. Dots show the locations of volcanoes known or believed to have erupted within the past 12,000 years. Each dot represents a single volcano or cluster of volcanoes. (After data of NOAA. Copyright © A. N. Strahler.)

LANDFORMS OF TECTONIC ACTIVITY

Recall from Chapter 11, our introduction to global plate tectonics, that there are two basic forms of tectonic activity: compressional and extensional (see Figure 11.10). Along converging lithospheric plate boundaries, tectonic activity is primarily compression. In subduction zones, sedimentary layers of the ocean floor are compressed within a trench as the descending plate forces them against the overlying plate. In continental collision, compression is of the severest kind. In zones of rifting of continental plates, however, the brittle continental crust is pulled apart and yields by faulting. This motion is extensional and produces faults. We begin our examination of landforms of tectonic activity with folding produced by compression.

Fold Belts

When continental collision begins to take place, broad wedges of strata of a passive continental margin come under strong forces of compression. The strata, which were originally more or less flat-lying, experience **folding** (Figure 11.11). The wavelike shapes imposed on the strata consist of alternating archlike upfolds, called *anticlines*, and troughlike downfolds, called *synclines*. Thus, the initial landform associated with an anticline is a broadly rounded mountain ridge, and the landform corresponding to a syncline is an elongate, open valley.

> **Compressional tectonic activity produces folding and overthrust faulting. Extensional tectonic activity produces normal faults, uplifted blocks, and rift valleys.**

An example of open folds of comparatively young geologic age that has long attracted the interest of geographers is the Jura Mountains of France and Switzerland. Figure 12.14 is a block diagram of a small portion of that fold belt. The rock strata are mostly limestone layers and were capable of being deformed by bending with little brittle fracturing. Folding occurred in late Cenozoic (Miocene) time. Notice that each mountain crest is associated with the axis of an anticline, while each valley lies over the axis of a syncline. Some of the anticlinal arches have been partially removed by erosion processes. The rock structure can be seen clearly in

(a)

(b)

Figure 12.14 Anticlines and synclines of the Jura Mountains, France and Switzerland. A cross section shows the folds and ground surface (*a*). The landscape developed on the folds is shown in the block diagram (*b*). (After E. Raisz.)

the walls of the winding gorge of a major river that crosses the area. The Jura folds lie just to the north of the main collision orogen of the Alps. Because of their location near a mountain mass, they are called *foreland folds*.

Faults and Fault Landforms

A **fault** in the brittle rocks of the earth's crust occurs when rocks suddenly yield to unequal stresses by fracturing. Faulting is accompanied by a displacement—a slipping motion—along the plane of breakage, or *fault plane*. Faults are often of great horizontal extent, so that the surface trace, or fault line, can sometimes be followed along the ground for many kilometers. Most major faults extend down into the crust for at least several kilometers.

Faulting occurs in sudden slippage movements that generate earthquakes. A single fault movement may result in a slippage of as little as a centimeter or as much as 15 m (about 50 ft). Successive movements may occur many years or decades apart, even several centuries apart. Over long time spans, the accumulated displacements can amount to tens or hundreds of kilometers. In some places, clearly recognizable sedimentary rock layers are offset on opposite sides of a fault, allowing the total amount of displacement to be measured accurately.

Normal Faults

One common type of fault associated with crustal rifting is the **normal fault** (Figure 12.15*a*). The plane of slippage, or fault plane, is steeply inclined. The crust on one side is raised, or upthrown, relative to the other, which is downthrown. A normal fault results in a steep, straight, clifflike feature called a *fault scarp* (Figure 12.15). Fault scarps range in height from a few meters to a few hundred meters (Figure 12.16). Their length is usually measurable in kilometers. In some cases they attain lengths as great as 300 km (about 200 mi).

Normal faults are not usually isolated features. Commonly, they occur in multiple arrangements, often as a set of parallel faults. This gives rise to a grain or pattern of rock structure and topography. A narrow block dropped down between two normal faults is a *graben* (Figure 12.17). A narrow block elevated between two normal faults is a *horst*. Grabens make conspicuous topographic trenches, with straight, parallel walls. Horsts make blocklike plateaus or mountains, often with a flat top but steep, straight sides.

In rifted zones of the continents, regions where normal faulting occurs on a grand scale, mountain masses called *block mountains* are produced. The up-faulted mountain blocks can be described as either tilted or lifted (Figure 12.18). A tilted block has one steep face—the

Figure 12.15, part (a): labels **Fault plane**, **Fault scarp**, **Upthrown block**, **Downthrown block**, **Normal fault**

Part (b): labels **Landsliding**, **Up**, **Down**, **Reverse fault**

Part (c): labels **Rift zone**, **Transcurrent fault**

Part (d): labels **Thrust plane**, **Overthrust fault**

Figure 12.15 Four types of faults.

Figure 12.16 This fault scarp was formed during the Hebgen Lake, Montana, earthquake of 1959. In a single instant, a displacement of 6 m (20 ft) took place on a normal fault.

(Figure 12.15*c*), or sometimes as *strike-slip faults.* In a transcurrent fault, the movement is predominantly horizontal. Where the land surface is nearly flat, no scarp, or a very low one at most, results. Only a thin fault line is traceable across the surface. In some places a narrow trench, or rift, marks the fault.

The best known of the active transcurrent faults is the great San Andreas fault, which can be followed for a distance of about 1000 km (about 600 mi) from the Gulf of California to Cape Mendocino, a location well north of

Figure 12.17 Initial landforms of normal faulting. A graben is a downdropped block, often forming a long, narrow valley. A horst is an upthrown block, forming a plateau, mesa, or mountain. (A. N. Strahler.)

Labels: **Graben**, **Horst**

fault scarp—and one gently sloping side. A lifted block, which is a type of horst, is bounded by steep fault scarps on both sides.

Transcurrent Faults

Recall that lithospheric plates slide past one another along major transform faults and that these features comprise one type of lithospheric plate boundary. Long before the principles of plate tectonics became known, geologists referred to such faults as **transcurrent faults**

Figure 12.18 Fault block mountains may be of tilted type (*left*) or lifted type (*right*). (After W. M. Davis.)

Figure 12.19 The San Andreas fault in southern California. The fault is marked by a narrow trough. The fault is slightly offset in the middle distance.

the San Francisco area, where it heads out to sea. It is a transform fault that marks the active boundary between the Pacific plate and the North American plate (see Figure 11.13). The Pacific plate is moving toward the northwest, which means that a great portion of the state of California and all of Lower (Baja) California is moving bodily northwest with respect to the North American mainland.

On a transcurrent fault, rock masses move in a horizontal direction past one another.

Throughout many kilometers of its length, the San Andreas fault appears as a straight, narrow scar. In some places this scar is a trenchlike feature, and elsewhere it is a low scarp (Figure 12.19). Frequently, a stream valley takes an abrupt jog—for example, first right, then left—when crossing the fault line. This offset in the stream's course shows that many meters of movement have occurred in fairly recent time.

Reverse and Overthrust Faults

In a *reverse fault*, the inclination of the fault plane is such that one side rides up over the other and a crustal shortening occurs (Figure 12.15*b*). Reverse faults produce fault scarps similar to those of normal faults, but the possibility of landsliding is greater because an overhanging scarp tends to be formed. The San Fernando, California, earthquake of 1971 was generated by slippage on a reverse fault.

The *low-angle overthrust fault* (Figure 12.15*d*) involves predominantly horizontal movement. One slice of rock rides over the adjacent ground surface. A thrust slice may be up to 50 km (30 mi) wide. The evolution of low-angle thrust faults was explained in Chapter 11 and illustrated in Figure 11.11.

The Rift Valley System of East Africa

Rifting of continental lithosphere is the very first stage in the splitting apart of a continent to form a new ocean basin. The process is beautifully illustrated by the East African rift valley system. This region has attracted the attention of geologists since the early 1900s. They gave the name *rift valley* to what is basically a graben, but with a more complex history that includes the building of volcanoes on the graben floor.

Figure 12.20 is a sketch map of the East African rift valley system. It is about 3000 km (1900 mi) long and extends from the Red Sea southward to the Zambezi River. Along this axis, the earth's crust is being lifted and spread apart in a long, ridgelike swell. The rift valley system consists of a number of grabenlike troughs. Each is a separate rift valley ranging in width from about 30 to 60 km (20 to 40 mi). As geologists noted in early field surveys of this system, the rift valleys are like keystone blocks of a masonry arch that have slipped down between neighboring blocks because the arch has spread apart somewhat. Thus, the floors of the rift valleys are above the elevation of most of the African continental surface. Major rivers and several long, deep lakes—Lake Nyasa and Lake Rudolph, for example—occupy some of the valley floors.

The sides of the rift valleys typically consist of multiple fault steps (Figure 12.21). Sediments derived from the high plateaus that form the flanks of troughs cover the floors of the valleys in thick layers. Two great stratovolcanoes have been built close to the rift valley east of Lake Victoria. One is Mount Kilimanjaro, whose summit rises to over 6000 m (about 19,000 ft). The other, Mount Kenya, is only a little lower and lies right on the equator.

Figure 12.20   A sketch map of the East African rift valley system and the Red Sea to the north.

Figure 12.21   The rift valley wall in Ethiopia. Multiple fault scarps give the landscape a stepped appearance.

EARTHQUAKES

You have probably seen television news accounts of disastrous earthquakes and their destructive effects. Californians know about severe earthquakes from first-hand experience, but several other areas in North America have also experienced strong earthquakes, and a few of these have been very severe. An **earthquake** is a motion of the ground surface, ranging from a faint tremor to a wild motion capable of shaking buildings apart.

The earthquake is a form of energy of wave motion transmitted through the surface layer of the earth. Waves move outward in widening circles from a point of sudden energy release, called the *focus*. Like ripples produced when a pebble is thrown into a quiet pond, these *seismic waves* gradually lose energy as they travel outward in all directions. (The term *seismic* means "pertaining to earthquakes.") SCIENCE NEWS 21

Most earthquakes are produced by sudden slip movements along faults. They occur when rock on both sides of the fault is slowly bent over many years by tectonic forces. Energy accumulates in the bent rock, just as it does in a bent archer's bow. When a critical point is reached, the strain is relieved by slippage on the fault, and the rocks on opposite sides of the fault move in different directions. A large quantity of energy is instantaneously released in the form of seismic waves, which shake the ground. In the case of a transcurrent fault, on which movement is in a horizontal direction, slow bending of the rock that precedes the shock takes place over many decades. Sometimes a slow, steady displacement known as *fault creep* occurs, which tends to reduce the accumulation of stored energy.

The devastating San Francisco earthquake of 1906 resulted from slippage along the San Andreas fault, which is dominantly a transcurrent fault. This fault also passes about 60 km (40 mi) inland of Los Angeles, placing the densely populated metropolitan Los Angeles region in great jeopardy. Associated with the San Andreas fault are several important parallel and branching transcurrent faults, all of which are capable of generating severe earthquakes.

Earthquakes can also be produced by volcanic activity, as when magma rises or recedes within a volcanic chamber.

The Richter scale measures the energy released by an earthquake.

A scale of earthquake magnitudes was devised in 1935 by the distinguished seismologist Charles F. Richter. Now called the *Richter scale*, it describes the quantity of energy released by a single earthquake. Scale numbers range from 0 to 9, but there is really no upper limit other than nature's own energy release limit. For each whole unit of increase (say, from 5.0 to 6.0), the quantity of energy release increases by a factor of 32. The Chilean earthquake of 1960 is the largest observed to date. It earned a rating of 8.3 on the Richter scale, but that value was later updated to 9.5. The great San Francisco earthquake of 1906 is now rated as magnitude 7.9.

Earthquakes and Plate Tectonics

Seismic activity—the repeated occurrence of earthquakes—occurs primarily near lithospheric plate boundaries. Figure 12.22, which shows the location of all earthquake centers during a typical seven-year period, clearly reveals this pattern. The greatest intensity of seismic activity is found along converging plate boundaries where oceanic plates are undergoing subduction. Strong pressures build up at the downslanting contact of the two plates, and these are relieved by sudden fault slippages that generate earthquakes of large magnitude. This mechanism explains the great earthquakes experienced in Japan, Alaska, Central America, Chile, and other narrow zones close to trenches and volcanic arcs of the Pacific Ocean basin.

Good examples can be cited from the Pacific coast of Mexico and Central America, where the subduction boundary of the Cocos Plate lies close to the shoreline. The great earthquake that devastated Mexico City in 1986 was centered in the deep trench offshore. Two great shocks in close succession, the first of magnitude 8.1 and the second of 7.5, damaged cities along the coasts of the Mexican states of Michoacán and Guerrero. Although Mexico City lies inland about 300 km (about 185 mi) distant from the earthquake epicenters, it experienced intense ground shaking of underlying saturated clay formations, with the resulting death toll of some 10,000 persons (Figure 12.23).

Transform boundaries that cut through the continental lithosphere are also sites of intense seismic activity, with moderate to strong earthquakes. The most familiar example is the San Andreas fault.

Spreading boundaries are the location of a third class of narrow zones of seismic activity related to lithospheric plates. Most of these boundaries are identified with the midoceanic ridge and its branches. For the most part, earthquakes in this class are limited to moderate intensities.

Earthquakes also occur at scattered locations over the continental plates, far from active plate boundaries. In many cases, no active fault is visible, and the geologic cause of the earthquake is uncertain. For example, the great New Madrid earthquake of 1811 was centered in the Mississippi River floodplain in Missouri. It produced

Figure 12.22 World map of earthquake center locations and centers of great earthquakes. Center locations of all earthquakes originating at depths of 0 to 100 km (62 mi) during a six-year period are shown by red dots. Each dot represents a single location or a cluster of centers. Black circles identify centers of earthquakes of Richter magnitude 8.0 or greater during an 80-year period. The map clearly shows the pattern of earthquakes occurring at subduction boundaries. (Compiled by A. N. Strahler from data of U.S. government. Copyright © A. N. Strahler.)

Figure 12.23 Earthquake devastation in Mexico City following the earthquake of September 1985.

three great shocks in close succession, rated from 8.1 to 8.3 on the Richter scale.

Seismic Sea Waves

An important environmental hazard often associated with a major earthquake centered on a subduction plate boundary is the *seismic sea wave*, or **tsunami**, as it is known to the Japanese. A train of these water waves is often generated in the ocean by a sudden movement of the sea floor at a point near the earthquake source. The waves travel over the ocean in ever-widening circles, but they are not perceptible at sea in deep water. (Seismic sea waves are sometimes referred to as "tidal waves," but since they have nothing to do with the tides, the name is quite misleading.)

When a tsunami arrives at a distant coastline, the effect is to cause a rise of water level. Normal wind-driven waves, superimposed on the heightened water level, attack places inland that are normally above their reach. For example, in this century several destructive seismic sea waves in the Pacific Ocean attacked ground as high as 10 m (30 ft) above normal high-tide level, causing widespread destruction and many deaths by drowning in low-lying coastal areas. It is thought that the coastal flooding that occurred in Japan in 1703, with an estimated loss of life of 100,000 persons, may have been caused by seismic sea waves. **SCIENCE NEWS** 22

Earthquakes along the San Andreas Fault

More than nine decades have passed since the great San Francisco earthquake of 1906 was generated by movement on the San Andreas fault. The maximum horizon-

Figure 12.24 A sketch map of the San Andreas fault and Transverse Ranges showing the locked sections of the fault alternating with active sections. (Based on data of the U.S. Geological Survey.)

tal displacement of the ground was about 6 m (20 ft). Since then, this sector of the fault has been locked—that is, the rocks on the two sides of the fault have been held together without sudden slippage (Figure 12.24). In the meantime, the two lithospheric plates that meet along the fault have been moving steadily with respect to one another. This means that a huge amount of unrelieved strain energy has already accumulated in the crustal rock on either side of the fault.

On October 17, 1989, the San Francisco Bay area was severely jolted by an earthquake with a Richter magnitude of 7.1. The earthquake's epicenter was located near Loma Prieta peak, about 80 km (50 mi) southeast of San Francisco, at a point only 12 km (7 mi) from the city of Santa Cruz, on Monterey Bay. The city of Santa Cruz suffered severe structural damage to older buildings. Destructive ground shaking in the distant San Francisco Bay area proved surprisingly severe. Buildings, bridges, and viaducts on landfills were particularly hard hit (Figure 12.25). Altogether, 62 lives were

Figure 12.26   This area within the city of Kobe, Japan, was reduced to rubble by the earthquake of January 1995 and the fires that followed in its aftermath.

Figure 12.25   A collapsed section of the double-decked Nimitz Freeway (Interstate 880) in Oakland, California. Shaking by the Loma Prieta earthquake caused the upper level of the freeway to collapse on the lower deck, crushing at least 39 persons in their cars.

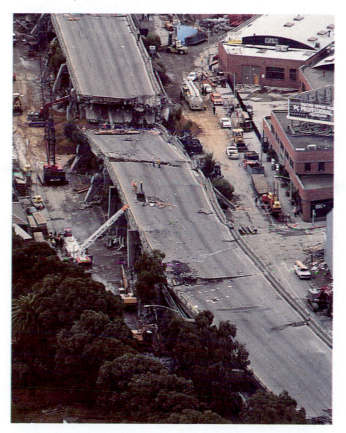

lost in this earthquake, and the damage was estimated to be about $6 billion. In comparison, the 1906 earthquake took a toll of 700 lives and property damage equivalent to 20 billion 1987 dollars.

The displacement that caused the Loma Prieta earthquake occurred deep beneath the surface not far from the San Andreas fault, which has not slipped since the great San Francisco earthquake of 1906. The slippage on the Loma Prieta fault amounted to about 1.8 m (6 ft) horizontally and 1.2 m (4 ft) vertically but did not break the ground surface above it. Geologists state that the Loma Prieta slippage, though near the San Andreas fault, probably has not relieved more than a small portion of the strain on the San Andreas. While the time of occurrence of another major earthquake in the San Francisco region cannot be predicted with precision, it is inevitable. **SCIENCE NEWS** 23, 24

Surprising as it may seem, you may need to look to Japan for a scenario that the citizens of the San Francisco Bay area could experience when a branch of the San Andreas fault lets go. The Hyogo-ken Nanbu earthquake that devastated the city of Kobe in January of 1995 (Figure 12.26) occurred on a short side branch of a major strike-slip fault quite similar tectonically to the San Andreas. A similar side branch of the main San Andreas fault, called the Hayward fault, runs through the East Bay region of San Francisco. Seismologists estimate that a slip on the Hayward fault would generate a

EYE ON THE ENVIRONMENT

The Northridge Earthquake of 1994

The January 17, 1994, Northridge earthquake struck a heavily populated suburb of Los Angeles, California, at 4:30 A.M., PST. Assigned a Richter scale rating of 6.7, it produced the strongest ground motions ever recorded in an urban setting in North America and the greatest financial losses from a natural disaster in the United States since the great San Francisco earthquake of 1906.

Northridge lies in an east-west trending fault-rich valley close to the base of the lofty San Gabriel Mountains. This range is an uplifted fault-block that has produced earthquakes at many locations along the base of its steep southern flank. One of these was the equally severe San Fernando earthquake of 1971, centered not far to the north in the same valley. These active earthquake faults belong to a tectonic belt known as the Transverse Ranges, which runs more or less east-west through the Los Angeles basin and heads out across the Pacific coastline beyond Santa Barbara (see Figure 12.24). Although the San Andreas fault lies not far distant to the north, these two fault systems have different trends and operate quite independently.

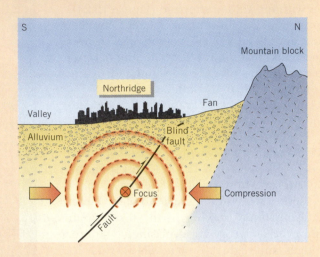

Geologic setting of the Northridge earthquake.

The Northridge and San Fernando earthquake faults are of the reverse type (see Figure 12.15*b*). Our sketch shows how it works. Large arrows pointing toward each other indicate that along the Transverse Ranges the earth's crust is under a strong, north-south, horizontal compressional force. The fault plane slants steeply upward. When the fault suddenly starts to slip at the focus, located about 20 km below the valley surface, the upper block delivers a powerful upward punch that sharply lifts the ground surface. As with a boxer's "uppercut," what counts is the acceleration of that first upward slip motion. Seismic waves also spread out radially from the focus, generating strong ground shaking with horizontal motions. Notice that the displacement along the earthquake fault did not extend upward to the ground surface—that is, it is a "blind fault." Nev-

quake of magnitude 7.0—about the same as the Kobe quake—and has a 28-percent probability of occurring by the year 2018. Although the nation of Japan prides itself on its earthquake preparedness, the Kobe catastrophe left 5000 dead and 300,000 homeless, and did property damage estimated to have been ten times that of the Northridge earthquake of 1994, which was the most costly earthquake in American history.

Along the southern California portion of the San Andreas fault, a recent estimate placed at about 50 percent the likelihood that a very large earthquake will occur within the next 30 years. In 1992 three severe earthquakes occurred in close succession along local

faults a short distance north of the San Andreas fault within the area labeled the southern active area on the map (Figure 12.24). The second of these, the Landers earthquake, a powerful 7.5 on the Richter scale, occurred on a strike-slip fault trending north-northwest. It caused a 80-km (50-mi) rupture across the Mojave Desert. These three events have led to speculation that the likelihood of a major slip on the nearby San Andreas fault in the near future has substantially increased.

For residents of the Los Angeles area, an additional serious threat lies in the large number of active faults close at hand. Movements on these local faults have produced more than 40 damaging earthquakes since ertheless, disturbances of the ground surface occurred over a wide area.

Northridge had more severe structural damage than San Fernando, although both were Richter 6.7 energy releases. In the Northridge case, the focus lay directly beneath the city, whereas the focus of the San Fernando earthquake was located some distance back under the mountain block. Northridge casualties included 33 persons dead as a direct

result of the earthquake, more than 7000 injuries treated at local hospitals, and over 20,000 persons left homeless. Total financial losses were estimated as high as $20 billion. Sections of three freeways were closed, including the busiest highway in the country, I-5.

In the wake of the severe structural damage of the 1971 San Fernando earthquake, a program of strengthening (retrofitting) of buildings, freeways, and dams had been in progress

for some two decades throughout the Los Angeles region by the time of the Northridge quake. While the retrofitting may have reduced some forms of damage, it was soon discovered that the long-established practice of making welded connections in the steel frames of buildings and freeway bridges failed badly during the quake. The welded joints proved brittle and easy to fracture. In contrast, the few modern buildings designed to move sidewise easily on a fixed base performed well and without damage. The implications of these revelations are obviously of enormous consequence to the residents of the California "Southland." Is it now too late to correct a monstrous design error? There will be many more earthquakes like those of Northridge and San Fernando, occurring unexpectedly at widely separated points within the Transverse Ranges.

Questions

1. Explain the geologic setting of the Northridge earthquake of 1994. What type of fault was involved? Where was it located? What type of motion occurred?

2. What discovery was made concerning construction practices as a result of the Northridge quake?

This building suffered severe structural damage in the Northridge earthquake of 1994.

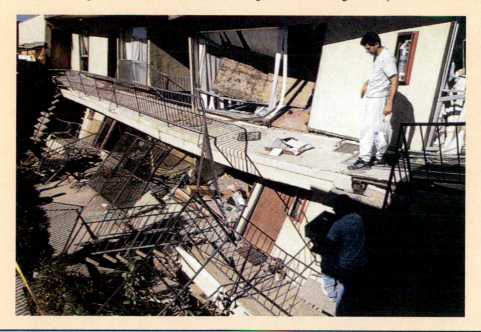

1800, including the Long Beach earthquakes of the 1930s and the San Fernando earthquake of 1971. The San Fernando earthquake measured 6.6 on the Richter scale and produced severe structural damage near the earthquake center. The brief but intense primary shock and aftershocks that followed damaged beyond repair many older structures built of unreinforced brick masonry. In 1987 an earthquake of magnitude 6.1 struck the vicinity of Pasadena and Whittier, located within about 20 km (12 mi) of downtown Los Angeles. Known as the Whittier Narrows earthquake, it was generated along a local fault system that had not previously shown significant seismic activity. The Northridge earth-

quake of 1994, described in *Eye on the Environment* • *The Northridge Earthquake of 1994*, measured 6.7 on the Richter scale and produced severe structural damage near the earthquake center.

A slip along the San Andreas fault, some 50 km (31 mi) to the north of the densely populated region of Los Angeles, will release an enormously larger quantity of energy than local earthquakes, such as the Northridge or San Fernando earthquakes. On the other hand, the destructive effects of a San Andreas earthquake in downtown Los Angeles will be somewhat moderated by the greater travel distance. Although the intensity of ground shaking might not be much different from that of the San Fernando earthquake, for example, it will last much longer and cover a much wider area of the Los Angeles region. The potential for damage and loss of life is enormous. **SCIENCE NEWS** 25

A LOOK AHEAD

In the last three chapters, we have surveyed the composition, structure, geologic activity, and initial landforms of the earth's crust. We began with a study of the rocks and minerals that make up the earth's crust and core. We saw how the rock cycle describes a continuous cycling and recycling of rocks and minerals that has occurred over some 3 billion years or more of geologic time. In the second chapter, we developed plate tectonics as the mechanism powering the rock cycle. The global pattern of plate tectonics also explains the geographical distribution of mountain ranges, ocean basins, and continental shields occurring on the earth's surface. In this chapter, we described the initial landforms that result directly from volcanic and tectonic activity, occurring primarily at the boundaries of spreading or colliding lithospheric plates.

With this survey of the earth's crust and the geologic processes that shape it now completed, we can turn to other landform-creating processes in the following chapters. First will be the processes of weathering, which breaks rock into small particles, and mass wasting, which moves them downhill as large and small masses under the influence of gravity. Then we will turn to running water in three chapters that first describe the behavior of rivers and streams, secondly their work in shaping landforms, and lastly how running water dissects rock layers to reveal their inner structures. In our two concluding chapters, we will examine landforms created by waves, wind, and glacial ice.

VOLCANIC AND TECTONIC LANDFORMS IN REVIEW

Landforms are the surface features of the land, and geomorphology is the scientific study of the processes that shape landforms. Initial landforms are shaped by volcanic and tectonic activity, while sequential landforms are sculpted by agents of denudation, including running water, waves, wind, and glacial ice.

Volcanoes are landforms marking the eruption of lava at the earth's surface. Stratovolcanoes, formed by the emission of thick, gassy, felsic lavas, have steep slopes and are prone to explosive eruptions that can form calderas. Most active stratovolcanoes lie along the Pacific rim, where subduction of oceanic lithospheric plates is occurring. At hotspots, rising mantle material provides mafic magma that erupts as basaltic lavas. Because these lavas are more fluid and contain little gas, they form broadly rounded shield volcanoes. Hotspots occurring beneath continental crust can also provide vast areas of flood basalts. Some basaltic volcanoes occur along the midoceanic ridge.

The two forms of tectonic activity are compressional and extensional. Compressional activity occurs at lithospheric plate collisions. At first, the compression produces folding—anticlines (upfolds) and synclines (downfolds). If compression continues, folds may be overturned and eventually reverse to produce overthrust faulting. Extensional activity occurs where lithospheric plates are spreading apart, generating normal faults. These can produce upthrown and down-dropped blocks that are sometimes as large as mountain ranges or rift valleys. Transcurrent faults occur where two rock masses move horizontally past each other.

Earthquakes occur when rock layers, bent by tectonic activity, suddenly fracture and move. The sudden motion at the fault produces earthquake waves that shake and move the ground surface in the adjacent region. The energy released by an earthquake is measured by the Richter scale. Large earthquakes occurring near developed areas can cause great damage. Most severe earthquakes occur near plate collision boundaries. The San Andreas fault is a major transcurrent fault located near two great urban areas—Los Angeles and San Francisco. The potential for a severe earthquake on this fault is high, and the probability of a major earth movement increases every year.

KEY TERMS

landforms	stratovolcano	normal fault
geomorphology	shield volcano	transcurrent fault
denudation	folding	earthquake
volcano	fault	tsunami

REVIEW QUESTIONS

1. Distinguish between initial and sequential landforms. How do they represent the balance of power between internal earth forces and external forces of denudation agents?
2. What is a stratovolcano? What is its characteristic shape, and why does that shape occur? Where do stratovolcanoes generally occur and why?
3. What is a shield volcano? How is it distinguished from a stratovolcano? Where are shield volcanoes found, and why? Give an example of a shield volcano. How are flood basalts related to shield volcanoes?
4. What types of landforms characterize foreland fold belts? Where do these belts occur, and how are they formed?
5. Briefly describe the rift valley system of East Africa as an example of normal faulting.
6. How does a transcurrent fault differ from a normal fault? What landforms are expected along a transcurrent fault? How are transcurrent faults related to plate tectonic movements?
7. What is an earthquake, and how does it arise? How are the locations of earthquakes related to plate tectonics?
8. Describe the tsunami, including its origin and effects.
9. Briefly summarize the geography and recent history of the San Andreas fault system in California. What are the prospects for future earthquakes along the San Andreas fault?

VISUALIZING EXERCISES

1. Sketch a cross section through a normal fault, labeling the fault plane, upthrown side, downthrown side, and fault scarp.
2. Sketch a cross section through a foreland fold belt showing rock layers in different colors or patterns. Label anticlines and synclines.

A DEEPER VIEW OF VOLCANIC AND TECTONIC LANDFORMS

1. Write a fictional news account of a volcanic eruption. Select a type of volcano—composite or shield—and a plausible location. Describe the eruption and its effects as it was witnessed by observers. Make up any details you need, but be sure they are scientifically correct.
2. How are mountains formed? Provide the plate tectonic setting for mountain formation, and then describe how specific types of mountain landforms arise.

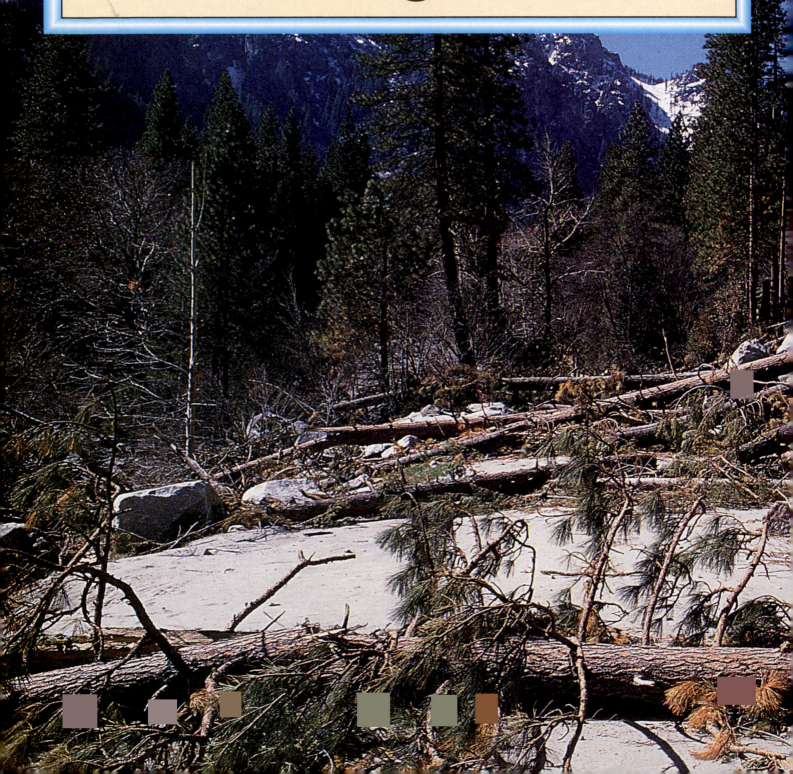

Chapter 13

Weathering and Mass Wasting

A rockslide blocks a road in Yosemite National Park with boulders and tree trunks.

PUTTING YOU IN THE PICTURE

The pull of gravity is relentless and unforgiving. It acts to move all things closer to the earth's center, and for objects on the earth's surface, that generally means downhill. If you are a skier, you may be grateful for that downhill pull (providing the ski lift is working to handle the uphill part of skiing). As a hiker, the downhill force makes the trek home from the summit a lot easier (if your knees and leg muscles are in good shape). But if you are an inhabitant of a peaceful valley below a steep mountain slope that suddenly gives way, you may not be so grateful for the force that has buried your home under tons of rocky debris.

Landslide, earthflow, mudflow, soil creep—these words describe the different ways that earth materials move downhill under the force of gravity. Sometimes the materials move very quickly, other times very slowly. Perhaps you've walked the sidewalks of a town or city built on a steep slope and seen retaining walls leaning over or even toppled by the slow downhill creep of soil. While annoying and expensive for a homeowner, a failing retaining wall is not likely to result in death or destruction. But a landslide, earthflow, or mudflow can easily take lives and destroy property.

In 1903, an enormous mass of limestone rock slid from the face of Turtle Mountain, near Frank, Alberta, into the valley below (see Figure 13.23). The volume of the landslide was estimated at about 27 million m³ (35 million yd³). It moved down through 900 m (3000 ft) of elevation at great speed and slid uphill on the opposite side of the valley to regain over 100 m (300 ft) of elevation. The rocky debris buried part of the town of Frank, Alberta, taking 70 lives.

A similar landslide in the Madison Canyon of Montana was triggered by the Hebgen Lake earthquake of 1959. About 28 million m³ (about 36 million yd³) of rock broke away from the south wall of the canyon of the Madison River and descended more than half a kilometer (0.3 mi) at a speed later estimated to have exceeded 45 m/s (about 100 mi/hr). The slide debris blocked the Madison River and a new lake quickly began to form. Within three weeks' time, it was about 60 m (about 200 ft) deep. (Our *Eye on the Environment* provides more information about this extreme event.)

Earthflows are slower but no less destructive. For residents of the Palos Verdes Hills, a coastal peninsula in Los Angeles County, the Portuguese Bend "landslide" was a nightmare that wouldn't go away. About 160 hectares (400 acres) of residential land settled and moved downslope some 20 m (70 ft) over a three-year period. Houses were slowly tilted, cracked, and broken apart as the land underneath them crept along a shale layer softened by water from the septic systems and abundant lawn-watering of the residential area.

The Nicolet earthflow of 1955 occurred when a clay layer beneath the city of Nicolet, Quebec, unexpectedly became liquid. A large chunk of the town quickly flowed into the Nicolet River. Only three lives were lost, but damage was in the millions.

Some large mudflows can follow stream and river courses for long distances. Immediately following the eruption of Mount St. Helens in 1980, a mudflow of fresh ash and melted snow hurtled down the Toutle River, destroying fields and forests. It traveled a total distance of more than 120 km (75 mi), eventually reaching the Columbia River.

Mass movements of earth under the influence of gravity are one of the two main subjects of this chapter. These movements create distinctive landforms of an obvious nature. However, the subject we will present first is weathering—a set of physical and chemical processes that break rocks apart to create large and small particles and to alter their mineral composition. Weathering prepares the way for mass movements, and it can also create its own distinctive landforms, as we will see.

Chapter Outline

Slopes and Regolith

Physical Weathering
Frost Action
Salt-Crystal Growth
Unloading
Other Physical Weathering Processes

Chemical Weathering and Its Landforms
Hydrolysis and Oxidation
Acid Action

Mass Wasting
Soil Creep
Earthflow
Mudflow
Landslide
Eye on the Environment • The Great Hebgen Lake Disaster

Induced Mass Wasting
Induced Earthflows
Scarification of the Land

WEATHERING AND MASS WASTING

Now that we have completed our study of the earth's crust—including its mineral composition, its moving lithospheric plates, and its tectonic and volcanic landforms—we can focus on the shallow life layer itself. At this sensitive interface, the external processes of denudation carve sequential landforms from the earth materials uplifted by the earth's internal processes. Our investigation of what happens to rock upon exposure at the surface began in Chapter 10 with a description of mineral alteration of rock and the production of sediment, which is an essential part of the cycle of rock transformation. In this chapter, we look further at the softening and breakup of rock and then at how the resulting particles move downhill under the force of gravity.

Weathering is the general term applied to the combined action of all processes that cause rock to disintegrate physically and decompose chemically because of exposure near the earth's surface. In Chapter 10, we described the processes of mineral alteration (oxidation, hydrolysis, and carbonic acid solution) that transform minerals from types that were stable when the rocks were formed to types that are now stable at surface temperatures and pressures. These processes are part of chemical weathering, in which rock particles are altered chemically at the surface. A second type of weathering is physical weathering, in which rock particles are fractured and broken apart. Weathering also leads to a number of distinctive landforms, which we will describe in this chapter.

This chapter also examines how gravity acts on rock fragments, created by weathering, to produce landforms. In this process, gravity induces the spontaneous downhill movement of soil, regolith, and rock, but without the action of moving water, air, or ice. The downhill movement is referred to as **mass wasting**. Movement of a mass of soil or rock to lower levels takes place when the internal strength declines to a critical point below which the force of gravity cannot be resisted. This failure of strength under the ever-present force of gravity takes many forms and scales, and we will see that human activity causes or aggravates several forms of mass wasting.

SLOPES AND REGOLITH

Mass wasting occurs on slopes. As used in physical geography, the term *slope* designates a small strip or patch of the land surface that is inclined from the horizontal. Thus, we speak of "mountain slopes," "hill slopes," or "valley-side slopes" to describe some of the inclined ground surfaces that we might encounter on a landscape. Slopes guide the downhill flow of surface water and fit together to form drainage systems within which surface-water flow converges into stream channels. Nearly all natural surfaces slope to some degree—very few are perfectly horizontal or vertical.

Wasting of slopes refers to the downhill movement of soil and regolith under the force of gravity.

As described in Chapter 9, most slopes are mantled with *regolith*, a surface layer of weathered rock particles. The regolith grades downward into solid, unaltered rock, known simply as **bedrock**. Regolith, in turn, provides the source for *sediment*, consisting of detached mineral particles that are transported and deposited in a fluid medium. This fluid may be water, air, or even glacial ice. As we saw in Chapter 9, both regolith and sediment comprise parent materials for the formation of soil, which is a surface layer capable of supporting the growth of plants.

Figure 13.1 shows a typical hill slope that forms one wall of the valley of a small stream. Soil and regolith

Figure 13.1 Soil, regolith, and outcrops on a hillslope. Alluvium, a form of transported regolith, lies in the floor of an adjacent stream valley. (Drawn by A. N. Strahler.)

blanket the bedrock, except in a few places where the bedrock is particularly hard and projects in the form of *outcrops. Residual regolith* is derived directly from the rock beneath and moves very slowly down the slope toward the stream. Beneath the valley bottom are layers of transported regolith, called **alluvium**, which is sediment transported and deposited by the stream. This sediment had its source in regolith prepared on hill slopes many kilometers or miles upstream. All accumulations of sediment on the land surface, whether deposited by streams, waves and currents, wind, or glacial ice, can be designated *transported regolith*, in contrast to residual regolith.

The thickness of soil and regolith is quite variable. Although the soil is rarely more than 1 or 2 m thick (about 3 to 6 ft), residual regolith on decayed and fragmented rock may extend down to depths of 5 to 100 m (about 15 to 300 ft), or more at some locations. On the other hand, soil or regolith, or both, may be missing. In some places, everything is stripped off down to the bedrock, which then appears at the surface as an outcrop. In other places, following cultivation or forest fires, the fertile soil is partly or entirely eroded away, and severe erosion exposes the regolith.

PHYSICAL WEATHERING

Physical weathering produces fine particles of regolith from massive rock by the action of forces strong enough to fracture the rock. The physical weathering processes discussed in this chapter include frost action, salt-crystal growth, unloading, and wedging by plant roots. The term *mechanical weathering* is also used for physical weathering.

Frost Action

One of the most important physical weathering processes in cold climates is *frost action*, the repeated growth and melting of ice crystals in the pore spaces of soil and in rock fractures. In contrast to most other liquids, water expands when it freezes. Perhaps you've accidentally left a bottle of water chilling in the freezer overnight and observed this phenomenon first hand, removing a mass of ice surrounded by broken glass the next morning. The expansion of water in freezing can rupture even extremely hard rocks. Frost action and ice-crystal growth produce a number of conspicuous effects and forms in all climates that have cold winters. Features caused by frost action and the buildup of ice below the surface are particularly visible in the tundra climate of arctic coastal fringes and islands, and above the timberline in high mountains.

Almost everywhere, bedrock is cut through by sys-

tems of fractures called *joints* (Figure 13.1). Rarely is igneous rock free of numerous joints that permit the entry of water. In sedimentary rocks, the planes of stratification, or bedding planes, comprise a natural set of planes of weakness cut at right angles by sets of joints. Obviously, comparatively weak stresses can separate joint blocks, whereas strong stresses are required to make fresh fractures through solid rock. The process of breaking up rock along joint and bedding planes can be called *joint block separation* (Figure 13.2).

As coarse-grained igneous rock becomes weakened by chemical decomposition, water is able to penetrate the contact surfaces between mineral grains. In this location, the water can freeze and exert strong forces to separate the grains. This form of breakup is termed *granular disintegration* (Figure 13.2). The end product is a fine gravel or coarse sand in which each grain consists of a single mineral particle separated from the others along its original crystal or grain boundaries.

When water freezes in bedrock joints and bedding planes, it expands and can split rocks apart.

The effects of frost action can be seen in all climates having a winter season with many alternations of freeze and thaw. Where bedrock is exposed on knolls and mountain summits, joint blocks are pried apart by water that freezes in joint cracks (Figure 13.3). Under the most favorable conditions, seen on high mountain summits and in the arctic tundra, large angular rock fragments accumulate in a layer that completely blankets the bedrock beneath.

Frost action on cliffs of bare rock in high mountains detaches rock fragments that fall to the cliff base. Where

Figure 13.2 Two common forms of bedrock disintegration. (Drawn by A. N. Strahler.)

Joint-block separation

Granular disintegration

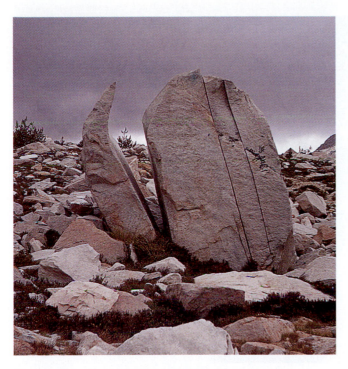

Figure 13.3 Frost-shattered blocks, Sierra Nevada, California. Ice-crystal growth within the joint planes of rock can cause the rock to split apart. This boulder is an example.

production of fragments is rapid, they accumulate to form *talus slopes.* Most cliffs are notched by narrow ravines that funnel the rock fragments into separate tracks. Each track, or chute, feeds a growing, conelike talus body. Talus cones are arranged side by side along the base of the cliff (Figure 13.4). Fresh talus slopes are unstable, so that the disturbance created by walking across the slope or dropping a large rock fragment from the cliff above will easily set off a sliding or rolling motion within the surface layer of fragments.

In fine-textured soils and sediments composed largely of silt and clay, freezing of soil water takes place in horizontal layers or lens-shaped bodies. As these ice layers thicken, the overlying soil layer is heaved upward. Prolonged frost heaving can produce minor irregularities and small mounds on the soil surface. Where a rock fragment lies at the surface, perpendicular ice needles can grow beneath the fragment and raise it above the surface (Figure 13.5). The same process acting on a rock fragment below the soil surface will eventually bring the fragment to the surface.

A related process of ice-crystal growth acts in coarse-textured regolith of the barren tundra to cause the

Figure 13.4 Talus cones near the shore of Lake Louise in the Canadian Rockies.

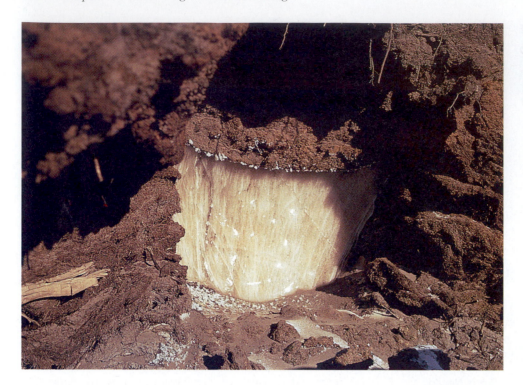

Figure 13.5 Needle ice growth. At night, water in the soil freezes at the surface, creating ice needles that can lift particles of soil or move larger stones.

coarsest fragments—pebbles and cobbles—to move horizontally and to become sorted out from the finer particles. This type of sorting produces ringlike arrangements of coarse fragments. Linked with adjacent rings, the result is a netlike pattern of *stone rings* (Figure 13.6).

In silty alluvium, such as that formed on river floodplains and delta plains in the arctic environment, ice

Figure 13.6 Sorted circles of gravel form a system of netlike stone rings on this nearly flat land surface where water drainage is poor. The circles in the foreground are 3 to 4 m (10 to 13 ft) across; the gravel ridges are 20 to 30 cm (8 to 12 in.) high. Broggerhalvoya, western Spitsbergen, latitude 78° N.

takes place in the porous outer zone of the sandstone, tiny crystals of salts such as halite (sodium chloride), calcite (calcium carbonate), or gypsum (calcium sulfate) are left behind. The growth force of these crystals produces grain-by-grain breakup of the sandstone, which crumbles into a sand and is swept away by wind and rain.

Especially susceptible are zones of rock lying close to the base of a cliff, because there the ground water seeps outward to reach the rock surface (Figure 13.8). In the southwestern United States, many of the deep niches or cavelike recesses formed in this way were occupied by Native Americans. Their cliff dwellings gave them protection from the elements and safety from armed attack (Figure 13.9).

Salt-crystal growth can break rocks apart grain by grain, producing niches, shallow caves, and rock arches.

Salt crystallization also damages masonry buildings, as well as concrete sidewalks and streets. Brick and concrete in contact with moist soil are highly susceptible to grain-by-grain disintegration from salt crystallization. The salt crystals can be seen as a soft, white, fibrous layer on damp basement floors and walls. The deicing salts spread on streets and highways can be quite destructive. Sodium chloride (rock salt), widely used for this purpose, is particularly damaging to concrete pavements and walks, curbstones, and other exposed masonry structures.

Figure 13.7   These polygons are formed by the growth of ice wedges. On the Alaskan north slope, near the border of Alaska and Yukon Territory, Canada.

Figure 13.8   In dry climates there is a slow seepage of water from the cliff base. Salt-crystal growth separates the grains of permeable sandstone, breaking them loose and creating a niche. (After A. N. Strahler.)

accumulates in vertical wedge-forms in deep cracks in the sediment. These ice wedges are interconnected into a system of polygons, called *ice-wedge polygons* (Figure 13.7). Ice wedges are thought to originate as shrinkage cracks formed during extreme winter cold. During the spring melt, surface water enters the cracks and becomes frozen, adding to the thickness of the wedges.

Salt-Crystal Growth

Closely related to the growth of ice crystals is the weathering process of rock disintegration by growth of salt crystals in rock pores. This process, *salt-crystal growth*, operates extensively in dry climates and is responsible for many of the niches, shallow caves, rock arches, and pits seen in sandstone formations. During long drought periods, ground water is drawn to the surface of the rock by capillary action. As evaporation of the water

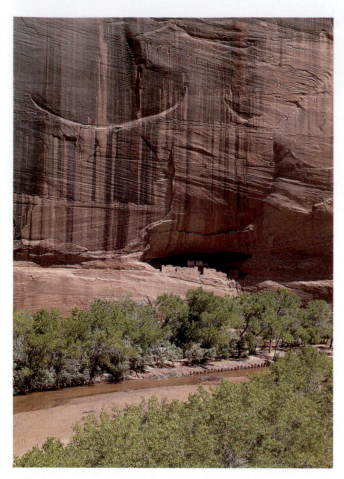

Figure 13.9 The White House Ruin, a former habitation of Native Americans, occupies a large niche in sandstone in the lower wall of Canyon de Chelly, Arizona.

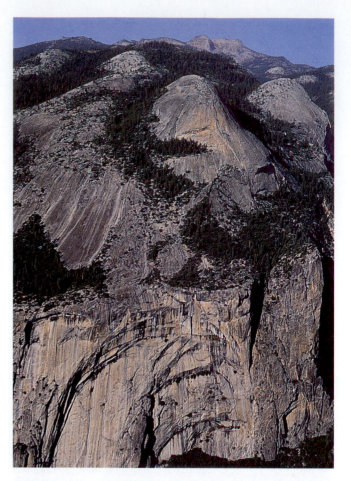

Figure 13.10 North Dome and Royal Arches, Yosemite National Park, California. Thick exfoliation rock layers, visible in the lower part of the photo, underlie the domed forms.

Unloading

A widespread process of rock disruption related to physical weathering results from *unloading*, the relief of confining pressure of overlying rock. *Exfoliation* is another term used for unloading. Unloading occurs as rock is brought near the surface by erosion of overlying layers. Rock formed at great depth beneath the earth's surface (particularly igneous and metamorphic rock) is in a slightly compressed state because of the confining pressure of overlying rock. As the rock above is slowly worn away, the pressure is reduced, and the rock expands slightly in volume. Expansion causes thick shells of rock to break free from the parent mass below. The new surfaces of fracture are a form of jointing called *sheeting structure*. The rock sheets show best in massive rocks such as granite and marble.

Where sheeting structure has formed over the top of a single large body of massive rock, an *exfoliation dome* is produced (Figure 13.10). Domes are among the largest of the landforms shaped primarily by weathering. In Yosemite Valley, California, where domes are spectacu-

larly displayed, the individual rock sheets may be as thick as 15 m (50 ft). (See *Eye on the Environment • Marvelous, Majestic, Monolithic Domes*, in Chapter 16.)

Other Physical Weathering Processes

Most rock-forming minerals expand when heated and contract when cooled. Where rock surfaces are exposed daily to the intense heating of the sun alternating with nightly cooling, the resulting expansion and contraction exerts powerful disruptive forces on the rock. Although first-hand evidence is lacking, it seems likely that daily temperature changes can cause the breakup of a surface layer of rock already weakened by other agents of weathering.

Another mechanism of rock breakup is the growth of plant roots, which can wedge joint blocks apart. You have probably observed concrete sidewalk blocks uplifted and fractured by the growth of tree roots. This process is also active when roots grow between rock layers or joint blocks. Even fine rootlets in joint fractures can cause the loosening of small rock scales and grains.

CHEMICAL WEATHERING AND ITS LANDFORMS

We investigated **chemical weathering** processes briefly in Chapter 10. Recall that the dominant processes of chemical change affecting silicate minerals are oxidation, hydrolysis, and acid action. Oxidation and hydrolysis are the chemical addition of oxygen and water, respectively. Acid action converts some minerals to soluble forms that are carried away by water. While feldspars and the mafic minerals are very susceptible to chemical weathering, quartz is a highly stable mineral, almost immune to decay.

Hydrolysis and Oxidation

Decomposition by hydrolysis and oxidation changes strong rock into very weak regolith. This change allows erosion to operate with great effectiveness wherever the regolith is exposed. Weakness of the regolith also makes it susceptible to natural forms of mass wasting. In warm, humid climates of the equatorial, tropical, and subtropical zones, hydrolysis and oxidation often result in the decay of igneous and metamorphic rocks to depths as great as 100 m (about 300 ft). To the construction engineer, deeply weathered rock is of major concern in the building of highways, dams, or other heavy structures. Although the regolith is soft and can often be removed by power shovels with little blasting, foundations on regolith can fail under heavy loads. This regolith also has undesirable plastic properties because of its high content of clay minerals.

> In chemical weathering, the minerals that make up rocks are chemically altered or dissolved.

The hydrolysis of exposed granite surfaces is accompanied by the grain-by-grain breakup of the rock. This process creates many interesting boulder and pinnacle forms by the rounding of angular joint blocks (Figure 13.11). These forms are particularly conspicuous in arid regions. In most deserts there is ample moisture for hydrolysis to act, given sufficient time. The products of grain-by-grain breakup form a fine desert gravel, which consists largely of quartz and partially decomposed feldspar crystals.

Acid Action

Chemical weathering is also produced by acid action, largely that of *carbonic acid*. This weak acid is formed when carbon dioxide dissolves in water. Rainwater, soil water, and stream water all normally contain dissolved carbon dioxide. Carbonic acid slowly dissolves some types of minerals. Carbonate sedimentary rocks, such as limestone and marble, are particularly susceptible to the acid action. In this process, the mineral calcium carbonate is dissolved and carried away in solution in stream water.

Carbonic acid reaction with limestone produces many interesting surface forms, mostly of small dimensions. Outcrops of limestone typically show cupping,

Figure 13.11 Large joint blocks of granite are gradually rounded into smooth forms by grain-by-grain disintegration in a desert environment. Alabama Hills, Owens Valley, California.

rilling, grooving, and fluting in intricate designs (Figure 13.12). In a few places, the scale of deep grooves and high wall-like rock fins reaches proportions that keep people and animals from passing through. Dissolution of limestone by carbonic acid in ground water can produce underground caverns as well as distinctive landscapes that are formed when underground caverns collapse. These landforms and landscapes will be described in Chapter 14.

In urban areas, air pollution by sulfur and nitrogen oxides commonly occurs. When these gases dissolve in rainwater, the result is acid precipitation. (See *Eye on the Environment • Air Pollution* in Chapter 4.) The acids rapidly dissolve limestone and chemically weather other types of building stones. The result can be very damaging to stone sculptures, building decorations, and tombstones (Figure 13.13).

In the wet low-latitude climates, mafic rock, particularly basaltic lava, undergoes rapid removal under attack by soil acids and produces landforms quite similar to those formed by carbonic acid on massive limestones in the moist climates of higher latitudes. The effects of solution removal of basaltic lava are displayed in spectacular grooves, fins, and spires on the walls of deep alcoves in part of the Hawaiian Islands (Figure 13.14).

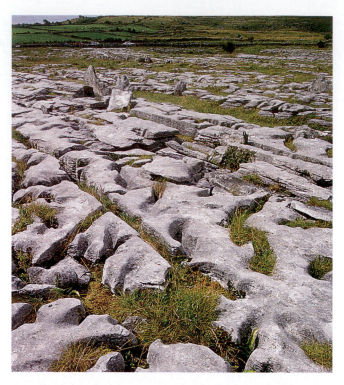

Figure 13.12 This outcrop of pure limestone shows rills and cups formed by solution weathering. County Clare, Ireland.

Figure 13.13 Weathered tombstones from a burying ground in Boston, Massachusetts. The marker on the left, carved in marble, has been strongly weathered, weakening the lettering. The marker on the right, made of slate, is much more resistant to erosion.

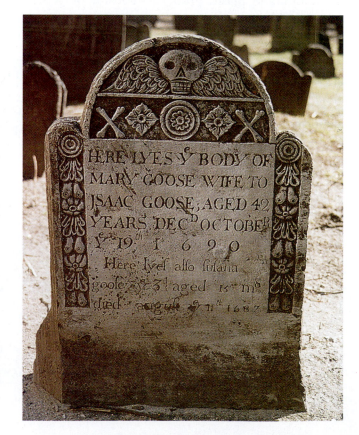

Figure 13.14 The steep walls of many narrow coastal ravines are deeply grooved by chemical weathering of the basaltic lava on the Napili Coast, Island of Hawaii, Hawaii.

MASS WASTING

With our discussion of weathering, we have described an array of processes that act to alter rock chemically and break it up into fragments. These rock fragments are subjected to gravity, running water, waves, wind, and the flow of glacial ice in landform-making processes. In the remainder of this chapter, we consider the first of these landform agents—gravity. We will return to the others in the following chapters.

Everywhere on the earth's surface, gravity pulls continuously downward on all materials. Bedrock is usually so strong and well supported that it remains fixed in place. However, when a mountain slope becomes too steep, bedrock masses can break free and fall or slide to new positions of rest. In cases where huge masses of bedrock are involved, the result can be catastrophic to towns and villages in the path of the slide. Such slides are a major environmental hazard in mountainous regions. They are but one form of mass wasting, which we defined earlier as spontaneous downhill movement of soil, rock, and regolith under the influence of gravity.

Because soil, regolith, and many forms of sediment are held together poorly, they are much more susceptible to movement under the force of gravity than hard, massive bedrock. On most slopes, at least a small amount of downhill movement is going on constantly. Although much of this motion is imperceptible, the regolith sometimes slides or flows rapidly.

The processes of mass wasting and the landforms they produce are extremely varied and tend to grade one into another. We have selected only a few of the most important forms of mass wasting to present in this chapter—they are shown in Figure 13.15. At the top of the diagram, three categories of information are pro-

Figure 13.15 Processes and forms of mass wasting.

Kinds of earth materials:	Rock (dry)	Regolith, soil, alluvium, clays + water	Water + sediment
Physical properties:	Hard, brittle, solid	Plastic substance	Fluid
Kinds of motion:	Falling, rolling, sliding	Flowage within the mass	Fluid flow

Forms of mass movement: Very slow → Very fast

ROCK CREEP TALUS CREEP · SOIL CREEP · SOLIFLUCTION · LANDSLIDES: BEDROCK SLUMP · ROCKSLIDE · EARTHFLOW (slump or flowage) · MUDFLOW · STREAM FLOW · ROCKFALL · ALPINE DEBRIS AVALANCHE · DEBRIS FLOOD vided, including the earth material involved, the properties of the material, and the kind of motion that occurs. Boxes in the body of the diagram contain the names of several forms of mass movements, arranged according to composition (left to right) and speed (top to bottom). As we describe each of these forms, this diagram will help you understand how it compares to the others.

Soil Creep

On almost any soil-covered slope, there is extremely slow downhill movement of soil and regolith, a process called **soil creep**. Figure 13.16 shows some of the evidence that soil creep is occurring. Joint blocks of distinctive rock types are found to be moved far downslope from the outcrop. In some layered rocks such as shales or slates, edges of the strata seem to "bend" in the downhill direction (Figure 13.17). This is not true plastic bending, but is the result of the slight movement of many rock pieces on small joint cracks. Creep causes fence posts and utility poles to lean downslope and even shift measurably out of line. Roadside retaining walls can buckle and break under the pressure of soil creep.

Soil creep refers to the slow downhill movement of soil and regolith.

What causes soil creep? Alternate drying and wetting of the soil, growth of frost needles, heating and cooling of the soil, trampling and burrowing by animals, and shaking by earthquakes all produce some disturbance of

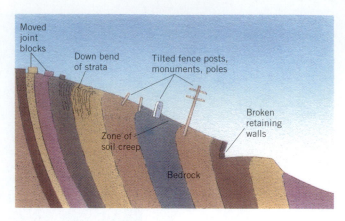

Figure 13.16 The slow, downhill creep of soil and regolith shows up in many ways on a hillside. (After C. F. S. Sharpe.)

the soil and regolith. Because gravity exerts a downhill pull on every such rearrangement, the particles very gradually work their way downslope.

Earthflow

In regions of humid climate, a mass of water-saturated soil, regolith, or weak shale may move down a steep slope during a period of a few hours in the form of an **earthflow**. Figure 13.18 is a sketch of an earthflow showing how the material slumps away from the top, leaving a steplike terrace bounded by a curved, wall-like scarp. The saturated material flows sluggishly to form a bulging toe.

Figure 13.17 Slow, downhill creep of regolith on this mountainside near Downieville, California, has caused vertical rock layers to seem to "bend over."

Figure 13.18 An earthflow with slump features well developed in the upper part. Flowage has produced a bulging toe. (After A. N. Strahler.)

Shallow earthflows, affecting only the soil and regolith, are common on sod-covered and forested slopes that have been saturated by heavy rains. An earthflow may affect a few square meters, or it may cover an area of several hectares (several acres). If the bedrock of a mountainous region is rich in clay (derived from shale or deeply weathered volcanic rocks), earthflows sometimes involve millions of metric tons of bedrock moving by plastic flowage like a great mass of thick mud.

Earthflows are a common cause of blockage of highways and railroad lines, usually during periods of heavy rains. Generally, the rate of flowage is slow, so that the flows are not a threat to life. Property damage to buildings, pavements, and utility lines is often severe where construction has taken place on unstable soil slopes.

A special variety of earthflow found in the arctic tundra is *solifluction* (from Latin words meaning "soil" and "flow"). In early summer, when thawing has penetrated the upper few tenths of meters, the soil is fully saturated. Soil water cannot escape downward because the frozen layer below will not allow water to drain through it. Flowing almost imperceptibly, this saturated soil forms solifluction terraces and lobes that give the tundra slope a stepped appearance (Figure 13.19).

Mudflow

One of the most spectacular forms of mass wasting and a potentially serious environmental hazard is the *mudflow*. This mud stream of fluid consistency pours swiftly down canyons in mountainous regions (Figure 13.20).

Figure 13.19 Solifluction lobes in the Richardson Mountains, Northwest Territories, Canada. Bulging masses of water-saturated regolith have slowly moved downslope, overriding the ground surface below while bearing intact their covers of plants and soil.

Figure 13.20 Thin, streamlike mudflows issue occasionally from canyon mouths in arid regions. The mud spreads out on the fan slopes below in long, narrow tongues. (After A. N. Strahler.)

In deserts, where vegetation does not protect the mountain soils, local thunderstorms produce rain much faster than it can be absorbed by the soil. As the water runs down the slopes, it forms a thin mud, which flows down to the canyon floors. Following stream courses, the mud continues to flow until it becomes so thick it must stop. Great boulders are carried along, buoyed up in the mud. Roads, bridges, and houses in the canyon floor are engulfed and destroyed. Where the mudflow emerges from the canyon and spreads across an alluvial fan, severe property damage and even loss of life may be the result (Figure 13.21).

As explained in Chapter 12, mudflows that occur on the slopes of erupting volcanoes are called *lahars.* Freshly fallen volcanic ash and dust are turned into mud by heavy rains or melting snows and flow down the slopes of the volcano. Herculaneum, a city at the base of Mount Vesuvius, was destroyed by a mudflow during the eruption of A.D.79. At the same time, the neighboring city of Pompeii was buried under volcanic ash.

Mudflows vary in consistency, from a mixture like concrete emerging from a mixing truck to that of a muddy river flow. The watery type of mudflow is called a *debris flood* or *debris flow* in the western United States, where it occurs commonly in Southern California with disastrous effects.

Landslide

A **landslide** is the rapid sliding of large masses of bedrock. Wherever mountain slopes are steep, there is a possibility of large, disastrous landslides (Figure 13.22). In Switzerland, Norway, or the Canadian Rockies, for example, villages built on the floors of steep-sided valleys have been destroyed by the sliding of millions of cubic meters of rock set loose without warning. Severe earthquakes in mountainous regions are a major cause of landslides and earthflows. *Eye on the Environment • The Great Hebgen Lake Disaster* presents the history of an earthquake-triggered landslide that dammed a river to create a new, natural lake.

In a landslide, a large mass of rock suddenly moves from a steep mountain slope to the valley below.

Aside from occasional local catastrophes, landslides have rather limited environmental influence because they occur only sporadically and usually in thinly populated mountainous regions. Small slides can, however, repeatedly block or break important mountain highways or railway lines.

Figure 13.21 This mudflow, carrying numerous large boulders, issued from a steep mountain canyon in the Wasatch Mountains, Utah.

Figure 13.22 A great landslide descended from the high mountain summit at the upper left. The rock became pulverized into loose debris, which rode across the valley floor at great speed on a cushion of trapped air. Turtle Mountain Slide, Frank, Alberta.

INDUCED MASS WASTING

Human activities can induce mass wasting in forms ranging from mudflow and earthflow to landslide. These activities include (1) piling up of waste soil and rock into unstable accumulations that can move spontaneously, (2) removal of support by undermining natural masses of soil, regolith, and bedrock, and (3) accelerating natural mass wasting processes. We can refer to mass movements produced by human activities as *induced mass wasting*.

In Los Angeles County, California, real estate development has been carried out on very steep hillsides and mountainsides by the process of bulldozing roads and homesites out of the deep regolith. The excavated regolith is pushed out into adjacent embankments where its instability poses a threat to slopes and stream channels below. When saturated by heavy winter rains, these embankments can give way, producing earthflows, mudflows, and debris floods that travel far down the canyon floors and spread out on the alluvial fan surfaces below, burying streets and yards in bouldery mud.

Many debris floods of this area are also produced by heavy rains falling on mountain slopes denuded of vegetation by fire in the preceding dry season. Some of these fires are set carelessly or deliberately by humans.

Induced Earthflows

Examples of both large and small earthflows induced or aggravated by human activities are found in the Palos Verdes Hills of Los Angeles County, California (Figure 13.23). These movements occur in shales that tend to become plastic when water is added. The upper part of the earthflow undergoes a subsiding motion with backward rotation of the down-sinking mass, as illustrated in Figure 13.18. The interior and lower parts of the mass move by slow flowage, and a toe of extruded flowage material may be formed.

Largest of the earthflows in the Palos Verdes Hill area was the Portuguese Bend "landslide" that affected an area of about 160 hectares (about 400 acres). It was caused by the slippage of sedimentary rock layers on an underlying layer of clay. The total movement over a three-year period was about 20 m (about 70 ft). Damage to residential and other structures totaled some $10 million. The most interesting observation, from the point of view of assessing human impact on the environment, is that the slide has been attributed by geologists to infiltration of water from septic tanks and from irrigation water applied to lawns and gardens. A discharge of over 115,000 liters (about 30,000 gallons) of water per day from some 150 homes is believed to have sufficiently weakened the clay layer to start and sustain the flowage.

EYE ON THE ENVIRONMENT

The Great Hebgen Lake Disaster

For some 200 vacationers camping near the Madison River in a deep canyon not far west of Yellowstone Park, the night of August 17, 1959, began quietly, with almost everyone safely bedded down in their tents or camping trailers. Up to a certain point, it was everything a great vacation should be—that point in time was 11:37 P.M., Mountain Standard Time. At precisely that instant, not one but four terrifying forms of disaster were set loose on the sleeping vacationers—earthquake, landslide, hurricane-force wind, and raging flood. The earthquake, which measured 7.1 on the Richter scale, was the cause of it all. The first shock, lasting several minutes, rocked the campers violently in their trailers and tents. Those who struggled to go outside could scarcely stand up, let alone run for safety.

Then came the landslide. A dentist and his wife watched through the window of their trailer as a mountain seemed to move across the canyon in front of them, trees flying from its surface like toothpicks in a gale. Then, as rocks began to bang against the sides and top of their trailer, they got out and raced for safer ground. Later, they found that the slide had stopped only about 25 m (80 ft) from the trailer. Pushed by the moving mountain was a vicious blast of wind. It swept upriver, tumbling trailers end over end.

Then came the flood. Two women schoolteachers, sleeping in their car only about 5 m (15 ft) from the river bank, awoke to the violent shaking of the earthquake. Like other campers, they first thought they had a marauding bear on their hands. Puzzled and frightened, they started the engine and headed the car for higher ground. As they did so, they were greeted by a great roar coming from the mountainside above and behind them. An instant later, the car was completely engulfed by a wall of water that surged up the river bank, then quickly drained back. With the screams of drowning campers in their ears, the two women managed to drive the car to safe ground, high above the river.

After the first surge of water, generated as the landslide mass hit the river, the river began a rapid rise. This rise was aided by great surges of water topping the Hebgen Dam, located upstream, as earthquake aftershocks rocked the water of Hebgen Lake back and forth along its length of about 50 km (30 mi). But, of course, in the darkness of night, the terrified victims of the flood had no idea of what was happening. In the panic that ensued, a 71-year-old man performed an almost unbelievable act of heroism to save his wife and himself from drowning. As the water rose inside their house trailer, he forced open the door, pulled his wife with him to the trailer roof, and then carried her up on the branches of a nearby pine tree. Here they were finally able to reach safety. The water had risen 10 m (about 30 ft) above ground level in just minutes.

The Madison Slide, as the huge earth movement was later named, had

Figure 13.23 This aerial view shows houses on Point Fermun, in Palos Verdes, California, disintegrating as they slide downward toward the sea.

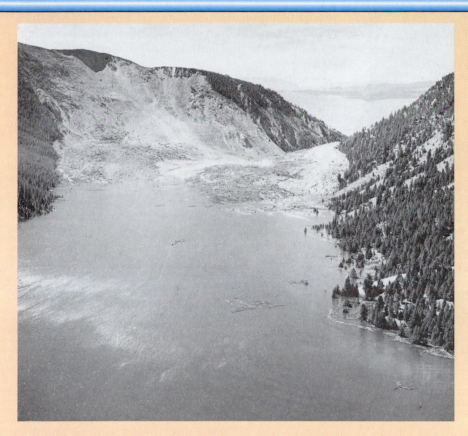

a bulk of 28 million m³ (37 million yd³) of rock (see photo). It consisted of a chunk of the south wall of the canyon, measuring over 600 m (about 2000 ft) in length and 300 m (about 1000 ft) in thickness. The mass descended over a half of a kilometer (about a third of a mile) to the Madison River, its speed estimated at more than 45 m/s (about 100 mi/hr). Pulverized into bouldery debris, the slide crossed the canyon floor, its momentum carrying it over 120 m (about 400 ft) in vertical distance up the opposite canyon wall. At least 26 persons died beneath the slide, and their bodies have never been recovered. Acting as a huge dam, the slide caused the Madison River to back up, forming a new lake. In three weeks' time, the lake was nearly 60 m (about 200 ft) deep. Today it is a permanent feature, named Earthquake Lake.

Question

Describe the formation of Earthquake Lake. What four forms of natural environmental disaster were involved?

Seen from the air a few days after the event, the Madison Slide forms a great dam of rubble across the Madison River Canyon.

Scarification of the Land

Industrial societies now possess enormous machine power and explosives capable of moving great masses of regolith and bedrock from one place to another. One use of this technology is the extraction of mineral resources. Another is the movement of earth in the construction of highway grades, airfields, building foundations, dams, canals, and various other large structures. Both activities involve removal of earth materials, a process that destroys the preexisting ecosystems and habitats of plants and animals. When the materials are used to build up new land on adjacent surfaces, ecosystems and habitats are also destroyed—by burial. What distinguishes artificial forms of mass wasting from the natural forms is that machinery is used to raise earth materials against the force of gravity. Explosives used in blasting can produce disruptive forces many times more powerful than the natural forces of physical weathering.

Scarification is a general term for excavations and other land disturbances produced to extract mineral resources. It includes the accumulation of rock waste as spoil or tailings. Among the forms of scarification are open-pit mines, strip mines, quarries for structural materials, borrow pits along highway grades, sand and gravel pits, clay pits, phosphate pits, scars from hydraulic mining, and stream gravel deposits reworked by dredging.

Strip mining is a particularly destructive scarification activity. Strip mining of coal is practiced where coal seams lie close to the surface or actually appear as outcroppings along hillsides (Figure 13.24). Here, earthmoving equipment removes the covering strata, or *overburden*, to bare the coal, which is lifted out by power shovels. The piled-up overburden is referred to as spoil. When saturated by heavy rains and melting snows, the spoil generates earthflows and mudflows that descend on houses, roads, and forest. The spoil also supplies sediment that clogs stream channels far down the valleys.

As a result of these problems, strip mining is under strict control in most locations. Mine operators are not permitted to create hazardous spoil slopes, and they

Figure 13.24 Contour strip mining near Lynch, Kentucky. A highway makes use of the winding bench at the base of the high rock wall.

must restore spoil banks and ridges to a natural condition. In this process, termed *reclamation*, topsoil is removed and saved as the mine is opened and worked. When spoil is returned to the mine cavity after extraction of the coal, it is covered with the topsoil and suitable plants are introduced. With good management and proper planning, the resulting reclaimed land becomes usable again. But there are also many reclamation failures in which the land has been damaged beyond repair.

Scarification is on the increase. Driven by an ever increasing human population, demands for coal and industrial minerals used in manufacturing and construction are on the rise. At the same time, as the richer and more readily available mineral deposits are consumed, industry turns to poorer and less easily accessible grades. As a result, the rate of scarification is further increased, making the success of reclamation efforts increasingly important.

A LOOK AHEAD

In this chapter, we have examined two related topics—weathering and mass wasting. In the weathering process, rock near the surface is broken up into smaller fragments and often altered in chemical composition. In the mass wasting process, weathered rock and soil move downhill in slow to sudden mass movements. The landforms of mass wasting are produced by gravity acting directly on soil and regolith. Gravity also powers another landform-producing agent—running water— which we take up in the next three chapters. The first chapter deals with water in the hydrologic cycle, in soil, and in streams. The second deals specifically with how streams and rivers erode regolith and deposit sediment to create landforms. The third describes how stream erosion strips away rock layers of different resistance, providing large landforms that reveal underlying rock structures.

WEATHERING AND MASS WASTING IN REVIEW

Weathering is the action of processes that cause rock to disintegrate and decompose because of surface exposure. Mass wasting is the spontaneous downhill motion of soil, rock, or regolith under gravity. Regolith, the surface layer of weathered rock particles, may be residual or transported.

Physical weathering produces regolith from solid rock by breaking bedrock into pieces. Frost action breaks rock apart by the repeated growth and melting of ice crystals in rock fractures and joints, as well as between individual mineral grains. In mountainous regions of vigorous frost, fields of angular blocks accumulate as felsenmeers. Slopes of rock fragments form talus cones. In soils and sediments, growing needle ice and ice lenses push rock and soil fragments upward. Salt-crystal growth in dry climates frees individual grains of rock and can damage brick and concrete. Unloading of the weight of overlying rock layers can cause some types of rock to expand and break loose into thick shells, producing exfoliation domes. Wedging by plant roots also forces rock masses apart.

Chemical weathering results from mineral alteration. Igneous and metamorphic rocks can decay to great depths through hydrolysis and oxidation, producing a regolith that is often rich in clay minerals. Carbonic acid reacts with and dissolves limestone. In warm, humid environments, basaltic lavas can also show features of solution weathering.

Soil creep is a process of mass wasting in which soil moves down slopes almost imperceptibly under the influence of gravity. In an earthflow, water-saturated soil or regolith slowly flows downhill. A special type of earthflow process, known as solifluction, occurs in the arctic spring when saturated soil flows downhill over a still-frozen base layer. A mudflow is much swifter than an earthflow. It follows stream courses, becoming thicker as it descends. A landslide is a rapid sliding of large masses of bedrock, sometimes triggered by an earthquake. Mass wasting can also be caused by removal of material in construction of roads or buildings, undermining the natural support of soil or regolith. Scarification of land by human activities, such as mining of coal or ores, can heap up soil and regolith into unstable masses that produce earthflows or mudflows. These actions are termed induced mass wasting.

KEY TERMS

weathering	physical weathering	mudflow
mass wasting	chemical weathering	landslide
bedrock	soil creep	scarification
alluvium	earthflow	

REVIEW QUESTIONS

1. What is meant by the term *weathering*? What types of weathering are recognized?
2. Define and contrast the terms *regolith*, *bedrock*, *sediment*, and *alluvium*.
3. How does frost action break up rock? Describe some landforms created by frost action and how they are formed.
4. How does salt-crystal growth break up rock? Give an example of a landform that arises from salt-crystal growth.
5. What is an exfoliation dome, and how does it arise? Provide an example.
6. Name three types of chemical weathering. Describe how limestone is often altered by a chemical weathering process.
7. Define mass wasting and identify the processes it includes.
8. What is soil creep, and how does it arise?
9. What is an earthflow? What features distinguish it as a landform? What special type of earthflow is encountered on the tundra?
10. Contrast earthflows and mudflows, providing an example of each.
11. Define the term *landslide*. How does a landslide differ from an earthflow?
12. Define and describe induced mass wasting. Provide some examples.
13. Explain the term *scarification*. Provide an example of an activity that produces scarification.

VISUALIZING EXERCISES

1. Sketch a cross section through a part of the landscape showing *regolith*, *bedrock*, *sediment*, and *alluvium* and label them on the sketch.
2. Identify and plot on Figure 13.15 the mass movement associated with each of the following locations: Turtle Mountain, Palos Verdes Hills, Nicolet, Mount St. Helens, Madison River, and Herculaneum. (*Hint*: Check *Putting You in the Picture*, *Eye on the Environment*, and chapter text.)

A DEEPER VIEW OF WEATHERING AND MASS WASTING

1. A landscape includes a range of lofty mountains elevated above a dry desert plain. Describe the processes of weathering and mass wasting that might be found on this landscape and identify their location.
2. Imagine yourself as the newly appointed director of public safety and disaster planning for your state. One of your first jobs is to identify locations where human populations are threatened by potential disasters, including those of mass wasting. Where would you look for mass wasting hazards and why? In preparing your answer, you may want to consult maps of your state.

Chapter 14

Fresh Water of the Continents

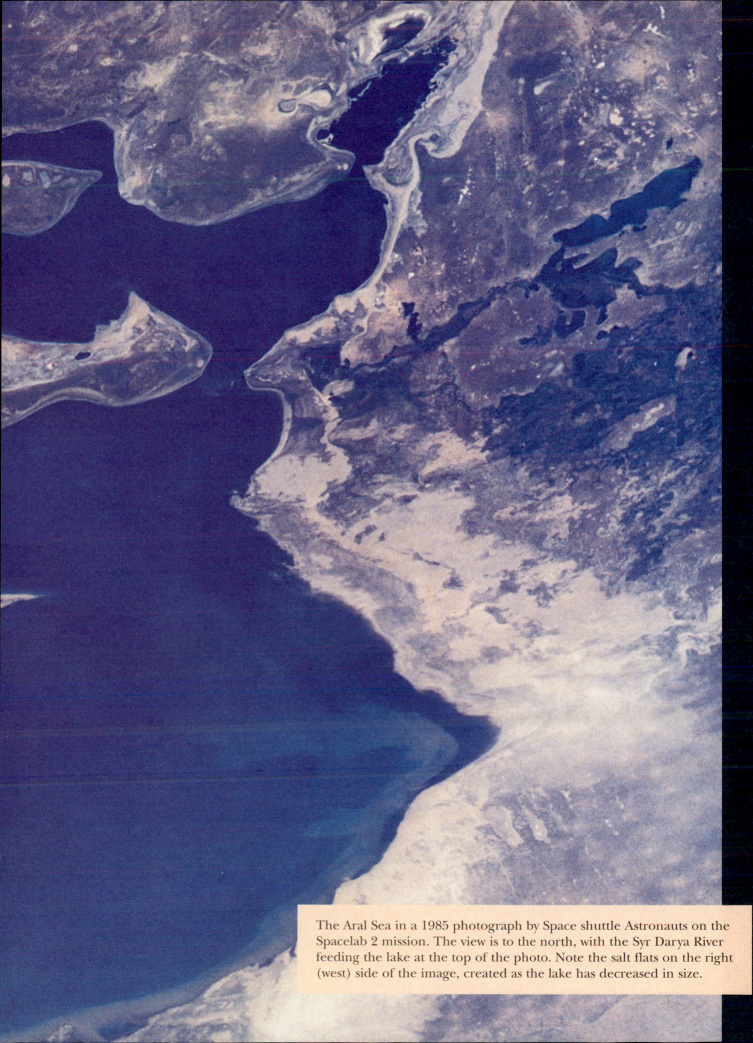

The Aral Sea in a 1985 photograph by Space shuttle Astronauts on the Spacelab 2 mission. The view is to the north, with the Syr Darya River feeding the lake at the top of the photo. Note the salt flats on the right (west) side of the image, created as the lake has decreased in size.

PUTTING YOU IN THE PICTURE

East of the Caspian Sea, astride the former Soviet republics of Kazakstan and Uzbekistan, lies an immense saline lake—the Aral Sea. Fed by meltwaters of high glaciers and snowfields in the lofty Hindu Kush, Pamir, and Tien Shan ranges, the lake endured through thousands of years as an oasis for terrestrial and aquatic wildlife deep in the heart of the central Asian desert.

But in the last 30 years, the Aral Sea, once larger than Lake Huron, has shrunk to a shadow of its former extent. The volume of its waters has decreased by 66 percent, and its salinity has increased from 1 percent to nearly 3 percent, making it almost as salty as sea water. Twenty of the 24 fish species native to the lake have disappeared. Its catch of commercial fish, which once supplied 10 percent of the total for the Soviet Union, has dwindled to zero. The deltas of the Amu Darya and Syr Darya rivers, which enter the south and east sides of the lake, were islands of great ecological diversity, teeming with fingerling fishes, birds, and their predators. Now only about half of the species of nesting birds remain. Many species of aquatic plants, shrubs, and grasses have vanished. Commercial hunting and trapping have almost ceased.

What caused this ecological catastrophe? The answer is simple—the lake's water supply was cut off. As an inland lake with no outlet, the Aral Sea receives water from the Amu Darya and Syr Darya, as well as a small amount from direct precipitation, but loses water by evaporation. Its gains balanced its losses, and, although these varied from year to year, the area, depth, and volume of the lake remained nearly constant until about 1960.

In the late 1950s, the Soviet government embarked on the first phases of a vast irrigation program, using water from the Amu Darya and Syr Darya for cotton cropping on the region's desert plain. The diversion of water soon became significant as more and more land came under irrigation. As a result, the influx fell to nearly zero by the early 1980s. The surface level of the Aral was sharply lowered and its area reduced. The sea became divided into two separate parts.

As the lake's shoreline receded, the exposed lakebed became encrusted with salts. The once-flourishing fishing port of Muynak became a ghost town, 50 km (30 mi) from the new lake shoreline. Strong winds now blow salt particles and mineral dusts in great clouds southwestward over the irrigated cotton fields and westward over grazing pastures. These salts—particularly the sodium chloride and sodium sulfate components—are toxic to plants. The salt dust permanently poisons the soil and can only be flushed away with more irrigation water.

The future of the lake appears grim indeed. With the present fresh water inflow nearly zero, the salinity of the remaining lake will soon be about the same as that of the ocean. Without the sacrifice of agricultural production for water to fill the lake, or the importation of more water from vast distances to the north, there is little that can be done to save the lake.

The Aral Sea is a small, but significant, part of the world's supply of water on the continents. This supply is largely fresh water found in lakes and streams, as well as in soils and deep underground reservoirs. Much of this water is evaporated from the oceans by solar energy, condensing or depositing over the continents to fall as precipitation and provide the fresh water on which terrestrial life depends. In this chapter, we trace the fate of precipitation as it sinks into the ground to recharge the stock of underground water, and as it runs off the land to feed the network of streams and rivers that blankets the landscape.

Chapter Outline

Ground Water
  The Water Table
  Aquifers
Limestone Solution by Ground Water
  Limestone Caverns
  Karst Landscapes
Problems of Ground Water Management
  Water Table Depletion
Eye on the Environment • Sinking Cities
  Contamination of Ground Water

Surface Water
  Overland Flow and Stream Flow
  Stream Discharge
  Drainage Systems
  River Floods
  Flood Prediction
  Lakes
  Saline Lakes and Salt Flats
  Desert Irrigation
  Pollution of Surface Water
Surface Water as a Natural Resource

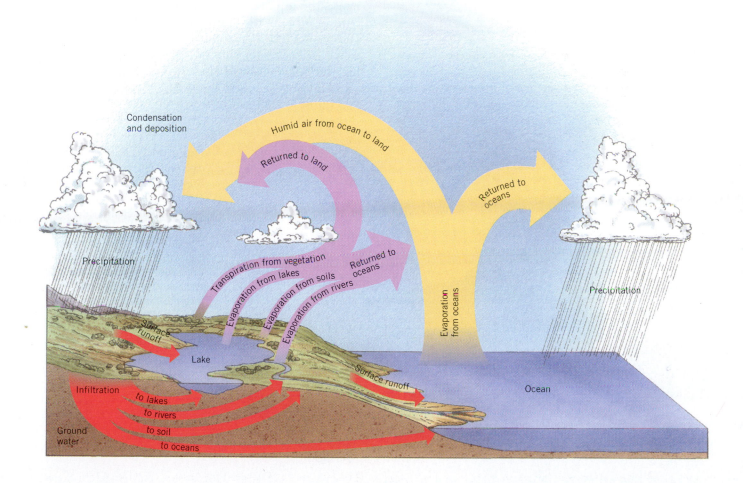

Chapter 14

FRESH WATER OF THE CONTINENTS

Before we continue with our survey of landforms and the processes that create them, we need to look more closely at the universal environmental agent that plays the dominant role in shaping the landscape—water. In this chapter we will focus first on two parts of the hydrologic cycle—the parts that involve water at the land surface and water that lies within the ground.

Recall from Chapter 4 that fresh water on the continents in surface and subsurface water constitutes only about 3 percent of the hydrosphere's total water. Most of this fresh water is locked into ice sheets and mountain glaciers. Ground water accounts for a little more than half of 1 percent, and fresh water in lakes, streams, and rivers constitutes only three-hundredths of 1 percent of the total water. Although very small proportion-

ally, the fresh, liquid water at the land surface is a vital part of the environment of plants and animals dwelling on continents.

In Chapter 4, we discussed atmospheric moisture and precipitation, describing the part of the hydrologic cycle in which water evaporates from ocean and land surfaces, condenses, or deposits, and then precipitates as rain or snow. What happens then to this precipitation? Figure 14.1 provides the answer. As the diagram shows, a portion of the precipitation returns directly to the atmosphere through evaporation from the soil. Another portion travels downward, moving through the soil under the force of gravity to become part of the underlying ground water body. Following underground flow paths, this subsurface water eventually emerges to become surface water, or it may emerge directly in the shore zone of the ocean. A third portion flows over the ground surface as runoff to lower levels. As it travels, the water flow becomes collected into streams, which eventually conduct the running water to the ocean.

In this chapter, we will trace the parts of the hydrologic cycle that include both the subsurface and surface

Figure 14.1 The hydrologic cycle traces the various paths of water from oceans, through the atmosphere, to land, and its return to oceans.

pathways of water flow. The study of these flows is part of the science of *hydrology*, which is the study of water as a complex but unified system on the earth.

Figure 14.2 shows what happens to water from precipitation as it first reaches the land surface. Most soil surfaces in their undisturbed, natural states are capable of absorbing the water from light or moderate rains by **infiltration**. In this process, water enters the small natural passageways between irregularly shaped soil particles, as well as the larger openings in the soil surface. These openings are formed by the borings of worms and animals, earth cracks produced by soil drying, cavities left from decay of plant roots, or spaces made by the growth and melting of frost crystals. A mat of decaying leaves and stems breaks the force of falling water drops and helps to keep these openings clear.

The precipitation that infiltrates the soil is temporarily held in the soil layer as soil water, occupying the *soil water belt* (Figure 14.2). Water within this belt can be returned to the surface and then to the atmosphere through a process that combines two components— evaporation and transpiration. As we explained in Chapter 9, these two forms of water vapor transport are combined in the term *evapotranspiration*.

Overland flow is a form of runoff in which water flows across the soil surface to reach stream channels.

When rain falls too rapidly to be passed downward through soil openings, **runoff** occurs and a surface water layer runs over the surface and down the direction of ground slope. This surface runoff is called **overland flow**. In periods of heavy, prolonged rain or rapid snowmelt, streams are fed directly by overland flow.

Overland flow also occurs when rainfall or snowmelt provides water to a soil that is already saturated. Since soil openings and pores are already filled, water can infiltrate the soil only as quickly as it can drain down to deeper layers. Under these conditions, nearly all of the precipitation will run off.

Runoff as overland flow moves surface particles from hills to valleys, and so it is an agent that shapes landforms. Because runoff supplies water to streams and rivers, it also allows rivers to cut canyons and gorges, and carry sediment to the ocean—but we are getting ahead of our story. We'll return to landforms carved by running water in Chapter 15.

GROUND WATER

As shown in Figure 14.2, water derived from precipitation can continue to flow downward beyond the soil water belt. This slow downward flow under the influence of gravity is termed *percolation*. Eventually, the percolating water reaches ground water. **Ground water** is the part of the subsurface water that fully saturates the pore spaces in bedrock, regolith, or soil, and so occupies the *saturated zone* (Figure 14.3). The **water table** marks the top of this zone. Above it is the *unsaturated zone*, in which water does not fully saturate the pores. Here, water is held by capillary tension—as thin films of water adhering to mineral surfaces. This zone also includes the soil water belt.

Ground water moves slowly in deep paths, eventually emerging by seepage into streams, ponds, lakes, and

Figure 14.2 Paths of precipitation falling on the land. Some is returned to the atmosphere through evapotranspiration. Some runs off the soil surface. The remainder sinks into the soil water belt, where it is accessible to plants. A portion of the infiltrating water passes through the soil water belt and percolates down to the ground water zone. (A. H. Strahler)

Figure 14.3 Zones of subsurface water.

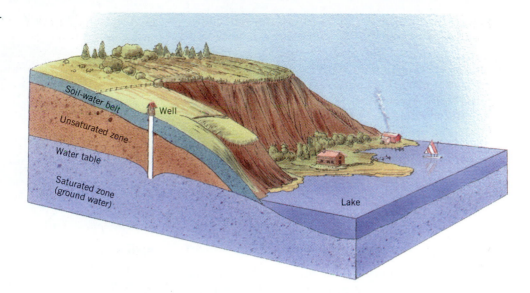

marshes. In these places the land surface dips below the water table. Streams that flow throughout the year—perennial streams—derive much of their water from ground water seepage.

The Water Table

Where there are many wells in an area, the position of the water table can be mapped in detail (Figure 14.4). This is done by plotting the water heights and noting the trend of change in elevation from one well to the other. The water table is highest under the highest areas of land surface—hilltops and divides. The water table declines in elevation toward the valleys, where it appears at the surface close to streams, lakes, or marshes.

The reason for this water table configuration is that water percolating down through the unsaturated zone tends to raise the water table, while seepage into lakes, streams, and marshes tends to draw off ground water and to lower its level. Because ground water moves extremely slowly, a difference in water table level is built up and maintained between high and low points on the water table. In periods of high precipitation, the water table rises under divide areas. In periods of water deficit, or during a drought, the water table falls (Figure 14.4).

The water table is highest under hilltops and divides. It slopes to intersect the surface at streams, lakes, or marshes.

Figure 14.4 The configuration of the water table surface conforms broadly with the land surface above it. It varies in response to prolonged wet and dry periods. Ground water flow paths circulate water to deep levels in a very slow motion and eventually feed streams by seepage.

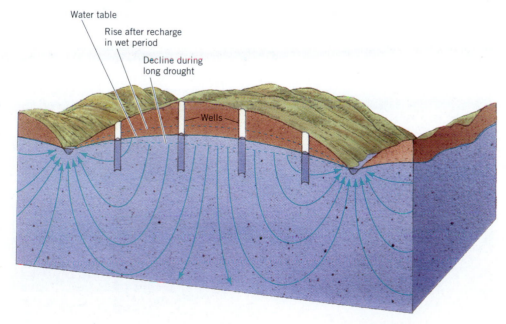

Figure 14.4 also shows paths of ground water flow. Water that enters the hillside midway between the hilltop and the stream flows rather directly toward the stream. Water reaching the water table midway between streams, however, flows almost straight down to great depths before recurving and rising upward again. Progress along these deep paths is incredibly slow, while flow near the surface is much faster. The most rapid flow is close to the stream, where the arrows converge. Over time, the level of the water table tends to remain stable, and the flow of water released to streams and lakes must balance the flow of water percolating down into the water table.

Aquifers

Sedimentary layers often exert a strong control over the storage and movement of ground water. For example, clean, well-sorted sand—such as that found in beaches, dunes, or stream alluvium—can hold an amount of ground water about equal to one-third of its bulk volume. A bed of sand or sandstone is thus often a good *aquifer*—that is, a layer of rock or sediment that contains abundant, freely flowing ground water. In contrast, beds of clay and shale are relatively impermeable and hold little free water. They are known as *aquicludes*. Figure 14.5 presents a situation in which a shale bed is overlain by a bed of sandstone. Ground water moves freely through the sandstone aquifer and across the top of the shale aquiclude, emerging in springs along the canyon shown at the right. Also shown is a small lens of shale, which accumulates a shallow *perched water table* above it.

When an aquifer is situated between two aquicludes, ground water in the aquifer may be under pressure and thus flow freely from a well. This type of self-flowing well is an **artesian well**. Figure 14.6 illustrates this situation. Since the elevation of the well head is below that of the range of hills feeding the aquifer, pressure causes water to rise in the well.

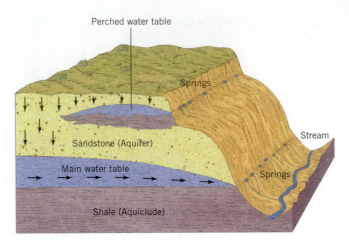

Figure 14.5 Ground water in horizontal strata. Water flows freely within the sandstone aquifer, but only very slowly in the shale aquiclude. A lens of shale creates a perched water table above the main water table. (Copyright © A. N. Strahler.)

LIMESTONE SOLUTION BY GROUND WATER

We saw in Chapter 13 that limestone at the surface is slowly dissolved by carbonic acid action in moist climates, producing lowland areas. The slow flow of ground water in the saturated zone can also dissolve limestone below the surface, producing deep underground caverns. These features can then collapse, producing a sinking of the ground above and the development of a unique type of landscape.

Limestone Caverns

You are probably familiar with such famous caverns as Mammoth Cave or Carlsbad Caverns. Millions of Americans have visited these natural wonders. *Limestone caverns* are interconnected subterranean cavities in bedrock formed by the corrosive action of circulating ground water on limestone. Figure 14.7 shows how cav-

Figure 14.6 An artesian well drilled into a dipping sandstone layer overlain by an impervious shale layer. (Drawn by Erwin Raisz. Copyright © A. N. Strahler.)

Figure 14.7 Cavern development in the ground water zone, followed by travertine deposition in the unsaturated zone. (Copyright © A. N. Strahler.)

erns develop. As shown in the upper diagram, the action of carbonic acid is particularly concentrated in the saturated zone just below the water table. This removal process forms many kinds of underground "landforms," such as tortuous tubes and tunnels, great open chambers, and tall chimneys. Subterranean streams can be found flowing in the lowermost tunnels, and these carry the products of solution to emerge along the banks of surface streams and rivers.

In a later stage, shown in the lower diagram, the stream has deepened its valley, and the water table has been correspondingly lowered to a new position. The cavern system previously formed is now in the unsaturated zone. Deposition of carbonate matter, known as *travertine*, takes place on exposed rock surfaces in the caverns. Encrustations of travertine take many beautiful forms—stalactites, stalagmites, columns, drip curtains, and terraces (Figure 14.8).

Figure 14.8 Travertine deposits in the Papoose Room of Carlsbad Caverns include stalactites (slender rods hanging from the ceiling) and sturdy columns.

Figure 14.9   Sinkholes in limestone, near Roswell, New Mexico.

Karst Landscapes

Where limestone solution is very active, we find a landscape with many unique landforms. This is especially true along the Dalmatian coastal area of Croatia, where the landscape is called *karst*. Geographers apply that term to the topography of any limestone area where sinkholes are numerous and small surface streams are nonexistent. A *sinkhole* is a surface depression in a region of cavernous limestone (Figure 14.9). Some sinkholes are filled with soil washed from nearby hillsides, while others are steep-sided, deep holes.

Development of a karst landscape is shown in Figure 14.10. In an early stage, funnel-like sinkholes are numerous. Later, the caverns collapse, leaving open, flat-floored valleys. Examples of some important regions of karst or karstlike topography are the Mammoth Cave region of Kentucky, the Yucatan Peninsula, and parts of Cuba and Puerto Rico. In regions such as southern China and west Malaysia, the karst landscape is dominated by steep-sided, conical limestone hills or towers, 100 to 500 m (about 300 to 1500 ft) high (Figure 14.11). These hills are often riddled with caverns and passageways.

Figure 14.10   Features of a karst landscape. (*a*) Rainfall enters the cavern system through sinkholes in the limestone. (*b*) Extensive collapse of caverns reveals surface streams flowing on shale beds beneath the limestone. Some parts of the flat-floored valleys can be cultivated. (Drawn by Erwin Raisz. Copyright © A. N. Strahler.)

Figure 14.11   Tower karst near Guilin (Kweilin), Guanxi Province, southern China at lat. 25° N. White limestone can be seen exposed in the nearly vertical sides of the towers.

PROBLEMS OF GROUND WATER MANAGEMENT

Rapid withdrawal of ground water has seriously affected the environment in many places. Increased urban populations and industrial developments require larger water supplies—needs that cannot always be met by constructing new surface-water reservoirs. To fill these needs, vast numbers of wells using powerful pumps draw great volumes of ground water to the surface, greatly altering nature's balance of ground water recharge and discharge.

In dry climates, agriculture is often heavily dependent on irrigation water from pumped wells—especially since major river systems are likely to be already fully utilized for irrigation. Wells are also convenient water sources. They can be drilled within the limits of a given agricultural or industrial property and can provide immediate supplies of water without any need to construct expensive canals or aqueducts.

In earlier times, the small well that supplied the domestic and livestock needs of a home or farmstead was actually dug by hand and sometimes lined with masonry. By contrast, a modern well supplying irriga-

tion and industrial water is drilled by powerful machinery that can bore a hole 40 cm (16 in.) or more in diameter to depths of 300 m (about 1000 ft) or more. Drilled wells are sealed off by metal casings that exclude impure near-surface water and prevent clogging of the tube by caving of the walls. Near the lower end of the hole, in the ground water zone, the casing is perforated to admit the water. The yields of single wells range from as low as a few hundred liters or gallons per day in a domestic well to many millions of liters or gallons per day for large industrial or irrigation wells.

Water Table Depletion

As water is pumped from a well, the level of water in the well drops. At the same time, the surrounding water table is lowered in the shape of a downward-pointing cone, termed the *cone of depression* (Figure 14.12). The difference in height between the cone tip and the original water table is the *drawdown*. The cone of depression may extend out as far as 16 km (10 mi) or more from a well where heavy pumping is continued. Where many wells are in operation, their intersecting cones will produce a general lowering of the water table.

EYE ON THE ENVIRONMENT

Sinking Cities

Ground water is a resource on which human civilization now depends to meet much of its demand for fresh water. However, it may take thousands of years to accumulate large reservoirs of underground water. When this water is removed rapidly by pumping, the rate of removal far exceeds the rate of recharge. The volume of water in the reservoir is reduced, and the water table falls. What are the effects of this change?

One important environmental effect of excessive ground water withdrawal is subsidence of the ground surface. Venice, Italy, provides a dramatic example of this side effect. Venice was built in the eleventh century A.D. on low-lying islands in a coastal lagoon, sheltered from the ocean by a barrier beach. Underlying the area are some 1000 m (about 3300 ft) of layers of sand, gravel, clay, and silt, with some layers of peat. Compaction of these soft layers has been going on gradually for centuries under the heavy load of city buildings. However, ground water withdrawal, which has been greatly accelerated in recent decades, has aggravated the condition.

Many ancient buildings in Venice now rest at lower levels and have suffered severe damage as a result of flooding during winter storms on the adjacent Adriatic Sea. Sea level is normally raised by the effects of coastal storms, and when high tides occur at the same time, water rises even higher. The problem of flooding during storms is aggravated by the fact that many of the canals of Venice receive raw sewage, so that the floodwater is contaminated.

Most of the subsidence in recent decades has been attributed to withdrawals of large amounts of ground water from industrial wells at Porto Marghere, the modern port of Venice, located a few kilometers distant on the mainland shore. This pumping has now been greatly curtailed, reducing the rate of subsidence to a very small natural rate (about 1 mm per year). However, the threat of flooding and damage to churches and other buildings of great historical value remains. Flood control now depends on the construction of seawalls and floodgates on the barrier beach that lies between Venice and the open ocean.

By the early 1980s, the port city of Bangkok, located on the Chao Phraya River a short distance from the Gulf of Thailand, had become the world's most rapidly sinking city as a result of massive ground water withdrawals from soft marine sediments beneath the city. During the 1980s, the rate of subsidence reached 14 cm (6 in.) per year, and major floods were frequent. Some reduction in the rate of ground water withdrawal has produced a modest decrease in the rate of subsidence, but as in the case of Venice, the flood danger remains.

Ground water withdrawal has affected several regions in California, where ground water for irrigation has been pumped from basins filled with alluvial (stream and lake-deposited) sediments. Water table levels in these basins have dropped over 30 m (98 ft), with a maximum drop of 120 to 150 m (394 to 492 ft) being recorded in one locality of California's San Joaquin Valley. In the Los Baños–Kettleman City area, ground subsidence of as much as 4.2 to 4.8 m (14 to 16 ft) was measured at some locations in a 35-year period.

Another important area of ground subsidence accompanying water withdrawal is beneath Houston, Texas, where the ground surface has subsided from 0.3 to 1 m (1 to 3 ft) in a metropolitan area 50 km (31 mi) across.

Perhaps the most celebrated case of ground subsidence is that of Mexico City where ground subsidence

---

Water table depletion often greatly exceeds recharge—the rate at which infiltrating water moves downward to the saturated zone. In arid regions, much of the ground water for irrigation is drawn from wells driven into thick aquifers of sand and gravel. These deposits are often recharged by the seasonal flow of streams that head high in adjacent mountains. Fanning out across the dry lowlands, the streams lose water, which sinks into the sands and gravels and eventually percolates to the water table below. The extraction of ground water by pumping can greatly exceed this recharge by stream flow, lowering the water table. Deeper wells and more powerful pumps are then required. The result is exhaustion of a natural resource that is not renewable except over very long periods of time.

Another byproduct of water table depletion is subsidence—a sinking of the land in response to the removal of water from underlying sediments. This problem has plagued a number of major cities that rely heavily on ground water wells for their water supplies. *Eye on the Environment • Sinking Cities* provides more information on this topic.

Contamination of Ground Water

Another major environmental problem related to ground water withdrawal is contamination of wells by

Land subsidence has subjected Venice to episodes of flooding by waters of the Adriatic Sea. Here, high water has flooded the Piazza San Marco and an outdoor café.

has ranged from about 4 to 7 m (13 to 23 ft). The subsidence has resulted from the withdrawal of ground water from an aquifer system beneath the city. The volume of clay beds overlying the aquifer has contracted greatly as water has been drained out. To combat the ground subsidence, recharge wells were drilled to inject water into the aquifer. In addition, new water supplies from sources outside the city area were developed to replace local ground water use. These efforts have stabilized the city center, but because Mexico City has no practical alternative to groud water pumping to provide water for its citizens, the subsidence continues.

Questions

1. Why has land subsidence occurred in Venice? What are the effects?

2. Identify four other locations that have suffered land subsidence caused by ground water withdrawal.

pollutants that infiltrate the ground and reach the water table. Both solid and liquid wastes are responsible. Disposal of solid wastes poses a major environmental problem in the United States because our advanced industrial economy provides an endless source of garbage and trash. Traditionally, these waste products were trucked to the town dump and burned there in continually smoldering fires that emitted foul smoke and gases. The partially consumed residual waste was then buried under earth.

In recent decades, a major effort has been made to improve solid-waste disposal methods. One method is high-temperature incineration, but it often leads to air pollution. Another is the sanitary landfill method in which waste is not allowed to burn. Instead, layers of waste are continually buried, usually by sand or clay available on the landfill site. The waste is thus situated in the unsaturated zone. Here it can react with rainwater that infiltrates the ground surface. This water picks up a wide variety of chemical compounds from the waste body and carries them down to the water table (Figure 14.13).

Once in the water table, the pollutants follow the flow paths of the ground water. As the arrows in the figure indicate, the polluted water may flow toward a supply well that draws in ground water from a large radius. Once the polluted water has reached the well, the water becomes unfit for human consumption.

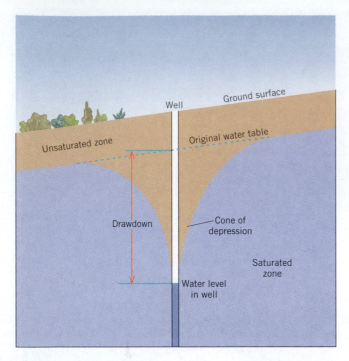

Figure 14.12 Drawdown and cone of depression in a pumped well. As the well draws water, the water table is depressed in a cone shape centered on the well. (A. N. Strahler)

Polluted water may also move toward a nearby valley, causing pollution of the stream flowing there (left side of Figure 14.13).

A common source of ground water contamination is the sanitary landfill.

Another source of contamination in coastal wells is *saltwater intrusion.* Since fresh water is less dense than saltwater, a coastal aquifer can be underlain by a layer of saltwater from the ocean. When the aquifer is depleted, the level of saltwater rises and eventually reaches the well from below, rendering the well unusable.

SURFACE WATER

So far, we have examined how water moves below the land surface. Now, we turn to tracing the flow paths of surplus water that runs off the land surface and ultimately reaches the sea. Here, we will be concerned primarily with rivers and streams. (In general usage, we speak of "rivers" as large watercourses and "streams" as smaller ones. However, the word "stream" is also used as a scientific term designating the channeled flow of surface water of any amount.)

Overland Flow and Stream Flow

As we saw earlier in this chapter, runoff that flows down the slopes of the land in broadly distributed sheets is overland flow. We can distinguish overland flow from *stream flow,* in which the water occupies a narrow channel confined by lateral banks. Overland flow can take several forms. It may be a continuous thin film, called *sheet flow,* where the soil or rock surface is smooth (Figure 14.14). Where the ground is rough or pitted, flow may take the form of a series of tiny rivulets connecting one water-filled hollow with another. On a grass-covered slope, overland flow is subdivided into countless tiny threads of water, passing around the stems. Even in a heavy and prolonged rain, you might not notice overland flow in progress on a sloping lawn. On heavily forested slopes, overland flow may pass entirely concealed beneath a thick mat of decaying leaves.

Overland flow eventually contributes to a stream, which is a much deeper, more concentrated form of runoff. We can define a **stream** as a long, narrow body of flowing water occupying a trenchlike depression, or channel, and moving to lower levels under the force of gravity. The **channel** of a stream is a narrow trough that contains the stream flow. It is shaped by the forces of flowing water to be most effective in moving the quantities of water and sediment supplied to the stream (Figure 14.15). A channel may be so narrow that you can easily jump across it, or, in the case of the Mississippi River, as wide as 1.5 km (about 1 mi) or more.

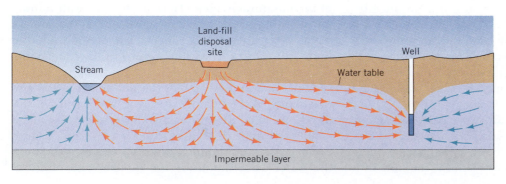

Figure 14.13 Polluted water, leached from a waste disposal site, moves toward a supply well (*right*) and a stream (*left*). (Copyright © A. N. Strahler.)

Figure 14.14 Overland flow taking the form of a thin sheet of water covers the nearly flat plain in the middle distance. This water is converging into stream flow in a narrow, steep-sided gully (*left*). The photograph was taken shortly after a summer thunderstorm had deluged the area. The locality, near Raton, New Mexico, shows steppe grassland vegetation.

As a stream flows under the influence of gravity, the water encounters resistance—a form of friction—with the channel walls. As a result, water close to the bed and banks moves slowly, while water in the central part of the flow moves faster. If the channel is straight and symmet-rical, the single line of maximum velocity is located in midstream. If the stream curves, the maximum velocity is found toward the bank on the outside of the curve.

Actually, the arrows in Figure 14.15 only show average velocity. In all but the most sluggish streams, the water is affected by *turbulence*, a system of countless eddies that are continually forming and dissolving. If we follow a particular water molecule, it will travel a highly irregular, corkscrew path as it is swept downstream. Motions include upward, downward, and sideward directions. Only if we measure the water velocity at a certain fixed point for a long period of time, say, several minutes, will the average motion at that point be downstream and in a line parallel with the surface and bed.

Stream Discharge

Stream flow at a given location is measured by its **discharge,** which is defined as the volume of water per unit time passing through a cross section of the stream at that location. It is measured in cubic meters (cubic feet) per second. The cross-sectional area and average velocity of a stream can change within a short distance, even though the stream discharge does not change (Figure 14.16). These changes occur because of changes in the *gradient*, or slope of the stream channel. The gradient is the rate of fall in elevation of the stream surface in the downstream direction (Figure 14.15).

When the gradient is steep, the force of gravity will act more strongly and flow velocity will be greater. When

Figure 14.15 Characteristics of stream flow. The velocity of flow is greatest in the middle (*a*) and at the top (*b*) of the stream. (*b*, Copyright © A. N. Strahler.)

Figure 14.16 In order to preserve a uniform discharge, a stream becomes wider and deeper in pools, where flow velocity is slower. In rapids, flow is swifter and the stream becomes narrower and shallower.

the stream channel has a gentle gradient, the velocity will be slower. However, in a short stretch of stream, the discharge will remain constant. As shown in Figure 14.16, this means that in stretches of rapids, where the stream flows swiftly, the stream channel will be shallow and narrow. In pools, where the stream flows more slowly, the stream channel will be wider and deeper to main-tain the same discharge. Sequences of pools and rapids can be found along streams of all sizes.

The discharges of streams and rivers change from day to day, and records of daily and flood discharges of major streams and rivers are important information. They are used in planning the development and distrib-ution of surface waters, as well as in designing flood-pro-tection structures and predicting floods as they progress down a river system. An important activity of the U.S. Geological Survey is the measurement, or gauging, of stream discharge in the United States. In cooperation with states and municipalities, this organization main-tains over 6000 gauging stations on principal streams and their tributaries.

The larger the cross section of a river, the lower will be its gradient.

Figure 14.17 is a map showing the relative discharge of major rivers of the United States. The mighty Mississippi with its tributaries dwarfs all other North America rivers. Two other discharges are of major pro-portions—the Columbia River, draining a large segment of the Rocky Mountains in southwestern Canada and the northwestern United States, and the Great Lakes,

Figure 14.17 This schematic map shows the relative magnitude of United States rivers. The width of the river as drawn is proportional to mean annual discharge. (After U.S. Geological Survey.)

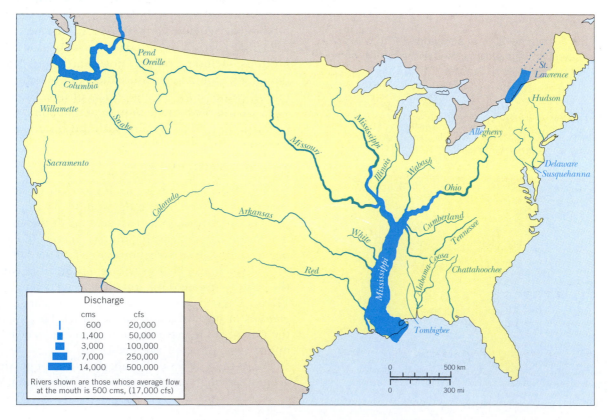

Figure 14.18 Channel network of a small drainage basin. Drainage divides mark the boundaries of stream basins. (Data of U.S. Geological Survey and Mark A. Melton.)

discharging through the St. Lawrence River. The Colorado River, a much smaller stream, crosses a vast semiarid and arid region in which little tributary flow is added to its snowmelt source high in the Rocky Mountains.

As the figure shows, the discharge of major rivers increases downstream as a natural consequence of the way streams and rivers combine to deliver runoff and sediment to the oceans. The gradient also changes in a downstream direction. The general rule is the larger the cross-sectional area of the stream, the lower the gradient. Great rivers, such as the Mississippi and Amazon, have gradients so low that they can be described as "flat." For example, the water surface of the lower Mississippi River falls in elevation about 3 cm for each kilometer of downstream distance (1.9 in. per mi).

Rivers with headwaters in high mountains have characteristics that are especially desirable for utilizing river flow as irrigation and for preventing floods. The higher ranges serve as snow storage areas, slowly releasing winter and spring precipitation through early or midsummer. As melting proceeds to successively higher levels, the meltwater is supplied to the river. In this way, a continuous river flow is maintained. Among the snow-fed rivers of the western United States are the Columbia, Snake, Missouri, Arkansas, and Colorado.

Drainage Systems

As runoff moves to lower and lower levels and eventually to the sea, it becomes organized into a **drainage system**. The system consists of a branched network of stream channels, as well as the sloping ground surfaces

that contribute overland flow to those channels. Between the channels on the ridges are *drainage divides*, which mark the boundary between slopes that contribute water to different streams or drainage systems. The entire system is bounded by a drainage divide that outlines a more-or-less pear-shaped **drainage basin**.

A drainage basin contains a branched network of stream channels and adjacent slopes.

A typical stream network within a drainage basin is shown in Figure 14.18. Each fingertip tributary receives runoff from a small area of land surface surrounding the channel. The entire surface within the outer divide of the drainage basin constitutes the **watershed** for overland flow. The drainage system provides a converging mechanism that funnels overland flow into streams and smaller streams into larger ones.

River Floods

You've probably been exposed to enough media coverage of river floods to have a good idea of what floodwaters look like and what kind of damage is caused by flood erosion and deposition of silt and clay. Even so, it is not easy to define the term **flood**. Perhaps it is enough to say that a flood condition exists when the discharge of a river cannot be accommodated within its normal channel. As a result, the water spreads over the adjoining ground, which is normally cropland or forest. Sometimes, however, the ground is occupied by houses, factories, or transportation corridors.

Most rivers of humid climates have a *floodplain*, a broad belt of low flat ground bordering the channel on one or both sides that is flooded by stream waters about once a year. This flood usually occurs in the season when abundant surface runoff combines with the effects of a high water table to supply more runoff than can be carried in the channel. This annual inundation is considered a flood, even though its occurrence is expected and does not prevent the cultivation of crops after the flood has subsided. The seasonal inundation does not interfere with the growth of dense forests, which are widely distributed over low, marshy floodplains in all humid regions of the world. Still higher discharges of water, the rare and disastrous floods that may occur as rarely as once in 30 or 50 years, inundate ground lying well above the floodplain (Figure 14.19).

For practical purposes, the National Weather Service, which provides a flood-warning service, designates a particular river surface height, or *stage*, at a given place as the *flood stage*. Above this critical level, inundation of the floodplain will occur.

Flood Prediction

How often can we expect a river to flood, and how high will the water be? The answer, of course, is different for each river and each location on it. Hydrologists have developed a graphical method to present the flood history of a river. Figure 14.20 shows a graph of the flood history of the Mississippi River at Vicksburg and explains the meaning of the bar symbols. The Mississippi responds largely to spring floods, producing a simple annual cycle.

Flood forecasts rely on precipitation patterns, past history, and present river levels to predict when and how high a flood will crest.

Flash floods are characteristic of streams draining small watersheds with steep slopes. When an intense rainfall event occurs, these streams rise very quickly to high water levels. The flood arrives as a swiftly moving wall of turbulent water, sweeping away buildings and vehicles in its path. In heavily forested watersheds, as in the Appalachians, the water carries little sediment load and is highly erosive. In arid western watersheds, great quantities of coarse rock debris are swept into the main channel and travel with the floodwaters, creating a mudflow (see Figure 13.21). Because flash floods often occur too quickly to warn affected populations, they can cause significant loss of life.

Figure 14.19 The Red River, near Fargo, North Dakota, in flood in April, 1997. The normal channel is marked by the sinuous bands of trees running across the lower part of the photo. Dikes have kept the water from flooding the town in the center of the image.

Figure 14.20 Maximum monthly stages of the Mississippi River at Vicksburg, Mississippi. The river shows an annual cycle of spring flooding, with lowest water occurring in the fall. (After National Weather Service.)

The National Weather Service operates a River and Flood Forecasting Service through 85 offices located at strategic points along major river systems of the United States. When a flood threatens, forecasters analyze precipitation patterns and the progress of high waters moving downstream. Examining the flood history of the rivers and streams concerned, they develop specific flood forecasts. These are delivered to communities within the associated district, which usually covers one or more large watersheds. Flood warnings are publicized by every possible means. Close cooperation is maintained with various agencies to plan evacuation of threatened areas and the removal or protection of property.

Although our awareness of floods is usually based on the damage to settlements and structures on floodplains caused by flooding, floods may also have beneficial effects. Overbank flooding serves to recharge ground water tables in flooded areas. Nutrient-rich silt and clay particles are deposited on some parts of the floodplain, enhancing the productivity of floodplain soils. Erosion and scouring with deposition of sand and gravel on other parts of the floodplain provides new land for successional plant species, thus preserving biodiversity. Most natural floodplain ecosystems are well-adjusted to flooding. The problems arise when a human population invests heavily in floodplain development without due regard for the occurrence of floods that are rare on the human life scale, but frequent on the time scale over which landforms and ecosystems evolve. **SCIENCE NEWS** 26

Lakes

A **lake** is a water body that has an upper surface exposed to the atmosphere and no appreciable gradient. The term *lake* includes a wide range of water bodies. Ponds (which are small, usually shallow water bodies), marshes, and swamps with standing water can all be included under the definition of a lake. Lakes receive water input from streams, overland flow, and ground water, and so are included as parts of drainage systems. Many lakes lose water at an outlet, where water drains over a dam (natural or constructed) to become an outflowing stream. Lakes also lose water by evaporation. Lakes, like streams, are landscape features but are not usually considered to be landforms.

Basins occupied by lakes show a wide range of origins as well as a vast range in dimensions. Lake basins, like stream channels, are true landforms. Basins are created by a number of geologic processes. For example, the tectonic process of crustal faulting creates many large, deep lakes. Lava flows often form a dam in a river valley, causing water to back up as a lake. Landslides suddenly create lakes, as we saw in the case of the Madison Slide (see *Eye on the Environment • The Great Hebgen Lake Disaster* in Chapter 13).

In moist climates, the water level of lakes and ponds coincides closely with the water table in the surrounding area. Seepage of ground water into the lake, as well as runoff from streams, maintains these free water surfaces permanently throughout the year. Freshwater marshes and swamps, in which water stands at or close to the

Glacial sand and gravel

Water table

Pond

Ground water zone

Well

Figure 14.21 Freshwater ponds in sandy glacial deposits on Cape Cod, Massachusetts. (Redrawn from *A Geologist's View of Cape Cod*, copyright © A. N. Strahler, 1966. Used by permission of Doubleday, a division of Bantam Doubleday Dell Publishing Group, Inc.)

ground surface over a broad area, also represent the appearance of the water table at the surface. Such areas of poor surface drainage are included under the name of *wetlands.*

Examples of freshwater lakes and ponds maintained by ground water seepage and runoff are widely distributed in glaciated regions of North America and Europe. Here, plains of glacial sand and gravel contain natural pits and hollows left by the melting of stagnant ice masses that were buried in the sand and gravel deposits (see Chapter 18). Figure 14.21 is a block diagram showing small freshwater ponds on Cape Cod. The surface elevation of these ponds coincides closely with the level of the surrounding water table.

An important point about lakes in general is that they are short-lived features on the geologic time scale. Lakes disappear by several processes that sometimes work in combination. First, lakes that have stream outlets will be gradually drained as the outlets are eroded to lower levels. Where a strong bedrock threshold underlies the outlet, erosion will be slow but nevertheless certain.

Second, lakes accumulate inorganic sediment carried by streams entering the lake and organic matter produced by plants and animals within the lake. Eventually, they fill up, forming a boggy wetland with little or no free water surface (Figure 14.22). The ultimate result is a bog with its surface close to the water table.

If climate changes and becomes more arid, lakes can shrink in size or dry up completely.

Third, lakes can also disappear when climate changes. If precipitation is reduced within a region, or temperatures and net radiation increase, evaporation can exceed input and the lake will dry up. Many former lakes of the southwestern United States flourished in moister periods of glacial advance during the Ice Age. Today, they have shrunk greatly or have disappeared entirely under the present arid regime.

Lakes are quite important from the human viewpoint. They are frequently used as sources of fresh water, and they also support ecosystems that provide

Figure 14.22 A freshwater pond in Wisconsin. Vegetation is slowly growing inward at the edges, and eventually the pond will become a bog supporting wet forest.

Figure 14.23 These salt encrustations at the edge of Great Salt Lake, Utah, were formed when the lake level dropped during a dry period.

food for humans. Where dammed to a high level above the outlet stream, they can provide hydroelectric power as well. Lakes and ponds are also important recreation sites and sources of natural beauty.

Where lakes are not naturally present in the valley bottoms of drainage systems, we create lakes as needed by placing dams across the stream channels. Many regions that formerly had almost no natural lakes are now abundantly supplied. Some are small ponds built to serve ranches and farms, while others cover hundreds of square kilometers. In some areas, the number of artificial lakes is large enough to have significant effects on the region's hydrologic cycle

Saline Lakes and Salt Flats

Lakes with no surface outlet are characteristic of arid regions. Here, the average rate of water loss by evaporation balances the average rate of stream inflow. If the rate of inflow increases, the lake level will rise. At the same time, the lake surface will increase in area, allowing a greater rate of evaporation. A new balance can then be achieved. Similarly, if the region becomes more arid, reducing input and increasing evaporation, the water level will fall to a lower level. The Aral Sea, which was the subject of our opening feature, is a prime example. However, in this case the lake has shrunk in response to the agricultural diversion of large amounts of water from the rivers that feed it.

Lakes without outlets often show salt buildup. Dissolved solids are brought into the lake by streams—usually streams that head in distant highlands where a water surplus exists. Since evaporation removes only pure water, the salts remain behind and the salinity of the water slowly increases (Figure 14.23). Salinity, or degree of "saltiness," refers to the abundance of certain common ions in the water. Eventually, salinity levels reach a point where salts are precipitated as solids, forming evaporites (see Chapter 10).

Sometimes the surfaces of such lakes lie below sea level. An example is the Dead Sea, with a surface elevation of −396 m (−1299 ft). The largest of all lakes, the Caspian Sea, has a surface elevation of −25 m (82 ft). Both of these large lakes are saline.

In regions where climatic conditions consistently favor evaporation over input, the lake may be absent. Instead, a shallow basin covered with salt deposits (see Figure 15.32), otherwise known as a *salt flat* or dry lake, will occur. In Chapter 15 we will describe dry lakebeds as landforms. On rare occasions, these flats are covered by a shallow layer of water, brought by flooding streams heading in adjacent highlands.

A number of desert salts are of economic value and have been profitably extracted from salt flats. In shallow coastal estuaries in the desert climate, sea salt for human consumption is commercially harvested by allowing it to evaporate in shallow basins. One well-known source of this salt is the Rann of Kutch, a coastal lowland in the tropical desert of westernmost India, close to Pakistan. Here the evaporation of shallow water of the Arabian Sea has long provided a major source of salt for inhabitants of the interior.

Desert Irrigation

Human interaction with the tropical desert environment is as old as civilization itself. Two of the earliest sites of civilization—Egypt and Mesopotamia—lie in the tropical deserts. Successful occupation of these deserts requires irrigation with large supplies of water from nondesert sources. For Egypt and Mesopotamia, the water sources of ancient times were the rivers that cross the desert but derive their flow from regions that have a water surplus. These are referred to as *exotic rivers*, because their flows are derived from an outside region.

Irrigation systems in arid lands divert the discharge of an exotic river such as the Nile, Indus, Jordan, or Colorado into a distribution system that allows the water to infiltrate the soil of areas under crop cultivation. Ultimately, such irrigation projects can suffer from two undesirable side effects: salinization and waterlogging of the soil.

Salinization occurs when salts build up in the soil to levels that inhibit plant growth. This happens because an irrigated area within a desert loses large amounts of soil water through evapotranspiration. Salts contained in the irrigation water remain in the soil and increase to high concentrations. Salinization may be prevented or cured by flushing the soil salts downward to lower levels by the use of more water. This remedy requires greater water use than for crop growth alone. In addition, new drainage systems must be installed to dispose of the excess saltwater.

Salinization and waterlogging are undesirable side effects of continued irrigation.

Waterlogging occurs when irrigation with large volumes of water causes a rise in the water table, bringing the zone of saturation close to the surface. Most food crops cannot grow in perpetually saturated soils. When the water table rises to the point at which upward movement under capillary action can bring water to the surface, evaporation is increased and salinization is intensified.

Agricultural areas of major salinization include the Indus River Valley in Pakistan, the Euphrates Valley in Syria, the Nile Delta of Egypt, and the wheat belt of western Australia. In the United States, extensive regions of heavily salinized agriculture are found in the San Joaquin and Imperial valleys of California. Other areas of salinization occur throughout the entire semiarid and arid regions of the western United States. In Chapter 9, *Eye on the Environment • Death of a Civilization* documents the effects of salinization and waterlogging on the civilizations of ancient Mesopotamia.

Pollution of Surface Water

Streams, lakes, bogs, and marshes are specialized habitats of plants and animals. Their ecosystems are particularly sensitive to changes induced by human activity in the water balance and in water chemistry. Our industrial society not only makes radical physical changes in water flow by construction of engineering works (dams, irrigation systems, canals, dredged channels), but also pollutes and contaminates our surface waters with a large variety of wastes.

Among the common chemical pollutants of both surface water and ground water are sulfate, chloride, sodium, nitrate, phosphate, and calcium ions. In urban and suburban areas, sources of these ions include sewage effluent, deicing salts, and lawn conditioners (lime and fertilizers). In agricultural regions, important sources of pollutant ions are fertilizers and livestock wastes.

Water pollutants include various types of common ions and salts, as well as heavy metals, organic compounds, and acids.

Phosphate and nitrate ions are plant nutrients and can lead to excessive growth of algae and other aquatic plants in streams and lakes. Applied to lakes, this process is known as *eutrophication*, which is often described as the "aging" of a lake. In eutrophication, the accumulation of nutrients stimulates plant growth, producing a large supply of dead organic matter in the lake. Microorganisms break down this organic matter but require oxygen in the process. However, oxygen dissolves only slightly in water, and so it is normally present only in low concentrations. The added burden of oxygen use by the decomposers reduces the oxygen level to the point where other organisms, such as desirable types of fish, cannot survive. After a few years of nutrient pollution, the lake can take on the characteristics of a shallow pond that results when a lake is slowly filled with sediment and organic matter over thousands of years by natural "aging" processes.

A particular form of chemical pollution of surface water goes under the name of *acid mine drainage*. It is an important form of environmental degradation in parts of Appalachia where abandoned coal mines and strip-mine workings are concentrated (Figure 14.24). Ground water emerges from abandoned mines and as soil water percolating through strip-mine waste banks. This water contains sulfuric acid and various salts of metals, particularly of iron. Acid of this origin in stream waters can have adverse effects on animal life. In sufficient concentrations, it is lethal to certain species of fish and has at times caused massive fish kills.

Some industrial plants dispose of toxic metals and organic compounds by discharging them directly into streams and lakes. Toxic metals, among them mercury, along with pesticides and a host of other industrial chemicals, are introduced in quantities that are locally damaging or lethal to plant and animal communities. Many towns and cities still discharge untreated or partly

Figure 14.24 This small stream is badly polluted by acid waters that have percolated through strip-mine waste.

treated sewage wastes, introducing live bacteria and viruses that are classed as biological pollutants. These can pose a threat to the health of humans and animals.

Thermal pollution is a term applied generally to the discharge of heat into the environment from combustion of fuels and from the conversion of nuclear energy into electric power. Thermal pollution of water takes the form of heavy discharges of heated water into streams, estuaries, and lakes. The thermal environmental impact may be quite drastic in a local area. With power plants and nuclear arms factories, there is also the possibility of contamination by radioactive substances. All told, there are many threats to the fresh water supply that human and natural ecosystems depend on.

SURFACE WATER AS A NATURAL RESOURCE

Fresh surface water is a basic natural resource essential to human agricultural and industrial activities. Runoff held in reservoirs behind dams provides water supplies for great urban centers, such as New York City and Los Angeles. When diverted from large rivers, it provides irrigation water for highly productive lowlands in arid lands, such as the Sacramento and San Joaquin valleys of California and the Nile Valley of Egypt. To these uses of runoff are added hydroelectric power, where the gradient of a river is steep, or routes of inland navigation, where the gradient is gentle.

Our heavily industrialized society requires enormous supplies of fresh water for its sustained operation. Urban dwellers consume water in their homes at rates of 150 to 400 liters (50 to 100 gallons) per person per day (Figure 14.25). Large quantities of water are used for cooling in air conditioning units and power plants.

Much of this water is obtained from surface water supplies, and the demand increases daily.

Unlike ground water, which represents a large water storage body, fresh surface water in the liquid state is stored only in small quantities. (An exception is the Great Lakes system.) Recall from Chapter 4 that the global quantity of available ground water is about 20 times as large as that stored in freshwater lakes and that the water held in streams is only about one one-hundredth of that in lakes. Because of small natural storage capacities, surface water can be drawn only at a rate comparable with its annual renewal through precipitation. Dams are built to develop useful storage capacity for runoff that would otherwise escape to the sea, but, once the reservoir has been filled, water use must be scaled to match the natural supply rate averaged over the year. Development of surface-water supplies brings

Figure 14.25 How water is used in an average home in Akron, Ohio. (U.S. Geological Survey.)

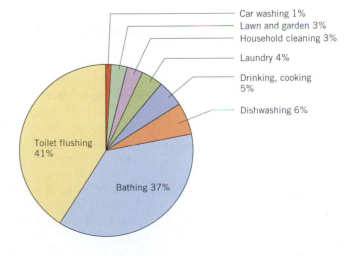

Car washing 1%
Lawn and garden 3%
Household cleaning 3%
Laundry 4%
Drinking, cooking 5%
Dishwashing 6%
Toilet flushing 41%
Bathing 37%

on many environmental changes, both physical and biological, and these must be taken into account in planning for future water developments.

A LOOK AHEAD

This chapter has focused on water, including the precipitation that runs off the land and flows into the sea or accumulates in inland basins. Think for a moment about the drainage system of a stream that conducts this flow of runoff. The smallest streams catch runoff from slopes, carrying the runoff into larger streams. The larger streams, in turn, receive runoff from their side slopes and also pass their flow on to still larger streams. However, this cannot happen unless the gradients of slopes and streams are adjusted so that water keeps flowing downhill. This means that the landscape is shaped and organized into landforms that are an essential part of the drainage system. The shaping of landforms within the drainage system occurs as running water erodes the landscape, which is the subject of the next chapter.

FRESH WATER OF THE CONTINENTS IN REVIEW

The fresh water of the lands accounts for only a small fraction of the earth's water. Since it is produced by precipitation over land, it depends on the continued operation of the hydrologic cycle for its existence. The soil layer plays a key role in determining the fate of precipitation by diverting it in three ways: to the atmosphere as evapotranspiration, to ground water through percolation, and to streams and rivers as runoff.

Ground water occupies the pore spaces in rock and regolith. The water table marks the upper surface of the saturated zone of ground water, where pores are completely filled with water. Ground water moves in slow paths deep underground, recharging rivers, streams, ponds, and lakes by upward seepage and thus contributing to runoff. Solution of limestone by ground water can produce limestone caverns and generate karst landscapes.

Wells draw down the water table and, in some regions, lower the water table more quickly than it can be recharged. Ground water contamination can occur when precipitation percolates through contaminated soils or waste materials. Landfills and dumps are common sources of ground water contaminants.

Runoff includes overland flow, moving as a sheet across the land surface, and flow in streams and rivers, which is confined to a channel. Rivers and streams are organized into a drainage network that moves runoff from slopes into channels, and from smaller channels into larger ones. The discharge of a stream measures the flow rate of water moving past a given location. Discharge increases downstream as tributary streams contribute additional runoff.

Floods occur when river discharge increases, and the flow can no longer be contained within the river's usual channel. Water spreads over the floodplain, inundating low fields and forests adjacent to the channel. When discharge is high, floodwaters can rise beyond normal levels to inundate nearby developed areas, causing damage and sometimes taking lives. Flash floods occur in small, steep watersheds and can be highly destructive.

Lakes are especially important parts of the drainage system because they are sources of fresh water. They are also used for recreation and, in many cases, can supply hydroelectric power. Where lakes occur in inland basins, they are often saline. When climate changes, such lakes can dry up, creating salt flats.

Irrigation is the diversion of fresh water from streams and rivers to supply the water needs of crops. In desert regions, where irrigation is most needed, salinization and waterlogging can occur, reducing productivity and eventually creating unusable land.

Water pollution arises from many sources, including industrial sites, sewage treatment plants, agricultural activities, mining, and processing of mineral deposits. Sulfate, nitrate, phosphate, chloride, sodium, and calcium ions are frequent contaminants. Toxic metals, pesticides, and industrial chemicals are also hazards.

Ground water and surface water resources are essential for human activities. Human civilization is dependent on abundant supplies of fresh water for many uses. But because the fresh water of the continents is such a small part of the global water pool, utilization of water resources takes careful planning and management.

KEY TERMS

infiltration	artesian well	drainage basin
runoff	stream	watershed
overland flow	channel	flood
ground water	discharge	lake
water table	drainage system	

REVIEW QUESTIONS

1. What happens to precipitation falling on soil? What processes are involved?
2. How and under what conditions does precipitation reach ground water?
3. How do caverns come to be formed in limestone? Describe the key features of a karst landscape.
4. How do wells affect the water table? What happens when pumping exceeds recharge?
5. How is ground water contaminated? Describe how a well might become contaminated by a nearby landfill dump.
6. Define discharge (of a stream) and the two quantities that determine it. How does discharge vary in a downstream direction? How does gradient vary in a downstream direction?
7. What is a drainage system? How are slopes and streams arranged in a drainage basin?
8. Define the term *flood*. What is the floodplain? What factors are used in forecasting floods?
9. How are lakes defined? What are some of their characteristics? What factors influence the size of lakes?
10. Describe some of the problems that can arise in long-continued irrigation of desert areas.
11. Identify common surface-water pollutants and their sources.
12. How is surface water utilized as a natural resource?

VISUALIZING EXERCISES

1. How and where does ground water flow? Sketch a diagram showing the flow paths of ground water.
2. Sketch a cross section through the land surface showing the position of the water table and indicating flow directions with arrows. Be sure to include a stream in your diagram. Label the saturated and unsaturated zones.
3. Why does water rise in an artesian well? Illustrate with a sketched cross-sectional diagram showing the aquifer, aquiclude, and the well.

A DEEPER VIEW OF FRESH WATER OF THE CONTINENTS

1. A thundershower causes heavy rain to fall in a small region near the headwaters of a major river system. Describe the flow paths of that water as it returns to the atmosphere and ocean. What human activities influence the flows? in what ways?
2. Imagine yourself a recently elected mayor of a small city located on the banks of a large river. What issues might you be concerned with that involve the river? In developing your answer, choose and specify some characteristics for this city—such as its population, its industries, its sewage systems, and the present uses of the river for water supply or recreation.

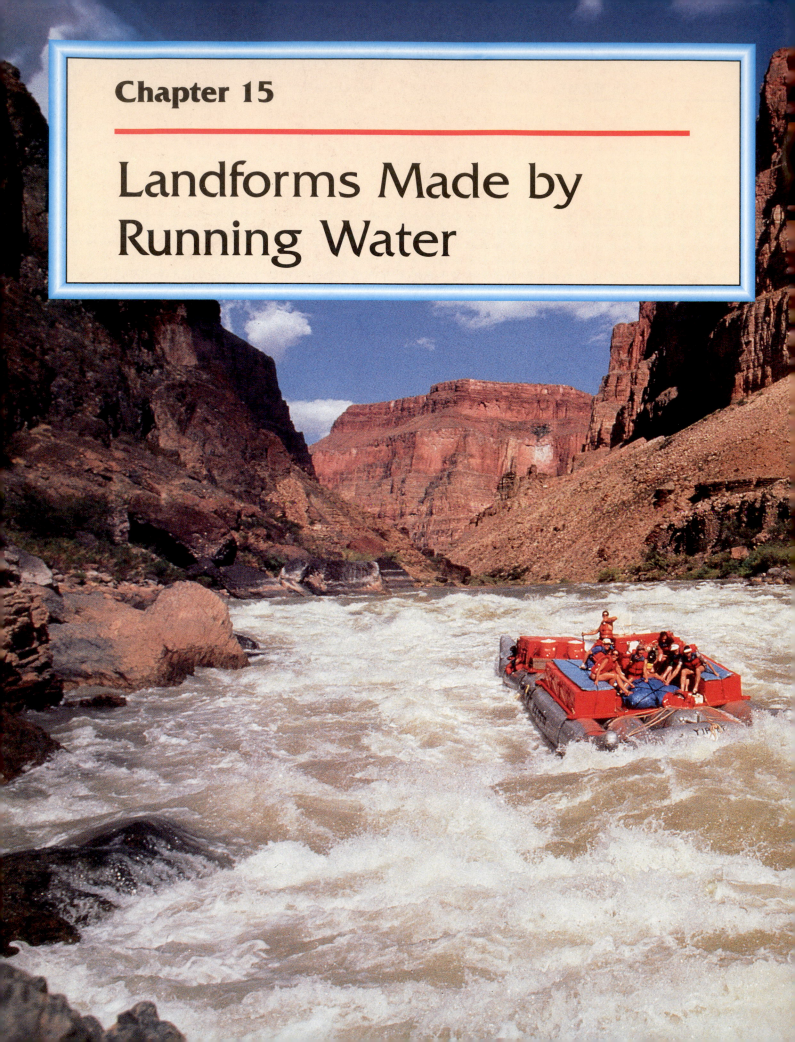

Chapter 15

Landforms Made by Running Water

White-water rafting on the Colorado River in Grand Canyon National Park, Arizona.

PUTTING YOU IN THE PICTURE

Have you ever experienced the power of a river first-hand? One of the best ways to learn about rivers is to ride one for a while. Picture yourself on a rafting trip down the Colorado River as you follow the narrative below.

The rickety school bus turns off the pavement and bumps down the dirt road toward the riverbank. With a lurch, it stops on a wide, gravel-covered spot next to the river, and you clamber off. Behind you is the ancient truck carrying the rafts, paddles, and gear that you and your group will need for your 12-day rafting trip through the majestic Grand Canyon of the Colorado River.

Soon, you are afloat, seven to a raft, slowly drifting with the current down the wide river. "Left side, forward!" "Right side, back!" The river guide barks the orders as you practice paddling together as a team. Sitting on the outer tube of the raft, you twist sideways, dipping your canoe paddle deeply into the blue-green water and pushing hard. This is real work! The river guide explains that safe passage through the hundred or more rapids you'll encounter on the trip will depend on how well you and your raftmates can move your craft across the swift currents of the mighty river.

Then, almost too quickly, you are at the first riffle. The raft glides easily through the rough water, and you think that perhaps this won't be so difficult after all. Then your river guide tells you

that before the day is done, you'll be shooting the Badger Creek and Soap Creek Rapids, which have claimed several lives over the years. Back to paddle exercises, this time in earnest!

As the days pass, you come to feel quite at home with the river. The thrill of wild white-water rides has alternated with the calm of slow drifting in the bright summer sun. Although you've taken foot trips to explore the side canyons and historic sites along the way, you've always returned to the river to continue your journey, and you feel a close bond with it. But this comfortable feeling about the river is due to change. Lava Falls—the last major obstacle on your river trip—lies just downriver. With its 11 m (37 ft) drop, it is the highest in the canyon.

A booming noise begins to build, telling you that the rapids are approaching. The smooth, swift tongue of water that leads into the rapids accelerates the raft. You and your crew paddle furiously across the current, aiming for the side of a huge drop that lies ahead. Suddenly,

you're falling over the edge and down into the froth below. Gliding and twisting, the raft splashes into the white water and is thrown violently to the side. Flying spray hits you, hard. The shock of the cold water on your face and arms makes you gasp for breath. The bow rises as the raft hits the huge standing wave on the downstream side of the hole.

"Pull! Pull!" The guide's urgent shout rises above the roar. Your paddle splits the water, backed by every ounce of strength you can muster. Perched on top of the standing wave, the raft seems for a sickening moment about to slip back into the pitching waters of the hole behind. But suddenly, the raft's bow drops and you're yanked forward as the raft falls into the trough beyond. Then it's a wild roller coaster ride through the haystacks of waves below. The raft rebounds to the right off a huge pillow of foam, then bounces into a side eddy, a patch of calmer water well below the roaring fall. You catch your breath, your heart pounding with the excitement, surrounded by white water on every side. What an amazing trip!

Rivers are powerful, indeed, and especially so in flood. From the physical geographer's perspective, running water is the most important of all the agents that create landforms. The processes of erosion, transportation, and deposition by fluvial systems, and the landforms they create, are the subjects of this chapter.

Chapter Outline

Fluvial Processes and Landforms
Erosional and Depositional Landforms
Slope Erosion
Accelerated Erosion
Sheet Erosion and Rilling
Colluvium and Alluvium
Slope Erosion in Semiarid and Arid Environments
The Work of Streams
Stream Erosion
Stream Transportation
Capacity of a Stream to Transport Load

Eye on the Environment•Battling the Mighty Mississippi
Stream Gradation
Landscape Evolution of a Graded Stream
Great Waterfalls
Evolution of a Fluvial Landscape
Aggradation and Alluvial Terraces
Alluvial Rivers and Their Floodplains
Entrenched Meanders
Fluvial Processes in an Arid Climate
Alluvial Fans
The Landscape of Mountainous Deserts

LANDFORMS MADE BY RUNNING WATER

Most of the world's land surface has been sculpted by running water. Flowing as a sheet across a land surface, running water picks up particles and moves them downslope into a stream channel. When rainfall is heavy, streams and rivers swell, lifting large volumes of sediment and carrying them downstream. In this way, running water erodes mountains and hills, carves valleys, and deposits sediment. This chapter describes the work of running water and the landforms it creates.

Running water is one of four flowing substances that erode, transport, and deposit mineral and organic matter. The other three are waves, glacial ice, and wind. These four fluid agents carry out the processes of *denudation*, which we discussed in Chapter 12. Recall that denudation is the total action of all processes by which the exposed rocks of the continents are worn away and the resulting sediments are transported to the sea or to closed inland basins.

Denudation is an overall lowering of the land surface. If left unchecked to operate over geologic time, denudation will reduce a continent to a nearly feature-less, sea-level surface. However, recall that plate tectonics has always kept continental crust elevated well above the ocean basins. The result is that running water, waves, glacial ice, and wind have always had plenty of raw material available to create the many landforms that we see around us.

Thanks to denudation processes and plate tectonics, the land environments of life are constantly changing, even as plants and animals undergo evolutionary development. Wind, water, waves, and ice produce and maintain a wide variety of landforms that are the habitats for evolving life-forms. In turn, the life-forms become adapted to those habitats and diversify to a degree that matches the diversity of the landforms themselves.

FLUVIAL PROCESSES AND LANDFORMS

Landforms shaped by running water are described as **fluvial landforms**. They are shaped by the **fluvial processes** of overland flow and stream flow, which we described in Chapter 14. Weathering and the slower forms of mass wasting (Chapter 13), such as soil creep, operate hand in hand with overland flow, providing the rock and mineral fragments that are carried into stream systems.

Fluvial landforms and fluvial processes dominate the continental land surfaces the world over. Throughout geologic history, glacial ice has been present only in continental areas located in mid- and high latitudes and in high mountains. Landforms made by wind action are found only in very small parts of the continental surfaces. And landforms made by waves and currents are restricted to the narrow contact zone between oceans and continents. That is why, in terms of area, the fluvial landforms dominate the environment of terrestrial life.

Areas of fluvial landforms are also exploited for agriculture. Except for areas in the northern hemisphere that were formerly occupied by glacial ice, most land areas used in crop cultivation or for grazing have been shaped by fluvial processes. For this reason, fluvial landforms are very important to the support of the human species.

Erosional and Depositional Landforms

All agents of denudation perform the geological activities of erosion, transportation, and deposition. Consequently, there are two major groups of landforms—erosional landforms and depositional landforms. When an initial landform, such as an uplifted crustal block, is created, it is attacked by the processes of denudation and especially by fluvial action. Valleys are formed where rock is eroded away by fluvial agents. Between the valleys are ridges, hills, or mountain summits, representing the remaining parts of the crustal block that are as yet uncarved by running water. These sequential landforms, shaped by progressive removal of the bedrock mass, are *erosional landforms*. Fragments of soil, regolith, and bedrock that are removed from the parent rock mass are transported and deposited elsewhere to make an entirely different set of surface features—the *depositional landforms*.

Figure 15.1 illustrates the two groups of landforms as produced by fluvial processes. The ravine, canyon, peak, spur, and col are erosional landforms. The fan, built of

Figure 15.1 Erosional and depositional landforms. (A. N. Strahler.)

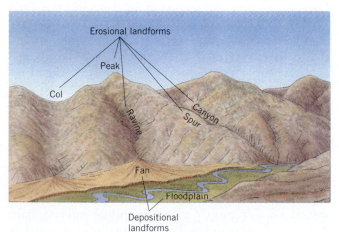

rock fragments below the mouth of the ravine, is a depositional landform. The floodplain, built of material transported by a stream, is also a depositional landform.

SLOPE EROSION

Fluvial action starts on the uplands as *soil erosion.* By exerting a dragging force over the soil surface, overland flow picks up particles of mineral matter ranging in size from fine colloidal clay to coarse sand or even gravel. The size grade selected depends on the speed of the current and the degree to which the particles are bound by plant rootlets or held down by a mat of leaves. Added to this solid matter is dissolved mineral matter in the form of ions produced by acid reactions or direct solution.

Erosional landforms are shaped by progressive removal of rock material. Depositional landforms are built of transported and deposited material.

This ongoing removal of soil is part of the natural geological process of denudation. It occurs everywhere that precipitation falls on land. Under stable natural conditions in a humid climate, the erosion rate is slow enough that a soil with distinct horizons is formed and maintained. Each year a small amount of soil is washed away, while a small amount of solid rock material becomes altered to new regolith and soil. These conditions also enable plant communities to maintain themselves in a stable equilibrium. Soil scientists refer to this state of activity as the *geologic norm.*

Accelerated Erosion

In contrast, the rate of soil erosion may be enormously speeded up by human activities or by rare natural erosional events to produce a state of *accelerated erosion.* What happens then is that the soil is removed much faster than it can be formed, and the uppermost soil horizons are progressively exposed. Accelerated erosion arises most commonly when the plant cover and the physical state of the ground surface change. Destruction of vegetation by the clearing of land for cultivation or by forest fires sets the stage for a series of drastic changes. No foliage remains to intercept rain, and the protection of a ground cover of fallen leaves and stems is removed. Consequently, the raindrops fall directly on the mineral soil.

The direct force of falling drops on bare soil causes a geyserlike splashing in which soil particles are lifted and then dropped into new positions. This process is termed *splash erosion* (Figure 15.2). Soil scientists estimate that a torrential rainstorm has the ability to disturb as much as 225 metric tons of soil per hectare (about 100 U.S. tons

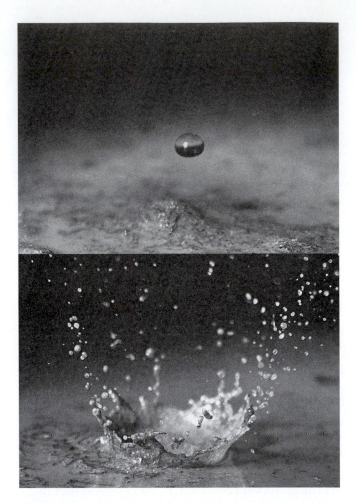

Figure 15.2 A large raindrop (*above*) lands on a wet soil surface, producing a miniature crater (*below*). Grains of clay and silt are thrown into the air and the soil surface is disturbed.

per acre). On a sloping ground surface, splash erosion shifts the soil slowly downhill. An even more important effect is to cause the soil surface to become much less able to absorb water. This occurs because the natural soil openings become sealed by particles shifted by raindrop splash. Reduced infiltration, in turn, permits a much greater depth of overland flow to occur from a given amount of rain. So, the rate of soil erosion is intensified.

Another effect of the destruction of vegetation is to reduce greatly the resistance of the ground surface to the force of erosion under overland flow. On a slope covered by grass sod, even a deep layer of overland flow causes little soil erosion. This protective action is present because the energy of the moving water is dissipated in friction with the grass stems, which are tough and elastic. On a heavily forested slope, the surface layer of leaves, twigs, roots, and even fallen tree trunks takes up the force of overland flow. Without such a cover, the eroding force is applied directly to the bare soil surface, easily dislodging the grains and sweeping them downslope.

Sheet Erosion and Rilling

Accelerated soil erosion is a constant problem in cultivated regions with a substantial water surplus. When the natural cover of forest or prairie grasslands is first removed and the soil is plowed for cultivation, little erosion will occur until the action of rain splash has broken down the soil aggregates and sealed the larger openings. Then, overland flow begins to remove the soil in rather uniform thin layers, a process termed *sheet erosion*. Because of seasonal cultivation, the effects of sheet erosion are often little noticed until the upper horizons of the soil are removed or greatly thinned.

Where land slopes are steep, runoff from torrential rains produces a more destructive activity, *rill erosion*, in which many closely spaced channels are cut into the soil and regolith. If the rills are not destroyed by soil tillage, they may soon begin to join together into still larger channels. These deepen rapidly and soon become *gullies*—steep-walled, canyonlike trenches whose upper ends grow progressively upslope (Figure 15.3). Ultimately, a rugged, barren topography results from accelerated soil erosion that is allowed to proceed unchecked.

Colluvium and Alluvium

With an increased depth of overland flow and no vegetation cover to absorb the eroding force, soil particles are easily picked up and moved downslope. Eventually, they reach the base of the slope, where the surface slope becomes more gentle and meets the valley bottom. There the particles come to rest and accumulate in a thickening layer termed *colluvium*. Because this deposit is built by overland flow, it has a sheetlike distribution and may be little noticed, except where it eventually buries fence posts or tree trunks.

Sediment deposited by streams is referred to as alluvium.

If not deposited as colluvium, sediment carried by overland flow eventually reaches a stream in the adjacent valley floor. Once in the stream, it is carried farther downvalley and may accumulate as alluvium in layers on the valley floor. The term **alluvium** is used to describe any stream-laid sediment deposit. In a region of accelerated erosion, deposition of alluvium can bury fertile floodplain soil under infertile, sandy layers. Coarse alluvium chokes the channels of small streams and can cause the water to flood broadly over the valley bottoms.

Slope Erosion in Semiarid and Arid Environments

Thus far, we have discussed slope erosion in moist climates with a natural vegetation of forest or a dense

Figure 15.3   Deep branching gullies have carved up an overgrazed pasture near Shawnee, Oklahoma. Check dams, located to left and right of the central gullied area, have halted the gullying by stopping overland flow before it becomes concentrated into channels.

prairie grassland. Conditions are quite different in a midlatitude semiarid climate with summer drought. Here, the natural plant cover consists of short-grass prairie (steppe). Although it is sparse and provides a rather poor ground cover of plant litter, the grass cover is normally strong enough that a slow pace of erosion can be sustained.

Much the same conditions are also found in the tropical savanna grasslands. In these semiarid environments, however, the natural equilibrium is highly sensitive and easily upset. Depletion of the plant cover by fires or the grazing of herds of domesticated animals can easily set off rapid erosion. These sensitive, marginal environments require cautious use, because they lack the potential to recover rapidly from accelerated erosion once it has begun.

Erosion at a very high rate by overland flow is actually a natural process in certain locations in semiarid and arid lands. Here, the erosion produces *badlands*. Badlands are underlain by clay formations, which are easily eroded by overland flow. Erosion rates are too fast to permit plants to take hold, and no soil can develop. A maze of small stream channels is developed, and ground slopes are very steep (Figure 15.4).

One well-known area of badlands in the semiarid short-grass prairie is the Big Badlands of South Dakota, along the White River. Badlands such as these are self-sustaining and have been in existence on continents throughout much of geologic time. Badlands can also result from poor agricultural processes, especially when the vegetation cover of clay formations is disturbed by plowing or overgrazing.

THE WORK OF STREAMS

The work of streams consists of three closely related activities—erosion, transportation, and deposition. **Stream erosion** is the progressive removal of mineral material from the floor and sides of the channel, whether bedrock or regolith. **Stream transportation** consists of movement of the eroded particles dragged over the stream bed, suspended in the body of the stream, or held in solution as ions. **Stream deposition** is the accumulation of transported particles on the stream bed and floodplain, or on the floor of a standing body of water into which the stream empties. Erosion cannot occur without some transportation taking place, and the transported particles must eventually come to rest. Thus, erosion, transportation, and deposition are simply three phases of a single activity.

Stream Erosion

Streams erode in various ways, depending on the nature of the channel materials and the tools with which the current is armed. The force of the flowing water not only sets up a dragging action on the bed and banks, but also causes particles to impact the bed and banks. Dragging and impact can easily erode alluvial materials, such as gravel, sand, silt, and clay. This form of erosion, called *hydraulic action*, can excavate enormous quantities in a short time. The undermining of the banks causes large masses of alluvium to slump into the river, where the particles are quickly separated and become part of the stream's load. This process of bank caving is an

Figure 15.4   Badlands at Zabriskie Point, Death Valley National Monument, California.

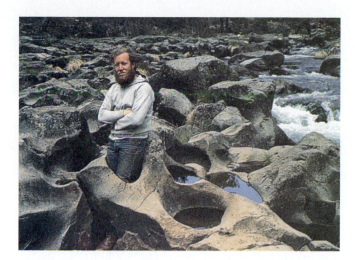

Figure 15.5 These potholes in lava bedrock attest to the abrasion that takes place on the bed of a swift mountain stream. McCloud River, California.

important source of sediment during high river stages and floods.

Where rock particles carried by the swift current strike against bedrock channel walls, chips of rock are detached. The large, strong fragments become rounded as they travel. The rolling of cobbles and boulders over the stream bed further crushes and grinds the smaller grains to produce a wide assortment of grain sizes. This process of mechanical wear is called *abrasion*. It is the principal means of erosion in bedrock that is too strong to be affected by simple hydraulic action. A striking example of abrasion is the erosion of a *pothole*. This occurs when a shallow depression in the bedrock of a stream bed acquires one or several grinding stones, which are spun around and around by the flowing water and carve a deep depression (Figure 15.5).

Finally, the chemical processes of rock weathering—acid reactions and solution—are effective in removing rock from the stream channel. The process is called *corrosion*. Effects of corrosion are conspicuous in limestone, which develops cupped and fluted surfaces.

Stream Transportation

The solid matter carried by a stream is the **stream load**. It is carried in three forms (Figure 15.6). *Dissolved matter* is transported invisibly in the form of chemical ions. All streams carry some dissolved ions resulting from mineral alteration. Sand, gravel, and larger particles move as *bed load* close to the channel floor by rolling or sliding. Clay and silt are carried in *suspension*—that is, they are held within the water by the upward elements of flow in turbulent eddies in the stream. This fraction of the transported matter is the *suspended load* (Figure 15.7). Of the three forms, suspended load is generally the largest.

A large river such as the Mississippi carries as much as 90 percent of its load in suspension. Most of the suspended load comes from its great western tributary, the Missouri River, which is fed from semiarid lands, including the Dakota Badlands. (Figure 14.10 shows the Mississippi River system and its major branches.)

The Yellow (Huang) River of China heads the world list in annual suspended sediment load, and its watershed sediment yield is one of the highest known for a large river basin. This is because much of its basin consists of cultivated upland surfaces of wind-deposited silt that is very easily eroded. (See Chapter 17 and Figure 17.35.) In addition, much of the river's upper watershed is in a semiarid climate with dry winters. Vegetation is sparse, and the runoff from heavy summer rains sweeps up a large amount of sediment.

Figure 15.6 Streams carry their load as dissolved, suspended, and bed loads. Suspended load is kept in suspension by turbulence. Bed load moves by sliding or rolling.

Capacity of a Stream to Transport Load

The maximum solid load of debris that can be carried by a stream at a given discharge is a measure of the *stream capacity*. This load is usually measured in units of metric tons per day passing downstream at a given location. Total solid load includes both bed load and suspended load.

A stream's capacity to carry suspended load increases sharply with an increase in the stream's velocity, because the swifter the current, the more intense is the turbulence. The capacity to move bed load also increases with velocity—the faster water motion produces a stronger dragging force against the bed. In fact, the capacity to move bed load increases according to the third to fourth power of the velocity. In other words, when a stream's velocity is doubled in flood, its ability to transport bed load is increased from eight to sixteen times. Thus, most of the conspicuous changes in the channel of a stream occur in a flood stage.

When water flow increases, a stream in a channel flowing over thick layers of silt, sand, and gravel will easily widen and deepen its channel. When the flow slackens, the stream will deposit material in the bed, filling the channel again. Where a stream flows in a channel of hard bedrock, the channel cannot be quickly deepened in response to rising waters and may change very little during a single flood. Such conditions exist in streams that occupy deep canyons and have steep gradients.

A major river in flood can move vast quantities of sediment and do enormous damage to settlements on near-by floodplains. *Eye on the Environment • Battling the Mighty Mississippi* documents some of the effects of the great Mississippi River Flood of 1993.

STREAM GRADATION

A main stream, fully developed within its drainage basin, has normally undergone thousands of years of adjustment of its channel. It can discharge not only the surplus runoff produced by the basin, but also the solid load that the tributary channels supply. This transport capability is related to the gradient of the channel (discussed in Chapter 14). The steeper the gradient, the higher the stream velocity and the greater the ability of the stream to carry sediment. The gradient of a stream adjusts over time to achieve an average balanced state of operation, year in and year out and from decade to decade. In this equilibrium condition, the stream is referred to as a **graded stream**.

> **Rivers carry most of their sediment load in suspension. Land cultivation can greatly increase that load by increasing soil erosion.**

To develop the concept of a graded stream system, we will investigate the changes that take place along a stretch of stream that is initially very poorly adjusted to the transport of its load. Such an ungraded channel is illustrated by an example in Figure 15.8. The side of the block shows a series of *stream profiles*—plots of elevation of the stream with distance from the sea. The starting profile (1) is produced by crustal uplift in a series of fault steps, bringing to view a land surface that was formerly beneath the ocean and exposing it to fluvial processes for the first time. Overland flow collects in shallow depressions, which fill and overflow from higher to lower levels. In this way, a through-flowing channel originates and begins to conduct runoff to the sea. The profile of this stream is far from smooth, containing many *nickpoints*, where slope changes abruptly. As time passes, the landscape is slowly eroded by fluvial action, and the stream profile is smoothed out into a uniform curve, shown as profile line 3. The profile has now been graded. From that point on into time, this *graded profile* is steadily lowered in elevation as the landscape is further eroded (curves 4 through 6).

Landscape Evolution of a Graded Stream

Early Stages

Figure 15.9 illustrates the stream gradation process in a series of block diagrams. In (*a*), we see *waterfalls* and *rapids*, which are simply portions of the channel with steep gradients. Flow velocity at these points is greatly

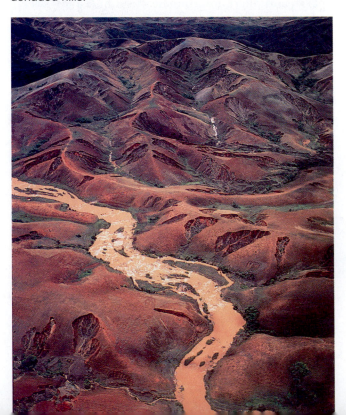

Figure 15.7 This turbulent stream in Madagascar carries a heavy load of suspended sediment. Gullying erodes the denuded hills.

Figure 15.8 Schematic diagram of gradation of a stream. Originally, the channel consists of a succession of lakes, falls, and rapids. (Copyright © A. N. Strahler.)

increased, and abrasion of bedrock is therefore most intense. As a result, the falls are cut back and the rapids are trenched. At the same time, the ponded stretches of the stream are first filled by sediment and are later lowered in level as the lake outlets are cut down. In time, the lakes disappear and the falls are transformed into rapids.

Figure 15.9 Evolution of a stream and its valley. (*a*) Stream established on a land surface dominated by landforms of recent tectonic activity. (*b*) Gradation in progress. The lakes and marshes drained. The gorge is deepening and tributary valleys are extending. (*c*) Graded profile attained. Floodplain development is beginning, and the widening of the valley is in progress. (*d*) Floodplain widened to accommodate meanders. Floodplains now extend up tributary valleys. (Drawn by E. Raisz. Copyright © A. N. Strahler.)

EYE ON THE ENVIRONMENT

Battling the Mighty Mississippi

The violent blast cut the still night air with a force that broke windows and set off car alarms on the far side of the mile-wide river. The muddy waters of the mighty Mississippi, thrown back at first by the force of the dynamite, surged forward through a new gap in the levee just upstream from Prairie du Rocher, Illinois. Fifteen minutes later, a second thundering explosion resounded across the width of the seething river, and floodwaters spilled through another new gash.

At 3:30 that morning, August 4, 1993, the commissioners of the Fort Chartres–Ivy Landing Drainage and Levee District had approved a risky plan to dynamite two new crevasses in the levee upstream from the historic village of Prairie du Rocher. They hoped that the new breaches would skim the top off the crest of high waters approaching the main levee that had so far held the town secure from flooding. The waters from the breached levee would take hostage the fields and farms of a large area of upriver floodplain in return for sparing the village. It wasn't long before their vote was affirmed by the two blasts that echoed for miles up and down the river.

In the eyes of the commissioners, Prairie du Rocher was a prize worth saving. The town traces its roots back to its founding as a French trading post in 1722 and considers itself to be the birthplace of Illinois. Its quaint, turn-of-the-century architecture and its reputation for bed-and-breakfast inns and fine French food made the small town famous in the region as an upscale tourist mecca.

The vote of the commissioners to dynamite the levee was against the advice of the U.S. Army Corps of Engineers, the agency charged with building and maintaining the levees, dams, reservoirs, and other flood control structures that normally keep the powerful Mississippi River shackled and controlled. The Corps' engineers were afraid that the shock wave from the blast, transmitted through the river, would damage the levee near the town, flooding the village instead of protecting it. The pressure wave might also cause the levee at the site to lose its strength, turning rapidly into a mixture of sediment and water in a process called liquefaction. The resulting break would be vast enough to send a huge volume of water through the levee that could swamp other levees and dikes and eventually attack the town from the rear. Even without liquefaction, the flooding could be severe enough to engulf areas that would have otherwise remained dry.

On August 3, one day earlier, the Corps had created its own breach in the upstream levee, dispatching a crane shovel on a barge to remove the top 1.2 m (4 ft) of a 120-m (400 ft) section of the levee. According to plan, an area of about 50 km² (20 mi²) was flooded sooner rather than later, thus helping to blunt the rapidly approaching flood crest. But the water continued to rise, relentlessly threatening the historic village and leading to the late-night meeting and vote by the district commissioners.

Fortunately, none of the catastrophes foretold by the Corps' engineers came about. Prairie du Rocher was saved from flooding, but whether the extreme measures adopted by the desperate commissioners made the difference in keeping the floodwaters from overtopping the levees is not really known.

The summer of 1993 was a time of extreme measures all along the upper reaches of the Mississippi. In June, unprecedented rainfall in the upper Mississippi basin totaled over 30 cm (12 in.) for large areas of the region. Southern Minnesota and western Wisconsin received even larger amounts. Monthly rainfall totals were the highest on many records dating back more than 100 years. The heavy rainfall continued into July, centered on Iowa, Illinois, and Missouri. Inundated by huge volumes of rain, the vast wet landscape came to resemble a sixth Great Lake when viewed from the air.

As the water drained from the landscape, so the Mississippi and its tributaries rose, creating a flood crest of swiftly moving turbid water that worked its way slowly downstream through July and early August. At nearly every community along the length of the Mississippi and Missouri rivers, citizens fought to raise and reinforce the levees that protected their lands, homes, and businesses from the ravages of the flood (see Figure 14.12). Where levees failed to keep them in check, the rivers filled their broad floodplains from bluff to bluff. By the time it was over, the waters had crested at 14.8 m (49.4 ft) at St. Louis, nearly 6 m (20 ft) above flood stage.

The Mississippi River Flood of 1993 was an environmental disaster of the first magnitude. An area twice the size of New Jersey was flooded, taking 50 lives and driving nearly 70,000 people from their homes. Property damage, including loss of agricultural crops, was estimated at $12 billion. Of the region's 1400 levees, at least 800 were breached or overtopped. Some flood

These two Landsat satellite images show St. Louis and vicinity during a normal year and during the flood of 1993. In these images, vegetation appears green, and urban areas appear in pink and purple tones. Some clouds appear in the 1993 image. The Mississippi River (*topmost*) joins the Missouri River to the north of the city. In 1993, the two rivers left their banks to spread into the bottomlands of their floodplains. The floodwaters crested at an even higher level at St. Louis on August 1, 1993.

engineers classed the event as a 500-year flood, meaning that only once in a 500-year interval is a flood of this magnitude likely to occur. However, this does not mean that a similar flood, or an even larger one, could not occur in the near future. For the people living on its banks and bottomlands, the mighty Mississippi is a sleeping giant that awakens with davastating consequences. Only time will tell when the giant is to awaken again.

Questions

1. Why were levees dynamited near Prairie du Rocher? Was the action justified?

2. What caused the Mississippi River Flood of 1993? What were some effects of the flood?

In the early stages of gradation and tributary extension, the capacity of the stream exceeds the load supplied to it, so that little or no alluvium accumulates in the channels. Abrasion continues to deepen the major channels, with the result that they come to occupy steep-walled *gorges* or *canyons* (Figure 15.10). Weathering and mass wasting of these rock walls contribute an increasing supply of rock debris to the channels. Also on the increase is debris shed from land surfaces that contribute overland flow to the newly developed branches.

> **Over time, a stream develops a graded profile in which the gradient is just sufficient to carry the average annual load of water and sediment produced by its drainage basin.**

The erosion of rapids reduces the gradient to a slope angle that more closely approximates the average gradient of that section of the stream (Figure 15.9*b*). At the same time, branches of the main stream are being extended into higher parts of the original land mass. These carve out many new small drainage basins. Thus, the original tectonic landscape is transformed into a complete fluvial landform system.

Attaining a Graded Condition

As landscape change continues, a gradual decrease in a stream's capacity to move bed load results from the gradual reduction in the channel gradient. Simultaneously, the load supplied to the stream from the entire upstream area is on the increase. So, a time will come when the supply of load exactly matches the stream's capacity to transport it. Figure 15.9*c* shows the landscape at this point in time. Now, all the major streams have achieved the graded condition and possess graded profiles that descend smoothly and uniformly.

The first indication that a stream has attained a graded condition is the beginning of floodplain development. On the outside of each bend, the channel shifts laterally (sidewise) to make a curve of larger radius (Figure 15.11). This forced change requires the stream to undercut and steepen the valley wall. On the inside of the bend, alluvium accumulates as a long, curving deposit of sediment—termed a *point bar*. Widening of the bar deposit produces a crescent-shaped area of low ground, which is the first stage in floodplain development. This stage is illustrated in Figure 15.9*c*. As lateral cutting continues, the floodplain strips are widened, and the channel develops sweeping bends. These winding river bends are called **alluvial meanders** (*d*). As they expand, the meanders also move downriver. By their action, the floodplain is widened into a continuous belt of flat land between steep valley walls. Figure 15.12 illustrates these changes.

Floodplain development reduces the frequency with which the river attacks and undermines the adjacent valley wall. Weathering, mass wasting, and overland flow can then act to reduce the steepness of the valley-side slopes (Figure 15.12). As a result, in a humid climate, the gorgelike aspect of the valley gradually disappears

Figure 15.10 The Grand Canyon of the Yellowstone River has been carved in volcanic rock (rhyolite) of the Yellowstone Plateau. Yellowstone National Park, Wyoming.

Figure 15.11 Idealized map and cross-profile of a meander bend of a large alluvial river, such as the lower Mississippi. Arrows show the position of the swiftest current. (Copyright © A. N. Strahler.)

Figure 15.12 Following stream gradation, the valley walls become gentler in slope, and the bedrock is covered by soil and weathered rock. (After W. M. Davis. Copyright © A. N. Strahler.)

and eventually gives way to an open valley with soil-covered slopes protected by a dense plant cover. The time required for the stream to reach a graded condition and erode a broad valley will be a few tens of millions of years.

Great Waterfalls

Although small, high waterfalls are common features of alpine mountains carved by glacial erosion (Chapter 18), large waterfalls on major rivers are comparatively rare the world over. Faulting and dislocation of large crustal blocks have caused spectacular waterfalls on several east African rivers (Figure 15.13). As we explained in Chapter 12, this is the Rift Valley region, where block faulting has been taking place.

Another class of large waterfalls involves new river channels resulting from glacial activity in the Pleistocene Epoch. Erosion and deposition by large moving ice sheets greatly disrupted drainage patterns in northern continental regions, creating lakes and causing river courses to be shifted to new locations. Niagara Falls is a good example. The outlet from Lake Erie into Lake Ontario, the Niagara River, is situated over a gently inclined layer of limestone, beneath which lies easily eroded shale (Figure 15.14). The fall is maintained by continual undermining of the limestone by erosion of the less resistant shale at the base of the fall (Figure 15.15). The height of the falls is now 52 m (171 ft), and its discharge is about 17,000 m³ per second (about 200,000 ft³/s). The drop of Niagara Falls is utilized for the production of hydroelectric power by the Niagara

Figure 15.13 Victoria Falls is located on the Zambesi River, which forms the border between Zambia and Zimbabwe in southern Africa. The water falls 128 m (420 ft) to the bottom of a deep, narrow gorge excavated along a fault line.

Figure 15.14 A bird's-eye view of the Niagara River with its falls and gorge carved in strata of the Niagara Escarpment. The view is toward the southwest from a point over Lake Ontario. (After a sketch by G. K. Gilbert. Redrawn from A. N. Strahler. Copyright © A. N. Strahler.)

Power Project. Water is withdrawn upstream from the falls and is carried in tunnels to generating plants located about 6 km (4 mi) downstream from the falls.

Dams and Resources

Because most large rivers of steep gradient do not have falls, dams are necessary to create the vertical drop required to spin the turbines of electric power generators. Hydroelectric power is cheap, nonpolluting, and renewable, and large dams provide fresh water for urban use and irrigation. However, lakes behind dams drown river valleys and can inundate the gorges, rapids, and waterfalls of major rivers. Scenic and recreational resources such as these can be lost. White-water boating and rafting, for example, are now popular sports on our wild rivers. The Grand Canyon of the Colorado River, probably more than any single product of fluvial processes, demonstrates the scenic and recreational value of a great river gorge.

Dam construction also destroys ecosystems adapted to the river environment. In addition, deposition of sediment behind the dam rapidly reduces the holding capacity of the lake and within a century or so may fill the lake basin. This reduces the ability of the lake to provide consistent supplies of water and hydroelectric power. It is small wonder, then, that new dam projects can meet with stiff opposition from concerned local citizens' groups and national environmental organizations.

Evolution of a Fluvial Landscape

With the concept of a graded stream in mind, this is a good place to examine some of the broader aspects of

fluvial denudation. Just as a stream undergoes a life cycle of change, an entire land mass has its stages of evolution.

Consider a landscape made up of many drainage basins and their branching stream networks. The region is rugged, with steep mountainsides and high, narrow crests (Figure 15.16a). The total volume of rock

Figure 15.15 Niagara Falls is formed where the river passes over the eroded edge of a massive limestone layer. Continual undermining of weak shales at the base keeps the fall steep. (Drawn by E. Raisz. Copyright © A. N. Strahler.)

lying above sea level and available for removal by fluvial action is called the *landmass*. Rock debris is transported out of each drainage basin by the main graded streams at the same average rate as the debris is being contributed from the land surfaces within the basin. Eventually, the export of debris lowers the land surface generally, and the average altitude of the land surface steadily declines. This decline must be accompanied by a reduction in the average gradients of all streams (*b*).

As time passes, the streams and valley-side slopes of the drainage basins undergo gradual change to lower gradients. In theory, the ultimate goal of the denudation process is to reduce the landmass to a featureless plain at sea level. In this process, a sea-level surface imagined to lie beneath the entire landmass represents the lower limiting level, or *base level*, of the fluvial denudation (labeled in Figures 15.8 and 15.16). But because the rate of denudation becomes progressively slower, the land surface approaches the base-level surface of zero elevation at a slower and slower pace. Under this scenario, the ultimate goal can never be reached. Instead, after the passage of some millions of years, the land surface is reduced to a gently rolling surface of low elevation, called a **peneplain** (*c*). You can think of this strange term as meaning an "almost-plain."

Production of a peneplain requires a high degree of crustal and sea-level stability for a period of many millions of years. One region that has been cited as a possible example of a contemporary peneplain is the Amazon–Orinoco basin of South America. This vast region is a stable continental shield of ancient rock.

Uplifted peneplains that are now high-standing land surfaces are numerous within the continental shields. These regions are characterized by an upland surface of nearly uniform elevation. The upland is trenched by stream valleys graded with respect to a lower base level.

If crustal uplift does not occur, land surfaces will eventually be worn down to a low, gently rolling surface, called a peneplain.

Figure 15.16*d* shows the peneplain of (*c*) uplifted to an elevation of several hundred meters. The base level is now far below the land surface, and a large thickness of

new landmass has been created. Soon, however, streams begin to trench the landmass and to carve deep, steep-walled valleys, shown in (*e*). This process is called *landmass rejuvenation*. Landscapes in various stages of rejuvenation are recognized throughout the continental shield area of North America. With the passage of many millions of years, the landscape will be carved into the rugged stage shown in (*a*), and the later stages of (*b*) and (*c*) will follow.

(*a*)

(*b*)

(*c*)

(*d*)

(*e*)

Figure 15.16 Reduction of a landmass by fluvial erosion. (*a*) In early stages, relief is great, slopes are steep, and the rate of erosion is rapid. (*b*) In an advanced stage, relief is greatly reduced, slopes are gentle, and rate of erosion is slow. Soils are thick over the broadly rounded hill summits. (*c*) After many millions of years of fluvial denudation, a peneplain is formed. Slopes are very gentle, and the landscape is an undulating plain. Floodplains are broad, and the stream gradients are extremely low. All of the land surface lies close to base level. (*d*) The peneplain is uplifted. (*e*) Streams trench a new system of deep valleys in the phase of landmass rejuvenation. (Drawn by A. N. Strahler.)

Aggradation and Alluvial Terraces

A graded stream, delicately adjusted to its supply of water and rock waste from upstream sources, is highly sensitive to changes in those inputs. Changes in climate or vegetation cover bring changes in discharge and load at downstream points, and these changes in turn require channel readjustments. One kind of change is the buildup of alluvium in the valley floors.

Consider first what happens when bed load increases, exceeding the transporting capacity of the stream. Along a section of channel where the excess load is introduced, the coarse sediment accumulates on the stream bed in the form of bars of sand, gravel, and pebbles. These deposits raise the elevation of the stream bed, a process called *aggradation*. As more bed materials accumulate, the stream channel gradient is steepened and flow velocity increases. This increase enables bed materials to be dragged downstream and spread over the channel floor at more and more distant downstream sections. In this way, sediment introduced at the head of a stream will be gradually spread along all the length of the stream.

Aggradation typically changes the channel cross section from a narrow and deep form to a wide and shallow one. Because bars are continually being formed, the flow is divided into multiple threads. These rejoin and subdivide repeatedly to give a typical *braided stream* (Figure 15.17). The coarse channel deposits spread across the former floodplain, burying fine-textured alluvium under the coarse material.

What processes cause alluvium to build up on valley floors and induce stream aggradation? One way is for alluvium to accumulate as a result of accelerated soil erosion, which we have already discussed. Other causes for aggradation are related to major changes in global climate, such as the onset of an ice age. Figure 15.17 illustrates one natural cause of aggradation—glacial activity—that has been of major importance in stream systems of North America and Eurasia during the recent Ice Age. In our photo example, a modern valley glacier has provided a large quantity of coarse rock debris at the head of the valley. Valley aggradation of a similar kind was widespread in a broad zone near the edges of the great ice sheets of the Ice Age. The accumulated alluvium filled most valleys to depths of several tens of meters. Figure 15.18*a* shows a valley filled in this manner by an aggrading stream. The case could represent any one of a large number of valleys in New England or the Middle West.

Suppose, next, that the source of bed load is cut off or greatly diminished. In the case illustrated in Figure 15.18, the ice sheets have disappeared from the distant headwater areas and, with them, the supplies of coarse rock debris. Reforestation of the landscape restores a protective cover to valley-side and hill slopes of the region, keeping coarse mineral particles from entering the stream through overland flow. Now the streams have abundant water discharges but little bed load. In other words, they are operating below their transporting capacity. The result is channel scour. The channel form becomes both deeper and narrower, and it also begins to develop meanders.

Alluvial terraces form when an aggrading river loses its sediment input and begins degrading its bed, leaving terraces behind as it cuts deeper into its sediment-filled valley.

Gradually, the stream profile level is lowered, in a process called *degradation*. Because the stream is very close to being in the graded condition at all times, its dominant activity is lateral (sidewise) cutting by growth of meander bends, as shown in Figure 15.18*b*. By this process the valley alluvium is gradually excavated and carried downstream, but it cannot all be removed because the channel encounters hard bedrock masses in many places. These obstructions prevent cutting away of more alluvium. Consequently, as shown in (*c*), step-like alluvial surfaces remain on both sides of the valley. The treads of these steps are called *alluvial terraces*.

Figure 15.17 The braided channel of the Chitina River, Wrangell Mountains, Alaska, shows many separate channels separating and converging on a floodplain filled with glacial debris.

(a)

Braided channel

Alluvium

Bedrock

(b)

Meander

Alluvial terrace

Bluff

(c)

Bedrock outcrops

Figure 15.18 Alluvial terraces form when a graded stream slowly cuts away the alluvial fill in its valley. (Drawn by A. N. Strahler.)

Alluvial terraces have always attracted human settlement because of the advantages over both the valley-bottom floodplain—which is subject to annual flooding—and the hill slopes beyond—which may be too steep and rocky to cultivate. Besides, terraces are easily tilled and make prime agricultural land (Figure 15.19). Towns are also easily laid out on the flat ground of a terrace. Roads and railroads can be constructed along the terrace surfaces with little difficulty.

Alluvial Rivers and Their Floodplains

We turn now to the graded river with its floodplain. As time passes, the floodplain is widened, so that broad areas of floodplain lie on both sides of the river channel. Civil engineers have given the name **alluvial river** to a large river of very low channel gradient. It flows on a thick floodplain accumulation of alluvium constructed by the river itself in earlier stages of its activity. Characteristically, an alluvial river experiences overbank floods each year or two. These occur during the season of large water surplus over the watershed. Overbank flooding of an alluvial river normally inundates part or all of a floodplain that is bounded on either side by rising steep slopes, called *bluffs*.

Typical landforms of an alluvial river and its floodplain are illustrated in Figure 15.20. Dominating the floodplain is the meandering river channel itself and abandoned stretches of former channels. Meanders develop narrow necks, which are cut through, thus shortening the river course and leaving a meander loop abandoned. This event is called a *cutoff*. It is quickly followed by deposition of silt and sand across the ends of the abandoned channel, producing an *oxbow lake*. The

Figure 15.19 Alluvial terraces of the Rakaia River gorge on the South Island of New Zealand. The flat terrace surface in the foreground is used as pasture for sheep. Two higher terrace levels can be seen at the left.

oxbow lake is gradually filled in with fine sediment brought in during high floods and with organic matter produced by aquatic plants. Eventually, the oxbows are converted into swamps or marshes, but their identity is retained indefinitely (Figure 15.21).

During periods of overbank flooding, when the entire floodplain is inundated, water spreads out from the main channel over adjacent floodplain deposits (Figure 15.22). As the spreading water slackens, sand and silt are deposited in a zone adjacent to the channel. The result is an embankment of higher land on either side of the channel known as a **natural levee**. Because deposition is heavier closest to the channel and decreases away from the channel, the levee surface slopes away from the channel (Figure 15.20). Between the levees and the bluffs is lower ground, called the *backswamp*. In Figure 15.12, the small settlements are located on the higher ground of the levee, next to the river, while the agricultural fields occupy the backswamp area.

Overbank flooding not only results in the deposition of a thin layer of silt on the floodplain, but also brings an infusion of dissolved mineral substances that enter the soil. As a result of the resupply of nutrients, floodplain soils retain their remarkable fertility, even though they are located in regions of rainfall surplus from which these nutrients are normally leached away.

Entrenched Meanders

What happens when a broadly meandering river is uplifted in the process of stream rejuvenation? The uplift increases the river's gradient, so that it cuts downward into the bedrock below. This forms a steep-walled inner gorge. On either side lies the former floodplain, now a flat terrace high above river level. Any river deposits left on the terrace are rapidly stripped off by runoff, since no new sediment is accumulating.

Rejuvenation may cause the meanders to become impressed into the bedrock and give the inner gorge a meandering pattern. These sinuous bends are termed *entrenched meanders* to distinguish them from the floodplain meanders of an alluvial river (Figure 15.23).

Although entrenched meanders are not free to shift about as floodplain meanders do, they can enlarge slowly so as to produce cutoffs. Cutoff of an entrenched meander leaves a high, round hill separated from the valley wall by the deep abandoned river channel and the shortened river course (Figure 15.23). As you might guess, such hills formed ideal natural fortifications. Many European fortresses of the Middle Ages were built on such cutoff meander spurs. Under unusual circumstances, where the bedrock includes a strong, massive sandstone formation, meander cutoff leaves a *natural bridge* formed by the narrow meander neck.

FLUVIAL PROCESSES IN AN ARID CLIMATE

Desert regions look strikingly different from humid regions, in both vegetation and landforms. Obviously, the lower precipitation makes the difference. Vegetation is sparse or absent, and land surfaces are

Figure 15.21 This aerial photo of the Mississippi River floodplain shows three oxbow lakes—former river bends that were cut off as the river shifted its course. Vegetation appears bright red in this color-infrared photo.

Figure 15.20 Floodplain landforms of an alluvial river. (Drawn by A. N. Strahler.)

Figure 15.22 A river floodplain in Bangladesh is largely under water during a 1973 flood. Villages occupy the higher ground of the natural levees bordering the river channel.

mantled with mineral material—sand, gravel, rock fragments, or bedrock itself.

Although deserts have low precipitation, rain falls in dry climates as well as in moist, and most landforms of desert regions are formed by running water. A particular locality in a dry desert may experience heavy rain only once in several years. But when rain does fall, stream channels carry water and perform important work as agents of erosion, transportation, and deposition. Fluvial processes are especially effective in shaping desert landforms because of the sparse vegetation cover. The few small plants that survive offer little or no protection to soil or bedrock. Without a thick vegetative cover to protect the ground and hold back the swift

Figure 15.23 Rejuvenation of a meandering stream has produced entrenched meanders. One meander neck has been cut through, forming a natural bridge. (Drawn by E. Raisz. Copyright © A. N. Strahler.)

High, round hill

Inner gorge

Entrenched meander

Former floodplain

Natural bridge

downslope flow of water, large quantities of coarse rock debris are swept into the streams. A dry channel is transformed in a few minutes into a raging flood of muddy water heavily charged with rock fragments (Figure 15.24).

Although rain is infrequent in desert environments, running water shapes desert landforms with great effectiveness because of the lack of vegetation cover.

An important contrast between regions of arid and humid climates lies in the way in which the water enters and leaves a stream channel (Figure 15.25). In a humid region (*a*) with a high water table sloping toward a stream channel, ground water moves steadily toward the channel and seeps into the stream bed, producing permanent (perennial) streams. In arid regions (*b*), the water table normally lies far below the channel floor. Where a stream flows across a plain of gravel and sand, water is lost from the channel by seepage. Loss of discharge by seepage and evaporation strongly depletes the flow of streams in alluvium-filled valleys of arid regions. As a result, aggradation occurs and braided channels are common. Streams of desert regions are often short and end in alluvial deposits or on the floors of shallow, dry lakes.

(a)

(b)

Figure 15.25   In humid regions (*a*), a stream channel receives ground water through seepage. In arid regions (*b*), stream water seeps out of the channel and into the water table below. (Copyright © A. N. Strahler.)

Alluvial Fans

One very common landform built by braided, aggrading streams is the **alluvial fan**, a low cone of alluvial sands and gravels resembling in outline an open

Figure 15.24   A flash flood has filled this desert channel in the Tucson Mountains of Arizona with raging, turbid waters. A distant thunderstorm produced the runoff.

Figure 15.26 Features of an alluvial fan. A cross section from the head of the canyon (A) to the playa (A') shows mudflow layers interbedded with sand layers, providing water (arrows) for a well in the fan. (Copyright © A. N. Strahler.)

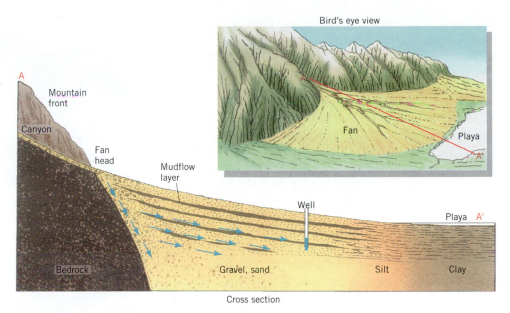

Japanese fan (Figure 15.26). The apex, or central point of the fan, lies at the mouth of a canyon or ravine. The fan is built out on an adjacent plain. Alluvial fans are of many sizes. In fact, some desert fans are many kilometers across (Figure 15.27).

Fans are built by streams carrying heavy loads of coarse rock waste from a mountain or an upland region. The braided channel shifts constantly, but its position is firmly fixed at the canyon mouth. The lower part of the channel, below the apex, sweeps back and forth. This activity accounts for the semicircular fan form and the downward slope in all radial directions away from the apex.

Large, complex alluvial fans also include mudflows (Figure 15.26). Mud layers are interbedded with sand and gravel layers. Water infiltrates the fan at its head,

Figure 15.27 Great alluvial fans extend out upon the floor of Death Valley. The canyons from which they originate have carved deeply into a great uplifted fault block.

making its way to lower levels along sand layers. The mudflow layers serve as barriers to ground water movement. The trapped ground water is under pressure from water higher in the fan apex. When a well is drilled into the lower slopes of the fan, water rises spontaneously as artesian flow. (See Chapter 14 and Figure 14.6.)

Alluvial fans are the primary sites of ground water reservoirs in the southwestern United States. In many fan areas, sustained heavy pumping of these reserves for irrigation has lowered the water table severely. The rate of recharge is extremely slow in comparison. At some locations, recharge is increased by building water-spreading structures and infiltrating basins on the fan surfaces. A serious side effect of excessive ground water withdrawal is subsidence (sinking) of the land surface. (See *Eye on the Environment • Sinking Cities*, in Chapter 14.)

The Landscape of Mountainous Deserts

Where tectonic activity has recently produced block faulting in an area of continental desert, the assemblage of fluvial landforms is particularly diverse. The basin-and-range region of the western United States is such an area. It includes large parts of Nevada and Utah, southeastern California, southern Arizona and New Mexico, and adjacent parts of Mexico. The uplifted and tilted blocks, separated by downdropped tectonic basins, provide an environment for the development of spectacular erosional and depositional fluvial landforms.

Figure 15.28 demonstrates some landscape features of mountainous deserts. Figure 15.28*a* shows two uplifted fault blocks with a downdropped block between them. Although denudation acts on the uplifted blocks as they are being raised, we have shown them as very little modified at the time tectonic activity has ceased. At first, the faces of the fault block are extremely steep. They are scored with deep ravines, and talus blocks form cones at the bases of the blocks.

Figure 15.28*b* shows a later stage of denudation in which streams have carved up the mountain blocks into a rugged landscape of deep canyons and high divides. Rock waste furnished by these steep mountain slopes is carried from the mouths of canyons to form numerous large alluvial fans. The fan deposits form a continuous apron extending far out into the basins.

Alluvial fans and playas are features of valleys in mountainous deserts.

In the centers of desert basins lie the saline lakes and dry lake basins mentioned in Chapter 14. Accumulation of fine sediment and precipitated salts produces an extremely flat basin floor, called a **playa** in the southwestern United States and in Mexico. Salt flats are found where an evaporite layer forms the surface. In some playas, shallow water forms a salt lake.

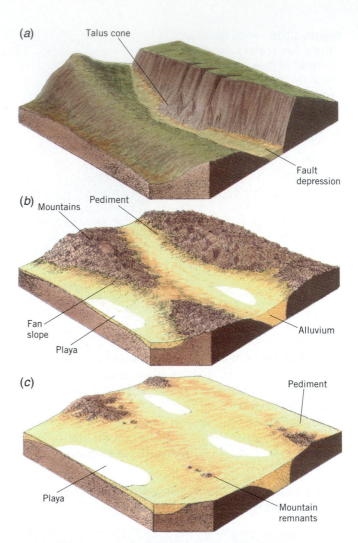

Figure 15.28 Idealized diagrams of landforms of the mountainous deserts of the southwestern United States. (*a*) Initial uplift of two blocks, with a downdropped valley between them. (*b*) Stage of rapid filling of tectonic basins with debris from high, rugged mountain blocks. (*c*) Advanced stage with small mountain remnants and broad playa and fan slopes. (Drawn by A. N. Strahler.)

Figure 15.29 is an air photograph of a mountainous desert landscape in the Death Valley region of eastern California. Three environmental zones can be seen in the photo. First is a region of rugged mountain masses dissected into canyons with steep rocky walls. At lower elevations, a zone of coalescing alluvial fans lines the mountain front. Last is the white playa occupying the central part of the basin.

In this type of mountainous desert, fluvial processes are limited to local transport of rock particles from a mountain range to the nearest adjacent basin, which receives all the sediment. The basin gradually fills as the mountains diminish in elevation. Each arid basin becomes a closed system as far as transport of sediment is involved. Only the hydrologic system is open, with water entering as precipitation and leaving as evaporation.

Figure 15.29 Racetrack Playa, a flat, white plain, is surrounded by alluvial fans and rugged mountains. This desert valley lies in the northern part of the Panamint Range, which is not far west of Death Valley, California. In the distance rises the steep eastern face of the Inyo Mountains, a great fault block.

As the desert mountain masses are lowered in height and reduced in extent, a gently sloping rock floor, called a *pediment*, develops close to the receding mountain front (Figure 15.30). As the remaining mountains shrink further in size, the pediment surfaces expand to form wide rock platforms thinly veneered with alluvium. This advanced stage is shown in Figure 15.28c. In the southwestern United States and northern Mexico, the graded slope from the mountain front to the playa is called a *bajada*. It includes both pediment and fan surfaces. The desert land surface produced in an advanced stage of fluvial activity is an undulating plain consisting largely of areas of pediment surrounded by areas of alluvial fan and playa surfaces.

A LOOK AHEAD

Running water, the primary agent of denudation, does not act equally on all types of rocks. Some rocks are more resistant to erosion, while others are less so. As a result, fluvial erosion can create unique and interesting landscapes that reveal rock structures of various kinds. This will be the subject of our next chapter.

Figure 15.30 A pediment formerly carved in granite at the foot of a desert mountain slope is now seen in profile as a straight, sloping line. Later erosion has exposed the bedrock, and weathering has formed the joint blocks into boulders (intermediate distance). Little Dragoon Mountains, near Adams Peak, about 20 km (about 12 mi) northeast of Benson, Arizona.

LANDFORMS MADE BY RUNNING WATER IN REVIEW

This chapter has covered the landforms and land-forming processes of running water, one of the four fluid agents of denudation. Like the other agents, running water erodes, transports, and deposits rock material, forming both erosional and depositional landforms.

The work of running water begins on slopes, producing colluvium where overland flow moves soil particles downslope, and producing alluvium when the particles enter stream channels and are later deposited. In most natural landscapes, soil erosion and soil formation rates are more or less equal, a condition known as the geologic norm. Badlands are an exception in which natural erosion rates are very high.

The work of streams includes stream erosion and stream transportation. Where stream channels are carved into soft materials, large amounts of sediment can be obtained by hydraulic action. Where stream channels flow on bedrock, channels are deepened only by the abrasion of bed and banks by mineral particles, large and small. Both the suspended load and bed load of rivers increase greatly as velocity increases. Velocity, in turn, depends on gradient.

Over time, streams tend to a graded condition in which their gradients are adjusted to move the average amount of water and sediment supplied to them by slopes. Lakes and waterfalls, created by tectonic, volcanic, or glacial activity, are short-lived events, geologically speaking, that give way to a smooth, graded stream profile. Grade is maintained as landscapes are eroded toward base level. If the land surface is tectonically stable for a very long period, a peneplain can form. When provided with a sudden inflow of rock material, as, for example, by glacial action, streams build up their beds by aggradation. When that inflow ceases, streams resume downcutting, leaving behind alluvial terraces.

Large rivers with low gradients that move large quantities of sediment are termed alluvial rivers. The meandering of these rivers forms cutoff meanders, oxbow lakes, and other typical landforms. Alluvial rivers are sites of intense human activity. Their fertile floodplains yield agricultural crops and provide easy transportation paths. When a region containing a meandering alluvial river is uplifted, the rejuvenation can produce entrenched meanders.

Although rainfall is scarce in deserts, running water is very effective there in producing fluvial landforms. Desert streams, subject to flash flooding, build alluvial fans at the mouths of canyons. Water sinks into the fan deposits, creating local ground water reservoirs. Eventually, desert mountains are worn down into gently sloping pediments. Fine sediments and salts, carried by streams, accumulate in playas, from which water evaporates, leaving sediment and salt behind.

KEY TERMS

fluvial landforms	stream deposition	alluvial river
fluvial processes	stream load	natural levee
alluvium	graded stream	alluvial fan
stream erosion	alluvial meanders	playa
stream transportation	peneplain	

REVIEW QUESTIONS

1. List and briefly identify the four flowing substances that serve as agents of denudation.
2. Describe the process of slope erosion. What is meant by the geologic norm?
3. Contrast the two terms *colluvium* and *alluvium*. Where on a landscape would you look to find each one?
4. What special conditions are required for badlands to form?
5. When and how does sheet erosion occur? How does it lead to rill erosion and gullying?

6. In what ways do streams erode their bed and banks?
7. What is stream load? Identify its three components. In what form do large rivers carry most of their load?
8. How is velocity related to the ability of a stream to move sediment downstream?
9. Describe the evolution of a fluvial landscape. What is meant by rejuvenation?
10. How does stream degradation produce alluvial terraces?
11. Define the term *alluvial river*. Identify some characteristic landforms of alluvial rivers. Why are alluvial rivers important to human civilization?
12. Why is fluvial action so effective in arid climates, considering that rainfall is scarce? How do streams in arid climates differ from streams of moist climates?
13. Describe the evolution of the landscape in a mountainous desert. Use the terms *alluvial fan*, *playa*, and *pediment* in your answer.

VISUALIZING EXERCISES

1. Compare erosional and depositional landforms. Sketch an example of each type.
2. What is a graded stream? Sketch the profile of a graded stream and compare it with the profile of a stream draining a recently uplifted set of landforms.
3. Sketch the floodplain of a graded, meandering river. Identify key landforms on the sketch. How do they form?

A DEEPER VIEW OF LANDFORMS MADE BY RUNNING WATER

1. A river originates high in the Rocky Mountains, crosses the high plains, flows through the agricultural regions of the Midwest, and finally reaches the sea. Describe the fluvial processes and landforms you might expect to find on a journey along the river from its headwaters to the ocean.
2. What would be the effects of climate change on a fluvial system? Choose either the effects of cooler temperatures and higher precipitation in a mountainous desert or warmer temperatures and lower precipitation in a humid agricultural region.

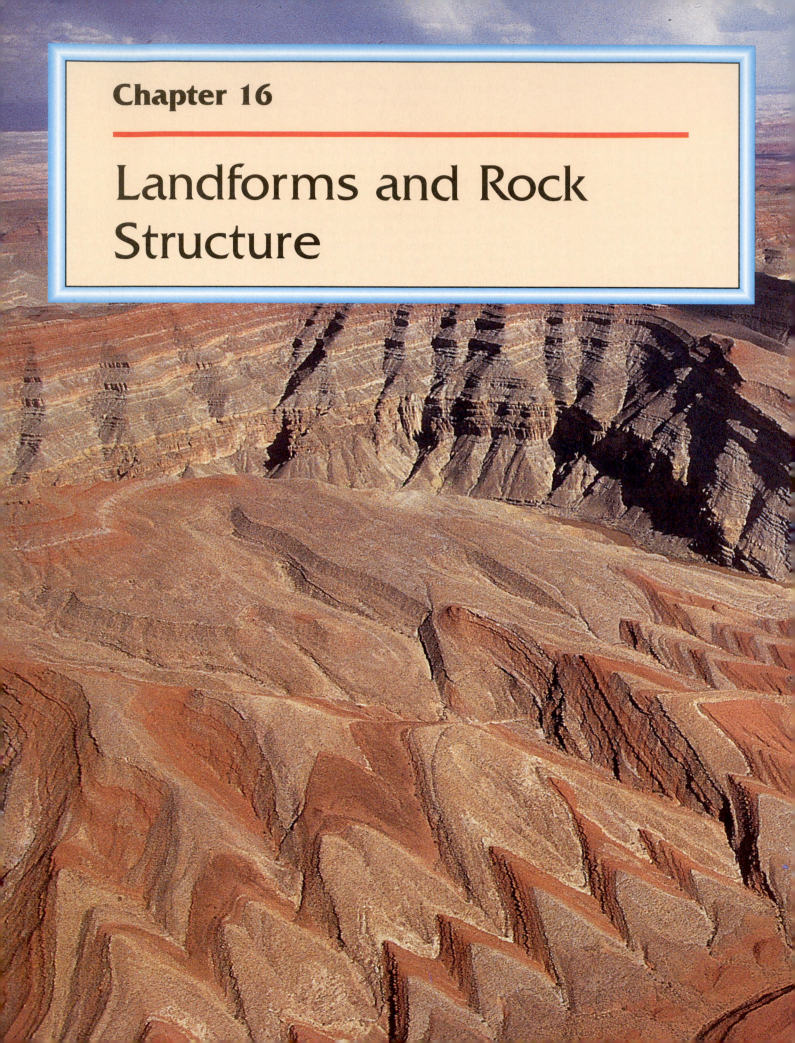

Chapter 16

Landforms and Rock Structure

An aerial view of the Rapley Monocline, on the Colorado Plateau. Near Mexican Hat, Utah.

PUTTING YOU IN THE PICTURE

An aerial view of the earth provides a unique perspective on the landscape. At the scale of a high-flying passenger aircraft, an overview emerges that brings out the broader patterns of landscape. In regions of diverse rock types, this big picture emphasizes the contrast between easily eroded rock formations and those that are more resistant. The central Rocky Mountains and Colorado Plateau are such regions. Imagine yourself on a flight from Denver to Las Vegas on a clear day in June. Here's what you might experience:

The force of the powerful jet engines pushes you deeply into your window seat as you roll down the runway. Quickly, the plane is airborne and heads westward toward the magnificent Rockies. The slope of the Front Range looms large, and as you head across the mountain barrier, the ground seems to rise up to meet you. Below is I-70, twisting and turning its way up the narrow valleys toward the summit. There's the entrance to the Eisenhower Tunnel! The tiny cars below disappear into the mountain slope for the long traverse beneath the mountain crest.

Looking south, you see the peaks of the Front Range spread out in a long line. The overlying sedimentary rocks have long been eroded away here in the center of this vast elongated dome, exposing the hard granite at the core. You easily identify Pikes Peak in the far distance, its summit silhouetted against the dark blue sky. Looking straight down, you see the gray of the ancient granite rock, mantling the slopes between the patches of snow still lying in the low places. Snowpacks in high valley heads below you hint at how the mountains must have looked when their glaciers slowly and majestically marched valleyward, carving broad, U-shaped valleys and leaving the ridges and piles of sediment now resting on the valley floors.

As the flight progresses, the landscape becomes drier, and the vast high desert of the Colorado Plateau replaces the forested lower slopes of the ranges. Below is a landscape of nearly flat-lying rocks. Long lines of cliffs, marking the edges of resistant rock beds, advance and recede as the aircraft moves steadily southwestward.

"Well, folks, you're in luck this afternoon." Suddenly, the voice of the captain cuts the low rushing sound that fills the airplane's cabin. "Air traffic control has vectored us to the south, so we'll be approaching Las Vegas from just over the North Rim of the Grand Canyon. The views should be spectacular!"

And spectacular they are. The gaping chasm cuts open the surrounding plateaus with a broad red gash. Within the canyon, the flat-lying rock formations form a succession of giant steps leading down to a narrow inner gorge that contains a blue-gray ribbon of water. The canyons of tributary creeks cut deeply into the rock sequence, creating wedding cakes and turrets of multileveled rock.

But it is the coloring of the rocks that is the most breathtaking. Buff and gray layers of limestone alternate with reds and purples of siltstones and shales. The green tones of vegetation overshadow the earth colors on the wider benches and platforms. The play of light and shade on the landscape etches the shapes in relief, enhancing their coloration and making the canyon's landforms look somehow even more grand and monumental.

Descending now, the airplane approaches Las Vegas. This flight will remain etched in your memory for years to come, not only because of its sheer scenic beauty, but also for the way in which the landscape revealed its inner structure to your educated eye.

This chapter focuses on the adjustment of landforms to the nature of the rock beneath them. As we will see, weak rocks tend to form valleys or lowlands, while strong rocks tend to stand out as highlands, ridges, or cliffs. Keep this principle in mind as we examine the relation between landforms and rock structure.

Chapter Outline

Rock Structure as a Landform Control
  Strike and Dip
Landforms of Horizontal Strata and Coastal Plains
  Arid Regions
  Drainage Patterns
  Coastal Plains
Landforms of Warped Rock Layers
  Sedimentary Domes
  Fold Belts

Landforms Developed on Other Landmass Types
  Erosion Forms on Fault Structures
  Metamorphic Belts
  Exposed Batholiths and Monadnocks
  Eye on the Environment • Marvelous, Majestic, Monolithic Domes
  Deeply Eroded Volcanoes

LANDFORMS AND ROCK STRUCTURE

Recall that the initial landforms created by volcanic and tectonic processes—volcanoes, lava flows, faults, and folds—were the subject of Chapter 12. Now that we have presented the role of fluvial processes in eroding and shaping the landscape, we can turn to the *sequential landforms* of volcanic and tectonic processes. These are the landforms of denudational processes that follow after initial landforms have been created.

Denudation acts to wear down all rock masses exposed at the land surface. However, different types of rocks are worn down at different rates. Some are easily eroded, while others are highly resistant to erosion. Generally, weak rocks will underlie valleys, and strong rocks will underlie hills, ridges, and uplands. This means that the pattern of landforms on a landscape can reveal the sequence of rock layers underneath it. The sequence, in turn, is determined by the type of rock structure present—for example, a series of folds, or perhaps a dome, in a set of rock layers. Thus, there is often a direct relationship between landforms and rock structure.

Over the world's vast land area, many types of rock and rock structures appear. Repeated episodes of uplift, followed by periods of denudation, can bring to the surface rocks and rock structures formed deep within the crust. In this way, even ancient mountain roots, formed during continental collisions, may eventually appear at the surface and be subjected to denudation. Batholiths and other plutons—igneous rock structures produced by magmas that cool at depth—can similarly be exposed.

As we will see in this chapter, rock structure controls not only the locations of uplands and lowlands, but also the placement of streams and the shapes and heights of the intervening divides. A distinctive assemblage of landforms and stream patterns develops on each of the major types of crustal structures. We begin with a discussion of the relative resistance of several common rock types to erosion.

ROCK STRUCTURE AS A LANDFORM CONTROL

As denudation takes place, landscape features develop according to patterns of bedrock composition and structure. Figure 16.1 shows an arrangement of five layers of sedimentary rock, together with a mass of much older igneous rock. The sediments were originally deposited in horizontal layers atop the igneous rock, but the block of igneous and sedimentary rock has been bent and tilted in an ancient orogeny. The diagram shows the usual landform shape of each rock type and whether it forms valleys or mountains.

> **Shale is a weak rock that forms valleys, while sandstone and conglomerate are strong rocks that form ridges.**

Shale is a weak rock that is easily eroded and forms the low valley floors of the region. Limestone, subjected to solution by carbonic acid in rainwater and surface water, also forms valleys in humid climates. In arid climates, on the other hand, limestone is a resistant rock and usually stands high to form ridges and cliffs. Sandstone and conglomerate are typically resistant to denudation, and they form ridges or uplands. As a group, the igneous rocks are also resistant—they typically form uplands or mountains rising above adjacent areas of shale and limestone. Although metamorphic rocks in general are more resistant to denudation, individual metamorphic rock types—marble or schist, for example—vary in their resistance.

Strike and Dip

The surfaces of most rock layers are not flat but are tipped away from the horizontal. In addition, internal planes—such as joints—occur within rock layers and are likely to be tilted. We need a system of geometry to enable us to measure and describe the position in which these natural planes are held and to indicate them on maps. Examples of such planes include the bedding layers of sedimentary strata, the sides of a dike, and the joints in granite.

The tilt and orientation of a natural rock plane are measured with reference to a horizontal plane. Figure 16.2 shows a water surface resting against tilted sedimentary strata. Since it is horizontal, we can use the

Figure 16.1 Landforms evolve through the slow erosional removal of weaker rock, leaving the more resistant rock standing as ridges or mountains. (Drawn by A. N. Strahler.)

Figure 16.2 Strike and dip. (Drawn by A. N. Strahler.)

water surface as the horizontal plane. The acute angle formed between the rock plane and the horizontal plane is termed the *dip*. The amount of dip is stated in degrees, ranging from 0 for a horizontal rock plane to 90 for a vertical rock plane. The dip of a rock formation is important information for such activities as planning highways, developing mines, and following aquifers.

Next, we examine the line of intersection between the inclined rock plane and the horizontal water plane. This horizontal line will have an orientation with respect to the compass. This orientation is termed the *strike*. In Figure 16.2, the strike is along a north-south line. Since the strike of a rock formation will determine the compass direction of its surface boundary with adjacent rock types, it is important in siting transportation corridors and in studies of surface water.

LANDFORMS OF HORIZONTAL STRATA AND COASTAL PLAINS

Many continental areas are covered by thick sequences of horizontal sedimentary rock layers. These sequences were deposited in shallow inland seas, created when tectonic processes caused ancient shields to subside below sea level. Following crustal uplift, often with little disturbance other than minor crustal warping or faulting, these areas became continental surfaces undergoing denudation.

Arid Regions

In arid climates, where vegetation is sparse and the action of overland flow especially effective, sharply defined landforms develop on horizontal sedimentary strata (Figure 16.3). The normal sequence of landforms is a sheer rock wall, called a *cliff*, which develops at the edge of a resistant rock layer. At the base of the cliff is an inclined slope, which flattens out into a plain beyond.

In these arid regions, erosion strips away successive rock layers, leaving behind a broad platform capped by hard rock layers. This platform is usually called a **plateau**. Cliffs retreat as near-vertical surfaces because the weak clay or shale formations exposed at the cliff base are rapidly washed away by storm runoff and channel erosion. When undermined, the rock in the upper cliff face repeatedly breaks away along vertical fractures.

Plateaus, mesas, and buttes are landforms of flat-lying erosion-resistant strata in arid regions.

Cliff retreat produces a **mesa**, which is a table-topped plateau bordered on all sides by cliffs (Figure 16.4). Mesas represent the remnants of a formerly extensive layer of resistant rock. As a mesa is reduced in area by retreat of the rimming cliffs, it maintains its flat top. Eventually, it becomes a small steep-sided hill known as a **butte**. Further erosion may produce a single, tall column before the landform is totally consumed.

In the walls of the great canyons of the Colorado Plateau region, these landforms are wonderfully dis-

Figure 16.3 In arid climates, distinctive erosional landforms develop in horizontal strata. (Drawn by A. N. Strahler.)

Figure 16.4 Bright Angel Canyon, Grand Canyon National Park, Arizona. Note the "stairstep" landscape, in which resistant sandstones and limestones form flat-topped benches and mesas, while weaker shales form slopes connecting them.

played. Figure 16.4 shows a view of the Grand Canyon. The flat surface of the canyon rim marks the edge of the vast plateau surrounding the canyon. Within the canyon, resistant rock layers crop out at different levels, producing a series of steps. Cliffs of resistant rock form the risers, while weaker rocks form the sloping treads below. Benches or platforms develop atop resistant strata, extending the treads and producing a series of levels that break the descent into the canyon. Small mesas and buttes remain within the canyon, isolated by erosion of surrounding rock strata.

Drainage Patterns

Imagine making a map of a region that showed only its stream channels. This map would show the stream pattern, or **drainage pattern**, of the region. The drainage pattern that develops on each land mass type is often related to the underlying rock type and structure.

Regions of horizontal strata often show broadly branching stream networks formed into a *dendritic drainage pattern*. Figure 16.5 is a map of the stream channels in a small area within the Appalachian Plateau region of Kentucky. The smaller streams in this pattern take a great variety of compass directions. Because the rock layers lie flat, no one direction is favored. Large

Figure 16.5 This dendritic drainage pattern is from an area of horizontal strata in the Appalachian Plateau of northern Kentucky. (Derived from data of the U.S. Geological Survey.)

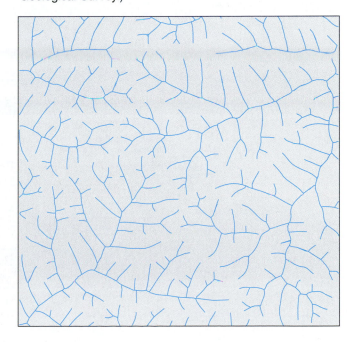

areas of uniform igneous rock also exhibit a dendritic drainage pattern, as we will see in a later section.

Coastal Plains

Coastal plains are found along passive continental margins that are largely free of tectonic activity. They are underlain by nearly horizontal strata that slope gently toward the ocean. Figure 16.6a shows a coastal zone that has recently emerged from beneath the sea. Formerly, this zone was a shallow continental shelf that accumulated successive layers of sediment brought from the land and distributed by currents. On the newly formed land surface, streams flow directly seaward, down the gentle slope. A stream of this origin is a *consequent stream*. It is a stream whose course is controlled by the initial slope of a new land surface. Consequent streams form on many kinds of initial landforms, such as volcanoes, fault blocks, or beds of drained lakes.

Coastal plains exhibit alternating belts of cuestas and lowlands.

In an advanced stage of coastal-plain development, a new series of streams and topographic features has developed (Figures 16.6b). Where more easily eroded strata (usually clay or shale) are exposed, denudation is rapid, making *lowlands*. Between them rise broad belts of hills called **cuestas**. Cuestas are commonly underlain by layers of sand, sandstone, limestone, or chalk that dip away from the cuesta at a low angle. The lowland lying between the area of older rock—the oldland—and the first cuesta is called the *inner lowland*.

Streams that develop along the trend of the lowlands, parallel with the shoreline, are known as *subsequent streams*. They take their position along a belt or zone of weak rock, and therefore follow closely the pattern of rock exposure. Subsequent streams occur in many regions, and we will mention them again in the discussion of folds, domes, and faults.

The drainage lines on a fully dissected coastal plain combine to form a *trellis (or lattice) drainage pattern*. In this type of pattern, a main stream has tributaries that are arranged at right angles. You can see this pattern in the streams of Figure 16.6b. The subsequent streams trend at about right angles to the consequent streams.

The coastal plain of the United States is a major geographical region, ranging in width from 160 to 500 km (about 100 to 300 mi) and extending for 3000 km (about 2000 mi) along the Atlantic and Gulf coasts. In Alabama and Mississippi (Figure 16.7), the coastal plain shows many of the features of Figure 16.6. In this region, the coastal plain sweeps far inland, as the oldland boundary shows. Cuestas and lowlands run in belts curving to follow the ancient coast. The cuestas, named "hills" or "ridges," are underlain by sandy formations and support pine forests. Limestone forms lowlands, such as the Black Belt in Alabama. This belt is named for its dark, fertile soils.

Figure 16.6 Development of a broad coastal plain. (*a*) Early stage—plain recently emerged. (*b*) Advanced stage—cuestas and lowlands developed. (Drawn by A. N. Strahler.)

Figure 16.7 The Alabama–Mississippi coastal plain is belted by a series of sandy cuestas and shale lowlands. (After A. K. Lobeck.)

LANDFORMS OF WARPED ROCK LAYERS

Although sedimentary rocks are horizontal or gently sloping in some regions, in other areas rock layers may be upwarped or downwarped. In some cases, these form domes and basins. In others, they form anticlines and synclines—wavelike forms of crests and troughs—as we noted in Chapter 12.

Sedimentary Domes

A distinctive land mass type is the **sedimentary dome**, a circular or oval structure in which strata have been forced upward into a domed shape. Sedimentary domes occur in various places within the covered shield areas of the continents. Igneous intrusions at great depth may have been responsible for some of these uplifts. For others, upthrusting on deep faults may have been the cause.

Erosion features of a sedimentary dome are illustrated in the two block diagrams of Figure 16.8. Strata are first removed from the summit region of the dome, exposing older strata beneath. Eroded edges of steeply dipping strata form sharp-crested sawtooth ridges called *hogbacks* (Figure 16.9). When the last of the strata have been removed, the ancient shield rock is exposed in the central core of the dome, which then develops a mountainous terrain.

The stream network on a deeply eroded dome shows dominant subsequent streams forming a circular system, called an *annular drainage pattern* (Figure 16.10). ("Annular" means "ring-shaped.") The shorter tributaries make a radial arrangement. The total pattern resembles a trellis pattern bent into a circular form.

The Black Hills Dome

A classic example of a large and rather complex sedimentary dome is the Black Hills dome of western South Dakota and eastern Wyoming (Figure 16.11). The Red

Figure 16.8 Erosion of sedimentary strata from the summit of a dome structure. (*a*) The strata are partially eroded, forming an encircling hogback ridge. (*b*) The strata are eroded from the center of dome, revealing a core of older igneous or metamorphic rock. (Drawn by A. N. Strahler.)

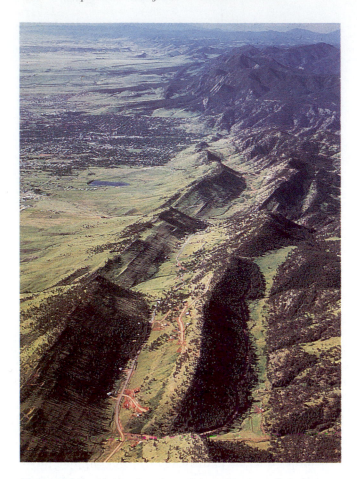

Figure 16.9 Hogbacks of sandstone lie along the eastern base of the Colorado Front Range, which is a large domed structure. The view is southward toward the city of Boulder. The rugged forested terrain in the distance is developed on igneous and metamorphic rock exposed in the core of the uplift.

Figure 16.10 The drainage pattern on an eroded dome combines annular and radial elements.

Valley is continuously developed around the entire dome and is underlain by a weak shale, which is easily washed away. On the outer side of the Red Valley is a high, sharp hogback of Dakota sandstone, known as Hogback Ridge, rising some 150 m (about 500 ft) above the level of the Red Valley. Farther out toward the margins of the dome, the strata are less steeply inclined and form a series of cuestas.

> **The Black Hills dome of South Dakota is a sedimentary dome that is eroded to produce hogback ridges, a circular valley, and a mountainous core of igneous and metamorphic rocks.**

The eastern central part of the Black Hills consists of a mountainous core of intrusive and metamorphic rocks. These mountains are richly forested, while the surrounding valleys are beautiful open parks. Thus, the region is very attractive as a summer resort area. In the northern part of the central core, there are valuable ore deposits. At Lead is the fabulous Homestake Mine, one

Figure 16.11 The Black Hills consist of a broad, flat-topped dome deeply eroded to expose a core of igneous and metamorphic rocks. (Drawn by A. N. Strahler.)

Figure 16.12   Stages in the erosional development of folded strata. (*a*) Erosion exposes a highly resistant layer of sandstone or quartzite, which controls much of the ridge-and-valley landscape. (*b*) Continued erosion partly removes the resistant formation but reveals another below it. (Drawn by A. N. Strahler.)

of the world's richest gold-producing mines. The west-central part of the Black Hills consists of a limestone plateau deeply carved by streams. The plateau represents one of the last remaining sedimentary rock layers to be stripped from the core of the dome

Fold Belts

According to the model of mountain making developed in Chapter 11, strata of the continental margins are deformed into folds along narrow belts during continental collision. As we explained in Chapter 12, a troughlike downbend of strata is a *syncline*, and the

archlike upbend next to it is an *anticline*. In a belt of folded strata, synclines alternate with anticlines, like a succession of wave troughs and wave crests on the ocean surface.

The two block diagrams of Figure 16.12 show some of the distinctive landforms resulting from fluvial denudation of a belt of folded strata. Deep erosion of these simple, open folds in this land mass type produces a **ridge-and-valley landscape**. Weaker formations such as shale and limestone are eroded away, leaving hard strata, such as sandstone or quartzite, to stand in bold relief as long, narrow ridges (Figure 16.13). Note that anticlines, or upwarps, are not always ridges. If a

Figure 16.13   Like an old wooden plank deeply etched by drifting sand to reveal the grain, the surface of south-central Pennsylvania shows zigzag ridges formed by bands of hard quartzite. Strata were crumpled into folds during a collision of continents that took place over 200 million years ago to produce the Appalachians. In this September image, the red color depicts vegetation cover, while the blue colors indicate mainly agricultural fields. The cloud bands are formed by condensation from passing aircraft. Note the shadows that the cloud bands cast on the ground.

Figure 16.14 A trellis drainage pattern on deeply eroded folds in the Central Appalachians.

resistant rock type at the center of the anticline is eroded through to reveal softer rocks underneath, an anticlinal valley may form. An *anticlinal valley* is shown in Figure 16.12*a*. A *synclinal* mountain is also possible. It occurs when a resistant rock type is exposed at the center of a syncline, and the rock stands up as a ridge (Figure 16.12*b*).

On eroded folds, the stream network is distorted into a trellis drainage pattern (Figure 16.14). The principal elements of this pattern are long, parallel subsequent streams occupying the narrow valleys of weak rock. In a few places, a larger stream cuts across a ridge in a deep, steep-walled *water gap*.

The folds illustrated in Figure 16.12 are continuous and even-crested. They produce ridges that are approximately parallel in trend and continue for great distances. In some fold regions, however, the folds are not continuous and level-crested. Instead, the fold crests rise or descend from place to place. A descending fold crest is said to "plunge." Plunging folds give rise to a zigzag line of ridges (Figure 16.15). Excellent examples of plunging folds can be seen in Figure 16.13.

LANDFORMS DEVELOPED ON OTHER LANDMASS TYPES

We now turn to the landforms developed on a few remaining landmass types. First are erosional forms that develop on a landmass exposed to faulting. Then we discuss the landforms that develop on metamorphic rocks and on exposed igneous batholiths. Last are the landforms of eroded volcanoes.

Erosion Forms on Fault Structures

In Chapter 12, we found that active normal faulting produces a sharp surface break called a *fault scarp* (see Figure 12.16). Repeated faulting may produce a great rock cliff hundreds of meters high. Erosion quickly modifies a fault scarp, but, because the fault plane extends hundreds of meters down into the bedrock, its effects on erosional landforms persist for long spans of geologic time. Figure 16.16 shows both the original fault scarp (above) and its later landform expression (below), known as a *fault-line scarp*. Even though the cover of sedimentary strata has been completely removed, exposing the ancient shield rock, the fault continues to produce a landform. Because the fault plane is a zone of weak rock that has been crushed during faulting, it is occupied by a subsequent stream. A scarp persists along the upthrown side. Such scarps on ancient fault lines are numerous in the exposed and covered continental shields.

Figure 16.17 shows some erosional features of a large, tilted fault block. The freshly uplifted block has a steep mountain face, but it is rapidly dissected into deep canyons. The upper mountain face becomes less steep

Figure 16.16 (*a*) A recently formed fault scarp. (*b*) Despite continental denudation over several million years, the fault plane causes a long, narrow valley bounded by a scarp. (Drawn by A. N. Strahler. Copyright © A. N. Strahler.)

Figure 16.15 Folds with crests that plunge downward give rise to zigzag ridges following erosion. (Drawn by Erwin Raisz. Copyright © by A. N. Strahler.)

Syncline   Anticline

(*a*)

(*b*)

Fault scarp

Fault-line scarp   Subsequent stream

as it is removed, and rock debris accumulates in the form of alluvial fans adjacent to the fault block. Vestiges of the fault plane are preserved in a line of triangular facets. Each facet represents the snubbed end of a ridge between two canyon mouths.

Metamorphic Belts

Where strata have been tightly folded and altered into metamorphic rocks during continental collision, denudation eventually develops a landscape with a strong grain of ridges and valleys. These features lack the sharpness and straightness of ridges and valleys in belts of open folds. Even so, the principle that governs the development of ridges and valleys in open folds applies to metamorphic landscapes as well—namely, that rocks resistant to denudation form highlands and ridges, while rocks that are more susceptible to denudation form lowlands and valleys.

Figure 16.18 shows typical erosional forms associated with parallel belts of metamorphic rocks, such as schist, slate, quartzite, and marble. Marble forms valleys, while slate and schist make hill belts. Quartzite stands out boldly and may produce conspicuous narrow hogbacks. Areas of gneiss form highlands.

Figure 16.18 Metamorphic rocks form long, narrow parallel belts of valleys and mountains. (Drawn by A. N. Strahler.)

Slate and marble are weak metamorphic rocks that underlie valleys. Schist, gneiss, and quartzite are more resistant and underlie uplands and ridges.

Parts of New England, particularly the Taconic and Green Mountains of New Hampshire and Vermont, illustrate the landforms eroded on an ancient metamorphic belt. The larger valleys trend north and south and are underlain by marble. These are flanked by ridges of gneiss, schist, slate, or quartzite.

Figure 16.17 The near-vertical faces of the foreground peaks in this photo mark a fault scarp along which the rocks have been uplifted by many successive motions. Talus slopes of rock debris line the base of the scarp. Eastern Sierra Nevada, California.

Exposed Batholiths and Monadnocks

Recall from Chapter 10 that batholiths—huge plutons of intrusive igneous rock—are formed deep below the earth's surface. Some are eventually uncovered by erosion and appear at the surface. Because batholiths are typically composed of resistant rock, they are eroded into hilly or mountainous uplands.

Granite batholiths are often found within the continental shields. A good example is the Idaho batholith, a granite mass exposed over an area of about 40,000 sq km (about 16,000 sq mi)—a region almost as large as New Hampshire and Vermont combined. Another example is the Sierra Nevada batholith of California, which makes up most of the high central part of that great mountain block.

A dendritic drainage pattern is often well developed on an eroded batholith (Figure 16.19). Note how similar this pattern is to that on horizontal strata (Figure 16.5).

Small bodies of granite, representing domelike projections of batholiths that lie below, are often found surrounded by ancient metamorphic rocks into which the granite was intruded. A good example is Stone Mountain, which rises above the Piedmont Upland of Georgia (Figure 16.20). Stone Mountain and other monolithic domes are unique landforms that are discussed further in *Eye on the Environment • Marvelous, Majestic, Monolithic Domes*.

The name **monadnock** has been given to an isolated mountain or hill (such as Stone Mountain) that rises conspicuously above a peneplain (Figure 16.21). A

Figure 16.19   This dendritic drainage pattern is developed on the dissected Idaho batholith.

monadnock develops because the rock within the monadnock is much more resistant to denudation processes than the bedrock of the surrounding region. The name is taken from Mount Monadnock in southern New Hampshire.

Deeply Eroded Volcanoes

Recall from Chapter 12 that volcanoes can be divided into two types—stratovolcanoes and shield volcanoes.

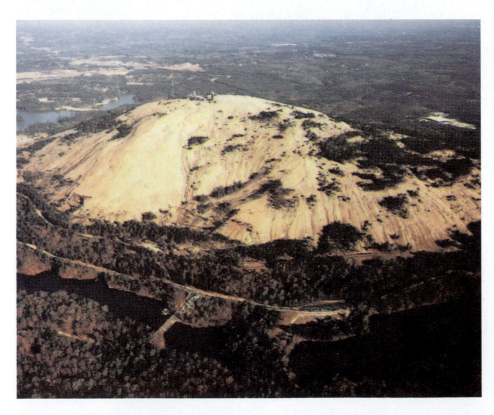

Figure 16.20   Stone Mountain, Georgia, is a striking erosion remnant, or monadnock, about 2.4 km (1.5 mi) long and rising 193 m (650 ft) above the surrounding Piedmont peneplain surface. The rock is granite, almost entirely free of joints, and has been rounded into a smooth dome by weathering processes.

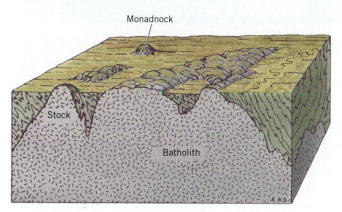

Figure 16.21 Batholiths appear at the land surface only after long-continued erosion has removed thousands of meters of overlying rocks. Small projections of the granite intrusion appear first and are surrounded by older rock. (Drawn by A. N. Strahler.)

Stratovolcanoes are typically constructed by explosive eruptions accumulating layers of thick, gummy andesite lava flows and airborne tephra, while shield volcanoes are built primarily of thin, runny flows of basalt. Thus, these two types of volcanoes produce different types of landforms as they erode away.

Figure 16.22 shows successive stages in the erosion of stratovolcanoes, lava flows, and a caldera. Shown in block (*a*) are active volcanoes in the process of building. These are initial landforms. Lava flows issuing from the volcanoes have spread down into a stream valley, following the downward grade of the valley and forming a lake behind the lava dam.

In block (*b*) some changes have taken place. The most conspicuous change is the destruction of the largest volcano to produce a caldera. A lake occupies the caldera, and a small cone has been built inside. One

Figure 16.22 Stages in the erosional development of stratovolcanoes and lava flows. (Drawn by Erwin Raisz. Copyright © by A. N. Strahler.)

EYE ON THE ENVIRONMENT

Marvelous, Majestic, Monolithic Domes

There is something awesome about a granite dome. Perhaps it's the way it projects above its surroundings, emerging abruptly from the ground and rising so steeply to its rounded summit. Perhaps it's the smooth, barren surface, hardly marred by a crack or crevice. Or maybe it's just the size—a huge, single, uniform object that dwarfs an observer like no other. In any event, these geological curiosities have aroused the interest of observers for centuries—perhaps even millennia.

Probably the most famous of all granite domes is Rio de Janeiro's Sugar Loaf, a monument that serves as a symbol for that cosmopolitan Brazilian city. Of course, Australia has its Ayres Rock, a gigantic rock mass in the desert of Northern Territory. But, alas, Ayres Rock is mere sandstone, not solid granite. Australia actually has some fine granite domes, located on the Eyre Peninsula of South Australia, but they have not achieved the international fame of Ayres Rock. Splendid granite domes also occur in the Nubian Desert of North Africa. Even Pasadena, California, has its own Little League contender in Eagle Rock, a small but prominent dome of massive conglomerate. These monolithic domes of uniform rock owe their rounded shapes to grain-by-grain disintegration of their rock surfaces. Because the domes lack the joints found in most other rocks, oxidation and hydrolysis act uniformly to weather the rock surface, producing a smooth, rounded form.

And what about the spectacular domes of Yosemite National Park in California's Sierra Nevada Range? Yosemite Valley has several splendid domes high up on either side of the deep canyon (see photo below). As we mentioned in Chapter 13, these are

These fine examples of monolithic granite domes are found in the Upper Yosemite Valley, Yosemite National Park, California.

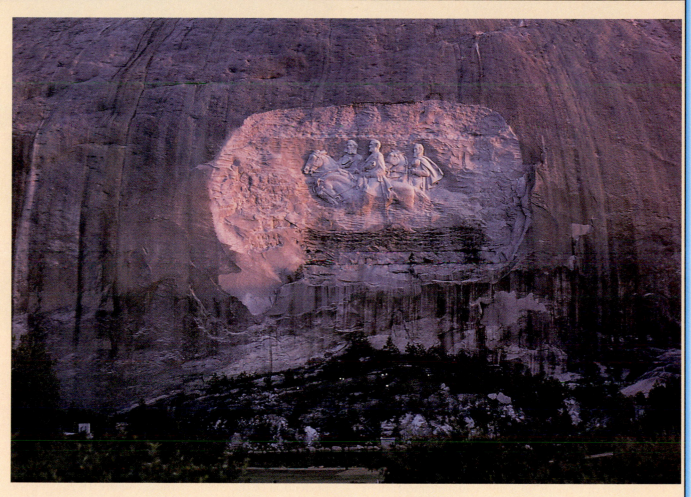

Monumental carvings honoring Lee, Jackson, and Davis. Stone Mountain, Georgia.

exfoliation domes, bearing thick rock shells. As the dome weathers, the outermost shells break apart and slide off as new ones are generated within. Beneath their shells, however, the Yosemite domes seem to be genuine monoliths.

Another fine granite dome is Stone Mountain, near Atlanta (Figure 16.20). Like Sugar Loaf, it is a uniform mass of granite seemingly without joints or fissures. It is the surface remains of an ancient, knoblike granite pluton intruded into older metamorphic rocks of the region. For millions of years, this region has been undergoing denudation, and most of it has been reduced to a rolling upland. Numerous monadnocks of quartzite with elongate, ridgelike forms rise above the upland, but Stone Mountain is unique in terms of its compact outline and smoothly rounded form.

Stone Mountain is unique in yet another way. Carved into its massive granite are huge equestrian sculptures of Robert E. Lee, Stonewall Jackson, and Jefferson Davis (see photo above). These were the artistic vision of Gutzon Borglum, the sculptor who carved the great busts of Washington, Jefferson, Lincoln, and Theodore Roosevelt in the granite of Mount Rushmore. Commissioned in the early 1920s to carve the sculptures by the Daughters of the Confederacy, he began the work with jack hammers and explosives, but in 1924 he resigned in a tiff with his sponsors and destroyed his plans. Eventually, two other sculptors completed his work, and despite long delays, it was dedicated in 1970.

Questions

1. What are the characteristics of domes? What process gives the dome its unique shape?

2. Briefly describe and locate two examples of granite domes.

Figure 16.23 Radial drainage patterns of volcanoes in the East Indies. The letter "C" shows the location of a crater.

Figure 16.24 Mount Shasta in the Cascade Range, northern California, is a partly dissected composite volcano. It has been eroded by streams and small alpine glaciers. Partway up the right-hand side is a more recent subsidiary volcanic cone, called Shastina, with a sharp crater rim.

of the other volcanoes, formed earlier, has become extinct. It has been dissected by streams and has lost its smooth conical form. Smaller, neighboring volcanoes are still active, and the contrast in form is marked.

The system of streams on a dissected volcano cone is a *radial drainage pattern*. Because these streams take their position on a slope of an initial land surface, they are of the consequent variety. It is often possible to recognize volcanoes from a drainage map alone because of the perfection of the radial pattern (Figure 16.23). A good example of a partly dissected volcano is Mount Shasta in California (Figure 16.24).

Figure 16.25 Ship Rock, New Mexico, is a volcanic neck enclosed by a weak shale formation. The peak rises about 520 m (about 1700 ft) above the surrounding plain. In the foreground and to the left, wall-like dikes extend far out from the central peak.

In Figure 16.22*c* all volcanoes are extinct and have been deeply eroded. The caldera lake has been drained, and the rim has been worn to a low, circular ridge. The lava flows have been able to resist erosion far better than the rock of the surrounding area and have come to stand as mesas high above the general level of the region.

Block (*d*) of Figure 16.22 shows an advanced stage of erosion of stratovolcanoes. All that remains now is a small, sharp peak, called a *volcanic neck*—which is the remains of lava that solidified in the pipe of the volcano. Radiating from it are wall-like dikes, formed of magma, which filled radial fractures around the base of the volcano. Perhaps the finest illustration of a volcanic neck with radial dikes is Ship Rock, New Mexico (Figure 16.25).

Shield volcanoes show erosion features that are quite different from those of stratovolcanoes. Figure 16.26*a*

Figure 16.27 Waimaea Canyon, over 760 m (2500 ft) deep, has been eroded into the flank of an extinct shield volcano on Kauai, Hawaii. Gently sloping layers of basaltic lava flows are exposed in the canyon walls.

shows the first stage of erosion of Hawaiian shield volcanoes, beginning in the upper diagram with the active volcano and its central depression. These are initial landforms. In (*b*), radial consequent streams cut deep canyons into the flanks of the extinct shield volcano, and these canyons are opened out into deep, steep-walled amphitheaters such as Waimea Canyon, on the island of Kauai (Figure 16.27). In the last stages (*c*), the original surface of the shield volcano is entirely obliterated, leaving a rugged mountain mass made up of sharp-crested divides and deep canyons.

Figure 16.26 Shield volcanoes in various stages of erosion make up the Hawaiian Islands. (*a*) Newly formed dome with central depression. (*b*) Early stage of erosion with deeply eroded valley heads. (*c*) Advanced erosion stage with steep slopes and mountainous relief. (Drawn by A. N. Strahler.)

(*a*)

0   8 km (5 mi)

(*b*)

(*c*)

A LOOK AHEAD

With this chapter, we conclude our series of three chapters on running water as a landform-making agent. Because the landscapes of most regions on the earth's land surface are produced by fluvial processes acting on differing rock types, running water is by far the most important agent in creating the variety of landforms that we see around us. The three remaining fluid agents of denudation are waves, wind, and glacial ice. These are the subjects of our final two chapters. In the first, we describe how waves shape the landforms of beaches and coasts, and how wind shapes the dunes of coasts and deserts. In the second, we relate how glacial ice scrapes, pushes, and deposits rock materials to form the distinctive landforms of glaciated regions.

LANDFORMS AND ROCK STRUCTURE IN REVIEW

This chapter has added a new dimension to the realm of landforms produced by fluvial denudation—the variety and complexity introduced by differences in rock composition and crustal structure. Because the various kinds of rocks offer different degrees of resistance to the forces of denudation, they exert a controlling influence on the shapes of landforms. We have seen, too, that streams carving up a landmass are controlled to a high degree by the structure of the rock on which they act. In this way, distinctive drainage patterns evolve.

In arid regions of horizontal strata, resistant rock layers produce vertical cliffs separated by gentler slopes on less resistant rocks. The drainage pattern is dendritic. Where strata are gently dipping, as on a coastal plain, the more resistant rock layers stand out as cuestas, interspersed with lowland valleys on weaker rocks. Consequent streams cut across the cuestas toward the ocean, and subsequent streams follow the valleys. This forms a trellis (lattice) drainage pattern.

Where rock layers are uparched into a dome, erosion produces a circular arrangement of rock layers outward from the center of the dome. Resistant strata form hogbacks, and weaker rocks form lowlands. Igneous rocks are often revealed in the center. The drainage pattern is annular. In fold belts, the sequence of synclines and anticlines brings a linear pattern of rock layers to the surface. Resistant strata form ridges, and weaker strata form valleys. Like the coastal plain, the drainage pattern will be trellised.

Faulting provides an initial surface along which rock layers are moved—the fault scarp. This feature can persist as a fault-line scarp long after the initial scarp is gone. In metamorphic belts, weak layers of shale, slate, and marble underlie valleys, while gneiss and schist form uplands. Quartzite stands out as a ridge-former. Exposed batholiths are often composed of uniform, resistant igneous rock. They erode to form a dendritic drainage pattern. Monadnocks of intrusive igneous rock stand up above a plain of weaker rocks.

Stratovolcanoes produce lava flows that initially follow valleys, but are highly resistant to erosion. After erosion of the surrounding area, they can remain as highlands, lava ridges, or mesas. At the last stages of erosion, all that remains of stratovolcanoes are necks and dikes. Hawaiian shield volcanoes are eroded by streams that form deeply carved valleys with steeply sloping heads. Eventually, these merge to produce a landscape of steep slopes and knifelike ridges.

KEY TERMS

plateau	coastal plain	scape
mesa	cuesta	monadnock
butte	sedimentary dome	
drainage pattern	ridge-and-valley land-	

REVIEW QUESTIONS

1. Why is there often a direct relationship between landforms and rock structure? How are geologic structures formed deep within the earth exposed at the surface?
2. Which of the following types of rocks—shale, limestone, sandstone, conglomerate, igneous rocks—tends to form lowlands? uplands?
3. How are the tilt and orientation of a natural rock plane measured?
4. How are coastal plains formed? Identify and describe two landforms found on coastal plains. What drainage pattern is typical of coastal plains?
5. What types of landforms are associated with sedimentary domes? How are they formed? What type of drainage pattern would you expect for a dome and why? Provide an example of an eroded sedimentary dome and describe it briefly.

6. How does a ridge-and-valley landscape arise? Explain the formation of the ridges and valleys. What type of drainage pattern is found on this landscape? Where in the United States can you find a ridge-and-valley landscape?
7. What type of landform(s) might you expect to find in the presence of a fault? Why?
8. Which of the following types of rocks—schist, slate, quartzite, marble, gneiss—tends to form lowlands? uplands?
9. Do shield volcanoes erode differently from stratovolcanoes? How so?

VISUALIZING EXERCISES

1. Identify and sketch a typical landform of flat-lying rock layers found in an arid region.
2. What types of landforms and drainage patterns develop on batholiths? Provide a sketch of a batholith. What is the difference between a batholith and a monadnock?
3. Describe and sketch the stages through which a landscape of stratovolcanoes is eroded. What distinctive landforms remain as the last remnants?

A DEEPER VIEW OF LANDFORMS AND ROCK STRUCTURE

1. Imagine the following sequence of sedimentary strata—sandstone, shale, limestone, and shale. What landforms would you expect to develop in this structure if the sequence of beds is (*a*) flat-lying in an arid landscape; (*b*) slightly tilted as in a coastal plain; (*c*) folded into a syncline and an anticline in a fold belt; (*d*) fractured and displaced by a normal fault? Use sketches in your answer.
2. A region of ancient mountain roots, now exposed at the surface, includes a central core of plutonic rocks surrounded on either side by belts of metamorphic rocks. What landforms would you expect for this landscape and why?

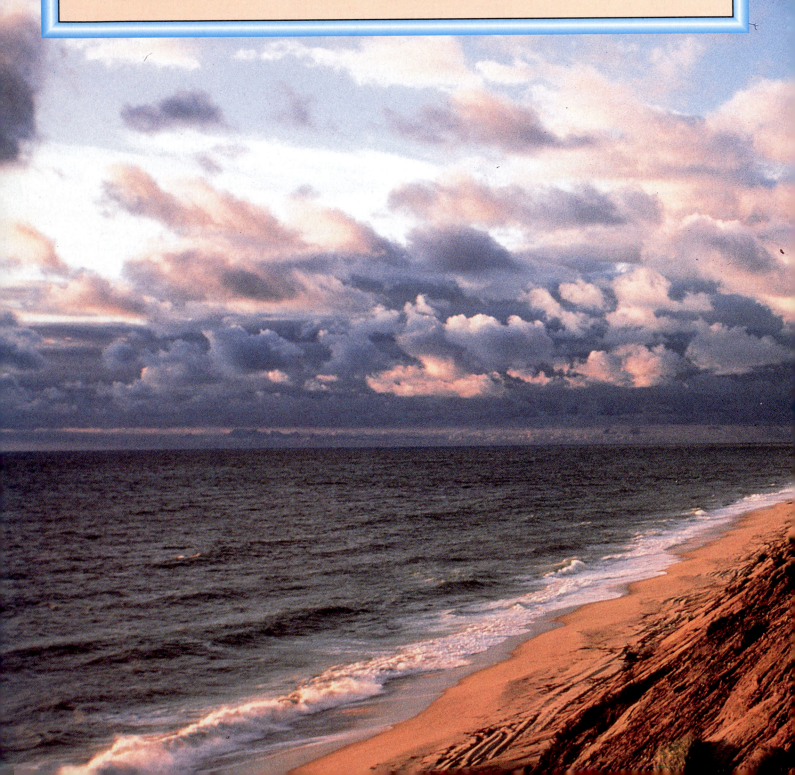

Landforms Made by Waves and Wind

An eroding bluff of Ice Age sediment at Sankaty Head Lighthouse, Nantucket Island, Massachusetts.

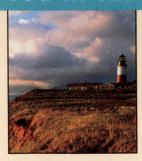

PUTTING YOU IN THE PICTURE

A coastal storm, occurring at the interface between land, ocean, and atmosphere, can be an awesome event. Lashed by unrelenting wind and rain, the surf attacks the beach, moving large quantities of sand offshore. At high tide, waves cut new paths through coastal dunes to inland bays. Shoreline bluffs are undercut and collapse, releasing new sediment to the shoreline. Picture yourself witnessing the power of a coastal storm as you read the account below.

You slow the car and carefully turn into a narrow, blacktopped road that leads to the lighthouse. The drive wasn't easy, with all those downed tree limbs and flooded spots along the way. What a way to end your Thanksgiving vacation! Before arriving at the cottage on Cape Cod, you had visions of long treks along the beach, inhaling the salt air and watching the sea birds rush to and fro with the swash of the breaking waves.

Soon after you arrived, however, the skies turned gray, and the wind and rain began—a "nor'easter" was in store. Last night the storm reached its peak, as rain fell in sheets and the wind howled without letup. This morning the rain stopped, and so you decided to venture out to the beach.

It's not long before the lighthouse comes into view, a wide, glass-topped column set against the low clouds of the dark sky. You enter the parking lot, which

is perched on top of the coastal bluff overlooking the beach. The far portion of the lot is blocked off with red- and white-striped sawhorses anchored by sandbags to keep them erect in the strong wind.

Stepping out of the car, you find that the force of the wind is strong, though not overwhelming. Beyond the sawhorses, you see that a large section of pavement has broken away and slid down the bluff to the beach below. The storm waves, striking the bluff at high tide, must have undermined the pavement, allowing it to fall into the breaker zone beneath. The wooden stairs down the bluff have fared somewhat better. The sand underneath the bottom part of the lowest flight has been undercut, but access to the beach is not difficult, and you descend. Fortunately, the tide is now out.

The beach looks totally different from the way it looked when you visited last summer. Then, the beach had a wide, flat upper portion, ideal for sunbathing, and sloped steeply in the breaker zone.

Now the beach is just a long, gentle slope stretching from the bluffs to the breakers. The waves are breaking much farther out, where the gray and white of the surf and foam converges with the gray and white of the sky. Rolls of frothing water move landward, eventually collapsing and running far up the low slope.

The roar of the surf is intense and continuous. The beach feels dense and hard. A few inches above the wet, smooth surface, sheets of wind-blown sand move downwind, braiding and separating into flowing, streamlike currents. You turn to look back at the bluff face. The freshly eroded layers of sand and silt are clear and distinct. Large, wave-washed lumps of sediment lie at the base of the bluff. It won't be long before the ocean's erosive power claims the lighthouse itself.

You linger for a time, enjoying the dark beauty of the scene, before heading for home. Although this trip to the beach wasn't exactly what you had in mind for the holiday, you are glad to have experienced first-hand the power of waves and wind.

Wind and waves are the primary agents that shape the shoreline. Unrelenting, they often work against human efforts to develop the coastline and preserve it in an unchanging configuration. Thus, human and natural forces often conflict in the coastal zone, as we will see in this chapter.

Chapter Outline

The Work of Waves
  Marine Cliffs
  Beaches
  Littoral Drift
  Littoral Drift and Shore Protection
Tidal Currents
  Tidal Current Deposits
Types of Coastlines
  Coastlines of Submergence
  Barrier-Island Coasts
  Delta and Volcano Coasts
  Coral-Reef Coasts

Raised Shorelines and Marine Terraces
Eye on the Environment • The Threat of Rising Sea Level from Global Warming
Wind Action
  Erosion by Wind
  Dust Storms
Sand Dunes
  Types of Sand Dunes
  Coastal Foredunes
Loess
  Induced Deflation

Chapter 17

LANDFORMS MADE BY WAVES AND WIND

In Chapter 5, we saw how the rotation of the earth on its axis, combined with unequal solar heating of the earth's surface, produces a system of global winds in which air moves as a fluid across the earth's surface. This motion of air produces a frictional drag that can move surface materials. When winds blow over broad expanses of water, waves are generated. These waves then expend their energy at coastlines, eroding rock, moving sediment, and creating landforms. When winds blow over expanses of soil or alluvium that are not protected by vegetation, fine particles can be picked up and carried long distances before coming to rest. Coarser particles, such as sand, can also be moved to build landforms, such as sand dunes. Because wave action on beaches provides abundant sand, we often find landforms of both wind and waves near coastlines.

This chapter describes the processes and landforms of erosion and deposition that result when wind power moves surface materials, either directly as wind or indirectly as waves. Unlike fluvial action or mass wasting, wind and breaking waves can move materials uphill, against the force of gravity. For this reason, wind power is a unique land-forming agent. Another agent that acts in coastal regions is the tide—the rhythmic rise and fall in sea level that is generated by the gravitational attraction of the sun and moon. This subject is also covered in our chapter.

Throughout this chapter we will use the term **shoreline** to mean the shifting line of contact between water and land. The broader term **coastline**, or simply *coast*, refers to a zone in which coastal processes operate or have a strong influence. The coastline includes the shallow water zone in which waves perform their work, as well as beaches and cliffs shaped by waves, and coastal dunes. Also often present are **bays**—bodies of water that are sheltered by the configuration of the coast from strong wave action. Where a river empties into an ocean bay, the bay is termed an **estuary**. In an estuary, fresh and ocean water mix, creating a unique habitat for many plants and animals which is neither fresh water nor ocean.

The continental shoreline is a unique, complex environmental zone of great importance. Here, the saltwater of the oceans receives fresh water from rivers and streams, and contacts the solid mineral base of the continents. Shallow waters and estuaries provide food resources of shellfish, finfish, and waterfowl. The shoreline also provides a base for ocean-going vessels that har-

vest marine food resources and transport goods between continents. In addition, the coastal zone is a recreational region, with sea breezes, bathing beaches, and opportunities for surfing, skin diving, sport fishing, and boating.

Although the presence of the ocean in the coastal zone provides unique opportunities, it also presents a unique set of restraints and hazards for humans and human development. Some coasts are more suitable for development than others. Where rocky and cliffed, the coast may not provide the sheltered harbors essential for marine commerce. Along other coasts, the enormous energy of storm waves can cut back the shore. Roads and buildings are undermined and eventually claimed by the sea. Storms can raise sea levels, flood low-lying areas, and elevate breaking waves to a level well above normal. Storm surges and seismic sea waves, which cause such inundations, have already been described in earlier chapters. For centuries, humanity has been at war with the sea, building fortresslike walls to keep out ocean waters. Sometimes the battle is won, as land under tidal waters is reclaimed to produce more crops for food and forage. Sometimes it is lost, as storm waves devastate a coastal community.

THE WORK OF WAVES

The most important agent shaping coastal landforms is wave action. The energy of waves is expended primarily in the constant churning of mineral particles and water as waves break at the shore. This churning erodes shoreline materials, moving the shoreline landward. But, as we will see shortly, the action of waves and currents can also move sediment along the shoreline for long distances. This activity can build beaches outward as well as form barrier islands just offshore.

Wave action is the most important agent in shaping shoreline landforms.

Waves travel across the deep ocean with little loss of energy. When waves reach shallow water, the drag of the bottom slows and steepens the wave until it falls and collapses as a *breaker* (Figure 17.1). Many tons of water surge forward, riding up the beach slope.

Where weak or soft materials—various kinds of regolith, such as alluvium—make up the coastline, the force of the forward-moving water alone easily cuts into the coastline. Here, erosion is rapid, and the shoreline recedes easily. Under these conditions, a steep bank—a *marine scarp*—is the typical coastal landform (Figure 17.2). It retreats steadily under attack of storm waves.

Figure 17.1 A breaking wave. As the wave approaches the beach (1–3), it steepens (4–5) and finally falls forward (6–7), rushing up the beach slope (8). (After W. M. Davis.)

Figure 17.2 Mohegan Bluffs, Block Island, Rhode Island. This marine scarp of Ice Age sediments is being rapidly eroded, threatening historic Southeast Lighthouse. In 1993 the lighthouse was moved to a safer spot 73.5 m (245 ft) away.

Marine Cliffs

Where a **marine cliff** lies within reach of the moving water, it is impacted with enormous force. Rock fragments of all sizes, from sand to cobbles, are carried by the surging water and thrust against bedrock of the cliff. The impact breaks away new rock fragments, and the cliff is undercut at the base. In this way, the cliff erodes shoreward, maintaining its form as it retreats. The retreat of a marine cliff formed of hard bedrock is exceedingly slow when judged in terms of a human life span.

Figure 17.3 illustrates some details of a typical marine cliff. A deep basal indentation, the *wave-cut notch*, marks the line of most intense wave erosion. The waves find points of weakness in the bedrock and penetrate deeply to form crevices and *sea caves*. Where a more resistant rock mass projects seaward, it may be cut through to form a picturesque *sea arch*. After an arch collapses, a rock column, known as a *stack*, remains. Eventually, the stack is toppled by wave action and is leveled. As the sea cliff retreats landward, continued wave abrasion forms an *abrasion platform* (Figure 17.4). This sloping rock floor continues to be eroded and widened by abrasion beneath the breakers. If a beach is present, it is little more than a thin layer of gravel and cobblestones atop the abrasion platform.

Beaches

Where sand is in abundant supply, it accumulates as a thick, wedge-shaped deposit, or **beach**. Beaches absorb the energy of breaking waves. During short periods of storm activity, the beach is cut back, and sand is carried offshore a short distance by the heavy wave action. However, the sand is slowly returned to the beach during long periods when waves are weak. In this way, a beach may retain a fairly stable but alternating configuration over many years' time.

Although beaches are most often formed from particles of fine to coarse quartz sand, some beaches are built from rounded pebbles or cobbles. Still others are formed from fragments of volcanic rock (Hawaii) or even shells (Florida).

Beaches are shaped by alternate landward and seaward currents of water generated by breaking waves

Figure 17.3 Landforms of sea cliffs. (Drawn by E. Raisz. Copyright © A. N. Strahler.)

(Figure 17.5). After a breaker has collapsed, a foamy, turbulent sheet of water rides up the beach slope. This *swash* is a powerful surge that causes a landward movement of sand and gravel on the beach. When the force of the swash has been spent against the slope of the beach, a return flow, or *backwash*, pours down the beach (Figure 17.6*a*). This "undercurrent" or "undertow," as it is popularly called, can be strong enough to sweep unwary bathers off their feet and carry them seaward beneath the next oncoming breaker.

Figure 17.4   A wave-cut abrasion platform on the central coast of California, Montaña d'Oro State Park.

Littoral Drift

The unceasing shifting of beach materials with swash and backwash of breaking waves also results in a sidewise movement known as *beach drift* (Figure 17.6a). Wave fronts usually approach the shore at less than a right angle, so that the swash and its burden of sand ride obliquely up the beach. After the wave has spent its energy, the backwash flows down the slope of the beach in the most direct downhill direction. The particles are dragged directly seaward and come to rest at a position to one side of the starting place. This movement is repeated with each wave and individual rock particles travel long distances along the shore. Multiplied many thousands of times to include the numberless particles of the beach, beach drift becomes a significant form of sediment transport.

Sediment in the shore zone is also moved along the beach in a related, but different, process. When waves approach a shoreline at an angle to the beach, a current is set up parallel to the shore in a direction away from the wind. This is known as a *longshore current* (Figure 17.6b). When wave and wind conditions are favorable, this current is capable of carrying sand along the sea bottom. The process is called *longshore drift*. Beach drift and longshore drift, acting together, move particles in the same direction for a given set of onshore winds. The total process is called **littoral drift** (Figure 17.6c). ("Littoral" means "pertaining to a coast or shore.")

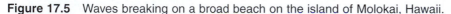

The swash and backwash of breakers attacking the shore at an angle creates beach drift, a process that moves sediment along the shoreline.

Littoral drift operates to shape shorelines in two quite different situations. Where the shoreline is straight or broadly curved for many kilometers at a stretch, littoral drift moves the sand along the beach in one direction for a given set of prevailing winds. This situation is shown in Figure 17.6c. Where a bay exists, the sand is carried out into open water as a long finger, or *sandspit* (Figure 17.7). As the sandspit grows, it forms a barrier, called a *bar*, across the mouth of the bay.

A second situation is shown in Figure 17.8. Here the coastline consists of prominent headlands, projecting seaward, and deep bays. Approaching wave fronts slow when the water becomes shallow, and this slowing effect causes the wave front to wrap around the headland. High, wave-cut cliffs develop shoreward of an abrasion platform. Sediment from the eroding cliffs is carried by littoral drift along the sides of the bay, converging on

Figure 17.5 Waves breaking on a broad beach on the island of Molokai, Hawaii.

Figure 17.6 How waves move sediment by littoral drift. (*a*) Swash and backwash move particles along the beach in beach drift. (*b*) Waves set up a longshore current that moves particles by longshore drift. (*c*) Littoral drift, produced by these two processes, creates a sandspit.

(*a*)

(*b*)

(*c*)

Figure 17.7 This white sandspit is growing in a direction toward the observer. Monomoy Point, Cape Cod, Massachusetts.

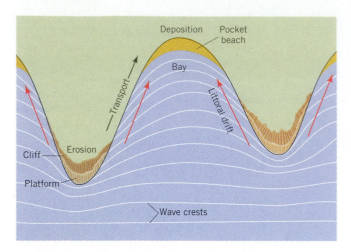

Figure 17.8 On an embayed coast, sediment is carried from eroding headlands to the bayheads, where pocket beaches are formed. (Copyright © A. N. Strahler.)

the head of the bay. The result is a crescent-shaped beach, often called a *pocket beach*.

Littoral Drift and Shore Protection

When sand arrives at a particular section of the beach more rapidly than it is carried away, the beach is widened and built oceanward. This change is called

progradation (building out). When sand leaves a section of beach more rapidly than it is brought in, the beach is narrowed and the shoreline moves landward. This change is called **retrogradation** (cutting back).

In some cases, the source of beach sand is sediment delivered to the coast by a river. Construction of dams far upstream on the river may drastically reduce the sediment load of the river, cutting off the source of sand for littoral drift. Retrogradation can then occur on a long stretch of shoreline.

Groins trap longshore drift and help prevent beach retrogradation.

Along stretches of shoreline affected by retrogradation, the beach may be seriously depleted or even entirely disappear, destroying valuable shore property. In some circumstances, structures can be installed that will cause progradation, and so build a broad, protective beach. This is done by installing groins at close intervals along the beach. A *groin* is simply a wall or embankment built at right angles to the shoreline. It may be constructed of large rock masses, concrete, or wooden pilings. The groins act to trap sediment moving along the shore as littoral drift.

The effect of groins is shown in Figure 17.9. The beach toward the back of the scene is protected by

Figure 17.9 Erosion by severe storms during the winter of 1993 has carved out an inlet in this barrier beach on the south shore of Long Island, New York. A system of groins (arrows) has trapped sand, protecting the far stretch of beach. In the foreground, the beach has receded well inland of the houses that were once located on its edge.

groins, while the beach in the foreground is not. The trapping action of the groins has starved the foreground beach of sand, narrowing the beach and leaving the houses unprotected. Storm waves have severely eroded the unprotected beach and cut an inlet across the beach.

TIDAL CURRENTS

Most marine coastlines are influenced by the *ocean tide*, a rhythmic rise and fall of sea level under the influence of changing attractive forces of moon and sun on the rotating earth. Where tides are great, the effects of changing water level and the tidal currents thus set in motion are of major importance in shaping coastal landforms.

The tidal rise and fall of water level is graphically represented by the *tide curve*. Figure 17.10 is a tide curve for Boston Harbor covering a day's time. The water reached its maximum height, or high water, at 3.6 m (11.8 ft) on the tide staff, and then fell to its minimum height, or low water, at 0.8 m (2.6 ft), occurring about $6\frac{1}{4}$ hours later. A second high water occurred about $12\frac{1}{2}$ hours after the previous high water, completing a single tidal cycle. In this example, the tidal range, or difference between heights of successive high and low waters, is 2.8 m (9.2 ft).

In bays and estuaries, the changing tide sets in motion currents of water known as *tidal currents*. The relationships between tidal currents and the tide curve are shown in Figure 17.11. When the tide begins to fall,

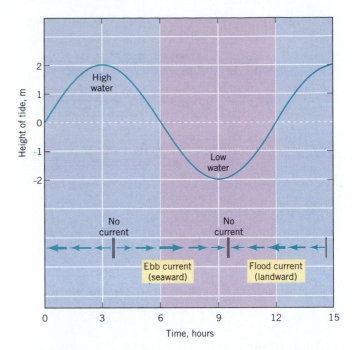

Figure 17.11 The ebb current flows seaward as the tide level falls. The flood current flows landward as the tide level rises. Tidal currents are strongest when water level is changing most rapidly, in the middle of the cycle.

an *ebb current* sets in. This flow ceases about the time when the tide is at its lowest point. As the tide begins to rise, a landward current, the *flood current*, begins to flow.

Tidal Current Deposits

Ebb and flood currents generated by tides perform several important functions along a shoreline. First, the currents that flow in and out of bays through narrow inlets are very swift and can scour the inlet strongly. This keeps the inlet open, despite the tendency of shore-drifting processes to close the inlet with sand.

Second, tidal currents carry large amounts of fine silt and clay in suspension. This fine sediment is derived from streams that enter the bays or from bottom muds agitated by storm wave action. It settles to the floors of the bays and estuaries, where it accumulates in layers and gradually fills the bays. Much organic matter is present in this sediment.

In time, tidal sediments fill the bays and produce mud flats, which are barren expanses of silt and clay. They are exposed at low tide but covered at high tide. Next, a growth of salt-tolerant plants takes hold on the mud flat. The plant stems trap more sediment, and the flat is built up to approximately the level of high tide, becoming a *salt marsh* (Figure 17.12). A thick layer of peat is eventually formed at the surface. Tidal currents maintain their flow through the salt marsh by means of a highly complex network of winding tidal streams.

Figure 17.10 Height of water at Boston Harbor measured every half hour.

Figure 17.12 A tideland salt marsh. Chincoteague National Wildlife Refuge, Virgina.

Salt marshes can be drained and used for agriculture. The salt marsh is first cut off from the sea by construction of an embankment of earth (a dike). Gates are installed in the dike to allow the freshwater drainage of the land to exit when the ocean's tide level is low. Gradually, the saltwater is excluded, and the soil water of the diked land becomes fresh. Such diked lands are intensively developed in Holland (polders) and southeast England (fenlands).

The surface of the reclaimed salt marsh subsides over many decades because of compaction of the underlying peat layers. In fact, the land surface can lie well below mean sea level. The threat of flooding by saltwater, when storm waves break the dikes, constantly hangs over the inhabitants of such low areas.

TYPES OF COASTLINES

The world's coastlines fall into a few distinctive types, each depending on the type of land mass against which the ocean water has come to rest. One group of coastline types derives its qualities from **submergence**, the partial drowning of a coast by a rise of sea level or a sinking of the crust. Another group derives its qualities from **emergence**, the exposure of submarine landforms by a falling of sea level or a rising of the crust. Another group of coastline types results when new land is built out into the ocean by volcanoes and lava flows, by the growth of river deltas, or by the growth of coral reefs.

Ria coasts and fiord coasts result from submergence of a land mass, while a barrier-island coast results from emergence.

A few important types of coastlines are illustrated in Figure 17.13. The first two are the result of submergence. The *ria coast* (*a*) is a deeply embayed coast resulting from submergence of a landmass dissected by streams. The *fiord coast* (*b*) is deeply indented by steep-walled fiords, which are submerged glacial troughs (discussed further in Chapter 18). The *barrier-island coast* (*c*) is associated with a recently emerged coastal plain. Here, the offshore slope is very gentle, and a barrier island of sand, lying a short distance from the coast, is created by wave action. Large rivers build elaborate deltas, producing *delta coasts* (*d*). The *volcano coast* (*e*) is formed by the eruption of volcanoes and lava flows, partly constructed below water level. Reef-building corals create new land and make a *coral-reef coast* (*f*). Downfaulting of the coastal margin of a continent can allow the shoreline to come to rest against a fault scarp, producing a *fault coast* (*g*).

Coastlines of Submergence

Coastlines of submergence include ria coasts and fiord coasts. The ria coast, which takes its name from the Spanish word for estuary, *ria*, has many offshore islands. A ria coast is formed when a rise of sea level or a crustal sinking (or both) brings the shoreline to rest against the sides of valleys previously carved by streams (Figure 17.13*a*). Wave attack forms cliffs on the exposed seaward sides of islands and headlands. Sediment produced by wave action accumulates in the form of beaches along the cliffed headlands and at the heads of bays. This sediment is carried by littoral drift and is often built into sandspits across bay mouths and as connecting links between islands and the mainland.

The fiord coast is similar to the ria coast. However, the submerged valleys were carved by flowing glaciers instead of streams. As a result, the valleys are deep, with straight, steep sides. Because sediment rapidly sinks into the deep water, beaches are rare. Fiords are described further in Chapter 18 (see Figure 18.8).

Many glaciated coastlines, such as those of New England and the Canadian Maritime Provinces, are coastlines of submergence. During the Ice Age, the weight of ice sheets depressed the crust of these regions, and they are still slowly rising now that the ice sheets are gone. Meanwhile, ocean waves and currents are rapidly eroding their rocky shorelines, creating bays, bars, and estuaries.

Barrier-Island Coasts

In contrast to the bold relief and deeply embayed outlines of coastlines of submergence are low-lying coasts

(*a*) Ria coast

(*b*) Fiord coast

(*c*) Barrier-island coast

(*d*) Delta coast

(*e*) Volcano coast (*left*)   (*f*) Coral-reef coast (*right*)

(*g*) Fault coast

Figure 17.13 Seven common kinds of coastlines. These examples have been selected to illustrate a wide range in coastal features. (Drawn by A. N. Strahler.)

from which the land slopes gently beneath the sea. The coastal plain of the Atlantic and Gulf coasts of the United States presents a particularly fine example of such a gently sloping surface. As we explained in Chapter 16, this coastal plain is a belt of relatively young sedimentary strata, formerly accumulated beneath the sea as deposits on the continental shelf. During the latter part of the Cenozoic Era and into recent time, the coastal plain emerged from the ocean as a result of repeated crustal uplifts.

Along much of the Atlantic Gulf coast there exist *barrier islands*, low ridges of sand built by waves and further increased in height by the growth of sand dunes (Figure 17.14). Behind the barrier island lies a *lagoon*. It is a broad expanse of shallow water in places largely filled with tidal deposits.

A characteristic feature of barrier islands is the presence of gaps, known as *tidal inlets*. Strong currents flow alternately landward and seaward through these gaps as

Figure 17.14 A barrier island is separated from the mainland by a wide lagoon. Sediments fill the lagoon, while dune ridges advance over the tidal flats. An inlet allows tidal flows to pass in and out of the lagoon. (Drawn by A. N. Strahler.)

Inlet

Tidal flats

Shallow lagoon

Dunes

Beach ridges

(a)

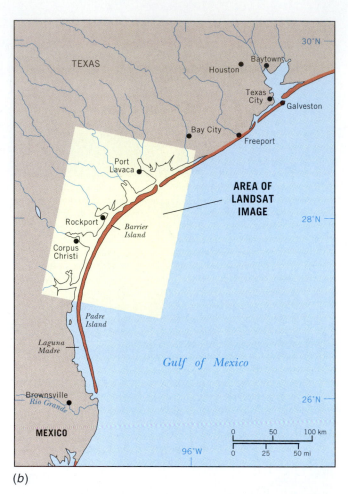

(b)

Figure 17.15 (*a*) A Landsat image of the Texas barrier-island coast. Red colors on the Landsat image indicate vegetation. Bright white areas are dunes or beach sand. Lighter blue tones mark the presence of sediment in water inside of the barrier beach. (*b*) Index map of the location.

the tide rises and falls. In heavy storms, the barrier island may be breached by new inlets (Figure 17.9). After that occurs, tidal current scour will tend to keep a new inlet open. In some cases, the inlet is closed later by shore drifting of beach sand.

Perhaps the finest example of a barrier-island coast is the Gulf coast of Texas (Figure 17.15). Here, the barrier island is unbroken for more than 160 km (100 mi) at a single stretch, and natural passes are few. The lagoon is 8 to 16 km (5 to 10 mi) wide. As is true of most barrier-island coasts, deep natural harbors are lacking. Extensive channel dredging is required. Ocean-going ships must enter and leave the lagoon through passes in the barrier island.

Delta and Volcano Coasts

The deposit of clay, silt, and sand made by a stream or river where it flows into a body of standing water is known as a **delta**. Deposition is caused by rapid reduction in velocity of the current as it pushes out into the standing water. Typically, the river channel divides and subdivides into lesser channels called *distributaries*. The coarser sand and silt particles settle out first, while the fine clays continue out farthest and eventually come to rest in fairly deep water. Contact of fresh with saltwater causes the finest clay particles to clot together and form larger particles that settle to the sea floor.

A delta results when a river empties into the ocean, depositing sediment that is worked by waves and currents.

Deltas show a wide variety of outlines. The Nile delta has the triangular shape of the Greek letter delta (Δ). From a spacecraft, this delta appears as a fertile plain bounded by desert land (Figure 17.16). In outline, it resembles an alluvial fan. The delta has two major distributaries, one of which terminates in the Damietta mouth and the other in the Rosetta mouth. The mouths project as cusps (tooth-forms) into the Mediterranean Sea. Sediment from the mouths has been carried along the shore by littoral drift to form curved barrier beaches.

Figure 17.16 A Landsat image of the Nile delta. The lush vegetation cover of the delta lands appears in deep red tones. Blue colors indicate fallow land. A few puffy clouds are present near the southeast edge of the delta.

The Mississippi delta has a different shape. Long, branching fingers grow far out into the Gulf of Mexico at the ends of the distributaries, giving the impression of a bird's foot. A satellite image of the delta, Figure 17.17, shows the great quantity of suspended sediment—clay and fine silt—being discharged by the river into the Gulf. It amounts to about 1 million metric tons (about 1.1 million English tons) per day.

Delta growth is often rapid, ranging from 3 m (about 10 ft) per year for the Nile to 60 m (about 200 ft) per year for the Mississippi delta. Some cities and towns that were at river mouths several hundred years ago are today several kilometers inland. An important engineering problem is to keep an open channel for ocean-going vessels that must enter the delta distributaries to reach port. The mouths of the Mississippi River delta distributaries, known locally as "passes," have been extended by the construction of jetties. The narrowed stream is forced to move faster and thus scours a deep channel suitable for the passage of large vessels.

A volcano coast is similar to a delta coast in that it marks the boundary between the ocean and newly created land. However, the new land consists of volcanic deposits—lava and ash—that flow from active volcanoes into the ocean. Low cliffs occur when wave action erodes the fresh deposits. Beaches are typically narrow, steep, and composed of fine particles of volcanic deposits.

Coral-Reef Coasts

Coral-reef coasts are unique in that the addition of new land is made by organisms—corals and algae. Growing together, these organisms secrete rocklike deposits of mineral carbonate, called **coral reefs**. As coral colonies die, new ones are built on them, accumulating as limestone. Coral fragments are torn free by wave attack, and the pulverized fragments accumulate as sand beaches.

Coral-reef coasts occur in warm, tropical and equatorial waters between the limits of lat. 30° N and 25° S. Water temperatures above 20°C (68°F) are necessary for

Figure 17.17 The Mississippi delta recorded from the Landsat orbiting satellite. The natural levees of the bird-foot delta appear as lacelike filaments in a great pool of turbid river water. New Orleans can be seen at the upper left, occupying the region between the natural levees of the Mississippi River and the southern shore of Lake Pontchartrain.

Figure 17.18   The Island of Moorea and its coral reefs, Society Islands, South Pacific Ocean. The island is a deeply dissected volcano with a history of submergence. Tahiti lies in the background.

dense growth of coral reefs. Reef corals live near the water surface. The sea water must be free of suspended sediment and well aerated for vigorous coral growth. For this reason, corals thrive in positions exposed to wave attack from the open sea. Because muddy water prevents coral growth, reefs are missing opposite the mouths of muddy streams. Coral reefs are remarkably flat on top. They are exposed at low tide and covered at high tide.

There are three distinctive types of coral reefs—fring-

ing reefs, barrier reefs, and atolls. *Fringing reefs* are built as platforms attached to shore (Figure 17.18). They are widest in front of headlands where wave attack is strongest. *Barrier reefs* lie out from shore and are separated from the mainland by a lagoon (Figure 17.18). Narrow gaps occur at intervals in barrier reefs. Through these openings, excess water from breaking waves is returned from the lagoon to the open sea.

Atolls are more or less circular coral reefs enclosing a lagoon but have no land inside (Figure 17.19). On large

Figure 17.19   A Pacific atoll. The blue waters of the lagoon are ringed by submerged coral reefs. Four patches of land, covered by dark vegetation, are also seen on the rim. Kayangel atoll, Republic of Belau, Micronesia.

atolls, parts of the reef have been built up by wave action and wind to form low island chains connected by the reef. Most atolls are built on a foundation of volcanic rock. This foundation is thought to have been a basaltic volcano, originally built on the deep ocean floor and rising above the ocean surface. After being eroded by streams and planed off by wave erosion, the extinct volcano slowly subsided, while the fringing coral reef continued to build upward.

Raised Shorelines and Marine Terraces

The active life of a shoreline is sometimes cut short by a sudden rise of the coast. When this tectonic event occurs, a *raised shoreline* is formed. If present, the marine cliff and abrasion platform are abruptly raised above the level of wave action. The former abrasion platform has now become a *marine terrace* (Figure 17.20). Of course, fluvial denudation acts to erode the terrace as soon as it is formed. The terrace may also undergo partial burial under alluvial fan deposits.

Raised shorelines are common along the continental and island coasts of the Pacific Ocean, because tectonic processes are active along these mountain and island arcs. Repeated uplifts result in a series of raised shorelines in a steplike arrangement. Fine examples of these multiple marine terraces are seen on the western

Figure 17.20 A raised shoreline becomes a cliff parallel with the newer, lower shoreline. The former abrasion platform is now a marine terrace. (Drawn by A. N. Strahler.)

slope of San Clemente Island, off the California coast (Figure 17.21).

Raised shorelines occur when sea level drops relative to land. What happens when sea level rises? The obvious answer is that old coastal features are drowned and new ones are formed at higher locations. A critical problem of global warming is the accompanying rise of sea level that is anticipated by the melting of glaciers and thermal expansion of sea water in response to warmer tem-

Figure 17.21 Marine terraces on the western slope of San Clemente Island, off the southern California coast. More than 20 terraces have been identified in this series. The highest has an elevation of about 400 m (about 1300 ft).

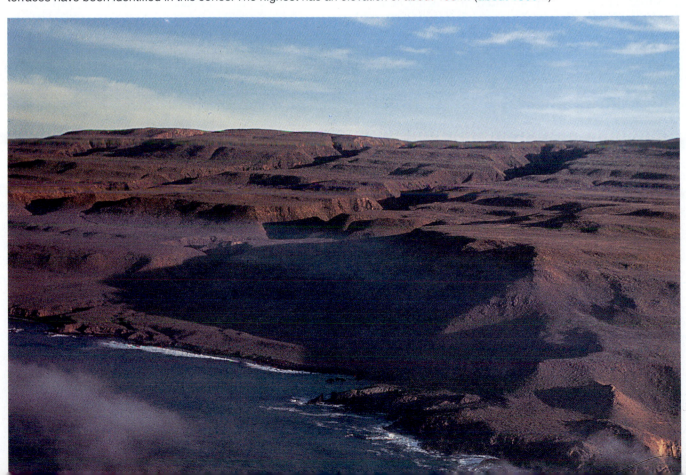

EYE ON THE ENVIRONMENT

Eye on the Environment • The Threat of Rising Sea Level from Global Warming

The earth's climate seems to be warming, probably in response to human activity. Have you ever thought about some of the possible effects of climatic warming? One of these effects is the rise of sea level. Imagine, for a moment, the effect of an increase of several meters. Low-lying areas—such as tidal estuaries and deltas—would be flooded. Coastal cities would have to be defended from wave attack. Barrier islands would be swept landward or overtopped. Atolls and other low islands would disappear. The threat of a rise in sea level is real, but like other possible impacts of global warming, it is far from certain how long it will take sea level to rise and how great the effect will be.

A review of the scientific principles involved here may be helpful. First, these changes in sea level are eustatic—that is, they refer to worldwide sea-level change. Local relative changes in sea level resulting from the uplift or downsinking of the earth's crust are tectonic effects and not included in eustatic change. Eustatic change can result from a change either in water volume of the world ocean (as by melting of glaciers) or in the water-holding capacity of the ocean basins. However, any change in the capacity of the ocean basins would be a result of plate tectonics and thus occur on a longer time scale than that involved in global warming.

Climate affects the water volume of the oceans in two ways. First, a volume increase results from water temperature increase. Global warming will raise the temperature of the uppermost layer of the oceans—at most a few hundred meters thick—and the resulting thermal expansion of that water will cause a small but significant eustatic rise.

Second, global warming may increase the rate of glacial melting, causing mountain glaciers and ice sheets to shrink in volume. Meltwater would then increase the volume of the world ocean. Keep in mind, however, that, by increasing precipitation, global warming might instead promote the growth of glaciers by increasing precipitation. Although it is not clear which process will dominate in the short term, we do know that large swings in world sea level have taken place throughout the past 3 million years as ice sheets have cyclically expanded and contracted. (In Chapter 18 we will return to a discussion of the mechanism of possible sea-level rise by ice melting.)

Has a slow eustatic rise of sea level accompanied the overall global temperature rise that started in the mid-1880s? Apparently so. Scientists examining tide gauge records have concluded that in the 80-year period 1900–1980, the total eustatic rise has been about 80 mm (about 3 in.). About half of this increase is attributed to thermal expansion. The average rate of rise was about 1.2 mm/yr (about 0.05 in./yr) through the early and middle 1900s. Recently, the rate has doubled to about 2.4 mm/yr (about 0.1 in./yr).

Projections of the rate of sea-level rise into the next century vary widely. In 1995, the Intergovernmental Panel on Climate Change, a United Nations-sponsored group of scientists, predicted a sea-level rise ranging form 15–90 cm (6–35 in.) by the year 2100. The impacts of such a rise, especially if toward the high end of the estimate, could be very costly for major coastal cities of the world. Large river deltas, including both the cities built on

peratures. The implications are discussed in *Eye on the Environment • The Threat of Rising Sea Level from Global Warming.*

WIND ACTION

Transportation and deposition of sand by wind is an important process in shaping certain coastal landforms. We have already mentioned coastal sand dunes, which are derived from beach sand. In the remainder of this chapter, we investigate the transport of mineral particles by wind and the shaping of dune forms. Our discussion also provides information about dune forms far from the coast—in desert environments, where the lack of vegetation cover allows dunes to develop if a source of abundant sand particles is present.

Wind action normally moves mineral particles only when they are in dry state and unprotected by a vegetation cover.

Wind blowing over the land surface is one of the active agents of landform development. Ordinarily, wind is not strong enough to dislodge mineral matter from the surfaces of unweathered rock, or from moist, clay-rich soils, or from soils bound by a dense plant cover. Instead, the action of wind in eroding and transporting sediment is limited to land surfaces where small mineral and organic particles are in a loose, dry state. These areas are typically deserts and semiarid lands (steppes). An exception is the coastal environment, where beaches provide abundant supplies of loose sand. In this environment, wind action shapes coastal dunes,

Coastal erosion has undermined the bluff beneath these two buildings, depositing them on the beach below.

quently with global warming, would shut down the city's subways, riverside highways, and La Guardia Airport, bringing the city's transportation system to a virtual halt. On Long Island, category 2 hurricanes would demolish beach communities and submerge Jones Beach, Fire Island, and the Hampton beaches. The south fork of the Island would be split into smaller sections. Water tables would rise, creating more frequent local flooding at locations far from the beach front. Although these impact studies are based on the high end of sea-level rise projections, they demonstrate some of the impacts that could be felt by low-lying coastal plains even with lesser rises.

It seems certain now that some degree of sea-level rise, brought on by global warming, is inevitable. The question now is, when and how much?

Questions

1. How might global climate change bring about a rise in sea level? What are the possible effects of such a rise?

2. Is a sea-level rise occurring at present? What are the projections for the future, and what are the implications of the projections?

them and their reclaimed tidal estuaries and marshes, would face catastrophic damage through storm surf and storm surges in concert with high tides. Along cliffed coasts under direct attack by waves, retrogradation would be intensified, resulting in massive losses of shorefront properties (see photo). Some island nations would lose most or all of their land. The Maldives, a chain of atolls in the Indian Ocean with a population of about a quarter of a million people, is an example.

Recent studies by the Environmental Defense Fund and the National Environmental Trust project how a rise of 1 m (3.3 ft) might affect the New York area. Flooding from severe storms, occurring more fre-

---

even where the climate is humid and the land surface inland from the coast is well protected by a plant cover.

Landforms shaped and sustained by wind erosion and deposition provide distinctive life environments. The ecosystems that occupy these habitats are often highly specialized. In semidesert climates with little soil water, the growth of plants can be limited due to fierce wind action. Wind action tends to rework the landforms, while plants tend to stabilize the landforms and protect them from wind action. The balance between these two forces can easily be upset by human activities, often with serious consequences of land degradation. Changes in climate, both natural and those induced by human activity, can also upset the balance. To understand these environmental changes, we need to acquire a working knowledge of the physical processes of wind action on the land surfaces.

Erosion by Wind

Wind performs two kinds of erosional work: abrasion and deflation. In the process of *wind abrasion*, wind drives sand and dust particles against an exposed rock or soil surface. This causes the surface to be worn away by the impact of the particles. Abrasion requires cutting tools—mineral particles that are carried by the wind.

The sandblasting action of wind abrasion against exposed rock surfaces is limited to the basal meter or two of a rock mass that rises above a flat plain. This height is the limit to which sand grains can rise high into the air. Wind abrasion produces pits, grooves, and hollows in the rock. Wooden utility poles on windswept sandy plains are quickly cut through at the base unless a protective metal sheathing or heap of large stones is placed around the base.

The removal of loose particles from the ground is termed **deflation**. Deflation acts on loose soil or sediment. Dry river courses, beaches, and areas of recently formed glacial deposits are susceptible to deflation. In dry climates, much of the ground surface is subject to deflation because the soil or rock is largely bare of vegetation.

Wind is selective in its deflational action. The finest particles, those of clay and silt sizes, are lifted and raised into the air—sometimes to a height of a thousand meters (about 3300 ft) or more. Sand grains are moved only when winds are at least moderately strong and usually travel within a meter or two (about 3 to 6 ft) of the ground. Gravel fragments and rounded pebbles can be rolled or pushed over flat ground by strong winds, but they do not travel far. They become easily lodged in hollows or between other large grains. Consequently, where a mixture of size of particles is present on the ground, the finer-sized particles are removed and the coarser particles remain behind.

A landform produced by deflation is a shallow depression called a **blowout**. The size of the depression may range from a few meters (10 to 20 ft) to a kilometer (0.6 mi) or more in diameter, although it is usually only a few meters deep. Blowouts form in plains regions of dry climate. Any small depression in the surface of the plain, especially where the grass cover has been broken or disturbed, can form a blowout. Rains fill the depression and create a shallow pond or lake. As the water evaporates, the mud bottom dries out and cracks, leaving small scales or pellets of dried mud. These particles are lifted out by the wind. In grazing lands, animals trample the margins of the depression into a mass of mud, breaking down the protective grass-root structure and facilitating removal of dried particles. In this way, the depression is enlarged. Blowouts are also found on rock surfaces where the rock is being disintegrated by weathering.

Deflation is also active in semidesert and desert regions. In the southwestern United States, playas often occupy large areas on the flat floors of tectonic basins. Deflation has reduced many playas several meters in elevation.

Rainbeat, overland flow, and deflation may be active for a long period on the gently sloping surface of a desert alluvial fan or alluvial terrace. These processes remove fine particles, leaving coarser, heavier materials behind. As a result, rock fragments ranging in size from pebbles to small boulders become concentrated into a surface layer known as a **desert pavement** (Figure 17.22). The large fragments become closely fitted together, concealing the smaller particles—grains of sand, silt, and clay—that lie beneath. The pavement acts as an armor that protects the finer particles from rapid removal by deflation. However, the pavement is easily disturbed by the wheels of trucks and motorcy-

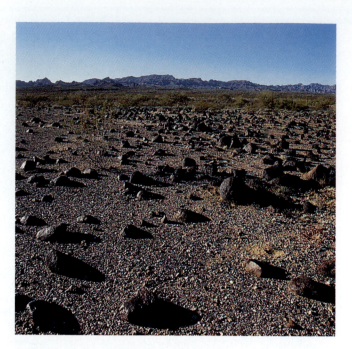

Figure 17.22 This desert pavement is formed of closely fitted rock fragments. Lying on the surface are fine examples of wind-faceted rocks, which attain their unusual shapes by long-continued sandblasting.

cles, exposing the finer particles and allowing severe deflation and water erosion to follow.

Dust Storms

Strong, turbulent winds blowing over barren surfaces lift great quantities of fine dust into the air, forming a dense, high cloud called a **dust storm**. In semiarid grasslands, a dust storm is generated where ground surfaces have been stripped of protective vegetation cover by cultivation or grazing. Strong winds cause soil particles and coarse sand grains to hop along the ground. This motion breaks down the soil particles and disturbs more soil. With each impact, fine dust is released that can be carried upward by turbulent winds.

A dust storm approaches as a dark cloud extending from the ground surface to heights of several thousand meters (Figure 17.23). Typically, the advancing cloud wall represents a rapidly moving cold front. Within the dust cloud there is deep gloom or even total darkness. Visibility is cut to a few meters, and a fine choking dust penetrates everywhere.

Estimates reveal that as much as 1000 metric tons of dust may be suspended in a cubic kilometer of air (about 4000 tons per cu mi). On this basis, a large dust storm can carry more than 100 million tons of dust—enough to make a hill 30 m (about 100 ft) high and 3 km (about 2 mi) across the base. Dust travels long distances in the air. Dust from a single desert storm can be traced as far as 4000 km (about 2500 mi).

Figure 17.23 An approaching dust storm in eastern Kenya.

SAND DUNES

A **sand dune** is any hill of loose sand shaped by the wind. Active dunes constantly change form under wind currents. Dunes form where there is a source of sand—for example, a sandstone formation that weathers easily to release individual grains, or perhaps a beach supplied with abundant sand from a nearby river mouth. Dunes must be free of a vegetation cover in order to form and move. They become inactive when stabilized by a vegetation cover, or when patterns of wind or sand sources change. Table 17.1 summarizes six important types of sand dunes.

Dune sand is most commonly composed of the mineral quartz, which is extremely hard and largely immune to chemical decay. Under the force of strong winds, the grains move in long, low leaps, bouncing after impact with other grains. Rebounding grains rarely rise more than half a centimeter above the dune surface. Grains struck by bouncing grains are pushed forward, and, in this way, the surface sand layer creeps downwind. This type of hopping, bouncing movement is termed *saltation*.

Types of Sand Dunes

One common type of sand dune is an isolated heap of free sand called a *barchan*, or *crescentic dune*. This type of dune has the outline of a crescent, and the points of the crescent are directed downwind (Figure 17.24). On the upwind side of the crest, the sand slope is gentle and smoothly rounded. On the downwind side of the dune,

Table 17.1 Types of Sand Dunes

Type	Shape	Description
Barchan dunes	Crescentric	Crescent-shaped, with points directed downwind. May occur as isolated dunes or as strings or swarms of dunes.
Transverse dunes	Wavelike ridges	Resembling storm waves on an ocean. Requires very abundant sand supply.
Star dune	Star with many points	A large hill of sand with radiating arms. Largely fixed in position.
Coastal blowout dunes	Horseshoe	Dunes moving away from a beach. May engulf and smother inland vegetation or buildings.
Parabolic dunes	Curving arc	Broad, low dune ridges without steep slip faces. May be elongated into hairpin shapes, with points directed upwind.
Longitudinal dunes	Parallel ridges	Long, narrow ridges parallel with wind direction. May cover vast areas of deserts.

Figure 17.24 Barchan dunes. The arrow indicates wind direction. (Drawn by A. N. Strahler.)

hill, rock, or clump of brush. Once a sufficient mass of sand has formed, it begins to move downwind, taking on the crescent form. For this reason, the dunes are usually arranged in chains extending downwind from the sand source.

In a barchan dune, the points of the crescent are directed downwind.

Where sand is so abundant that it completely covers the solid ground, dunes take the form of wavelike ridges separated by troughlike furrows. These dunes are called *transverse dunes* because, like ocean waves, their crests trend at right angles to the direction of the dominant wind (Figure 17.26). The entire area may be called a *sand sea*, because it resembles a storm-tossed sea suddenly frozen to immobility. The sand ridges have sharp crests and are asymmetrical, the gentle slope being on the windward and the steep slip face on the lee side. Deep depressions lie between the dune ridges. Sand seas require enormous quantities of sand, supplied by material weathered from sandstone formations or from sands in nearby alluvial plains. Transverse dune belts also form adjacent to beaches that supply abundant sand and have strong onshore winds.

Wind is a major agent of landscape development in the Sahara Desert. Enormous quantities of reddish dune sand have been derived from weathering of sand-

within the crescent, is a steep dune slope, the *slip face*. This face maintains an angle of about 35° from the horizontal (Figure 17.25). Sand grains slide down the steep face after being blown over the sharp crest. When a strong wind is blowing, the flying sand makes a visible cloud at the crest.

Barchan dunes usually rest on a flat, pebble-covered ground surface. The life of a barchan dune may begin as a sand drift in the lee of some obstacle, such as a small

Figure 17.25 This aerial view shows a large barchan dune moving from right to left. At its apex is a smaller barchan dune that is overtaking it. Salton Sea region, California.

Figure 17.26   Transverse dunes of a sand sea, near Yuma, Arizona. Isolated crescent dunes can be seen at the right. Prevailing winds are from the left.

stone formations. The sand is formed into a great sand sea, called an *erg*. Elsewhere, there are vast flat-surfaced sheets of sand that are armored by a layer of pebbles forming a desert pavement. A surface of this kind in the Sahara is called a *reg*.

Some of the Saharan dunes are elaborate in shape. For example, the *star dune* (heaped dune) is a large hill of sand whose base resembles a many-pointed star in plan (Figure 17.27). Radial ridges of sand rise toward the dune center and culminate in a sharp peak as high as 100 m (about 300 ft) or more above the base. The Arabian star dunes remain fixed in position and have served for centuries as reliable landmarks for desert travelers. The star dune also occurs in the deserts of the border region between the United States and Mexico.

Another group of dunes belongs to a family in which the curve of the dune crest is bowed outward in the downwind direction. (This curvature is the opposite of the barchan dune.) These are termed *parabolic dunes*. A common type of parabolic dune is the *coastal blowout dune*, formed adjacent to beaches. Here, large supplies of sand are available, and the sand is blown landward by prevailing winds (Figure 17.28*a*). A saucer-shaped depression is formed by deflation, and the sand is heaped in a curving ridge resembling a horseshoe in plan. On the landward side is a steep slip face that advances over the lower ground and buries forests,

Figure 17.27   Star dunes of the Libyan Desert, seen from an altitude of 10 km (6 mi). Dune peaks rise to heights of 100 to 200 m (about 300 to 600 ft) or more above the intervening level ground.

Figure 17.28 Three types of parabolic dunes. The prevailing wind direction is the same for all three types. (*a*) Coastal blowout dunes. (*b*) Parabolic dunes on a semiarid plain. (*c*) Parabolic dunes drawn out into hairpin forms. (Drawn by A. N. Strahler.)

(*a*)

(*b*)

(*c*)

Figure 17.29 This coastal blowout dune is advancing over a coniferous forest, with the slip face gradually burying the tree trunks. Pacific coast, near Florence, Oregon.

killing the trees (Figure 17.29). Coastal blowout dunes are well displayed along the southern and eastern shore of Lake Michigan. Dunes of the southern shore have been protected for public use as the Indiana Dunes State Park.

On semiarid plains, where vegetation is sparse and winds are strong, groups of parabolic blowout dunes develop to the lee of shallow deflation hollows (Figure 17.28*b*). Sand is caught by low bushes and accumulates on a broad, low ridge. These dunes have no steep slip faces and may remain relatively immobile. In some cases, the dune ridge migrates downwind, drawing the dune into a long, narrow form with parallel sides resembling a hairpin in outline (Figure 17.28*c*).

Another class of dunes, described as *longitudinal dunes*, consists of long, narrow ridges oriented parallel with the direction of the prevailing wind (Figure 17.30).

Figure 17.30 Longitudinal dunes run parallel to the direction of the wind. (Drawn by A. N. Strahler.)

Figure 17.31 Longitudinal sand dunes in the southern Arabian Peninsula trend northeast-southwest, parallel with the prevailing northeast winds. In the lower right-hand part of the frame is a dissected plateau of the flat-lying sedimentary strata.

These dune ridges may be many kilometers long and cover vast areas of tropical and subtropical deserts in Africa and Australia (Figure 17.31).

Coastal Foredunes

Landward of sand beaches, we usually find a narrow belt of dunes in the form of irregularly shaped hills and depressions. These are the *foredunes*. They normally bear a cover of beachgrass and a few other species of plants capable of survival in the severe environment (Figure 17.32). On coastal foredunes, the sparse cover of beachgrass and other small plants acts as a baffle to trap sand moving landward from the adjacent beach. As a result, the foredune ridge is built upward to become a barrier standing several meters above high-tide level.

Foredunes form a protective barrier for tidal lands that often lie on the landward side of a beach ridge or barrier island. In a severe storm, the swash of storm waves cuts away the upper part of the beach. Although the foredune barrier may then be attacked by wave action and partly cut away, it will not usually yield. Between storms, the beach is rebuilt, and, in due time, wind action restores the dune ridge, if plants are maintained.

Foredunes form a protective barrier against storm wave action that keeps waves from overwashing a beach ridge or barrier island.

If the plant cover of the dune ridge is reduced by vehicular and foot traffic, a blowout will rapidly develop. The new cavity can extend as a trench across the dune ridge. With the onset of a storm that brings high water levels and intense wave action, swash is funneled through the gap and spreads out on the tidal marsh or tidal lagoon behind the ridge. Sand swept through the gap is spread over the tidal deposits. If eroded, the gap can become a new tidal inlet for ocean water to reach the bay beyond the beach. For many coastal communities of the eastern United States seaboard, the breaching of a dune ridge with its accompanying overwash may bring unwanted change to the tidal marsh or estuary.

Where land, rather than tidal bays, lies behind the beach, blowout dunes can advance on forests, roads, buildings, and agricultural lands. In the Landes region of the southwestern coast of France, advancing blowout dunes have overwhelmed houses and churches and have even caused entire towns to be abandoned.

Figure 17.32 Beachgrass thriving on coastal foredunes has trapped drifting sand to produce a dune ridge. Queen's County, Prince Edward Island, Canada.

LOESS

In several large midlatitude areas of the world, the surface is covered by deposits of wind-transported silt, which has settled out from dust storms over many thousands of years. This material is known as **loess**. (The pronunciation of this German word is somewhere between "lerse" and "luss.") It generally has a uniform yellowish to buff color and lacks any visible layering (Figure 17.33). Loess tends to break away along vertical cliffs wherever it is exposed by the cutting of a stream or grading of a roadway. It is also very easily eroded by running water and is subject to rapid gullying when the vegetation cover that protects it is broken.

The thickest deposits of loess are in northern China, where a layer over 30 m (about 100 ft) thick is common and a maximum thickness of 100 m (about 300 ft) has been measured. It covers many hundreds of square kilometers and appears to have been brought as dust from the interior of Asia. Because it is easily excavated, loess has been widely used for cave dwellings both in China and in Central Europe. Extensive loess deposits are also found in the United States, Central Europe, Central Asia, and Argentina, where they often form productive, but erosion-prone, agricultural soils.

In the United States, thick loess deposits lie in the Missouri–Mississippi Valley (Figure 17.34). Large areas of the prairie plains region of Indiana, Illinois, Iowa,

Figure 17.33 This thick layer of loess in New Zealand was deposited during the Ice Age. Loess has excellent cohesion and often forms vertical faces as it wastes away.

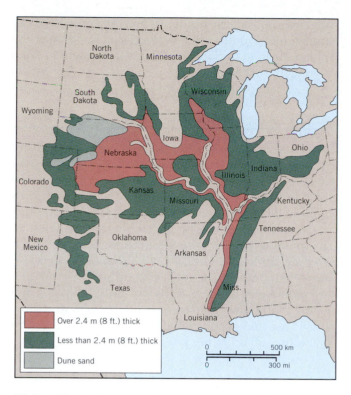

Figure 17.34 Map of loess distribution in the central United States. (Data from *Map of Pleistocene Eolian Deposits of the United States*, Geological Society of America.)

Missouri, Nebraska, and Kansas are underlain by loess ranging in thickness from 1 to 30 m (about 3 to 100 ft). Extensive deposits also occur in Tennessee and Mississippi in areas bordering the lower Mississippi River floodplain. Still other loess deposits are in the

Palouse region of northeast Washington and western Idaho.

The American and European loess deposits are directly related to the continental glaciers of the Pleistocene Epoch. At the time when the ice covered much of North America and Europe, a generally dry winter climate prevailed in the land bordering the ice sheets. Strong winds blew southward and eastward over the bare ground, picking up silt from the floodplains of braided streams that discharged the meltwater from the ice. This dust settled on the ground between streams, gradually building up a smooth, level surface. The loess is particularly thick along the eastern sides of the valleys because of prevailing westerly winds. It is well exposed along the bluffs of most streams flowing through these regions today.

Loess is a deposit of wind-blown silt that may be as thick as 30 m (about 100 ft) in some regions of North America.

The thick loess deposit covering a large area of north-central China in the province of Shanxi and adjacent provinces poses a difficult problem of severe soil erosion. Although the loess is capable of standing in vertical walls, it also succumbs to deep gullying during the period of torrential summer rains. From the steep walls of these great scars, fine sediment is swept into streams and carried into tributaries of the Huang He (Yellow River). The Chinese government has implemented an intensive program of slope stabilization by using artificial contour terraces (seen in Figure 17.35) in combina-

Figure 17.35 The loess plateau of northern China. A succession of contour terraces, cut into thick loess and covered by tree plantings, is designed to prevent deep gullying and loss of soil. Arched entrances to cave dwellings can be seen at lower left and lower right. Grain fields on the flat valley floor occupy surfaces of thick sediment trapped behind dams. Shanxi Province, near Xian, People's Republic of China.

tion with tree planting. Valley bottoms have been dammed so as to trap the silt to form flat patches of land suitable for cultivation.

Induced Deflation

Induced deflation is a frequent occurrence when short-grass prairie in a semiarid region is cultivated without irrigation. Plowing disturbs the natural soil surface and grass cover, and in drought years, when vegetation dies out, the unprotected soil is easily eroded by wind action. Much of the Great Plains region of the United States has suffered such deflation and experienced dust storms generated by turbulent winds. Strong cold fronts frequently sweep over this area and lift dust high into the troposphere at times when soil moisture is low. The "Dust Bowl" of the 1930s is an example (see Chapter 9).

Human activities in very dry, hot deserts contribute measurably to the raising of high dust clouds. In the desert of northwest India and Pakistan (the Thar Desert bordering the Indus River), the continued trampling of fine-textured soils by hooves of grazing animals and by human feet produces a dust cloud that hangs over the region for long periods. It extends to a height of 9 km (about 30,000 ft).

A LOOK AHEAD

One final agent remains to be examined on the list of active agents of erosion, transportation, and deposition—glacial ice. Compared to wind and water, glacial ice moves much more slowly but is far steadier in its motion. Like a vast conveyor belt, glacial ice moves sediment forward relentlessly, depositing the sediment at the ice margin, where the ice melts. By plowing its way over the landscape, glacial ice also shapes the local terrain—bulldozing loose rock from hillsides and plastering sediments underneath its vast bulk. This slow but steady action is very different from that of water, wind, and waves, and produces a distinctive set of landforms that is the subject of our last text chapter.

LANDFORMS MADE BY WAVES AND WIND IN REVIEW

This chapter has described the landforms of waves and wind. Waves act at the shoreline—the boundary between water and land. Waves expend their energy as breakers, which erode hard rock into marine cliffs and create marine scarps in softer materials.

Beaches, usually formed of sand, are shaped by the swash and backwash of waves, which continually work and rework beach sediment. Wave action produces littoral drift, which moves sediment parallel to the beach. This sediment accumulates in bars and sandspits, which further extend the beach. Depending on the nature of longshore currents and the availability of sediment, shorelines can experience progradation or retrogradation.

Tidal forces cause sea level to rise and fall rhythmically, and this change of level produces tidal currents in bays and estuaries. Tidal flows redistribute fine sediments within bays and estuaries, which can accumulate with the help of vegetation to form salt marshes. These are sometimes reclaimed to form new agricultural land.

Coastlines of submergence result when coastal lands sink below sea level or sea level rises rapidly. Scenic ria and fiord coasts are examples. Coastlines of emergence include barrier-island coasts and delta coasts. Coral-reef coasts occur in regions of warm tropical and equatorial waters. Along some coasts, rapid uplift has occurred, creating raised shorelines and marine terraces.

Wind is a landform-creating agent that acts by moving sediment. Deflation occurs when wind removes mineral particles—especially clay and silt, which can be carried long distances. Deflation creates blowouts in semidesert regions and lowers playa surfaces in deserts. In arid regions, deflation produces dust storms.

Sand dunes form when a source, such as a sandstone outcrop or a beach, provides abundant sand that can be moved by wind action. Barchan dunes are arranged individually or in chains leading away from the sand source. Transverse dunes form a sand sea of frozen "wave" forms arranged perpendicular to the wind direction. Parabolic dunes are arc-shaped—coastal blowout dunes are an example. Longitudinal dunes parallel the wind direction and cover vast desert areas. Coastal foredunes are stabilized by dune grass and help protect the coat against storm wave action.

Loess is a surface deposit of fine, wind-transported silt. It can be quite thick, and it typically forms vertical banks. Loess is very easily eroded by water and wind. In eastern Asia, the silt forming the loess was transported by winds from extensive interior deserts located to the north and west. In Europe and North America, the silt was derived from fresh glacial deposits during the Pleistocene Epoch. Human activities can hasten the action of deflation by breaking protective surface covers of vegetation and desert pavement.

KEY TERMS

shoreline	progradation	deflation
coastline	retrogradation	blowout
bay	submergence	desert pavement
estuary	emergence	dust storm
marine cliff	delta	sand dune
beach	coral reef	loess
littoral drift	atoll	

REVIEW QUESTIONS

1. What is the energy source for wind and wave action?
2. What landforms can be found in areas where bedrock meets the sea?
3. What is littoral drift, and how is it produced by wave action?
4. Identify progradation and retrogradation. How can human activity influence retrogradation?
5. How are salt marshes formed? How can they be reclaimed for agricultural use?
6. What key features identify a coastline of submergence? Identify and compare the two types of coastlines of submergence.
7. Under what conditions do barrier-island coasts form? What are the typical features of this type of coastline?
8. What conditions are necessary for the development of coral reefs? Identify three types of coral-reef coastlines.
9. How are marine terraces formed?
10. What is deflation, and what landforms does it produce? What role does the dust storm play in deflation?
11. How do sand dunes form? Describe and compare barchan dunes, transverse dunes, star dunes, coastal blowout dunes, parabolic dunes, and longitudinal dunes.
12. What is the role of coastal dunes in beach preservation? How are coastal dunes influenced by human activity? What problems can result?
13. Define the term *loess*. What is the source of loess and how are loess deposits formed?

VISUALIZING EXERCISES

1. Describe the features of delta coasts and their formation. Sketch and compare the shapes of the Mississippi and Nile deltas.
2. Take a piece of paper and let it represent a map with winds coming from the north, at the top of the page. Then sketch the shapes of the following types of dunes: barchan, transverse, parabolic, and longitudinal.

A DEEPER VIEW OF LANDFORMS MADE BY WAVES AND WIND

1. Consult an atlas to identify a good example of each of the following types of coastlines: ria coast, fiord coast, barrier-island coast, delta coast, coral-reef coast, and fault coast. For each example, provide a brief description of the key features you used to identify the coastline type.
2. Wind action moves sand close to the ground in a bouncing motion, whereas silt and clay are lifted and carried longer distances. Compare landforms and deposits that result from wind transportation of sand with those that result from wind transportation of silt and finer particles.

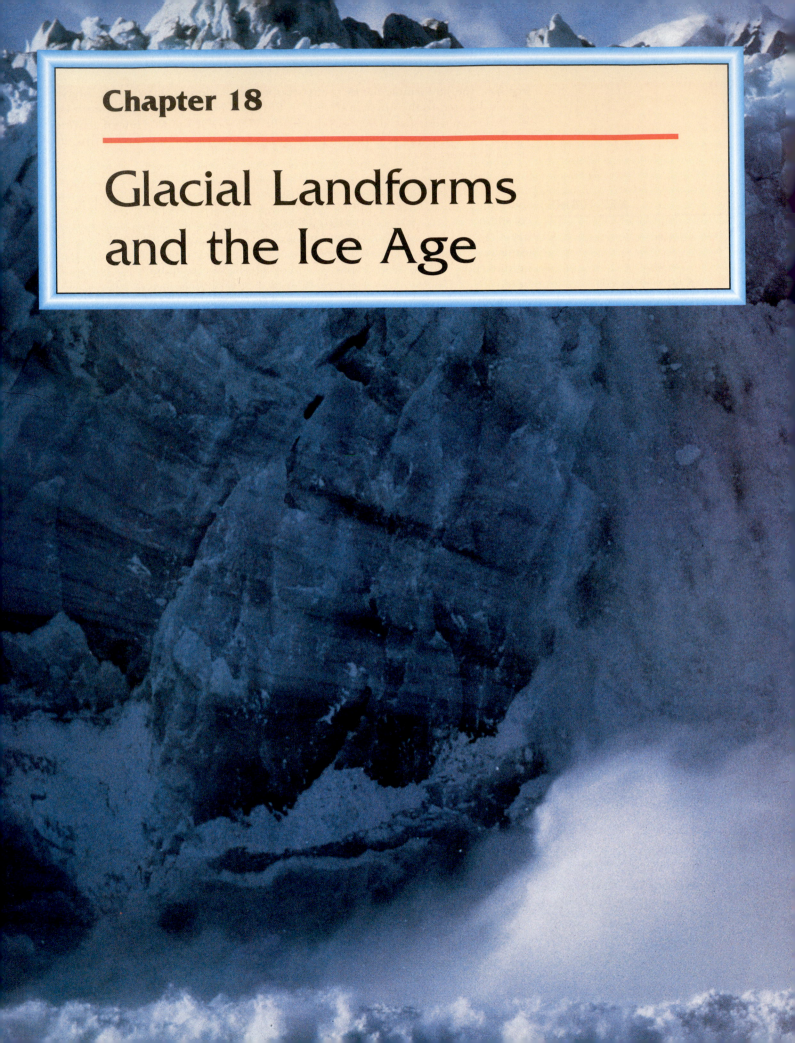

Chapter 18

Glacial Landforms and the Ice Age

A huge block of ice falls from the front of a tidewater glacier, hurling tons of seawater high in the air. Wrangell–St. Elias National Park, Alaska

Glacial ice, flowing slowly and majestically under the influence of gravity, is a powerful geomorphic agent that creates many types of distinctive landforms. Perhaps the best way to appreciate the power of glaciers is to visit one. Imagine yourself on an ocean cruise along the inland passage from Vancouver to Juneau with the ultimate destination of an Alaskan tidewater glacier. Here's what you might experience:

The huge cruise ship moves silently and effortlessly between the two steep, forest-covered slopes. The green walls of spruce and hemlock glide by slowly, the spires of the forest giants fading into the low ceiling of gray clouds overhead. Tiny droplets of drizzle cool your face as you lean against the rail, observing the somber scene. The trees seem almost close enough to reach out and touch. Below you, the ship's bow parts the gray-green water with little resistance, providing only a low rushing noise that barely breaks the stillness. Ahead, the sea and forest merge, fade, and then blend into the clouds and fog.

So this is Alaska! The travel brochure showed your ship cruising in a crystal-blue channel past magnificent vistas of snow-capped peaks, but so far the trip has provided only low clouds, fog, and conifer-clad slopes. The weather may soon change, however. The forecast calls for slow clearing later in the day, about the time that your ship will reach Glacier

Bay. This steep-walled fiord is a drowned valley that was once filled with a huge tongue of glacial ice. Now, however, it is filled only with sea water and small drifting ice fragments released by the many streams of glacial ice that enter the bay along its 100 km (60 mi) length.

Soon you notice a change in the ship's rhythm. A graceful curving turn, and you leave Icy Strait to enter the mouth of Glacier Bay. Miraculously, the overcast skies have yielded to a blue and white patchwork of sky and cloud, and the sun has appeared. Your ship moves slowly down the narrow passage leading to the Grand Pacific Glacier. To either side are low, rocky peaks, sculpted by the ice into stark angular shapes composed of points and edges. Ahead are the high peaks of the St. Elias Mountains, entirely capped with white.

As the glacier comes into view, the ship slows further and approaches the steep ice face carefully. It is fractured into tall, vertical columns with steplike horizontal facets at the top. The color of the glacial ice is what strikes you most. Beneath a top white layer of decaying ice

and snow, the ice is a bright greenish-blue. The color is so intense that the ice seems to glow from within. Above the clear aqua of the ice are the grays and browns of the rocks, the white of the snow-capped peaks, and the blue and white patchwork of sky. Below is the gray-green of the water, dotted with white ice fragments of many shapes and sizes.

The ice is also noisy. It groans, screeches, clicks, and pops. A loud crack startles you, then reverberates against the walls of the fiord. A monumental column of ice breaks free and falls, as if in slow motion. A great splash wells upward as it breaks the surface of the water. Next to the huge ice front, the bulk of the cruise ship seems very small. Suddenly, you connect the force of the moving ice with the shaping of the vast fiords that have contained your small ship on its passage. The power that carved these huge, water-filled canyons is awesome indeed.

If you live in a region that was covered by ice during the last Ice Age, the evidence of glaciation is all around. From ponds formed by the melting of buried ice blocks to the great hills of sediments dumped by glacial ice sheets at their margins, the action of glaciers has drastically modified much of the landscape of northern regions. In this chapter, you will learn about how glaciers create unique landforms like these.

Chapter Outline

Glaciers
Alpine Glaciers
  Landforms Made by Alpine Glaciers
  Glacial Troughs and Fiords
Ice Sheets of the Present
Sea Ice and Icebergs
The Ice Age
  Glaciation During the Ice Age
Landforms Made by Ice Sheets
  Erosion by Ice Sheets

  Deposits Left by Ice Sheets
  Environmental Aspects of Glacial Deposits
Investigating the Ice Age
  Possible Causes of the Late-Cenozoic Ice Age
  Possible Causes of Glaciation Cycles
Holocene Environments
Eye on the Environment • Ice Sheets and Global Warming

Chapter 18

GLACIAL LANDFORMS AND THE ICE AGE

In this chapter we turn to the last of the active agents that create landforms—glacial ice. Not long ago, during the Ice Age, much of northern North America and Eurasia was covered by massive sheets of glacial ice. As a result, glacial ice has played a dominant role in shaping landforms of large areas in midlatitude and subarctic zones. Glacial ice still exists today in two great accumulations of continental dimensions—the Greenland and Antarctic Ice Sheets—and in many smaller masses in high mountains.

The glacial ice sheets of Greenland and Antarctica strongly influence the radiation and heat balance of the globe. Because of their whiteness, they reflect much of the solar radiation they receive. Their intensely cold surface air temperatures contrast with temperatures at more equatorial latitudes. This temperature difference helps drive the system of meridional heat transport that we described in Chapters 2 and 4. In addition, these enormous ice accumulations represent water in storage in the solid state and are a major component of the global water balance. When the volume of glacial ice increases, as during an ice age, sea levels must fall. When ice sheets melt away, sea level rises. Today's coastal environments evolved during the rising sea level that followed the melting of the last ice sheets of the Ice Age.

GLACIERS

Most of us know ice only as a brittle, crystalline solid because we are accustomed to seeing it in small quantities. Where a great thickness of ice exists, the pressure on the ice at the bottom makes the ice lose its rigidity. That is, the ice becomes plastic. This allows the ice mass to flow in response to gravity, slowly spreading out over a larger area or moving downhill. On steep mountain slopes, the ice can also move by sliding. Movement is the key characteristic of a *glacier*, defined as any large natural accumulation of land ice affected by present or past motion.

Glacial ice accumulates when the average snowfall of the winter exceeds the amount of snow that is lost in summer by ablation. The term **ablation** means the loss of snow and ice by evaporation and melting. When winter snowfall exceeds summer ablation, a layer of snow is added each year to what has already accumulated. As the snow compacts by surface melting and refreezing, it turns into a granular ice and is then compressed by overlying layers into hard crystalline ice. When the ice mass is so thick that the lower layers become plastic, outward or downhill flow starts, and the ice mass is now an active glacier.

Glacial ice forms where temperatures are low and snowfall is high. These conditions can occur both at high elevations and at high latitudes. In mountains, glacial ice can form even in tropical and equatorial zones if the elevation is high enough to keep average annual temperatures below freezing. Orographic precipitation encourages the growth of glacial ice. In high mountains, glaciers flow from small high-elevation collecting grounds down to lower elevations, where temperatures are warmer. Here the ice disappears by ablation. Typically, mountain glaciers are long and narrow because they occupy former stream valleys. These **alpine glaciers** are a distinctive type.

Alpine glaciers and ice sheets are two forms of moving glacial ice.

In arctic and polar regions, prevailing temperatures are low enough that snow can accumulate over broad areas, eventually forming a vast layer of glacial ice. Accumulation starts on uplands that intercept heavy snowfall. The uplands become buried under enormous volumes of ice, which can reach a thickness of several thousand meters. The ice then spreads outward, over surrounding lowlands, and covers all landforms it encounters. This extensive type of ice mass is called an **ice sheet**. As already noted, ice sheets exist today in Greenland and Antarctica.

Glacial ice normally contains abundant rock fragments ranging from huge angular boulders to pulverized rock flour. Some of this material is eroded from the rock floor on which the ice moves. In alpine glaciers, rock debris is also derived from material that slides or falls from valley walls onto the ice.

Glaciers can erode and deposit great quantities of sediment. *Glacial abrasion* is a glacial erosion process caused by rock fragments that are held within the ice and scrape and grind against bedrock (Figure 18.1). Erosion also occurs by *plucking*, as moving ice lifts out blocks of bedrock that have been loosened by the freezing and expansion of water in joint fractures. Abrasion and plucking act to smooth the bed of a glacier as glacial flow continues through time. Rock debris brought into a glacier is eventually deposited at the lower end of a glacier, where the ice melts. Both erosion and deposition result in distinctive glacial landforms.

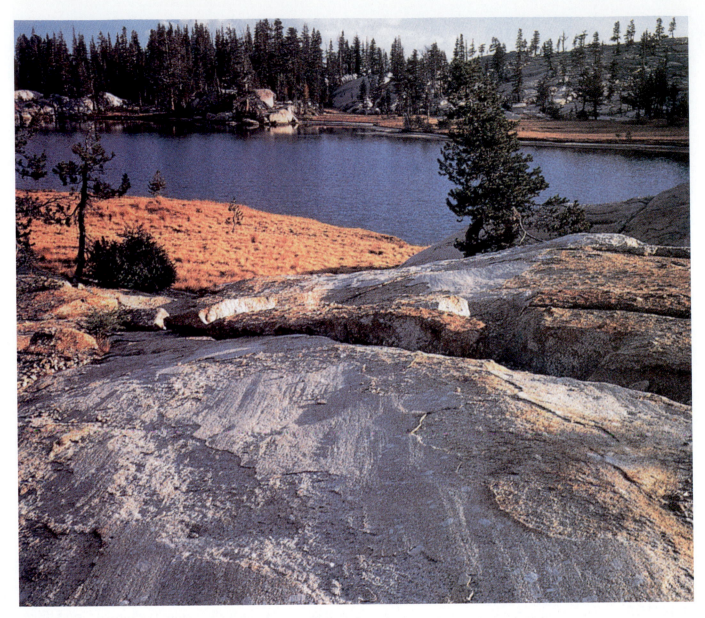

Figure 18.1 This grooved and polished surface, now partly eroded, marks the former path of glacial ice. Cathedral Lakes, Yosemite National Park, California.

ALPINE GLACIERS

Figure 18.2 illustrates a number of features of alpine glaciers. The illustration shows a simple glacier occupying a sloping valley between steep rock walls. Snow collects at the upper end in a bowl-shaped bedrock depression, the **cirque**. The upper end lies in a zone of accumulation. Layers of snow in the process of compaction and recrystallization are called *firn*.

The smooth firn field is slightly bowl-shaped in profile. Flowage in the glacial ice beneath the firn carries the ice downvalley and out of the cirque. The rate of ice flow is accelerated at a steep rock step, where deep *crevasses* (gaping fractures) mark an ice fall. The lower part of the glacier lies in the zone of ablation. In this area, the rate of ice wastage is rapid, and old ice is exposed at the glacier surface. This surface may be quite rough, with deep crevasses. At its lower end, or terminus, the glacier carries abundant rock debris. As the downward-flowing ice melts, the debris accumulates.

Although the uppermost layer of a glacier is brittle and fractures readily into crevasses, the ice beneath behaves as a plastic substance and moves by slow flowage (Figure 18.3). Like stream flow, glacier flow is most rapid far from the glacier's bed—near the midline and toward the top of the glacier's surface. Alpine glaciers also move by basal sliding. In this process, the ice slides downhill, lubricated by meltwater and mud at its base.

Figure 18.2 Cross section of an alpine glacier. Ice accumulates in the glacial cirque, then flows downhill, abrading and plucking the bedrock. Glacial debris accumulates at the glacier terminus. (After A. N. Strahler.)

A glacier establishes a dynamic balance in which the rate of accumulation at the upper end balances the rate of ablation at the lower end. This balance is easily upset by changes in the average annual rates of accumulation or ablation, causing the glacier's terminus to move forward or melt back.

Glacial flow is usually very slow. It amounts to a few centimeters per day for large ice sheets and the more sluggish alpine glaciers, but as fast as several meters per day for an active alpine glacier. However, some alpine glaciers experience episodes of very rapid movement, termed *surges*. A surging glacier may travel downvalley at speeds of more than 60 m (about 200 ft) per day for several months. The reasons for surging are not well understood, but probably involve mechanisms that increase the amount of meltwater beneath the ice, enhancing basal sliding. Most glaciers do not experience surging.

Landforms Made by Alpine Glaciers

Landforms made by alpine glaciers are shown in a series of diagrams in Figure 18.4. Mountains are eroded and shaped by glaciers, and after the glaciers melt, the remaining landforms are exposed to view. Diagram *a* shows a region sculptured entirely by weathering, mass wasting, and streams. The mountains have a smooth, rounded appearance. Soil and regolith are thick.

Imagine now that a climatic change results in the accumulation of snow in the heads of the higher valleys. An early stage of glaciation is shown at the right side of diagram *b*, where snow is collecting and cirques are being carved by the grinding motion of the ice. Deepening of the cirques is aided by intensive frost shat-

Figure 18.3 Motion of glacial ice. Ice moves most rapidly on the glacier's surface at its midline. Movement is slowest near the bed, where the ice contacts bedrock or sediment.

(a)

(b)

(c)

Figure 18.4 Landforms produced by alpine glaciers. (*a*) Before glaciation sets in, the region has smoothly rounded divides and narrow, V-shaped stream valleys. (*b*) After glaciation has been in progress for thousands of years, new erosional forms are developed. (*c*) With the disappearance of the ice, a system of glacial troughs is exposed. (Drawn by A. N. Strahler.

tering of the bedrock near the masses of compacted snow. At a later stage (center), glaciers have filled the valleys and are integrated into a system of tributaries that feed a trunk glacier. Tributary glaciers join the main glacier smoothly. The cirques grow steadily larger. Their rough, steep walls soon replace the smooth, rounded slopes of the original mountain mass. Where two cirque walls intersect from opposite sides, a jagged, knifelike ridge, called an *arête*, is formed. Where three or more cirques grow together, a sharp-pointed peak, or *horn*, is formed. Where opposed cirques have intersected deeply, a pass or notch, called a *col*, is formed. On the left side of the diagram, a portion of the landscape is drawn as unglaciated. This portion retains its rounded forms and V-shaped valleys.

A ridge or pile of rock debris left by glacial action that marks the edge of a glacier is termed a **moraine**. A *lateral moraine* is a debris ridge formed along the edge of the ice adjacent to the trough wall (Figure 18.2). Where two ice streams join, this marginal debris is dragged along to form a narrow band riding on the ice in midstream (Figures 18.4*b*, 18.5), called a *medial moraine*. At the terminus of a glacier, rock debris accumulates in a *terminal moraine*, an embankment curving across the valley floor and bending upvalley along each wall of the trough (Figure 18.6).

Glacial Troughs and Fiords

Glacier flow constantly deepens and widens its rock channel, so that after the ice has finally melted, a deep, steep-walled **glacial trough** remains (Figure 18.4*c*). The

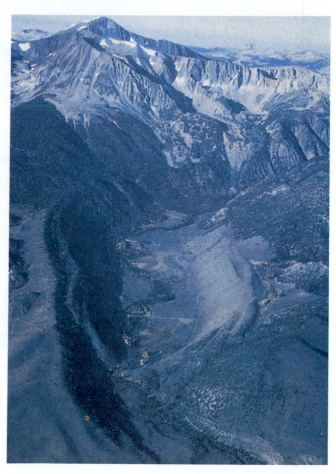

Figure 18.6 A terminal moraine, shaped like the bow of a great canoe, lies at the mouth of a deep glacial trough on the east face of the Sierra Nevada. Cirques can be seen in the distance. Near Lee Vining, California.

Figure 18.5 Aerial view of the Grand Plateau Glacier, St. Elias Mountains, Glacier Bay National Park. The dark stripes are medial moraines.

Figure 18.7 This U-shaped glacial trough in the Beartooth Plateau, near Red Lodge, Montana, was carved by an alpine glacier during the Ice Age. Talus cones have been built out from the steep trough walls.

trough typically has a U-shape in cross-profile (Figure 18.7). Tributary glaciers also carve U-shaped troughs, but they are smaller in cross section and less deeply eroded by their smaller glaciers. Because the floors of these troughs lie high above the level of the main trough, they are called *hanging valleys* (Figure 18.4c). After glaciation, a stream will often occupy the abandoned valley, providing a scenic waterfall that cascades over the lip of the hanging valley to the main trough below. High up in the smaller troughs, the bedrock is unevenly excavated, so that the floors of troughs and cirques contain rock basins and rock steps. The rock basins are occupied by small lakes, called *tarns* (Figure 18.5c). Major troughs sometimes hold large, elongated trough lakes.

Cirques are rounded rock basins that contain the heads of alpine glaciers. They form high on glaciated peaks.

When the floor of a trough open to the sea lies below sea level, the sea water enters as the ice front recedes. The result is a deep, narrow estuary known as a **fiord** (Figure 18.8). Fiords are opening up today along the Alaskan coast, where some glaciers are melting back rapidly and ocean waters are filling their troughs. Fiords are found largely along mountainous west coasts between lat. 50° and 70° N and S. On these coasts, glaciers were nourished by heavy orographic snowfall, associated with the marine west-coast climate ⑧.

Figure 18.8 Geirangerfjord, Norway, is a deeply carved glacial trough occupied by an arm of the sea.

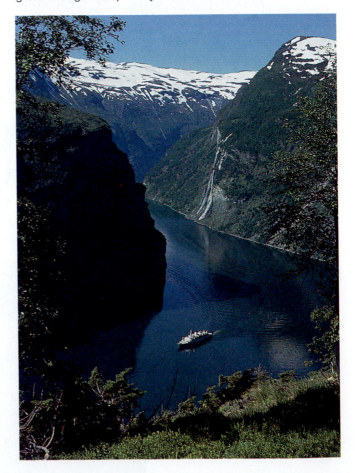

ICE SHEETS OF THE PRESENT

In contrast to alpine glaciers are the enormous ice sheets of Antarctica and Greenland. These are huge plates of ice, thousands of meters thick in the central areas, resting on land masses of subcontinental size. The Greenland Ice Sheet has an area of 1.7 million km² (about 670,000 mi²) and occupies about seven-eighths of the entire island of Greenland (Figure 18.9). Only a narrow, mountainous coastal strip of land is exposed.

Figure 18.9 The Greenland Ice Sheet. (Based on data of R. F. Flint, *Glacial and Pleistocene Geology*, John Wiley & Sons, New York.)

Figure 18.10 The Antarctic Ice Sheet and its ice shelves. (Based on data of American Geophysical Union.

The Antarctic Ice Sheet covers about 13 million km² (about 5 million mi²) (Figure 18.10). Both ice sheets are developed on large, elevated land masses in high latitudes. No ice sheet exists near the north pole, which is positioned in the vast Arctic Ocean. Ice there occurs only as floating sea ice.

The surface of the Greenland Ice Sheet has the form of a very broad, smooth dome. Underneath the ice sheet, the rock floor lies near or slightly below sea level under the central region but is higher near the edges. The Antarctic Ice Sheet is thicker than the Greenland Ice Sheet—as much as 4000 m (about 13,000 ft) at maximum. At some locations, ice sheets extend long tongues, called outlet glaciers, to reach the sea at the heads of fiords (Figure 18.11). An important glacial feature of Antarctica is the presence of great plates of floating glacial ice, called *ice shelves* (Figure 18.10). Ice shelves are fed by the ice sheet, but they also accumulate new ice through the compaction of snow.

Figure 18.11   This outlet glacier on the coast of Antarctica moves ice rapidly away from the interior to the ocean. Ambers Island, Antarctic Peninsula.

SEA ICE AND ICEBERGS

Free-floating ice on the sea surface is of two types—sea ice and icebergs. *Sea ice* is formed by direct freezing of ocean water. The surface zone of sea ice is composed of fresh water, while the deeper ice is salty. Sea ice does not exceed 5 m (15 ft) in thickness. *Pack ice* is sea ice that completely covers the sea surface (Figure 18.12). Under the forces of wind and currents, pack ice breaks up into individual patches called ice floes. The narrow strips of open water between such floes are known as *leads*. Where ice floes are forcibly brought together by winds, the ice margins buckle and turn upward into pressure ridges that often resemble walls of ice. Travel on foot across the polar sea ice is extremely difficult because of such obstacles.

Icebergs are bodies of land ice that have broken free from glaciers terminating in the ocean (Figure 18.13). An iceberg may be as thick as several hundred meters. Because it is only slightly less dense than sea water, the iceberg floats very low in the water, with about five-sixths of its bulk submerged. The ice is composed of fresh water since it is formed from compacted and recrystallized snow.

In the northern hemisphere, icebergs are derived largely from outlet glaciers of the Greenland Ice Sheet. They drift slowly south with the Labrador and Greenland currents and may find their way into the North Atlantic in the vicinity of the Grand Banks of Newfoundland. They are irregular in shape and present rather peaked outlines above water. In contrast, antarctic icebergs are commonly tabular in form, with flat tops and steep clifflike sides (Figure 18.13). This shape is not surprising, since these tabular bergs are parts of ice shelves. A large tabular berg of the Antarctic may be tens of kilometers broad and over 600 m (about 2000 ft) thick, with an ice wall rising 100 m (about 300 ft) above sea level.

THE ICE AGE

The period during which continental ice sheets grow and spread outward over vast areas is known as a **glaciation**. Glaciation is associated with a general cooling of average air temperatures over the regions where the ice sheets originate. At the same time, ample snowfall must persist over the growth areas to allow the ice masses to build in volume.

When the climate warms or snowfall decreases, ice sheets become thinner and cover less area. Eventually, the ice sheets may melt completely. This period is called a *deglaciation*. Following a deglaciation, but preceding the next glaciation, is a period in which a mild climate prevails—an **interglaciation**. The last complete interglaciation began about 140,000 years ago and ended between 120,000 and 110,000 years ago. A succession of alternating glaciations and interglaciations, spanning a period of 1 to 10 million years or more, constitutes an *ice age*.

An ice age includes cycles of glaciation, deglaciation, and interglaciation.

Throughout the past 3 million years or so, the earth has been experiencing the **Late-Cenozoic Ice Age** (or, simply, the **Ice Age**). As you may recall from Chapter 11, the Cenozoic Era has seven epochs (see Table 10.4). The Ice Age falls within the last three epochs: Pliocene, Pleistocene, and Holocene. These three epochs comprise only a small fraction—about one-twelfth—of the total duration of the Cenozoic Era.

During the first half of this century, most geologists associated the Ice Age with the Pleistocene Epoch, which began about 1.6 million years ago. However, new evidence obtained from deep-sea sediments shows that the glaciations of the Ice Age actually began in late Pliocene time, perhaps 2.5 to 3.0 million years ago.

At present, we are within an interglaciation of the Late-Cenozoic Ice Age, following a deglaciation that set in quite rapidly about 15,000 years ago. In the preceding glaciation, called the *Wisconsinan Glaciation*, ice sheets covered much of North America and Europe, as well as parts of northern Asia and southern South America. The maximum ice advance of the Wisconsinan Glaciation was reached about 18,000 years ago.

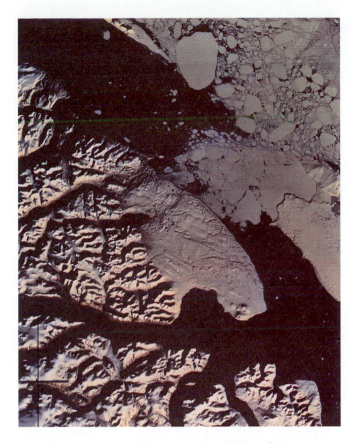

Figure 18.12 A Landsat image of a portion of the Canadian arctic archipelago. An ice cap on a landmass is visible in the center of the photo, with exposed mountainous ridges on the uncovered portions of the landmass. A branching glacial trough, now a water-filled fiord, is seen in the lower part of the image. In the upper left are huge chunks of free-floating sea ice.

Figure 18.13 An iceberg off the coast of Antarctica near Elephant Island.

Glaciation During the Ice Age

Figures 18.14 and 18.15 show the maximum extent to which North America and Europe were covered during the last advance of the ice. Most of Canada was engulfed by the vast Laurentide Ice Sheet. It spread south into the United States, covering most of the land lying north of the Missouri and Ohio rivers, as well as northern Pennsylvania and all of New York and New England. Alpine glaciers of the western ranges coalesced into a single ice sheet that spread to the Pacific shores and met the Laurentide sheet on the east. Notice that an area in southwestern Wisconsin escaped inundation. Known as the Driftless Area, it was apparently bypassed by glacial lobes moving on either side.

In Europe, the Scandinavian Ice Sheet centered on the Baltic Sea, covering the Scandinavian countries. It spread south into central Germany and far eastward to cover much of Russia. In north-central Siberia, large ice caps formed over the northern Ural Mountains and highland areas farther east. Ice from these centers grew into a large sheet covering much of central Siberia. The European Alps were capped by enlarged alpine glaciers. The British Isles were mostly covered by a small ice sheet that had several centers on highland areas and spread outward to coalesce with the Scandinavian Ice Sheet.

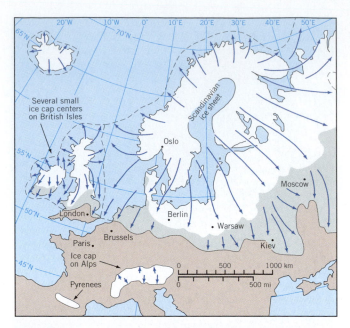

Figure 18.15 The Scandinavian Ice Sheet dominated northern Europe during the last glaciation. As noted in Figure 18.14, the present coastline is far inland from the coastline that prevailed during glaciations. (Based on data of R. F. Flint, *Glacial and Pleistocene Geology*, John Wiley & Sons, New York.)

Figure 18.14 Continental ice sheets of the last glaciation in North America at their maximum extent reached as far south as the present Ohio and Missouri rivers. Note that during glaciations sea level was much lower. The present coastline is shown for reference only. (Based on data of R. F. Flint, *Glacial and Pleistocene Geology*, John Wiley & Sons, New York.)

At the maximum spread of these ice sheets, sea level was as much as 125 m (410 ft) lower than today, exposing large areas of the continental shelf on both sides of the Atlantic basin. The shelf supported a vegetated landscape populated with animal life, including Pleistocene elephants (mastodons and mammoths). The drawdown of sea level explains why the ice sheets shown on our maps extend far out into what is now the open ocean.

Although the largest ice sheets covered North America and Eurasia, South America also developed an ice sheet.

South America, too, had an ice sheet. It grew from ice caps on the southern Andes Range south of about latitude 40° S and spread westward to the Pacific shore, as well as eastward, to cover a broad belt of Patagonia. It covered all of Tierra del Fuego, the southern tip of the continent. The South Island of New Zealand, which today has a high spine of alpine mountains with small relict glaciers, developed a massive ice cap in late Pleistocene time. All high mountain areas of the world underwent greatly intensified alpine glaciation at the time of maximum ice sheet advance. Today, most remaining alpine glaciers are small ones. In less favorable locations, the Ice Age alpine glaciers are entirely gone.

LANDFORMS MADE BY ICE SHEETS

Landforms made by the last ice advance and recession are very fresh in appearance and show little modification by erosion processes. It is to these landforms that we now turn our attention.

Erosion by Ice Sheets

Like alpine glaciers, ice sheets are highly effective eroding agents. The slowly moving ice scrapes and grinds away much solid bedrock, leaving behind smoothly rounded rock masses that often bear grooves and scratches trending in the general direction of ice movement (see Figure 18.1). Sometimes the ice polishes the rock to a smooth, shining surface. The evidence of ice abrasion is common throughout glaciated regions of North America and may be seen on almost any hard rock surface that is freshly exposed. Conspicuous knobs of solid bedrock shaped by the moving ice are also common features (Figure 18.16). The side from which the ice approached is usually smoothly rounded. The lee side, where the ice plucked out angular joint blocks, is irregular and blocky.

The ice sheets also excavated enormous amounts of rock at locations where the bedrock was weak and the flow of ice was channeled by the presence of a valley trending in the direction of ice flow. Under these conditions, the ice sheet behaved like a valley glacier, scooping out a deep, U-shaped trough. The Finger Lakes of western New York State are fine examples (Figure 18.17). Here, a set of former stream valleys lay largely parallel to the southward spread of the ice, and a set of long, deep basins was eroded. Blocked at their north ends by glacial debris, the basins now hold lakes. Many hundreds of lake basins were created by glacial erosion and deposition over the glaciated portion of North America.

Deposits Left by Ice Sheets

The term **glacial drift** includes all varieties of rock debris deposited in close association with glaciers. Drift is of two major types. *Stratified drift* consists of layers of

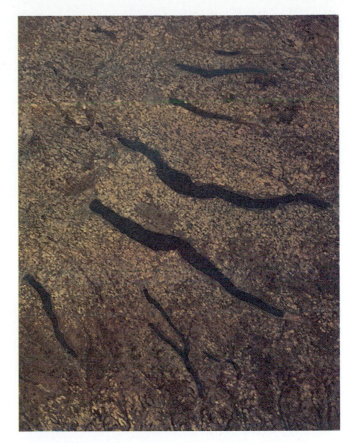

Figure 18.17 The Finger Lakes region of New York is shown in this photo taken by astronauts aboard the Space Shuttle. The lakes occupy valleys that were eroded and deepened by glacial ice.

sorted and stratified clays, silts, sands, or gravels. These materials were deposited by meltwater streams or in bodies of water adjacent to the ice. **Till** is an unstratified mixture of rock fragments, ranging in size from clay to boulders, that is deposited directly from the ice without water transport. Where till forms a thin, more or less even cover, it is referred to as *ground moraine*.

Over those parts of North America formerly covered by late-Cenozoic ice sheets, glacial drift thickness averages from 6 m (about 20 ft) over mountainous terrain, such as New England, to 15 m (about 50 ft) and more

Figure 18.16 A glacially abraded rock knob. Glacial action abrades the rock into a smooth form as it approaches the rock summit, then plucks bedrock blocks from the lee side, producing a steep, rocky slope. (Copyright © A. N. Strahler.)

over the lowlands of the north-central United States. Over Iowa, drift thickness is from 40 to 60 m (about 150 to 200 ft), and over Illinois it averages more than 30 m (about 100 ft). In some places where deep stream valleys existed prior to glacial advance, as in parts of Ohio, drift is much thicker.

To understand the form and composition of deposits left by ice sheets, it will help to examine the conditions prevailing at the time of the ice sheet's existence. Figure 18.18*a* shows a region partly covered by an ice sheet with a stationary front edge. This condition occurs when the rate of ice ablation balances the amount of ice brought forward by spreading of the ice sheet. Although the ice fronts of the Ice Age advanced and receded in many minor and major fluctuations, there were long periods when the front was essentially stable and thick deposits of drift accumulated.

Moraines

An ice sheet is like a huge conveyor belt for sediment. Anything carried on the belt is dumped off at the end as the ice evaporates or melts. If not constantly removed, sediment will pile up in increasing quantities. Glacial till that accumulates at the immediate ice edge forms an irregular heap of rubble—the *terminal moraine.* After the ice has disappeared (Figure 18.18*b*), the moraine appears as a belt of knobby hills interspersed with basin-like hollows, or kettles, some of which hold small lakes. The name *knob-and-kettle* is often applied to morainal belts.

Terminal moraines form great curving patterns. The outward curvature is southward and indicates that the ice advanced as a series of great *ice lobes*, each with a curved front (Figure 18.19). Where two lobes come

Figure 18.18 Marginal landforms of continental ice sheets. (*a*) With the ice front stabilized and the ice in a wasting, stagnant condition, various depositional features are built by meltwater. (*b*) The ice has wasted completely away, exposing a variety of new landforms made under the ice. (Drawn by A. N. Strahler.)

together, the moraines curve back and fuse together into a single *interlobate moraine* pointed northward. In its general recession accompanying disappearance, the ice front paused for some time along a number of positions, causing morainal belts similar to the terminal moraine belt to be formed. These belts are known as *recessional moraines* (Figure 18.18*b*). They run roughly parallel with the terminal moraine but are often thin and discontinuous.

Moraines are built of rock debris that is deposited at the edge of a melting glacier.

Outwash and Eskers

Figure 18.18 shows a smooth, sloping plain lying in front of the ice margin. This is the *outwash plain*, formed of stratified drift left by braided streams issuing from the ice. The plain is built of layer upon layer of sands and gravels.

Large streams carrying meltwater issue from tunnels in the ice. These form when the ice front stops moving for many kilometers back from the front. After the ice has gone, the position of a former ice tunnel is marked by a long, sinuous ridge of sediment known as an *esker* (Figure 18.20). The esker is the deposit of sand and gravel laid on the floor of the former ice tunnel. After the ice has melted away, only the stream bed deposit remains, forming a ridge. Many eskers are several kilometers long.

Drumlins and Till Plains

Another common glacial form is the *drumlin*, a smoothly rounded, oval hill resembling the bowl of an inverted teaspoon. It consists in most cases of glacial till (Figure 18.21). Drumlins invariably lie in a zone behind the terminal moraine. They commonly occur in groups or swarms and may number in the hundreds. The long axis of each drumlin parallels the direction of ice movement. The origin of drumlins is not well understood. They seem to have been formed under moving ice by a plastering action in which layer upon layer of bouldery clay was spread on the drumlin.

Between moraines, the surface overridden by the ice is covered by glacial till. This cover is often inconspicuous since it forms no prominent landscape feature. The till layer may be thick and may obscure, or entirely bury, the hills and valleys that existed before glaciation. Where thick and smoothly spread, the layer forms a level *till plain*. Plains of this origin are widespread throughout the central lowlands of the United States and southern Canada.

Marginal Lakes and Their Deposits

When the ice advances toward higher ground, valleys that may have opened out northward are blocked by ice.

Figure 18.19 Moraine belts of the north-central United States have a curving pattern left by ice lobes. Some regions of interlobate moraines are shown by the color overlay. (Based on data of R. F. Flint and others, *Glacial Map of North America*, Geological Society of America.)

Under such conditions, marginal glacial lakes form along the ice front (see Figure 18.18*a*). These lakes overflow along the lowest available channel between the ice and the rising ground slope, or over a low pass along a divide. Streams of meltwater from the ice build *glacial deltas* into these marginal lakes.

When the ice withers away, the lakes are drained, leaving a flat floor of fine clay and silt exposed. Glacial lake plains often contain extensive areas of marshland. The deltas are now curiously isolated, flat-topped landforms known as *delta kames*, composed of well-washed and well-sorted sands and gravels (Figure 18.8*b*).

Pluvial Lakes

During the Ice Age, some regions south of the glacial border experienced a cooler, moister climate. In the western United States, closed basins filled with water, forming pluvial lakes. The largest of these, glacial Lake Bonneville, was about the size of Lake Michigan and occupied a vast area of western Utah. With the warmer and drier climate of the present interglacial period,

Figure 18.20 The curving ridge of sand and gravel in this photo is an esker, marking the bed of a river of meltwater flowing underneath a continental ice sheet near its margin. Kettle-Moraine State Park, Wisconsin.

Figure 18.21 This small drumlin, located south of Sodus, New York, shows a tapered form from upper right to lower left, indicating that the ice moved in that direction (north to south).

these lakes shrank greatly in volume. Lake Bonneville became the present-day Great Salt Lake. Many other lakes dried up completely, forming desert playas. The history of these pluvial lakes is known from their ancient shorelines, some as high as 300 m (about 1000 ft) above present levels.

Environmental Aspects of Glacial Deposits

Because much of Europe and North America was glaciated by the Pleistocene ice sheets, landforms associated with the ice are of major environmental importance. Agricultural influences of glaciation are both favorable and unfavorable, depending on preglacial topography and the degree and nature of ice erosion and deposition.

In hilly or mountainous regions, such as New England, the glacial till is thinly distributed and extremely stony. Till cultivation is difficult because of countless boulders and cobbles in the clay soil. Till accumulations on steep mountain or roadside slopes are subject to mass movement as earthflows. Movement occurs when clay in the till becomes weakened after absorbing water from melting snows and spring rains. Along moraine belts, the steep slopes, the irregularity of knob-and-kettle topography, and the abundance of boulders conspired to prevent crop cultivation but invited use as pasture.

Flat till plains, outwash plains, and lake plains, on the other hand, can sometimes provide very productive agricultural land. Bordering the Great Lakes, fertile soils have formed on till plains and on exposed lakebeds. This fertility is enhanced by a blanket of wind-deposited silt (loess) that covers these plains (Chapter 17).

Stratified drift deposits are of great commercial value. The sands and gravels of outwash plains, deltas, and eskers provide necessary materials for both concrete manufacture and highway construction. Where it is thick, stratified drift forms an excellent aquifer and is a major source of ground water supplies.

INVESTIGATING THE ICE AGE

A great scientific breakthrough in the study of Ice Age glacial history came in the 1960s. First, scientists learned how to measure the absolute age of certain types of water-laid sediments by means of ancient magnetism. The earth's magnetic field experienced many sudden reversals of polarity in Cenozoic time, and the absolute ages of these reversals have been firmly established. Second, techniques were developed to take long sample cores of undisturbed fine-textured sediments from the deep ocean floor. Within each core, scientists could determine the age of sediment layers at various control points by identifying magnetic polarity reversals. By further studying the composition and chemistry of the layers within the core, a record of ancient temperature cycles in the air and ocean could be established.

Deep-sea cores reveal a long history of alternating glaciations and interglaciations going back at least as far as 2 million years and possibly 3 million years before present. The cores show that in Late-Cenozoic time more than 30 glaciations occurred, spaced at time intervals of about 90,000 years. How much longer this sequence will continue into the future is not known, but perhaps for 1 or 2 million years, or even longer.

Possible Causes of the Late-Cenozoic Ice Age

What caused the earth to enter into an ice age with its numerous cycles of glaciation and interglaciation? Three causes seem possible. First is a change in the placement of continents on the earth's surface through plate tectonic activity. Second is an increase in the number and severity of volcanic eruptions. Third is a reduction in the sun's energy output.

Increased volcanic activity during the Cenozoic Era may have helped to trigger the Ice Age.

Perhaps the answer lies in plate tectonics, through the motions of lithospheric plates following the breakup of Pangea. Recall from Chapter 11 that in Permian time, only the northern tip of the Eurasian continent projected into the polar zone. But as the Atlantic basin opened up, North America moved westward and poleward to a position opposite Eurasia, while Greenland took up a position between North America and Europe.

The effect of these plate motions was to bring an enormous landmass area to a high latitude and to surround a polar ocean with land. Because the flow of warm ocean currents into the polar ocean was greatly reduced, or at times totally cut off, this arrangement was favorable to the growth of ice sheets. The polar ocean was ice-covered much of the time and average air temperatures in high latitudes were at times lowered enough to allow ice sheets to grow on the encircling continents. Furthermore, Antarctica moved southward during the breakup of Pangea and took up a position over the south pole. In that location, it was ideally situated to develop a large ice sheet. Some scientists have also proposed that the uplift of the Himalayan Plateau, a result of the collision of the Austral-Indian and Eurasian plates, could have modified weather patterns sufficiently to trigger the Ice Age.

The second geological mechanism suggested as a basic cause of the Ice Age is increased volcanic activity on a global scope in Late-Cenozoic time. Volcanic eruptions produce dust veils that linger in the stratosphere and reduce the intensity of solar radiation reaching the ground (Chapter 2). Temporary cooling of near-surface air temperatures follows such eruptions. Although the geologic record shows periods of high levels of volcanic activity in the Miocene and Pliocene epochs, their role in initiating the Ice Age has not yet been convincingly demonstrated. **SCIENCE NEWS** 27

Another possible cause of the Ice Age is a slow decrease in the sun's energy output over the last several million years, perhaps as part of a cycle of slow increase and decrease over many million years' duration. As yet, data are insufficient to identify this mechanism as a possible basic cause. However, research on this topic is being stepped up as new knowledge of the sun is acquired from satellites that probe the sun's atmosphere and map its surface features.

Possible Causes of Glaciation Cycles

What timing and triggering mechanisms are responsible for the many cycles of glaciation and interglaciation that the earth is experiencing during the present Ice Age? Although many causes for glacial cycles have been proposed, we will limit our discussion here to one major contender called the **astronomical hypothesis**. It has been under consideration for about 40 years and is based on well-established motions of the earth in its orbit around the sun.

Two cyclic factors are involved—the changing distance between earth and sun, and the changing angle of tilt of the earth's axis of rotation. As to the first factor, recall from Chapter 1 that the earth's orbit is not perfectly circular but is slightly elliptical. The shape of the ellipse varies slowly through time. As a result, the distance that separates the earth and sun at summer solstice, June 21, undergoes a cyclic variation. A single cycle lasts 21,000 years. Second, we take into account that the earth's axial tilt (now $23\frac{1}{2}°$) undergoes a 40,000-year cycle of change in which the tilt angle may increase to as much as about 24° and decrease to as little as 22°. If we pick a point on the earth at, say, lat. 65° N, we can calculate the cycle of total change in the intensity of incoming solar radiation (insolation) on a day in summer (June 21), resulting from changing earth–sun distance and the cycle of axial tilt.

The mechanism that links ice sheet growth and decline with cycles of insolation is complex and not well understood.

The changing insolation, plotted against time on a graph, yields a series of peaks and valleys (see the graph in the *Eye on the Environment • Earth–Sun Cycles and Ice Ages*, Chapter 1). The major peaks are repeated at intervals of about 80,000 to 90,000 years. Each peak seems to have triggered a deglaciation, thereby ending a glaciation and bringing on an interglaciation. During the longer periods of reduced insolation, ice sheets formed and grew in volume. However, the glacial events may have lagged behind the insolation peaks and valleys by some thousands of years.

EYE ON THE ENVIRONMENT

Ice Sheets and Global Warming

What effects will global warming have on the earth's ice sheets? The Antarctic Ice Sheet holds 91 percent of the earth's ice. If it were to melt entirely, the earth's mean sea level would rise by about 40 m (about 200 ft). Additional water could be released by the melting of the Greenland Ice Sheet, which holds most of the remaining volume of land ice. Although global warming will accelerate glacial melting and thinning of the ice at the edges of ice sheets, increased precipitation, held by warmer air over the ice sheets, will likely produce a net growth in ice sheet thickness. Thus, it seems most unlikely that global warming will melt these ice sheets completely, unless the warming is much greater than antici-pated. However, there may be other factors at work than snow buildup and marginal melting in the case of indi-vidual ice sheets.

In 1998 NASA researchers repeat-ed measurements of the height of the southern half of the Greenland Ice Sheet that were originally made in 1993 using a laser altimeter mounted on an aircraft. They discovered that marginal thinning had reduced the volume of the ice sheet by about 42 km³ (about 10 mi³) in that five-year period. The thinning was primarily on the eastern side of the Ice Sheet. On the western side, melting was more in balance with new snow accu-mulation. The loss of this ice volume is not large enough to have much effect on global sea level. But a sub-stantial flow of fresh meltwater into the ocean could influence the ther-mohaline circulation of the northern Atlantic (see Chapter 3), triggering an ocean cycle of several centuries' duration that would bring very cold winter weather to Europe.

At the other pole, recent attention has focused on the Western Antarctic Ice Sheet, shown in the map below. Much of this vast expanse of Antarctic ice is "grounded" on a bedrock base that is well below sea level—that is, a large portion of the ice sheet rests directly on deep bedrock without sea water underneath. Attached to the grounded ice sheet are the Ross, Ronne, and Filchner ice shelves, which are underlain by sea water. They are also grounded at some points where the shelves enclose islands and overlay higher parts of undersea topography. At present, the grounded ice shelves act to hold back the flow of the main part of the ice sheet.

Geophysicists regard this as an unstable situation. A rapid melting or deterioriation of the ice shelves, per-haps in response to global warming, would release the back pressure on the main part of the ice sheet, which would then move forward and thin rapidly. With a reduced thickness, reduced pressure at the bottom of the ice would allow sea water to enter, and soon most of the sheet would be float-ing in ocean water. The added bulk would then raise sea level by as much as 6 m (about 20 ft).

A key feature of the Western Antarctic Ice Sheet is the presence of

A map of Antarctica showing the West-ern Antarctic Ice Sheet. (Adapted with permission from C.R. Bentley, *Science* Vol. 275, p. 1077. Copyright © American Assoc-iation for the Advancement of Science.)

The actual mechanisms by which insolation changes cause ice sheets to grow or to disappear are unclear. The entire subject is so complex that it is difficult for even a research scientist to grasp fully. Interactions between the atmosphere, the oceans, and the continental sur-faces (including the ice sheets) are numerous and close-ly interrelated with many threads of cause and effect. Changes that occur in one earth realm are fed back to the other realms in a most complex manner.

HOLOCENE ENVIRONMENTS

The elapsed time span of about 10,000 years since the Wisconsinan Glaciation ended is called the *Holocene Epoch*. It began with a rapid warming of ocean surface temperatures. Continental climate zones then quickly shifted poleward, and plants soon became reestablished in glaciated areas.

A satellite image of the Palmer Peninsula acquired on February 22, 1995, showing the Larsen Ice Shelf. Land areas are outlined in white. A large iceberg (A, center of image) has broken away from the shelf. Above (north of) the berg is a region of disintegrated ice (B) that has produced a swarm of many smaller bergs. (The black line crossing the image diagonally is a data gap.) (National Snow and Ice Data Center.)

formation of an iceberg of this size is not unusual.

Scientists hypothesized that the unusual disintegration of the Larsen Ice Shelf was caused by extreme surface melting during several warm seasons of the 1990s and by a regional climatic warming of about 2.5°C (4.5°F) observed over the past 50 years at a nearby station. However, other measurements show that the temperature of the Antarctic continent as a whole has not increased significantly since the mid-1960s. The Larsen Ice Shelf is different from the larger Ross and Ronne ice shelves in that it is not fed primarily by glacial ice, but rather by snowfall. As a result, it is thinner and likely to be more sensitive to small changes in climate.

The possible effects of climate change on global ice sheets are far from certain. We hope that human-induced global warming will result in slow, progressive changes to which human civilization can readily adapt.

Questions

1. What would be the effect on sea level if climatic warming melted the Antarctic Ice Sheet? What process seems to be offsetting the rate of melting of ice sheets as climate warms?

2. Why is the Western Antarctic Ice Sheet regarded as unstable? What is the role of ice streams in maintaining the present size of the ice sheet?

3. What recent events concerning the Larsen Ice Shelf have been attributed to climatic warming?

a number of ice streams within the sheet within which the ice flows at a much more rapid rate. These ice streams blend gradually into the ice shelves, providing a main source of their supply of ice. The flow of the ice streams is essential to maintaining the ice shelves. If these streams slowed, the ice shelf might retreat, which could release the unstable western ice sheet, producing catastrophic flooding. Although recent studies have suggested that the total flow of these ice streams has been quite constant for at least the last 1500 years, new evidence suggests that there has been at least one collapse of the West Antarctic Ice Sheet during an interglacial period. Sediments extracted from below the ice sheet show the presence of marine organisms that indicate freely circulating ocean water at some time within the last two million years. Moreover, temperatures at the time of the collapse are inferred to be not much greater than those of today.

Although it is not certain what the fate of climatic warming may be on the Antarctic Ice Sheet as a whole, satellite images acquired in 1995 showed that a large region of the Larsen Ice Shelf (Figure 18.10), about 2000 km² (about 770 mi²), had disintegrated into small icebergs during a storm (Figure 18.22). This unusual event was accompanied by the breaking off of a very large iceberg—nearly the size of the state of Rhode Island—that was strikingly visible on satellite images. Although more spectacular,

Three major climatic periods occurred during the Holocene Epoch leading up to the last 2000 years. These periods are inferred from studies of changes in vegetation cover types as observed in fossil pollen and spores preserved in glacial bogs. The earliest of the three is known as the Boreal stage and was characterized by boreal forest vegetation in midlatitude regions. There followed a general warming until the Atlantic stage, with temperatures somewhat warmer than today,

was reached about 8000 years ago (−8000 years). Next came a period of temperatures that were below average, the Subboreal stage. This stage spanned the age range −5000 to −2000 years.

Through the availability of historical records and of more detailed evidence, the climate of the past 2000 years can be described on a finer time scale. A secondary warm period occurred in the period A.D. 1000 to 1200 (−1000 to −800 years). This warm episode was followed by the Little Ice Age, A.D. 1450–1850 (−550 to −150 years). During the latter, valley glaciers made new advances and extended to lower elevations. More recent temperature cycles are discussed in Chapter 3. SCIENCE NEWS 28, 29

Within the last century, global temperatures have been slowly warming, as we noted in Chapter 3. What will be the effect of global warming on the existing continental ice sheets of Greenland and Antarctica? *Eye On The Environment • Ice Sheets and Global Warming* discusses the possibilities.

A LOOK AHEAD

The group of chapters we have now completed has reviewed landform-making processes that operate on the surface of the continents. These have ranged from mass wasting to the erosion and deposition brought about by glacial ice. The great variety and complexity of landforms we have described are not difficult to understand when each agent of denudation is examined in turn.

Human influence on landforms is felt most strongly on surfaces of fluvial denudation because of the severity of surface changes caused by agriculture and urbanization. Landforms shaped by wind and by waves and currents are also highly sensitive to changes induced by human activity. Only glaciers maintain their integrity and are thus far largely undisturbed by human activity. Perhaps even this last realm of nature's superiority will eventually fall prey to human interference through climate changes induced by industrial activity.

Our closing chapter, the Epilogue, dramatizes four areas of ongoing conflict between the human species and its natural environment—sea-level rise from climatic warming, population pressures, rainforest clearing, and development in earthquake-prone areas. As the human presence continues to impact our planet's natural systems, such conflicts will become increasingly important to resolve.

GLACIAL LANDFORMS AND THE ICE AGE IN REVIEW

Glaciers form when snow accumulates to a great depth, creating a mass of ice that is plastic in lower layers and flows outward or downhill from a center in response to gravity. As they move, glaciers can deeply erode bedrock by abrasion and plucking. The eroded fragments, incorporated into the flowing ice, leave depositional landforms when the ice melts.

Alpine glaciers develop in cirques in high mountain locations. Alpine glaciers flow downvalley on steep slopes, picking up rock debris and depositing it in lateral and terminal moraines. Through erosion, glaciers carve U-shaped glacial troughs that are distinctive features of glaciated mountain regions. They become fiords if later submerged by rising sea level.

Ice sheets are huge plates of ice that cover vast areas. They are present today in Greenland and Antarctica. The Antarctic Ice Sheet includes ice shelves—great plates of floating glacial ice. Icebergs form when glacial ice flowing into an ocean breaks into great chunks and floats free. Sea ice, which is much thinner and more continuous, is formed by direct freezing of ocean water and accumulation of snow on the sea-ice surface.

An ice age includes alternating periods of glaciation, deglaciation, and interglaciation. During the past 2 to 3 million years, the earth has experienced the Late-Cenozoic Ice Age. During this ice age, continental ice sheets have grown and melted as many as 30 times. The most recent glaciation is the Wisconsinan Glaciation, in which ice sheets covered much of North America and Europe, as well as parts of northern Asia, southern South America, and New Zealand.

Moving ice sheets create many types of landforms. Bedrock is grooved and scratched. Where rocks are weak, long valleys can be excavated to depths of hundreds of meters. The melting of glacial ice deposits glacial drift, which may be stratified by water flow or deposited directly as till. Moraines accumulate at ice edges. Outwash plains are built up by meltwater streams. Tunnels within the ice leave streambed deposits as eskers. Till may be spread smooth and thick under an ice sheet, leaving a till plain. This may be studded with elongated till mounds, termed drumlins. Meltwater streams build deltas into lakes formed at the ice margin and line lake bottoms with clay and silt. When the lakes drain, these features remain.

Several factors have been proposed to explain the cause of present ice age glaciations and interglaciations. These include ongoing change in the global position of continents, an increase in volcanism, and a reduction in the sun's energy output. Individual cycles of glaciation seem strongly related to small, cyclic changes in earth–sun distance and axial tilt.

KEY TERMS

ablation	glacial trough	glacial drift
alpine glacier	fiord	till
ice sheet	glaciation	astronomical hypothesis
cirque	interglaciation	
moraine	Late-Cenozoic Ice Age	

REVIEW QUESTIONS

1. How does a glacier form? What factors are important? Why does a glacier move?
2. Distinguish between alpine glaciers and ice sheets.
3. What is a glacial trough and how is it formed? What is its basic shape? In what ways can a glacial trough appear after glaciation is over?
4. Where are ice sheets present today? How thick are they?
5. Contrast sea ice and icebergs, including the processes by which they form.
6. Identify the Late-Cenozoic Ice Age. When did it begin? What was the last glaciation in this cycle? When will it end?
7. What areas were covered with ice sheets by the last glaciation? How was sea level affected?
8. What are moraines? How are they formed? What types of moraines are there?
9. Identify the landforms and deposits associated with stream action at or near the front of an ice sheet.
10. Identify the landforms and deposits associated with deposition underneath a moving ice sheet.
11. Identify the landforms and deposits associated with lakes that form at ice sheet margins.
12. How have environments changed during the Holocene Epoch? What periods are recognized, and what are their characteristics?

VISUALIZING EXERCISES

1. What are some typical features of an alpine glacier? Sketch a cross section along the length of an alpine glacier and label it.
2. Refer to Figure 18.10, which shows the Antarctic continent and its ice cap. Identify the Ross, Filchner, and Larsen ice shelves. Use the scale to measure the approximate area of each in square kilometers or miles. Consulting an atlas or almanac, identify the state or province of the United States or Canada that is nearest in area to each.

A DEEPER VIEW OF GLACIAL LANDFORMS AND THE ICE AGE

1. Imagine that you are planning a car trip to the Canadian Rockies. What glacial landforms would you expect to find there? Where would you look for them?
2. At some time during the latter part of the Pliocene Epoch, the earth entered an ice age. Describe the nature of this ice age and the cycles that occur within it. What explanations are proposed for causing an ice age and its cycles of glaciation? What climatic cycles have been observed since the last ice sheets retreated?

EPILOGUE

News from the Future

Our text began with a Prologue describing a trip you might take as a lunar astronaut, traveling home from the moon to earth in 2050. What would life be like in the middle of the twenty-first century? Imagine turning on a futuristic television set and watching the news for June 26, 2050. It might go something like this.

"Good evening, America, and welcome to the Nightly News. This is Dan Rogers in Washington. Tonight's big stories are from the environment. First, we'll go to Galveston, where the story is the flooding from the first storm of the season, Hurricane Alma. Here's our correspondent, Lorraine Jackson, on the scene. Lorraine, what's it like there now?"

"Thanks, Dan. The storm has slackened off in the last few hours, but the system of dams and pumps they have here has worked, at least for now, to keep the floodwaters out of the center of the city. As you can see, the rain is still falling and the winds are still strong. But the real problem with this storm has not been the wind and rain. Alma is one of the weaker storms to reach the Texas coast in the

past several hurricane seasons. The problem is with the ocean flooding that this coastal outpost endures when a storm like Alma passes close by.

"The ocean dikes have been taking a terrible beating from the wave action.

Here's some footage of the waves breaking on the dikes that we shot earlier. All that wave water spilling over the dikes has to be pumped out, and fast. We understand from the local authorities that several smaller dikes west of the city were

Climatic change may raise sea level, enhancing the ability of hurricanes to destroy coastal settlements. Here, Hurricane Alicia lashes the Texas coast at Galveston.

A crowded street in Hong Kong during the lunch hour.

breached this morning, adding to the floodwaters. Apparently, several subdivisions out there are well under water now. There was a scare when the power went out earlier in the day, and one of the huge backup generators supplying power to the pumps refused to start. But the other generators are keeping up with the load of the pumps."

"When the citizens of this Texas city decided 30 years ago to fight against the rise in sea level rather than move inland like most other beach communities, they knew that the battle against the ocean would be costly and could eventually be a losing one. But for now, the valiant city of Galveston seem to be holding its own, Dan."

"Lorraine, what about the death toll?"

"So far, Dan, the authorities have not reported any deaths from the storm, although there were some injuries in the evacuation. We had a report that one of the huge buses they use for evacuations hit the rear end of another bus, but only the driver and two passengers were injured. Apparently, the accident didn't slow down the river of huge vehicles streaming across the viaduct to the mainland. As I reported this morning, last night's evacuation went pretty smoothly. The residents here have been through so many evacuations in the last few years that they are all familiar with the emergency routine."

"Thanks, Lorraine. That is relatively good news."

"Our next story marks a momentous event for the human race—the birth of the 7 billionth citizen of earth. For the story, we go to Beijing, to our correspondent Li Feng Sun. Where are you, Feng Sun?"

"Dan, I'm here in the corridor outside the maternity ward in this small birthing center in a suburb west of Beijing, waiting word about the birth of Zhu Peng, destined to be the world's 7 billionth citizen. Of course, we're not precisely certain that baby Zhu Peng will be the world's 7 billionth citizen, but scientists have predicted that in precisely four minutes and fourteen seconds, the earth's population will reach 7 billion. They also tell us that the 7 billionth citizen is most likely to be Chinese.

"Economists and population planners have been marking the event as a real milestone. Before the turn of the century, they projected that the world's population would be over 9 billion by 2050. You may remember some years ago the many predictions that the plans for sustainable

As progressively larger areas of the earth's surface are developed, the potential for damage from extreme natural events, such as earthquakes, increases. This interchange on the Golden State Freeway, Los Angeles, was destroyed in the Whittier Narrows Earthquake of 1987.

world development would crash along with populations of Africa and Southeast Asia, which were decimated by the AIDS epidemic. China was only little impacted because AIDS spread quite slowly here, and because of China's crash program to inoculate its entire population with the AIDS vaccine shortly after it was developed. That left this huge country to continue its slow population growth coupled with rapid economic growth fueled by this country's vast natural and human resources. The result is the success story we're all familiar with—China's economy surpassing the economies of Europe, the Americas, and finally the East Asian trading block of Japan, Taiwan, Korea and Singapore."

"Feng Sun, I hear some chanting in the background. Are those demonstrators?"

"That's right, Dan. The militant arm of the Chinese Green party has mounted a demonstration here. In fact, we have one of the leaders here, Huang Chung, ready to talk with us. Huang Chung, what are the demonstrators here for?"

"Well, Feng Sun, we're protesting this event because we don't think the world needs any more people. Every consumer requires more food and uses more resources. We've already cut down almost half of the world's forests. Global warming has made the deserts spread and our cropland become less productive. We all know that global warming was generated by consumers demanding products and lifestyles that depend on fossil fuel...."

"Feng Sun, I hate to break in, but we've got an interrupt here. Apparently, there's been a breakthrough on the biodiversity talks at the U.N. For the story, let's go to Maria Remarquez in New York. Maria?"

"Hi, Dan. We've just gotten word here that the rainforest nations will back down from their threat to withdraw from the biodiversity treaty. Apparently, the U.S., the European Union, and Chinese-East Asian bloc have agreed to make some of the payments demanded for the use of the rainforest genetic materials that the rainforest nations contend were stolen from them by biotechnology companies.

"Here to tell us about it is Professor Kate Carver of New York's City University. Professor Carver, why is this agreement important?"

"Well, Maria, because with this agreement in place, we can go ahead with the plan to set up rainforest preservation zones, perhaps as early as next year. The rainforest ecosystems are our most complex ecological communities, and they hold the greatest diversity of life of all habitats on earth. They are a reservoir of genetic material that is irreplaceable, and they need to be protected as vast areas forever. Without these genetic reserves, we'd be missing many of our cures for diseases ranging from cancer to AIDS.

"You know, Maria, I might also point out that this agreement is good for the environment in another way. The money will be used by the rainforest nations to stimulate the use of nonpolluting technologies for sustainable development. That is, the funds will be helping these nations promote more use of solar power, wind power, and methane gas production. In this way, they will reduce their reliance on old-fashioned fossil fuels and with it the release of carbon dioxide that produces climate change."

"Thanks for that perspective, Professor. It looks like a win for the environment here at the United Nations this evening, Dan."

"And thanks for that fine story, Maria. For our next story, remember that it was exactly one month ago when the disastrous El Cajon earthquake struck Los Angeles. We go now to Pasadena for a report on the cleanup and rebuilding. Our reporter with the story is Keith Mancini. Keith?"

"It was early on a Sunday morning just one month ago when Angelinos were thrown from their beds by the violent shaking that devastated so much of Los Angeles. It was the largest earthquake to hit Southern California in more than a century, measuring magnitude 8.1 on the Richter Scale, flattening thousands of buildings, and rendering thousands more unusable. The death toll, estimated at over 16,000 with more than 100,000 injured, was also by far the largest in this century for an American natural disaster.

"Today Los Angeles is still digging out. The magnitude of the task ahead is mind-boggling, Dan. The electric freeway system is virtually closed for all but a few segments. Nearly every viaduct and bridge experienced some damage, and engineers are struggling to inspect them all for safety and set the priorities for repair. Most of the city's skyscrapers are still unoccupied, since structural repairs are expected to take many more months to complete. In the hardest-hit areas containing old masonry buildings, almost nothing was left standing. Now that the searches for casualties are over, bulldozers and dump trucks are removing the debris.

"In the Long Beach area, the cleanup is beginning from the fires that started in the oil processing plants and spread to the chemical factories. The toxic compounds carried by the smoke and ash contaminated a huge area. It has taken several weeks for the National Guard to replace their cordon of troops and vehicles with the fencing and barbed wire that will keep looters out while the area is being detoxified, which is expected to take six to nine months.

"Meanwhile, homeless residents are flocking to the hundreds of cities constructed of portable geodomes now being set up by the National Guard. The demand increases daily as troops and local authorities evict inhabitants still trying to live in the dangerous, crumbling buildings. The geodome cities at least provide shelter, hot food, and sanitation for the thousands of displaced residents, luxuries that are not easy to come by here.

"But the amazing thing, Dan, is the dedication of these residents to rebuilding their city. Everywhere you go in this vast metropolis, there is activity. People are pulling together to rebuild their homes, their businesses, and their lives. There's a real 'can-do' spirit here, and L.A. should emerge from this tragedy all the stronger for it."

"Thanks, Keith, for that fine report. It won't be easy for the folks in Los Angeles in the weeks and months ahead, but at least they know that the rest of the nation is pulling for them.

"And, that's it for the top news stories today. Stay with us for the local report, coming up after these messages."

PERSPECTIVE ON THE NEWS FROM THE FUTURE

This news report dramatizes some of the present concerns about the environment that have arisen in our survey of physical geography. The first news story calls attention to the sea-level rise that is predicted to accompany global warming. The latest predictions are that by the year 2100 global temperature will increase by 1°C to 3.5°C (1.8°F to 6.3°F), with a best estimate of 2°C (3.6°F). Sea level is expected to rise by 0.15 to 0.9 m (0.5 to 3.0 ft), with a best estimate of 0.48 m (1.6 ft).

The problem with rising sea level is not so much with day-to-day levels, but rather the levels that occur when tides combine with storms to produce intense wave action on low-lying coasts. The Republic of Maldives, which is a nation of 1190 small islands, would be entirely submerged by a rise of 2 m (6.6 ft). Even with a rise of only 1 m (3.3 ft), the country could be wiped out by a severe storm surge.

Sea-level rise is just one of the possible effects of climatic warming. Even though the extent of warming is uncertain, scientists have concluded that warming would bring not only warmer temperatures, but also more global precipitation. However, this precipitation increase will not be evenly distributed. Computer models predict that high latitudes will experience more precipitation, while midlatitude continental interiors will experience drier summers. These types of changes would have important effects on crop growth, shifting the varieties and types of crops grown within large regions. Increased irrigation could also be needed.

Because the effects of climatic warming could be so far-reaching, scientists have managed to alert the public about the danger of continued increases of CO2 and other gases that contribute to global warming. At the 1992 Earth Summit in Rio de Janeiro, most of the world's nations agreed on a treaty to reduce human impact on climate. The industrialized countries pledged to reduce the levels of greenhouse gases they release, thus slowing the rapid increases in atmospheric CO2, methane, chlorofluorocarbons, and nitrous oxide (N_2O) that have been measured. At subsequent meetings in Kyoto, Japan, and Buenos Aires, Argentina, specific reduction targets were hammered out and agreed upon. Although the goals will be difficult to achieve, the momentum is strongly toward reducing greenhouse emissions significantly.

Another action of the Rio Earth Summit was the adoption of a treaty on biodiversity, which is the subject of the third news story. In the treaty, nations pledged to protect the genetic resources in their plant and animal

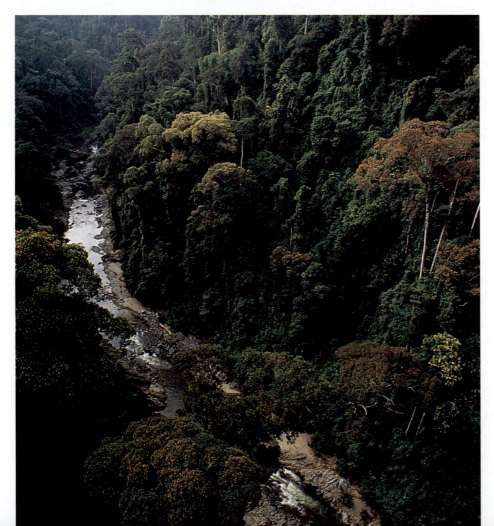

High biodiversity is a feature of equatorial rainforests, such as this example in the Datum River Gorge, Sabah, Borneo.

populations by shielding them from extinction. Forest preservation was addressed as a need as well. The low-latitude rainforests cover only about 7 percent of the earth's surface, but they contain over half of the world's species. They are especially rich in insect species and flowering plants. Plants of these regions are a particularly valuable resource. They have provided most of the species on which humans rely for food. Many highly effective medical drugs are produced from these plants.

Not only are low-latitude rainforests valuable resources for human exploitation, but they are also very complex ecosystems that have evolved over millions of years. Many environmentalists believe that humans have a responsibility to preserve our planet's natural systems and maintain their diversity. It is encouraging that the world's nations have recognized the importance of ecological resources such as the low-latitude rainforests and are working toward their preservation.

The last future news story, about an earthquake in Los Angeles, demonstrates the impact of natural disasters on human activities. The story is only slightly exaggerated—an earthquake of magnitude 8.1 on the San Andreas fault would most certainly have devastating effects on a nearby city the size of Los Angeles.

By nature, the human species has a short-term outlook, preferring to prepare for dangers that are likely to occur within a time span of a few years to a few decades. Many natural processes work on far longer time scales, however. Although a major earthquake on the San Andreas fault may not happen more than once every few generations, there were probably tens of thousands of severe earthquakes on that fault in the Los Angeles region during the Ice Age, with hundreds of these occurring since the last advance of glacial ice. A major flood on the Mississippi River, like that of 1993, is a still more likely event. Although infrequent in human terms, these rare events have shaped much of the earth's landscape. For the geologist or physical geographer trained in earth processes, the evidence of these past events is everywhere.

Earthquakes, floods, and tropical storms are not new phenomena. What is new is that human civilization has come to dominate most areas of the planet, so that these natural occurrences now have major impacts on human societies in the regions in which they occur. Taken in the aggregate, they become a factor of global change.

The event of the second future news story, which dramatizes the birth of the 7 billionth citizen of earth, probably has the most important implications. It is actually quite an optimistic scenario, for present estimates of the world population in 2050 are closer to 9.5 billion. The story refers to future population crashes in Africa and Southeast Asia produced by the AIDS epidemic that now rages unchecked throughout the world, and advances the idea that China will be impacted less severely than other developing nations. Whether this future will come to pass is uncertain. But in any event,

the human population will continue to increase until either it regulates itself or destroys itself.

Why is population growth so important? Humans are the most powerful force transforming the surface of the earth today. About 15 to 20 percent of the world's forest area has been cleared since human civilization began to exploit forest resources. In the last 300 years, humans have increased the area of cropland by 450 percent. Populations of the largest urban areas have increased 25 times in that same time span. These changes have directly influenced the global energy balance, the hydrologic cycle, and the dynamics of the atmosphere, with effects that are unknown in the long term.

Recognizing these trends, the nations of the world have come together in an effort to promote sustainable development. Under this policy, the needs and aspirations of present and future generations are met without impacting the ability of future generations to satisfy their needs. The idea is to focus on social and economic progress that is compatible with the environment. The concept draws on the premise that human knowledge, sophisticated technology, and access to resources are now so well developed that they can be used to allow developing countries to increase the well-being of their populations in a way that does not damage the environment. In this effort, much help and cooperation is required from developed nations, not only to provide the funds to stimulate sustainable development, but also to provide the technological base that will be required.

There are now some hopeful signs that the long-continued increase of human population will slow. Reports in 1996 revealed for the first time that fertility rates had declined in all regions of the world. A survey of population experts predicted that the earth's population will peak at around 11 billion by 2075 and remain level or decrease slightly from 2075 to 2100. Other studies of the economic impact of population growth have noted that technological innovation seems to naturally lead to increasing use of technologies that have a smaller environmental impact. Although the citizens of underdeveloped countries will want and need to advance their standard of living, their consumption will perhaps be less damaging to the environment. These recent trends in population and resource utilization can be expected to ease the transition from environmental exploitation to sustainable development in many world regions.

The social and economic aspects of continued population growth and policies such as sustainable development are far beyond the scope of physical geography, but they are concerns of which every responsible citizen on our planet must be aware. This book has provided a background and context on the workings of the earth and the environment that we, as authors, hope will be valuable to you, its readers, as you face the challenges of world citizenship in the twenty-first century.

More About Maps

Maps play an essential role in the study of physical geography, because much of the information content of geography is stored and displayed on maps. Map literacy—the ability to read and understand what a map shows—is a basic requirement for day-to-day functioning in our society. Maps appear in almost every issue of a newspaper and in nearly every TV newscast. Most people routinely use highway maps and street maps. The purpose of this appendix is to supplement the map-reading skills you have already acquired with some additional information on the subject of cartography, the science of maps and their construction. In understanding the following material, it will be helpful to review the section "Map Projections" in Chapter 1 of the text.

MAP PROJECTIONS

Recall from Chapter 1 that a *map projection* is an orderly system of parallels and meridians used as a base on which to draw a map on a flat surface. A projection is needed because the earth's surface is not flat but rather is curved in a shape that is very close to the surface of a sphere. All map projections misstate the shape of the earth is some way. It's simply impossible to transform a spherical surface to a flat (planar) surface without some violation of the true surface as a result of cutting, stretching, or otherwise distorting the information that lies on the sphere.

Perhaps the simplest of all map projections is a grid of perfect squares. In this simple map, horizontal lines are parallels, and vertical lines are meridians. This grid is often used in modern computer-generated world maps displaying data that consist of a single number for each square, representing, for example, 1, 5, or 10 degrees of latitude and longitude. A grid of this kind can show the true spacing (approximately) of the parallels, but it fails to show how the meridians converge toward the two poles. This grid fails dismally in high latitudes, and the map usually has to be terminated at about 70° to 80° north and south.

Early attempts to find satisfactory map projections made use of a simple concept. Imagine the spherical earth grid as a cage of wires (a kind of bird cage). A tiny light source is placed at the center of the sphere, and the image of the wire grid is cast on a surface outside the sphere. This situation is like a reading lamp with a lampshade. Basically, three kinds of "lampshades" can be used, as shown in Figure A1.1.

First is a flat paper disk balanced on the north pole. The shadow of the wire grid on this plane surface will appear as a combination of concentric circles (parallels) and radial straight lines (meridians). Here we have a polar-centered projection. Second is a cone of paper resting point-up on the wire grid. The cone can be slit down the side, unrolled, and laid flat to produce a map that is some part of a full circle. This is called a conic projection. Parallels are arcs of circles, and meridians are radiating straight lines. Third, a cylinder of paper can be wrapped around the wire sphere so as to be touching all around the equator. When slit down the side along a meridian, the cylinder can be unrolled to produce a cylindrical projection, which is a true rectangular grid.

Take note that none of these three projection methods can show the entire earth grid, no matter how large a sheet of paper is used to receive the image. Obviously, if the entire earth grid is to be shown, some quite different system must be devised. Many such alternative solutions have been proposed. In Chapter 1, we described

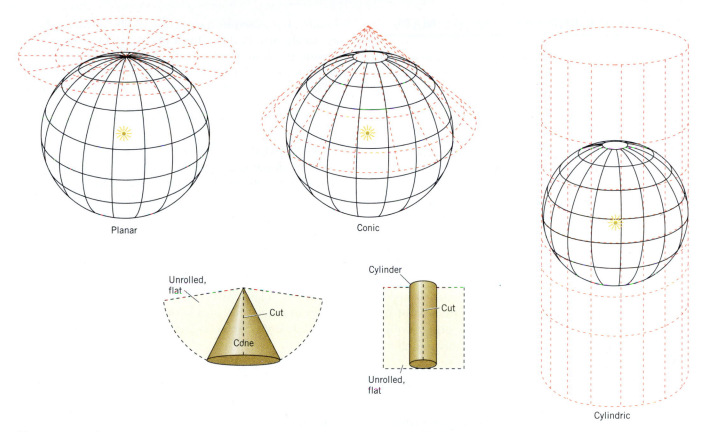

Figure A1.1 Simple ways to generate map projections. Rays from a central light source cast shadows of the spherical geographic grid on target screens. The conical and cylindrical screens can be unrolled to become flat maps. (A. N. Strahler.)

three types of projections used throughout the book—the *polar projection*; the *Mercator projection*, which is a cylindrical projection; and the *Goode projection*, which uses a special mathematical principle.

SCALES OF GLOBES AND MAPS

All globes and maps depict the earth's features in much smaller size than the true features they represent. Globes are intended in principle to be perfect scale models of the earth itself, differing from the earth only in size. The *scale* of a globe is therefore the ratio between the size of the globe and the size of the earth, where "size" is some measure of length or distance (but not of area or volume).

Take, for example, a globe 20 cm (about 8 in.) in diameter, representing the earth, which has a diameter of about 13,000 km. The scale of the globe is the ratio between 20 cm and 13,000 km. Dividing 13,000 by 20, we see that this ratio reduces to a scale stated as follows: One centimeter on the globe represents 650 kilometers on the earth. This relationship holds true for distances between any two points on a globe.

Scale is more usefully stated as a simple fraction, termed the *fractional scale*, or *representative fraction*. It can be obtained by reducing both earth and globe distances to the same unit of measure, which in this case is centimeters. (There are 100,000 centimeters in one kilometer.) The advantage of the representative fraction is that it is entirely free of any specified units of measure, such as the foot, mile, meter, or kilometer. Persons of any nationality can understand the fraction, regardless of their language or units of measure.

Being a true-scale model of the earth, a globe has a constant scale everywhere on its surface, but this is not true of a map projection drawn on a flat surface. In flattening the curved surface of the sphere to conform to a plane surface, all map projections stretch the earth's surface in a nonuniform manner, so that the map scale changes from place to place. So we can't say about any world map: "Everywhere on this map the scale is 1:65,000,000." It is, however, possible to select a meridian or parallel—the equator, for example—for which a fractional scale can be given, relating the map to the globe it represents. For example, in Figure 1.8, the scale of the Mercator projection along its equator is about 1:325,000,000.

SMALL-SCALE AND LARGE-SCALE MAPS

From maps on the global scale, showing the geographic grid of the entire earth or a full hemisphere, we turn to maps that show only small sections of the earth's surface. These maps of large scale are capable of carrying the enormous amount of geographic information that is available and must be shown in a convenient, effective manner. For practical reasons, maps are printed on sheets of paper usually less than a meter (3 ft) wide, as in the case of the ordinary highway map or navigation chart. Bound books of maps—atlases, that is—consist of pages that are usually no larger than 30 by 40 cm (about 12 by 16 in.), whereas maps found in textbooks and scientific journals are even smaller.

Actually, it is not simply the size of map sheet or page that determines how much information a map can carry. Instead, it is the scale on which the surface is depicted. Keep in mind that the relative magnitude of two different map scales is determined according to which representative fraction is the larger and which the smaller. For example, a scale of 1:10,000 is twice as large as a scale of 1:20,000. Don't make the mistake of supposing that the fraction with the larger denominator represents the larger scale. If in doubt, ask yourself, "Which fraction is the larger: 1/4 or 1/2?" and apply this to the scales you are comparing.

In the case of globes and maps of the entire earth, the fractional scale may range from 1:100,000,000 to 1:10,000,000; these are *small-scale maps*. *Large-scale maps* have fractional scales generally greater than 1:100,000. Stated in words, this fraction means: "One centimeter on the map represents one-hundred-thousand centimeters on the ground," or given that 100,000 cm equals 1 km, "One centimeter on the map represents one kilometer on the ground." Most large-scale maps carry a *graphic scale*, which is a line marked off into units representing kilometers or miles. Figure A1.2 shows a portion of a large-scale map on which sample graphic scales in miles, feet, and kilometers are superimposed. Graphic scales make it easy to measure ground distances.

Figure A1.2 A portion of a modern, large-scale map for which three graphic scales have been provided. (U.S. Geological Survey.)

INFORMATIONAL CONTENT OF MAPS

Up to this point, our investigation of cartography has yielded a geographic grid with only the barest outlines of the continents. Information conveyed by a map projection grid system is limited to one category only: absolute location of points on the earth's surface. What other categories of information can be shown on the flat map sheet? The number of categories is almost infinite, since the map is capable of carrying information about anything that occupies a particular location at a specified time. For the needs of physical geography, which is focused primarily on the physical environment of life in the surface layer, the list is greatly reduced, but still enormously long.

Multipurpose maps show a number of different types of information. A simple example is shown in Figure A1.3. Map sheets published by national governments are usually multipurpose maps. Using a great variety of symbols, patterns, and colors, these maps carry a high information content available on demand to the general user, who may select and use only one information category, or *theme*, at any one time. A real example of a multipurpose map is shown on the endpapers of this book, located on the inside of the back cover. It is a portion of a U.S. Geological Survey topographic quadrangle map for San Rafael, California.

In contrast to the multipurpose map is the *thematic map*, which shows only one type of information. Some examples are distributions of the human population, boundaries of election districts, geologic age of exposed rock, elevation of the land surface above sea level, number of days of frost in the year, and speed of winds in the upper atmosphere.

MAP SYMBOLS

Symbols on maps associate information with points, lines, and areas. To express points as symbols, they can be renamed "dots." Broadly defined, a dot can be any small device to show point location. It might be a closed circle, open circle, letter, numeral, or an icon of the object it represents (see "church with tower" in Figure A1.3). A line can vary in width and can be single or double. The line can also consist of a string of dots or dashes. A specific area of surface can be referred to as a "patch." The patch can be shown simply by a line marking its edge, or it can be depicted by a distinctive pattern or a solid color. Patterns are highly varied. Some consist of tiny dots and others of parallel, intersecting, or wavy lines.

A map consisting of dots, lines, and patches can carry a great deal of information, as Figure A1.3 shows. In this case, the map uses two kinds of dot symbols (both symbolic of churches), three kinds of line symbols, and three kinds of patch symbols. Altogether, eight types of information are offered. Line symbols freely cross patches, and dots can appear within patches. Two different kinds of patches can overlap.

The relation of map symbols to map scale is of prime importance in cartography. Maps of very large scale, along with architectural and engineering plans, can show objects to their true outline form. As map scale is decreased, representation becomes more and more generalized. In physical geography an excellent example is the depiction of a river, such as the lower Mississippi. Figure A1.4 shows the river channel at three scales, starting with a detailed plan, progressing to a double-line analog that generalizes the channel form, and ending with a single-line symbol. As generalization develops, the details of the river banks and channel bends are simplified as well. The level of depiction of fine details is described by the term *resolution*. Maps of large scale have much greater resolving power than maps of small scale.

Figure A1.3 Multipurpose map of an imaginary area with 10 villages. (After J. P. Cole and C.A.M. King, *Quantitative Geography*, copyright © John Wiley & Sons, London. Used by permission.)

(a)

(b)

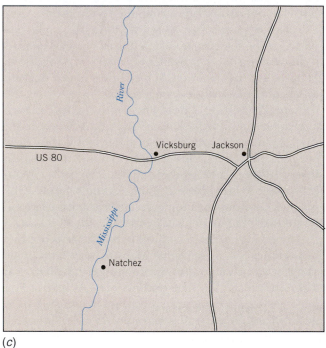

(c)

Figure A1.4 Maps of the Mississippi River on three scales. (*a*) 1:20,000. Channel contours give depth below mean water level. (*b*) 1:250,000. Waterline is shown only to depict channel. (*c*) 1:3,000,000. Channel is shown as a solid line symbol. (Maps slightly enlarged for reproduction.) (Modified from U.S. Army Corps of Engineers.)

PRESENTING NUMERICAL DATA ON THEMATIC MAPS

How are numerical data shown on maps? In physical geography, much of the information collected about particular areas is in the form of numbers. The numbers might represent readings taken from a scientific instrument at various places throughout the study area. A simple example is the collection of weather data, such as air temperature, air pressure, wind speed, and amount of rainfall. Another example is the set of measurements of the elevation of a land surface, given in meters above sea level. Another category of information consists merely of the presence or absence of a quantity or attribute. In such cases, we can simply place a dot to mean "present," so that when entries are completed, we have before us a field of scattered dots (Figure A1.5).

In some scientific programs, measurements are taken

Table A1.1 Examples of Isopleths

Name of Isopleth	Greek Root	Property Described	Examples in Figures
Isobar	*baros*, weight	Barometric pressure	5.17
Isotherm	*therme*, heat	Temperature of air, water, or soil	3.18
Isotach	*tachos*, swift	Fluid velocity	5.22
Isohyet	*hyetos*, rain	Precipitation	4.18
Isohypse	*hypso*, height	Elevation	18.9, 18.10

uniformly, for example, at the centers of grid squares laid over a map. For many classes of data, however, the locations of the observation points are predetermined by a fixed and nonuniform set of observing stations. For example, data of weather and climate are collected at stations typically located at airports. Whatever the sampling method used, we end up with an array of numbers and dots indicating their location on the field of the base map.

While the numbers and locations may be accurate, it may be difficult to see the spatial pattern present in the data being displayed. For this reason, cartographers often simplify arrays of point values into isopleth maps.

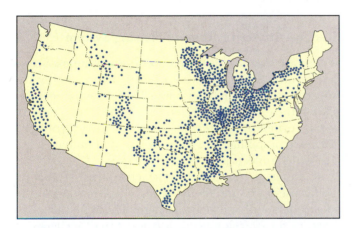

Figure A1.5 A dot map showing the distribution of soils of the order Alfisols in the United States. (From P. Gersmehl, *Annals of the Assoc. of Amer. Geographers*, vol. 67. Copyright © Association of American Geographers. Used by permission.)

An *isopleth* is a line of equal value (from the Greek *isos*, "equal," and *plethos*, "fullness" or "quantity"). Figure 3.17 shows how an isopleth map is constructed for temperature data. In this case, the isopleth is an *isotherm*, or line of constant temperature. In drawing an isopleth, the line is routed among the points in a way that best indicates a uniform value, given the observations at hand.

Isopleth maps are important in various branches of physical geography. Table A1.1 gives a partial list of isopleths of various kinds used in the earth sciences, together with their special names and the kinds of information they display. Examples are cited from our text.

A special kind of isopleth, the *topographic contour* (or isohypse), is shown on the maps in Figures A1.2, A1.4(*a*) and in the rear endpapers. Topographic contours show the configuration of land surface features, such as hills, valleys, and basins. Topographic map reading is a skill that is especially useful for outdoor activities, such as camping and hiking. The *Exercise Manual* that accompanies this book provides a concluding section on U.S. topographic maps that describes how they are constructed and how to read them.

Maps are especially valuable tools for the geographer. In this appendix we have touched briefly on the art and science of cartography—a very rich field that deserves study in its own right by serious students of geography. If you are interested in maps, learning more about cartography will be quite rewarding.

Remote Sensing and Geographic Information Systems

Aprimary activity of geographers is the acquisition, processing, and display of *spatial data*—pieces of information that are in some way associated with a specific location or area of the earth's surface. Two important areas of scientific and technical development have greatly extended the ability of geographers to acquire, examine, and manipulate spatial data in recent years—**remote sensing** and **geographic information systems**. Utilizing the power of computers to acquire and process large volumes of spatial data, these technologies have allowed geographers, geologists, geophysicists, ecologists, planners, landscape architects, and others to develop applications of spatial data processing ranging from planning land subdivisions on the fringes of suburbia to monitoring the deforestation of the Amazon basin. This appendix provides a brief introduction to some of the key concepts in these two dynamic areas of the geographic sciences.

REMOTE SENSING

Remote sensing refers to gathering information from great distances and over broad areas, usually through instruments mounted on aircraft or orbiting space vehicles that are known collectively as *remote sensors*. These instruments measure electromagnetic radiation coming from the earth's surface and atmosphere as probed from the aircraft or spacecraft platform. As we saw in Chapter 2, all substances, whether naturally occurring or synthetic, are capable of reflecting, absorbing, and emitting electromagnetic energy. The spatial data acquired by remote sensors are typically displayed as images—photographs or similar depictions on a computer screen or color printer—but are often processed further to provide many other types of outputs, as we will see in the examples below.

Radar—An Active Remote Sensing System

There are two classes of remote sensor systems: active and passive. *Active systems* use a beam of wave energy as a source, sending the beam toward an object or surface. Part of the energy is reflected back to the source, where it is recorded by a detector. A simple analogy would be the use of a spotlight on a dark night to illuminate a target, which reflects light back to the eye.

Radar is an example of an active sensing system. Radar uses radiation from the *microwave* portion of the electromagnetic spectrum, so named because the waves have a short wavelength with respect to other types of radio waves. Radar systems emit short pulses of microwave radiation and then "listen" for a returning microwave echo. Many radar systems can penetrate clouds to provide images of the earth's surface in any weather. At shorter wavelengths, however, microwaves can be scattered by water droplets and produce a return signal sensed by the radar apparatus. This effect is the basis for weather radars, which can detect rain and hail and are used in local weather forecasting.

Figure A2.1 shows a radar image of the folded Appalachian Mountains in south-central Pennsylvania. It is produced by an airborne radar instrument that sends pulses of radio waves downward and sideward as the airplane flies forward. Surfaces oriented most near-

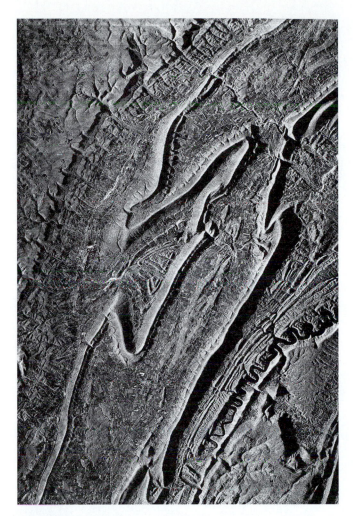

Figure A2.1 Radar image of a portion of the folded Appalachians in south-central Pennsylvania. The area shown is about 40 km (25 mi) wide. Compare this with Figure 16.15, which is a Landsat image of this region. (SAR image courtesy of Intera Technologies Corporation, Calgary, Alberta, Canada.)

ly at right angles to the slanting radar beam will return the strongest echo and therefore appear lightest in tone. In contrast, those surfaces facing away from the beam will appear darkest. The effect is to produce an image resembling a shaded relief map. Various types of surfaces, such as forest, rangeland, and agricultural fields, can also be identified by variations in image tone and pattern.

Passive Remote Sensing Systems

Passive systems measure radiant energy that is reflected or emitted by an object. Reflected energy falls mostly in

the visible light and near-infrared regions, while emitted energy lies in the longer thermal infrared region. The most familiar instrument of the passive type is the camera, which uses film that is sensitive to the light energy reflected from the scene. Aerial photographs, taken by downward-pointing cameras from aircraft, have been in wide use since before World War II. Commonly, the field of one photograph overlaps the next along the plane's flight path, so that the photographs can be viewed stereoscopically for a three-dimensional effect. Because of its high resolution (degree of sharpness), aerial photography remains one of the most valuable of the older remote-sensing techniques.

Photography can also extend into infrared wavelengths. Within the infrared spectrum, there is a region in the near-infrared, immediately adjacent to the visible red region, in which reflected rays can be recorded by cameras with suitable film and filter combinations. One example is color infrared film (Figure A2.2). In this type of film, the red color is produced as a response to infrared light, the green color is produced by red light, and the blue color by green light. Because healthy, growing vegetation reflects much more strongly in the near-infrared than in the red or green regions of the spectrum, vegetation has a characteristic red appearance. Figure 15.21, which shows a portion of the Mississippi River and its floodplain, is an example of an air photo taken with color-infrared film.

Satellite images are often processed to give the appearance of color-infrared film. Examples are Figures 16.13 and 17.15. On these images, agricultural crops appear as color shades ranging from pink to orange-red to deep red. Mature crops and dried vegetation (dormant grasses) appear yellow or brown. Urbanized areas typically appear in tones of blue and gray. Shallow water areas appear blue; deep water appears dark blue to blue-black.

Photography has been extended to greater distances through the use of cameras on orbiting space vehicles. A recent Space Shuttle flight included a specially constructed large-format camera, designed to produce very large, very detailed transparencies of the earth's surface suitable for precise topographic mapping. An excellent example is shown in Figure A2.3.

Scanning Systems

Passive remote sensing also includes images acquired by scanning systems, which may be mounted in aircraft or orbiting space vehicles. *Scanning* is the process of receiv-

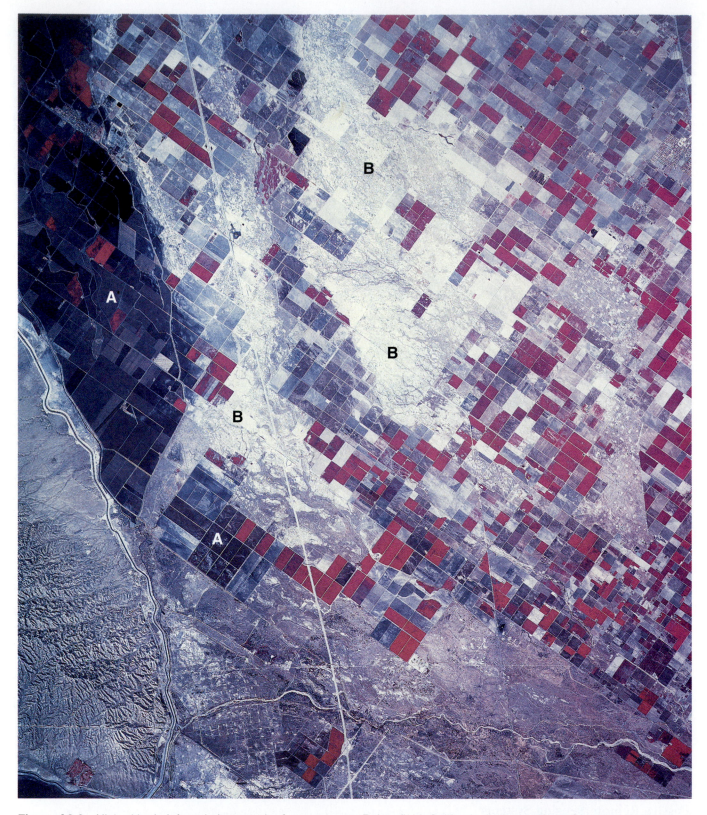

Figure A2.2 High-altitude infrared photograph of an area near Bakersfield, California, in the southern San Joaquin Valley, taken by a NASA U-2 aircraft flying at approximately 18 km (60,000 ft). Photos such as this can be used to study problems associated with agriculture. Problems affecting crop yields arise in (A) perched ground water areas, which appear dark, and (B) areas of high soil salinity, which appear light. The various red tones are associated with different types of crops. (Courtesy NASA, compiled and annotated by John E. Estes and Leslie W. Senger.)

Figure A2.3 This color infrared photograph was taken with a large-format camera carried on a flight of the NASA Space Shuttle. It gives an extremely detailed picture of the terrain and is suitable for precision mapping. The area shown, about 125 km in width, includes the Sulaiman Range in West Pakistan (upper left). The Indus and Sutlej rivers, fed by snowmelt from the distant Hindu Kush and Himalaya Ranges, cross the scene flowing from northeast to southwest. A mottled pattern of green fields (red) interspersed with barren patches of saline soil (white) covers much of the lower fourth of the area. (NASA)

Figure A2.4 Multispectral scanning from aircraft. As the aircraft flies forward, the scanner sweeps side to side. The result is a digital image covering the overflight area. (A. H. Strahler.)

ing information instantaneously from only a very small portion of the area being imaged (Figure A2.4). The scanning instrument senses a very small field of view that runs rapidly across the ground scene. Light from the field of view is focused on a detector that measures its intensity at very short time intervals and records the intensities as a series of numbers that can be processed by a computer. Later, the computer reconstructs an image of the ground scene from the measurements acquired by the scanning system.

Most scanning systems in common use are *multispectral scanners*. These devices measure brightness in several wavelength regions simultaneously. An example is the Multispectral Scanning System (MSS) used aboard the Landsat series of earth-observing satellites. This instrument simultaneously collects reflectance data in four spectral bands. Two are bands in the visible light spectrum (one green, one red), and two in the infrared region. A successor to this system, the Landsat Thematic Mapper (TM), collects data in seven spectral bands. The text contains a number of fine examples of Landsat images. (See Figures 14.19, 16.13, 17.15, 17.16, 17.17, 17.31, and 18.12.)

Another type of scanner provides images from the *thermal infrared* portion of the spectrum, which is sensitive to the temperature of the objects in the scene. Since warmer objects emit more infrared radiation than do cooler ones, the warmer objects will appear lighter on thermal infrared imagery. An example is shown in Figure 2.3. Here, a color spectrum has been used so that the warmest surfaces are shown in red, while cooler surfaces appear in blue, violet, or black.

Orbiting Earth Satellites

With the development of orbiting earth satellites carrying remote sensing systems, remote sensing has expanded into a major branch of geographic research, going far beyond the limitations of conventional aerial photography. One reason for this advance is that most satellite remote sensors are scanners that provide data instead of photographs. These data can be processed and enhanced by computers much more effectively than photographic images. Another reason is the ability of orbiting satellites to monitor nearly all the earth's surface. As a result, global and regional studies that could not have been carried out with air photographs have become possible.

How does a satellite sensor cover the entire earth? Nearly all satellites used for remote sensing are placed in a *sun-synchronous orbit*. In this orbit, all the satellite images of a location are acquired at about the same time of day. For the Landsat series, this is normally between 9:30 and 10:30 A.M. local time. This means that images collected on different dates can easily be compared, since they are all illuminated in a similar way by the sun. The orbit is also designed so that images are acquired from all locations on the globe before the orbit repeats. This can take from two to twenty-seven days, depending on how large an area is covered by each image.

Although in the past NASA has dominated the field of earth-observing satellites with its Landsat series, new satellites have been developed and launched by other nations. An example is the French satellite system, SPOT (an acronym for the French title *Système Probatoire d'Observation de la Terre*). The first SPOT satellite was launched in 1986, and the series continues to the present. Figure A2.5 shows a SPOT color image of the New York Harbor area. The resolution, which is finer than that of the Landsat series, shows how much small detail can be seen with new-generation sensing systems. In the 1998 time frame and beyond, several commercial firms are expected to produce and deliver satellite images with spatial resolutions as fine as 1 m (about 3 ft). These will be close to high-altitude aerial photography in quality.

Many technical advances will continue to be made in remote sensing. An international effort, led by NASA, is developing the *Earth Observing System*, which includes many satellites and sensing instruments to be used to study global change during the next few decades. Using these instruments, research in physical geography will be significantly enhanced. Remote sensing will be extended not only into remote areas of the earth that have never been mapped in detail, but also to global studies of the earth as a whole, utilizing the vantage point of earth orbit as never before in the history of science.

Figure A2.5 Digital image of New York Harbor, acquired by the SPOT satellite in June 1986. The data are presented as a color infrared picture. The tip of Manhattan is at the top center; the blocky, dark structures at the tip are Wall Street skyscrapers and their shadows. Just below the tip is Governor's Island, a Coast Guard military reservation, and immediately to the west (left) is small Liberty Island, home of the Statue of Liberty. The long container-loading piers of Port Newark are visible on the left, as are piers on the Brooklyn shoreline to the right. At the center bottom, the Verrazano Narrows Bridge connects Brooklyn with Staten Island. The fine detail of the street grid pattern, and even of the wakes of ships in the harbor, shows the fine resolution of the SPOT's imaging instrument. (Copyright © 1993 by CNES; Courtesy SPOT Image Corporation, Reston, Virginia.)

GEOGRAPHIC INFORMATION SYSTEMS

Another new area of technical development in the processing of spatial data is that of *geographic information systems*. What is a geographic information system, or *GIS*?

In essence, it is a system for acquiring, processing, storing, querying, creating, and displaying spatial data.

A simple example of a GIS is a map overlay system. Imagine that a planner is deciding how to divide a tract of land into building lots. Appropriate inputs would include a topographic map, giving contour lines of equal elevation; a vegetation map, showing the type of existing vegetation cover; a map of existing roads and trails; a map of streams and watercourses; a map of wetlands areas; a map of utility corridors crossing the area (power transmission lines and gas pipelines); and so forth. From these, the planner would locate the positions of roads and bridges; identify steep slopes, utility corridors, wetlands, and regions near streams as conservation areas; and then lay out individual lots having a minimum area of land suitable for building, access to roads and utilities, and other necessary attributes.

To do this, the planner needs to be able to overlay the various maps. Normally, this would be done by tracing them onto clear layers, then stacking the layers as needed while drawing the subdivision plans on the top layer. However, some problems may be encountered in this process. One common one is that the maps may not be drawn to the same scale or compiled on the same map projection. Another is that sometimes some maps need further work, as, for example, to identify areas of steep slopes given a map of topographic contours. Modern, computer-based geographic information systems are designed to solve these problems and allow ready manipulation of spatial data for diverse applications.

Spatial Objects in Geographic Information Systems

Geographic information systems are designed to manipulate spatial objects. A *spatial object* is a geographic area to which some information is attached. This information may be as simple as a place name or as complicated as a large data table with many types of information. A spatial object will normally have a boundary, which is described by a *polygon* that outlines the object. Some spatial objects are illustrated in Figure A2.6.

Spatial objects also include points and lines. A *point* can be thought of as a special type of spatial object with a location, but no area. A *line* is also a spatial object with no area, but it has two points associated with it, one for each end of the line. These special points are often referred to as *nodes*. Normally, a line is straight, but it can also be defined as a smooth curve having a certain shape. If the two nodes marking the ends of the line are differentiated as starting and ending, then the line has a direction. If the line has a direction, then its two sides can be distinguished. This allows information to be attached to each side—for example, labels differentiating land from water. Lines connect to other lines when they share a common node. A series of connected lines that form a closed chain is a polygon.

By defining spatial objects in this way, computer-based geographic information systems allow easy

Figure A2.6 Some different types of spatial objects.

manipulation of the objects and permit many different types of operations to compare objects and generate new objects. As an example, suppose we have a GIS data layer composed of conservation land in a region represented as polygons and another layer containing the location of preexisting water wells as points within the region (Figure A2.7). It is very simple to use the GIS to identify the wells that are on conservation land. Or, the conservation polygons containing wells may also be identified and may even be output as a new data layer. By comparing the conservation layer with a road network layer portrayed as a series of lines, we could identify the conservation polygons containing roads. We could also compare the conservation layer to a layer of polygons showing vegetation type, and tabulate the amount of conservation land in forest, grassland, brush, and so forth. We could even calculate distance zones around a spatial object, for example, to create a map of buffer zones that are located within, say, 100 meters of conservation land. Many other possible mannipulations exist.

Key Elements of a GIS

A geographic information system has five elements: data acquisition, preprocessing, data management, data manipulation and analysis, and product generation. Each is a component or process needed to ensure the functioning of the system as a whole. In the *data acquisition* process, data are gathered together for the particular application. These may include maps, air photos, tabular data, and other forms as well. In *preprocessing*, the assembled spatial data are converted to forms that can be ingested by the GIS to produce data layers of spatial objects and their associated information. The *data management* component creates, stores, retrieves, and modifies data layers and spatial objects. It is essential to proper functioning of all parts of the GIS. The *manipulation and analysis* component is the real workhorse of

Figure A2.7 Data layers in a simple geographic information system.

the GIS. Utilizing this component, the user asks and answers questions about spatial data and creates new data layers of derived information. The last component of the GIS, *product generation*, produces output products in the form of maps, graphics, tabulations, statistical reports and the like that are the end products desired by the users. Taken together, these components provide a system that can serve many geographic applications at many scales.

Many new and exciting areas of geographic research are associated with geographic information systems, ranging from algorithms for manipulation of spatial data to the modeling of spatial processes using a GIS. An especially interesting area is understanding how out-

puts are affected by errors and uncertainty in spatial data inputs and how this information can be communicated effectively to users.

Remote sensing and geographic information systems are two new, emerging fields of geographic research and application. Both rely heavily on computers and computer technology. Given the rate at which computers become ever more powerful as the technology improves, we can expect great strides in these fields in future years.

The Canadian System of Soil Classification*

The formation of the National Soil Survey Committee of Canada in 1940 was a milestone in the development of soil classification and of pedology generally in Canada. Prior to that time, Canada used an American system based on classification of soil horizons. Although the Canadian experience showed that the concept of zonal soils was useful in the western plains, it proved less applicable in eastern Canada where parent materials and relief factors had a dominant influence on soil properties and development in many areas.

Canadian pedologists observed closely the evolution of the U.S. Comprehensive Soil Classification System (CSCS) during the 1950s and 1960s and ultimately adopted several important features of that system. Nevertheless, the special needs of a workable Canadian national system required that a completely independent classification system be established. Because Canada lies entirely in a latitude zone poleward of the 40th parallel, there was no need to incorporate those soil orders found only in lower latitudes. Furthermore, the vast expanse of Canadian territory lying within the boreal forest and tundra climates necessitated the recognition at the highest taxonomic level (the order) of soils of cold regions that appear only as suborders and even as great groups in the CSCS—for example, the Cryaquents and Cryorthents that occupy much of the soils map of northern Canada above the 50th parallel (Figure 19.10).

The overall philosophy of the Canadian system is pragmatic: the aim is to organize knowledge of soils in a reasonable and usable way. The system is a natural or taxonomic one in which the classes (taxa) are based on properties of the soils themselves and not on interpretations of the soils for various uses. Thus, the taxa are concepts based on generalization of properties of real bodies of soils rather than idealized concepts of the kinds of soils that would result from the action of presumed genetic processes. In this respect, the philosophy agrees with that used in the CSCS. Although taxa in the Canadian system are defined on the basis of actual soil properties that can be observed and measured, the system has a genetic bias in that properties or combinations of properties that reflect genesis are favored in distinguishing among the higher taxa. Thus, the soils brought together under a single soil order are seen as the product of a similar set of dominant soil-forming processes resulting from broadly similar climatic conditions.

The Canadian system recognizes the *pedon* as the basic unit of soils; it is defined as in the CSCS. Major mineral horizons of the soil (A, B, C) are defined in much the same way as in the U.S. system. Thus, the Canadian system of soil taxonomy is more closely related to the U.S. system than to any other. Both are hierarchical, and the taxa are defined on the basis of measurable soil properties. However, they differ in several respects. The Canadian system is designed to classify only the soils that occur in Canada and is not a comprehensive system. The U.S. system includes the suborder, a taxon not recognized in the Canadian system. Because 90 percent of the area of Canada is not likely to be cultivated, the Canadian system does not recognize as diagnostic those horizons strongly affected by plowing and application of soil conditioners and fertilizers.

---

*Throughout this section numerous sentences and phrases are taken verbatim or paraphrased from the following work: Canada Soil Survey Committee, *The Canadian System of Soil Classification*, Research Branch, Canada Department of Agriculture, Publication 1646, 1978. Table A3.1 and Figure A3.2 are also compiled from this source.

SOIL HORIZONS AND OTHER LAYERS

The definitions of classes in the Canadian system are based mainly on kinds, degrees of development, and sequence of soil horizons and other layers in pedons. The major mineral horizons are A, B, and C. The major organic horizons are L, F, and H, which are mainly forest litter at various stages of decomposition, and O, which is derived mainly from bog, marsh, or swamp vegetation. Subdivisions of horizons are labeled by adding lowercase suffixes to the major horizon symbols—for example, Ah or Ae.

Besides the horizons, nonsoil layers are recognized. Two such layers are R, rock, and W, water. Lower mineral layers not affected by pedogenic processes are also identified. In organic soils, layers are described as tiers.

The principal mineral horizons, A, B, and C, are defined as follows:

A   Mineral horizon found at or near the surface in the zone of leaching or eluviation of materials in solution or suspension, or of maximum *in situ* accumulation of organic matter, or both.

B   Mineral horizon characterized by enrichment in organic matter sesquioxides, or clay; or by the development of soil structure; or by change of color denoting hydrolysis, reduction, or oxidation.

C   Mineral horizon comparatively unaffected by the pedogenic processes operative in A and B horizons. The processes of gleying and the accumulation of calcium and magnesium and more soluble salts can occur in this horizon.

Lowercase suffixes, used to designate subdivisions of horizons, are shown in Table A3.1.

Table A3.1   Subhorizons and Organic Horizons of the Canadian System of Soil Classification

Subhorizons (Lowercase Suffixes)

b   Buried soil horizon.

c   Cemented (irreversible) pedogenic horizon.

ca   Horizon of secondary carbonate enrichment in which the concentration of lime exceeds that in the unenriched parent material.

e   Horizon characterized by the eluviation of clay, Fe, Al, or organic matter alone or in combination.

f   Horizon enriched with amorphous material, principally Al and Fe combined with organic matter; reddish near upper boundary, becoming yellower at depth.

g   Horizon characterized by gray colors, or prominent mottling, or both, indicative of permanent or intense reduction.

h   Horizon enriched with organic matter.

j   Used as a modifier of suffixes e, f, g, n, and t to denote an expression of, but failure to meet, the specified limits of a suffix it modifies.

k   Denotes presence of carbonate as indicated by visible effervescence when dilute HCl is added.

m   Horizon slightly altered by hydrolysis, oxidation, or solution, or all three to give a change in color or structure, or both.

n   Horizon in which the ratio of exchangeable Ca to exchangeable Na is 10 or less. It must also have the following distinctive morphological characteristics: prismatic or columnar structure, dark coatings on ped surfaces, and hard to very hard consistence when dry.

p   Horizon disturbed by human activities such as cultivation, logging, and habitation.

s   Horizon of salts, including gypsum, which may be detected as crystals or veins or as surface crusts of salt crystals.

sa   Horizon with secondary enrichment of salts more soluble than Ca and Mg carbonates; the concentration of salts exceeds that in the unenriched parent material.

t   Illuvial horizon enriched with silicate clay.

u   Horizon that is markedly disrupted by physical or faunal processes other than cryoturbation.

x   Horizon of fragipan character. (A fragipan is a loamy subsurface horizon of high bulk density and very low organic matter content. When dry it is hard and seems to be cemented.)

y   Horizon affected by cryoturbation as manifested by disrupted or broken horizons, incorporation of materials from other horizons, and mechanical sorting.

z   A frozen layer.

Organic Horizons

O   Organic horizon developed mainly from mosses, rushes, and woody materials.

L   Organic horizon characterized by an accumulation of organic matter derived mainly from leaves, twigs, and woody materials in which the organic structures are easily discernible.

F   Same as L, above, except that original structures are difficult to recognize.

H   Organic horizon characterized by decomposed organic matter in which the original structures are indiscernible.

SOIL ORDERS OF THE CANADIAN SYSTEM

Nine soil orders make up the highest taxon of the Canadian System of Soil Classification. Listed in alphabetical order, they are as follows.

Brunisolic   Gleysolic   Podzolic

Chernozemic   Luvisolic   Regosolic

Cryosolic   Organic   Solonetzic

Table A3.2 lists the great groups within each order.

Brunisolic Order

The central concept of the *Brunisolic order* is that of soils under forest having brownish-colored Bm horizons. Most Brunisolic soils are well to imperfectly drained. They occur in a wide range of climatic and vegetative environments, including boreal forest; mixed forest,

Table A3.2   Great Groups of the Canadian Soil Classification System

Order	Great Group
Brunisolic	Melanic Brunisol
	Eutric Brunisol
	Sombric Brunisol
	Dystric Brunisol
Chernozemic	Brown
	Dark Brown
	Black
	Dark Gray
Cryosolic	Turbic Cryosol
	Static Cryosol
	Organic Cryosol
Gleysolic	Humic Gleysol
	Gleysol
	Luvic Gleysol
Luvisolic	Gray Brown Luvisol
	Gray Luvisol
Organic	Fibrisol
	Mesisol
	Humisol
	Folisol
Podzolic	Humic Podzol
	Ferro-Humic Podzol
	Humo-Ferric Podzol
Regosolic	Regosol
	Humic Regosol
Solonetzic	Solentz
	Solodized Solonetz
	Solod

shrubs, and grass; and heath and tundra. As compared with the Chernozemic soils, the Brunisolic soils show a weak B horizon of accumulation attributable to their moister environment. Brunisolic soils lack the diagnostic podzolic B horizon of the Podzolic soils, in which accumulation in the B horizon is strongly developed. The Melanic Brunisol shown in profile in Figure A3.1 can be found in the St. Lawrence Lowlands, surrounded by Podzolic soils (Figure A3.2).

Chernozemic Order

The general concept of the *Chernozemic order* is that of well to imperfectly drained soils having surface horizons darkened by the accumulation of organic matter from the decomposition of xerophytic or mesophytic grasses and forms representative of grassland communities or of grassland–forest communities with associated shrubs and forbs. The major area of Chernozemic soils is the cool, subarid to subhumid interior plains of western Canada. Most Chernozemic soils are frozen during some period each winter, and the soil is dry at some period each summer. The mean annual temperature is higher than 0°C and usually less than 5.5°C. The associated climate is typically the semiarid (steppe) variety of the dry midlatitude climate ⑨ (Figure A3.2).

Essential to the definition of soils of the Chernozemic order is that they must have an A horizon (typically, Ah) in which organic matter has accumulated, and they must meet several other requirements. The A horizon is at least 10 cm thick; its color is dark brown to black. It usually has sufficiently good structure that it is neither massive and hard nor single-grained when dry. The profile shown in Figure A3.1 is that of the Orthic subgroup of the Brown great group; it shows a Bm horizon that is typically of prismatic structure. The C horizon is one of lime accumulation (Cca). Clearly, the Chernozemic soils can be closely correlated with the Mollisols of the CSCS.

Cryosolic Order

Soils of the *Cryosolic order* occupy much of the northern third of Canada where permafrost remains close to the surface of both mineral and organic deposits. Cryosolic soils predominate north of the tree line, are common in the subarctic forest area in fine-textured soils, and extend into the boreal forest in some organic materials and into some alpine areas of mountainous regions. Cryoturbation (intense disturbance by freeze–thaw activity) of these soils is common, and it may be indicated by patterned ground features such as sorted and nonsorted nets, circles, polygons, stripes, and earth hummocks.

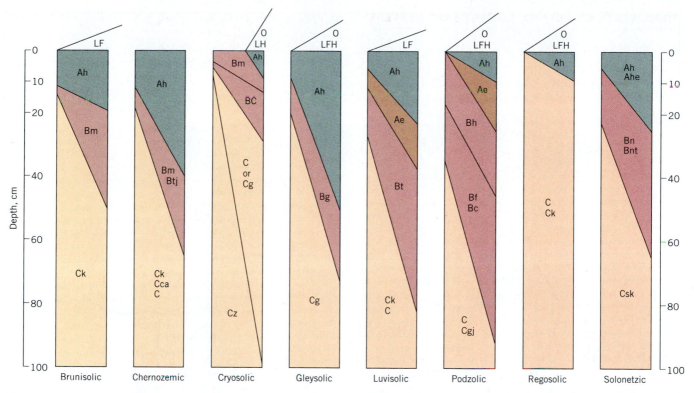

Figure A3.1 Representative schematic profiles of eight of the nine orders of the Canadian system of soil classification. Slanting lines show the range in depth and thickness of each horizon. (The horizon planes are actually approximately horizontal within the pedon.) See Table A3.1 for an explanation of symbols. (From Canada Soil Survey Committee, Research Branch, Canada Department of Agriculture, 1978.)

Cryosolic soils are found in either mineral or organic materials that have permafrost either within 1 m of the surface or within 2 m if more than one-third of the pedon has been strongly cryoturbated, as indicated by disrupted, mixed, or broken horizons. The profile shown in Figure A3.1 is that of the Orthic subgroup of the Static Cryosol great group. Note the presence of organic L, H, and O surface horizons and the thin Ah horizon. The Cryosolic soils are closely correlated with the Cryaquepts of the CSCS.

Gleysolic Order

Soils of the *Gleysolic order* have features indicating periodic or prolonged saturation with water and reducing conditions. They commonly occur in patchy association with other soils in the landscape. Gleysolic soils are usually associated with either a high ground water table at some period of the year or temporary saturation above a relatively impermeable layer. Some Gleysolic soils may be submerged under shallow water throughout the year. The profile shown in Figure A3.1 is that of the Gleysol great group. It has a thick Ah horizon. The underlying Bg horizon is grayish and shows mottling typical of reducing conditions.

Luvisolic Order

Soils of the *Luvisolic order* generally have light-colored, eluvial horizons (Ae), and they have illuvial B horizons (Bt) in which silicate clay has accumulated. These soils develop characteristically in well to imperfectly drained sites, in sandy loam to clay base-saturated parent materials under frost vegetation in subhumid to humid, mild to very cold climates. The genesis of Luvisolic soils is thought to involve the suspension of clay in the soil solution near the soil surface, downward movement of the suspended clay with the soil solution, and deposition of the translocated clay at a depth where downward motion of the soil solution ceases or becomes very slow. The representative profile shown in Figure A3.1 is that of the Orthic subgroup of the Gray Brown Luvisol great group.

Luvisolic soils occur from the southern extremity of Ontario to the zone of permafrost and from Newfoundland to British Columbia. The largest area of these soils are Gray Luvisols occurring in the central to northern interior plains under deciduous, mixed, and coniferous forest. In this location, they appear to correlate with the Boralfs of the Alfisol order in the CSCS. Gray-Brown Luvisolic soils of southern Ontario would correlate with the suborder of Udalfs.

Figure A3.2   Generalized map of soil regions of Canada. (Courtesy of Land Resources Research Institute, Agriculture Canada.) (Illustration is taken from *Fundamentals of Soil Science*, 7th ed., by Henry D. Foth, John Wiley & Sons.)

1. Podzols, Brunisols
2. Luvisols, Brunisols, rock
3. Brunisols, rock
4. Luvisols, Organics
5. Chernozemic soils
6. Luvisols
7. Brunisols, Gleysols
8. Organics, Cryosols
9. Podzols, Brunisols, Organics
10. Brunisols, Cryosols, rock
11. Cryosols, rock, ice
12. Mainly rock

Approximate Equivalents
in Soil Taxonomy

Brunisol–Inceptisol
Chernozemic–Mollisol
Cryosol–Inceptisols
  with permafrost

Gleysol–Inceptisol, wet
Luvisol–Alfisol
Organic–Histosol
Podzol–Spodosol

Organic Order

Soils of the *Organic order* are composed largely of organic materials. They include most of the soils commonly known as peat, muck, or bog soils. Organic soils contain 17 percent or more organic carbon (30 percent organic matter) by weight. Most Organic soils are saturated with water for prolonged periods. They occur widely in poorly and very poorly drained depressions and level areas in regions of subhumid to humid climate and are derived from vegetation that grows in such sites. However, one group of Organic soils consists of leaf litter overlying rock or fragmental material; soils of this group may occur on steep slopes and may rarely be saturated with water. (No profile of the Organic soils is shown in Figure A3.1.) Organic soils can be correlated with the Histosols of the CSCS.

Podzolic Order

Soils of the *Podzolic order* have B horizons in which the dominant accumulation product is amorphous material composed mainly of humified organic matter in varying degrees with Al and Fe. Typically, Podzolic soils occur in coarse- to medium-textured, acid parent materials, under forest and heath vegetation in cool to very cold humid to very humid climates. Podzolic soils can usually be readily recognized in the field. Generally, they have organic surface horizons that are commonly L, F, and H. Most Podzolic soils have a reddish brown to black B horizon (Bh) with an abrupt upper boundary. The profile shown in Figure A3.1 is that of the Orthic subgroup of the Humic Podzol great group. The Podzolic soils correspond closely to the Spodosols (Orthods) of the CSCS.

Regosolic Order

Regosolic soils have weakly developed horizons. The lack of development of genetic horizons may be due to any number of factors: youthfulness of the parent material, for example, recent alluvium; instability of the material, for example, colluvium on slopes subject to mass wasting; nature of the material, for example, nearly pure quartz sand; climate, for example, dry cold conditions. Regosolic soils are generally rapidly to imperfectly drained. They occur in a wide range of vegetation and climates. The profile shown in Figure A3.1 is that of the Orthic subgroup of the Regosol great group. It has only a thin humic A horizon (Ah) and a surface horizon of organic materials. Regosolic soils correspond with the Entisols of the CSCS.

Solonetzic Order

Soils of the *Solonetzic order* have B horizons that are very hard when dry and swell to a sticky mass of very low permeability when wet. Typically, the Solonetzic B horizon has prismatic or columnar macrostructure that breaks into hard to extremely hard, blocky peds with dark coatings. Solonetzic soils occur on saline parent materials in some areas of the semiarid to subhumid interior plains in association with Chernozemic soils and to a lesser extent with Luvisolic and Gleysolic soils. Most Solonetzic soils are associated with a vegetative cover of grasses and forbs. The profile shown in Figure A3.1 is that of the Brown subgroup of the Solonetz great group.

Solonetzic soils are thought to have developed from parent materials that were more or less uniformly salinized with salts high in sodium. Leaching of salts by descending rainwater presumably mobilizes the sodium-saturated colloids. The colloids are apparently carried downward and deposited in the B horizon. Further leaching results in depletion of alkali cations in the A horizon, which becomes acidic, and a platy Ahe horizon usually develops. The underlying Solonetzic B horizon (Bn, Bnt) usually consists of darkly stained, fused, intact columnar peds. This stage is followed by the structural breakdown of the upper part of the B horizon and eventually by its complete destruction in the most advanced stage, known as solodization. Solonetzic soils are correlated with the suborder of Argids in the order of Aridisols under the CSCS.

GLOSSARY

This glossary contains definitions of terms shown in the text in italics or boldface. Terms that are *italicized* within the definitions will be found as individual entries elsewhere in the glossary.

A horizon *mineral* horizon of the *soil,* overlying the *E* and *B* horizons.

ablation a wastage of glacial ice by both *melting* and *evaporation.*

abrasion erosion of *bedrock* of a *stream channel* by impact of particles carried in a *stream* and by rolling of larger *rock* fragments over the stream bed; abrasion is also an activity of glacial ice, waves, and *wind.*

abrasion platform sloping, nearly flat *bedrock* surface extending out from the foot of a *marine cliff* under the shallow water of the breaker zone.

absorption of radiation transfer of *electromagnetic energy* into heat *energy* within a *gas* or *liquid* through which the *radiation* is passing or at the surface of a *solid* struck by the radiation.

abyssal plain large expanse of very smooth, flat ocean floor found at depths of 4600 to 5500 m (about 15,000 to 18,000 ft).

accelerated erosion *soil erosion* occurring at a rate much faster than *soil horizons* can be formed from the parent *regolith.*

accretion of lithosphere production of new *oceanic lithosphere* at an active *spreading plate boundary* by the rise and solidification of *magma* of basaltic composition.

acid deposition the deposition of acid raindrops or dry acidic dust particles on vegetation and ground surfaces.

acid mine drainage sulfuric acid effluent from *coal* mines, mine tailings, or spoil ridges made by *strip mining.*

acid rain rainwater having an abnormally low *pH,* between 2 and 5, as a result of air pollution by sulfur *oxides* and nitrogen oxides.

active continental margins continental margins that coincide with tectonically active plate boundaries. (See also *continental margins, passive continental margins.*)

active systems *remote sensing* systems that emit a beam of wave *energy* at a source and measure the intensity of that *energy* reflected back to the source.

adiabatic lapse rate (See *dry adiabatic lapse rate, wet adiabatic lapse rate.*)

adiabatic process change of temperature within a *gas* because of compression or expansion, without gain or loss of heat from the outside.

advection fog *fog* produced by *condensation* within a moist basal air layer moving over a cold land or water surface.

aerosols tiny dust particles present in the *atmosphere,* so small and light that the slightest movements of air keep them aloft.

aggradation raising of *stream channel* altitude by continued *deposition* of *bed load.*

air a mixture of gases that surrounds the earth.

air mass extensive body of air within which upward gradients of temperature and moisture are fairly uniform over a large area.

air pollutant an unwanted substance injected into the *atmosphere* from the earth's surface by either natural or human activities; includes *aerosols, gases,* and *particulates.*

air temperature temperature of air, normally observed by a *thermometer* under standard conditions of shelter and height above the ground.

albedo percentage of *electromagnetic energy* reflected from a surface.

albic horizon pale, often sandy *soil horizon* from which *clay* and free iron *oxides* have been removed; found in the profile of the *Spodosols.*

Alfisols *soil order* consisting of *soils* of humid and subhumid climates, with high *base status* and an *argillic horizon.*

alluvial fan gently sloping, conical accumulation of coarse *alluvium* deposited by a *braided stream* undergoing *aggradation* below the point of emergence of the channel from a narrow *gorge* or *canyon.*

alluvial meanders sinuous bends of a *graded stream* flowing in the alluvial deposit of a *floodplain.*

alluvial river *stream* of low *gradient* flowing upon thick deposits of *alluvium* and experiencing approximately annual overbank flooding of the adjacent *floodplain.*

alluvial terrace benchlike landform carved in *alluvium* by a *stream* during *degradation.*

alluvium any stream-laid *sediment* deposit found in a *stream channel* and in low parts of a stream valley subject to flooding.

alpine chains high mountain ranges that are narrow belts of *tectonic activity* severely deformed by *folding* and thrusting in comparatively recent geologic time.

alpine glacier long, narrow, mountain *glacier* on a steep downgrade, occupying the floor of a troughlike valley.

alpine tundra a plant *formation class* within the *tundra biome*, found at high altitudes above the limit of *tree* growth.

amphibole group *silicate minerals* rich in calcium, magnesium, and iron, dark in color, high in *density*, and classed as *mafic minerals*.

andesite *extrusive igneous rock* of *diorite* composition, dominated by *plagioclase feldspar*; the extrusive equivalent of diorite.

Andisols a *soil order* that includes *soils* formed on volcanic ash; often enriched by organic matter, yielding a dark soil color.

anemometer *weather* instrument used to indicate *wind* speed.

angle of repose natural surface inclination (*dip*) of a *slope* consisting of loose, coarse, well-sorted *rock* or *mineral* fragments; for example, the *slip face* of a *sand dune*, a *talus slope*, or the sides of a *cinder cone*.

annuals plants that live only a single growing season, passing the unfavorable season as a seed or spore.

annular drainage pattern a *stream* network dominated by concentric (ringlike) major *subsequent streams*.

antarctic circle *parallel of latitude* at $66\frac{1}{2}°$S.

antarctic front zone frontal zone of interaction between antarctic *air masses* and polar air masses.

antarctic zone *latitude* zone in the latitude range 60° to 75° S (more or less), centered on the *antarctic circle*, and lying between the *subantarctic zone* and the *polar zone*.

anticlinal valley valley eroded in weak *strata* along the central line or axis of an eroded *anticline*.

anticline upfold of *strata* or other layered *rock* in an archlike structure; a class of *folds*. (See also *syncline*.)

anticyclone center of high *atmospheric pressure*.

aphelion point on the earth's elliptical orbit at which the earth is farthest from the sun.

aquatic ecosystem *ecosystem* of a *lake*, bog, pond, river, *estuary*, or other body of water.

Aquepts *suborder* of the *soil order Inceptisols*; includes Inceptisols of wet places, seasonally saturated with water.

aquiclude *rock mass* or layer that impedes or prevents the movement of *ground water*.

aquifer *rock* mass or layer that readily transmits and holds *ground water*.

arc curved line that forms a portion of a circle.

arctic circle *parallel of latitude* at $66\frac{1}{2}°$N.

arctic front zone frontal zone of interaction between arctic *air masses* and polar air masses.

arctic tundra a plant *formation class* within the *tundra biome*, consisting of low, mostly herbaceous plants, but with some very small stunted *trees*, associated with the *tundra climate* ⑫.

arctic zone *latitude* zone in the latitude range 60° to 75° N (more or less), centered about on the *arctic circle*, and lying between the *subarctic zone* and the *polar zone*.

arête sharp, knifelike divide or crest formed between two *cirques* by alpine glaciation.

argillic horizon *soil horizon*, usually the *B horizon*, in which *clay minerals* have accumulated by *illuviation*.

arid (dry climate subtype) subtype of the dry climates that is extremely dry and supports little or no vegetation cover.

Aridisols *soil order* consisting of soils of dry climates, with or without *argillic horizons*, and with accumulations of *carbonates* or soluble salts.

artesian well drilled well in which water rises under hydraulic pressure above the level of the surrounding *water table* and may reach the surface.

aspect compass orientation of a *slope* as an inclined element of the ground surface.

asthenosphere soft layer of the upper *mantle*, beneath the rigid *lithosphere*.

astronomical hypothesis explanation for *glaciations* and *interglaciations* making use of cyclic variations in the form of solar *energy* received at the earth's surface.

atmosphere envelope of gases surrounding the earth, held by *gravity*.

atmospheric pressure pressure exerted by the *atmosphere* because of the force of *gravity* acting on the overlying column of air.

atoll circular or closed-loop *coral reef* enclosing an open *lagoon* with no island inside.

atomic mass number total number of protons and neutrons within the nucleus of an atom.

atomic number number of protons within the nucleus of an atom; determines element name and chemical properties of the atom.

autumnal equinox *equinox* occurring on September 22 or 23.

available water capacity amount of water in soil that is available to plants; difference between *soil storage capacity* and *wilting point*.

axial rift narrow, trenchlike depression situated along the center line of the *midoceanic ridge* and identified with active sea-floor spreading.

axis of rotation center line around which a body revolves, as the earth's axis of rotation.

B horizon mineral *soil horizon* located beneath the A *horizon*, and usually characterized by a gain of *mineral matter* (such as *clay minerals* and oxides of aluminum and iron) and organic matter (*humus*).

backswamp area of low, swampy ground on the *floodplain* of an *alluvial river* between the *natural levee* and the *bluffs*.

backwash return flow of *swash* water under the influence of *gravity*.

badlands rugged land surface of steep *slopes*, resembling miniature mountains, developed on weak *clay* formations or clay-rich *regolith* by fluvial erosion too rapid to permit plant growth and soil formation.

bajada graded slope of *alluvial fan* and *pediment* surfaces between a mountain front and a playa in the arid landscapes of the southwestern United States and northern Mexico.

bar low ridge of *sand* built above water level across the mouth of a *bay* or in shallow water paralleling the shoreline; may also refer to embankment of sand or gravel on the floor of a *stream channel*.

bar (pressure) unit of pressure or *capillary tension* equal to 105 Pa (*pascals*); approximately equal to the pressure of the earth's *atmosphere* at sea level.

barchan dune *sand dune* of crescentic base outline with a sharp crest and a steep lee *slip face*, with crescent points (horns) pointing downwind.

barometer instrument for measurement of *atmospheric pressure*.

barrier island long, narrow island, built largely of beach *sand* and dune sand, parallel with the mainland and separated from it by a *lagoon*.

barrier reef *coral reef* separated from mainland *shoreline* by a *lagoon*.

barrier-island coast *coastline* with broad zone of shallow water offshore (a *lagoon*) shut off from the ocean by a *barrier island*.

basalt *extrusive igneous rock* of *gabbro* composition; occurs as *lava*.

base level lower limiting surface or level that can ultimately be attained by a *stream* under conditions of stability of the earth's crust and sea level; an imaginary surface equivalent to sea level projected inland.

base status of soils quality of a *soil* as measured by the presence or absence of *clay minerals* capable of holding large numbers of *bases*. Soils of high *base status* are rich in base-holding *clay minerals*; soils of low *base status* are deficient in such minerals.

bases certain positively charged *ions* in the *soil* that are also plant nutrients; the most important are calcium, magnesium, potassium, and sodium.

batholith large, deep-seated body of *intrusive igneous rock*, usually with an area of surface exposure greater than 100 km^2 (40 mi^2).

bauxite mixture of several *clay minerals*, consisting largely of aluminum *oxide* and water with impurities; a principal ore of aluminum.

bay a body of water sheltered from strong wave action by the configuration of the *coast*.

beach thick, wedge-shaped accumulation of *sand, gravel,* or cobbles in the zone of breaking waves.

beach drift transport of *sand* on a beach parallel with a *shoreline* by a succession of landward and seaward water movements at times when *swash* approaches obliquely.

bearing direction angle between a line of interest and a reference line, which is usually a line pointing north.

bed load that portion of the *stream load* moving close to the stream bed by rolling and sliding.

bedrock solid *rock* in place with respect to the surrounding and underlying rock and relatively unchanged by *weathering processes*.

biomass dry weight of living organic matter in an *ecosystem* within a designated surface area; units are kilograms of organic matter per square meter.

biome largest recognizable subdivision of *terrestrial ecosystems*, including the total assemblage of plant and animal life interacting within the *life layer*.

biosphere all living organisms of the earth and the environments with which they interact.

block mountains class of mountains produced by block faulting and usually bounded by *normal faults*.

block separation separation of individual joint blocks during the process of *physical weathering*.

blowout shallow depression produced by continued *deflation*.

bluffs steeply rising ground slopes marking the outer limits of a *floodplain*.

bog a shallow depression filled with organic matter, for example, a glacial *lake* or pond basin filled with *peat*.

Boralfs *suborder* of the *soil order Alfisols*; includes Alfisols of *boreal forests* or high mountains.

boreal forest variety of *needleleaf forest* found in the *boreal forest climate* (11) regions of North America and Eurasia.

boreal forest climate (11) cold *climate* of the *subarctic zone* in the northern *hemisphere* with long, extremely severe winters and several months of zero *potential evapotranspiration* (*water need*).

Borolls *suborder* of the *soil order Mollisols*; includes Mollisols of cold-winter semiarid plains (*steppes*) or high mountains.

braided stream *stream* with shallow channel in coarse *alluvium* carrying multiple threads of fast flow that subdivide and rejoin repeatedly and continually shift in position.

breaker sudden collapse of a steepened water wave as it approaches the *shoreline*.

broadleaf deciduous forest *forest* type consisting of broadleaf *deciduous trees* and found in the *moist subtropical climate* ⑥ and in parts of the *marine west-coast climate* ⑧. (See also *midlatitude deciduous forest*.)

broadleaf evergreen forest *forest* type consisting of broadleaf *evergreen trees* and found in the *wet equatorial* and tropical climates. (See also *low-latitude rainforest*.)

Brunisolic order a class of *forest soils* in the Canadian Soil Classification System with brownish *B horizon*.

butte prominent, steep-sided hill or peak, often representing the final remnant of a resistant layer in a region of flat-lying *strata*.

C horizon *soil horizon* comprised of *sediment* or *regolith* that is the *parent material* of the A, E, and B horizons above.

calcification accumulation of *calcium carbonate* in a soil, usually occurring in the B horizon, or sometimes in the C horizon.

calcite mineral having the composition *calcium carbonate*.

calcium carbonate compound consisting of calcium (Ca) and carbonate (CO_3) *ions*, formula $CaCO_3$, occurring naturally as the *mineral calcite*.

caldera large, steep-sided circular depression resulting from the explosion and subsidence of a *stratovolcano*.

canyon (See *gorge*.)

capillary tension a cohesive force among surface molecules of a *liquid* that gives a droplet its rounded shape.

carbohydrate class of organic compounds consisting of the elements carbon, hydrogen, and oxygen.

carbon dioxide the chemical compound CO_2, formed by the union of two atoms of oxygen and one atom of carbon; normally a *gas* present in low concentration in the *atmosphere*.

carbonates (carbonate minerals, carbonate rocks) *minerals* that are carbonate compounds of calcium or magnesium or both, that is, *calcium carbonate* or magnesium carbonate. (See also *calcite*.)

carbonic acid a weak acid created when CO_2 *gas* dissolves in water.

carbonic acid action chemical reaction of *carbonic acid* in rainwater, *soil water*, and *ground water* with *minerals;* most strongly affects carbonate minerals and *rocks*, such as limestone and marble; an activity of *chemical weathering*.

carnivore animal that consumes other animals as a source of metabolic *energy*.

cartography the science and art of making maps.

Celsius scale temperature scale in which the *freezing* point of water is 0°, and the boiling point is 100°.

Cenozoic Era last (youngest) of the *eras* of geologic time.

channel (See *stream channel*.)

chaparral sclerophyll scrub and dwarf *forest* plant *formation class* found throughout the coastal mountain ranges and hills of central and southern California.

chemical energy *energy* stored within an organic molecule and capable of being transformed into *heat* during metabolism.

chemical weathering chemical change in *rock*-forming minerals through exposure to atmospheric conditions in the presence of water; mainly involving *oxidation, hydrolysis, carbonic acid action*, or direct solution.

chemically precipitated sediment *sediment* consisting of *mineral matter* precipitated from a water solution in which the matter has been transported in the dissolved state as *ions*.

chernozem type of *soil order* closely equivalent to *Mollisol;* an order of the Canadian Soil Classification System.

Chernozemic order a class of grassland *soils* in the Canadian Soil Classification System with a thick *A horizon* rich in organic matter.

chert *sedimentary rock* composed largely of silicon dioxide and various impurities, in the form of nodules and layers, often occurring with *limestone* layers.

chinook wind a *local wind* occurring at certain times to the lee of the Rocky Mountains; a very dry wind with a high capacity to evaporate *snow*.

chlorofluorocarbons (CFCs) synthetic chemical compounds containing chlorine, fluorine, and carbon atoms that are widely used as coolant fluids in refrigeration systems.

cinder cone conical hill built of coarse *tephra* ejected from a narrow volcanic vent; a type of small *volcano*.

circle of illumination great circle that divides the globe at all times into a sunlit *hemisphere* and a shadowed hemisphere.

circum-Pacific belt chains of *andesite volcanoes* making up mountain belts and *island arcs* surrounding the Pacific Ocean basin.

cirque bowl-shaped depression carved in *rock* by glacial processes and holding the *firn* of the upper end of an *alpine glacier*.

clastic sediment *sediment* consisting of particles broken away physically from a parent *rock* source.

clay sediment *sediment* consisting of particles smaller than 0.004 mm in diameter.

clay minerals class of *minerals* produced by alteration of *silicate minerals*, having plastic properties when moist.

claystone *sedimentary rock* formed by lithification of *clay* and lacking *fissile* structure.

cliff sheer, near-vertical *rock* wall formed from flat-lying resistant layered *rocks*, usually *sandstone, limestone*, or *lava* flows; may refer to any near-vertical rock wall. (See also *marine cliff*.)

climate generalized statement of the prevailing *weather* conditions at a given place, based on statistics of a long period of record and including mean values, departures from those means, and the probabilities associated with those departures.

climax stable community of plants and animals reached at the end point of *ecological succession*.

climograph a graph on which two or more climatic variables, such as monthly mean temperature and monthly mean *precipitation*, are plotted for each month of the year.

cloud forest a type of low evergreen rainforest that occurs high on mountain slopes, where *clouds* and *fog* are frequent.

clouds dense concentrations of suspended water or ice particles in the diameter range 20 to 50 µm. (See *cumuliform clouds, stratiform clouds*.)

coal *rock* consisting of hydrocarbon compounds, formed of compacted, lithified, and altered accumulations of plant remains (*peat*).

coarse textured (rock) having *mineral* crystals sufficiently large that they are at least visible to the naked eye or with low magnification.

coast (See *coastline*.)

coastal blowout dune high *sand dune* of the *parabolic dunes* class formed adjacent to a beach, usually with a deep *deflation* hollow (*blowout*) enclosed within the dune ridge.

coastal forest subtype of *needleleaf evergreen forest* found in the humid coastal zone of the northwestern United States and western Canada.

coastal plain a coastal belt that has emerged from beneath the sea as a former *continental shelf*, underlain by *strata* with gentle *dip* seaward.

coastline (coast) zone in which coastal processes operate or have a strong influence.

col natural pass or low notch in an *arête* between opposed *cirques*.

cold front moving weather *front* along which a cold *air mass* moves underneath a warm air mass, causing the latter to be lifted.

cold-blooded a characteristic of animals with body temperatures that follow the environment; includes fish, reptiles, and invertebrates.

cold-core ring circular eddy of cold water, surrounded by warm water and lying adjacent to a warm, poleward-moving *ocean current*, such as the Gulf Stream. (See also *warm-core ring*.)

colloids particles of extremely small size, capable of remaining indefinitely in suspension in water; may be mineral or organic in nature.

colluvium deposit of *sediment* or *rock* particles accumulating from overland flow at the base of a *slope* and originating from higher slopes where *sheet erosion* is in progress. (See also *alluvium*.)

composite volcano (See s*tratovolcano*.)

compression (tectonic) squeezing together, as horizontal compression of crustal layers by *tectonic* processes.

condensation process of change of matter in the gaseous state (*water vapor*) to the *liquid* state (liquid water) or *solid* state (ice).

condensation nucleus a tiny bit of solid matter (*aerosol*) in the *atmosphere* on which *water vapor* condenses to form a tiny water droplet.

conduction of heat transmission of *sensible heat* through matter by transfer of *energy* from one atom or molecule to the next in the direction of decreasing temperature.

cone of depression conical configuration of the lowered *water table* around a well from which water is being rapidly withdrawn.

conglomerate a *sedimentary rock* composed of pebbles in a matrix of finer *rock* particles.

consequent stream *stream* that takes its course down the slope of an *initial landform*, such as a newly emerged *coastal plain* or a *volcano*.

consumers animals in the *food chain* that live on organic matter formed by *primary producers* or by other consumers. (See also *primary consumers, secondary consumers*.)

consumption (of a lithospheric plate) destruction or disappearance of a subducting *lithospheric plate* in the *asthenosphere*, in part by *melting* of the upper surface, but largely by softening because of heating to the temperature of the surrounding *mantle rock*.

continental collision event in *plate tectonics* in which *subduction* brings two segments of the *continental lithosphere* into contact, leading to formation of a *continental suture*.

continental crust crust of the continents, of felsic composition in the upper part; thicker and less dense than *oceanic crust*.

continental drift hypothesis, introduced by Alfred Wegener and others early in the 1900s, of the breakup of a parent continent, *Pangea*, starting near the close of the *Mesozoic Era*, and resulting in the present arrangement of *continental shields* and intervening *ocean-basin floors*.

continental lithosphere *lithosphere* bearing *continental crust* of *felsic igneous rock*.

continental margins (1) topographic: one of three major divisions of the ocean basins, being the zones directly adjacent to the continent and including the *continental shelf, continental slope*, and *continental rise*; (2) tectonic: marginal belt of *continental crust* and *lithosphere* that is in contact with *oceanic crust* and *lithosphere*, with or without an active plate boundary being present at the contact. (See also *active continental margins, passive continental margins*.)

continental rise gently sloping seafloor lying at the foot of the *continental slope* and leading gradually into the *abyssal plain*.

continental rupture crustal spreading apart affecting the *continental lithosphere*, so as to cause a *rift valley* to appear and to widen, eventually creating a new belt of *oceanic lithosphere*.

continental scale scale of observation at which we recognize continents and other large earth surface features, such as ocean currents.

continental shelf shallow, gently sloping belt of seafloor adjacent to the continental shoreline and terminating at its outer edge in the *continental slope*.

continental shields ancient crustal *rock* masses of the continents, largely *igneous rock* and *metamorphic rock*, and mostly of *Precambrian age*.

continental slope steeply descending belt of seafloor between the *continental shelf* and the *continental rise*.

continental suture long, narrow zone of crustal deformation, including underthrusting and intense *folding*, produced by a *continental collision*. Examples: Himalayan Range, European Alps.

convection (atmospheric) air motion consisting of strong updrafts taking place within a *convection cell*.

convection cell individual column of strong updrafts produced by atmospheric *convection*.

convection loop circuit of moving *fluid*, such as *air* or water, created by unequal heating of the fluid.

convectional precipitation a form of *precipitation* induced when warm, moist air is heated at the ground surface, rises, cools, and condenses to form water droplets, raindrops, and eventually, rainfall.

converging boundary boundary between two crustal plates along which *subduction* is occurring and *lithosphere* is being consumed.

coral reef rocklike accumulation of *carbonates* secreted by corals and algae in shallow water along a marine *shoreline*.

coral-reef coast *coast* built out by accumulations of *limestone* in *coral reefs*.

core of earth spherical central mass of the earth composed largely of iron and consisting of an outer *liquid* zone and an interior *solid* zone.

Coriolis effect effect of the earth's *rotation* tending to turn the direction of motion of any object or *fluid* toward the right in the northern *hemisphere* and to the left in the southern hemisphere.

corrosion erosion of *bedrock* of a *stream channel* (or other *rock* surface) by chemical reactions between solutions in stream water and *mineral* surfaces.

counterradiation *longwave radiation* of *atmosphere* directed downward to the earth's surface.

covered shields areas of *continental shields* in which the ancient *rocks* are covered beneath a thin layer of sedimentary *strata*.

crater central summit depression associated with the principal vent of a *volcano*.

crescentic dune (See *barchan dune*.)

crevasse gaping crack in the brittle surface ice of a *glacier*.

crude oil liquid fraction of *petroleum*.

crust of earth outermost solid shell or layer of the earth, composed largely of *silicate minerals*.

Cryaquepts great group within the soil *suborder* of *Aquepts*; includes Aquepts of cold climate regions and particularly the *tundra climate* (12).

Cryosolic order a class of *soils* in the Canadian Soil Classification System associated with strong *frost action* and underlying *permafrost*.

cuesta *erosional landform* developed on resistant *strata* having low to moderate *dip* and taking the form of an asymmetrical low ridge or hill belt with one side a steep slope and the other a gentle slope; usually associated with a *coastal plain*.

cumuliform clouds *clouds* of globular shape, often with extended vertical development.

cumulonimbus cloud large, dense *cumuliform cloud* yielding *precipitation*.

cumulus *cloud* type consisting of low-lying, white cloud masses of globular shape well separated from one another.

cutoff cutting-through of a narrow neck of land, so as to bypass the *stream flow* in an *alluvial meander* and cause it to be abandoned.

cycle of rock change total cycle of changes in which *rock* of any one of the three major rock classes—*igneous rock, sedimentary rock, metamorphic rock*—is transformed into rock of one of the other classes.

cyclone center of low *atmospheric pressure*. (See also *tropical cyclone, wave cyclone*.)

cyclonic precipitation a form of *precipitation* that occurs as warm moist air is lifted by air motion occurring in a *cyclone*.

cyclonic storm intense *weather* disturbance within a moving *cyclone* generating strong *winds*, cloudiness, and *precipitation*.

data acquisition component component of a *geographic information system* in which data are gathered together for input to the system.

data management component component of a *geographic information system* that creates, stores, retrieves, and modifies data layers and *spatial objects*.

daylight saving time time system under which time is advanced by one hour with respect to the *standard time* of the prevailing *standard meridian*.

debris flood (debris flow) streamlike flow of muddy water heavily charged with *sediment* of a wide range of size grades, including boulders, generated by sporadic torrential rains on steep mountain watersheds.

decalcification removal of *calcium carbonate* from a *soil horizon* or horizons as *carbonic acid* reacts with *carbonate mineral matter*.

December solstice (See *winter solstice*.)

deciduous plant *tree* or *shrub* that sheds its leaves seasonally.

declination of sun *latitude* at which the sun is directly overhead; varies from $-23\frac{1}{2}°$ ($23\frac{1}{2}°$ S lat.) to $+23\frac{1}{2}°$ ($23\frac{1}{2}°$ N lat.)

decomposers organisms that feed on dead organisms from all levels of the *food chain*; most are microorganisms and bacteria that feed on decaying organic matter.

deep sea cone a fan-shaped accumulation of undersea *sediment* on the *continental rise* produced by sediment-rich currents flowing down the *continental slope*.

deflation lifting and transport in *turbulent suspension* by *wind* of loose particles of *soil* or *regolith* from dry ground surfaces.

deglaciation widespread recession of *ice sheets* during a period of warming global *climate*, leading to an interglaciation. (See also *glaciation, interglaciation*.)

degradation lowering or downcutting of a *stream channel* by *stream erosion* in *alluvium* or *bedrock*.

degree of arc measurement of the angle associated with an *arc*, in degrees.

delta *sediment* deposit built by a *stream* entering a body of standing water and formed of the *stream load*.

delta coast *coast* bordered by a *delta*.

delta kame flat-topped hill of *stratified drift* representing a glacial *delta* constructed adjacent to an *ice sheet* in a marginal glacial lake.

dendritic drainage pattern *drainage pattern* of treelike branched form, in which the smaller *streams* take a wide variety of directions and show no parallelism or dominant trend.

density of matter quantity of mass per unit of volume, stated in kg/m².

denudation total action of all processes whereby the exposed *rocks* of the continents are worn down and the resulting *sediments* are transported to the sea by the *fluid agents*; also includes *weathering* and *mass wasting*.

deposition (atmosphere) the change of state of a substance from a *gas* (*water vapor*) to a *solid* (ice); in the science of *meteorology*, the term sublimation is used to describe both this process and the change of state from solid to vapor. (See also *sublimation*.)

deposition (of sediment) (See *stream deposition*.)

depositional landform *landform* made by *deposition* of *sediment*.

desert biome *biome* of the dry *climates* consisting of thinly dispersed plants that may be *shrubs*, grasses, or perennial *herbs*, but lacking in *trees*.

desert pavement surface layer of closely fitted pebbles or coarse *sand* from which finer particles have been removed.

desertification (See *land degradation*.)

detritus decaying organic matter on which *decomposers* feed.

dewpoint lapse rate rate at which the dewpoint of an air mass decreases with elevation; typical value is 1.8°C/1000 m (1.0°F/1000 ft).

dewpoint temperature temperature of an *air mass* at which the air holds its full capacity of *water vapor*.

diagnostic horizons *soil horizons*, rigorously defined, that are used as diagnostic criteria in classifying *soils*.

diffuse radiation solar *radiation* that has been *scattered* (deflected or reflected) by minute dust particles or cloud particles in the *atmosphere*.

diffuse reflection solar *radiation scattered* back to space by the earth's *atmosphere*.

digital image numeric representation of a picture consisting of a collection of numeric brightness values (pixels) arrayed in a fine grid pattern.

dike thin layer of *intrusive igneous rock*, often near-vertical or with steep *dip*, occupying a widened fracture in the surrounding *rock* and typically cutting across older rock planes.

diorite *intrusive igneous rock* consisting dominantly of *plagioclase feldspar* and *pyroxene*; a *felsic igneous rock*.

dip acute angle between an inclined natural *rock* plane or surface and an imaginary horizontal plane of reference; always measured perpendicular to the *strike*. Also a verb, meaning to incline toward.

discharge volume of flow moving through a given cross section of a *stream* in a given unit of time; commonly given in cubic meters (cubic feet) per second.

dissolved matter in *stream transportation*, a portion of the *stream load* that is carried as chemical *ions* in solution.

distributary branching *stream channel* that crosses a delta to discharge into open water.

diurnal adjective meaning "daily."

doldrums belt of calms and variable *winds* occurring at times along the *equatorial trough*.

dolomite carbonate *mineral* or *sedimentary rock* having the composition calcium magnesium carbonate.

dome (See *sedimentary dome*.)

drainage basin total land surface occupied by a *drainage system*, bounded by a *drainage divide* or watershed.

drainage divide imaginary line following a crest of high land such that overland flow on opposite sides of the line enters different *streams*.

drainage pattern the plan of a network of interconnected *stream channels*.

drainage system a branched network of *stream channels* and adjacent land *slopes*, bounded by a *drainage divide* and converging to a single channel at the outlet.

drainage winds *winds*, usually cold, that flow from higher to lower regions under the direct influence of *gravity*.

drawdown (of a well) difference in height between base of *cone of depression* and original *water table* surface.

drought occurrence of substantially lower-than-average *precipitation* in a season that normally has ample precipitation for the support of food-producing plants.

drumlin hill of glacial *till*, oval or elliptical in basal outline and with smoothly rounded summit, formed by plastering of till beneath moving, debris-laden glacial ice.

dry adiabatic lapse rate rate at which rising air is cooled by expansion when no *condensation* is occurring; 10°C per 1000 m (5.5°F per 1000 ft).

dry desert plant *formation class* in the *desert biome* consisting of widely dispersed xerophytic plants that may be small, hardleaved or spiny *shrubs*, succulent plants (cacti), or hard grasses.

dry midlatitude climate ⑨ dry climate of the *midlatitude zone* with a strong annual cycle of *potential evapotranspiration* (*water need*) and cold winters.

dry subtropical climate ⑤ dry *climate* of the *subtropical zone*, transitional between the *dry tropical climate* ④ and the *dry midlatitude climate* ⑨.

dry tropical climate ④ dry *climate* of the *tropical zone* with large total annual *potential evapotranspiration* (*water need*).

dune (See *sand dune*.)

dust storm heavy concentration of dust in a turbulent *air mass*, often associated with a *cold front*.

E horizon *mineral* horizon lying below the *A horizon* and characterized by the loss of *clay minerals* and *oxides* of iron and aluminum; it may show a concentration of *quartz* grains and is often pale in color.

earth's crust (See *crust of earth*.)

earthflow moderately rapid downhill flowage of masses of water-saturated *soil*, *regolith*, or weak *shale*, typically forming a steplike terrace at the top and a bulging toe at the base.

earthquake a trembling or shaking of the ground produced by the passage of *seismic waves*.

easterly wave weak, slowly moving trough of low pressure within the belt of *tropical easterlies*; causes a *weather* disturbance with rain showers.

ebb current oceanward flow of *tidal current* in a *bay* or tidal stream.

ecological succession time-succession (sequence) of distinctive plant and animal communities occurring within a given area of newly formed land or land cleared of plant cover by burning, clear cutting, or other agents.

ecology science of interactions between life forms and their environment; the science of *ecosystems*.

ecosystem group of organisms and the environment with which the organisms interact.

El Niño episodic cessation of the typical *upwelling* of cold deep water off the coast of Peru; literally, "The Christ Child," for its occurrence in the Christmas season once every few years.

electromagnetic radiation (electromagnetic energy) wavelike form of *energy* radiated by any substance possessing heat; it travels through space at the speed of light.

electromagnetic spectrum the total *wavelength* range of *electromagnetic energy*.

eluviation soil-forming process consisting of the downward transport of fine particles, particularly *soil colloids* (both mineral and organic), carrying them out of an upper *soil horizon*.

emergence exposure of submarine landforms by a lowering of sea level or a rise of the *crust*, or both.

energy the capacity to do work, that is, to bring about a change in the state or motion of matter.

energy balance (global) balance between *shortwave* solar *radiation* received by the earth–*atmosphere* system and radiation lost to space by shortwave reflection and *longwave radiation* from the earth–atmosphere system.

energy balance (of a surface) balance between the flows of *energy* reaching a surface and the flows of energy leaving it.

Entisols *soil order* consisting of mineral soils lacking *soil horizons* that would persist after normal plowing.

entrenched meanders winding, sinuous valley produced by *degradation* of a *stream* with trenching into the *bedrock* by downcutting.

environmental temperature lapse rate rate of temperature decrease upward through the *troposphere*; standard value is 6.4 C°/km (3 F°/1000 ft).

epipedon *soil horizon* that forms at the surface.

epiphytes plants that live above ground level out of contact with the soil, usually growing on the limbs of *trees* or *shrubs*; also called air plants.

epoch a subdivision of geologic time.

equator *parallel of latitude* occupying a position midway between the earth's poles of *rotation*; the largest of the parallels, designated as *latitude 0°*.

equatorial current westward-flowing *ocean current* in the belt of the *trade winds*.

equatorial easterlies upper-level easterly air flow over the *equatorial zone*.

equatorial rainforest plant *formation class* within the *forest biome*, consisting of tall, closely set broadleaf *trees* of evergreen or semideciduous habit.

equatorial trough atmospheric *low-pressure trough* centered more or less over the *equator* and situated between the two belts of *trade winds*.

equatorial zone *latitude* zone lying between lat. 10° S and 10° N (more or less) and centered on the *equator*.

equinox instant in time when the *subsolar point* falls on the earth's *equator* and the *circle of illumination* passes through both poles; *vernal equinox* occurs on March 20 or 21, *autumnal equinox* on September 22 or 23.

era major subdivision of geologic time consisting of a number of geologic periods; the three eras following *Precambrian* time are *Paleozoic*, *Mesozoic*, and *Cenozoic*.

erg large expanse of active *sand dunes* in the Sahara Desert of North Africa.

erosional landforms class of the *sequential landforms* shaped by the removal of *regolith* or *bedrock* by agents of erosion. Examples: *gorge*, glacial *cirque*, marine *cliff*.

esker narrow, often sinuous embankment of coarse gravel and boulders deposited in the bed of a meltwater *stream* enclosed in a tunnel within stagnant ice of an *ice sheet*.

estuary *bay* that receives fresh water from a river mouth and saltwater from the ocean.

Eurasian-Indonesian belt *mountain arc* system extending from southern Europe across southern Asia and Indonesia.

eutrophication excessive growth of algae and other related organisms in a *stream* or *lake* as a result of the input of large amounts of nutrient *ions*, especially phosphate and nitrate.

evaporation process in which water in *liquid* state or *solid* state passes into the vapor state.

evaporites class of *chemically precipitated sediment* and *sedimentary rock* composed of soluble salts deposited from saltwater bodies.

evapotranspiration combined water loss to the *atmosphere* by *evaporation* from the *soil* and *transpiration* from plants.

evergreen plant *tree* or *shrub* that holds most of its green leaves throughout the year.

exfoliation development of sheets, shells, or scales of *rock* on exposed *bedrock outcrops* or boulders, usually in concentric spheriodal arrangments. (See also *exfoliation dome*, *unloading*.)

exfoliation dome smoothly rounded *rock* knob or hilltop bearing rock sheets or shells produced by spontaneous expansion accompanying *unloading*.

exotic river *stream* that flows across a region of dry *climate* and derives its *discharge* from adjacent uplands where a *water surplus* exists.

exotic species an imported plant that is not native to a region in which it is currently found.

exposed shields areas of *continental shields* in which the ancient basement *rock*, usually of *Precambrian* age, is exposed to the surface.

extension (tectonic) drawing apart of crustal layers by *tectonic activity* resulting in *faulting*.

extrusion release of molten *rock magma* at the surface, as in a flow of *lava* or shower of volcanic ash.

extrusive igneous rock *rock* produced by the solidification of *lava* or ejected fragments of *igneous rock* (*tephra*).

Fahrenheit scale temperature scale in which the *freezing* point of water is 32° and the boiling point 212°.

fallout *gravity* fall of atmospheric *particulates* reaching the ground.

fault sharp break in *rock* with a displacement (slippage) of the block on one side with respect to an adjacent block. (See *normal fault, overthrust fault, strike-slip fault, transform fault*.)

fault coast *coast* formed when a *shoreline* comes to rest against a *fault scarp*.

fault creep more or less continuous slippage on a *fault plane*, relieving some of the accumulated strain.

fault plane surface of slippage between two earth blocks moving relative to each other during faulting.

fault scarp clifflike surface feature produced by faulting and exposing the *fault plane*, commonly associated with a *normal fault*.

fault-line scarp erosion scarp developed on an inactive *fault* line.

feldspar group of *silicate minerals* consisting of silicate of aluminum and one or more of the metals potassium, sodium, or calcium. (See also *plagioclase feldspar, potash feldspar*.)

felsenmeer expanse of large blocks of *rock* produced by *joint* block separation and shattering by *frost* action at high altitudes or in high latitudes; from the German for "rock sea."

felsic igneous rock *igneous rock* dominantly composed of *felsic minerals*.

felsic minerals (felsic mineral group) *quartz* and *feldspars* treated as a *mineral* group of light color and relatively low density. (See also *mafic minerals*.)

fine textured (rock) having *mineral* crystals too small to be seen by eye or with low magnification.

fiord narrow, deep ocean embayment partially filling a *glacial trough*.

fiord coast deeply embayed, rugged *coast* formed by partial *submergence* of *glacial troughs*.

firn granular old *snow* forming a surface layer in the zone of accumulation of a *glacier*.

fissile adjective describing a *rock*, usually *shale*, that readily splits up into small flakes or scales.

flash flood *flood* occurring on a small *watershed* with steep slopes that produces a wave of turbulent water sweeping downstream.

flood *stream flow* at a stream *stage* so high that it cannot be accommodated within the *stream channel* and must spread over the banks to inundate the adjacent *floodplain*.

flood basalts large-scale outpourings of basalt *lava* to produce thick accumulations of *basalt* over large areas.

flood current landward flow of a *tidal current*.

flood stage designated stream-surface level for a particular point on a *stream*, higher than which overbank flooding may be expected.

floodplain belt of low, flat ground, present on one or both sides of a *stream channel*, subject to inundation by a *flood* about once annually and underlain by *alluvium*.

fluid substance that flows readily when subjected to unbalanced stresses; may exist as a *gas* or a *liquid*.

fluid agents *fluids* that erode, transport, and deposit *mineral* matter and organic matter; they are running water, waves and currents, glacial ice, and *wind*.

fluvial landforms *landforms* shaped by running water.

fluvial processes geomorphic processes in which running water is the dominant *fluid* agent, acting as *overland flow* and *stream flow*.

fog *cloud* layer in contact with land or sea surface, or very close to that surface. (See also *advection fog, radiation fog*.)

folding process by which *folds* are produced; a form of *tectonic activity*.

folds wavelike corrugations of *strata* (or other layered rock masses) as a result of crustal *compression*.

food chain (food web) organization of an *ecosystem* into steps or levels through which *energy* flows as the organisms at each level consume energy stored in the bodies of organisms of the next lower level.

forb broad-leaved *herb*, as distinguished from the grasses.

foredunes ridge of irregular *sand dunes* typically found adjacent to *beaches* on low-lying *coasts* and bearing a partial cover of plants.

foreland folds *folds* produced by *continental collision* in *strata* of a *passive continental margin*.

forest assemblage of *trees* growing close together, their crowns forming a layer of foliage that largely shades the ground.

forest biome *biome* that includes all regions of *forest* over the lands of the earth.

formation classes subdivisions within a *biome* based on the size, shape, and structure of the plants that dominate the vegetation.

fossil fuels naturally occurring hydrocarbon compounds that represent the altered remains of organic materials enclosed in *rock*; examples are *coal, petroleum* (*crude oil*), and *natural gas*.

fractional scale (See *scale fraction*.)

freezing change from *liquid* state to *solid* state accompanied by release of *latent heat*, becoming *sensible heat*.

fringing reef *coral reef* directly attached to land with no intervening *lagoon* of open water.

front surface of contact between two unlike *air masses*. (See *cold front, occluded front, polar front, warm front*.)

frost action *rock* breakup by forces accompanying the *freezing* of water.

gabbro *intrusive igneous rock* consisting largely of pyroxene and *plagioclase feldspar*, with variable amounts of *olivine*; a *mafic igneous rock*.

gas (gaseous state) *fluid* of very low density (as compared with a *liquid* of the same chemical composition) that expands to fill uniformly any small container and is readily compressed.

geographic grid complete network of parallels and meridians on the surface of the globe, used to fix the locations of surface points.

geographic information system (GIS) a system for acquiring, processing, storing, querying, creating, and displaying *spatial data*; normally computer-based.

geologic norm stable natural condition in a moist *climate* in which slow *soil erosion* is paced by maintenance of *soil horizons* bearing a plant community in an equilibrium state.

geology science of the solid earth, including the earth's origin and history, materials comprising the earth, and the processes acting within the earth and on its surface.

geomorphology science of *landforms*, including their history and processes of origin.

geostrophic wind *wind* at high levels above the earth's surface blowing parallel with a system of straight, parallel *isobars*.

geyser periodic jetlike emission of hot water and steam from a narrow vent at a geothermal locality.

glacial abrasion *abrasion* by a moving *glacier* of the *bedrock* floor beneath it.

glacial delta *delta* built by meltwater *streams* of a *glacier* into standing water of a marginal glacial *lake*.

glacial drift general term for all varieties and forms of *rock* debris deposited in close association with *ice sheets* of the *Ice Age*.

glacial plucking removal of masses of *bedrock* from beneath an *alpine glacier* or *ice sheets* as ice moves forward suddenly.

glacial trough deep, steep-sided *rock* trench of U-shaped cross section formed by *alpine glacier* erosion.

glaciation (1) general term for the total process of glacier growth and *landform* modification by *glaciers*; (2) single episode or time period in which *ice sheets* formed, spread, and disappeared.

glacier large natural accumulation of land ice affected by present or past flowage. (See *alpine glacier*.)

Gleysolic order a class of *soils* in the Canadian Soil Classification System characterized by indicators of periodic or prolonged water saturation.

gneiss variety of *metamorphic rock* showing banding and commonly rich in *quartz* and *feldspar*.

Gondwana a *supercontinent* of the Permian *Period* including much of the regions that are now South America, Africa, Antarctica, Australia, New Zealand, Madagascar, and peninsular India.

Goode projection an equal-area *map projection*, often used to display areal thematic information, such as *climate* or *soil* type.

gorge (canyon) steep-sided *bedrock* valley with a narrow floor limited to the width of a *stream channel*.

graben trenchlike depression representing the surface of a crustal block dropped down between two opposed, infacing *normal faults*. (See *rift valley*.)

graded profile smoothly descending profile displayed by a *graded stream*.

graded stream *stream* (or *stream channel*) with *stream gradient* so adjusted as to achieve a balanced state in which average *bed load* transport is matched to average bed load input; an average condition over periods of many years' duration.

gradient degree of *slope*, as the gradient of a river or a flowing *glacier*.

granite *intrusive igneous rock* consisting largely of *quartz, potash feldspar*, and *plagioclase feldspar*, with minor amounts of biotite and hornblende; a *felsic igneous rock*.

granitic rock general term for *rock* of the upper layer of the *continental crust*, composed largely of *felsic igneous* and *metamorphic rock*; *rock* of composition similar to that of *granite*.

granular disintegration grain-by-grain breakup of the outer surface of coarse-grained *rock*, yielding *sand* and gravel and leaving behind rounded boulders.

graphic scale map scale as shown by a line divided into equal parts.

grassland biome *biome* consisting largely or entirely of *herbs*, which may include grasses, grasslike plants, and *forbs*.

gravity gravitational attraction of the earth upon any small mass near the earth's surface.

gravity gliding the sliding of a *thrust sheet* away from the center of an *orogen* under the force of *gravity*.

great circle circle formed by passing a plane through the exact center of a perfect sphere; the largest circle that can be drawn on the surface of a sphere.

greenhouse effect accumulation of heat in the lower *atmosphere* through the absorption of *longwave radiation* from the earth's surface.

greenhouse gases atmospheric gases such as CO_2 and *cholorofluorocarbons (CFCs)* that absorb outgoing *longwave radiation*, contributing to the *greenhouse effect*.

groin wall or embankment built out into the water at right angles to the *shoreline*.

ground moraine *moraine* formed of *till* distributed beneath a large expanse of land surface covered at one time by an *ice sheet*.

ground water *subsurface water* occupying the *saturated zone* and moving under the force of *gravity*.

gullies deep, V-shaped trenches carved by newly formed *streams* in rapid headward growth during advanced stages of *accelerated soil erosion*.

guyot sunken remnant of a volcanic island.

gyres large, circular *ocean current* systems centered on the oceanic subtropical *high-pressure cells*.

habitat subdivision of the plant environment having a certain combination of *slope*, drainage, *soil* type, and other controlling physical factors.

Hadley cell atmospheric circulation cell in low latitudes involving rising air over the *equatorial trough* and sinking air over the *subtropical high-pressure belts*.

hail form of *precipitation* consisting of pellets or spheres of ice with a concentric layered structure.

hanging valley stream valley that has been truncated by marine erosion so as to appear in cross section in a *marine cliff*, or truncated by glacial erosion so as to appear in cross section in the upper wall of a *glacial trough*.

haze minor concentration of *pollutants* or natural forms of *aerosols* in the *atmosphere* causing a reduction in visibility.

heat (See *latent heat, sensible heat*.)

heat island persistent region of higher air temperatures centered over a city.

hemisphere half of a sphere; that portion of the earth's surface found between the *equator* and a pole.

herb a tender plant, lacking a woody stem, usually small or low; may be annual or perennial.

herbivore organism that consumes plants as a source of metabolic *energy*.

high base status (See *base status of soils*.)

high-latitude climates group of *climates* in the *subarctic zone*, *arctic zone*, and *polar zone*, dominated by arctic *air masses* and polar air masses.

high-level temperature inversion condition in which a high-level layer of warm air overlies a layer of cooler air, reversing the normal trend of cooling with altitude.

high-pressure cell center of high barometric pressure; an *anticyclone*.

Histosols *soil order* consisting of *soils* with a thick upper layer of organic matter.

hogbacks sharp-crested, often sawtooth ridges formed of the upturned *edge* of a resistant *rock* layer of *sandstone*, *limestone*, or *lava*.

Holocene Epoch last *epoch* of geologic time, commencing about 10,000 years ago; it followed the *Pleistocene Epoch* and includes the present.

horn sharp, steep-sided mountain peak formed by the intersection of headwalls of glacial *cirques*.

horse latitudes *subtropical high-pressure belt* of the North Atlantic Ocean, coincident with the central region of the Azores high; a belt of weak, variable *winds* and frequent calms.

horst crustal block uplifted between two normal *faults*.

hot springs springs discharging heated *ground water* at a temperature close to the boiling point; found in geothermal areas and thought to be related to a *magma* body at depth.

hotspot center of intrusive *igneous* and *volcanic* activity thought to be located over a rising *mantle plume*.

human habitat the lands of the earth that support human life.

human-influenced vegetation vegetation that has been influenced in some way by human activity, for example, through cultivation, grazing, timber cutting, or urbanization.

humidity general term for the amount of *water vapor* present in the air. (See also *relative humidity, specific humidity*.)

humification *pedogenic process* of transformation of plant tissues into *humus*.

humus dark brown to black organic matter on or in the *soil*, consisting of fragmented plant tissues partly digested by organisms.

hurricane *tropical cyclone* of the western North Atlantic and Caribbean Sea.

hydraulic action *stream erosion* by impact force of the flowing water on the bed and banks of the *stream channel*.

hydrologic cycle total plan of movement, exchange, and storage of the earth's free water in *gaseous, liquid*, and *solid* state.

hydrology science of the earth's water and its motions through the *hydrologic cycle*.

hydrolysis chemical union of water molecules with *minerals* to form different, more stable mineral compounds.

hydrosphere total water realm of the earth's surface zone, including the oceans, surface waters of the lands, *ground water*, and water held in the *atmosphere*.

ice age span of geologic time, usually on the order of 1 to 3 million years, or longer, in which glaciations alternate with

interglaciations repeatedly in rhythm with cyclic global *climate* changes. (See also *glaciation, interglaciation*.)

Ice Age (Late-Cenozoic Ice Age) the present *ice age*, which began in late Pliocene time, perhaps 2.5 to 3 million years ago.

ice lobes (glacial lobes) broad, tonguelike extensions of an *ice sheet* resulting from more rapid ice motion where terrain is more favorable.

ice sheet large thick plate of glacial ice moving outward in all directions from a central region of accumulation.

ice-sheet climate ⑬ severely cold *climate*, found on the Greenland and Antarctic *ice sheets*, with *potential evapotranspiration* (*water need*) effectively zero throughout the year.

ice shelf thick plate of floating glacial ice attached to an *ice sheet* and fed by the ice sheet and by *snow* accumulation.

ice storm occurrence of heavy glaze of ice on solid surfaces.

ice wedge vertical, wall-like body of ground ice, often tapering downward, occupying a shrinkage crack in *silt* of *permafrost* areas.

ice-wedge polygons polygonal networks of *ice wedges*.

iceberg mass of glacial ice floating in the ocean, derived from a *glacier* that extends into tidal water.

igneous rock *rock* solidified from a high-temperature molten state; rock formed by cooling of *magma*. (See also *extrusive igneous rock, felsic igneous rock, intrusive igneous rock, mafic igneous rock, ultramafic igneous rock*.)

illuviation accumulation in a lower *soil horizon* (typically, the B horizon) of materials brought down from a higher horizon; a soil-forming process.

Inceptisols *soil order* consisting of soils having weakly developed *soil horizons* and containing weatherable minerals.

induced deflation loss of *soil* by *wind* erosion that is triggered by human activity such as cultivation or overgrazing.

induced mass wasting *mass wasting* that is induced by human activity, such as creation of waste *soil* and *rock* piles or undercutting of *slopes* in construction.

infiltration absorption and downward movement of *precipitation* into the *soil* and *regolith*.

infrared imagery images formed by *infrared radiation* emanating from the ground surface as recorded by a remote sensor.

infrared radiation *electromagnetic energy* in the *wavelength* range of 0.7 to about 200 μm.

initial landforms *landforms* produced directly by internal earth processes of *volcanism* and *tectonic activity*. Examples: *volcano, fault scarp*.

inner lowland on a *coastal plain*, a shallow valley lying between the first *cuesta* and the area of older *rock* (oldland).

insolation interception of solar *energy* (*shortwave radiation*) by an exposed surface under conditions of no *atmosphere* and a spherical earth.

inspiral horizontal inward spiral or motion, such as that found in a *cyclone*.

interglaciation within an *ice age*, a time interval of mild global *climate* in which continental *ice sheets* were largely absent or were limited to the Greenland and Antarctic ice sheets; the interval between two glaciations. (See also *deglaciation, glaciation*.)

interlobate moraine *moraine* formed between two adjacent lobes of an *ice sheet*.

International Date Line the 180° *meridian of longitude*, together with deviations east and west of that meridian, forming the time boundary between adjacent *standard time zones* that are 12 hours fast and 12 hours slow with respect to Greenwich standard time.

interrupted projection projection subdivided into a number of sectors (gores), each of which is centered on a different central *meridian*.

intertropical convergence zone (ITC) zone of convergence of *air masses* of *tropical easterlies* (*trade winds*) along the axis of the *equatorial trough*.

intrusion body of *igneous rock* injected as *magma* into preexisting crustal *rock*. Examples: *dike* or *sill*.

intrusive igneous rock *igneous rock* body produced by solidification of *magma* beneath the surface, surrounded by pre-existing *rock*.

inversion (See *temperature inversion*.)

ion atom or group of atoms bearing an electrical charge as the result of a gain or loss of one or more electrons.

island arcs curved lines of volcanic islands associated with active *subduction zones* along the boundaries of *lithospheric plates*.

isobars lines on a map passing through all points having the same *atmospheric pressure*.

isohyet line on a map drawn through all points having the same numerical value of *precipitation*.

isopleth line on a map or globe drawn through all points having the same value of a selected property or entity.

isotherm line on a map drawn through all points having the same air temperature.

isotope form of an element with a unique *atomic mass number*.

jet stream high-speed air flow in narrow bands within the *upper-air westerlies* and along certain other global *latitude* zones at high levels.

joints fractures within *bedrock*, usually occurring in parallel and intersecting sets of planes.

June solstice (See *summer solstice*.)

karst landscape or topography dominated by surface features of *limestone* solution and underlain by a *limestone cavern* system.

Kelvin scale (K) temperature scale on which the starting point is absolute zero, equivalent to −273°C.

knob and kettle terrain of numerous small knobs of *glacial drift* and deep depressions usually situated along the *moraine* belt of a former *ice sheet*.

La Niña enhancement of *upwelling* of cold deep water off the coast of Peru; opposite condition to *El Niño*.

lagoon shallow body of open water lying between a *barrier island* or a *barrier reef* and the mainland.

lahar rapid downslope or downvalley movement of a tongue-like mass of water-saturated *tephra* (volcanic ash) originating high up on a steep-sided volcanic cone; a variety of *mudflow*.

lake terrestrial body of standing water surrounded by land or glacial ice.

land breeze local *wind* blowing from land to water during the night.

land degradation *degradation* of the quality of plant cover and *soil* as a result of overuse by humans and their domesticated animals, especially during periods of *drought*.

land mass large area of *continental crust* lying above sea level (base level) and thus available for removal by *denudation*.

land mass rejuvenation episode of rapid fluvial *denudation* set off by a rapid crustal rise, increasing the available *land mass*.

landforms configurations of the land surface taking distinctive forms and produced by natural processes. Examples: hill, valley, plateau. (See also *depositional landforms, erosional landforms, initial landforms, sequential landforms*.)

landslide rapid sliding of large masses of *bedrock* on steep mountain slopes or from high *cliffs*.

lapse rate rate at which temperature decreases with increasing altitude. (See also *dry adiabatic lapse rate, environmental temperature lapse rate, wet adiabatic lapse rate*.)

large-scale map map with *fractional scale* greater than 1:100,000; usually shows a small area.

Late-Cenozoic Ice Age the series of *glaciations, deglaciations,* and *interglaciations* experienced during the late *Cenozoic Era*.

latent heat heat absorbed and held in storage in a *gas* or *liquid* during the processes of *evaporation*, or *melting*, or *sublimation*; distinguished from *sensible heat*.

latent heat transfer flow of *latent heat* that results when water absorbs heat to change from a *liquid* or *solid* to a *gas* and then later releases that heat to new surroundings by *condensation* or *deposition*.

lateral moraine *moraine* forming an embankment between the ice of an *alpine glacier* and the adjacent valley wall.

laterite rocklike layer rich in *sequioxides* and iron, including the minerals *bauxite* and *limonite*, found in low *latitudes* in association with *Ultisols* and *Oxisols*.

latitude arc of a *meridian* between the *equator* and a given point on the globe.

lattice drainage pattern (See *trellis drainage pattern*.)

Laurasia a *supercontinent* of the Permian *Period* including much of the regions that are now North America and western Eurasia.

lava *magma* emerging on the earth's solid surface, exposed to air or water.

leaching *pedogenic process* in which material is lost from the *soil* by downward washing out and removal by percolating surplus *soil water*.

leads narrow strips of open ocean water between ice floes.

level of condensation elevation at which an upward-moving parcel of moist air cools to the *dewpoint* and *condensation* begins to occur.

liana woody vine supported on the trunk or branches of a *tree*.

lichens plant forms in which algae and fungi live together (in a symbiotic relationship) to create a single structure; they typically form tough, leathery coatings or crusts attached to *rocks* and *tree* trunks.

life cycle continuous progression of stages in a growth or development process, such as that of a living organism.

life-form characteristic physical structure, size, and shape of a plant or of an assemblage of plants.

life layer shallow surface zone containing the *biosphere*; a zone of interaction between *atmosphere* and land surface, and between atmosphere and ocean surface.

life zones series of vegetation zones describing vegetation types that are encountered with increasing elevation, especially in the southwestern United States.

lifting condensation level elevation at which *condensation* occurs when a parcel of moist air moves upward.

limestone nonclastic *sedimentary rock* in which *calcite* is the predominant *mineral*, and with varying minor amounts of other *minerals* and *clay*.

limestone caverns interconnected subterranean cavities formed in *limestone* by *carbonic acid action* occurring in slowly moving *ground water*.

limonite mineral or group of *minerals* consisting largely of iron *oxide* and water, produced by *chemical weathering* of other iron-bearing minerals.

line type of *spatial object* in a *geographic information system* that has starting and ending *nodes*; may be directional.

liquid *fluid* that maintains a free upper surface and is only very slightly compressible, as compared with a *gas*.

lithosphere (1) general term for the entire solid earth realm; (2) in *plate tectonics*, the strong, brittle outermost *rock* layer lying above the *asthenosphere*.

lithospheric plate segment of *lithosphere* moving as a unit, in contact with adjacent lithospheric plates along plate boundaries.

littoral drift transport of *sediment* parallel with the *shoreline* by the combined action of *beach drift* and *longshore current* transport.

loam soil-texture class in which no one of the three size grades (*sand, silt, clay*) dominates over the other two.

local winds general term for *winds* generated as direct or immediate effects of the local terrain.

loess accumulation of yellowish to buff-colored, fine-grained *sediment*, largely of *silt* grade, upon upland surfaces after transport in the air in *turbulent suspension* (i.e., carried in a *dust storm*).

longitude *arc* of a *parallel* between the *prime meridian* and a given point on the globe.

longitudinal dunes class of *sand dunes* in which the dune ridges are oriented parallel with the prevailing *wind*.

longshore current current in the breaker zone, running parallel with the *shoreline* and set up by the oblique approach of waves.

longshore drift *littoral drift* caused by action of a *longshore current*.

longwave radiation *electromagnetic energy* emitted by the earth, largely in the range from 3 to 50 μm.

low base status (See *base status of soils*.)

low-angle overthrust fault *overthrust fault* in which the *fault plane* or fault surface has a low angle of *dip* or may be horizontal.

low-latitude climates group of *climates* of the *equatorial zone* and *tropical zone* dominated by the *subtropical high-pressure belt* and the *equatorial trough*.

low-latitude rainforest evergreen *broadleaf forest* of the *wet equatorial* and tropical climate zones.

low-latitude rainforest environment low-latitude environment of warm temperatures and abundant *precipitation* that characterizes rainforest in the *wet equatorial* ① and *monsoon and trade-wind coastal* ② climates.

low-level temperature inversion atmospheric condition in which temperature near the ground increases, rather than decreases, with elevation.

low-pressure trough zone of low pressure between two *anticyclones*.

lowlands broad, open valleys between two *cuestas* of a *coastal plain*. (The term may refer to any low areas of land surface.)

Luvisolic order a class of *forest soils* in the Canadian Soil Classification System in which the *B horizon* accumulates *clay*.

mafic igneous rock *igneous rock* dominantly composed of *mafic minerals*.

mafic minerals (mafic mineral group) *minerals*, largely *silicate minerals*, rich in magnesium and iron, dark in color, and of relatively great density.

magma mobile, high-temperature molten state of *rock*, usually of *silicate mineral* composition and with dissolved *gases*.

manipulation and analysis component component of a *geographic information system* that responds to spatial queries and creates new data layers.

mantle *rock* layer or shell of the earth beneath the *crust* and surrounding the *core*, composed of *ultramafic igneous rock* of *silicate mineral* composition.

mantle plume a columnlike rising of heated *mantle rock*, thought to be the cause of a *hotspot* in the overlying *lithospheric plate*.

map projection any orderly system of parallels and meridians drawn on a flat surface to represent the earth's curved surface.

marble variety of *metamorphic rock* derived from *limestone* or *dolomite* by recrystallization under pressure.

marine cliff *rock* cliff shaped and maintained by the undermining action of breaking waves.

marine scarp steep seaward *slope* in poorly consolidated *alluvium, glacial drift*, or other forms of *regolith*, produced along a *coastline* by the undermining action of waves.

marine terrace former *abrasion platform* elevated to become a steplike coastal *landform*.

marine west-coast climate ⑧ cool, moist *climate* of west coasts in the *midlatitude zone*, usually with a substantial annual *water surplus* and a distinct winter *precipitation* maximum.

mass number (See *atomic mass number*.)

mass wasting spontaneous downhill movement of *soil, regolith*, and *bedrock* under the influence of *gravity*, rather than by the action of *fluid* agents.

mean annual temperature mean of daily air temperature means for a given year or succession of years.

mean daily temperature sum of daily maximum and minimum air temperature readings divided by two.

mean monthly temperature mean of daily air temperature means for a given calendar month.

meanders (See *alluvial meanders*.)

mechanical weathering (See *physical weathering*.)

medial moraine long, narrow deposit of fragments on the surface of a *glacier*; created by the merging of *lateral moraines* when two glaciers join into a single stream of ice flow.

Mediterranean climate ⑦ *climate* type of the *subtropical zone*, characterized by the alternation of a very dry summer and a mild, rainy winter.

melting change from *solid* state to *liquid* state, accompanied by absorption of *sensible heat* to become *latent heat*.

Mercator projection conformal *map projection* with horizontal parallels and vertical meridians and with map scale rapidly increasing with increase in *latitude*.

mercury barometer *barometer* using the Torricelli principle, in which *atmospheric pressure* counterbalances a column of mercury in a tube.

meridian of longitude north-south line on the surface of the global *oblate ellipsoid*, connecting the *north pole* and *south pole*.

mesa table-topped *plateau* of comparatively small extent bounded by *cliffs* and occurring in a region of flat-lying *strata*.

Mesozoic Era second of three geologic *eras* following *Precambrian time*.

metamorphic rock *rock* altered in physical structure or chemical (*mineral*) composition by action of heat, pressure, *shearing* stress, or infusion of elements, all taking place at substantial depth beneath the surface.

meteorology science of the *atmosphere*, particularly the physics of the lower or inner atmosphere.

mica group aluminum-silicate *mineral* group of complex chemical formula having perfect cleavage into thin sheets.

micrometer metric unit of length equal to one-millionth of a meter (0.000001 m); abbreviated μm.

microwaves waves of the *electromagnetic radiation* spectrum in the *wavelength* band from about 0.03 cm to about 1 cm (0.01 to 0.4 in.).

midlatitude climates group of *climates* of the *midlatitude zone* and s*ubtropical zone*, located in the *polar front* zone and dominated by both tropical *air masses* and polar air masses.

midlatitude deciduous forest plant *formation class* within the *forest biome* dominated by tall, broadleaf *deciduous trees*, found mostly in the *moist continental climate* ⑩ and *marine west-coast climate* ⑧.

midlatitude zones latitude zones occupying the *latitude* range 35° to 55° N and S (more or less) and lying between the *subtropical zones* and the *subarctic (subantarctic) zones*.

midoceanic ridge one of three major divisions of the ocean basins, being the central belt of submarine mountain topography with a characteristic *axial rift*.

millibar unit of *atmospheric pressure* equal to one-thousandth of a bar. (See also *bar*.)

mineral naturally occurring inorganic substance, usually having a definite chemical composition and a characteristic atomic structure. (See also *felsic minerals, mafic minerals, silicate minerals*.)

mineral alteration chemical change of *minerals* to more stable compounds upon exposure to conditions at the earth's surface; same as *chemical weathering*.

mineral matter (soils) component of *soil* consisting of weathered or unweathered mineral grains.

mineral oxides (soils) secondary *minerals* found in *soils* in which original minerals have been altered by chemical combination with oxygen.

minute (of arc) 1/60 of a degree.

mistral local drainage wind of cold air affecting the Rhone Valley of southern France.

Moho contact surface between the earth's *crust* and *mantle*, a contraction of Mohorovičić, the name of the seismologist who discovered this feature.

moist continental climate ⑩ moist *climate* of the *midlatitude zone* with strongly defined winter and summer seasons, adequate *precipitation* throughout the year, and a substantial annual *water surplus*.

moist subtropical climate ⑥ moist *climate* of the *subtropical zone*, characterized by a moderate to large annual *water surplus* and a strongly seasonal cycle of *potential evapotranspiration* (*water need*).

mollic epipedon relatively thick, dark-colored surface *soil horizon*, containing substantial amounts of organic matter (*humus*) and usually rich in *bases*.

Mollisols *soil order* consisting of *soils* with a *mollic* horizon and high *base status*.

monadnock prominent, isolated mountain or large hill rising conspicuously above a surrounding *peneplain* and composed of a *rock* more resistant than that underlying the peneplain; a *landform* of *denudation* in moist *climates*.

monsoon and trade-wind coastal climate ② moist *climate* of low *latitudes* showing a strong rainfall peak in the season of high sun and a short period of reduced rainfall.

monsoon forest *formation class* within the *forest biome* consisting in part of deciduous *trees* adapted to a long dry season in the *wet-dry tropical climate* ③.

monsoon system system of low-level *winds* blowing into a continent in summer and out of it in winter, controlled by *atmospheric pressure* systems developed seasonally over the continent.

montane forest plant *formation class* of the *forest biome* found in cool upland environments of the *tropical zone* and *equatorial zone*.

moraine accumulation of rock debris carried by an *alpine glacier* or an *ice sheet* and deposited by the ice to become a *depositional landform*. (See also *lateral moraine, terminal moraine*.)

mountain arc curving section of an *alpine chain* occurring on a *converging boundary* between two crustal plates.

mountain roots erosional remnants of deep portions of ancient *continental sutures* that were once *alpine chains*.

mountain winds daytime movements of air up the *gradient* of valleys and mountain slopes; alternating with nocturnal *valley winds*.

mucks organic *soils* largely composed of fine, black, sticky organic matter.

mud *sediment* consisting of a mixture of *clay* and *silt* with water, often with minor amounts of *sand* and sometimes with organic matter.

mudflow a form of *mass wasting* consisting of the downslope flowage of a mixture of water and *mineral* fragments (*soil, regolith*, disintegrated *bedrock*), usually following a natural drainage line or *stream channel*.

mudstone *sedimentary rock* formed by the lithification of *mud*.

multipurpose map map containing several different types of information.

multispectral image image consisting of two or more images, each of which is taken from a different portion of the spectrum (e.g., blue, green, red, infrared).

multispectral scanner *remote sensing* instrument, flown on an aircraft or spacecraft, that simultaneously collects multiple *digital images* (*multispectral images*) of the ground. Typically, images are collected in four or more spectral bands.

nappe overturned recumbent *fold* of *strata*, usually associated with *thrust sheets* in a collision *orogen*.

natural bridge natural *rock* arch spanning a *stream channel*, formed by cutoff of an *entrenched meander* bend.

natural gas naturally occurring mixture of hydrocarbon compounds (principally methane) in the gaseous state held within certain porous *rocks*.

natural levee belt of higher ground paralleling a meandering *alluvial river* on both sides of the *stream channel* and built up by *deposition* of fine *sediment* during periods of overbank flooding.

natural vegetation stable, mature plant cover characteristic of a given area of land surface largely free from the influences and impacts of human activities.

needleleaf evergreen forest needleleaf *forest* composed of evergreen *tree* species, such as spruce, fir, and pine.

needleleaf forest plant *formation class* within the *forest biome*, consisting largely of needleleaf *trees*. (See also *boreal forest*.)

net primary production rate at which *carbohydrate* is accumulated in the tissues of plants within a given *ecosystem*; units are kilograms of dry organic matter per year per square meter of surface area.

net radiation difference in intensity between all incoming *energy* (positive quantity) and all outgoing *energy* (negative quantity) carried by both *shortwave radiation* and *longwave radiation*.

nickpoint location in a *stream profile* at which the stream slope changes abruptly.

nitrogen fixation chemical process of conversion of *gaseous* molecular nitrogen of the *atmosphere* by certain microorganisms into compounds or *ions* that can be directly utilized by plants.

node point marking the end of a *line* or the intersection of lines as *spatial objects* in a *geographic information system*.

noon (See *solar noon*.)

noon angle (of the sun) angle of the sun above the horizon at its highest point during the day.

normal fault variety of *fault* in which the *fault plane* inclines (*dips*) toward the downthrown block and a major component of the motion is vertical.

north pole point at which the northern end of the earth's *axis of rotation* intersects the earth's surface.

northeast trade winds surface *winds* of low *latitudes* that blow steadily from the northeast. (See also *trade winds*.)

O$_i$ horizon surface *soil horizon* containing decaying organic matter that is recognizable as leaves, twigs, or other organic structures.

O$_a$ horizon *soil horizon* below the O$_i$ *horizon* containing decaying organic matter that is too decomposed to recognize as specific plant parts, such as leaves or twigs.

oblate ellipsoid geometric solid resembling a flattened sphere, with polar axis shorter than the equatorial diameter.

occluded front *weather front* along which a moving *cold front* has overtaken a *warm front*, forcing the warm *air mass* aloft.

ocean basin floors one of the major divisions of the ocean basins, comprising the deep portions consisting of *abyssal plains* and low hills.

ocean current persistent, dominantly horizontal flow of ocean water.

ocean tide periodic rise and fall of the ocean level induced by gravitational attraction between the earth and moon in combination with earth *rotation*.

oceanic crust *crust* of basaltic composition beneath the ocean floors, capping *oceanic lithosphere*. (See also *continental crust*.)

oceanic lithosphere *lithosphere* bearing *oceanic crust*.

oceanic trench narrow, deep depression in the seafloor representing the line of *subduction* of an oceanic *lithospheric plate* beneath the margin of a continental lithospheric plate; often associated with an *island arc*.

olivine *silicate mineral* with magnesium and iron but no aluminum, usually olive-green or grayish-green; a *mafic mineral*.

omnivore animal that consumes both plant and animal tissue as a source of metabolic *energy*.

organic matter (soils) material in *soil* that was originally produced by plants or animals and has been subjected to decay.

Organic order a class of *soils* in the Canadian Soil Classification System that is composed largely of organic materials.

organic sediment *sediment* consisting of the organic remains of plants or animals.

orogen the mass of tectonically deformed *rocks* and related *igneous rocks* produced during an *orogeny*.

orogeny major episode of *tectonic activity* resulting in *strata* being deformed by folding and faulting.

orographic pertaining to mountains.

orographic precipitation *precipitation* induced by the forced rise of moist air over a mountain barrier.

outcrop surface exposure of *bedrock*.

outspiral horizontal outward spiral or motion, such as that found in an *anticyclone*.

outwash glacial deposit of stratified drift left by *braided streams* issuing from the front of a *glacier*.

outwash plain flat, gently sloping plain built up of *sand* and *gravel* by the *aggradation* of meltwater *streams* in front of the margin of an *ice sheet*.

overburden *strata* overlying a layer or *stratum* of interest, as overburden above a *coal* seam.

overland flow motion of a surface layer of water over a sloping ground surface at times when the *infiltration* rate is exceeded by the *precipitation* rate; a form of *runoff*.

overthrust fault *fault* characterized by the overriding of one crustal block (or *thrust sheet*) over another along a gently inclined *fault plane*; associated with crustal *compression*.

oxbow lake crescent-shaped lake representing the abandoned channel left by the *cutoff* of an *alluvial meander*.

oxidation chemical union of free oxygen with metallic elements in *minerals*.

oxide chemical compound containing oxygen; in *soils*, iron oxides and aluminum oxides are examples.

Oxisols *soil order* consisting of very old, highly weathered *soils* of low *latitudes*, with an oxic horizon and *low base status*.

ozone a form of oxygen with a molecule consisting of three atoms of oxygen, O$_3$.

ozone layer layer in the *stratosphere*, mostly in the altitude range 20 to 35 km (about 12 to 31 mi), in which a concentration of *ozone* is produced by the action of solar *ultraviolet radiation*.

pack ice floating *sea ice* that completely covers the sea surface.

Paleozoic Era first of three geologic *eras* comprising all geologic time younger than *Precambrian time*.

Pangea hypothetical parent continent, enduring until near the close of the *Mesozoic Era*, consisting of the *continental shields* of *Laurasia* and *Gondwana* joined into a single unit.

parabolic dunes isolated *low sand dunes* of parabolic outline, with points directed into the prevailing *wind*.

parallel of latitude east-west circle on the earth's surface, lying in a plane parallel with the *equator* and at right angles to the *axis of rotation.*

parent material inorganic, *mineral* base from which the *soil* is formed; usually consists of *regolith.*

particulates *solid* and *liquid* particles capable of being suspended for long periods in the *atmosphere.*

pascal metric unit of pressure, defined as a force of one newton per square meter (1 N/m^2); symbol, Pa; $100 \text{ Pa} = 1$ mb, $10^5 \text{ Pa} = 1$ bar.

passive continental margins continental margins lacking active plate boundaries at the contact of *continental crust* with *oceanic crust.* A passive margin thus lies within a single *lithospheric plate.* Example: Atlantic continental margin of North America. (See also *active continental margins, continental margins.*)

passive systems electromagnetic *remote sensing systems* that measure radiant *energy* reflected or emitted by an object or surface.

peat partially decomposed, compacted accumulation of plant remains occurring in a *bog* environment.

ped individual natural *soil* aggregate.

pediment gently sloping, rock-floored land surface found at the base of a mountain mass or *cliff* in an arid region.

pedogenic processes group of recognized basic soil-forming processes, mostly involving the gain, loss, *translocation,* or transformation of materials within the *soil* body.

pedology science of the *soil* as a natural surface layer capable of supporting living plants; synonymous with *soil science.*

pedon soil column extending down from the surface to reach a lower limit in some form of *regolith* or *bedrock.*

peneplain land surface of low elevation and slight relief produced in the late stages of *denudation* of a *land mass.*

perched water table *water table* of a lens or layer of *ground water* formed over an *aquiclude* and occupying a position above the main water table.

percolation slow, downward flow of water by *gravity* through *soil* and subsurface layers toward the *water table.*

perennials plants that live for more than one growing season.

peridotite *ultramafic igneous rock,* consisting largely of *olivine* and *pyroxene,* typically occurring as a *pluton;* also thought to compose much of the upper *mantle.*

perihelion point on the earth's elliptical orbit at which the earth is nearest to the sun.

period of geologic time time subdivision of the *era,* each ranging in duration between about 35 and 70 million years.

permafrost condition of permanently frozen water in the *soil, regolith,* and *bedrock* in cold *climates* of subarctic and arctic regions.

petroleum (crude oil) natural *liquid* mixture of many complex hydrocarbon compounds of organic origin, found in accumulations (oil pools) within certain *sedimentary rocks.*

pH measure of the concentration of hydrogen *ions* in a solution. (The number represents the logarithm to the base 10 of the reciprocal of the weight in grams of hydrogen ions per liter of water.) Acid solutions have pH values less than 6, and basic solutions have pH values greater than 6.

photoperiod duration of daylight on a given day at a given *latitude.*

photosynthesis production of *carbohydrate* by the union of water with *carbon dioxide* while absorbing light *energy.*

physical geography the study and synthesis of selected subject areas from the natural sciences—especially atmospheric science, *hydrology,* physical oceanography, *geology, geomorphology, soil science,* and plant ecology—in order to gain a complete picture of the physical environment of humans and to examine the interactions of humans with that environment.

physical weathering breakup of massive *rock* (*bedrock*) into small particles through the action of physical forces acting at or near the earth's surface. (See also *weathering.*)

phytoplankton microscopic floating plants found largely in the uppermost layer of ocean water.

pioneer plants plants that first invade an environment of new land or a *soil* that has been cleared of vegetation cover; often these are annual *herbs.*

plagioclase feldspar aluminum-silicate *mineral* with sodium or calcium, or both.

plane of the ecliptic imaginary plane in which the earth's orbit lies.

plant nutrients *ions* or chemical compounds that are needed for plant growth.

plate tectonics theory of *tectonic activity* dealing with *lithospheric plates* and their activity.

plateau upland surface, more or less flat and horizontal, upheld by resistant beds of *sedimentary rock* or *lava* flows and bounded by a steep *cliff.*

playa flat land surface underlain by fine *sediment* or *evaporite minerals* deposited from shallow *lake* waters in a dry *climate* in the floor of a closed topographic depression.

Pleistocene Epoch epoch of the *Cenozoic Era,* comprising the latter part of the *Ice Age,* it preceded the *Holocene Epoch.*

plinthite iron-rich concentrations present in some kinds of *soils* in deeper *soil horizons* and capable of hardening into rocklike material with repeated wetting and drying.

plucking (See *glacial plucking.*)

pluton any body of *intrusive igneous rock* that has solidified below the surface, enclosed in preexisting *rock.*

pocket beach *beach* of crescentic outline located at a *bay* head.

podzol type of *soil order* closely equivalent to *Spodosol;* an order of the Canadian Soil Classification System.

Podzolic order a class of *forest* and heath *soils* in the Canadian Soil Classification System in which an amorphous material of humified organic matter with Al and Fe accumulates.

point *spatial object* in a *geographic information system* with no area.

point bar deposit of coarse *bed-load alluvium* accumulated on the inside of a growing *alluvial meander.*

polar easterlies system of easterly surface *winds* at high *latitude,* best developed in the southern *hemisphere* over Antarctica.

polar front *front* lying between cold polar *air masses* and warm tropical air masses, often situated along a *jet stream* within the *upper-air westerlies.*

polar front jet stream *jet stream* found along the *polar front,* where cold polar air and warm tropical air are in contact.

polar high persistent low-level center of high *atmospheric pressure* located over the *polar zone* of Antarctica.

polar outbreak tongue of cold polar air, preceded by a *cold front,* penetrating far into the *tropical zone* and often reaching the *equatorial zone*; it brings rain squalls and unusual cold.

polar projection *map projection* centered on earth's *north pole* or *south pole*.

polar zones *latitude* zones lying between 75° and 90° N and S.

poleward heat transport movement of heat from equatorial and tropical regions toward the poles, occurring as *latent* and *sensible heat transfer*.

pollutants in air pollution studies, foreign matter injected into the lower *atmosphere* as *particulates* or as chemical pollutant *gases*.

pollution dome broad, low dome-shaped layer of polluted air, formed over an urban area at times when *winds* are weak or calm prevails.

pollution plume (1) the trace or path of pollutant substances, moving along the flow paths of *ground water*; (2) trail of polluted air carried downwind from a pollution source by strong *winds*.

polygon type of *spatial object* in a *geographic information system* with a closed chain of connected *lines* surrounding an area.

potash feldspar aluminum-silicate *mineral* with potassium as the dominant metal.

potential evapotranspiration (water need) ideal or hypothetical rate of *evapotranspiration* estimated to occur from a complete canopy of green foliage of growing plants continuously supplied with all the *soil water* they can use; a real condition reached in those situations where *precipitation* is sufficiently great or irrigation water is supplied in sufficient amounts.

pothole cylindrical cavity in hard *bedrock* of a *stream channel* produced by *abrasion* of a rounded rock fragment rotating within the cavity.

prairie plant *formation class* of the *grassland biome*, consisting of dominant tall grasses and subdominant *forbs*, widespread in subhumid continental *climate* regions of the *subtropical zone* and *midlatitude zone*. (See also *short-grass prairie*, *tall-grass prairie*.)

Precambrian time all of geologic time older than the beginning of the Cambrian Period, that is, older than 600 million years.

precipitation particles of *liquid* water or ice that fall from the *atmosphere* and may reach the ground. (See also *convectional precipitation, cyclonic precipitation, orographic precipitation*.)

preprocessing component component of a *geographic information system* that prepares data for entry to the system.

pressure gradient change of *atmospheric pressure* measured along a line at right angles to the *isobars*.

pressure gradient force force acting horizontally, tending to move air in the direction of lower *atmospheric pressure*.

prevailing westerly winds (westerlies) surface *winds* blowing from a generally westerly direction in the *midlatitude zone*, but varying greatly in direction and intensity.

primary consumers organisms at the lowest level of the *food chain* that ingest *primary producers* or *decomposers* as their *energy* source.

primary minerals in *pedology* (*soil science*), the original, unaltered *silicate minerals* of *igneous rocks* and *metamorphic rocks*.

primary producers organisms that use light *energy* to convert *carbon dioxide* and water to *carbohydrates* through the process of *photosynthesis*.

prime meridian reference meridian of zero *longitude*; universally accepted as the Greenwich meridian.

product generation component component of a *geographic information system* that provides output products such as maps, images, or tabular reports.

progradation shoreward building of a *beach, bar*, or *sandspit* by addition of coarse *sediment* carried by *littoral drift* or brought from deeper water offshore.

pyroxene group complex aluminum-silicate *minerals* rich in calcium, magnesium, and iron, dark in color, high in density, classed as *mafic minerals*.

quartz *mineral* of silicon dioxide composition.

quartzite *metamorphic rock* consisting largely of the *mineral quartz*.

radar an active *remote sensing* system in which a pulse of *radiation* is emitted by an instrument, and the strength of the echo of the pulse is recorded.

radial drainage pattern stream pattern consisting of *streams* radiating outward from a central peak or highland, such as a *sedimentary dome* or a *volcano*.

radiation (See *electromagnetic radiation*.)

radiation balance condition of balance between incoming *energy* of solar *shortwave radiation* and outgoing *longwave radiation* emitted by the earth into space.

radiation fog *fog* produced by radiational cooling of the basal air layer.

radioactive decay spontaneous change in the nucleus of an atom that leads to the emission of matter and *energy*.

radiogenic heat heat from the earth's interior that is slowly released by the *radioactive decay* of unstable forms of certain chemical elements.

rain form of *precipitation* consisting of falling water drops, usually 0.5 mm or larger in diameter.

rain gauge instrument used to measure the amount of *rain* that has fallen.

rain-green vegetation vegetation that puts out green foliage in the wet season, but becomes largely dormant in the dry season; found in the *tropical zone*, it includes the *savanna biome* and *monsoon forest*.

rainshadow belt of *arid climate* to lee of a mountain barrier, produced as a result of adiabatic warming of descending air.

raised shoreline former *shoreline* lifted above the limit of wave action; also called an elevated shoreline.

rapids steep-*gradient* reaches of a *stream channel* in which *stream* velocity is high.

recessional moraine *moraine* produced at the ice margin during a temporary halt in the recessional phase of *deglaciation*.

reclamation in *strip mining*, the restoration of a suitable surface vegetation cover after a working portion of the mine is closed.

recumbent overturned, as a folded sequence of *rock* layers in which the *folds* are doubled back upon themselves.

reflection outward scattering of *radiation* toward space by the *atmosphere* or earth's surface.

reg desert surface armored with a pebble layer, resulting from long-continued *deflation*; found in the Sahara Desert of North Africa.

regional scale the scale of observation at which subcontinental regions are discernible.

regolith layer of *mineral* particles overlying the *bedrock*; may be derived by *weathering* of underlying bedrock or be transported from other locations by *fluid* agents. (See also *residual regolith, transported regolith*.)

Regosolic order a class of *soils* in the Canadian Soil Classification System that exhibits weakly developed *horizons*.

relative humidity ratio of *water vapor* present in the air to the maximum quantity possible for *saturated air* at the same temperature.

remote sensing measurement of some property of an object or surface by means other than direct contact; usually refers to the gathering of scientific information about the earth's surface from great heights and over broad areas, using instruments mounted on aircraft or orbiting space vehicles.

remote sensor instrument or device measuring *electromagnetic radiation* reflected or emitted from a target body.

removal in *soil science*, the set of processes that result in the removal of material from a *soil horizon*, such as surface erosion or *leaching*.

representative fraction (R.F.) (See *scale fraction*.)

residual regolith *regolith* formed in place by alteration of *the bedrock* directly beneath it.

resolution on a map, power to resolve small objects present on the ground.

respiration the *oxidation* of organic compounds by organisms that powers bodily functions.

retrogradation cutting back (retreat) of a *shoreline, beach, marine cliff*, or *marine scarp* by wave action.

reverse fault type of *fault* in which one fault block rides up over the other on a steep *fault plane*.

revolution motion of a planet in its orbit around the sun, or of a planetary satellite around a planet.

rhyolite *extrusive igneous rock* of *granite* composition; it occurs as *lava* or *tephra*.

ria coastal embayment or *estuary*.

ria coast deeply embayed *coast* formed by partial *submergence* of a *land mass* previously shaped by fluvial *denudation*.

Richter scale scale of magnitude numbers describing the quantity of *energy* released by an *earthquake*.

ridge-and-valley landscape assemblage of *landforms* developed by *denudation* of a system of open *folds* of *strata* and consisting of long, narrow ridges and valleys arranged in parallel or zigzag patterns.

rift valley trenchlike valley with steep, parallel sides; essentially a *graben* between two *normal faults*; associated with crustal spreading.

rill erosion form of *accelerated erosion* in which numerous, closely spaced miniature channels (rills) are scored into the surface of exposed *soil* or *regolith*.

rock natural aggregate of *minerals* in the *solid* state; usually hard and consisting of one, two, or more mineral varieties.

Rossby waves horizontal undulations in the flow path of the *upper-air westerlies*; also known as upper-air waves.

rotation spinning of an object around an axis.

runoff flow of water from continents to oceans by way of *stream flow* and *ground water* flow; a term in the water balance of the *hydrologic cycle*. In a more restricted sense, runoff refers to surface flow by *overland flow* and channel flow.

Sahel (Sahelian zones) belt of *wet-dry tropical* ③ and *semiarid dry tropical* ④ climate in Africa in which *precipitation* is highly variable from year to year.

salic horizon *soil horizon* enriched by soluble salts.

salinization *precipitation* of soluble salts within the *soil*.

salt flat shallow basin covered with salt deposits formed when *stream* input to the basin is subjected to severe *evaporation*; may also form by evaporation of a saline *lake* when *climate* changes.

salt marsh *peat*-covered expanse of *sediment* built up to the level of high *tide* over a previously formed tidal *mud* flat.

salt-crystal growth a form of *weathering* in which *rock* is disintegrated by the expansive pressure of growing salt crystals during dry weather periods when *evaporation* is rapid.

saltation leaping, impacting, and rebounding of sand grains transported over a *sand* or pebble surface by *wind*.

saltwater intrusion movement of saltwater into a coastal well from below; usually produced by overpumping.

sand *sediment* particles between 0.06 and 2 mm (0.002 and 0.078 in.) in diameter.

sand dune hill or ridge of loose, well-sorted *sand* shaped by *wind* and usually capable of downwind motion.

sand sea field of *transverse dunes*.

sandspit narrow, fingerlike embankment of *sand* constructed by *littoral drift* into the open water of a *bay*.

sandstone variety of *sedimentary rock* consisting largely of *mineral* particles of *sand* grade size.

Santa Ana easterly *wind*, often hot and dry, that blows from the interior desert region of southern California and passes over the coastal mountain ranges to reach the Pacific Ocean.

saturated air air holding the maximum possible quantity of *water vapor* at a given temperature and pressure.

saturated zone zone beneath the land surface in which all pores of the *bedrock* or *regolith* are filled with *ground water*.

saturation condition in which air holds the maximum possible quantity of *water vapor* at a given temperature and pressure.

savanna a vegetation cover of widely spaced *trees* with a grassland beneath.

savanna biome *biome* that consists of a combination of *trees* and grassland in various proportions.

savanna woodland plant *formation class* of the *savanna biome* consisting of a *woodland* of widely spaced *trees* and a grass layer, found throughout the *wet-dry tropical climate* ③ regions in a belt adjacent to the *monsoon forest* and *low-latitude rainforest*.

scale fraction ratio that relates distance on the earth's surface to distance on a map or surface of a globe.

scale of globe ratio of size of a globe to size of the earth, where size is expressed by a measure of length or distance.

scale of map ratio of distance between two points on a map and the same two points on the ground.

scanning systems *remote sensing* systems that make use of a scanning beam to generate images over the frame of surveillance.

scarification general term for artificial excavations and other land disturbances produced for purposes of extracting or processing *mineral* resources.

scattering turning aside by reflection of solar *shortwave radiation* by *gas* molecules of the *atmosphere*.

schist foliated *metamorphic rock* in which mica flakes are typically found oriented parallel with foliation surfaces.

sclerophyll forest plant *formation class* of the *forest biome*, consisting of low sclerophyll *trees*, and often including sclerophyll woodland or *scrub,* associated with regions of *Mediterranean climate* ⑦.

sclerophyll woodland plant *formation class* of the *forest biome* composed of widely spaced sclerophyll *trees* and *shrubs.*

sclerophylls hardleaved evergreen *trees* and *shrubs* capable of enduring a long dry season.

scoria *lava* or *tephra* containing numerous cavities produced by expanding gases during cooling.

scrub plant *formation class* or subclass consisting of shrubs and having a canopy coverage of about 50 percent.

sea arch archlike *landform* of a rocky, cliffed coast created when waves erode through a narrow headland from both sides.

sea breeze local *wind* blowing from sea to land during the day.

sea cave cave near the base of a *marine cliff,* eroded by breaking waves.

sea fog *fog* layer formed at sea when warm moist air passes over a cool ocean current and is chilled to the *condensation point.*

sea ice floating ice of the oceans formed by direct *freezing* of ocean water.

second of arc 1/60 of a minute, or 1/3600 of a degree.

secondary consumers animals that feed on *primary consumers.*

secondary minerals in *soil science, minerals* that are stable in the surface environment, derived by *mineral alteration* of the *primary minerals.*

sediment finely divided *mineral matter* and organic matter derived directly or indirectly from preexisting rock and from life processes. (See also *chemically precipitated sediment, organic sediment.*)

sedimentary dome up-arched *strata* forming a circular structure with domed summit and flanks with moderate to steep outward *dip.*

sedimentary rock *rock* formed from accumulation of *sediment.*

seismic sea wave (tsunami) train of sea waves set off by an *earthquake* (or other sea-floor disturbance) traveling over the ocean surface.

seismic waves waves propagated through the solid earth during an *earthquake* by faulting or other crustal disturbance.

semiarid (dry climate subtype) subtype of the dry *climates* exhibiting a short wet season supporting the growth of grasses and *annual* plants.

semidesert plant *formation class* of the *desert biome,* consisting of xerophytic *shrub* vegetation with a poorly developed herbaceous lower layer; subtypes are semidesert scrub and *woodland.*

sensible heat heat measurable by a *thermometer.*

sensible heat transfer flow of heat from one substance to another by direct contact.

sequential landforms *landforms* produced by external earth processes in the total activity of *denudation.* Examples: *alluvial fan, floodplain, gorge.*

shale fissile, *sedimentary rock* of mud or *clay* composition, showing lamination.

shearing (of rock) slipping motion between very thin *rock* layers, like a deck of cards fanned with the sweep of a palm.

sheet erosion type of *accelerated soil erosion* in which thin layers of *soil* are removed without formation of rills or *gullies.*

sheet flow overland flow taking the form of a continuous thin film of water over a smooth surface of *soil, regolith,* or *rock.*

sheeting structure thick, subparallel layers of massive *bedrock* formed by spontaneous expansion accompanying *unloading.*

shield volcano low, often large, domelike accumulation of *basalt lava* flows emerging from long radial fissures on flanks.

shoreline shifting line of contact between water and land.

short-grass prairie plant *formation class* in the *grassland biome* consisting of short grasses sparsely distributed in clumps and bunches and some *shrubs,* widespread in areas of semi-arid *climate* in continental interiors of North America and Eurasia; also called *steppe.*

shortwave infrared *infrared radiation* with wavelengths shorter than 3 μm.

shortwave radiation *electromagnetic energy* in the range from 0.2 to 3 μm, including most of the *energy* spectrum of solar *radiation.*

shrubs woody *perennial* plants, usually small or low, with several low-branching stems and a foliage mass close to the ground.

silica silicon dioxide in any of several *mineral* forms.

silicate minerals (silicates) *minerals* containing silicon and oxygen atoms, linked in the crystal space lattice in units of four oxygen atoms to each silicon atom.

sill *intrusive igneous rock* in the form of a plate where magma was forced into a natural parting in the *bedrock,* such as a bedding surface in a sequence of *sedimentary rocks.*

silt *sediment* particles between 0.004 and 0.06 mm (0.0002 and 0.0024 in.) in diameter.

sinkhole surface depression in *limestone,* leading down into *limestone caverns.*

slash-and-burn agricultural system, practiced in the *low-latitude rainforest,* in which small areas are cleared and the *trees* burned, forming plots that can be cultivated for brief periods.

slate compact, fine-grained variety of *metamorphic rock,* derived from *shale,* showing well-developed cleavage.

sleet form of *precipitation* consisting of ice pellets, which may be frozen raindrops.

sling psychrometer apparatus for measuring humidity consisting of a wet-bulb thermometer and a dry-bulb thermometer.

slip face steep face of an active *sand dune,* receiving *sand* by *saltation* over the dune crest and repeatedly sliding because of oversteepening.

slope (1) degree of inclination from the horizontal of an element of ground surface, analogous to *dip* in the geologic sense; (2) any portion or element of the earth's solid surface; (3) verb meaning "to incline."

small-scale map map with *fractional scale* of less than 1:100,000; usually shows a large area.

smog mixture of *aerosols* and chemical *pollutants* in the lower *atmosphere,* usually found over urban areas.

snow form of *precipitation* consisting of ice particles.

soil natural terrestrial surface layer containing living matter and supporting or capable of supporting plants.

soil colloids *mineral* particles of extremely small size, capable of remaining suspended indefinitely in water; typically, they have the form of thin plates or scales.

soil creep extremely slow downhill movement of *soil* and *regolith* as a result of continued agitation and disturbance of the particles by such activities as *frost action*, temperature changes, or wetting and drying of the soil.

soil enrichment additions of materials to the *soil* body; one of the *pedogenic processes*.

soil erosion erosional removal of material from the *soil* surface.

soil horizon distinctive layer of the *soil*, more or less horizontal, set apart from other soil zones or layers by differences in physical and chemical composition, organic content, structure, or a combination of those properties, produced by soil-forming processes.

soil orders those 11 *soil* classes forming the highest category in the classification of soils.

soil profile display of *soil horizons* on the face of a freshly cut vertical exposure through the *soil*.

soil science (See *pedology*.)

soil structure presence, size, and form of aggregations (lumps or clusters) of *soil* particles.

soil texture descriptive property of the *mineral* portion of the *soil* based on varying proportions of *sand*, *silt*, and *clay*.

soil water water held in the *soil* and available to plants through their root systems; a form of *subsurface water*.

soil-water belt *soil* layer from which plants draw *soil water*.

soil-water recharge restoring of depleted *soil water* by *infiltration* of *precipitation*.

solar constant intensity of solar *radiation* falling on a unit area of surface held at right angles to the sun's rays at a point outside the earth's *atmosphere*, equal to an *energy* flow of about 1400 W/m^2.

solar day average time required for the earth to complete one *rotation* with respect to the sun; time elapsed between one *solar noon* and the next, averaged over the period of one year.

solar noon instant at which the *subsolar point* crosses the *meridian of longitude* of a given point on the earth; instant at which the sun's shadow points exactly due north or due south at a given location.

solids substances in the *solid* state; they resist changes in shape and volume, are usually capable of withstanding large unbalanced forces without yielding, but will ultimately yield by sudden breakage.

solifluction tundra (arctic) variety of *earthflow* in which the saturated thawed layer over *permafrost* flows slowly downhill to produce multiple terraces and solifluction lobes.

Solonetzic order a class of *soils* in the Canadian Soil Classification System with a *B horizon* of sticky *clay* that dries to a very hard condition.

sorting separation of one grade size of *sediment* particles from another by the action of currents of air or water.

source region extensive land or ocean surface over which an *air mass* derives its temperature and moisture characteristics.

south pole point at which the southern end of the earth's *axis of rotation* intersects the earth's surface.

southeast trade winds surface *winds* of low *latitudes* that blow steadily from the southeast. (See also *trade winds*.)

Southern Oscillation episodic reversal of prevailing barometric pressure differences between two regions, one centered on Darwin, Australia, in the eastern Indian Ocean, and the other on Tahiti in the western Pacific Ocean; a precursor to the occurrence of an El Niño event. (See also *El Niño*.)

southern pine forest subtype of *needleleaf forest* dominated by pines and occurring in the *moist subtropical climate* ⑥ .

spatial data information associated with a specific location or area of the earth's surface.

spatial object a geographic area, *line*, or *point* to which information is attached.

specific heat property of a substance that governs the temperature change of the substance with a given input of heat *energy*.

specific humidity mass of *water vapor* contained in a unit mass of air.

spit (See *sandspit*.)

splash erosion *soil erosion* caused by direct impact of falling raindrops on a wet surface of *soil* or *regolith*.

spodic horizon *soil horizon* containing precipitated amorphous materials composed of organic matter and *oxides* of aluminum, with or without iron.

Spodosols *soil order* consisting of *soils* with a *spodic horizon*, an *albic horizon*, with low *base status*, and lacking in *carbonate* materials.

spreading plate boundary *lithospheric plate* boundary along which two plates of *oceanic lithosphere* are undergoing separation, while at the same time new lithosphere is being formed by *accretion*. (See also *transform plate boundary*.)

stable air mass *air mass* in which the *environmental temperature lapse rate* is less than the *dry adiabatic lapse rate*, inhibiting *convectional* uplift and mixing.

stack (marine) isolated columnar mass of *bedrock* left standing in front of a retreating *marine cliff*.

stage height of the surface of a river above its bed or a fixed level near the bed.

standard meridians *standard time* meridians separated by 15° of *longitude* and having values that are multiples of 15°. (In some cases meridians are used that are multiples of $7\frac{1}{2}$s.)

standard time system time system based on the local time of a *standard meridian* and applied to belts of *longitude* extending $7\frac{1}{2}$s° (more or less) on either side of that meridian.

standard time zone zone of the earth in which all inhabitants keep the same time, which is that of a *standard meridian* within the zone.

star dune large, isolated *sand dune* with radial ridges culminating in a peaked summit; found in the deserts of North Africa and the Arabian Peninsula.

steppe (See *short-grass prairie*.)

steppe climate (See *semiarid [dry climate subtype]*.)

stone polygons (See *stone rings*.)

stone rings linked ringlike ridges of cobbles or boulders lying at the surface of the ground in arctic and alpine tundra regions.

storage capacity maximum capacity of *soil* to hold water against the pull of *gravity*.

storm surge rapid rise of coastal water level accompanying the onshore arrival of a *tropical cyclone*.

strata layers of *sediment* or *sedimentary rock* in which individual beds are separated from one another along bedding planes.

stratified drift *glacial drift* made up of sorted and layered *clay*, *silt*, *sand*, or gravel deposited from meltwater in *stream channels*, or in marginal *lakes* close to the ice front.

stratiform clouds *clouds* of layered, blanketlike form.

stratosphere layer of *atmosphere* lying directly above the *troposphere*.

stratovolcano *volcano* constructed of multiple layers of *lava* and *tephra* (volcanic ash).

stratus *cloud* type of the low-height family formed into a dense, dark gray layer.

stream long, narrow body of flowing water occupying a *stream channel* and moving to lower levels under the force of gravity. (See *consequent stream, graded stream, subsequent stream.*)

stream capacity maximum *stream load* of *solid* matter that can be carried by a *stream* for a given *discharge.*

stream channel long, narrow, troughlike depression occupied and shaped by a *stream* moving to progressively lower levels.

stream deposition accumulation of transported particles on a *stream* bed, upon the adjacent *floodplain*, or in a body of standing water.

stream erosion progressive removal of *mineral* particles from the floor or sides of a *stream channel* by drag force of the moving water, or by *abrasion*, or by *corrosion.*

stream flow water flow in a *stream channel*; same as channel flow.

stream gradient rate of descent to lower elevations along the length of a *stream channel*, stated in m/km, ft/mi, degrees, or percent.

stream load *solid* matter carried by a *stream* in dissolved form (as *ions*), in *turbulent suspension*, and as *bed load.*

stream profile a graph of the elevation of a *stream* plotted against its distance downstream.

stream transportation downvalley movement of eroded particles in a *stream channel* in solution, in *turbulent suspension*, or as *bed load.*

strike compass direction of the line of intersection of an inclined *rock* plane and a horizontal plane of reference. (See also *dip.*)

strike-slip fault variety of *fault* on which the motion is dominantly horizontal along a near-vertical *fault plane.*

strip mining mining method in which *overburden* is first removed from a seam of *coal*, or a sedimentary ore, allowing the coal or ore to be extracted.

subantarctic zone *latitude* zone lying between lat. 55° and 60° S (more or less) and occupying a region between the *midlatitude zone* and the *antarctic zone.*

subarctic zone *latitude* zone between lat. 55° and 60° N (more or less), occupying a region between the *midlatitude zone* and the *arctic zone.*

subduction descent of the downbent edge of a *lithospheric plate* into the *asthenosphere* so as to pass beneath the edge of the adjoining plate.

sublimation process of change of ice (*solid* state) to *water vapor* (*gaseous* state); in *meteorology*, sublimation also refers to the change of state from water vapor (*liquid*) to ice (solid), which is referred to as *deposition* in this text.

submergence inundation or partial drowning of a former land surface by a rise of sea level or a sinking of the *crust*, or both.

suborder a unit of *soil* classification representing a subdivision of the *soil order.*

subsequent stream *stream* that develops its course by *stream erosion* along a band or belt of weaker *rock.*

subsolar point point on the earth's surface at which solar rays are perpendicular to the surface.

subsurface water water of the lands held in *soil, regolith*, or *bedrock* below the surface.

subtropical broadleaf evergreen forest a formation class of the *forest biome* composed of broadleaf evergreen *trees*; occurs primarily in the regions of the *moist subtropical climate* ⑥ .

subtropical evergreen forest a subdivision of the *forest biome* composed of both broadleaf and needleleaf evergreen *trees.*

subtropical high-pressure belts belts of persistent high *atmospheric pressure* trending east-west and centered about on lat. 30° N and S.

subtropical jet stream *jet stream* of westerly *winds* forming at the *tropopause*, just above the *Hadley cell.*

subtropical needleleaf evergreen forest a *formation class* of the *forest biome* composed of needleleaf evergreen *trees* occurring in the *moist subtropical climate* ⑥ of the southeastern United States; also referred to as the *southern pine forest.*

subtropical zones *latitude* zones occupying the region of lat. 25° to 35° N and S (more or less) and lying between the *tropical zones* and the *midlatitude zones.*

summer monsoon inflow of maritime air at low levels from the Indian Ocean toward the Asiatic low-pressure center in the season of high sun; associated with the rainy season of the *wet-dry tropical climate* ③ and the Asiatic monsoon climate.

summer solstice solstice occurring on June 21 or 22, when the *subsolar point* is located at $23\frac{1}{2}°$ N.

sun-synchronous orbit satellite orbit in which the orbital plane remains fixed in position with respect to the sun.

supercontinent single world continent, formed when *plate tectonic* motions move continents together into a single, large land mass. (See also *Pangea.*)

supercooled water water existing in the *liquid* state at a temperature lower than the normal *freezing* point.

surface water water of the lands flowing freely (as *streams*) or impounded (as ponds, *lakes*, marshes).

surges episodes of very rapid downvalley movement within an *alpine glacier.*

suspended load that part of the *stream load* carried in *turbulent suspension.*

suspension (See *turbulent suspension.*)

suture (See *continental suture.*)

swash surge of water up the *beach* slope (landward) following collapse of a breaker.

synclinal mountain steep-sided ridge or elongate mountain developed by erosion of a *syncline.*

syncline downfold of *strata* (or other layered *rock*) in a troughlike structure; a class of *folds*. (See also *anticline.*)

taiga plant *formation class* consisting of *woodland* with low, widely spaced *trees* and a ground cover of *lichens* and mosses, found along the northern fringes of the region of *boreal forest climate* (11); also called cold woodland.

tall-grass prairie a *formation class* of the *grassland biome* that consists of tall grasses with broad-leaved *herbs.*

talus accumulation of loose *rock* fragments derived by the fall of *rock* from a *cliff.*

talus slope slope formed of *talus.*

tarn small *lake* occupying a *rock* basin in a *cirque* of *glacial trough.*

tectonic activity process of bending (folding) and breaking (faulting) of crustal mountains, concentrated on or near active *lithospheric plate* boundaries.

tectonics branch of *geology* relating to tectonic activity and the features it produces. (See also *plate tectonics, tectonic activity*.)

temperature gradient rate of temperature change along a selected line or direction.

temperature inversion upward reversal of the normal *environmental temperature lapse rate*, so that the air temperature increases upward. (See also *high-level temperature inversion, low-level temperature inversion*.)

temperature regime distinctive type of annual temperature cycle.

tephra collective term for all size grades of solid *igneous* rock particles blown out under *gas* pressure from a volcanic vent.

terminal moraine *moraine* deposited as an embankment at the terminus of an *alpine glacier* or at the leading edge of an *ice sheet*.

terrestrial ecosystems *ecosystems* of land plants and animals found on upland surfaces of the continents.

thematic map map showing a single type of information.

theme category or class of information displayed on a map.

thermal infrared a portion of the *infrared radiation wavelength* band, from approximately 3 to 20 μm, in which objects at temperatures encountered on the earth's surface (including fires) emit *electromagnetic radiation*.

thermal pollution form of water pollution in which heated water is discharged into a *stream* or *lake* from the cooling system of a power plant or other industrial heat source.

thermistor electronic device that measures (air) temperature.

thermohaline circulation global scale circulation pattern of ocean waters involving both deep and surface *ocean currents*; driven by sinking of cold, dense ocean water at the northern edges of the North Atlantic Ocean.

thermometer instrument measuring temperature.

thermometer shelter louvered wooden cabinet of standard construction used to hold *thermometers* and other weather-monitoring equipment.

thorntree semidesert plant *formation class* within the *desert biome* consisting of sparse, xerophytic, often thorny, *trees* and *shrubs*.

thorntree–tall-grass savanna plant *formation class*, transitional between the *savanna biome* and the *grassland biome*, consisting of widely scattered *trees* in an open grassland.

thrust sheet sheetlike mass of *rock* moving forward over a *low-angle overthrust fault*.

thunderstorm intense, local convectional storm associated with a *cumulonimbus cloud* and yielding heavy *precipitation*, also with lightning and thunder, and sometimes the fall of *hail*.

tidal current current set in motion by the *ocean tide*.

tidal inlet narrow opening in a *barrier island* or baymouth *bar* through which *tidal currents* flow.

tide (See *ocean tide*.)

tide curve graphical presentation of the rhythmic rise and fall of ocean water because of *ocean tides*.

till heterogeneous mixture of *rock* fragments ranging in size from *clay* to boulders, deposited beneath moving glacial ice or directly from the *melting* in place of stagnant glacial ice.

till plain undulating, plainlike land surface underlain by glacial *till*.

time zones zones or belts of given east-west (*longitudinal*) extent within which *standard time* is applied according to a uniform system.

topographic contour *isopleth* of uniform elevation appearing on a map.

tornado small, very intense *wind* vortex with extremely low air pressure in its center, formed beneath a dense *cumulonimbus cloud* in proximity to a *cold front*.

trade winds (trades) surface *winds* in low *latitudes*, representing the low-level air flow within the *tropical easterlies*.

transcurrent fault *fault* on which the relative motion is dominantly horizontal, in the direction of the *strike* of the fault; also called a *strike-slip fault*.

transform fault special case of a *strike-slip fault* making up the boundary of two moving *lithospheric plates*; usually found along an offset of the *midoceanic ridge* where sea-floor spreading is in progress.

transform plate boundary *lithospheric plate* boundary along which two plates are in contact on a *transform fault*; the relative motion is that of a *strike-slip fault*.

transform scar linear topographic feature of the ocean floor taking the form of an irregular scarp or ridge and originating at the offset *axial rift* of the *midoceanic ridge*; it represents a former *transform fault* but is no longer a plate boundary.

transformation (soils) a class of soil-forming processes that transform materials within the *soil* body; examples include *mineral alteration* and *humification*.

translocation a *soil*-forming process in which materials are moved within the soil body, usually from one horizon to another.

transpiration evaporative loss of water to the *atmosphere* from leaf pores of plants.

transportation (See *stream transportation*.)

transported regolith *regolith* formed of *mineral* matter carried by *fluid* agents from a distant source and deposited on the *bedrock* or on older regolith. Examples: floodplain silt, lake clay, beach sand.

transverse dunes field of wavelike *sand dunes* with crests running at right angles to the direction of the prevailing *wind*.

traveling anticyclone center of high pressure and outspiraling *winds* that travels over the earth's surface; often associated with clear, dry *weather*.

traveling cyclone center of low pressure and inspiraling *winds* that travels over the earth's surface; includes *wave cyclones*, *tropical cyclones*, and *tornadoes*.

travertine *carbonate mineral matter*, usually *calcite*, accumulating on *limestone cavern* surfaces situated in the *unsaturated zone*.

tree large, erect woody *perennial* plant typically having a single main trunk, few branches in the lower part, and a branching crown.

trellis drainage pattern *drainage pattern* characterized by a dominant parallel set of major *subsequent streams*, joined at right angles by numerous short tributaries; typical of *coastal plains* and belts of eroded *folds*.

tropic of cancer *parallel of latitude* at $23\frac{1}{2}°$ N.

tropic of capricorn *parallel of latitude* at $23\frac{1}{2}°$ S.

tropical cyclone intense *traveling cyclone* of tropical and subtropical *latitudes*, accompanied by high winds and heavy rainfall.

tropical easterlies low-latitude *wind* system of persistent air flow from east to west between the two *subtropical high-pressure belts*.

tropical easterly jet stream upper-air *jet stream* of seasonal occurrence, running east to west at very high altitudes over Southeast Asia.

tropical high-pressure belt a high-pressure belt occurring in tropical *latitudes* at a high level in the *troposphere*; extends downward and poleward to form the *subtropical high-pressure belt*, located at the surface.

tropical zones *latitude* zones centered on the *tropic of cancer* and the *tropic of capricorn*, within the latitude ranges 10° to 25° N and 10° to 25° S, respectively.

tropical-zone rainforest plant *formation class* within the *forest biome* similar to *equatorial rainforest*, but occurring farther poleward in tropical regions.

tropopause boundary between *troposphere* and *stratosphere*.

troposphere lowermost layer of the *atmosphere* in which air temperature falls steadily with increasing altitude.

tsunami (See *seismic sea wave.*)

tundra biome *biome* of the cold regions of *arctic tundra* and *alpine tundra*, consisting of grasses, grasslike plants, flowering *herbs*, dwarf *shrubs*, mosses, and *lichens*.

tundra climate ⑫ cold *climate* of the *arctic zone* with eight or more consecutive months of zero *potential evapotranspiration* (*water need*).

tundra soils soils of the arctic *tundra climate* ⑫ regions.

turbulence in *fluid* flow, the motion of individual water particles in complex eddies, superimposed on the average downstream flow path.

turbulent flow mode of *fluid* flow in which individual fluid particles (molecules) move in complex eddies, superimposed on the average downstream flow path.

turbulent suspension *stream transportation* in which particles of *sediment* are held in the body of the *stream* by turbulent eddies. (Also applies to *wind* transportation.)

typhoon *tropical cyclone* of the western North Pacific and coastal waters of Southeast Asia.

Udalfs suborder of the *soil order Alfisols*; includes Alfisols of moist regions, usually in the *midlatitude zone*, with deciduous *forest* as the natural vegetation.

Udolls suborder of the *soil order Mollisols*; includes Mollisols of the moist *soil-water* regime in the *midlatitude zone* and with no horizon of *calcium carbonate* accumulation.

Ultisols *soil order* consisting of *soils* of warm soil temperatures with an *argillic horizon* and low *base status*.

ultramafic igneous rock *igneous rock* composed almost entirely of *mafic minerals*, usually *olivine* or *pyroxene group*.

ultraviolet radiation *electromagnetic energy* in the *wavelength* range of 0.2 to 0.4 μm.

unloading process of removal of overlying *rock* load from *bedrock* by processes of *denudation*, accompanied by expansion and often leading to the development of *sheeting structure*.

unsaturated zone *subsurface water* zone in which pores are not fully saturated, except at times when *infiltration* is very rapid; lies above the *saturated zone*.

unstable air air with substantial content of *water vapor*, capable of breaking into spontaneous convectional activity leading to the development of heavy showers and *thunderstorms*.

upper-air westerlies system of westerly *winds* in the upper *atmosphere* over middle and high *latitudes*.

upwelling upward motion of cold, nutrient-rich ocean waters, often associated with cool equatorward currents occurring along *continental margins*.

Ustalfs suborder of the *soil order Alfisols*; includes Alfisols of semiarid and seasonally dry *climates* in which the *soil* is dry for a long period in most years.

Ustolls suborder of the *soil order Mollisols*; includes Mollisols of the semiarid *climate* in the *midlatitude zone*, with a horizon of *calcium carbonate* accumulation.

valley winds air movement at night down the *gradient* of valleys and the enclosing mountainsides; alternating with daytime *mountain winds*.

veins small, irregular, branching network of *intrusive rock* within a preexisting *rock* mass.

vernal equinox *equinox* occurring on March 20 or 21, when the *subsolar point* is at the *equator*.

Vertisols *soil order* consisting of *soils* of the *subtropical zone* and the *tropical zone* with high *clay* content, developing deep, wide cracks when dry, and showing evidence of movement between aggregates.

visible light *electromagnetic energy* in the *wavelength* range of 0.4 to 0.7 μm.

volcanic bombs boulder-sized, semisolid masses of *lava* that are ejected from an erupting *volcano*.

volcanic neck isolated, narrow steep-sided peak formed by erosion of *igneous rock* previously solidified in the feeder pipe of an extinct *volcano*.

volcanism general term for *volcano* building and related forms of *extrusive igneous* activity.

volcano conical, circular structure built by accumulation of *lava* flows and *tephra*. (See also *shield volcano, stratovolcano*.)

volcano coast *coast* formed by *volcanoes* and *lava* flows built partly below and partly above sea level.

warm front moving *weather front* along which a warm *air mass* slides up over a cold air mass, leading to production of *stratiform clouds* and *precipitation*.

warm-core ring circular eddy of warm water, surrounded by cold water and lying adjacent to a warm, poleward moving ocean current, such as the Gulf Stream. (See also *cold-core ring*.)

washout downsweeping of atmospheric *particulates* by *precipitation*.

watershed land surface within a *drainage basin* that sheds water by *overland flow*.

water gap narrow transverse *gorge* cut across a narrow ridge by a *stream*, usually in a region of eroded *folds*.

water need (See *potential evapotranspiration*.)

water surplus water disposed of by *runoff* or percolation to the ground water zone after the *storage capacity* of the *soil* is full.

water table upper boundary surface of the *saturated zone*; the upper limit of the *ground water* body.

water vapor the gaseous state of water.

waterfall abrupt descent of a *stream* over a *bedrock* downstep in the *stream channel*.

waterlogging rise of a *water table* in *alluvium* to bring the zone of saturation into the root zone of plants.

watt (W) unit of power equal to the quantity of work done at the rate of one joule per second (1 J/s).

wave cyclone traveling, vortexlike *cyclone* involving interaction of cold and warm *air masses* along sharply defined *fronts.*

wave-cut notch *rock* recess at the base of a *marine cliff* where wave impact is concentrated.

wavelength distance separating one wave crest from the next in any uniform succession of traveling waves.

weak equatorial low weak, slowly moving low-pressure center (*cyclone*) accompanied by numerous convectional showers and *thunderstorms;* it forms close to the *intertropical convergence zone* in the rainy season, or *summer monsoon.*

weather physical state of the *atmosphere* at a given time and place.

weather system organized state of the *atmosphere* associated with a characteristic *weather* pattern, such as a *cyclone* or *anticyclone.*

weathering total of all processes acting at or near the earth's surface to cause physical disruption and chemical decomposition of *rock.* (See *chemical weathering, physical weathering.*)

west-wind drift ocean drift current moving eastward in zone of *prevailing westerlies.*

westerlies (See *prevailing westerly winds, upper-air westerlies.*)

wet adiabatic lapse rate reduced *adiabatic lapse rate* when *condensation* is taking place in rising air; value ranges between 4 and 9°C per 1000 m (2.2 and 4.9°F per 1000 ft).

wet equatorial climate ① moist *climate* of the *equatorial zone* with a large annual *water surplus,* and with uniformly warm temperatures throughout the year.

wetlands land areas of poor surface drainage, such as marshes and swamps.

wet-dry tropical climate ③ *climate* of the *tropical zone* characterized by a very wet season alternating with a very dry season.

wilting point quantity of stored *soil water,* less than which the foliage of plants not adapted to *drought* will wilt.

wind air motion, dominantly horizontal relative to the earth's surface.

wind abrasion mechanical wearing action of *wind*-driven *mineral* particles striking exposed rock surfaces.

wind vane *weather* instrument used to indicate *wind* direction.

winter monsoon outflow of continental air at low levels from the Siberian high, passing over Southeast Asia as a dry, cool northerly *wind.*

winter solstice solstice occurring on December 21 or 22, when the *subsolar point* is at $23\frac{1}{2}°$ S.

Wisconsinan Glaciation last *glaciation* of the *Pleistocene Epoch.*

woodland plant *formation class,* transitional between *forest biome* and *savanna biome,* consisting of widely spaced *trees* with canopy coverage between 25 and 60 percent.

Xeralfs suborder of the *soil order Alfisols;* includes Alfisols of the *Mediterranean climate* ⑦.

xeric animal animal adapted to dry *climate* conditions.

Xerolls suborder of the *soil order Mollisols;* includes Mollisols of the *Mediterranean climate* ⑦.

xerophytes plants adapted to a dry environment.

PHOTO CREDITS

Title Page

©Jess Stock/Tony Stone Images/ New York, Inc.

Prologue

Opener: Courtesy NOAA. **Figure P.A:** Courtesy NASA. **Figure P.B:** ©Earth Satellite Corporation. **Figure P.C:** ©Earth Satellite Corporation/Science Photo Library /Photo Researchers. **Figure P.D:** Courtesy NASA.

Chapter 1

Opener & Page 12: ©Ron Watts/Black Star. **Figure 1.1:** ©William James Warren/West Light. **Figure 1.6:** ©National Maritime Museum, Greenwich, London.

Chapter 2

Opener & Page 30: ©Bill Ross/West Light. **Figure 2.3:** Courtesy Daedalus Enterprises, Inc. & National Geographic Magazine. **Page 41:** Courtesy NASA.

Chapter 3

Opener & Page 50: ©Ron Watts/West Light. **Figure 3.2a:** ©A. N. Strahler. **Figure 3.2b:** ©Steve McCutcheon. **Figure 3.8:** ©George Wuerthner. **Figure 3.11:** ©Christi Carter/Grant Heilman Photography. **Figure 3.22:** ©Shawn Henry/SABA.

Chapter 4

Opener & Page 76: ©Mike Magnuson/Tony Stone Images/ New York, Inc. **Figure 4.8:** ©A. N. Strahler. **Figure 4.12a:** ©Darryl Torckler/Tony Stone Images/ New York, Inc. **Figure 4.12b:** ©Kees van den Berg/Photo Researchers. **Figure 4.12c:** ©Trevor Mein/Tony Stone Images/ New York, Inc. **Figure 4.12d:** ©Larry Ulrich/Tony Stone Images/ New York, Inc. **Figure 4.13:** ©Ron Sanford/Tony Stone Images/ New York, Inc. **Figure 4.14:** ©Larry West/Bruce Coleman, Inc. **Figure 4.15:** ©A. N. Strahler. **Page 89 (left):** Data received by the Canada Centre for Remote Sensing. Image provided courtesy of RADARSAT International.. **Page 89 (right):** ©Paul Barton/The Stock Market. **Page 91:** © John S. Shelton. **Page 94:** ©Jim Strawser/Grant Heilman Photography. **Figure 4.20:** ©Keith Kent/Peter Arnold, Inc. **Figure 4.22:** ©Nuridsany et Perennou/Photo Researchers. **Figure 4.23:** ©John H. Hoffman/Bruce Coleman, Inc.

Chapter 5

Opener & Page 106: ©Joe Sohm/Photo Researchers. **Figure 5.15:** ©P. F. Bentley/Black Star. **Figure 5.21:** Courtesy NASA. **Figure 5.23:** Courtesy Otis B. Brown, Robert Evans, and M. Carle, University of Miami, Rosenstiel School of Marine and Atmospheric Science, Florida, and NOAA Satellite Data Services Division. **Figure 5.24:** Courtesy Otis B. Brown, University of Miami, and Gene Carl Feldman, NASA/Goddard Space Flight Center.

Chapter 6

Opener & Page 134: Courtesy NASA Goddard Space Flight Center/A.F. Hasler, H. Pierce, K. Palaniappan, and M.

Manvin. **Figure 6.11:** ©Keith Brewster/ Weatherstock. **Figure 6.13:** ©Alex MacLean/ Landslides. **Page 148:** ©Robert Holland/The Image Bank. **Page 149:** ©Ted Levin/Earth Scenes.

Chapter 7

Opener & Page 154: ©Chris Brown/SABA. **Figure 7.10:** ©Will & Deni McIntyre/Photo Researchers. **Figure 7.13:** ©Boyd Norton. **Page 173:** ©Alain Nogues/Sygma. **Figure 7.16:** ©Frans Lanting/Minden Pictures, Inc. **Figure 7.19:** ©Fred Hirschmann. **Figure 7.24:** ©Tom McHugh/Photo Researchers. **Figure 7.26:** ©David Muench Photography. **Figure 7.30:** ©Porterfield/Chickering/Photo Researchers. **Figure 7.36:** ©Warren Garst/Tom Stack & Associates.

Chapter 8

Opener & Page 196: ©Philip Condit II/Tony Stone Images/ New York, Inc. **Figure 8.3:** Courtesy NOAA/NES-DIS/NCDC/SDSD. **Figure 8.5:** ©Kenneth W. Fink/Photo Researchers. **Figure 8.6:** ©Animals Animals. **Figure 8.7:** ©Barbara Gerlach/Dembinsky Photo Associates. **Figure 8.8:** ©Josef Muench. **Figure 8.9:** ©F. Stuart Westmorland/Photo Researchers. **Figure 8.12:** ©Steve McCutcheon. **Figure 8.14:** ©M. Freeman/Bruce Coleman, Inc. **Figure 8.16:** ©Ferrero/Labat/Auscape International. **Figure 8.17:** ©Tom Bean. **Figure 8.19:** © John S. Shelton. **Page 215:** ©Jacques Jangoux/Peter Arnold, Inc. **Figure 8.22:** ©Kenneth Murray/Photo Researchers. **Figure 8.24:** ©Jake Rajs. **Figure 8.26:** ©Michael Townsend/Tony Stone Images/ Seattle. **Figure 8.27a:** ©A. N. Strahler. **Figure 8.27b:** ©Alan H. Strahler. **Figure 8.29:** ©A. N. Strahler. **Figure 8.30:** ©David Muench Photography. **Figure 8.31:** ©Annie Griffiths Belt/DRK Photo. **Figure 8.32:** ©Joe Viesti/Viesti Associates, Inc. **Figure 8.35:** ©Brian A. Vikander. **Figure 8.36 & 8.37:** ©A. N. Strahler.

Chapter 9

Opener & Page 238: Kevin Fleming/©CORBIS. **Figure 9.4:** Harry Gold, Courtesy R. B. Krone, San Francisco District Corps. of Engineers, U. S. Army. **Figure 9.5:** ©R. Schaetzl. **Figure 9.9:** ©A. N. Strahler. **Figure 9.11:** ©Henry D. Foth. **Figure 9.12:** ©Alan H. Strahler. **Figure 9.13:** USDA/ Soil Conservation Service. **Figure 9.14:** ©Henry D. Foth. **Figure 9.15:** ©William E. Ferguson. **Figure 9.18:** ©R. Schaetzl. **Figure 9.19:** ©Henry D. Foth. **Figure 9.20:** ©Ric Ergenbright/Tony Stone Images/ Seattle. **Page 259:** ©Erich Lessing/Art Resource. **Figure 9.21:** ©Robin White/Fotolex Associates. **Figure 9.22:** Courtesy USDA/ Soil Conservation Service. **Figure 9.23:** Ned L. Reglein. **Figure 9.24:** ©Henry D. Foth.

Chapter 10

Opener & Page 266: ©Gregory Dimijian/Photo Researchers. **Figure 10.2:** ©Roberto de Gugliemo/Science Photo Library /Photo Researchers. **Figure 10.6 & 10.7:** Ward's Natural Science Establishment. **Figure 10.8:** ©Travelpix/FPG

International. **Figure 10.9:** Ward's Natural Science Establishment. **Figure 10.11:** ©A. N. Strahler. **Page 274:** Courtesy of Richard S. Williams, Jr., U. S. Geological Survey. **Figure 10.12:** ©Carr Clifton/Minden Pictures, Inc. **Figure 10.13, 10.14 & 10.15:** ©A. N. Strahler. **Figure 10.16:** ©Victor Englebert/Photo Researchers. **Figure 10.17:** ©Charles R. Belinky/Photo Researchers. **Figure 10.18:** ©A. N. Strahler. **Figure 10.19:** ©Freeman Patterson/Masterfile.

Chapter 11:

Opener & Page 288: Courtesy NASA. **Figure 11.9:** Courtesy David T. Sandwell, Scripps Institution of Oceanography and Walter H. F. Smith, National Oceanic & Atmospheric Administration.

Chapter 12

Opener & Page 314: ©Galen Rowell/ Mountain Light Photography, Inc. **Figure 12.3:** ©James Mason/Black Star. **Figure 12.4:** ©K. Hamdorf/Auscape International. **Figure 12.5:** ©Greg Vaughn/Tom Stack & Associates. **Figure 12.6:** ©Werner Stoy/Camera Hawaii. **Figure 12.9 & 12.10:** ©A. N. Strahler. **Figure 12.11:** ©Larry Ulrich. **Figure 12.12:** ©Steve Vidler/Leo de Wys, Inc. **Figure 12.16:** ©A. N. Strahler. **Figure 12.19:** ©James Balog/Black Star. **Figure 12.21:** ©Georg Gerster/Comstock, Inc. **Figure 12.23:** ©Exelsior/Sipa Press. **Figure 12.25:** ©G. Hall/Woodfin Camp & Associates. **Figure 12.26:** ©Vladimir Sichov/Sipa Press. **Page 333:** ©Rick Rickman/Matrix International, Inc.

Chapter 13

Opener & Page 338: ©Galen Rowell/ Mountain Light Photography, Inc. **Figure 13.3:** ©Susan Rayfield/Photo Researchers. **Figure 13.4:** ©A. N. Strahler. **Figure 13.5:** ©Steve McCutcheon. **Figure 13.6:** Bernard Hallet, Periglacial Laboratory, Quaternary Research Center. **Figure 13.7:** ©Stephen J. Krasemann/DRK Photo. **Figure 13.9:** ©Kunio Owaki/The Stock Market. **Figure 13.10:** ©George Wuerthner. **Figure 13.11:** ©David Muench Photography. **Figure 13.12:** ©Tom Bean. **Figure 13.13:** ©Alan H. Strahler. **Figure 13.14:** ©Douglas Peebles Photography. **Figure 13.17:** ©Mark A. Melton. **Figure 13.19:** ©Steve McCutcheon/Alaska Pictorial Service. **Figure 13.21:** Orlo E. Childs. **Figure 13.22:** Courtesy NOAA. **Figure 13.23:** Courtesy Los Angeles County Department of Public Works. **Page 353:** ©A. N. Strahler. **Figure 13.24:** ©Bill Davis/Black Star.

Chapter 14

Opener & Page 358: Courtesy NASA. **Figure 14.8:** ©Laurence Parent. **Figure 14.9:** © John S. Shelton. **Figure 14.11:** ©Bruno Barbey/Magnum Photos, Inc. **Page 367:** ©M. Smith/Sipa Press. **Figure 14.14:** ©Mark A. Melton. **Figure 14.19:** ©Dan Koeck/Gamma Liaison. **Figure 14.22:** ©Tom Bean/DRK Photo. **Figure 14.23:** ©Tom Till Photography. **Figure 14.24:** ©National Audubon Society/Photo Researchers.

Chapter 15

Opener & Page 382: ©John Beatty/Tony Stone Images/ New York, Inc. **Figure 15.2:** Official U.S. Navy photograph. **Figure 15.3:** ©Mark A. Melton. **Figure 15.4:** ©Joe Englander/Viesti Associates, Inc. **Figure 15.5:** ©Alan H. Strahler. **Figure 15.7:** ©Frans Lanting/Minden Pictures, Inc. **Page 391:** ©Earth Observation Satellite Company. **Figure 15.10:** ©Dick Dietrich/Dietrich Photography. **Figure 15.13:** ©H. J. de Blij.

Figure 15.15: ©(photo credit).**Figure 15.17:** ©Tom Bean. **Figure 15.19:** ©F. Kenneth Hare.**Figure 15.21:** ©NASA/Science Source/Photo Researchers. **Figure 15.22:** ©Albert Moldvay/Eriako Associates. **Figure 15.24:** ©T. A. Wiewandt/DRK Photo. **Figure 15.27:** ©Mark A. Melton. **Figure 15.29:** © John S. Shelton. **Figure 15.30:** ©Mark A. Melton.

Chapter 16

Opener & Page 408: ©Tom Bean/DRK Photo. **Figure 16.4:** ©Larry Ulrich/DRK Photo. **Figure 16.9:** © John S. Shelton. **Figure 16.13:** ©Earth Satellite Corporation. **Figure 16.17:** ©J. A. Kraulis/Masterfile. **Figure 16.20:** ©Landis Aerial Photo. **Page 420:** ©Carr Clifton. **Page 421:** ©Nancy S. Dituri/Envision. **Figure 16.24:** © John S. Shelton. **Figure 16.25:** ©Alex McLean/Landslides. **Figure 16.27:** ©Larry Ulrich.

Chapter 17

Opener & Page 428: ©Jake Rajs/Tony Stone Images/ New York, Inc. **Figure 17.2:** ©Alex S. MacLean. **Figure 17.4:** ©A. N. Strahler. **Figure 17.5:** ©Michael Yamashita. **Figure 17.7:** ©Steve Dunnell/The Image Bank. **Figure 17.9:** ©Vic De Lucia/NYT Pictures. **Figure 17.12:** ©David Muench Photography. **Figure 17.15a:** Courtesy NASA. **Figure 17.16:** ©Earth Satellite Corporation. **Figure 17.17:** Courtesy NASA. **Figure 17.18:** ©David Hiser/Photographers Aspen. **Figure 17.19:** ©Douglas Faulkner/Photo Researchers. **Figure 17.21:** © John S. Shelton. **Page 443:** ©Stephen Rose/Gamma Liaison. **Figure 17.22:** ©Tom Bean. **Figure 17.23:** ©M. J. Coe/Animals Animals/Earth Scenes. **Figure 17.25 & 17.26:** ©John S. Shelton. **Figure 17.27:** ©Aero Service Corporation, Litton Industries. **Figure 17.29:** ©William E. Ferguson. **Figure 17.31:** ©ERIM, Ann Arbor, Michigan. **Figure 17.32:** ©J.A. Kraulis/Masterfile. **Figure 17.33:** ©G. R. Roberts. **Figure 17.35:** ©Alan H. Strahler.

Chapter 18

Opener & Page 456: ©Tom Bean/Tony Stone Images/ New York, Inc. **Figure 18.1:** ©Carr Clifton/Tony Stone Images/ Seattle. **Figure 18.5:** ©Fred Hirschmann Wilderness Photography. **Figure 18.6:** © John S. Shelton. **Figure 18.7:** ©Alan H. Strahler. **Figure 18.8:** ©Floyd L. Norgaard/Ric Ergenbright Photography. **Figure 18.11:** ©D. Parer & E. Parer–Cook/Auscape International Pty. Ltd.. **Figure 18.12:** Courtesy NASA. **Figure 18.13:** ©Wolfgang Kaehler. **Figure 18.17:** Courtesy NASA. **Figure 18.20:** ©C. Wolinsky/Stock, Boston. **Figure 18.21:** ©A. N. Strahler. **Page 473:** Courtesy National Snow and Ice Data Center.

Epilogue

Opener: ©Russell Gordon/Odyssey Productions. **Figure EP.1:** ©Alicia Hurr/ Weatherstock. **Figure EP.2:** ©Tom McHugh/Photo Researchers. **Figure EP.3:** ©Frans Lanting/Minden Pictures, Inc.

Appendix 2

Figure A2.1: SAR image courtesy of Intera Technologies Corporation, Calgary, Alberta, Canada. **Figure A2.2:** Courtesy NASA, compiled and annotated by John E. Estes and Leslie W. Senger. **Figure A2.3:** ©Photo provided by CAPS, Inc./Syed Sikander/Chicago Aerial Photo Services, Inc. **Figure A2.5:** Copyright ©1993 by CNES; Courtesy SPOT Image Corporation, Reston, Virginia.

INDEX

Ablation, 457
Abrasion, 386
Abrasion platforms, 430
Absorption, 40–41
Abyssal plains, 294
Accretion, plate tectonics, 300
Acid deposition, 92-94
Acid mine drainage, 376
Acid rain, 92–94
Active continental margins, 295
Active remote sensing systems, 488–489
Adiabatic processes, 82–84
Advection fog, 86
Aerial photography, 489, *490–491*
Aerosols, 57
    as pollutants, 88
African plate, 303
African Sahel, 171, 172–173
Aggradation, 396–397
Agriculture, 206. *See also* Soils,
    glaciation effects, 470
    old–field succession, 204
    and water table depletion, 365–66
A horizon, 245
AIDS, long-term impact, 481
Air. *See* Atmosphere
Air masses, 135–139
    classification, 135–136
    fronts, 137–139, 140
    North American, 137
Air pollution, 88–94. *See also* Acid rain
    and clouds and fog, 84
    control, 94
    fallout and washout, 90
    and ozone layer, 39
    pollution dome, 91, 92
    relative proportion of different pollutants,
        in U.S., 88
Air temperature, 51
    annual cycles, 60–63
    annual range, 68–69
    atmospheric temperature structure, 56–59
    daily cycle, 53–55
    land vs. water, 62–63
    distance to ground and, 54
    elevation variation, 56–58
    global patterns, 63–68
    global warming, 68–72
    high-latitude climates, 187–192
    and humidity, 79–80
    land-water contrasts, 61–63
    lapse rate, 56, 57
    low-latitude climates, 163–175

mean annual temperature: 1856–1996, 70
    measurement, 51–52
    midlatitude climates, 176–187
    mountain environments, 57–58
    temperature inversion, 58–60
    urban vs. rural surface characteristics, 55–56
    and vegetation, 203–204
    world maps, *65–66*
Albedo, 42
Albic soil horizon, 255
Aleutian Low, 142
Alfisols, 249, *250, 254–255*
Alluvial fans, 401–402
Alluvial meanders, 392
Alluvial rivers, 398
Alluvial terraces, 397
Alluvium, 340, 385
Alpine chains, 292
Alpine folds, 299
Alpine glaciers, 457, *458–462*
Alpine tundra, 190, 229
Alteration, of rocks, 273
Altocumulus clouds, *85*
Altostratus clouds, *85*
Aluminum, in earth's crust, 267
Aluminum oxides, in soil, 243, 253
Amazon rainforest, 169, *214–215*
Amazon River, 371
American plate, 301
Amphiboles, 270
Andesite, *272*
Andes Mountains, 319
Andisols, 249, *251*
Anemometers, 109
Animals, and ecosystem
    boreal needleleaf forest, 219
    deciduous forest, 218–219
    desert, 228
    grassland, 225–227
    light and, 204
    plant cover composition and, 206
    rainforest, 211
    savanna, 224
    temperature and, 203–204
    tundra, 229–230
    water needs of, 201–203
Annual cycles, air temperature, and global
    warming, 68, 69
Annual plants, 207
Antarctic air mass, 136
Antarctic circle, 25
Antarctic front zone, 163
Antarctic Ice Sheet, 472, 473

Antarctic plate, 303
Antarctic zone, 37
Anticlinal valleys, 416
Anticlines, 323
Anticyclones, 112, 135
    traveling, 135
Anvil head clouds, 85, *99*
Aphelion, 22
Appalachian mountains, 294
Aquatic ecosystem, 297
    biomass of, 199
Aquept soils, 258
Aral Sea, *356–357*, 375
Arc minutes, 16
Arc seconds, 16
Arctic air mass, 136, 137
Arctic circle, 25
Arctic-front zone, 163
Arctic tundra, 229
Arctic zone, 37
Arêtes, 461
Argillic soil horizons, 252
Arid climates, 160. *See also* Desert biomes
    fluvial processes, 399–403
    landforms of, 410–411
    vegetation adaptions to, 201–202
Arid deserts. *See* Dry deserts
Aridisols, 251, *251,* 259–260
Aspect, of ground surface, 246
Asthenosphere, 291
Astronomical hypothesis, of glaciation cycles,
    26, 471
Atmosphere, 7
    composition, 38
    heat/moisture transport, 128–129
    humidity, 70–82
    insolation losses, 39–41
    longwave radiation from, 44
    structure of, 56–57
    temperature inversion, 58–60
    temperature profile, 56–59
Atmospheric pressure, 107–108
    day-to-day variation, 108
    global pressure patterns and winds, 113–118
    gradients, and wind, 109 (*see also* Winds)
    measurement, 107–108
    sea level, 107, 108
    units of, 108
    variation with altitude, 108
Atolls, 440–441
Australian time zones, 21
Austral-Indian plate, 303
Autumnal equinox, 23

Axial rift, 294
Ayres Rock, 420
Azores High, 116

Backwash, 431, 433
Badlands, 385–386
Bangkok, subsidence in, 366
Barchan dunes, 445–446
Barometers, 107–108
Bar (pressure unit), 108
Barrier-island coasts, 436–438
Barrier reefs, 440
Bars, 432
Basalt, 272
Bases, in soil, 241
Base status, of soil, 243
Batholiths, 272
    exposed, 418
Bauxite, 243
Bays, 429
Beach drift, 432, 433
Beaches, 430–43
    progradation and retrogradation, 434–435
Bed load, 387
Bedrock, 240, 339
    disintegration, 340
Benguela Current, 123
Berlin Climate Convention (1995), 72
Biodiversity, *480*
Biogeography, 197
Biomass, 199
    energy conversion, 200
Biomes, 197, 208–209
    desert biome, 208, 227–228
    forest biome, 208, 209–223
    formation classes 209
    grassland biome, 208, 224–225, 227
    savanna biome, 208, 223–224
    tundra biome, 208, 229
Biosphere, 7, 197
    and energy flow in ecosystems, 197–198
Biotite mica, 269
Black Belt (Alabama), 412
Black Hills dome, 413–415
Block mountains, 324–325, *326*
Blowout, 444
Bluffs, 398
Bogs, *256*, 257
Boralf soils, 254
Boreal forest environment, 188–189
Boreal forests, 219–220
Boreal stage, 473
Boroll soils, 256
Boundaries, tectonic plates, 300
Braided streams, 396
Breakers, 429
Bright Angel Canyon, *411*
Broadleaf evergreen forest, 210
Brunisolic soil order (Canadian system), 498
Brushfires, 119, 223
B soil horizon, 244, 245, 254
Buttes, 410

Calcification, 246
Calcite. *See* Calcium carbonate
Calcium, in earth's crust, 267
Calcium Carbonate, 278
    translocation, 246
Caldera, 318
Caledonides, 294
California Current, 123
Canadian soil classification system, 498–501

Canopy, rainforest, 209–2210
Canyons, 389, 392, 411
Capillary tension, 243
Carbonate rocks, 278
Carbon cycle, and thermohaline circulation, 128
Carbon dioxide
    in atmosphere, 38
    and El Niño, 125
    emissions targets, 72
    and global warming, 71
Carbonic acid, 273
Carbonic acid action, 345–346
Carbon monoxide, as pollutant, 88
Carnivores, 198
Cartography, 17, 482
Cascade Mountains, 319
Caspian Sea, 375
Caverns
    formation, 346
    sea caves, 430
Celsius temperature scale, 51, 52
Cementation, in clastic sediments, 277
Cenozoic era, 282, 291
CFCs, and ozone layer, 40–41
Chalk, *279*
Channel, 268–269
Chaparral, 223
Chemically-precipitated sediment, 276
Chernozemic soil order (Canadian system), 498
Chernozems, 256
Chert, 278
Cherty limestone, 278
Chilean earthquake (1960), 328
China, time zones, 20
Chinook winds, 119
Chlorofluorocarbons (CFCs), and ozone layer, 40–41
Cindercones, 321–322
Circle of illumination, 24, 25
Circum-Pacific belt, 292
    volcanism, 319
Cirques, 460, 461, 462
Cirrocumulus clouds, *85*
Cirrostratus clouds, *85*
Cirrus clouds, *85*
    in hurricanes, 146
Clastic sedimentary rocks, 276–278
Clastic sediments, 276
Clay, in soil, 240
Clay minerals, 243, 273
Clay-rich loams, 240
Claystone, 277
Clear cutting, 206, 214
Cliffs, 410–411
    marine, 430
Climates, 155
    change
        global warming and, 71–72
        thermohaline circulation and, 128
    classification, 157–163
    front zones, 158, 163
    high-latitude group, 163, 187–192
        boreal forest, 187–188, 199
        climographs, 187
        ice sheet, 188, 191–192
        tundra, 188, 189–190, 199
        Köppen system, 159–162
    low-latitude group, 158, 161–175
        climographs for, 165

dry tropical, 166, 171, 174–175, 199
    monsoon and trade wind coastal, 166, 167–168, 199
    wet-dry tropical, 166, 169–171, 172–173, 199, 202
    wet equatorial, 166–167, 199
midlatitude group, 176–187
    climographs, 176
    dry midlatitude, 177, 184–185, 199
    dry subtropical, 177–178, 199
    marine west coast, 177, 183, 199
    Mediterranean, 177, 180–183, 199, 202
    moist continental, 177, 185–187, 199, 202
    moist subtropical, 177, 179–180, 199
    net productivity of, by type, 199
    undifferentiated highland, 164
    and vegetation, 202, 203
    world map of, 164–165
Climatic factors, and net primary productivity, 199
Climatic gradients, and vegetation types, 231–233
Climax, of econogical succession, 205
Climographs, 163
Cloud forest, 212
Cloud forms, 84, 85
Clouds, 84–86. *See also* Precipitation
    condensation nucleus, 84
    formation, 97
    and fronts, 139
    in hurricanes, 146
Coal, 277, 280
Coarse-textured rocks, 271
Coastal aquifers, 368
Coastal blowout dunes, 445, 448
Coastal foredunes, 449
Coastal forests, 220
Coastal plains, 412
Coastlines, 429
    coast types, 436–441
    submergence/emergence, 436
Cocos Plate, 328
Col, 461
Cold-blooded animals, 204
Cold fronts, 137–139, 139
Colluvium, 385
Colorado Front Range, *414*
Columbia Plateau, 320–321
Composite volcanoes, 316. *See also* Stratovolcanoes
Compression, tectonic process, 298, 299
Condensation, 77
    level of, and adiabatic process, 83
Condensation nucleus, 84
Conduction, 39, 51
Cone of depression, 365, 368
Conglomerate, 277–278, *278*
Consequent streams, 412
Consumers, 198, 199
    primary, 198
    secondary, 198
Consumption, plate tectonics, 300
Contamination, of ground water, 366–368
Continental antarctic air mass, 136, 137
Continental arctic air mass, 136, 137
Continental colisions, 304–305
Continental crust, 290
Continental drift, 308–309. *See also* Plate tectonics
Continental lithosphere, 299

Continental margins, 295
Continental polar air mass, 136, 137
Continental rise, 295
Continental rupture, 305
Continental shelf, 295
Continental shields, 292–293
Continental slope, 126, 295
Continental sutures, 304
Continental tropical air mass, 136, 137
Continents, relief features, 291–294
Convection, 96–97
    and sensible heat transfer, 39
Convectional precipitation, 95, 96–101
    unstable air and, 99–101
Convection cells, 96–101
Converging boundaries, 300
Coral-reef coasts, 436, 439–441
Coral reefs, 439–441
Core, of the earth, 289–290
Coriolis effect, 14
    equatorial regions, 145
    and winds, 110–111
Corrosion, 386
Counterradiation, 42
Covered shields, 293
Cranberry bogs, 257
Crater, of volcanoes, 316
Crater Lake, 318, 322
Crescentic dunes, 445–446
Crevasses, 458
Crude oil, 280
Crust, of the earth, 267, 290
Cryaquept soils, 261
Cryosolic soil order (Canadian system),
    498–499
C soil horizon, 244, 245
Cuestas, 412, 413
Cumuliform clouds, 84, *85*
Cumulonimbus clouds, 84, *85*, *96*, 138
    in hurricanes, 146
    and tornados, 144
Cumulus clouds, 84, *85*
    and cold fronts, 139
    formation, 96
Cutoffs, 398
Cyclone families, 142
Cyclones, 111–112, 135
    Simpsons-Saffir Scale of Intensity, 146
    traveling (*see* Traveling cyclones)
    tropical, 145–150
Cyclone tracks, 142
Cyclonic precipitaion, 94, 135
Cyclonic storms, 135

Daily cycles, air temperature, 53–56
    land vs. water, 61–62
Dams, 394–395
Data acquisition, 494
Data management, 494
Data manipulation and analysis, 494–495
Daylight, and ecosystems, 204
Daylight saving time, 21–22
Day-night cycle, 204
Dead Sea, 375
Debris flow, 350
Decalcification, 246
Deciduous forest, 218–220
Deciduous plants, 202. *See also* Forests,
    midlatitude deciduous
Declination, of sun, 25
Decomposers, 198, 199

Dediduous plants, 202
Deep-sea cones, 305
Deesertification, 172–173
Deflation, 444
    induced, 452
Deforestation, 214–215
Deglaciation, 464
Degradation, 397
Degrees of arc, 15
Delta coasts, 436, 437, 438–439
Delta kames, 469
Dendritic drainage pattern, 411
Density, 270
Denudation, 315, 395–399
    and rock structure, 409–410
Deposition, 77
Depositional landforms, 383–384
Desert biome, 208, 227–228
    and animals, 228
    and organisms, 202
Desert climates. *See* Arid climates; Dry desert
Desertification, 172–173
Desert pavements, 441
Deserts, 171, 174
    irrigation, 376
    mountainous, 402–403
Detritus, 198
Dew-point temperature, 80, 81
Diffuse radiation, 40
Dikes, 272, *273*, 436
Diorite, 269, 270
Dip, 409–410
Discharge, or streams, 369–371
Distributaries, 438
Diurnal rhythm, 14
Doldrums, 112, 113
Dolomite, 278
Drainage basins, 371
Drainage divides, 371
Drainage patterns, 411, 412, 422
Drainage winds, 118
Drawdown, 365
Drought, 172–173
Drumlins, 468, 469, 470
Dry adiabatic rate, 83
Dry climates (Köppen system), 159
    and organisms, 201–202
Dry desert, 228
Dust storms, 444, *445*

Eagle Rock, 420
Earth. *See also* Continents; Oceans
    albedo, 42
    asthenosphere, 291
    atmospheric composition, 38
    axis
        and insolation by latitude, 36–37
        tilt, 13, 23
    core, 289–290
    crust, 287, 290
    global energy flows/balance, 31, 39–45, *43*
    land/ocean percentage, 291
    lithosphere (*see* Lithosphere)
    longwave radiation from, 34
    mantle, 289, 290
    as oblate ellipsoid, 14
    radiation balance, 34–37
    revolution, 12, 22–25
    rotation, 12, 13, 14–15 (*see also* Coriolis
        effect)
    shape and curvature, 13–14

    structure, 289–290
    sun cycles, and Ice Ages, 26
    water balance, 78, 79
Earthflow, 348–349
    induced, 351–352
Earth Orbiting System, 493
Earthquake Lake, 353
Earthquakes, 327–334. *See also* San Francisco
        earthquake; and otherspecific earthquakes
    activity map, 329
    California, 327–328, 330–334
    focus, 327
    Richter scale, 328
    seismic sea waves from, 330
    tectonic activity and, 328–330
Earth Summit. *See* Rio de Janeiro Earth Summit
East African rift valley, 322, 327
Easterlies, 113
    equatorial, 119–120
Easterly saves, 145
Ebb current, 435
Ecliptic, 23
Ecological succession, 205–206
Ecology, 197
Ecosystems, 197–233. *See also* Biomes
    aquatic, 197
    biomass of, 199
    defined, 197
    energy flow in, 197–199
    net primary production, 199–200
    organisms
        and habitat in, 200–201
        and temperature in, 203–204
        and water needs in, 201–203
    terrestrial, 197
Electromagnetic radiation, 31–34
    longave, 33–34, 44
    shortwave, 33
Electromagnetic spectrum, 32–33, *33*
Elephant grass, 223
Elevation, air temperature variation, 56–57, 58
El Niño, 124–125
Eluviation, 245
Emperor Seamounts, 320
Energy, renewable, 200
Energy balance, of the earth, 31, 39–45
Energy flow, in ecosystems, 197–206
Entisols, 249, 257–258
Entrenches meanders, 398–399
Environmental temperature lapse rate, 56
Environments
    boreal forest, 188–189
    rainforest, 168–169
    subtropical desert, 178
    tundra, 190–191
Epipedon, 248
Epiphytes, 211
Epochs, geologic, 291
Equator, 15
Equatorial air mass, 136
Equatorial easterlies, 119, 120
Equatorial lows, weak, 145
Equatorial trough, 112, 113
Equatorial zone, 37
Equinoxes, 23–24, 25, 35
Eras, geologic, 291
Ergs, 447
Erosion
    ice sheets, 467
    slopes, 384–386
    stream work, 386–388
    wind, 442–444

Erosional landforms, 383
Eskers, 468, 469
E soil horizon, 244, 245
Estuaries, 429
Eurasian-Indonesian mountain belt, 292
Eurasian plate, 301–303
Europe, and thermohaline circulation, 129
Eustatic change, 442
Eutrophication, 376
Evaporation, 51, 77
  and global water balance, 79
Evaporites, 279
Evapotranspiration, 423, 360
Everglades, 149
Evergreen plants, 202
Exfoliation domes, 344
Exotic rivers, 376
Exposed shields, 293
Extension, tectionic process, 298
Extrusion, 271
Extrusive igneous rocks, 271–272
Eye, of hurricane, 146

Fahrenheit temperature scale, 41, 42
Fallout, air pollutants, 90
Farming. *See* Agriculture
Fault coasts, 436, 437
Fault creep, 327
Fault-line scarps, 416
Faults, 298, 299, 300
  and earthquakes, 328
  landforms, 324–326
  normal, 324–325
  overthrust, 325, 326
  reverse, 324, 326
  transcurrent (strike-slip), 325
Fault scarps, 324, 325, 416
Feldspars, 268
Felsic minerals, 270
Fenlands, 436
Fesic igneous rocks, 270
Filchner ice shelf, 472
Fine-textured rocks, 271
Finger Lakes, 467
Fiord coasts, 436, 437
Fiords, 462
Fires
  brushfries, 118, 119
  in savanna woodlands, 223
Firn, 458
Fissile rocks, 278
Flash floods, 372–373, 400
Flood basalts, 320–321
Flood current (tides), 435
Floodplains, 258, 372
  alluvial rivers, 398–399
  development, 373, 389, 392
Floods, 371–373
Flood stages, 372
Fluvial landforms, 383–395
  depositional, 383–384
  erosional, 383
  evolution, 395–399
  mountainous deserts, 402–403
Fluvial processes, 383
  arid climates, 399–403
  slope erosion, 384–386
  stream erosion and transportation, 386–388
Fly ash, 88
Focus, earthquakes, 327
Fog, 86
  types of, 85

Fold belts, 415–415
Folding processes, 299
Folds, 299
Foliation, in schist, 281
Food web or chain, 197–199
  consumers in, 198, 199
  decomposers in, 198, 199
  energy flow in, 199
  primary producers in, 197–198
  and species population size, 198–199
  in tundra, 230
Forbs, 224
Foredunes, 449
Foreland folds, 324
Forest biome, 208, 209–223
Forests, 207. *See also* Rainforests
  acid deposition damage, 93–94
  animal composition, 206
  biomass, 199
  boreal, 216, 219–220, *220*
  coastal, 220
  cloud, 212
  diseases, 206
  midlatitude deciduous, 216–219
  monsoon, 213–214
  mountain, 212
  needleleaf, 219–220
  oak-hickory climax, 206
  sclerophyll, 222–223
  subtropical evergreen, 214–216
  succession, 205
Formation classes, of biomes, 209
Fossil fuels, 280
  combustion, and air pollution, 88
Fractional scale, of maps, 483
Fracture zones, 306
Freezing, 77
Freshwater ecosystems, 197
  biomass of, 199
Fringing reefs, 440
Fronts, 136–139, 140
Frost, 58–59, 77
Frost action, 340–343

Gabbro, 270–271
Gases
  in atmosphere, 38
  as pollutants, 88
Geneva global warming treaty (1997), 72
Geographic grid, 15–16. *See also* Map ,
  projections
Geographic information systems (GIS),
  488, 493–495
  key elements of, 494–495
  and spatial objects, 494
Geologic norm, 384
Geologic time, 282, 291
Geology, 267
Geomorphology, 315
Geostrophic wind, 119
Geysers, 322
Glacial ablation, 457
Glacial deltas, 468, 469
Glacial drift, 467
Glacial till, 467, 468, 469
Glaciation, 26, 464, 465. *See also* Ice Ages
  and solar cycles, 26
Glaciation cycles, 464–465, 471–472
Glaciers, 457
  alpines, 457, 458–462
  flow rate, 459
  ice sheets (*see* Ice Ages; Ice sheets)

Gleysolic soil order (Canadian system),
  498, 499
Global warming, 68–72
  future impact, 70–72, 480–481
  ice sheets and, 472–473
  sea level increase and, 442–443, 480
  temperature record for, 70
  thermohaline circulation and, 128
Global water balance, 78
Globes, scale, 483–484
Gneiss, 281
Gondwana, 306–307
Goode projection, 18–19, 483
Gorges, 389
Graben, 324, 325
Graded streams, 389, 392
Gradient, of stream channel, 369–370, 371
Grand Canyon, *278, 392*, 394, 411
Granite, 270
Granite domes, 420–421
Graniteic rock, 290
Granular disintegration, 340
Graphic scales, 484
Grassland biome, 208, 224–225
  animals of, 225–227
  biomass of, 199
Great circle, 18
Great realms, of the earth, 7–8
Great Salt Lake, 470
Greenhouse effect, 42
  and global warming, 68–70 (*see also* Global
    warming)
  water's role, 146–147
Greenhouse gases, 68, 71, 480
  and Rio de Janeiro Earth Summit, 72
Greenland Ice Sheet, 463, 472
Green Mountains, 417
Greenwich, England, 12, 16, 21
  meridian, 20
Groins, 434–435
Ground moraine, 467
Ground subsidence, 366–367
Ground water, 78, 360–362, 366–368
  and lakes, 373
  water table, 360–362, 365–366
Ground water management, 365–368
  contamination, 366–367
Gulf Stream, 122, 123, 126
Gullies, 385
Guyot, 320
Gyres, 122

Habitats, 200, 201
Hadley cells, 112, 113, 116, 128–129
Hail, 87, 100–101
Halite, 279
Hanging valleys, 460, 462
Hawaiian High, 116
Hawaiian islands, 319
Hawaiian seamount chain, 320
Hawaiian volcanoes, 319
Hayward fault, 331–332
Haze, 89
Heaped dunes, 447
Heat islands, 55
Heat transfer, global, 39–45, 128–129
Hebgen Lake, 352–353
Heimaey volcano, 273, 274–275
Hemispheres, 16
Herbivores, 198, 211, 219
Herbs, 207
Hibernation, 204

High-latitude climate group, 164–165, 187–192, 188 *table*
  boreal forest, 187–189
  climographs, 187
  ice sheet, 188, 191–192
  tundra, 188, 189–191
High-level temperature inversion, 91
High-level winds, 119–121
High-mountain environments, 57–58
Histosols, 249, 251
Hogbacks, 413, *414*
Hoh rainforest, *184*
Hole, in ozone layer, 40–41
Holocene epoch, 465, 472–474
Homestake Mine, 414–415
Horns, 460, 461
Horse latitudes, 113
Horst, 324
Hotspot, 319–320
Hot springs, 322
Human, period of existence, 282
Human-influenced vegetation, 206. *See also*
  Agriculture; Vegetation
Humboldt Current, 123, 124
  and La Niña, 125–127
Humidity, 79–80
  dew-point temperature, 81
  measurement, 81–82
  pole-to-pole profiles, *80*
  relative, 79–82
  specific, 79–80
Humification, 246
Humus, 239
Hurricane Andrew, 148–149
Hurricane Mitch, 147–148
Hurricanes (tropical cyclones), 145–150, *476*
Hydraulic action, 386
Hydrocarbons, in sedimentary rocks, 280
Hydroelectric dams, 394–395
Hydrologic cycle, 78–79, 359–360
  global water balance and, 79
Hydrology, 360
Hydrolysis, 273, 345
Hydrosphere, 7, 77–79
  types of water in: "reservoirs" of the ,
    hydrosphere, 78

Ice Ages, 464–466
  astronomical hypothesis, 26, 471–472
  cycle of, 404–405
  Earth-Sun cycles and, 26
  Late-Cenozoic Ice Age, 465–466
  causes of, 471
  Little Ice Age, 474
  study of, 470–471
Icebergs, 464, 473
Ice-crystal growth, 340–342
Iceland, volcanism, 320
Icelandic Low, 142
Ice lobes, 468, 469
Ice sheets, 457
  and global warming, 472–473
  landform development by, 467–470
  sea and pack ice, 464
Ice shelves, 463
  and global warming, 472
Ice storms, 87
Ice streams, and volcanoes, 472–473
Ice-wedge polygons, 343
Idaho batholith, 418
Igneous rocks, 268–273, 269 *table*

alteration of, 273
  extrusive vs. instrusive, 269, 271–273
  felsic, 270
  mafic, 271
  resistance to denudation, 409
  ultramafic, 271
Illuviation, 245
Inceptisols, 249, 258
Indiana Dunes State Park, 448
Indian Peninsula, 307
Induced deflation, 452
Induced earthflows, 351–352
Induced mass wasting, 351–354
Infiltration, 360
Infrared aerial photography, 489, *490–491*
Infrared radiation, 33
Initial landforms, 315
Inner lowlands, 412
Insolation, 34–35, 36
  annual, by latitude, 36–37
  atmospheric losses, 39–41
  daily through year, 36
  daily cycle, 53
Inspiral winds, 111, 112
Interglaciation, 465
Interlobate moraines, 469
International Date Line, 21
International Panel on Climate Change, 71
Interrupted projections, 18–19
Intertropical convergence zone, 112, 116–117, 128, 129
Intrusion, 271
Intrusive igneous rocks, 271–272
Ions, in soil, 241
Iron
  in core, 290
  in earth's crust, 267
Iron oxides, in soil, 243, 253
Irrigation, 376
Island arcs, 292
Isobars, 109
Isohyets, 96, 155
Isopleths, 487
Isotherms, 63, 487
  seasonal migration, 67
Isotopes, 306

Jet streams, 120, 121
Joint block separation, 340
Joints, 340, 409
Juan de Fuca plate, 301
Jura Mountains, 323–324

Kalahari desert, 175
Kamchatka Current, 133
Karst, 364, *365*
Knob-and-kettle landform, 468
Kobe, Japan, earthquake (1995), 331
Köppen-Geiger climate system, 159–162, 163
Krakatoa, 318
Kuroshio Current, 122
Kyoto Protocal treaty, 72, 480

Labrador Current, 123, 126
Lagoons, 437
Lahar, 350
Lake Bonneville, 469–470
Lakes, 373–375
  acid deposition in, 93
  created by landslides, 352–353, 373
  disappearance, 374–375
  eutrophication, 376
  saline lakes and salt flats, 375

Land
  air temperature, vs. water, 61–63
  breezes, 109–110
  degradation, 172–173
Landfills, and ground water contamination, 366–368
Landforms, 315–316
  arid climates, 410–411
  coastal
    coastal foredunes, 449
    coast types, 435–441
    wave action, 429–435
  coastal plains, 412
  eroded volcanoes, 419–420
  exposed batholiths and monad
    nocks, 418, 420–421
  fault structure erosion, 416–417
  fluvial (*see* Fluvial landforms)
  and glaciers, 457, 459–462
  horizontal rock layers, 410–412
  ice sheets and, 467–470
  initial, 315
  limestone regions: caverns and Karst, 364, *365*
  metamorphic belts, 417
  and rock structure, 409–410
  sequential, 315–316, 409
  tectonic, 323–327
  volcanic, 316–322
  warped rock layers, 413–416
  wave action, 429–4435
  wind action, 442–435
  sand dunes, 445–449
Landmass, 395
Landmass rejuvenation, 396
Landsat, 492
Landslides, 347, 350–351
  and faulting, 325
  Madison Slide disaster, 342–343
Land subsidence, 352–353
La Niña, 124–127
Lapse rate, 56
Large-scale maps, 484
Larsen Ice Shelf, 473
Late-Cenozoic Ice Age, 465–466, 471
Latent heat, 39, 77
Latent heat transfer, 39, 44
Lateral moraines, 460, 461
Laterite, 253, *253*
Latitude, 15–16
  insolation variation with latitude, 36
Latitude zones, 37
Laurasia, 306–307
Laurel forest, 215
Laurentide Ice Sheet, 466
Lava, 271–272, 314
  controlling flows, 274–275
Leaching, soil, 245
Leads, 464
Levees, natural, 398
Level of condensation, 83
Lianas, 207, 211
Lichens, 207
Life layer, 7–8
Life zones, 231
Light, and ecosystems, 204
Lightning, 101
Lime, and soil acidity, 242
Limestone, 275, *279*
  carbonic acid action on, 345–346
  resistance to denudation, 409

Limestone caverns, 362–364
  formation, 345–346
Limonite, 243
Lines, 494
Lithosphere, 7, 29–291. *See also* Plate tectonics
  continental, 295
  crosssection, 303
  global plate system, 300–303, *301–302*
  movement across asthenosphere, 291
  oceanic, 299
  rocks in, 290
Lithospheric plates, 291
Little Ice Age, 474
Littoral drift, 432–435
Loams, 240, 241, 243
Local winds, 118
Loess, 450–452
Loma Prieta earthquake (1989), 331
Long Beach earthquakes (1930s), 333
Longitude, 15, 16–17
Longshore current, 432, 433
Longshore drift, 432
Longwave radiation, 33
  from atmosphere, 43, 44
  from earth, 34
Low-angle overthrust faults, 326
Lowlands, 412
Low-latitude climate group, 158, 161 *table*,
  163–175
  climographs for, 165
  dry tropical, 166, 171, 174–175
  monsoon and trade wind coastal, 165, 166,
    167–168
  wet-dry tropical, 164, 166, 169–171
  wet equatorial, 164, 166, 167
Low-latitude rainforest, 209–212
  environment, 168–169
  exploitation, 212, 214–215
Low-level temperature inversion, 90
Luvisolic soil order (Canadian system), 498,
  499–500

Mackerel sky, *85*
Madison Slide, 352–353
Mafic igneous rocks, 271
Mafic minerals, 270
Magma, 270, 271
Magnesium, in earth's crust, 267
Mammoth Cave, 362, 364
Mantle, of earth, 289, 290
Map projections, 17–19, 482–483
  Goode projection, 18–19, 483
  interrupted, 18–19
  mercator projection, 17–18, 483
  polar projection, 17, 483
  scale fraction, 17
Maps, 483–487
  data representaion on, 485–487
  informational content, 485
  multipurpose, 485
  resolution, 485
  scale, 483–484
  thematic, 486–487
Map symbols, 485
Marble, 281
Marginal lakes, 468, 469
Marine cliffs, 430
Marine ecosystems, 197
Marine scarps, 429, 430
Marine terraces, 441
Maritime equatorial air mass, 136–137
Maritime polar air mass, 136, 137

Maritime tropical air mass, 136–137
Marshes, 373–374
Mass wasting, 347–350
  induced, 351–354
Meanders, 389, 392, 398–399
Medial moraines, 460, 461
Melting, 77
Mercator projection, 17–18, 483
Mercury barometer, 107–108
Meridians, 15
  prime, 16
  and time zones, 20–21
Mesas, 410
Mesozoic era, 282, 291
Metamorphic belts, 417
Metamorphic rocks, 268, 280–281
  resistance to denudation, 409
Methane, 280
Mexico City
  earthquake (1985), *329*
  earthquake (1986), 328
  subsidence in, 367
Micas, 269, 270
Micrometer, 31
Microwave radar systems, 488–489
Midlatitude climate group, 158, 176–187,
    177 *table*
  climographs, 176
  dry midlatitude, 177, 184–185
  dry subtropical, 177, 178
  marine west coast, 177, 183
  Mediterranean, 177, 180–182
  moist continental, 177, 185, 187
  moist subtropical, 177, 179–180
Midlatitude deciduous forests, 216–218
Midlatitude zones, 37
Mid-oceanic ridge, 294
Milankovitch, curve, 26
Mild, humid climates (Köppen system), 159
Millibar (pressure unit), 108
Mineral alteration, 242, 273
Mineral oxides, 243
Minerals, 267–268. *See also* specific minerals and
    mineral classes
  in soil, 242–243
Mineral soil horizons, 244, 245
Mining
  acid mine drainage, 376
  scarification and, 353–354
Minutes of arc, 16
Mississippi delta, 439
Mississippi River, 368, 370–371, 387
  flooding, 371, 373, 390–391
Mistral, 119
Moho, 290
Mojave desert, 178, 179, 228
Mollic epipedon horizon, 255
Mollisols, *251*, 255–256
Monadnocks, 418
Monsoon forest, 213–214
Monsoons, 116–117
Montaine forest, 212
Montserrat, 318
Moraines, 460, 461, 467, 468–469
Moss, 207
Mountain arcs, 292
Mountain environments, 57–58
Mountainous deserts, 402–403
Mountain roots, 293–294
Mountains, and orographic
    precipitation, 95–96
Mountain winds, 118
Mount Fujiyama, 316

Mount Hekla, 320
Mount Hood, 316
Mount Kenya, 327
Mount Kilimanjaro, 327
Mount Mayon, 316, *317*
Mount Pelée, 318
Mount Pinatubo eruption, 40, 70, 125
Mount Shasta, *422*
Mount St. Helens, 316, *317*
Mucks, 257
Mud, 278
Mudflow, 345–346
Mudstone, 277, 278
Multipurpose maps, 485
Multispectral scanners, 492

Nappes, 299, 304
National Acid Precipitation Assessment
    Program (NAPAP), 94
National Weather Service River and
    Flood Forecasting Service, 373
Natural bridges, 399
Natural gas, 280
Natural levees, 398
Natural vegetation, 206. *See also* Vegetation
  world regions of, *208–209*
Needleleaf forests, 219–220
Net primary production, 199–200
  remote sensing to monitor, *200*
Net radiation, 44
  annual cycle, 60–61
  daily cycle, 53–54
New Madrid earthquake (1811–1812), 328
Niagra Falls, 394
Nile delta, 438
Nimbostratus clouds, *85*
Nitric oxide emissions, and acid deposition, 92
Nitrogen, in atmosphere, 38
Nitrogen oxides, as pollutants, 88, 90
Nodes, 494
Normal faults, 324–325
North American air masses, 137
North Atlantic
  freshwater inputs and, 128
  and thermohaline circulation, 127, 129
North Atlantic Drift, 123
Northeast trade winds, 113
Northern hemisphere, 16
North pole, 14
Northridge earthquake (1994), 332–333

Obsidīian, *222*
Occluded fronts, 139
Ocean basins
  floor, 294—295, *296–297*
  formation, 305
  relief features, 294–295
Ocean currents, 122–124
  and carbon cycle, 128
  deep, and thermohaline circulation,
    127–128
  role in global heat transport, 129
  surface, and El Niño, 124–125
  and winds, 122–124
Ocean ecosystems, biomass of, 199
Oceanic crust, 290
Oceanic lithosphere, 299
Oceanic trenches, 292, 295
Oceans, 77
  air temperature, vs. land, 61–63
  level increase from global warming, 442,
    443, 480

Ocean tides, 15, 435, 436
Ocean waves, 429–435
Oil (petroleum), 280
Old-field succession, 205
Olivine, 270
Omnivores, 198
Organic sediments, 276
Organic soil horizons, 244
Organic soil order (Canadian system), 498, 500
Organisms, in ecosystem, 200–206
  ,habitat, 200–201
  temperature and, 203–204
  water needs, 201–202
Orogens, 299, 304–395
Orogeny, 299, 304–395
Orographic precipitation, 95–96, 155–156
O soil horizons, 244
Outcrops, 340
Outspiral winds, 112
Outwash, 468, 469
Outwash plains, 468, 469
Overland flow, 360
  vs. stream flow, 368–369
Overthrust faults, 298, 299, 326
Oxbow lakes, 298
Oxidation, 273, 345
Oxisols, 249, *250, 252,* 253
Oxygen, in atmosphere, 38
Ozone, 38, 40
  in upper atmosphere, 38–39
Ozone layer, 38–39, 40–41

Pack ice, 464
Paleogeography, 306
Paleozoic era, 282, 291
Pangaea, 306, 307, 308
Panthalassa, 306
Parabolic dunes, 445, 447
Parallels, 15
Parent material, of soil, 240
Particulates, 88
Passive continental margins, 294
Passive remote sensing systems, 489
Peat, 257, 280
Peat bogs, 257
Pediments, 402–403
Pedology, 239
Peds, 242
Peneplains, 394–396
Percolation, 360
Perennial plants, 206–207
Peridotite, 271
Perihelion, 22
Periods, geologic, 291
Permafrost, 190, *261*
Petroleum, 280
pH, 92, 241
Photoperiod, 204
Photosynthesis, 198
  and temperature, 199
  and water, 199
Physical geography
  defined, 2
  great realms of earth, 7–8
  and human habitat, 8–9
  overview of earth, 2–6
Pioneer plants, 205
Placific plate, 301, 302
Plagioclase feldspar, 268
Plant nutrients, 241
Plants. *See* Vegetation

Plateaus, 410
Plate tectonics, 298–308
  boundaries, 300
  continent history: continental drift, 306–309
  deep-sea cones, 395, 305
  faults, 299, 300
  landforms, 324–326
  global plate system, 300–303, *301–302*
  subduction, 300, 303–304
    and volcanism, 319
  tectonic processes, 298–306
    continental collisions, 304–305
    continental rupture, 305
    energy source, 306
    folding, 299, 323–324
    orogeny, 299, 204–205
    transform scars, 306
Playas, 402, *403*
Pleistocene epoch, 282, 465
Plinthite, 253
Pliocene epoch, 282, 465
Ploar-front jet stream, 121
Plucking, 457, 459
Plunging folds, 416
Plutons, 272
Pluvial lakes, 469–470
Pocket beaches, 434
Podzolic soil order (Canadian system),498, 501
Podzols, 255
Point bars, 392
Points, 494
Polar air mass, 136
Polar climates (Köppen system), 159
Polar easterlies, 113
Polar front, 1113, 137
Polar front zone, 158, 163
Polar outbreaks, 113
Polar projection, 17, 483
Polders, 436
Poleward heat transport, 45, 128–129
Pollution
  air (*see* Air pollution)
  thermal, 377
  water (*see* Water pollution)
Pollution dome, 91, 92
Polygon, 494
Ponds, 373, 374, 375
Population growth, future impact, 481
Population sizes, 199
Portuguese Bend "landslide," 351
Potash feldspar, 268
Potassium, in earth's crust, 267
Potholes, 386
Prairie, 185, 186, 224, 225
Precambrian time, 282, 291
Precipitation, 57, 86–96. *See also* Acid rain
  adiabatic process and, 80–84
  convectional precipitation, 95, 96–101
    unstable air and, 99–101
  cyclonic precipitation, 94
  fate of, 79
  formation of, 86
  fronts and, 139
  global water balance and, 78–79
  high-latitude climates, 188 *table*
  low-latitude climates, 166 *table*
  mean annual global patterns, 155–157
  measurement, 87
  midlatitude climates, 176–177
  net primary production and, 199
  orographic precipitation, 95–96

processes, 95–99
runoff paths, 360 (*see also* Ground water; Surface water)
Preprocessing, 494
Pressure, atmospheric. *See* Atmospheric pressure
Pressure gradient force, 109
Pressure gradients, 109
Prevailing westerlies, 113
Primary minerals, 242
Primary producers, 197
Prime meridian, 16
Product generation, 495
Prograduation, 434–435
Psychrometer, sling, 81–82
Pyroxenes, 270

"Quaternary Period," 282
Quartz, 267, 268, 273
Quartzite, 281

Racetrack Playa, *403*
Radar systems, 488–489
Radia drainage pattern, 422
Radiation, electromagnetic. *See also* Electrogmagnetic radiation; Solar radiation
Radiation balance, of earth, 34–37
Radiation fog, 86
Radiative heat transfer, 39–45
  net radiation, 44–45
Radioactive decay, as power source for plate tectonics, 306
Radiogenic heat, 306
Rainfall, 89. *See also* Acid rain; Precipitatation
  acidity, across North America, 93
  formation, 86
  global patterns, 155–157
  hurricane, 146
  net primary production and, 199
  snowfall equivalent, 87
Rainforest environment, 168–169
Rainforests. *See also* Forests
  exploitation, 214–215
  low-latitude, 209–212
  profusion of species in, 211
Rain gauges, 87
Rain-green vegetation, 171
Rainshadow, 96
Raised shorelines, 441–442
Rann of Kutch, 374
Rapids, 289
Reclamation, 354
Recumbent folds, 299
Red Sea, 305
Reefs, 439–441
Reflection, 40
Regolith, 240, 399. *See also* Weathering
  transported and residual, 340
Regosolic soil order (Canadian system), 498, 501
Relative humidity, 81–82
Remote sensing, 488–492
  active, 488
  net primary productivity monitoring, 199, *200*
  passive, 489
  scanning systems, 489–492
Removal processes, soil, 245
Representative fraction, of maps, 483
Residual regolith, 340
Resolution, of maps, 485
Respiration, 198

Retrogradation, 434–435
Reverse faults, 326
Revolution, of the earth, 22–24
Rhyolite, 271–272
Ria coasts, 436, 437
Richter scale, 328
Ridge-and-valley landscape, 415–416
Rift valleys, 305, 327
Rill erosion, 385
"Ring of fire," 322
Rio de Janiero Earth Summit (1992), 72, 480
Rivers, 368–373. *See also* Streams
    alluvial, 398–399
    discharge, 369–371
    exotic, 376
    floods, 371–373
Rocks, 267–268
    cycle of change, 268, 282–284
    dip of formation, 410
    igneous, 268–273
    lithosphere, 291
    metamorphic, 268, 280–281
    mineral alteration, 273
    sedimentary, 268, 273–280
Rock structure, 409–412
Ronne Ice Shelf, 472
Rossby waves, 120, 128–129
Ross Ice Shelf, 472, 473
Rotation of earth, 14–15
Runoff, 79, 360, 373
    and global water balance, 79
Rural surface characteristics, and air
    temperature, 55
Russia, time zones in, 20

Sahara Desert, 446–447
Sahel, 172–173
St. Lawrence River, 370
Salic horizon, 246, *260*
Saline lakes, 375
Salinization, 246, 258–259, 376
Salt-crystal growth, 343
Salt flats, 375
Salt marshes, 435–436
Saltwater intrusion, 368
San Andreas Fault, 325, *326*, 330– 334
San Clemente Island, 441
Sand, in soil, 240–241
Sandbars, 432
Sand dunes, 445
    coastal foredunes, 449
    types, 445–449, 445 *table*
Sand seas, 446
Sandspits, 431, 432, *433*
Sandstone, 277–278
    resistance to denudation, 409
    weathering formations, 343
Sandy loams, 240
San Fernando earthquake (1971), 333–334
San Franciso earthquake (Loma Prieta,
    1989), 331
Santa Ana wind, 119
Satellite remote sensing, *491, 492*
Saturated zone, 360
Savanna biome, 208, 223–224
Savannas, 171
    slope erosion, 385
Savanna woodland, 223–224
Scale, of globes and maps, 483–484
Scale fraction, 17
Scandinavian Ice Sheet, 466

Scarification, 353–354
Scarps
    fault and fault–line, 324, 325, 416
    marine, 129, *430*
Scattering, of radiation, 40
Schist, 281
Sclerophyll forests, 222–223
Sclerophylls, 202
Sclerophyll woodland, 222
Scoria, 2271, *272*
Scotia plate, 301, 302
Sea arches, 430
Sea breezes, 109–110
Sea caves, 430
Seafloor spreading, 320
Sea fog, 86
Sea ice, 464
Sea level, increase from global warming,
    442–443, 480
Seamounts, 320
Secondary minerals, 242–243
Secons of arc, 16
Sedimentary domes, 413–415
Sedimentary rocks, 268, 273–280
    chemically precipitated, 278–279
    clastic, 276–278
    hydrocarbons in, 280
Sediments, 276, 339
    chemically–precipitated, 276
    clastic, 276–277
    oceanic, 295
    organic, 276
Seismic sea waves, 330
Seismic waves, 327
Semiarid regions, 175
Semiarid (steppe) climates (Köppen system),
    160
Semidesert, 227–228
Sensible heat, 39
Sensible heat transfer, 39, 44
Sequential landforms, 315–316, 409
Serengeti Plains, *171*
Sesquioxides, in soil, 243
Shale, 277
    resistance to denudation, 409
Shearing, in rocks, 281
Sheet erosion, 385
Sheet flow, 368
Sheeting structure, 344
Shield volcanoes, 319–320
    erosion, 423
Ship Rock, *422,* 423
Shorelines, 429
    raised, 441
Short–grass prairie, 185, 225
Shortwave infrared, 33
Shortwave radiation, 33
Shrubs, 207, *223*
Siberian High, 116
Sierra Nevada batholith, 418
Silica, 277
Silicate minerals, 268–270
Sills, 272
Silt
    accumulation on agricultural land,259
    in soil, 240
Silty loams, 240, 241
Simpsons-Saffir Scale of Tropical Cyclone
    Intensity, 146
Sinkholes, 364
Slash-and-burn, 214, 253

Slate, 281
Slaughter Stone, 12
Sleet, 87
Sling psychrometer, 81–82
Slip face, dunes, 446
Slope, of ground, 246
Slopes, 339
    erosion, 384–386
Small-scale maps, 484
Smog, 89
Snow. *See also* Precipitation
    formation, 86–87
    global patterns, 155–157
    measurement, 87
    rainfall equivalent, 87
Snowy-forest climates (Köppen system), 159
Sodium, in earth's crust, 267
Soil acidity and alkalinity, 241
Soil colloids, 240, 241
Soil creep, 348
Soil erosion, 384–386
    stream work, 386–388
Soil horizons, 244
    Canadian system, 496–497
Soil orders, 247–249, *249*, *250*
    Andisols, 249, 258–259
    Aridisols, 251, 260
    Canadian system, 497–501
    Entisols, 249, 257–258
    Histisols, 249, *251*
    Inceptisols, 249, 258
    Mollisols, *251,* 255–256
    Oxisols, 249, *250,* 252
    soil profiles: photographs of, 250–251
    Spodosols, 250, 255
    suborders, 247 *table,* 248–249
    summary of, 247 *table*
    tundra soils, 261
    Ultisols, 249, *250,* 252–253
    Vertisols, 249, *250,* 254
    world distribution map, 248–249
Soil profiles, 244
Soils, 239–240
    acidity/alkalinity, 241
    base status, 243
    color, 240
    enrichment, 245
    formation processes, 245–247
    mass wasting and, 339
    minererals in, 242–243
    moisture, 243
    storage capacity, 243, 244
    surface configuration and, 246
    temperature, 246
    texture, 240–241
Soil solum, 244
Soil structure, 242
Soil water, 78
Soil water belt, 360, 361
Soil water recharge, 243
Solar constant, 33–34
Solar day, 14
Solar energy, and food chain, 198
Solar noon, 19
Solar radiation, 32–34
    and earth's radiation balance, 34–37
Solifluction, 349
Solonetzic soil order (Canadian system), 496,
    501
Solstices, 12, 23, 24
    conditions, 25

Sonoran desert, 17i, 228
Sorting, in clastic sediments, 277
Source regions, 135
    and climate types, 158, 163
Southeast trade winds, 113
Southern hemisphere, 16
Southern Oscillation, 124
South pole, 14
Spatial data, 488
Spatial object, 494
Species diversity. *See* Animals; Vegetation
Specific heat, 62
Specific humidity, 79–80
Sphagnum peat, 257
Splash erosion, 384
Spodic soil horizon, 255
Spodosols, *250*, 255
SPOT satellite, 492, *493*
Spreading boundaries, 300
Stable air, 91
Stalactites, 363
Stalagmites, 363
Standard time system, 20
Star dunes, 447
Steppe climates (Köppen system), 160
Steppes, 174, 224
    slope erosion, 385–386
Stonehenge, 12
Stone Mountain, 418, 421
Stone rings, 342
Storage capacity, soils, 243, 244
Storm surge, 147
Strata, 276
Stratiform clouds, 84, *85*
    in hurricanes, 146
Stratocumulus clouds, *85*
Stratosphere, 57
    ozone in, 38–39
Stratovolcanoes, 316–318, 327
    erosion, 418–423
Stratus clouds, 84, *85* and warm fronts, 139
Stream capacity, 387–388
Stream deposition, 386
Stream erosion, 386
Stream flow, vs. overland flow, 368–369
Stream load, 387
Stream profiles, 388
Streams, 368. *See also* Rivers
    aggradation, 396–397
    arid regions, 400
    braided, 396
    consequent, 412
    degradation, 397
    discharge, 369–371
    gradation, 388–399
        landscape evolution in, 389–393
    load-carrying capacity, 387–388
    rejuvenation, 398–399
    subsequent, 412
Stream transportation, 386–388
Strike, 409–410
Strike-slip fault, 325
Strip mining, 353
Subantarctic zone, 37
Subarctic zone, 37
Subboreal stage, 473
Subduction, 300, 303–304
Sublimation, 77
Subsequent streams, 412
Subsidence, 366–367
Subsolar point, 24, 25

Subsurface, 78
Subtropical desert environment, 178
Subtropical evergreen forests, 214–215, 216
Subtropical high-pressure belts, 112, 113, 116
Subtropical jet stream, 121
Subtropical zones, 37
Suction vortex, 149
Sugar Loaf, 420
Sulfur dioxide emissions, and acid deposition, 92–93
Sulfur oxides, as pollutants, 88, 89–90
Sumerian civilization, death of, 258–259
Summer monsoons, 117
Summer solstice, 23, 36
Sun
    angle in sky, 34–36
    cycles of energy output, and glaciation, 26, 471
    movements of, 12
Sun-synchronous orbit, 492
Supercooling, 84
Surface water, 78, 368–377
    desert irrigation, 376
    drainage systems 371
    floods, 371–373
    lakes, 373–375
    as natural resource, 377–378
    pollution, 376–377
    stream flow, 368–371
Surges, 459
Suspended load, 387
Swamps, 373–374
Swash, 431, 433
Synclinal mountains, 416
Synclines, 323, 415

Taiga, 219
Tall-grass prairie, 186, 224
Talus slopes, 341
Tarns, 460, 462
Tectonic activity, 291, 298–308
Tectonics, 298. *See also* Plate tectonics
Temperature, 51. *See also* Air temperature
    organisms and, 203–204
    net primary production and, 199
    radiation wavelength and, 31–32
Temperature gradients, 64
Temperature inversion, 58–60
    and smog, 90–91
Temperature scales, 51, 52
Tephra, 274, 316
Terminal moraines, 416
Terraces
    alluvial, 397
    marine, 441
Terrestrial ecosystems, 197. *See also* Biomes
"Tertiary Period," 282
Thematic maps, 486–487
Thermal erosion, 191
Thermal infrared radiation, 33
Thermal infrared remote sensing, 492
Thermal pollution, 377
Thermistors, 52
Thermohaline circulation, 127–128, 129
Thermometers, 41
Thermometer shelter, 51
Thorntree semidesert, 228
Thorntree-tall grass savanna, 223
Thrust sheets, 299
Thunderstroms, 97, 99–101
    and cold fronts, 138

Tidal currents, 435–436
Tidal inlets, 437–438
Tidal waves, 330
Tide curve, 435
Tides, 14, 435
Till, 467, 469
Tilt, of earth's axis, 23
Time systems, 19–22. *See also* Geologic time
    daylight saving time, 21–22
    standard time, 20
Time zones, 20
    U.S. and Canadian, 20, 21
    world, 20–21, 22
Topographic contours, 487
Tornados, 135, 143–144
    incidence and occurrence, 143, 144
Trade winds, 113
Transcurrent faults, 325
Transformation, in soil, 246
Transform boundaries, 300
Transform faults, 300
Transform scars, 306
Translocation processes, soils, 245
Transpiration, 55
    and photosynthesis, 199
Transported regolith, 340
Transverse dunes, 445, 446
Traveling anticyclones, 135
    tornados, 135, 143–145
    tropical cyclones: hurricanes and typhoons, 145–147
    wave cyclones, 135, 138–141
Travertine, 363
Trees, 202, 206–207
Trellis drainage pattern, 412
Trenches (ocean), 292, 295
Tropical air mass, 136
Tropical cyclones, 145–150
    impacts of, 147–49
    naming of, 147
    variable activity of, 147–50
Tropical easterly jet stream, 121
Tropical high-pressure belt, 119
Tropical rainy climates (Köppen system), 159
Tropical storms, 150
Tropical zones, 37
Tropic of cancer, 25
Tropic of capricorn, 24
Tropopause, 57
Troposphere, 47
    winds in, 119–121
Truncated spurs, 460
Tsunami, 330
Tundra, 190
Tundra biome, 208, 229, *230*
    animals, 229–230
Tundra environment, 190–191
Tundra soils, 261
Turbulence, 369
Typhoons, 145–147

Udalf soils, 254
Udoll soils, 256
Ultisols, 249, *250*, 252–253, *252*
Ultramafic igneous rocks, 271
Ultraviolet radiation, 32
    and ozone layer, 39
Unloading, 344
Unsaturated zone, 360
Unstable air, 97–101
Updrafts, 97–100

Upper-air westerlies, 119–120
Upwelling, 123
Ural mountains, 305
Urban heat islands, 55–56
Urban surface characteristics, and air temperature, 55
Ustoll soils, 256

Valley winds, 118
Vegetation. *See also* Biomes; Ecosystems
    altitude zones, 230–231
    basic types and structure, 206–207
    climatic gradients, 231–233
    dry conditions and, 201–203
    ecological succession, 205–206
    food web and, 197–198
    habitats, 200, 201
    human-influenced, 205, 206
    natural, 206
        world regions of, *208–209*
    rain-green, 171
    role in rock breakup, 344
    soil quality and, 239
    temperature effects, 203–204
    transplantation, 206
    water requirements, 201–202
    wind effects, 205
Venice, subsidence in, 366
Vernal equinox, 23
Vertisols, 247, *250*, 254
Victoria Falls, *393*
Visible light, 33
Volcanic activity, global pattern, 322, 323
Volcanic bombs, 316
Volcanic eruptions, 274–275
Volcanic necks, *422*, 423
Volcanism, 291, 316–322
    flood basalts, 320–321
    geysers, 322
    greenhouse effect/global warming and, 70
    hot springs, 322
    subduction arcs and, 319
Volcano coasts, 436, 437, 439
Volcanoes, 272, 316
    caldera, 318
    cinder cones, 321–322
    crater, 316
    erosion, 419, 422–423
    hotspots, 319–320

ice streams and, 472–473
shield volcanoes, 319
types of, 316–320

Waimaea Canyon, 423
Warm-blooded animals, 204
Warm fronts, 138–139
Washout, air pollutants, 90
Wasting. *See* Mass wasting
Water. *See also* Precipitation available capacity,
    and soil storage capacity, 243
    availability of, and net photosynthesis, 199
    global balance, 78, 79
    hydrologic cycle, 78, 359–360
    need and organisms, 201–203
    role in global heat transfer, 77
    states of, 77
    transitions of, 77
    types of, in hydrosphere, 78
Waterfalls, 389, 393–394
Water gaps, 416
Waterlogging, of agricultural land, 258–259, 376
Water pollution
    ground water, 366–368
    surface water, 376–377
Water table, 360–362
    depletion, 365–366
    and water level of lakes, 373
Water vapor
    adiabatic process and, 82–84
    in atmosphere, 38, 80
    humidity (*see also* Humidity saturation)
Watt, 34
Wave action, 429–435
Wave–cut notches, 430
Wave cyclones, 139–142
Wavelength, 31
Waves, 429
Weak equatorial lows, 145
Weathering, 239–240, 339
    chemical, 345–346
    igneous rock alteration and, 273
    physical, 340–344
Weather systems, 135. *See also* Air masses
    easterly waves, 145
    wave cyclones, 135, 139–142
    weak equatorial lows, 145
Wegener, Alfred, 308–309

Wells
    contamination, 366–368
    water table depletion, 365–366
Westerlies, 113
    upper-air, 119, 120
Western Antarctic Ice Sheet, 472
West-wind drift, 122–123
Wet adiabatic lapse rate, 83–84
Wet adiabatic rate, 83–84
Wetlands, 374
Whittier Narrows earthquake (1987), 333, *478*
Wildfires, 119
Wilting point, 243
Wind erosion, 443–444
    dust storms, 444
Winds
    and ecosystem, 205
    Coriolis effect and, 110–112
    cyclones and anticyclones, 111–112
    direction designations, 109
    and global pressure patterns, 113–118
    higher latitude winds, 117–118
    high level, 119–121
        jet streams, 121–122
    on ideal earth: no suface feature
        effects, 112–113
    intertropical convergence zone, 112, 113,
        116–117, 128
    local, 118–119
    major surface winds, 112–113
    measurement, 109
    and ocean currents, 122–124
    pressure gradients and, 109
    sea vs. land breezes, 109–110
    subtropical high-pressure belts, 112–113,
        115
Wind vanes, 109
Winter monsoon, 117
Winter solstice, 23, 24, 35, 36
Wisconsinan Glaciation, 465
Woodland, 207

Xeralf soils, 255
Xeric animals, 202–203
Xeroll soils, 256
Xerophytes, 201
Xerophytes, 201–202

Yellow (Huang) River, 387

CONVERSION FACTORS

Metric to English

Metric Measure	Multiply by*	English Measure
LENGTH		
Millimeters (mm)	0.0394	Inches (in.)
Centimeters (cm)	0.394	Inches (in.)
Meters (m)	3.28	Feet (ft)
Kilometers (km)	0.621	Miles (mi)
AREA		
Square centimeters (cm^2)	0.155	Square inches (in^2)
Square meters (m^2)	10.8	Square feet (ft^2)
Square meters (m^2)	1.12	Square yards (yd^2)
Square kilometers (km^2)	0.386	Square miles (mi^2)
Hectares (ha)	2.47	Acres
VOLUME		
Cubic centimeters (cm^3)	0.0610	Cubic inches (in^3)
Cubic meters (m^3)	35.3	Cubic feet (ft^3)
Cubic meters (m^3)	1.31	Cubic yards (yd^3)
Milliliters(ml)	0.0338	Fluid ounces (fl oz)
Liters (l)	1.06	Quarts (qt)
Liters (l)	0.264	Gallons (gal)
MASS		
Grams (g)	0.0353	Ounces (oz)
Kilograms (kg)	2.20	Pounds (lb)
Kilograms (kg)	0.00110	Tons (2000 lb)
Tonnes (t)	1.10	Tons (2000 lb)

English to Metric

English Measure	Multiply by*	Metric Measure
LENGTH		
Inches (in.)	2.54	Centimeters (cm)
Feet (ft)	0.305	Meters (m)
Yards (yd)	0.914	Meters (m)
Miles (mi)	1.61	Kilometers (km)
AREA		
Square inches (in^2)	6.45	Square centimeters (cm^2)
Square feet (ft^2)	0.0929	Square meters (m^2)
Square yards (yd^2)	0.836	Square meters (m^2)
Square miles (mi^2)	2.59	Square kilometers (km^2)
Acres	0.405	Hectares (ha)
VOLUME		
Cubic inches (in^3)	16.4	Cubic centimeters (cm^3)
Cubic feet (ft^3)	0.0283	Cubic meters (m^3)
Cubic yards (yd^3)	0.765	Cubic meters (m^3)
Fluid ounces (fl oz)	29.6	Milliliters (ml)
Pints (pt)	0.473	Liters (l)
Quarts (qt)	0.946	Liters (l)
Gallons (gal)	3.79	Liters (l)
MASS		
Ounces (oz)	28.4	Grams (g)
Pounds (lb)	0.454	Kilograms (kg)
Tons (2000 lb)	907	Kilograms (kg)
Tons (2000 lb)	0.907	Tonnes (t)

*Conversion factors shown to 3 decimal-digit precision.

Below Sea Level — over 8000 · 8000 · 6000 · 4000 · 2000 · 1000 · 500 · 200 · 0 · 200 · 500 · 1000 · 2000 · 3500 · 5000 · over 5000 — **Above Sea Level**

RELIEF OF THE SURFACE OF THE EARTH

Maps of the continental surfaces and ocean floors (above and left) are based on the Mercator projection. Relief is shown by plastic shading with illumination from a simulated sun 20° above the western horizon. Depth or elevation zones, labeled in meters on the color scale, are shown by varying the hue and saturation of colors. The hemispheric images (right) use the same color system.

These maps were produced from digital data on a 5-minute by 5-minute latitude-longitude grid, representing the best available information from the following sources:

United States, Central America, Southern Canada, Western Europe, Japan: *U.S. Defense Mapping Agency.* ▪ Australia, New Zealand: *Branch of Mineral Resources CSIRO (Australia), Department of Scientific and Industrial Research (New Zealand).* ▪ Rest of the world land masses: *Fleet Numerical Oceanography Center (U.S. Navy) from 10-minute grid.* ▪ Oceanic areas: *U.S. Naval Oceanographic Office.*

Digital imagery by Dr. Peter W. Sloss, NOAA-National Geophysical Data Center, Boulder, Colorado, with image processing equipment at the National Snow and Ice Data Center, Boulder, Colorado, 1986. Data used in producing these images are available from the National Geophysical Data Center, Code E/GC, 325 Broadway, Boulder, Colorado, 80303-3328.

Hemispheric Images

TOPOGRAPHIC MAP SYMBOLS

VARIATIONS WILL BE FOUND ON OLDER MAPS

Hard surface, heavy duty road, four or more lanes	
Hard surface, heavy duty road, two or three lanes	
Hard surface, medium duty road, four or more lanes	
Hard surface, medium duty road, two or three lanes	
Improved light duty road	
Unimproved dirt road and trail	
Dual highway, dividing strip 25 feet or less	
Dual highway, dividing strip exceeding 25 feet	
Road under construction	

Railroad, single track and multiple track	
Railroads in juxtaposition	
Narrow gage, single track and multiple track	
Railroad in street and carline	
Bridge road and railroad	
Drawbridge, road and railroad	
Footbridge	
Tunnel, road and railroad	
Overpass and underpass	
Important small masonry or earth dam	
Dam with lock	
Dam with road	
Canal with lock	

Buildings (dwelling, place of employment, etc.)	
School, church, and cemetery	Cem
Buildings (barn, warehouse, etc.)	
Power transmission line	
Telephone line, pipeline, etc. (labeled as to type)	
Wells other than water (labeled as to type)	o Oil o Gas
Tanks; oil, water, etc. (labeled as to type)	• ● ◍ Water
Located or landmark object; windmill	o
Open pit, mine, or quarry; prospect	× x
Shaft and tunnel entrance	▪ Y

Horizontal and vertical control station:

Tablet, spirit level elevation	BM△ 5653
Other recoverable mark, spirit level elevation	△ 5455
Horizontal control station: tablet, vertical angle elevation	VABM△ 9519
Any recoverable mark, vertical angle or checked elevation	△3775
Vertical control station: tablet, spirit level elevation	BM × 957
Other recoverable mark, spirit level elevation	× 954
Checked spot elevation	×4675
Unchecked spot elevation and water elevation	× 5657 870

Boundary, national	
State	
County, parish, municipio	
Civil township, precinct, town, barrio	
Incorporated city, village, town, hamlet	
Reservation, national or state	
Small park, cemetery, airport, etc.	
Land grant	
Township or range line, United States land survey	
Township or range line, approximate location	
Section line, United States land survey	
Section line, approximate location	
Township line, not United States land survey	
Section line, not United States land survey	
Section corner, found and indicated	+ +
Boundary monument: land grant and other	□ □
United States mineral or location monument	▲

Index contour	Intermediate contour	
Supplementary contour	Depression contours	
Fill	Cut	
Levee	Levee with road	
Mine dump	Wash	
Tailings	Tailings pond	
Strip mine	Distorted surface	
Sand area	Gravel beach	

Perennial streams	Intermittent streams	
Elevated aqueduct	Aqueduct tunnel	
Water well and spring	Disappearing stream	
Small rapids	Small falls	
Large rapids	Large falls	
Intermittent lake	Dry lake	
Foreshore flat	Rock or coral reef	
Sounding, depth curve	10	Piling or dolphin
Exposed wreck	Sunken wreck	
Rock, bare or awash; dangerous to navigation		

Marsh (swamp)	Submerged marsh	
Wooded marsh	Mangrove	
Woods or brushwood	Orchard	
Vineyard	Scrub	
Inundation area	Urban area	

SCALE 1:24 000

CONTOUR INTERVAL 40 FEET
DATUM IS MEAN SEA LEVEL

SAN RAFAEL, CALIF.
NE/4 MT. TAMALPAIS 15' QUADRANGLE
N 3752.5—W 12230/7.5

QUADRANGLE LOCATION

SAUSALITO (CITY HALL) 1.9 MI.
SAN FRANCISCO (CIVIC CENTER) 9.8 MI.

INTERIOR—GEOLOGICAL SURVEY,
WASHINGTON, D. C.

INTRODUCING PHYSICAL GEOGRAPHY

GEOGRAPHY

Second Edition
Updated and Upgraded

Alan Strahler
Boston University

Arthur Strahler
Columbia University (emeritus)

WILEY

John Wiley & Sons, Inc.

New York • Chichester • Weinheim • Brisbane • Singapore • Toronto

ACQUISITIONS EDITOR	Nanette Kauffman
DEVELOPMENTAL EDITOR	Barbara Heaney
MARKETING MANAGER	Susan Lyons
PRODUCTION EDITOR	Sandra Russell
DESIGNER	Harold Nolan
PHOTO EDITOR	Lisa Gee
ILLUSTRATION EDITOR	Anna Melhorn/Sandra Rigby
ILLUSTRATIONS	Precision Graphics/GeoSystems/J/B Woolsey Associates
COVER PHOTOGRAPHS	Clockwise from top:
	Pete Turner/The Image Bank
	John Kelly/The Image Bank
	Michael Melford/The Image Bank
	Jess Stock/Tony Stone Images/New York, Inc.
	Dugald Bremner/Tony Stone Images/New York, Inc.
	Greg Vaughn/Tony Stone Images/New York, Inc.

This book was set in New Baskerville by HRS Electronic Text Management and Hermitage Publishing Services and printed and bound by Von Hoffmann Press, Inc. The cover was printed by The Lehigh Press, Inc. The color separations were prepared by Lehigh Press Colortronics.

This book is printed on acid-free paper. ∞

The paper in this book was manufactured by a mill whose forest management programs include sustained yield harvesting of its timberlands. Sustained yield harvesting principles ensure that the number of trees cut each year does not exceed the amount of new growth.

Library of Congress Cataloging in Publication Data:
Strahler, Alan H.
    Introducing physical geography / Alan Strahler, Arthur Strahler.
    ——2nd ed./Updated and Upgraded
          p.   cm.
    Includes index.
    ISBN 0-471-37293-5 (pbk. : alk. paper)
    1. Physical geography.   I. Strahler, Arthur Newell, 1918– .
II. Title.

GB54.5.S77	1999	97-13431
550—dc21		CIP

Printed in the United States of America

10 9 8 7 6 5 4 3 2 1

Brief Contents

🌐 *Eye on the Environment* feature

Prologue
INTRODUCING EARTH   2

CHAPTER 1   OUR PLACE IN THE SUN   10
🌐 Earth-Sun Cycles and Ice Ages

CHAPTER 2   THE EARTH'S GLOBAL ENERGY BALANCE   28
🌐 The Ozone Layer—Shield to Life

CHAPTER 3   AIR TEMPERATURE   48
🌐 Carbon Dioxide—On the Increase

CHAPTER 4   ATMOSPHERIC MOISTURE AND PRECIPITATION   74
🌐 Air Pollution

CHAPTER 5   WINDS AND GLOBAL CIRCULATION   104
🌐 El Niño

CHAPTER 6   WEATHER SYSTEMS   132
🌐 Hurricane Andrew—Killer Cyclone

CHAPTER 7   GLOBAL CLIMATES   152
🌐 Drought and Land Degradation in the African Sahel

CHAPTER 8   GLOBAL ECOSYSTEMS   194
🌐 Exploitation of the Low-Latitude Rainforest Ecosystem

CHAPTER 9   GLOBAL SOILS   236
🌐 Death of a Civilization

CHAPTER 10   EARTH MATERIALS   264
🌐 Battling Iceland's Heimay Volcano

CHAPTER 11   THE LITHOSPHERE AND PLATE TECTONICS   286
🌐 Alfred Wegener and Continental Drift

CHAPTER 12   VOLCANIC AND TECTONIC LANDFORMS   312
🌐 The Northridge Earthquake of 1994

CHAPTER 13   WEATHERING AND MASS WASTING   336
🌐 The Great Hegben Lake Disaster

CHAPTER 14   FRESH WATER OF THE CONTINENTS   356
🌐 Sinking Cities

CHAPTER 15   LANDFORMS MADE BY RUNNING WATER   380
🌐 Battling the Mighty Mississippi

CHAPTER 16   LANDFORMS AND ROCK STRUCTURE   408
🌐 Marvelous, Majestic, Monolithic Domes

CHAPTER 17   LANDFORMS MADE BY WAVES AND WIND   426
🌐 The Threat of Rising Sea Level from Global Warming

CHAPTER 18   GLACIAL LANDFORMS AND THE ICE AGE   454
🌐 Ice Sheets and Global Warming

Epilogue
NEWS FROM THE FUTURE   476

Appendix I   More about Maps   482

Appendix II   Remote Sensing for Physical Geography   488

Appendix III   The Canadian System of Soil Classification   496

Contents

| **Prologue** | **INTRODUCING EARTH   2** |

What is Physical Geography All About?   7
The Physical Geography of Our Planet   7
   The Four Great Realms   7
   The Life Layer   8
Scale   8
Physical Geography and the Human
   Habitat   8

| **Chapter 1** | **OUR PLACE IN THE SUN   13** |

The Shape of the Earth   13
Earth Rotation   14
   Direction of Rotation   14
   Environmental Effects of Earth
    Rotation   14
The Geographic Grid   15
   Parallels and Meridians   15
   Latitude and Longitude   15
   Map Projections   17
   Polar Projections   17
   Mercator Projection   17
   Goode Projection   18

Global Time   19
   Standard Time   20
   World Time Zones   20
   International Date Line   21
   Daylight Saving Time   21
The Earth's Revolution Around the Sun   22
   Tilt of the Earth's Axis   23
   Solstice and Equinox   23
   Equinox Conditions   23
   Solstice Conditions   25
⊕ *Eye on the Environment • Earth-Sun Cycles and*
Ice Ages   26

| **Chapter 2** | **THE EARTH'S GLOBAL ENERGY BALANCE   31** |

Electromagnetic Radiation   31
   Radiation and Temperature   31
   Solar Radiation   32
   Characteristics of Solar Energy   33
   Longwave Radiation from the Earth   34
   The Global Radiation Balance   34
Insolation over the Globe   34
   The Path of the Sun in the Sky   35
   Insolation Through the Year   36
   Annual Insolation by Latitude   36
World Latitude Zones   37
Composition of the Atmosphere   38
   Ozone in the Upper Atmosphere   38

⊕ *Eye on the Environment • The Ozone Layer—*
Shield to Life   40
Sensible Heat and Latent Heat
  Transfer   39
The Global Energy System   39
   Insolation Losses in the Atmosphere   39
   Albedo   42
   Counterradiation and the Greenhouse
    Effect   42
   Global Energy Budgets of the Atmosphere and
    Surface   43
Net Radiation, Latitude, and the Energy
  Balance   44

Chapter 3

AIR TEMPERATURE AND AIR TEMPERATURE CYCLES   51

Measurement of Air Temperature   51
The Daily Cycle of Air Temperature   53
   Daily Insolation and Net Radiation   53
   Daily Temperature   54
   Temperatures Close to the Ground   54
   Urban and Rural Temperature
     Contrasts   55
   The Urban Heat Island   55
Temperature Structure of the
  Atmosphere   56
   Troposphere   57
   Stratosphere   57
   High-Mountain Environments   57
   Temperature Inversion and Frost   58
The Annual Cycle of Air Temperature   60

   Net Radiation and Temperature   60
   Land and Water Contrasts   61
World Patterns of Air Temperature   63
   Factors Controlling Air Temperature
     Patterns   64
   World Air Temperature Patterns for January
     and July   64
   The Annual Range of Air Temperatures   67
Global Warming and the Greenhouse
  Effect   68
⊕ ***Eye on the Environment • Carbon Dioxide—On***
  the Increase   69
   The Temperature Record   70
   Future Scenarios   70

Chapter 4

ATMOSPHERIC MOISTURE AND PRECIPITATION   77

Water—The Global Perspective   77
   The Hydrosphere and the Hydrologic Cycle   77
The Global Water Balance   79
Humidity   79
   Specific Humidity   79
   Relative Humidity   81
The Adiabatic Process   82
   Dry Adiabatic Rate   82
   Wet Adiabatic Rate   83

Clouds   83
   Cloud Forms   84
   Fog   84
Precipitation   86
⊕ ***Eye on the Environment • Air Pollution   88***
   Precipitation Processes   95
   Orographic Precipitation   95
   Convectional Precipitation   96
   Thunderstorms   99

Chapter 5

WINDS AND GLOBAL CIRCULATION   107

Atmospheric Pressure   107
   Measuring Atmospheric Pressure   107
   How Air Pressure Changes with Altitude   108
Winds and Pressure Gradients   109
   Sea and Land Breezes   109
   The Coriolis Effect and Winds   110
   Cyclones and Anticyclones   111
   Surface Winds on an Ideal Earth   112
Global Wind and Pressure Patterns   113
   Subtropical High-Pressure Belts   116
   The ITC and the Monsoon Circulation   116
   Wind and Pressure Features of Higher
     Latitudes   117

Local Winds   118
Winds Aloft   119
   Global Circulation at Upper Levels   119
   Rossby Waves   120
   Jet Streams   121
Ocean Currents   122
   Surface Currents   122
⊕ ***Eye on the Environment • El Niño   124***
   Deep Currents and Thermohaline
     Circulation   127
Poleward Transport of Heat and
  Moisture   128

Chapter 6   WEATHER SYSTEMS   135

Traveling Cyclones and Anticyclones   135
    Air Masses   *135*
    North American Air Masses   *137*
    Cold, Warm, and Occluded Fronts   *137*
    Wave Cyclones   *139*
    Weather Changes Within a Wave
      Cyclone   *140*
    Cyclone Tracks and Cyclone Families   *142*
    The Tornado   *143*
Tropical and Equatorial Weather
   Systems   145

    Easterly Waves and Weak Equatorial
      Lows   *145*
    Tropical Cyclones   *145*
    Impacts of Tropical Cyclones   *147*
⊕ **Eye on the Environment • Hurricane Andrew—**
    Killer Cyclone   **148**
Cloud Cover, Precipitation, and Global
   Warming   150

Chapter 7   GLOBAL CLIMATES   155

Global Precipitation   155
Climate Classification   157
Special Supplement • The Köppen Climate
System   **159**
Low-Latitude Climates   163
    The Wet Equatorial Climate ①   *164*
    The Monsoon and Trade-Wind Coastal
      Climate ②   *167*
    The Wet-Dry Tropical Climate ③   *169*
    The Dry Tropical Climate ④   *171*
⊕ **Eye on the Environment • Drought and Land**
   Degradation in the African Sahel   **172**

Midlatitude Climates   176
    The Dry Subtropical Climate ⑤   *177*
    The Moist Subtropical Climate ⑥   *179*
    The Mediterranean Climate ⑦   *180*
    The Marine West-Coast Climate ⑧   *183*
    The Dry Midlatitude Climate ⑨   *184*
    The Moist Continental Climate ⑩   *185*
High-Latitude Climates   187
    The Boreal Forest Climate ⑪   *187*
    The Tundra Climate ⑫   *189*
    The Ice Sheet Climate ⑬   *191*

Chapter 8   GLOBAL ECOSYSTEMS   197

Energy Flow in Ecosystems   197
    The Food Web   *197*
    Biomass, Net Primary Production, and
      Climate   *199*
Organisms and Environment   200
    Habitat   *200*
    Organisms and Water Need   *201*
    Organisms and Temperature   *203*
    Other Climatic Factors   *204*
    Ecological Succession   *205*
Natural Vegetation   206

    Structure and Life-Form of Plants   *206*
Terrestrial Ecosystems—The Biomes   208
    Forest Biome   *209*
⊕ **Eye on the Environment • Exploitation of the**
   Low-Latitude Rainforest Ecosystem   **214**
    Savanna Biome   *223*
    Grassland Biome   *224*
    Desert Biome   *227*
    Tundra Biome   *229*
    Altitude Zones of Vegetation   *230*
    Climatic Gradients and Vegetation Types   *231*

Chapter 9 GLOBAL SOILS   239

The Nature of the Soil   239
  Soil Color and Texture   240
  Soil Colloids   241
  Soil Acidity and Alkalinity   241
  Soil Structure   242
  Minerals of the Soil   242
  Soil Moisture   243
Soil Development   244

Soil Horizons   244
Soil-Forming Processes   245
Soil Temperature   246
The Global Scope of Soils   247
  Soil Orders   249
  🌐 Eye on the Environment • Death of a
    Civilization   258
  Desert and Tundra Soils   259

Chapter 10 EARTH MATERIALS   267

The Crust and Its Composition   267
  Rocks and Minerals   267
Igneous Rocks   268
  Common Igneous Rocks   270
  Intrusive and Extrusive Igneous Rocks   271
  Chemical Alteration of Igneous Rocks   273
Sediments and Sedimentary Rocks   273
  🌐 Eye on the Environment • Battling Iceland's
    Heimaey Volcano   274

Clastic Sedimentary Rocks   276
Chemically Precipitated Sedimentary
  Rocks   278
Hydrocarbon Compounds in Sedimentary
  Rocks   280
Metamorphic Rocks   280
The Geologic Time Scale   282
The Cycle of Rock Change   282

Chapter 11 THE LITHOSPHERE AND PLATE TECTONICS   289

The Structure of the Earth   289
  The Earth's Interior   289
  The Lithosphere and Asthenosphere   290
Major Relief Features of the Earth's
  Surface   291
  Relief Features of the Continents   291
  Relief Features of the Ocean Basins   294
Plate Tectonics   298
  Tectonic Processes   298
  Plate Motions and Interactions   299

The Global System of Lithospheric Plates   300
Subduction Tectonics   303
Orogens and Collisions   304
Continental Rupture and New Ocean
  Basins   305
The Power Source for Plate Movements   306
Continents of the Past   306
  🌐 Eye on the Environment • Alfred Wegener and
    Continental Drift   308

Chapter 12 VOLCANIC AND TECTONIC LANDFORMS   315

Landforms   315
Volcanic Activity   316
  Stratovolcanoes   316
  Shield Volcanoes   319
  Volcanic Activity over the Globe   322
Landforms of Tectonic Activity   323
  Fold Belts   323
  Faults and Fault Landforms   324

The Rift Valley System of East Africa   327
Earthquakes   328
  Earthquakes and Plate Tectonics   328
  Seismic Sea Waves   330
  Earthquakes Along the San Andreas
    Fault   330
  🌐 Eye on the Environment • The Northridge
    Earthquake of 1994   332

Chapter 13   WEATHERING AND MASS WASTING   339

Slopes and Regolith   339
Physical Weathering   340
   Frost Action   340
   Salt-Crystal Growth   343
   Unloading   344
   Other Physical Weathering Processes   344
Chemical Weathering and Its
   Landforms   345
   Hydrolysis and Oxidation   345
   Acid Action   345

Mass Wasting   347
   Soil Creep   348
   Earthflow   348
   Mudflow   349
   Landslide   350
Induced Mass Wasting   351
   Induced Earthflows   351
   Eye on the Environment • The Great Hebgen
   Lake Disaster   352
   Scarification of the Land   353

Chapter 14   FRESH WATER OF THE CONTINENTS   359

Ground Water   360
   The Water Table   361
   Aquifers   362
Limestone Solution by Ground Water   362
   Limestone Caverns   362
   Karst Landscapes   364
Problems of Ground Water
   Management   365
   Water Table Depletion   365
   Eye on the Environment • Sinking Cities   366
   Contamination of Ground Water   366

Surface Water   368
   Overland Flow and Stream Flow   368
   Stream Discharge   369
   Drainage Systems   371
   River Floods   371
   Flood Prediction   372
   Lakes   373
   Saline Lakes and Salt Flats   375
   Desert Irrigation   376
   Pollution of Surface Water   376
Surface Water as a Natural Resource   377

Chapter 15   LANDFORMS MADE BY RUNNING WATER   383

Fluvial Processes and Landforms   383
   Erosional and Depositional Landforms   383
Slope Erosion   384
   Accelerated Erosion   384
   Sheet Erosion and Rilling   385
   Colluvium and Alluvium   385
   Slope Erosion in Semiarid and Arid
     Environments   385
The Work of Streams   386
   Stream Erosion   386
   Stream Transportation   387
   Capacity of a Stream to Transport Load   387
Stream Gradation   388

Landscape Evolution of a Graded
   Stream   389
   Eye on the Environment • Battling the Mighty
   Mississippi   390
   Great Waterfalls   393
   Evolution of a Fluvial Landscape   394
   Aggradation and Alluvial Terraces   396
   Alluvial Rivers and Their Floodplains   397
   Entrenched Meanders   398
Fluvial Processes in an Arid Climate   398
   Alluvial Fans   400
The Landscape of Mountainous
   Deserts   402

Chapter 16   # LANDFORMS AND ROCK STRUCTURE   409

Rock Structure as a Landform Control   409
   Strike and Dip   409
Landforms of Horizontal Strata and
   Coastal Plains   410
   Arid Regions   410
   Drainage Patterns   411
   Coastal Plains   412
Landforms of Warped Rock Layers   413
   Sedimentary Domes   413

Fold Belts   415
Landforms Developed on Other Landmass
   Types   416
   Erosion Forms on Fault Structures   416
   Metamorphic Belts   417
   Exposed Batholiths and Monadnocks   418
   Deeply Eroded Volcanoes   418
   🌐 *Eye on the Environment • Marvelous, Majestic,
   Monolithic Domes*   420

Chapter 17   # LANDFORMS MADE BY WAVES AND WIND   429

The Work of Waves   429
   Marine Cliffs   430
   Beaches   430
   Littoral Drift   432
   Littoral Drift and Shore Protection   434
Tidal Currents   435
   Tidal Current Deposits   435
Types of Coastlines   436
   Coastlines of Submergence   436
   Barrier-Island Coasts   436
   Delta and Volcano Coasts   438
   Coral-Reef Coasts   439

Raised Shorelines and Marine Terraces   441
🌐 *Eye on the Environment • The Threat of Rising
   Sea Level from Global Warming*   442
Wind Action   442
   Erosion by Wind   443
   Dust Storms   444
Sand Dunes   445
   Types of Sand Dunes   445
   Coastal Foredunes   449
Loess   450
   Induced Deflation   452

Chapter 18   # GLACIAL LANDFORMS AND THE ICE AGE   457

Glaciers   457
Alpine Glaciers   458
   Landforms Made by Alpine Glaciers   459
   Glacial Troughs and Fiords   461
Ice Sheets of the Present   463
Sea Ice and Icebergs   464
The Ice Age   465
   Glaciation During the Ice Age   466
Landforms Made by Ice Sheets   467
   Erosion by Ice Sheets   467

Deposits Left by Ice Sheets   467
*Environmental Aspects of Glacial
   Deposits*   470
Investigating the Ice Age   470
   *Possible Causes of the Late Cenozoic
   Ice Age*   471
   Possible Causes of Glaciation Cycles   471
🌐 *Eye on the Environment • Ice Sheets and Global
   Warming*   472
Holocene Environments   472

Epilogue   # NEWS FROM THE FUTURE   476

Perspective on the News   476

Appendices

I. More About Maps 482
II. Remote Sensing and Geographic
   Information Systems 488
III. The Canadian System of Soil
    Classification 496

GLOSSARY 502

PHOTO CREDITS 525

INDEX 527

CONVERSION FACTORS 537

RELIEF SURFACE OF THE EARTH 538

TOPOGRAPHIC MAP SYMBOLS 540

SCIENCE NEWS 544

VISUALIZATION 2.0 CD-ROM, GETTING STARTED 586

SCIENCE NEWS

Number	Title		Text Key*
1.	Greenhouse Warming Hurts Arctic Ozone	544	Chapter 2, p. 39
2.	Antarctic Ozone Hole Reaches Record Size	545	Chapter 2, p. 39
3.	Ozone Killer Confounds Expectations	546	Chapter 2, p. 39
4.	1998: Warmest Year Of Past Millennium	546	Chapter 3, p. 69
5.	Earth's Temperature Shot Skyward In 1998	546	Chapter 3, p. 69
6.	Recent Years Are Warmest Since 1400	547	Chapter 3, p. 69
7.	Continents Growing Wetter As Globe Warms	548	Chapter 3, p. 70
8.	Heat And Humidity Getting Worse	548	Chapter 3, p. 70
9.	Nations Draft Kyoto Climate Treaty	548	Chapter 3, p. 72
10.	China's Air Pollution Chokes Crop Growth	550	Chapter 4, p. 88
11.	How Inhaled Dust Harms The Lungs	550	Chapter 4, p. 88
12.	Spying On El Niño: The Struggle To Predict The Pacific Prankster	551	Chapter 5, p. 127
13.	La Niña Readies To Steal El Niño's Thunder	554	Chapter 5, p. 127
14.	Oklahoma Tornado Set Wind Record	554	Chapter 6, p. 144
15.	Diversity In Tropical Forest Edges	556	Chapter 8, p. 211
16.	Amazon Forests Caught In Fiery Feedback	556	Chapter 8, p. 212
17.	Hidden Threats Take Toll In Amazon	557	Chapter 8, p. 212
18.	Global Graveyard: New Images Of Earth's Interior Reveal The Fate Of Old Ocean Floor	558	Chapter 11, p. 300
19.	Eruption On Montserrat Gaining Strength	560	Chapter 12, p. 318
20.	Inside Old Faithful: Scientists Look Down The Throat Of A Geyser	561	Chapter 12, p. 322
21.	Seattle's Soft Spot Boosts Power Of Quakes	562	Chapter 12, p. 328
22.	Waves Of Death: Why The New Guinea Tsunami Carries Bad News For North America	562	Chapter 12, p. 330
23.	California Shakes Most Often In September	565	Chapter 12, p. 331
24.	California Quake Deficit Fades	565	Chapter 12, p. 331
25.	Racing The Waves: Seismologists Try To Catch Quake Tremors Quickly Enough To Save Lives	566	Chapter 12, p. 334
26.	Natural-Disaster Policies Need Shaking Up	569	Chapter 14, p. 373
27.	The Big Chill: Does Dust Drive Earth's Ice Ages?	569	Chapter 18, p. 471
28.	Signs Of Unstable Ice In Antarctica	572	Chapter 18, p. 474
29.	Sea Change In The Arctic: An Oceanful of Clues Points To Climatic Warming In The Far North	572	Chapter 18, p. 474

Preface

The world around us is changing quickly, and so is our understanding of how it works. In the four years since the manuscript for the first edition of *Introducing Physical Geography* was completed, the component sciences of physical geography have moved forward significantly. Especially relevant is the progress in achieving a better understanding of the relationships among the many components of the global climate system, as well as a new realization of the importance of human impact on climate. Moreover, new environmental events, from earthquakes to hurricanes, have captured the public's attention and can now provide an updated context for demonstrating the principles of physical geography to beginning students and relating them to their life experiences. Accordingly, the primary goal of our Second Edition is to incorporate these new developments in science as well as appropriate references to recent, newsworthy environmental events.

The second goal of our revision of *Introducing Physical Geography* is to enhance our learning environment by making the text more effective in communicating the concepts of physical geography. To accomplish this goal, we use a visualization strategy. Today's students are accustomed to learning visually and interactively, and our text will be more effective if it helps students visualize the concepts that they are learning. Our visualization strategy is described more fully in a following section.

CONTENT

Introducing Physical Geography has been written for a one-semester or one-quarter course in which a majority of students are nonmajors fulfilling a college science requirement. These students need to learn how to reason logically, to understand the nature of cause and effect, and to become acquainted with fundamental scientific principles. Physical geography can easily provide the context for this learning, and it has the added advantage of being relevant to many aspects of students' everyday experiences. For this audience, the appropriate focus is on the broad concepts of physical geography and their application, and this is the focus we have adopted for this book.

In designing the content of *Introducing Physical Geography*, we have emphasized the standard curriculum for physical geography that has evolved over the years. The sequence of topics is deliberately conventional and eminently teachable. However, as experienced instructors of physical geography, we know that the conventional curriculum is difficult to cover in a one-semester or one-quarter course. Every instructor has to come to grips with this fact. Our approach in this book has been to provide a balanced level of coverage of the field, so that when an instructor chooses to emphasize a particular area of science—climatology or geomorphology, for example—a firm foundation is there. If a topic is deemphasized, the text will provide a basic level of understanding that can stand without lecture support.

Physical geography is a broad field that draws from many component sciences. In developing the topical coverage for *Introducing Physical Geography*, we have had to make difficult decisions about what to include and what to omit. We have constantly asked, "Is this topic really necessary? Do other topics depend on it? Is it essential for understanding the big picture? Will it be useful and important to a nonscience major a year from now?" In producing a book of reasonable size and length, we have not had the luxury of including topics and concepts that are secondary to the book's main mission.

Second Edition Updated and Upgraded Changes

In preparing the second edition, faculty and student reviewers greatly helped us to fine tune our content and presentation. Some of the more significant changes made in response to reviewers are:

- Revising and editing chapter opening features (*Putting You in the Picture* essays) to be more focused and consistent from chapter to chapter.

- Moving text coverage of air pollution into a multipage *Eye on the Environment* feature within Chapter 4 and adding a new *Eye on the Environment* feature on Hurricane Andrew in Chapter 6.

- Enlarging global maps in Chapters 7, 8, and 9 to show more detail.

- Broadening Chapter 8 to focus on global ecosystems, extending our treatment of global vegetation to include basic concepts of plant and animal ecology. Biome descriptions now include examples of typical or characteristic fauna.

- Revising Chapters 10 and 11 to provide a better flow of topics in basic geology and plate tectonics.

- Adding new material on hotspots, sea floor spreading, and shield volcanoes to Chapter 12, accompanied by a new global map of volcano locations.

- Improving the treatment of aquifers and salt water intrusion in Chapter 14.

- Moving karst and cavern material from Chapter 16 to Chapter 14.

- Improving the art program by making changes throughout the text, including the addition of two new photos of cloud types to Chapter 4, a new mass wasting process diagram to Chapter 13, and a new figure on types of coastlines to Chapter 17. We have also sharpened block diagrams in Chapters 15 and 16.

In addition to these changes, we have also incorporated the latest knowledge in a number of rapidly moving areas of science by

- Updating ozone layer status in the *Eye on the Environment* feature of Chapter 2.

- Updating global temperatures and projections of global warming in Chapter 3 and enhancing material on variability in climate as a result of climate change.

- Updating coverage of CO_2 and other greenhouse gases vis a vis Kyoto and Buenos Aires meetings in 1997 and 1998.

- Updating sea level rise projections in Chapter 3.

- Adding coverage of thermohaline ocean current circulation with text and a new figure in Chapter 5.

- Coverage of El Niño and La Niña conditions in 1997 through 1999.

- Coverage of Hurricane Mitch in Chapter 6.

- Updating material on the stability of the Western Antarctic Ice Sheet in Chapter 18.

- Updating material on environmental issues and events since the Rio Earth Summit in the Epilogue.

Science News

The Environmental Update Edition of *Introducing Physical Geography*, published in 1996, added an appendix of Science News articles selected for relevance to text topics, and we have retained this popular feature. For the Updated and Upgraded 2nd Edition, we have selected an entirely new set of 29 features from its 1997, 1998 and 1999 issues. The selections are primarily designed to provide the latest information on global change topics, but include items from ecology, plate tectonics, and the human environment as well. They are specifically keyed to the running text, using an icon and number in the style shown here: SCIENCE NEWS 11

THE LEARNING ENVIRONMENT

Structure

Students need orientation as they read a textbook. Where are we in the scheme of this book? What topics will be covered next? Why are they important? How do they relate to topics that have already been presented? Forty years of classroom lecturing have convinced us of the importance of helping students understand the structure of a course as they learn, and we have carried this idea forward in our Second Edition.

We begin the book with a Prologue, in which the student becomes an astronaut stationed on the moon and embarks on a journey home to earth. On this journey, the student sees the planet at ever-increasing scales, first as a whole, then as a realm of the four great spheres, then as regions differentiated by climate, vegetation, and soils, and finally as a collection of individual landforms as the return spacecraft proceeds to a touchdown. The journey is both a metaphor and model for the learning to come.

We continue the involvement of the student as we open each chapter with a feature entitled *Putting You in the Picture*. Rather than simply preview the chapter's contents, this essay personalizes the chapter with an experience or topic that directly involves the student. For example, in Chapter 6 the student reads an account of experiencing Hurricane Andrew as it devastated the South Florida coast in 1992. In Chapter 15, the student is introduced to the erosive power of rivers by riding a raft down the Colorado River through the Grand Canyon. Most of these essays are written in the second person to involve the student more directly.

Communicating the connectivity between topics within and across chapters is essential in guiding the student through the subject matter of physical geography. The first few paragraphs of each chapter are specially written to provide an orientation to the material to come, and transitional paragraphs continue within the text whenever an important new subject begins. To enhance the forward linkages, we provide a brief final section, *A Look Ahead*, that puts the present chapter in perspective and leads into the next one.

Students enjoy learning that their new knowledge is important and useful. To help reinforce text material and provide relevance, we include a short *Eye on the Environment* feature within each chapter. These essays focus the concepts of the chapter on a specific environmental topic. Many are concerned with global climate and climatic change, which is a frequent concern of news and documentary coverage on television and appears often in the popular press.

The book concludes with an Epilogue that raises the environmental implications of the physical geography that the student has learned. In an extended script for a news broadcast from the mid-twenty-first century, students picture how life might change in the future as human impacts continue to reshape our planet. The news broadcast leads into a discussion of the most serious environmental issues facing the human species and the prognosis for the earth's immediate future.

Learning Tools

To help students master text material, we provide a number of learning tools. To build the necessary vocabulary, we have identified the dozen or so key terms that are most important in each chapter with the use of **boldface**. Less important terms are identified in *italics*. All terms are defined in our *Glossary*, which contains over 1000 individual entries and includes many terms that are part of a basic science vocabulary.

As another aid to learning, we have embedded in the text a number of brief key statements that focus on important concepts. They are set apart from the text in a larger type face and marked by color bars for easy identification.

Our suite of end-of-chapter materials is also designed to reinforce key concepts and ideas of the chapter. A summary, *In Review*, covers all the major concepts of the chapter. The list of *Key Terms* gathers together the critical vocabulary of the chapter. *Review Questions* are designed as oral or written exercises that require description or explanation of important ideas and concepts. *Visualizing Exercises* utilize sketching or graphing as a way of focusing attention on key illustrations. Also provided are *A Deeper View* questions that require more synthesis or the reorganization of knowledge in a new context.

VISUALIZATION

Our visualization strategy focuses on helping students form lasting visual images of the processes of physical geography. The strategy involves three components: illustrations, word pictures, and a CD-ROM.

Illustrations and Word Pictures

The art program for *Introducing Physical Geography*, entirely new in the first edition, provides scientific accuracy with visual appeal. It captures the essence of key concepts using both accurate and effective graphics and interesting and attractive photos.

Another important part of the visualization theme is building word pictures that will stay in students' minds,

serving as milestones along a journey of discovery. Clear word pictures must be painted to lay out and connect scientific ideas and images and place them in the context of their linkages to other processes and concepts. This involves not only accurate descriptions of processes within the main body of the text, but also virtual images in which the students experience the processes directly. These are present throughout the book, from the astronaut's journey from moon to earth in our Prologue, through the *Putting You in the Picture* essays that open each chapter, to the future news broadcast in our *Epilogue*. Throughout, the idea is to create striking images that will personalize the processes students are studying and connect them to their own lives.

Upgraded Visualization 2.0 CD-ROM

A vital part of our visualization strategy is to provide a CD-ROM as an integral part of the text. Using this device, we can extend the learning from the textbook into a multimedia environment, drawing on motion, color, and sound to illustrate key concepts and processes. By publishing this book with an integral CD-ROM, we ensure that every student has access to the complete array of learning tools that we have developed for physical geography. Version 2.0 of the Visualization CD-ROM includes four main types of information:

- 13 original animations that enable visualization of key concepts and processes.

- 11 minutes of film and video sequences to show landscape processes in action.

- Over 300 captioned photos from the authors' collections.

- Over 300 graphic art illustrations taken from the text.

By viewing the animations, students can visualize processes that are difficult to describe in words or show with fixed art. Examples include: the revolution of the earth around the sun and the seasons; the Coriolis effect; wave cyclone development; adiabatic heating and cooling; and margins of tectonic plates. A special animation provides an overlay tool for students to inspect and compare patterns of climate, vegetation, and soils for continental-sized regions.

Film and video sequences illustrate the processes of physical geography in action. Examples include: waterfalls; wave action and coastal landforms; volcanoes and their formation; landscapes of plate tectonic margins; and convectional precipitation.

The photos are color slides from the authors' personal collections, each with a caption explaining its content. For ease of use, the slides are keyed to the chapters to which they are most directly related. The graphic art from the text is also presented with captions and figure numbers.

To link the CD-ROM with the text, small icons like the one at the end of this paragraph are placed with the running text and illustration captions in the book. 📹

The CD comes complete with viewing software for PCs and Macintoshes and also includes a searchable keyword index tool to help guide users to the materials illustrating specific concepts. Besides providing a student supplement, the CD-ROM is designed to be used by the instructor in computer video-equipped classrooms as a teaching tool.

SUPPLEMENTS

On-Line Study Guide. Students will be able to interact over the Internet with a study guide at any time during the term with the On-Line Study Companion. Each chapter includes practice quizzes; e-mail functionality for sending work to instructors, teaching assistants, and peers; tutorials for selected topics; hot links to current World Wide Web sites; and other Internet services, all designed to make the On-Line Student Companion both a solid study resource and a perfect component of any distance learning or Web-delivered course. The address is: http://www.wiley.com/college/strahler.

Exercise Manual. Written by the textbook authors, Arthur and Alan Strahler, the Exercise Manual builds on the students' knowledge of physical geography developed through class lecture and textbook study and puts this knowledge to use by solving problems based on course topics. It provides all data, figures, and maps needed for the exercises, which are keyed into the parent textbook. An instructor's version of the Exercise Manual provides the answers to all the exercises.

Instructor's Web Resource. Instructors adopting *Introducing Physical Geography* will be issued a password that will provide access to a wide range of instructional resources on a protected Wiley Web site. These include test banks, chapter outlines, and a Supplements Integrator, a guide to using the Web for active learning in the physical geography course with a chapter-by-chapter list of well-maintained URLs hot-linked to Web sites. The Supplements Integrator cross references for each chapter the resources to which the instructor has access from across our multimedia offerings, including the animations, movies, photos, and drawings on the Visualization CD-ROM; the overhead transparency set; the slide set; and the On-Line Student Companion. The dynamic nature of this site also allows the publisher to post current events, maps, and other news and developments related to physical geography, from Wiley and the authors. The address is: http://www.wiley.com/college/strahler.

Full-color Overhead Transparencies. Over 115 full-color transparencies from the text are provided in a form suit-able for projection in the classroom. These illustrations will also be available as slides.

Supplementary Slide Set. Approximately 150 slides of graphic art and photos, some from the authors' personal collections, are available.

ANOTHER TITLE BY THE AUTHORS

Physical Geography: Science and Systems of the Human Environment (Wiley, 1997) provides a more thorough overview of physical geography, emphasizing a systems approach. The organization is slightly different, with global climate expanded to three chapters and with soils and two biogeography chapters concluding the text rather than immediately following global climates. Systems are threaded throughout, with special systems boxes providing examples of the systems approach. The book also includes a sequence of optional quantitative boxes that supplement important concepts with simple formulas and calculations.

ACKNOWLEDGMENTS

The preparation of *Introducing Physical Geography*, Second Edition was greatly aided by the reviewers who read and evaluated the various parts of the manuscript. They include

Tanya Allison, Montgomery College
Joseph M. Ashley, Montana State University
Roger Balm, Rutgers State University
David Bixler, Chaffey Community College
Kenneth L. Bowden, Northern Illinois University
David Butler, University of North Carolina, Chapel Hill
Les Dean, Riverside Community College
Christopher H. Exline, University of Nevada
Jerry Green, Miami University
Clarence Head, University of Central Florida
Peter W. Knightes, Central Texas College
Christopher S. Larsen, State University of New York, Buffalo
James F. Petersen, Southwest Texas State University
Clinton Rowe, University of Nebraska-Lincoln
Carol Shears, Ball State University
Carol I. Zinser, State University of New York, Plattsburgh

We were delighted by the comments we received from our student reviewers at the University of Central Florida: Kevin Chibar, Bryan M. Ciervo, Gerrit A. Lemmen, and Scott Vanderbeck.

We also thank the reviewers who read and evaluated the manuscript for the first edition:

Ted Alsop, Utah State University
Kevin Anderson, Augustana College
Joseph M. Ashley, Montana State University

George Aspbury, Illinois State

David Castillon, Southwest Missouri State University

Randall Cerveny, Arizona State University

Richard Crooker, Kutztown University

Don Gary, Nicholls State University

Roland L. Grant, Eastern Montana College

Maurice Hackett, Southwest Missouri State University

Richard Hackett, Oklahoma State University

Vance Halliday, University of Wisconsin

Peter Herron, University of Calgary

Julie Laity, California State University, Northridge

Jeffrey Lee, Texas Tech

Richard Mackinnon, Allan Hancock College

Andrew Marcus, Montana State University

Bernard McGonigle, Community College of Philadelphia

Marilyn Raphael, University of California at Los Angeles

Vatche Tchakerian, Texas A&M University

John Watkins, University of Kentucky

Jan Westerik, Concord College

Kenneth White, Nicholls State University

Thomas Wikle, Oklahoma State

For the Upgraded Visualization 2.0 CD-ROM we depended greatly on the thorough reviews provided by the following instructors and students. We believe their suggestions are well represented in the new version.

Dan Blair, University of Winnipeg

Hobart King, Mansfield University

Donald L. Morgan, Brigham Young University

Buffy Quinn, Syracuse University

Sam M. W. Scripter, University of Idaho

We thank David McCracken for his creative and painstaking work on the CD-ROM storyboards, Jerry Shue of The Pennsylvania State University for helping us develop the concepts for the animations and movies, Mary Lee and Sid Nolan of Academic Media Network for their work on the CD-ROM videos, Ed Burke of I-Media for putting our CD-ROM ideas into action, and Fred Schroyer for his comparative analysis.

It is with particular pleasure that we thank the staff at Wiley for their long hours of careful work in the preparation and production of our *Second Edition*. They include our editor, Nanette Kauffman; our developmental editors Barbara Heaney and Pui Szeto; our new-media editor, David Blair; our photo editor, Kim Khatchatourian; our designer, Harry Nolan; our illustration editor, Edward Starr; our marketing manager, Sue Lyons; and our editorial assistant Karen Ayoub. Special thanks go to our production editor, Sandra Russell, for keeping us all on track during the complicated job of turning words and sketches into books and CD-ROMs.

Alan Strahler
Cambridge, Massachusetts

Arthur Strahler
Santa Barbara, California

September 14, 1999

About the Authors

ABOUT THE AUTHORS

Alan Strahler (b. 1943) received his B.A. degree in 1964 and his Ph.D degree in 1969 from The Johns Hopkins University, Department of Geography and Environmental Engineering. He has held academic positions at the University of Virginia, the University of California at Santa Barbara, and Hunter College of the City University of New York, and is now Professor of Geography at Boston University. With Arthur Strahler, he is a coauthor of seven textbook titles with six revised editions on physical geography and environmental science. He has also published over 90 articles in the refered scientific literature, largely on the theory of remote sensing of vegetation. He has also contributed to the fields of plant geography, forest ecology, and quantitative methods. His work has been supported by over $4 million in grant and contract funds, primarily from NASA. In 1993, he was awarded the Association of American Geographers/Remote Sensing Specialty Group Medal for Outstanding Contributions to Remote Sensing.

Arthur Strahler (b. 1918) received his B.A. degree in 1938 from the College of Wooster, Ohio, and his Ph.D. degree in geology from Columbia University in 1944. He was appointed to the Columbia University faculty in 1941, serving as Professor of Geomorphology from 1958 to 1967 and as Chair of the Department of Geology from 1959 to 1962. He was elected as a Fellow of both the Geological Society of America and the Association of American Geographers for his pioneering contributions to quantitative and dynamic geomorphology, contained in over 30 major papers in leading scientific journals. He is the author or coauthor with Alan Strahler of 16 textbook titles with 12 revised editions in physical geography, environmental science, the earth sciences, and geology. His most recent new title, Understanding Science: An Introduction to Concepts and Issues, published by Prometheus Books in 1992, has been widely reviewed as an introductory college text on the philosophy of science.

To the Student

As authors, we are very concerned about what students think about our book. In preparing this revision, we have received a generous amount of student input that has improved our text. As a result, we feel that the Second Edition of *Introducing Physical Geography* is even more accessible to both students and instructors. We invite you to share your comments and suggestions with us for the next edition. Please feel free to write to us care of Geography Editor, John Wiley & Sons, 605 Third Avenue, New York, NY 10158 or send us an e-mail at nKauffma@wiley.com

The following quotations from four students at Central Florida University highlight some of the strengths of this edition, as well as give advice to you and other new students on how best to study from this text. We hope you find them interesting and helpful.

Kevin Chibar, animation major
Bryan Ciervo, pre-pharmacy major
Gerrit Lemmen, business management major
Scott Vanderbeck, civil engineering major

What are the strengths of this text? How did it help you understand difficult concepts?

Kevin: It is an easy-to-read book…and allows good understanding. The most difficult things to visualize in the course have been wind circulation and pressure patterns, including cyclones and anticyclones. The text helped me to understand them.

Bryan: I really like the way that the authors have so many pictures because it gives you a real look at what they're talking about. I found all the new information [on biogeography] more interesting.

Gerrit: The illustrations make it easier to process the information. Soils were difficult to understand, [but] the diagrams were very descriptive and clear. I like the added examples of ecosystems.

Scott: *Introducing Physical Geography* is extremely thorough and informative. The atmosphere and weather systems were difficult topics; Strahler & Strahler provided me with very strong text as well as diagrams that allowed me to visualize how these systems work. The CD-ROM would be extremely valuable [because] it would allow a more visual and interactive way of learning.

Have you lived through any of the phenomena described in this text, and did this text help you understand it better?

Kevin: Yes, Hurricane Andrew. The text did help me to understand…how and under what conditions a hurricane is formed.

Bryan: I lived through Hurricane Andrew. Very scary and emotional to talk about. Before, I had no clue [about hurricanes]. Now I can hold an interesting conversation about the hurricane.

Gerrit: Yes, sinkholes, a hurricane, and undertow at the beach. My experience did help me grasp concepts more easily.

Scott: Yes, several tornadoes and Hurricane Andrew. I had the chance to witness firsthand how powerful hurricanes are. *Introducing Physical Geography* supplied me with a great deal of information concerning hurricanes…such as their formation and the factors that contribute to their strengthening.

What advice would you give to students who will be studying from this text in the next few years?

Kevin: They should allow themselves enough time to read and understand the topics. I would suggest that they answer the questions at the end. Diagrams and photos are always helpful, especially with difficult concepts. The sentences that are "blocked off" from the rest of the paragraph [key statements between blue bars] basically summarize the information, while alerting students to important points.

Bryan: The numerous photos and drawings help you understand the concepts.

Gerrit: Read the entire chapter through and then go back and highlight the key ideas.

Scott: Read the text prior to class lectures and discussion. Use class time to gain a better understanding of material you may not have fully understood. The photos and drawings definitely help to visualize the concepts.

PROLOGUE

Introducing Earth

What Is Physical Geography All About?

Physical geography is the study of the surface of our home planet, Earth. It is not just a catalog of mountains and rivers, but rather a branch of science that investigates how and why the surface changes from day to day, from year to year, and over millions of years. For the physical geographer, the surface includes four parts. First is the land surface itself, which is covered by soil and vegetation. Second is the upper layer of the earth's crust, which provides rock materials that can be shaped into landforms by running water, wind, waves, and ice. Third is the lower layer of the atmosphere, in which clouds and weather occur. Last is the upper layer of the oceans, where ocean waters influence land near the coast.

Perhaps the easiest way to introduce physical geography is to examine the features of our planet—first as a whole and then by zooming in on continents, regions, and, finally, local areas. But to make it more interesting, imagine that the year is 2050 and that you are an astronaut on a lunar base. Relax and follow along as you move from the lunar base to a space station orbiting the earth, and finally to the descent and landing of a space shuttle carrying you safely home.

It's eight hundred hours, on December 17, Houston time, and your duty watch is just beginning. Taking your place at the lab's environmental control desk, you check the oxygen and carbon dioxide balances and review the pressure and water vapor concentration log from the previous watch. All is well. You rise from your desk and turn. A small push on your right foot produces a giant but graceful step. You are standing in front of the viewport, looking out across the moonscape. Not much scenery here at Lunar Base Alpha, at least not when you're in the middle of the long lunar night. But in the black sky above, there is a truly majestic sight—a huge blue, green, and brown disk veiled by swirls and films of white. It is Earth, your home.

From your lunar viewpoint, the earth's environmental realms are easily identified by their colors. The blue of the world oceans—the hydrosphere—*dominates the area of the disk. Rich browns characterize the* lithosphere—*the solid rocky portion of the planet—where it rises above the seas. Vast white cloud expanses form and fade in the gaseous portion of the planet—the* atmosphere. *The lush greens of plants are the visible portion of the* biosphere, *the realm of life on earth, which dominates the* life layer—*the interface between air, land, and ocean.*

The right-hand edge of the disk is in shadow, and you see the dramatic contrast between the day and night portions of the globe. The land and water masses of the illuminated side bask in the sunlight, and here the temperature of the life layer is increasing. On the shadowed side of the planet, however, the heat of the oceans, atmosphere, and continents is continuously being lost—radiated into outer space. Without sunlight, the earth's surface cools.

Over the next few hours, you notice the planet's slow rotation. The continent of South America glides across the center of the disk to its edge, then disappears. Australia emerges from behind the curtain of night and makes its way out onto the illuminated portion of the disk. As you watch, each land and water mass slowly takes its turn in the sun, absorbing and reflecting its quota of solar energy.

A Meteosat photo of earth showing cloud and water vapor patterns. Taken from geostationary orbit.

SCIENCE NEWS®

1. Greenhouse Warming Hurts Arctic Ozone 544
2. Antarctic Ozone Hole Reaches Record Size 545
3. Ozone Killer Confounds Expectations 546
4. 1998: Warmest Year Of Past Millennium 546
5. Earth's Temperature Shot Skyward In 1998 546
6. Recent Years Are Warmest Since 1400 547
7. Continents Growing Wetter As Globe Warms 548
8. Heat And Humidity Getting Worse 548
9. Nations Draft Kyoto Climate Treaty 548
10. China's Air Pollution Chokes Crop Growth 550
11. How Inhaled Dust Harms The Lungs 550
12. Spying On El Niño: The Struggle To Predict The Pacific Prankster 551
13. La Niña Readies To Steal El Niño's Thunder 554
14. Oklahoma Tornado Set Wind Record 554
15. Diversity In Tropical Forest Edges 556
16. Amazon Forests Caught In Fiery Feedback 556
17. Hidden Threats Take Toll In Amazon 557
18. Global Graveyard: New Images Of Earth's Interior Reveal The Fate Of Old Ocean Floor 558
19. Eruption On Montserrat Gaining Strength 560
20. Inside Old Faithful: Scientists Look Down The Throat Of A Geyser 561
21. Seattle's Soft Spot Boosts Power Of Quakes 562
22. Waves Of Death: Why The New Guinea Tsunami Carries Bad News For North America 562
23. California Shakes Most Often In September 565
24. California Quake Deficit Fades 565
25. Racing The Waves: Seismologists Try To Catch Quake Tremors Quickly Enough To Save Lives 566
26. Natural-Disaster Policies Need Shaking Up 569
27. The Big Chill: Does Dust Drive Earth's Ice Ages? 569
28. Signs Of Unstable Ice In Antarctica 572
29. Sea Change In The Arctic: An Oceanful of Clues Points To Climatic Warming In The Far North 572

Greenhouse Warming Hurts Arctic Ozone

SCIENCE NEWS, VOL. 153   APRIL 11, 1998

Although nations have started weaning themselves off ozone-depleting chemicals, a seemingly unrelated problem threatens to foil their efforts. Greenhouse warming in coming decades will alter global winds in ways that significantly slow the recovery of the ailing ozone shield, according to computer climate simulations by NASA scientists.

"Our research says that it won't get better starting in a couple of years [as predicted], but it will actually continue to get worse for the next 10 or 15 years or so," says Drew T. Shindell of the NASA Goddard Institute for Space Studies in New York City. "It is an interesting example of how global warming is such a big effect that it can have all sorts of unforeseen implications," he adds. Shindell's group reports its findings in the April 9 NATURE.

The NASA scientists explored the subtle influences of greenhouse warming by running experiments on a climate model that divides the atmosphere into boxes measuring 8° by 10°, stacked in 23 vertical layers. Shindell and his colleagues built in simplified ozone chemistry—a first for computer climate modelers.

Ozone destruction depends on cold conditions in the stratosphere, and Shindell's computer simulations showed two ways in which greenhouse warming abets this process by cooling the lower stratosphere.

First, the trapping of heat near the ground blocks radiation from reaching the stratosphere—a straightforward effect well known to scientists. A more potent and unanticipated cooling influence comes from the way that greenhouse gases alter winds in the lowest layer of air, the troposphere. Warming in the tropics strengthens the jet streams circling the globe in the midlatitudes, says Shindell. These enhanced winds prevent disturbances from migrating toward the Arctic and Antarctic, effectively sealing off the polar skies and allowing them to cool.

In the model, the lower temperatures in winter helped bring about massive ozone loss in the Arctic from 2010 to 2019, even as the instrument of the

Predicted Arctic ozone loss in March for the years 2010 through 2019.

destruction—chlorine pollution—became scarcer in the skies.

Because computer simulations represent simplified versions of the real atmosphere, researchers find it difficult to judge the reliability of model forecasts. Yet recent history has added credence to the NASA study, according to Shindell. During the 1990s, the Arctic stratosphere has remained more stable and colder than during the 1980s, a pattern similar to that exhibited by the model. Individual years vary tremendously, however, and the winter of 1998 in fact saw only meager ozone loss in the Arctic, says Shindell.

Other research groups are reaching conclusions similar to NASA's as they start to examine the interplay between global warming and ozone chemistry, says Neal Butchart of the United Kingdom Meteorological Office in Bracknell. In unfinished studies, both its model and a French one show the polar stratospheres growing more isolated and colder. "This looks like it could be a fairly robust feature in the models," says Butchart.

The details of the winds in the NASA model do not resemble those in other computer models, however, including one at the National Oceanic and Atmospheric Administration's Fluid Dynamics Laboratory in Princeton, N.J. The lab's director, Jerry D. Mahlman, argues that the NASA model splits the atmosphere too coarsely to investigate the intricacies of how winds will respond to global warming. "Pieces of their work are almost certainly correct, and other pieces are highly questionable."

The NASA study, he adds, didn't include several factors that will further delay ozone recovery. "The problem is even scarier than this paper suggests," says Mahlman.   —R. Monastersky

Drug prevents some cancer, poses risks

For "the first time in history," evidence indicates "that breast cancer can not only be treated but also prevented," says Bernard Fisher of Allegheny University of the Health Sciences in Pittsburgh.

Fisher is referring to the finding that women at especially high risk of breast cancer appear to halve their chances of developing the malignancy by taking tamoxifen, a synthetic hormone. These results, reported Monday, come from a federally funded trial directed by Fisher.

For 25 years, tamoxifen has been used to limit breast cancer recurrence in women treated for the disease. In 1992, the National Cancer Institute (NCI) in Bethesda, Md., launched a trial to see whether the drug could also prevent the scourge in women who had never had the disease. Half of some 13,000 volunteers received the drug; half got an inactive substance, or placebo.

While 154 of the women taking the placebo developed invasive breast cancer during the trial, only 85 women who were taking tamoxifen did.

Hailing the trial as a "historic" success, NCI Director Richard D. Klausner also noted that the drug poses serious risks. Among the women taking tamoxifen, a known carcinogen (SN: 3/2/96, p. 132), 33 developed uterine cancer, compared to 14 in the placebo group. The 17 pulmonary blood clots among tamoxifen users were almost triple the number in the placebo group and included two deaths. Clearly, Klausner said, "this drug is far from ideal," and the need to evaluate its benefits and risks individually for each patient "cannot be emphasized enough."

The increased risks showed up only in women over 50, notes NCI's Leslie Ford. Moreover, "none of the risks were unanticipated, and none were higher than expected" (SN: 5/9/92, p. 309).

Because two earlier studies suggested that taking tamoxifen for many years increases cancer rates and mortality (SN: 12/9/95, p. 391), NCI recommends that breast cancer survivors take it for no more than 5 years. Klausner said no one knows whether cancerfree women will need to adhere to the same limit.

"I'm glad they found a very substantial benefit in [breast] cancer prevention, because the drug's adverse effects in healthy women are very troubling," says breast cancer epidemiologist Trudy Bush of the University of Maryland in Baltimore.

Adds Cynthia Pearson of the National Women's Health Network in Washington, D.C., it's too early to know whether the tamoxifen group's lower cancer rate represents "a true prevention of disease versus just a delay."
                                                       —J. Raloff

Antarctic ozone hole reaches record size

SCIENCE NEWS, VOL. 154   OCTOBER 17, 1998

Chemical pollutants combined with unusually low temperatures high in the atmosphere to gnaw a record-breaking hole in Earth's protective ozone layer this year, according to measurements made in the frigid skies over Antarctica.

The stratosphere above both poles has grown colder in recent years for reasons not clear to researchers. The recent Antarctica temperatures, the lowest in 2 decades of measurements, raise concern that the ozone layer will not heal as quickly as scientists had predicted, even though nations are curbing the use of ozone-depleting chemicals.

"It may take longer to recover," says Paul Newman of NASA's Goddard Space Flight Center in Greenbelt, Md.

One fear among scientists is that greenhouse gases could be playing a role in lowering stratospheric temperatures. Carbon dioxide and other forms of pollution trap heat in the lower atmosphere, but they cool off the stratosphere—the layer between 10 and 50 kilometers in altitude.

The ozone hole above Antarctica starts forming in September, when springtime sunlight returns to the polar skies. The light energizes chlorine- and bromine-containing chemicals that break apart ozone molecules in the stratosphere. A key part of the chemical chain reaction takes place on the surface of frozen cloud particles, so cold temperatures worsen ozone destruction.

According to measurements by a NASA satellite, the ozone hole grew to 27.3 million square kilometers on September 19, larger than the North American Continent. The biggest previous ozone hole reached 26 million km² in 1996.

The satellite also showed that ozone concentrations in the worst section of the hole bottomed out at 90 Dobson units, only one-third of what should normally be there this time of year.

Balloon measurements over the South Pole recorded a value of 92 Dobson units, in good agreement with the satellite measurements, says David J. Hofmann of the National Oceanic and Atmospheric Administration in Boulder, Colo.

The cold temperatures this year helped the ozone hole reach new heights, according to Hofmann. "We saw some ozone loss all the way up to 24 km, which is higher than usual." Normally, temperatures are too warm at that altitude to allow the formation of frozen cloud particles.

Scientists trace some of the stratospheric cooling in recent years to the loss of ozone molecules, which absorb sunlight and heat up the surrounding air. But this process cannot explain the extremely low temperatures detected in August and September above Antarctica, before much sunlight had returned to the polar skies, says William J. Randel of the National Center for Atmospheric Research in Boulder.

One cause could be natural weather conditions in the lower atmos-

NASA

Dark-blue ring (arrow) marks the boundary of the ozone hole in this satellite image. Colors indicate different ozone concentrations.

Medical Nobel prize says yes to NO

In the 1860s, Alfred Nobel invented dynamite, an explosive using the volatile compound nitroglycerin. Yet, when a doctor prescribed nitroglycerin to treat his heart disease, Nobel refused the treatment, believing it unlikely to work. Ironically, research that has helped explain how nitroglycerin can ease chest pains has now earned one of the famous prizes established by Nobel's riches.

Robert F. Furchgott of the State University of New York Health Science Center in Brooklyn, Ferid Murad of the University of Texas Medical School in Houston, and Louis J. Ignarro of the University of California School of Medicine in Los Angeles share the 1998 Nobel Prize in Physiology or Medicine for "discoveries concerning nitric oxide as a signalling molecule in the cardiovascular system," the Karolinska Institute in Stockholm announced this week.

In 1977, Murad found that nitroglycerin and related drugs induce nitric oxide (NO) formation and that this gas relaxes the muscle cells that narrow and dilate blood vessels. Around 1980, Furchgott found that endothelial cells, which line the interior of blood vessels, produce a signaling molecule that causes nearby muscle cells to relax, thus regulating blood pressure.

Scientists didn't initially connect the two discoveries. Like other so-called free radicals, the highly reactive nitric oxide, a common air pollutant, was largely thought a destructive molecule not normally employed by the body. But as Furchgott and Ignarro, working independently, analyzed the nature of the endothelium's relaxing factor, the gas became a prime candidate. At a conference in 1986, each laid out a compelling case that nitric oxide was indeed the factor.

"This was bold, to say the least, because there was no reason to suspect that there was machinery for making nitric oxide" in the human body, notes Jack R. Lancaster Jr. of the Louisiana State University Medical Center in New Orleans, an editor of the journal NITRIC OXIDE.

Research into nitric oxide has blossomed since. Scientists have found the enzymes with which cells make the short-lived gas and have shown that it plays a role in activities as diverse as memory formation, tumor suppression, and immunity. Some brain cells communicate using the gas, and immune cells let loose bursts of nitric oxide to kill infectious organisms or cancer cells. By increasing blood flow, nitric oxide even plays a role in penile erections; the celebrated drug Viagra amplifies the actions of the gas. Physicians are also exploiting the effects of nitric oxide in several other ways. They administer it to premature infants to stimulate blood flow to underdeveloped lungs, for example.

"The NO field has outgrown its original discovery," says Jonathan S. Stamler of the Duke University Medical Center in Durham, N.C., who studies nitric oxide's interactions with hemoglobin (SN: 3/23/96, p. 180). "The impact has been truly remarkable. The ramifications apply to every organ system and cellular response. You name it . . . and nitric oxide has been implicated."

Considering that too much nitric oxide can damage cells, "it's astounding nature chose it to perform so many functions," adds Lancaster. —*J. Travis* phere, which can sometimes send pressure disturbances rippling up into the stratosphere. These so-called planetary waves warm the polar stratosphere and slow ozone destruction. In recent years, however, few planetary waves have buffeted the Arctic and Antarctica during the critical season of springtime ozone loss.

Earlier this year, a computer model suggested that greenhouse warming would reduce the number of planetary waves hitting the Arctic and Antarctica (SN: 4/11/98, p. 228). Other computer models have come to the opposite conclusion, raising questions about the validity of this prediction, says Newman.

The recent cooling, he says, "increases our worry about this potential problem. But we certainly can't say that this is evidence for it." — *R. Monastersky*

Ozone killer confounds expectations   3

SCIENCE NEWS, VOL. 155   APRIL 3, 1999

Atmospheric concentrations of an ozone-destroying compound have risen at a surprising rate over the past decade, according to Australian scientists. These findings suggest that the protective ozone layer in Earth's stratosphere will take longer to recover than previously expected.

Measurements made in Tasmania show an increase in the concentration of halon-1211, a chemical used in fire extinguishers. Researchers had expected emissions of this compound to peak in 1988 and eventually decline, as industrialized nations phased it out in accordance with the Montreal Protocol of 1987. Since 1988, however, emissions have increased by 25 percent, estimates the Commonwealth Scientific and Industrial Research Organisation (CSIRO) in Melbourne.

Developing nations, particularly China, are most likely causing the rise in concentrations, says Paul Fraser of CSIRO. The Montreal Protocol allows developing nations to produce halon-1211 until the year 2010. According to Fraser, the additional emissions will delay the anticipated recovery of the ozone layer by up to 10 years.

Stephen A. Montzka of the National Oceanic and Atmospheric Administration in Boulder, Colo., and his colleagues have also detected increases in halon-1211 concentrations, although they calculate that emissions have remained constant.   *—R.M.*

1998: Warmest year of past millennium   4

SCIENCE NEWS, VOL. 155   MARCH 20, 1999

The current millennium is ending not with a bang but a boil. A study of climate reaching back to the time of William the Conqueror indicates that recent temperatures are the highest of the past 1,000 years.

The new analysis, published in the March 15 GEOPHYSICAL RESEARCH LETTERS, combines information recorded in tree rings and glacial ice. Michael E. Mann of the University of Massachusetts at Amherst and his colleagues calibrated these natural thermometers against modern instruments, which enabled them to make a Northern Hemisphere temperature reconstruction of unprecedented length.

In a previous analysis of the past 6 centuries, the team had found no decade warmer than the 1990s (SN: 5/9/98, p. 303). Still, some scientists

Black line charts temperatures relative to millennial average. Yellow denotes range of potential error in the measurements.

had wondered about earlier times. In the first 3 centuries of the millennium, a time known as the medieval warm period, Europe was notably balmy. This span may now require a different name. While the North Atlantic basked in warmth, the high temperatures did not spread across the Northern Hemisphere, according to the new report.

Although there are large uncertainties in the temperature estimates for the earliest part of the record, the current warmth is unprecedented, says Mann. "The numbers for 1998 are so unusual. They push it up to the ballpark of 99 percent certainty that 1998 is the warmest year in the reconstruction," says Mann. It's also very likely that the 1990s are the warmest decade of this millennium, he adds.

Such strong statements, however, get a cool reception from tree-ring researcher Gordon Jacoby of Lamont-Doherty Earth Observatory in Palisades, N.Y. "There's so much uncertainty in the chronology," says Jacoby. He calls the study "a good first effort, but there's a lot more work to be done."   *—R.M.*

Earth's temperature shot skyward in 1998   5

SCIENCE NEWS, VOL. 155   JANUARY 2, 1999

Global temperatures in 1998 shattered the record high mark, making last year the warmest since at least 1860, and possibly since the end of the last millennium. El Niño deserves part of the credit, say climate scientists, but some researchers also see signs that people are helping to push temperatures into uncharted territory.

The World Meteorological Organization (WMO) in Geneva announced last month that the mean surface temperature of the globe in 1998 reached 0.58°C above a baseline average for the period from 1961 through

1990. For climatologists, who worry about global changes in hundredths of a degree, last year's warmth stands out like a Himalayan peak.

"It's quite large. It represents several days' lengthening of the growing season," says David Parker of the United Kingdom Meteorological Office's Hadley Centre in Bracknell. The Hadley Centre and the University of East Anglia in Norwich supplied much of the analysis in the WMO statement, which includes data through mid-December. The British groups combine surface-temperature measurements made at more than 1,000 land meteorological stations with readings of sea-surface temperature from almost 2,000 ships and buoys.

A separate analysis, completed by NASA's Goddard Institute for Space Studies in New York, also has 1998 setting a temperature record by a wide margin, says James E. Hansen of the institute.

Climate researchers trace part of the heat to the El Niño ocean warming, which first started developing in mid-1997 in the Pacific. El Niño faded in May 1998, turning the tropical eastern Pacific cool, but temperatures remained elevated in many other ocean regions. In particular, the year brought "unprecedented warmth" to the Indian Ocean, according to WMO.

All the continents baked in 1998 except for northern parts of Europe and Asia. The southern United States faced extreme heat and drought during spring and summer. In central Russia, a June hot spell killed more than 100 people and fed large fires, WMO reports.

With El Niño now only a memory in the Pacific, Parker expects the globe to cool off in 1999, although probably not back to the 1961 to 1990 baseline.

Red hues denote warm regions in 1998.

The globe has warmed markedly during the past decade, so much so that 7 of the 10 warmest years on record occurred after 1990. All of the top 10 postdate 1983.

The recent warmth amplifies concerns that greenhouse gases are turning up Earth's thermostat, according to some researchers. A United Nations consensus panel announced in 1995 that the balance of evidence suggests people are influencing climate. Now, says Parker, "the balance is tipping a bit further in that direction."

Greenhouse skeptics point out that the lower atmosphere up to 7 kilometers has not warmed over the last 20 years, since satellites started making measurements. These readings showed substantial warming in early 1998, but atmospheric temperatures later fell back to the 20-year average, says John Christy of the University of Alabama in Huntsville, who analyzes the satellite data.

Still, people are most concerned about Earth's surface, which has warmed by almost 0.7°C since the end of the 19th century, according to WMO. The hot spell of the past two decades may be unprecedented in the last 1,200 years, according to Jonathan T. Overpeck of the National Oceanic and Atmospheric Administration in Boulder, Colo., who discussed historical climate data last month at a meeting of the American Geophysical Union in San Francisco.

When Overpeck compiled work by scientists who have examined tree rings, glaciers, and sediments from lakes and oceans, he found no evidence for the existence of a global warm spell during the Medieval period—a time that climatologists once regarded as universally balmy. While Europe and Greenland were warm during this phase, South America, Antarctica, and Australia were not. Overall, he says, that time was not as warm as today.   —R. Monastersky

Recent years are warmest since 1400

6

SCIENCE NEWS, VOL. 153   MAY 9, 1998

Temperatures in the Northern Hemisphere have climbed higher in the last decade than at any other time in the past 6 centuries, according to a comprehensive study that pooled many different climate records. The results add weight to reports linking the recent warming to greenhouse gas pollution.

Starting in the 1980s, scientists who analyze thermometer records discovered that the world has warmed since the late 1800s. The Intergovernmental Panel on Climate Change concluded in 1995 that people have caused part of this climate shift, but questions remain about how much of the warming stems from long-term natural variations. To answer this question, researchers need centuries' worth of climate informa-tion—much more than most thermometer records can offer.

Accordingly, Michael E. Mann of the University of Massachusetts in Amherst and his colleagues examined studies of surrogate measures of temperature. These include tree rings, ice cores, coral, and historical weather records. By combining such proxy measurements with a few long-term thermometer records, the researchers were able to compile a 600-year-long temperature record for the Northern Hemisphere, which they describe in the April 23 NATURE.

In their analysis, which carries a 99.7 percent level of certainty, temperatures in 1990, 1995, and 1997 exceeded all others since 1400. They concluded that volcanic eruptions and solar variations strongly influenced temperatures before 1900. Since then, warming from greenhouse gas pollution has taken the dominant role.

Other climate researchers caution that significant uncertainties plague any study of temperatures reaching so far back. "Even with all the uncertainties," says Keith R. Briffa of the University of East Anglia in Norwich, England, "the best interpretation of the data, until proven differently, is that we are seeing an unprecedented change, over quite a few hundred years, occurring in the 20th century."

Briffa and East Anglia's Phil Jones have conducted separate studies of records since 1400. Both show unusual warming in the 20th century.

—R.M.

Continents growing wetter as globe warms

SCIENCE NEWS, VOL. 152   NOVEMBER 29, 1997

A study of global precipitation records going back to 1900 shows that the United States and other land areas in the midlatitudes have grown wetter, whereas the tropics have generally dried out, reports Aiguo Dai of the National Center for Atmospheric Research in Boulder, Colo. Such changes offer a taste of the kinds of precipitation shifts predicted to accompany greenhouse warming, the researchers say.

Rain and snow patterns profoundly influence social well-being by controlling factors ranging from food production to the frequency of floods. Concern about the potential for changes in precipitation underlie the international climate treaty negotiations taking place next week in Kyoto, Japan.

"Precipitation is what we really need to address. A lot of natural disasters are related more to precipitation than to temperature," says Inez Y. Fung of the University of Victoria, British Columbia. Dai, Fung, and Anthony D. Del Genio of NASA's Goddard Institute for Space Studies in New York report their findings in the November Journal of Climate.

Dai and his colleagues examined monthly precipitation records from 5,328 meteorological stations for the period 1900 through 1988. Using a statistical technique that teases out independent patterns in the data, they minimized errors introduced when instruments were changed or stations moved.

"This is by far the most comprehensive compilation of precipitation data. It is a much larger and cleaner data set than has existed before," says Fung.

On a global average, precipitation onto land areas increased through the first half of the century but has decreased since the mid-1970s, resulting in a small net gain. Much of the recent drop comes from a prolonged dry period in the Sahel region just south of the Sahara Desert.

The biggest increases have occurred in mid- and high-latitude areas, such as North America, northern Eurasia, Australia, and Argentina. Over the same period, the Philippines, Malaysia, and the Mediterranean lands have dried out.

This pattern matches the general picture emerging from computer climate models, say the researchers. In simulations of greenhouse warming, increased air temperatures cause more water to evaporate from Earth's surface. In the tropics, much of this additional moisture remains in the atmosphere because warmer air can more readily hold extra water vapor. In the middle and high latitudes, the vapor-storing capacity of the atmosphere also rises but not enough to hold most of the additional moisture, which drops as precipitation.

Despite the broad agreement between models and observations, Henry F. Diaz of the National Oceanic and Atmospheric Administration in Boulder cautions against concluding that greenhouse gases have caused this century's precipitation changes.

In fact, he notes, early models called for greenhouse warming to increase the frequency of midcontinent droughts in North America, whereas U.S. data show an increase in precipitation.

—*R. Monastersky*

Heat and humidity getting worse

SCIENCE NEWS, VOL. 155   JANUARY 16, 1999

People who loathe the withering days of summer will get positively steamed at the results of a new meteorological study. Records of temperature and humidity across the United States indicate that sweltering days have grown more common since 1949, a trend that increases the risk of heat-related problems affecting the elderly and other vulnerable people.

U.S. meteorologists had previously documented a long-term rise in average summertime temperatures, particularly at night. In the new study, Dian J. Gaffen and Rebecca J. Ross, both of the National Oceanic and Atmospheric Administration in Silver Spring, Md., searched for changes in the apparent temperature—an index that combines heat and humidity measurements.

Analyzing data from 113 U.S. weather stations, the two researchers determined how often the apparent temperature reached extreme values during the years 1949 through 1995. To qualify as extreme, the temperature and humidity had to exceed a threshold value—defined by the top 15 percent of the worst days in all the Julys and Augusts during that period at each station. Medical statistics indicate that heat-related deaths start to increase when temperatures top these thresholds.

Over the period of study, U.S. stations showed a significant increase in the number of extreme days, with the most dramatic changes occurring at night. Across the stations, the average number of sultry nights went from 8 per year in the early 1950s to 16 per year in the 1990s. The trends were strongest in the East and West, Gaffen and Ross report in the Dec. 10, 1998 NATURE.

The researchers also found significant increases in the number of heat waves—defined as 3 or more days in a row when the apparent temperature exceeded the threshold.

Although the scientists did not seek the cause of these changes, Gaffen says that "what we're seeing is generally consistent with what we would expect from an enhanced greenhouse effect."

—*R.M.*

Nations Draft Kyoto Climate Treaty

SCIENCE NEWS, VOL. 152   DECEMBER 20 & 27, 1997

It was a hard-won agreement, but after 10 days that ended last week in a marathon dusk-to-dawn negotiating frenzy, representatives of 160 nations finally thrashed out compromise wording for a new treaty to protect the global environment.

At its heart, the Kyoto Protocol, named for the Japanese city in which it was completed, seeks to rein in human activities that emit greenhouse gases. Atmospheric concentrations of the targeted pollutants, some of which have been growing since the dawn of Western industrialization, have reached a level that worries scientists. Theory, computer models, and climate-monitoring data all suggest that these gases are at or near the point where they could provoke a global warming that would shift patterns of rainfall, raise sea levels, and increase weather variability.

The European Union had wanted to limit the new curbs to just carbon dioxide, methane, and nitrous oxide. The United States succeeded in adding limits on three more—hydrofluorocarbons, perfluorocarbons, and sulfur hexafluoride. Collectively, the industrial countries agreed to drop emissions of these pollutants to 5.2 percent below their 1990 levels for the first three gases and to 5.2 percent below a later baseline for the three fluorocarbons. The deadline for achieving these cuts is 2012.

Individually, the nations must achieve differential limits. The European Union, the United States, and Japan agreed to cutbacks of 8 percent, 7 percent, and 6 percent, respectively. The Russian Federation, Ukraine, and New Zealand must return to 1990 levels, while Australia and Iceland have to stabilize their releases at 8 and 10 percent, respectively, above 1990 levels.

The treaty divides industrial countries into two "bubbles." Within a particular bubble, nations that achieve more than their required greenhouse gas reductions may trade or sell pollution rights—not to exceed their excess—to another country (SN: 5/24/97, p. 320).

Emissions credits for preserving forests or planting new ones were another hotly contested U.S. provision that won adoption. As trees grow, they soak up and retain large quantities of carbon dioxide (SN: 12/24&31/88, p. 411). The protocol allows industrial nations to receive some emissions credit for any land forested since 1990, but they must increase their reductions to account for any land deforested since then.

How much credit forests will receive was not settled. It's one of the unresolved issues to be ironed out next November when treaty negotiators reconvene in Buenos Aires.

While the new treaty would impose binding emissions limits on 39 nations, 121 less economically developed nations escaped regulation during this round of talks. Many of these are already emitting large amounts of greenhouse gases, and those emissions are expected to grow rapidly.

The U.S. Senate has pledged not to adopt a treaty that does not require "meaningful participation" by developing countries. So "what we have here is not ratifiable by the Senate," says Sen. John Kerry (D-Mass.), though the United States may begin abiding by the treaty even before it is ratified. In any case, he calls the treaty a "very significant first step."

Gail McDonald of the Global Climate Coalition in Washington, D.C., which represents many fossil fuel and other industry groups, expresses concern that the

Forest preservation has become part of the emissions equation.

U.S. commitment may require something close to "cutting [U.S.] energy use by a third in 10 years." Big industry won't be the only one having to make those cuts, she says, "but you and me and our thermostats."

Moreover, she contends, by agreeing to those limits now, "the U.S. has lost a lot of its bargaining power" in compelling developing countries to curb their emissions or in working out details on issues such as emissions trading.
—*J. Raloff*

Electric diode tunes in to plastic

Many polymers are called plastics because of their mechanical flexibility, but these materials are chemically versatile as well. Now, a researcher has used a polymer to remake a basic electronic component—a diode. By adjusting the electrochemical properties of the polymer, he makes the diode show a wide range of current-carrying capacities.

The device "puts a new tool in the toolbox of electrical engineers," says Mark C. Lonergan of the University of Oregon in Eugene. He reports his finding in the Dec. 19 SCIENCE.

Conventional diodes join a semiconductor and a metal, creating a junction that allows current to flow in only one direction. The new diode uses indium phosphide as the semiconductor and replaces the metal with polypyrrole, a polymer that conducts electricity (SN: 8/23/97, p. 119). A gold mesh embedded in the polypyrrole acts as the electric contact.

Lonergan adjusts polypyrrole's electrochemical properties by essentially making it part of a battery. After attaching electrodes to the gold mesh, he immerses the device in a liquid source of ions. By applying a small control voltage, Lonergan can "tune" the diode by adding or removing electrons from the polymer. This process changes the diode's current-carrying capacity.

"It combines in a single device a collection of properties that you would normally have to get from a whole series of separate semiconductor-metal interfaces," Lonergan says.

The tunability of the diode makes it a good platform for exploring the basic "technical aspects of metal-semiconductor contacts," says J. Campbell Scott of the IBM Almaden Research Center in San Jose, Calif. Scott doubts that the polymer diode will ever replace the conventional ones used today, but he says the new device opens up possibilities for future applications.

Polymer diodes could be used as tunable sensors with a range of sensitivities, Scott suggests. They could also form the heart of multiple-state computer memories, says Lonergan. Instead of adopting just two states, as in standard binary computer logic, a tunable diode could exist in several different states.

For practical applications, the diode needs several technical improvements, such as an increased electric switching speed and replacement of the liquid ion source with a solid. Lonergan is also planning to conduct more tests on the stability of the device, since frequent electric switching quickly degrades conducting polymers.
—*C. Wu*

China's air pollution chokes crop growth

SCIENCE NEWS, VOL. 155 MARCH 27, 1999

With China's economy and population expanding at breakneck speed, its infamous air-pollution problems have started to claim new victims. Data collected at rural Chinese sites indicate that ozone pollution has already reached the point where it's harming agriculture—raising questions about whether the world's fastest developing country will be able to feed itself in the next century.

"There has been a lot of discussion of whether China can meet its future food demands, but there has been very little discussion of how regional environmental degradation is affecting agriculture there," says William L. Chameides of the Georgia Institute of Technology in Atlanta, who led the study.

When ozone, a highly reactive gas, builds up in the lowest layer of the atmosphere, it harms plants and animals. The pollutant forms when hydrocarbons and nitrogen oxides mingle in the air and get cooked by sunlight. Both these ingredients come from the incomplete combustion of fossil fuels. Hydrocarbons also are released by vegetation.

In a collaboration with Chinese researchers, Chameides and his U.S. team analyzed ozone measurements made at four rural sites in China during 1994 and 1995. These are the first systematic ozone readings taken outside cities in China, according to the researchers.

They found the highest pollutant concentrations at two places: an agricultural town called Linan, near the coastal city of Hangzhou, and a part of Hong Kong Island upwind of the city. At both spots, ozone often exceeded 60 parts per billion, an approximate threshold for harmful effects. Experiments in the United States and elsewhere have shown that the range of ozone exposure at these two Chinese sites is enough to reduce crop yields by 10 percent or more, the scientists report in the April 1 GEOPHYSICAL RESEARCH LETTERS.

The scientists were surprised to find the highest ozone readings in fall, winter, and spring—the reverse of the U.S. pattern. In China, they say, summer storms are strong enough to clear out polluted air from the coastal sites where the measurements were made.

The highest ozone readings in the study occurred in Linan, which sits in the Yangtze River Delta, one of China's most important agricultural regions. The timing of the air-pollution problems threatens the delta's winter wheat crop, which supplies 20 percent of China's total wheat harvest, say the researchers.

Chameides' team used a computer model to see how far the ozone problem might extend beyond the four measured spots. In the simulation, harmful concentrations of ozone blanketed much of eastern China.

The Chinese pollution measured is "what you'd see at a rural U.S. site in summer, maybe a little bit dirtier," says Jennifer A. Logan, an atmospheric chemist at Harvard University.

Agricultural studies in the United States suggest that ozone pollution reduces U.S. crop yields by $5 billion to $10 billion annually, says Ellis B. Cowling of North Carolina State University in Raleigh. —R. Monastersky

How Inhaled Dust Harms the Lungs

SCIENCE NEWS, VOL. 153 JANUARY 31, 1998

Seven years ago, an Environmental Protection Agency statistician stunned researchers studying the effects of air pollution on health when he reported analyses indicating that as many as 60,000 U.S. residents die each year from breathing federally allowed concentrations of airborne dust (SN: 4/6/91, p. 212). This and subsequent studies figured prominently in EPA's decision last year to ratchet down the permitted concentration of breathable particles in urban air—and in human airways (SN: 7/5/97, p. 6).

At the time, many industrialists argued that they shouldn't have to pay for better pollution control because science had yet to suggest a plausible biological mechanism by which breathing low concentrations of urban dust might sicken or kill people.

Now, scientists at the University of Texas Houston Health Science Center describe how they uncovered what they think may be one of the basic elements of that toxicity.

On the alert for foreign debris, a community of white blood cells known as alveolar macrophages patrols small airways of the lung. When these cells encounter suspicious material, they

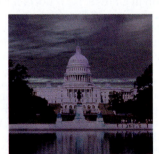

Studies of dust from Washington, D.C., and St. Louis suggest how particles in air damage the lungs.

identify it and send out a chemical clarion call to rally the immune system cells best suited to disabling and disposing of such matter.

The trick is to recruit only as many troops as are needed. If they call in too many, the lung can sustain inflammatory damage from friendly fire. Alongside the small troop of macrophages that stimulates defense measures, a larger squadron of macrophages halts immune activity when it threatens the host.

Andrij Holian and his coworkers in Houston have found that people with healthy lungs normally have 10 times as many suppressor macrophages as stimulatory ones. In people with asthma and other chronic lung diseases—who face an increased risk of respiratory disease from inhaling urban dust—that ratio may be only 3 to 1. The reason for the difference is not known.

In a report to published in the March ENVIRONMENTAL HEALTH PERSPECTIVES, Holian's team describes test-tube studies of human alveolar macrophages. The macrophages showed no response to ash collected from the Mount St. Helen's eruption. However, when exposed to airborne dust from St. Louis and Washington, D.C., most of the suppressor macrophages underwent apoptosis, or cellular suicide, while the stimulatory ones survived unaffected. Ash from burned residual oil, a viscous boiler fuel, proved even more potent at triggering suppressor cell suicides.

If this test-tube system models what's actually happening in the human lung, Holian told SCIENCE NEWS, the different responses of the two classes of lung macrophages could result in an overly aggressive immune response to normal triggering events. Indeed, he says, it would be the first step in a cascade that can end in inflammatory lung injury. "We may one day be able to target this upstream event and prevent that injury."

"This is, I think, an important contribution to the overall story," says Daniel L. Costa of EPA's pulmonary toxicology branch in Research Triangle Park, N.C.

Studies by EPA suggest that certain metals—especially iron, vanadium, nickel, and copper— in smoke from combustion of fossil fuels trigger particularly aggressive inflammatory responses by lung cells. Costa says these metals play a "preeminent" role in the toxicity of airborne particulates. When EPA researchers removed the metals, they also removed the toxicity, he says. Moreover, he notes, these metals tend to reside on the smallest water-soluble particles in urban air—the fraction targeted for more aggressive controls under the new rules.

John Vandenberg, assistant director of EPA's National Health and Environmental Effects Research Laboratory in Research Triangle Park, says Holian's results are "a nice complement to our studies." —J. Raloff

Uric acid linked to multiple sclerosis

Mice paralyzed by a disease resembling multiple sclerosis can walk again after receiving daily injections of uric acid, a compound that occurs naturally in the body.

Treatment lessened symptoms and lengthened survival, report researchers from Thomas Jefferson University in Philadelphia. Sixty percent of untreated animals died within 4 weeks of the disease's onset, but almost 90 percent of the mice treated with the highest doses of uric acid were alive 8 weeks later.

"Before the treatment kicked in, the mice had total hind limb paralysis," says immunologist D. Craig Hooper, a coauthor of the study published in the Jan. 20 PROCEEDINGS OF THE NATIONAL ACADEMY OF SCIENCES. "When we stopped the treatment, they had a little limp."

The mice were given experimental allergic encephalomyelitis (EAE), a mouse disease that causes chronic bouts of paralysis much like multiple sclerosis in humans (SN: 12/6/97, p. 356). Hooper and his colleagues had reported earlier that they could prevent most EAE symptoms by giving mice uric acid before inducing the disease.

Now, the researchers report that uric acid treatment can reverse, not just prevent, the progression of EAE.

The team also found lower amounts of uric acid in 46 people with multiple sclerosis than in 46 people with other neurological diseases. In a review of 20 million Medicare and Medicaid patient records, the team discovered almost no overlap between multiple sclerosis and gout, a disease caused by excess uric acid. The combined evidence suggests a relationship between low uric acid concentrations and multiple sclerosis in people, Hooper says.

In both multiple sclerosis and EAE, the immune system seems to attack the myelin sheath that insulates the nerves of the brain and spinal cord. Uric acid inactivates peroxynitrite, a compound generated by the immune system to combat invading viruses and bacteria. Thus a deficiency in uric acid might lead to attacks on normal tissue.

However, the idea that low uric acid concentrations cause multiple sclerosis is controversial, says neuroimmunologist Henry F. McFarland of the National Institute of Neurological Disorders and Stroke in Bethesda, Md. He adds that accurately inferring disease causes from patient records is hard, particularly for a difficult-to-diagnose disease like multiple sclerosis.

Stephen Reingold, a neurophysiologist at the National Multiple Sclerosis Society in New York, cautions that "the vast majority of things that work in EAE prove not very promising in multiple sclerosis." Nevertheless, he adds, "this is an intriguing, provocative finding. It needs to be followed up." —M. Jensen

Spying on El Niño

The struggle to predict the Pacific prankster

By RICHARD MONASTERSKY

SCIENCE NEWS, VOL. 152 OCTOBER 25, 1997

In Indonesia and Malaysia, abnormally dry skies have spurred on epic wildfires, blanketing the region with a smoky haze. In California, homeowners are nailing down new roofs as emergency officials prepare for potentially torrential winter rains with accompanying floods and landslides. In the tropical Atlantic Ocean, waters have remained remarkably calm this summer, spawning far fewer hurricanes than normal.

Meanwhile, in the middle of the Pacific Ocean, the source of all this unusual weather grows stronger by the week.

The El Niño—the rogue warming of Pacific waters—has come roaring onto the scene this year and has already started turning conditions across the world topsy-turvy. First detected in April, it has gathered steam far faster than anticipated and has now spread to an area 1.5 times the size of the contiguous United States.

The vast pool of warm water, up to 3.5°C above normal, is poised to surpass the greatest El Niño of the century, which wreaked havoc in 1982 and 1983. Although the damage from that event is impossible to quantify exactly, climate experts often quote a toll of $8 billion worldwide exacted by floods, storms, fires, and droughts.

Seventeen years ago, climate scientists had no reliable tools to provide advance warning of the El Niño. In fact, there were so few observational buoys in the Pacific that researchers had trouble recognizing the phenomenon even after it had started. That glaring gap in knowledge triggered a decade-long research effort called the Tropical Oceans and Global Atmosphere program, aimed at learning how to predict the Pacific's moods and their effects on weather outside the tropics.

Now, the world is reaping the fruits of that work. As early as the end of last year, sophisticated computer models started hinting that the Pacific would brew up an El Niño in 1997. Then, in the first 3 months of this year, satellite sensors and a network of buoys strung across the Pacific began to pick up signs of the warming. Forecasters with the National Oceanic and Atmospheric Administration (NOAA) issued an advisory in April, alerting the world to the possibility of a major El Niño.

These apparent successes, however, belie some major problems plaguing efforts to forecast El Niño. While several models correctly predicted that the Pacific would warm this year, all missed the magnitude and timing of the temperature increase. Some of them disagree on whether the El Niño will persist through next winter, a critical factor in determining how U.S. weather will fare. Perhaps most worrisome to researchers, the model with the longest track record failed miserably this year.

"This is the really scary part of the story because we don't know why that model failed," says Mojib Latif of the Max Planck Institute for Meteorology in Hamburg, Germany.

The earliest prediction schemes for El Niño, devised in the late 1970s and early 1980s, worked somewhat like the mind of a novice stock investor: They operated on the often flawed assumption that past performance guarantees future returns.

The early models did not seek to simulate the actual forces of nature that drive Pacific warmings. Instead, they were statistical models. They would take the temperatures and winds in the Pacific and then search past records to find comparable situations.

In the mid-1980s, climate scientists started trying to capture the physics of the El Niño process inside computer models. They devised cartoon versions of the ocean to predict how heat shuffles

El Niño warmings (red) have hit more frequently than La Niña coolings (purple) since 1976. Climate scientists wonder whether the change is part of a natural cycle or linked to greenhouse warming.

around the Pacific basin. These oceans-in-a-box have gradually gotten more complicated and are now coupled to computer models of the atmosphere. Simulated winds in the atmosphere stir up virtual ocean currents, which in turn alter the pressure patterns in the atmosphere that then feed back on the ocean, and so on.

The coupling of air and water is important because scientists have realized that El Niño ocean warmings are only one-half of a climatic marriage. The other partner, called the Southern Oscillation, is a see-saw variation in Pacific atmospheric pressure.

Under typical conditions, a high pressure system sits in the central Pacific near Tahiti and low pressure hangs over Indonesia, driving winds westward along the equator. These winds push surface waters to the west and pull up deep, nutrient-rich water that keeps the eastern Pacific cool and thriving with life. In the western Pacific, the same winds pile up warm water into a giant pool centered around Indonesia.

According to current theory, an El Niño develops when broad so-called

Kelvin waves sweep toward the Americas and shut down the upwelling of cold water in the east. The sea surface warms in this region. The winds blowing westward weaken. The warmth normally contained near Indonesia spreads toward the center of the ocean, carrying with it the rain storms that would traditionally hit Australia and nations of the western Pacific, says Ants Leetmaa, who heads NOAA's Climate Prediction Center (CPC) in Camp Springs, Md.

As Kelvin waves and other types of waves slosh back and forth across the Pacific, they can trigger a chilly phenomenon called La Niña. In many ways the opposite of El Niño, La Niña can also disrupt weather: It was implicated in the devastating 1988 drought in the central United States and the forest fires in the Southwest.

The El Niño–Southern Oscillation (ENSO), as the ocean-atmosphere pattern is called, constantly cycles from warm to normal to a cold La Niña and back, although the variation is highly erratic. The job of forecasters is to predict when the Pacific see-saw will shift and how far it will tip toward the extremes. On average, El Niño comes once every 4 years. Sometimes, though, the ocean can go 7 years without a warm phase. At other times, the Pacific can pull out of an El Niño, skip a cold phase, and fall right back into another warm episode the following year, as happened twice in the 1990s.

Currently, a dozen forecasting teams around the world are employing various types of models to keep tabs on the many moods of the Pacific. For most of them, the 1997 El Niño could be regarded as a coming of age.

The CPC ocean-atmosphere model, the most complex model currently in routine operation, fared particularly well this year, says Leetmaa. It started portending a 1997 El Niño as early as November 1996.

A much simpler model, built by researchers from the Scripps Institution of Oceanography in La Jolla, Calif., and the Max Planck Institute, also started picking up signs early on, says Latif. Unlike the CPC model, Latif's has a purely statistical atmosphere that relies on past weather patterns rather than simulating the underlying physics of jet streams and pressure patterns in the air.

Most surprising to the community of ENSO researchers, the longest-running and one of the most successful models went belly-up this year. Operated at Lamont-Doherty Earth Observatory in Palisades, N.Y., this computer program

The Atlantic's climate beast

In recent years, researchers have recognized that El Niño is only one of the goblins lurking in the climate system. For residents of Europe and the East Coast of the United States, a pattern called the North Atlantic Oscillation plays a bigger role in determining winter weather.

Under typical conditions in the North Atlantic, a high-pressure system sits over the Azores and a low-pressure one over Iceland. When that Icelandic pressure rises and the Azores' drops, Arctic air invades the northeastern United States. Storms that are normally bound for England head instead toward Spain. Northern Europe gets

extremely cold, as it has been in the last two winters.

Although meteorologists have long known of the North Atlantic Oscillation, oceanographers are only now finding links between the air pressure and currents in the ocean, giving them the hope of one day being able to forecast the North Atlantic's moods. At present, however, climate scientists lack a firm understanding of what drives the weather pattern, making it impossible to predict how it will behave this winter, says Gerald D. Bell of NOAA's Climate Prediction Center.

—R.M.

is a cartoon version of the ocean, but its simplicity and speed have brought good results in the last decade, says Lamont-Doherty's Mark Cane. The model accurately called for warmings in 1986 and 1992, and it picked up the 1988 La Niña.

Cane and others are struggling to make sense of why his model forecast a cold Pacific this year. The current whopper of a warming should have advertised itself more than most. "It's not obvious to me what went wrong," says Cane.

At a detailed level, even the best models face problems that threaten their accuracy, especially for predictions more than 6 months ahead. No model called for such an intense 1997 El Niño or one that developed so early in the year.

Looking into the future, there is uncertainty about how the El Niño will behave over the winter. "Models disagree on when it will peak and when it will go away. The impact that it will have on weather depends critically on having this event continue in winter and into the spring. Some models have it peak in the fall, in which case its impact will be less," says Leetmaa.

Part of the problem is that the models are still under development and are notoriously poor at predicting major transitions, such as when El Niño will die off, says Gerald D. Bell, a CPC forecaster. "Obviously, there's room for improvement. These models are in their infancy compared to where they will be in 10 to 15 years."

For Bell and his colleagues, climate forecasting is still an art that requires evaluating many different sources of information, only one of which is model predictions. The meteorologists rely heavily on actual measurements of the winds, pressure patterns, storm positions, and water temperatures in the Pacific. These facts, combined with a knowledge of past events, help the forecasters decide when to trust the models and when to regard them with suspicion.

For now, forecasters will not go out on a limb with only a model for support. "You can't put out an El Niño advisory based on just a model forecast because they're just not that reliable. You wait until there are at least El Niño–type conditions developing out there in the ocean to put the advisory out," says Bell.

For instance, when the CPC model began calling for a 1997 El Niño last November, forecasters had some doubts. In 1996, the model ran warmer than the actual Pacific Ocean, which stayed slightly cooler than normal. Bell and his coworkers began to trust the model only after they saw temperatures in the equatorial ocean warming markedly in the first quarter of 1997.

Even with substantial improvements, though, models will never make perfect climate forecasts. A heavy dose of chaos infects the streams of air and water coursing around the planet, explains Latif.

"There was an enthusiasm that went through the community a few years ago when we initially realized that we can predict El Niño. People overdid it a bit and were too optimistic. I think the claims that were made that ENSO may be predictable a few years in advance are probably not correct," he says. He estimates that unstable fluctuations of the ocean and atmosphere limit the predictability of ENSO to about 1 year.

For meteorologists, forecasting El Niño is only half the battle. Once they predict how the ocean will behave a half year hence, they must look into how something in the middle of the Pacific will affect the weather thousands of kilometers away, where more people live.

El Niño has such a long reach because it coaxes thunderstorms from the vicinity of Indonesia out into the middle of the Pacific. Like giant boulders in a river, these towering clouds

Expected effects of a typical El Niño.

redirect the atmospheric jet streams, shifting the normal track of storms over North and South America.

In the tropics, El Niño has an obvious influence. It tends to bring droughts to Indonesia, Australia, and northeast Brazil, as it inundates coastal Peru and Ecuador.

Meteorologists are now gaining insight into how the equatorial waters of the Pacific skew weather outside the tropics. Robert E. Livezey of the CPC has combed through 102 years' worth of U.S. temperature and precipitation records to document long-distance links between Pacific warmth and the weather in regions of each U.S. state.

The knowledge of these links, coupled with the record El Niño, has boosted CPC meteorologists' confidence in their ability to predict U.S. weather for this winter. "These factors have led us to make stronger forecast statements than we've ever done before," says Livezey.

The certainty shows up in CPC's map predicting U.S. precipitation for January through March 1998. The center has had to add a new hue to its maps to reflect the extremely high odds of dry conditions in the Ohio River Valley and of enhanced precipitation in parts of Texas and South Florida. The prediction also calls for a wet Southwest and Gulf Coast. The northern states will bask in unusual warmth, say the forecasts.

In the past, the weird weather spawned by strong El Niños has taken a heavy toll on people around the globe. Aside from direct effects, such as desiccated crops and flooded cities, the extreme conditions have also indirectly fostered the spread of diseases. Researchers have linked past occurrences of El Niño to outbreaks of cholera in Peru, malaria in Colombia, and hantavirus in the U.S. Southwest.

This year, however, the timely forecasts have provided advance warning of many of the upcoming disruptions. Last month, Antonio D. Moura of the newly created International Research Institute for Climate Prediction at Lamont-

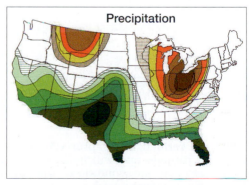

U.S. forecasts for January through March 1998. The temperature map shows warmer-than-average conditions in red shades. In the precipitation map, orange and brown depict below-average precipitation, whereas green and blue denote wet conditions.

Doherty spent 2 hours briefing the president of Brazil on how El Niño and Atlantic Ocean temperatures could spur a drought in the northeast part of that nation. At the same time, U.S. meteorologists met with representatives of southern African nations in Zimbabwe to discuss the potential problems associated with El Niño. This month and next, researchers will carry their message to Peru and Uruguay.

More than in the past, politicians and emergency planners seem to be listening. "I can see that in many regions, governments are starting to worry about this," says Moura. "They are beginning to believe the forecasts and explore what is their best strategy for decreasing the negative impacts."

La Niña readies to steal El Niño's thunder 13

SCIENCE NEWS, VOL. 154 JULY 4, 1998

Water temperatures in parts of the equatorial Pacific Ocean have dropped precipitously in the last 2 months, chilling El Niño's fever and setting the stage for the arrival of another climatic hooligan: La Niña.

Meteorologists were divided earlier this year on whether an episode of La Niña cooling would follow El Niño's demise, but the recent Pacific shift has brought consensus. "All of the forecasts are consistently indicating that we will have a La Niña shortly and it will continue over the next winter," says Vernon E. Kousky of the National Oceanic and Atmospheric Administration (NOAA) in Camp Springs, Md.

La Niña and its sibling, El Niño, are opposite extremes of a Pacific pendulum that causes the equatorial waters to swing from warm to cold and back again. During El Niño, warmth normally centered over Indonesia spreads eastward across the entire Pacific, dragging with it towering thunderheads that pump heat into the atmosphere. The storms redirect jet streams and turn typical weather patterns upside down around much of the globe.

During La Niña, the equatorial Pacific waters turn cold except around Indonesia, drawing thunderstorms back to that part of the ocean basin. The Pacific jet stream weakens, and wintertime weather in the United States turns much more mercurial. "The jet stream tends to undulate more. During our winter, the jet stream may be way north in Alaska and then come diving down south in certain periods. This will be much more variable in the sense that we can go from one month to the next and have a dramatically different pattern," says Kousky.

Measurements of water temperature, taken by buoys and satellites, show that a full-fledged La Niña has not yet developed. Between early May and early June, the equatorial temperatures fell markedly from 3°C above normal to 3°C below normal around 130°W, but the cooling stayed confined and has not spread westward. Moreover, warm remnants of El Niño linger to the north and south of the equator, helping to suppress rainfall needed to quench the fires raging in Mexico.

Meteorologists will be watching in coming months to see whether water temperatures drop further to the west, near the International Date Line. That's a key spot affecting how the weather evolves downstream in North America, says Arthur J. Miller of the Scripps Institution of Oceanography in La Jolla, Calif.

Cool water (shown in blue) has emerged in parts of the equatorial Pacific. Meteorologists predict further cooling leading to the development of La Niña.

The recent El Niño tamed the winter weather in the United States, bringing mild temperatures to much of the country. Researchers say it is hard to forecast what kind of weather will arrive this winter as the Pacific enters its cold phase. "La Niña conditions are much more difficult to predict. It's not as simple as the reverse of El Niño," says Miller.

La Niña tends to bring warmer-than-normal winter weather to the southeastern United States while the Northwest gets colder than normal, according to NOAA. The Pacific Northwest generally gets wet weather in late fall and early winter during La Niña, whereas the Southwest often dries out. "There is a strong potential for drought in the Southwest," says Kousky

—*R. Monastersky*

Oklahoma Tornado Sets Wind Record 14

SCIENCE NEWS, VOL. 155 MAY 15, 1999

Tornado-chasing scientists in Oklahoma last week measured wind speeds of 318 miles per hour, the highest ever documented on Earth. These radar observations, which still must be verified, also captured the birth of a tornado and will help researchers unravel the conditions that spawn such killer storms.

"It's really fascinating watching this tornado come together. That was the first time we were positioned so well with the radar to get these data," says Joshua Wurman, an atmospheric scientist at the University of Oklahoma in

Doppler radar picture of one of the Oklahoma tornadoes on May 3.

Norman. Wurman and his team used two Doppler radars mounted on separate trucks to study the tornadoes that ravaged the Oklahoma City area on May 3.

The Doppler radars, cousins of police radar guns, gauge wind speed by shooting pulses of microwaves that reflect off rain, dust, and other objects in the air (SN: 6/22/96, p. 388). The new wind speed measurement by one of Wurman's radars shatters the previous record of 286 mph, taken during a 1991 tornado by atmospheric scientist Howard B. Bluestein of the University of Oklahoma. He had used a more primitive Doppler radar.

"If this measurement holds up, it will be a remarkable thing," says Joseph H. Golden, a research meteorologist with the National Oceanic and Atmospheric Administration in Silver Spring, Md. He cautions, though, that "everybody has to be careful until people have a chance to sift through these data."

The Doppler radars measure winds more than 65 feet above the ground, making it difficult to compare these data with the Fujita scale, used to assess tornado damage to houses and other objects on the ground, says Wurman. The strongest documented tornadoes, designated F-5, carry frame houses considerable distances, send automobiles flying, and strip the bark off trees.

The late Tetsuya Theodore Fujita, who designed the scale, estimated that F-5 tornadoes have wind speeds of 261 to 318 mph just above the surface. Meteorologists, however, consider these numbers only rough guides. "The F scale has never been well cali-

Depression, sadness yield brain link

Major depression includes not only feelings of intense sadness and despair but also a wandering, indecisive train of thought. Distinct brain areas involved in emotion and attention together foster both depression and ordinary bouts of sadness, according to a new study.

Specific changes in these neural regions accompany recovery from major depression, whether achieved with an antidepressant drug or placebo pills, says a team of neuroscientists led by Helen S. Mayberg of the Rotman Research Institute in Toronto.

"The negative influence of depressed mood on attention is probably due to functional connections between these two brain regions," Mayberg says. "Successful treatment, including placebo use, alters those connections."

Mayberg's group first took positron emission tomography (PET) scans of eight women resting and after recalling a sad personal experience. The PET scans measured blood-flow changes in their brains, an indirect sign of boosts or drops in brain-cell activity. None of the women or their family members had been diagnosed with mood disorders.

In a second trial, the researchers took PET scans of eight men before and after successful treatment for major depression. Over 6 weeks, four men had improved after taking the antidepressant drug fluoxetine (Prozac); the rest had rallied in response to pills that they thought might be antidepressants but that contained no active ingredients.

When the women recalled sad experiences, blood-flow surged in a pair of the inner brain structures that regulate emotional responses and declined in two parts of the brain's outer layer previously linked to attention, the team reports in the May AMERICAN JOURNAL OF PSYCHIATRY. The men who had recovered from depression showed unusually high activity in the two attention areas and low activity in the emotion areas.

Before recovery, however, the depressed men had exhibited marked overactivity in only one of the two emotion-related areas characteristic of brief sadness in the women. Further research will examine more closely the processes that occur during depression, Mayberg says.

The findings suggest that a brain circuit incorporating emotion and attention "offers a plausible converging point" for antidepressant effects of drugs and psychotherapy, remark psychiatrist Charles B. Nemeroff of Emory University School of Medicine in Atlanta and his coworkers in an accompanying editorial.

Researchers haven't yet explored brain function in the substantial minority of depressed people who don't benefit from treatment, Mayberg adds. —B. Bower

brated, particularly at the F-4 and F-5 damage categories," says Golden.

Last week's record winds will help scientists probe the speed limits of the atmosphere, where friction and other forces put the brakes on flowing air, says Wurman. "There have always been questions about how high wind speeds could get."

Before recording the top wind speed, Wurman's team captured the birth of this violent tornado, which eventually passed through Oklahoma City. The researchers observed the storm for 8 minutes before the tornado's birth and then documented the first 6 minutes of its life.

Tornadoes develop when winds high above the ground flow at a different speed than winds near the surface, causing the air to roll along a horizontal axis, like the wheels of a car. If a thunderstorm is brewing at the same time, updrafts and downdrafts can tilt the rolling winds so that they spin around a vertical axis, setting a broad region of the storm slowly rotating. Meteorologists call such a storm a mesocyclone.

"What's not well understood is how the mesocyclone then spawns a tornado," says Wurman.

The data collected last week are expected to shed some light on that process and others, as different teams analyze their gigabytes of measurements. Bluestein and his coworkers studied some of the same Oklahoma storms using a Doppler radar with much higher resolution, designed by Andrew L. Pazmany of the University of Massachusetts at Amherst. "These are the highest-resolution images anyone's ever gotten inside a tornado," says Bluestein.

—R. Monastersky

Diversity in tropical forest edges

SCIENCE NEWS, VOL. 152   AUGUST 9, 1997

Interesting things happen at the edges, whether of continental plates or cultural periods. In ecology, too, the edges of habitats—called ecotones—are happening places. If they're large enough, they can be cradles of biological diversity, researchers report in the June 20 SCIENCE.

Biologist Thomas B. Smith of San Francisco State University and his colleagues studied a dozen populations of a small bird, the little greenbul (*Andropadus virens*), in Cameroon. Half of the populations lived in the dense tropical rain forest, while the other half came from the extensive and patchy edge habitat that separates the forest from the savanna. Although all of the birds were genetically similar and could interbreed, those from the edge habitat differed physically from the forest birds. They weighed more and had longer wings, for example.

The researchers maintain that these differences—as significant as the differences between some species—are good evidence of natural selection linked to the habitats the bird populations occupy. Longer wings might offer a faster means of escape in the edge habitat, where predators may be a greater problem. These habitats, though relatively neglected by conservationists, may therefore "be integral to the production and maintenance of biodiversity in tropical rain forests," the researchers say.   —C.M.

. . . and tropical forest soils

Tropical forests are famous for their biological diversity, much of it undescribed, but the real mother lode of new tropical species is underfoot, not overhead, according to an analysis of soil microbes from the eastern Amazon Basin in Brazil.

Rather than growing and then counting species and numbers of bacteria from a soil sample in the lab, agronomists James Borneman and Eric W. Triplett of the University of Wisconsin-Madison extracted DNA from samples of forested and deforested soil and analyzed 100 of the sequences they obtained.

"The microbial diversity [from this snapshot] is immense," the team reports in the July Applied and Environmental Microbiology. None of the sequences were duplicates, and none had been reported previously. Most of the sequences were related to those of bacteria such as Clostridium and Planctomyces, but 18 could not be assigned to any known bacterial group. Most of the novel sequences, presumably representing unknown microbes, came from the forested soils. Two sequences belong to microorganisms from the ancient and little-known domain of archaea (SN: 8/24/96, p. 116).
                      —C.M.

Amazon forests caught in fiery feedback

SCIENCE NEWS, VOL. 154   OCTOBER 3, 1998

One little fire inching through a tropical forest may not kill much. Yet it triggers a vicious cycle—fires preparing the way for bigger fires—that could ultimately turn Amazon jungles into savannas, according to new research.

During a typical 16-day dry spell, only some 5 percent of an intact rain forest dries out enough to catch fire, says Mark A. Cochrane of the Woods Hole (Mass.) Research Center. But even a small fire can sufficiently tatter the shade canopy —and leave behind enough extra debris for fuel—to render some 50 percent of that forest vulnerable to a second, more destructive blaze during a subsequent dry spell. As fires recur, virtually all the forest becomes susceptible, report Cochrane and Mark D. Schulze of Pennsylvania State University in State College.

Brazil's Tailândia region in Pará has already slipped into this fiery feedback loop, observe Cochrane and Schulze in the October CONSERVATION BIOLOGY.

A decade ago, Tailândia was the new Amazonian frontier. Settlers moved in, and accidental fires became common. Now, forests there that have previously caught fire reburn about every 3 years, too quickly to allow regeneration. Historically, the time between forest fires was at least 400 years.

"Fire is burning everything and everyone," Cochrane says. On a data-gathering trip last December, he found that fire had destroyed even the Brazilian forest service's sustainable management plot.

The first fire that attacks an intact Amazon forest looks "unimpressive," admits Cochrane. Most of the time, the flames spread as a thin ribbon barely ankle-high, creeping perhaps 100 meters a day. These fires take the night off, winking out around 5 p.m. and reigniting from smoldering sparks when the next day heats up around 10 a.m. They kill thin-skinned young trees but typically leave 90 percent of the forest's biomass alive.

A year after such a fire has worked through a forest, however, the tree canopy provides only 60 percent shade instead of its former 85 to 95 percent, the researchers report. Trees no longer create as much moist cover as they used to, and the next fire starts more easily, this time burning some 40 percent of the biomass. Unlike the first fire, the second one kills big trees as often as little ones.

Longtime tropical fire watcher Christopher Uhl, also from Penn State, comments in the same journal issue that "fire adds a whole new dimension to tropical disturbance ecology." Long gone are the days when researchers observed that Amazon jungles didn't burn. Uhl once sheltered part of the forest floor from rain for more than a month but couldn't get a blaze going. Now Uhl sees fire as a huge force for change in rain forests. "Even for those

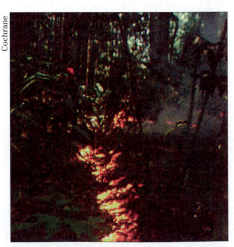

Even a sleepy, ankle-high blaze can trigger a vicious cycle of fires in Amazon forests.

Probing the heart of extragalactic jets

Astronomers have come a little closer to finding the origin of the twin jets that shoot out from the brightest objects in the cosmos. Some 10 percent of quasars and active galaxies emit such jets, which extend in opposite directions up to thousands of light-years.

By studying the radio emissions from the jets squirted out by the distant quasar 3C279, researchers have gathered the first compelling evidence that the jets consist mainly of electrons and their antiparticles, positrons. This composition hints at the processes taking place deep within the hearts of active galaxies, where the jets are produced.

Any bright galactic core not powered by starlight is known as an active galactic nucleus. The most luminous of these are called quasars, and they are thought to be fueled by massive black holes. The new observations point to interactions between subatomic particles just outside a black hole as the source of the jets.

John F.C. Wardle and his colleagues at Brandeis University in Waltham, Mass., used the Very Long Baseline Array, a network of 10 radio telescopes, to examine the radio waves emitted by the jets in 3C279. They measured the polarization, the direction in which the electric field of the radio waves oscillates.

In the Oct. 1 NATURE, Wardle's team reports finding circularly polarized radiation from the jets, meaning that the angle at which the electric field oscillates is not fixed but rotates around the wave's direction of travel. Earlier observations indicated that most of the radio emission comes from high-speed electrons accelerated by a magnetic field. Scientists had also determined that there must be roughly equal numbers of positive and negative charges in the jets, but they didn't know whether the positive charges were protons or positrons.

The pattern of polarization found by Wardle's team indicates that the electrons' positively charged companions are mostly positrons. The team has found similar results in jets from three other active galactic nuclei.

There are two ways to make beams of positrons and electrons, says Thomas W. Jones of the University of Minnesota in Minneapolis. In one scenario, material falling toward the black hole generates high-energy photons, which collide to produce pairs of electrons and positrons. These particles in turn radiate more photons, which create more electron-positron pairs. One drawback to this story line, says Jones, is that if the radiation generated by infalling material is too intense, it may act as a brake on the electrons and positrons, preventing them from attaining the high speeds observed in the jets.

A more likely possibility, he notes, is that energetic protons, too massive to be slowed by radiation, collide with either each other or photons. These reactions generate subatomic particles called pions, which can decay into electrons and positrons. The reaction also generates a small number of protons, consistent with the observation that most but not all the positively charged particles are positrons.

—R. Cowen

---

species that survive, these grand fires might be among the largest biological selection events in modern history."

The feedback effect does not surprise Norman L. Christensen, dean of the Nicholas School of the Environment at Duke University in Durham, N.C. "The pattern is similar to what we see in some of the coastal forests in the Pacific Northwest," where ecosystems have not evolved to cope with frequent blazes.

In other temperate forests, however, small fires have the opposite effect and reduce the chance of future blazes, notes James K. Agee of the University of Washington in Seattle. Little fires lap up dead leaves and branches, preventing fuel from building up. Adaptations like thick-er bark protect trees. Forest managers now set these so-called prescribed burns as preemptive housekeeping blazes.

"My take-home message is that we wouldn't want to take those temperate [forest] ideas and try to apply them too strongly to the tropics," Agee says. "Ecology is really a science of place."

—S. Milius

Hidden Threats Take Toll in Amazon

SCIENCE NEWS, VOL. 155 APRIL 10, 1999

Studies of deforestation in the Brazilian Amazon have overlooked more than half of the actual rainforest damage occurring each year, reports a team of U.S. and Brazilian researchers.

Past assessments of the Amazon have used satellite images to tally deforested areas, where farmers and ranchers have clear-cut and burned all trees. Such work has not addressed logging, which removes only selected trees, or surface fires that burn down individual trees but do not denude the forest.

The new analysis, published in the April 8 NATURE, indicates that logging and these fires degrade a larger area of forest than does strict deforesta-tion—especially in dry years.

"This will really change quite fundamentally the way we think about deforestation," says study leader Daniel C. Nepstad of the Woods Hole (Mass.) Research Center and the Amazon Institute for Environmental Research in Belém, Brazil. "It shows that the effects of people on forests can be much greater than we thought."

Other researchers question the exact numbers in the study but agree with the conclusion. "Our surveys do not consider impoverishment of the forest by logging," says Diógenes S. Alves of Brazil's National Institute for Space Research in São José dos Campos. "This is a serious issue."

Satellite surveys have difficulty picking out effects of logging and fires because they remain obvious only for a year, then new growth obscures the damage, says Nepstad. To estimate the impact of logging, he and his colleagues interviewed 1,393 managers of sawmills and obtained records of wood harvests for 1996 and 1997.

The team tested the accuracy of the sawmill data by measuring wood harvesting at 22 square-kilometer patches of forest. "What the loggers were telling us was right on the money," says Nepstad.

To estimate the effects of recent fires, the researchers interviewed 202 landholders whose properties total

Standard studies of satellite data (left) distinguish only between totally deforested (yellow) areas and primary forest (black). A new study (right) shows that much of what had been considered primary forest has been partially logged (green) or burned (red).

9,200 km². They checked this information against satellite data, which showed that the owners correctly gauged the number of fires but under-estimated the area of burning by 43 percent.

Nepstad and his coworkers calculate that logging damaged 10,000 to 15,000 km² of undisturbed forest each year in 1996 and 1997. This represents 50 to 90 percent as much land as that totally deforested during 1996. The fire estimates are less precise but suggest that fires harmed an area roughly equal to the area of deforestation.

Logging and fire injuries can have far-reaching effects. By taking down trees, logging and fires open up the canopy, dry the forest floor, and increase the risk of fire. On a global scale, such forest degradation releases large amounts of carbon dioxide that are not currently included in global computations.

The combined effect of logging and fires is "a potentially large and significant phenomenon," says David Skole, a geographer at Michigan State University in East Lansing who uses satellite data to track deforestation. The new study, however, "doesn't provide us with direct measurements of the magnitude of this effect," he adds.

"It's very important to identify the size of the problem," says Eduardo Martins, president of the Brazilian Environment Institute in Brasília. Logging is often the first step toward total deforestation, he says.

—R. Monastersky

Global Graveyard

New images of Earth's interior reveal the fate of old ocean floor

By RICHARD MONASTERSKY

SCIENCE NEWS, VOL. 152   JULY 19, 1997

In their effort to unravel the secrets of Earth's interior, geoscientists have been stumbling about, a bit like Mr. Magoo, the notoriously near-sighted cartoon character. Researchers have spent decades trying different tricks for peering into the planet, but they have been unable to see clearly enough to resolve some of the most basic issues about this hidden realm.

One particularly vexing question is what happens to oceanic rock after a few hundred million years. The planet is 4.5 billion years old and new ocean floor forms continuously, but almost all of the ocean crust over 200 million years old has vanished, sinking into Earth's interior through a process called subduction.

Just where this rock ends up when it disappears from view has important implications, both for basic knowledge about the planet and for improved insight into the driving force behind earthquakes and volcanoes. "This is absolutely fundamental for our understanding of how the Earth works. It just changes your whole idea of Earth history," says Guy Masters, a seismologist at the Scripps Institution of Oceanography in La Jolla, Calif.

Masters first became interested in subduction while in graduate school in 1975. At the time, scientists were trying

Two visions of Earth: Similar pictures of the mantle beneath North America emerge from independent tomographic studies of secondary seismic waves (top) and primary seismic waves (bottom). These vertical slices through the planet show what appears to be cold ocean rock (blue) plunging toward the bottom of the lower mantle.

to decide whether the subducting ocean rock remained trapped near Earth's surface or sank all the way to the core. That same debate continues today. "It was a major controversy then, and it's a bit depressing that 22 years later, we're still arguing the same issue," he says.

His spirits are getting a lift, however, from two new studies poised to end the dispute. After spending years sifting through thousands of earthquake records, independent teams of seismologists have produced some of the most detailed images yet of the deep mantle, the great rocky layer between Earth's metallic core and its outer shell, called the lithosphere.

Their analyses have convinced many geoscientists that subducting ocean crust sinks extremely deep into the planet, thereby helping to mix the entire mantle, as if stirring a pot of soup. This process of mixing, or convection, provides the power for pushing Earth's surface plates around and bears ultimate responsibility for earthquakes and volcanic eruptions.

The mantle debate has persisted for so long because different lines of evidence have pushed scientists toward opposite conclusions. Previous studies of earthquake waves convinced many seismologists that the mantle mixes from top to bottom as one big layer.

Geochemists, however, came to believe in a two-layered mantle. They drew their conclusions from volcanic rocks, which have risen from Earth's upper mantle and provide a glimpse into what goes on inside the planet. The chemistry of these rocks suggests that the mantle segregates itself into almost completely separate upper and lower regions. The presumed dividing line lies at a depth of 660 kilometers, where descending seismic waves speed up, indicating a change in the character of the rock at that level.

According to geochemists, ocean floor sinks only to the bottom of the upper mantle rather than dropping all the way to the base of the lower mantle, about 2,900 km below the surface.

In the split-level scheme of geochemists, the lower mantle remains essentially cut off from Earth's surface. This isolation has kept the deep rock in a pristine state, so it retains the primordial elements that Earth had in its infancy. In contrast, the upper mantle is like chewing gum that has lost its taste. The upper mantle is missing many of its primordial elements because rock from this region melted to form the crust early in its history, concentrating many of these elements at the surface.

Seismologists have tried to solve the mantle puzzle by using earthquake waves to produce images of the deep planet. Their technique is similar to computerized axial tomography, or CAT scanning, which uses X rays to make three-dimensional images of a person's internal anatomy.

In the past, pictures from seismic tomography were too fuzzy to reveal relatively thin features like subducting layers of ocean floor. The studies now emerging "give a much clearer image of Earth's interior than was previously possible," says Rob D. van der Hilst, a seismologist at the Massachusetts Institute of Technology. "On a few basic issues, I think we now have a definite answer."

In the April 10 NATURE, van der Hilst and his colleagues present seismic images showing slabs of subducting ocean floor diving down to the very bottom of the mantle—implying substantial mixing between the upper and lower portions of the mantle. Similar pictures appear in a separate study by Stephen P. Grand of the University of Texas in Austin, who discusses his research in the April GSA TODAY, a publication of the Geological Society of America.

"These are completely independent lines of research and completely different data sets, and they produce very similar images. The convergence of results is so convincing to everybody. By and large, we seem to reach a consensus that slabs do penetrate into the lower mantle," says van der Hilst, who collaborated with Sri Widiyantoro from the Australian National University in Canberra and E.

Geochemist Claude Allègre proposes that the mantle remained segregated into upper and lower sections (top figure) for most of Earth's history. In the last 500 million years (lower figure), the two-layer system has started breaking down, allowing mixing between the upper and lower mantles.

Robert Engdahl of the U.S. Geological Survey in Denver.

To make their images, van der Hilst and his colleagues spent 5 years poring over records of 100,000 earthquakes from the last 3 decades. The scientists studied primary waves, or P-waves, which pulse through the planet vibrating molecules forward and backward, like sound speeding through air. These waves arrive at seismic stations before secondary waves, or S-waves, which shake molecules from side to side.

The researchers examined how long it took P-waves from individual earthquakes to reach stations around the world. Because they know, in general, how quickly P-waves travel through normal mantle rock, the scientists could detect unusual features in the mantle by looking for waves that took too long to arrive or arrived too quickly. Computers processed millions of these data points to create a three-dimensional image showing precisely where in the mantle the waves sped up or slowed down.

With this technique, van der Hilst's group found two unusually fast corridors of rock resembling broad ramps slicing into the planet. One is diving beneath North and South America, and the other is sinking under the southern margin of Asia and Europe. These fast regions, the scientists conclude, are slabs of ancient ocean floor subducting quite deep in the mantle, to at least about 1,300 km below the surface and in some cases much deeper. The ocean rock is colder than the surrounding mantle rock, so seismic waves speed up as they pass through the descending slabs.

Grand took another approach, analyzing S-waves, which can echo off interior boundaries such as the junction between the core and the mantle or the mantle and the crust. He looked at only 250 earthquakes but analyzed each one in much more detail. Despite the different tactics, Grand found the same two long slabs descending beneath the Americas and beneath Europe and Asia.

The correspondence is striking, he says, because there has been little agreement in the past between tomographic studies relying on S-waves and those using P-waves.

Like fossils of extinct life, the cold slabs sitting deep in the planet are reminders of a period in Earth's history long past.

The feature beneath Europe and Asia, propose Grand and van der Hilst, is the floor of the Tethys, a sea that separated India and Africa from lands to the north. As continental movement closed the Tethys, the ocean floor disappeared into the mantle, forming the broad ramp visible in the seismic images. The Mediterranean, Black, Caspian, and Aral Seas are tiny remnants of the much larger Tethys.

The slab beneath the Americas represents a large section of the Pacific Ocean floor, called the Farallon plate, which once bordered western North and South America. Over the last 100 million years, much of the Farallon plate has slid eastward under the continents.

Strangely, neither of the new reports shows deep slabs where geophysicists most expected them: in the northwest Pacific. This hotbed of geologic activity is home to large earthquakes and major volcanic eruptions, all fueled by subducting ocean floor. From previous studies, geophysicists knew that the ocean slabs are sinking into the upper mantle beneath Japan, eastern Siberia, and the Aleutian Islands. The new tomographic studies, if accurate, show that these slabs do not venture from the upper mantle into the lower mantle.

The reason for this paradox may be that geoscientists are looking at the wrong moment in Earth's history. In the past, speculates van der Hilst, slabs in the western Pacific did sink all the way into the lower mantle. This process stopped temporarily about 40 million years ago, when movement of the Pacific seafloor plates changed direction, altering how the slabs descend and making it harder for them to break through the 660-km-deep barrier between upper and lower mantle, he says.

In support of this scenario, van der Hilst points to regions of cold rock sitting below Asia at a depth of 1,800 km. These lumps, he suggests, are remnants of the ancient Pacific floor that subducted into the lower mantle before the plate

rearrangement occurred. Eventually, the oceanic plates of the northwest Pacific will resume their trip into the lower mantle, he suggests.

Geophysicists like van der Hilst and Grand acknowledge that they must refine their studies further to produce sharper tomographic images of the mantle. Even so, the new studies have won some converts among geochemists who formerly dismissed the fuzzy seismic pictures.

One such geochemist, Claude J. Allègre of the Institute for Physics of the Globe in Paris, has an idea that reconciles the volcanic rock studies with the new seismic images. Allègre explains that the two types of evidence need not agree, because they record different time periods. The rock geochemistry gives a picture of the entire history of mantle evolution over the last 4.5 billion years, whereas seismic images provide a snapshot of today and the last few hundred million years.

Linking the two lines of evidence,

Allègre proposes that the pattern of mantle convection has changed during Earth's history. For the first 4 billion years, the mantle convected in two separate layers, thereby keeping the deep mantle relatively pristine. Over that time, though, Earth's interior cooled enough to alter the convection process. About 500 million years ago, the two-layer system started to break down, and it is now evolving toward a type of convection that mixes the mantle as one layer, he hypothesizes in the July 24 EARTH AND PLANETARY SCIENCE LETTERS.

Computer simulations support this concept, says Allègre, because they show that a two-layer convection system will switch over to a one-layer system as it cools.

"There is no conflict between the geochemistry and the geophysics," says Allègre, offering an olive branch.

It may take years, though, for the two opposing disciplines to break down old barriers and, like the mantle, blend into one.

Eruption on Montserrat gaining strength

SCIENCE NEWS, VOL. 152   AUGUST 16, 1997

A 5-day-long series of explosive eruptions assaulted the tiny Caribbean island of Montserrat last week, piling more misery on an already embattled population. The recent outbursts marked the strongest activity to date in a volcanic crisis that has dragged on for 2 years.

"The recent increase in explosions obviously raises the level of concern over where this eruption is going to go in the future," says Willy Aspinall, a scientist with the British Geological Survey who worked at the Montserrat Volcano Observatory this summer.

The Soufrière Hills Volcano in the southern region of Montserrat, a British territory, came to life with several small eruptions in July 1995 and has since grown gradually more violent. Authorities on the island eventually evacuated the capital, Plymouth, and other cities in the south. On June 25, 1997, the volcano claimed its first victims, at least 10 people who had ventured into the evacuated zone and were engulfed by pyroclastic flows—superheated gas and ash that can exceed speeds of 100 kilometers per hour.

Margaret Mangan, a volcanologist

with the U.S. Geological Survey in Hawaii, was working at the Montserrat Volcano Observatory at the time. "I think that people just had no idea of the type of fury that can come down the mountain. For many folks, June 25 was really an awakening to the dangers. In the past, my sense was that it was considered a nuisance, a pain in the neck, but not necessarily a personal threat. After the 25th, it became a lot more real and a lot more frightening for people."

Scientists are monitoring the volcano from a command center set up a few kilometers northwest of the volcano. Seismometers provide some warning of impending eruptions by picking up the tremors produced as molten rock rises through the volcano. Recent eruptions have destroyed the tiltmeters that had been gauging how much the mountain swells and deforms. The observatory is staffed by a rotating cast of researchers from the University of the West Indies in Trinidad and from the British Geological Survey. Several scientists from the USGS have also served at the observatory throughout the last 2 years.

The mountain's unrest increased

this summer, with periods of explosive eruptions in early July and again in early August that destroyed Plymouth. The series of blasts last week came like clockwork almost every 10 to 12 hours, occasionally showering rocks and ash onto the northern end of the island where Montserrat's remaining residents have sought refuge. Before the eruptions, the island had a population of more than 11,000 people. Scientists estimate that 7,000 of them have left.

A central mountain provides some protection for the northern end of Montserrat, but the events of recent weeks are causing authorities to reassess residents' safety, says Richard P. Hoblitt of the USGS in Vancouver, Wash., who has served at the Montserrat observatory.

"The shift to more explosive behavior has raised the concern that this thing could have a much larger event that could endanger greater parts of the island," he says. "People are getting increasingly pushed farther and farther into the north into less commodious circumstances. It's a small island to be living with an active volcano."

—*R. Monastersky*

Inside Old Faithful
Scientists look down the throat of a geyser

By SID PERKINS

SCIENCE NEWS, VOL. 152   OCTOBER 11, 1997

When the world's most famous geyser opened up and said "Aaaah" for geologists, there were surprises in store.

Under the supervision of National Park Service personnel, researchers lowered several instruments, including a video camera, into Yellowstone's Old Faithful to observe conditions there. Their findings, published in the October GEOLOGY, provide the first detailed look inside a geyser that is unmatched in its combination of size, frequency, and regularity of eruptions.

James A. Westphal, a geologist at the California Institute of Technology in Pasadena, says many of the team's discoveries confirmed previous suspicions, but there were also a few unexpected findings—for example, the eruption cycle behaves more simply than the scientists had surmised.

In April and October 1984, during several eruption cycles, the researchers measured temperature and pressure every 5 seconds at eight different depths along the uppermost 21.7 meters of the geyser, its only accessible portion. Susan W. Kieffer, a coauthor of the report, says these first data were so complicated that they did not match any established theory of geyser behavior.

Says Westphal, "That's when she [Kieffer] said to me, 'You know, I'd sure like to see what it looks like in there.'"

In 1992, when miniature video cameras became widely available, the scientists returned to Yellowstone. They insulated a 2-inch video camera and lowered it into Old Faithful to map the vent and see what happens there between eruptions. "The vent had a much more complicated shape than we expect-

Scientists used images from a video camera to generate the first detailed map of the upper 14.3 m of Old Faithful's vent.

ed," Westphal says.

Kieffer had assumed that the vent would be a smooth vertical tube, but the camera showed that it is actually a crack running in an east-west direction that extends downward at least 14.3 m. In some portions of this fissure, the walls lie beyond the 1.8-m field of view of the camera, but other regions are extremely narrow. In one spot, at a depth of about 6.8 m, the walls close to within 15 centimeters of each other—a feature critical to explaining the geyser's behavior.

The camera also revealed that the wall of the vent is riddled with cracks and that water enters continuously at several different depths. Cool water from the local water table flows into the fissure 5.5 m and 7.5 m underground, and superheated water and steam shoot into the vent 14.3 m down. Spikes in temperature readings at the beginning of an eruption—some of them reaching nearly 130°C—suggest that water and steam also surge into the fissure from geothermal sources at depths below 21.7 m, says Westphal.

Old Faithful blows its top every 79 minutes, on average, with intervals between eruptions ranging from 45 to 105 minutes, depending on the amount of boiling water left in the fissure when the geyser runs out of steam. "There's no real pattern, except that a short eruption is always followed by a long one," Westphal says. "The geyser pretty much keeps its own schedule."

Temperature and pressure measurements show that for the first 20 or 30 seconds of each 3- to 5-minute eruption, steam and boiling water rocket through the narrowest part of the fissure at the speed of sound. During that time, the 15-cm-wide slot behaves just like the throat of a wind tunnel, limiting the geyser's discharge rate. When the pressure driving the eruption falls below a critical value, the outflow slows and the

Westphal

A look downward at the narrowest portion of the vent, where the walls come within 15 cm of each other. The north wall of the fissure is at the top of photo.

geyser's plume, which can reach a height of 55 m, begins to shrink.

"I was assuming a simple geometry for the fissure and expected a complicated physical explanation for the geyser's behavior," Kieffer says. "The video camera showed us it was just the reverse."

The researchers observed the water as it boiled, churned, and surged through the fissure. The behavior of the bubbles in the water confirmed their suspicion that carbon dioxide is present in the geyser's water, although the power behind the eruption comes from steam.

Moreover, Westphal was surprised to find that even though groundwater constantly replenishes the pool of liquid in the fissure, the water will sometimes recede to the unobservable lower levels of the fissure for minutes at a time. What's happening in the deeper recesses of Old Faithful remains a mystery, he says.

The forces of nature that give life to the geyser also contain the agents of its ultimate demise. The superheated water that constantly recharges the fissure is saturated with minerals, which precipitate onto the walls of the vent as silica. In a geological equivalent of hardening of the arteries, Old Faithful will eventually become clogged and die, a fate Westphal says has already befallen several nearby geysers.

Mouth of Geyser

North →

Ductus

SCIENCE NEWS, VOL. 155   MAY 8, 1999

Seattle enjoys a buzzing reputation for its coffee culture, but the city faces the threat of a caffeine-free jolt in the future. Seismologists this week report that the thick foundation of sediments underlying Seattle is less rigid than previously thought and will amplify shaking during earthquakes.

Much of the new information about the city's seismic risk comes from a large project launched in March 1998 to explore the hidden geologic structure of the region surrounding Seattle and Vancouver, British Columbia. The $3 million effort sent ships with airguns throughout Puget Sound and connecting waterways, where they fired 33,000 volleys of sound through the water and into Earth's crust.

Seismologists recorded the vibrations that bounced off buried structures or bent back upwards by refraction. Recordings of these waves provided the first clear picture of the Seattle Basin—a 10-kilometer-deep oblong bowl filled with sediments. As the vibrations passed through the basin beneath Seattle, they traveled much more slowly than expected, says project leader Michael Fisher of the U.S. Geological Survey in Menlo Park, Calif.

"We're finding that the velocity of sound in these sedimentary rocks is very low, which means, probably, that the sediment is going to shake severely during a major earthquake. . . . So it's probably heightened the threat to the city of Seattle," says Fisher. He discussed the findings at a meeting of the Seismological Society of America in Seattle.

The low velocities indicate that the sediments beneath Seattle are softer than expected. At a depth of 5 km, sound waves rippled at a speed of 4 km per second, roughly four-fifths of the typical value seen in sedimentary basins in California, says Fisher.

When seismic waves from an earthquake travel upward and enter the basin, the sediments intensify the shaking. "Think of it like you have a plate with a chunk of Jell-O on it and you shake it. The thing wobbles all over the place, and you wouldn't want a house on that," he says.

One possible source for an earthquake in this region is the Seattle fault, a major fracture in Earth's crust that cuts in an east-west direction across the southern part of the city. Geologists discovered it only 7 years ago, when they found evidence of a strong earthquake about 1,100 years ago that ripped the southern edge of the city, where the Kingdome arena now stands.

At the meeting, researchers showed that the Seattle fault is not dead. Craig Weaver of the USGS in Seattle discussed a magnitude 4.9 quake that struck west of Puget Sound in June 1997. Weaver and his colleagues linked the quake to a recently identified strand of the Seattle fault, indicating that the fault is more active than thought.

"It's getting to be compelling evidence that this is a major actor here in Puget Sound," he says. At present, though, researchers can't say how often destructive earthquakes emanate from the fault.

When the fault does unleash a strong shock, the shape of the basin could give Seattle a double shot of trouble, according to new studies. In a computer simulation of a magnitude 6.5 earthquake deep down on the Seattle fault, two different sets of seismic waves ganged up on the city, reports Arben Pitarka of URS Greiner Woodward Clyde, a geological consulting firm in Pasadena, Calif.

As seismic waves radiated up and out from the fault, some bent when they entered the edge of the basin and then traveled along the surface toward downtown Seattle. Other waves took a more direct route. When these two sets collided, they intensified the shaking in some regions—an effect seen in Kobe, Japan, in 1995. Pitarka cautions, however, that this simulation relies on assumptions about Seattle's underground geology, which scientists are only starting to map out.   —R. Monastersky

Waves of Death

22

Why the New Guinea tsunami carries bad news for North America

By RICHARD MONASTERSKY

SCIENCE NEWS, VOL. 154   OCTOBER 3, 1998

When a devastating tsunami hits anywhere in the world, Costas Synolakis gets on the phone and starts arranging flights to the afflicted region. A hydrodynamic engineer, Synolakis has specialized in studying these giant sea waves, and nature has provided plenty for him to study in recent years. Between 1992 and 1997, Pacific earthquakes spawned nine tsunamis that claimed more than 1,800 lives. Yet these events didn't prepare him for the scale of the disaster in Papua New Guinea, where three mountainous waves pounded the northern coastline on July 17 and carried away at least 2,500 people.

"We were in a state of shock," says Synolakis, a researcher at the University of Southern California in Los Angeles and co-leader of a science team that visited Papua New Guinea in early August. "It was really something we had not seen before. It was sort of a new threshold in terms of what a wave can do."

Debris hanging from the tops of palm trees indicated that the waves reached heights of 14 meters, taller than a four-story building and more than twice the estimate reported by news agencies immediately after the catastrophe. In surveys of past disasters, Synolakis and his team had never before witnessed evidence of so much water flowing over a shoreline. "This is about double the worst overland flow that we had seen before," he says.

All this from an otherwise ordinary earthquake. The jolt that preceded the tsunami measured 7.1 on the moment magnitude scale, meaning it was a strong but fairly common quake. Tremors of at least this size strike somewhere on the globe about every 3 weeks. The United States has suffered three in the past 7 years.

During their survey, Synolakis and his colleagues grew convinced of something they had suspected even before reaching Papua New Guinea. "Once we were there, we realized the earthquake itself could not have generated such a large wave." Their favored hypothesis now is that the earthquake triggered an underwater landslide that in turn generated the giant waves—a possibility that raises disturbing questions about the tsunami hazard elsewhere.

The coastline that stretches from

The tsunami hit hardest in the villages of Arop and Warupu, and it claimed lives in the Malol villages.

California to the Pacific Northwest and southern Alaska has many similarities to the shores of Papua New Guinea, say tsunami researchers. The west coast of North America is laced with small coastal faults capable of producing moderate quakes and underwater slides in the steep-walled canyons offshore. "These [small faults] haven't been taken seriously as tsunami-generating sources before. We have to take a much closer look now," says Synolakis.

Of all natural disasters, tsunamis may rank as the most poorly understood by the general public. Media reports often describe them as tidal waves, although they bear no relation to tides. They are often portrayed as simply scaled-up versions of a breaking wave, similar to the famous painting by the Japanese artist Hokusai that appears on almost all tsunami web sites. But the truth lies far from that image.

Survivors of the Papua New Guinea disaster describe the tsunami as a wall of water barreling toward shore, according to the researchers. Unlike a normal wave with its prominent crest, the tsunami was more like a plateau of water, averaging 10 m high and extending 4 to 5 kilometers from front to back, says Synolakis. The largest wave swept over the shore at speeds of up to 20 km per hour for more than a minute, before draining away in preparation for the next. Imagine a mountain of water, appearing out of nowhere and plowing across the beach as if surging from a collapsed dam.

In their preliminary report to the National Science Foundation, which funded their postdisaster survey, Synolakis and his team say that the first of the three large waves arrived 5 to 10 minutes after the initial earthquake, with the others following several minutes later. The tsunami struck an extremely limited region extending over only 20 km of coastline, unlike other recent tsunamis that spread devastation over hundreds of kilometers.

The waves happened to hit a particularly vulnerable section of the north coast, where a narrow strip of land separates the ocean from a brackish body of water called Sissano Lagoon. The lagoon itself formed during a tsunami in 1907, providing a warning that went unheeded.

With Sissano Lagoon blocking the route inland, the families of fishermen living on the sand bar had no way to escape the waves. The water washed away all evidence of the houses on this 300-meter-wide coastal strip, except for some bent 2-by-4 stilts that people had used to raise their homes off the sand.

Evidence collected along the coast confirms the initial seismological reports of the quake's moderate size, says team member Emile A. Okal, a seismologist with Northwestern University in Evanston, Ill. People who lived only 15 km from the most devastated area told the investigators that the shaking was not particularly intense in their village. "There were no houses destroyed by the earthquake. There were very few objects thrown down," says Okal.

Seismologists have noted that the earthquake was of slightly longer duration than typical shocks (SN: 8/1/98, p. 69), but this factor would not account for the large size of the tsunami, says Okal. Instead, investigators have had to search for other factors.

During its survey of the coastline, the team discovered a cliff with evidence of a fresh avalanche, which local residents attributed to the quake. Okal and his colleagues speculate that a similar, but much larger, slide happened underwater. The seafloor off the north coast plunges steeply into a submarine trench, and the sediments piled on this slope may easily slump downhill when shaken.

That would fit with the distribution of damage along the shore. "It is known that slumps can create extremely devastating tsunamis and extremely localized ones," says Okal.

There may have been a second factor adding to the instability of the slope, suggest the researchers. Sediments along the margin of continents often contain a frozen substance called methane hydrate, which can form slick surfaces buried beneath the seafloor. "Nobody knows what happens if gas hydrates get shaken by earthquakes," says Okal.

In other locations—off the coast of Norway and the southeastern United States—oceanographers have found scars of slumps that originated in regions replete with methane hydrate deposits (SN: 11/9/96, p. 298). If similar deposits exist along the coast of Papua New Guinea, they could have helped the sediments slip downslope after the initial earthquake, the researchers suggest.

At present, Okal calls the case for slumping circumstantial. To test the hypothesis, the scientists would need an oceanographic ship to scan the seafloor with sonar, looking for direct evidence of large, fresh slides. U.S. officials say that they are trying to obtain funds for such a survey, but so far none have surfaced.

The seafloor off Papua New Guinea reminds some researchers of other areas notorious in the tsunami field. In 1992, a strong and extremely slow earthquake off the coast of Nicaragua sparked a massive tsunami that killed 170 people. A seafloor survey conducted afterward showed scars from slides in a region known to be rich with methane hydrate, says Eddie Bernard, who is director of the Pacific Marine Environmental Laboratory in Seattle and coordinates the National Tsunami Hazard Mitigation Program.

Although scientists have known that subsea landslides can cause trouble, this issue has received scant attention in tsunami research. The Papua New Guinea disaster may provide a stark lesson on the dangers of slumping. "I think what this has done to the U.S. program is given it more emphasis to examine the slump-generated tsunami problem," says Bernard.

For decades, the United States has been playing catch-up with tsunamis, taking protective steps only in the wake of a disaster. In 1946, an earthquake in the Aleutian Islands set off a tsunami that crossed the Pacific and killed 159 people in Hawaii. In response, the National Oceanic and Atmospheric Administration (NOAA) opened a tsunami warning center in Hawaii.

In 1964, a giant Alaskan earthquake spawned a tsunami that took 111 lives in Alaska, California, and Oregon. Following that tragedy, NOAA established a second tsunami center in Alaska.

These U.S. installations have focused on providing warnings of tsunamis that are triggered by massive earthquakes and then cross the ocean to threaten far-off coasts. For years, these so-called distant tsunamis were seen as the biggest threat to the U.S. shores, says Synolakis.

In the mid-1980s, however, geologists working in the Pacific Northwest started discovering evidence of a home-grown hazard. Bays and wetlands along the coast of Washington and Oregon preserve signs of monstrous tsunamis generated by a giant offshore fault called the Cascadia subduction zone. This structure roared to life most recently in 1700, drowning large sections of the Pacific Northwest coast with a tsunami that also crashed into the coast of Japan, where it was recorded in official documents (SN: 4/8/95, p. 223).

As the Cascadia evidence accumulated in the 1990s, the Pacific rim suffered a string of devastating tsunamis, all generated by strong earthquakes relatively near to shore. The combination forced U.S. researchers to start looking in their own backyards.

Now, they must also worry about smaller threats. "The main lesson [from Papua New Guinea] is that small, local

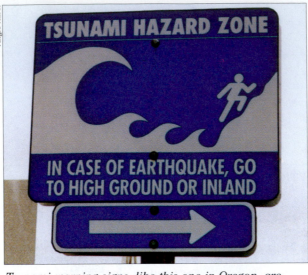

Tsunami warning signs, like this one in Oregon, are going up along the coastline of the Pacific Northwest. If a giant earthquake strikes the region, as happened in 1700, enormous waves could batter the shores for 8 hours.

faults have a much greater potential for tsunami generation than we had thought earlier," says Synolakis.

"This is something we're really worried about in California," he adds. Coastal regions such as Santa Barbara and Palos Verdes, near Los Angeles, face the threat of magnitude 7 quakes from small faults that run underwater, a situation reminiscent of Papua New Guinea.

Adding to the similarities, the seafloor off the coast of southern California drops into steep submarine canyons that are prone to landslides, says Bernard. Indeed, he notes that a highly localized tsunami struck Santa Barbara after an earthquake there in 1812. Although records from that time are sparse, written accounts report that the waves flooded a Spanish mission located 15 feet above sea level.

Okal draws parallels between Papua New Guinea and the Strait of Georgia, which runs between Vancouver Island and the mainland. The Fraser River has built up a large deposit of sediments in the strait that could slump in an earthquake. "We should target areas where sediments are piling up quickly and could be unstable," he says.

The Papua New Guinea disaster comes as the United States is waging an unprecedented campaign to reduce the hazard of tsunamis. The federal government in 1996 teamed up with Hawaii, Alaska, California, Oregon, and Washington to launch the "first systematic attempt to focus on the tsunami problem," says NOAA's Frank González who sits on the steering group of the mitigation program.

The team has taken a three-pronged approach. First, researchers such as

Synolakis are using computer models to produce inundation maps showing credible worst-case scenarios.

Second, local officials and police use these maps as part of a public education process aimed at teaching communities what to do in case of a tsunami.

Finally, the program seeks to develop sensors that would sit on the ocean floor and transmit warnings of an approaching tsunami. Located in the deep reaches of the ocean, such instruments would help primarily in cases of distant tsunamis, in which officials often have hours to prepare for the waves' arrival.

In the case of earthquakes close to a coastline, the tsunami hits the shore in only a few minutes and the deep-ocean warning system offers no aid. The mitigation program is therefore teaching local residents to rely on warning signs provided by nature. The message is fairly simple, says González: "If you can see the ocean and you feel the earthquake, run like hell away from the water."

Even just 2 minutes of warning gives people enough time to flee a few hundred meters inland, which can carry them beyond the strongest pulse of the tsunami. "It's irrefutable that you can save lives by this kind of training," says González.

More than a thousand residents of Okushiri, Japan, can testify to that statement. In 1993, a powerful earthquake, magnitude 7.8, hit the coast of this island at 10:00 p.m. Within about 6 minutes, the tsunami crashed into shore and killed 200 people. A total of 1,200 people escaped the waves, however, when they ran inland after feeling the shaking, says Bernard.

In its educational literature, NOAA advises coastal residents to evacuate if the shaking is strong enough that people must hold onto something to keep from falling. Yet the tremors in Papua New Guinea never grew that intense, according to estimates made by Okal and his colleagues.

Interviews with witnesses have provided some other warning signs that coastal residents can look for. People in Papua New Guinea saw the ocean withdraw before the first wave hit, an observation consistent with reports from several other recent tsunamis. Moreover, witnesses said they could hear the wave coming.

"If you hear a large roar, especially at night when you can't see the ocean, that's an indication a tsunami is approaching. You don't want to wait to see this thing. You want to be moving," says Bernard.

California shakes most often in September

23

SCIENCE NEWS, VOL. 152   DECEMBER 13, 1997

Seismologists have discovered a potential link between earthquakes and the weather—a connection that researchers had dismissed for decades.

A study of earthquakes in California and neighboring states indicates that fault activity in the last 5 years has followed a yearly pattern. The predominantly tiny quakes have occurred most frequently in September and least often in April, report Shang Xing Gao and Paul G. Silver of the Carnegie Institution of Washington (D.C.). After examining many potential causes of the cycle, Gao and Silver found that atmospheric pressure appears to be the controlling factor, they reported in San Francisco this week at a meeting of the American Geophysical Union.

When air pressure remains low, as it does during the warmth of summer, it lessens the weight of the atmosphere pressing on the ground, explains Gao. "This reduces the friction on rocks, so earthquakes can occur more easily," he says.

Some Californians have long asserted that earthquakes tend to follow certain weather patterns, but seismologists had been unable to find reliable connections, says Thomas V. McEvilly, a seismologist at the University of California, Berkeley.

Gao and Silver examined all earthquakes that occurred in more than a decade within 1,000 kilometers of Parkfield, Calif., a tiny town on the San Andreas fault. They found no obvious annual pattern in records before 1992. The yearly cycle first appeared in the data following June 28, 1992, when a magnitude 7.3 earthquake struck the Mojave Desert near Landers, Calif.

The Landers tremor may have sensitized certain faults across the western United States, enabling them to respond more readily to changes in atmospheric pressure, the researchers suggest.

"That makes sense," comments Don L. Anderson, a seismologist at the California Institute of Technology in Pasadena. The Landers earthquake, he notes, surprised seismologists by triggering aftershocks across a wide swath of the western United States, much farther than researchers had previously witnessed. Faults in northern California, Nevada, and even Yellowstone National Park in Wyoming awoke in the hours after the Landers quake.

Landers' far-flung aftershocks occurred mostly in volcanic regions, where magma and heated groundwater lie near Earth's surface. These may be weak areas of the crust that are unusually susceptible to changes in atmospheric stress, suggest Gao and Silver. If so, the sensitivity of these faults is waning; the seasonal variation in quake numbers has diminished over the last 5 years.

Other areas, such as the San Andreas fault, showed no heightened sensitivity following the Landers quake.

The researchers examined other potential causes of the seismic variations, such as changes in precipitation or seasonal variation in mining companies' blasting. The pattern of atmospheric pressure fluctuations best matched the rise and fall in quake numbers.

The new observations are important because they also offer new insight into the forces needed to generate aftershocks in the wake of a major quake. Seismologists have assumed that faults would not budge unless stresses jumped or dropped by more than 100 millibars, yet the variations in atmospheric pressure are only one-tenth that figure, say Gao and Silver.

—R. Monastersky

Western U.S. earthquakes have exhibited a yearly cycle since the Landers jolt.

California's quake deficit fades

SCIENCE NEWS, VOL. 153   MARCH 21, 1998

As Californians recover from El Niño–driven storms, they can take comfort from the news that the ground underfoot is less fickle than it once seemed.

In 1995, a panel of seismologists concluded that Southern California had experienced too few earthquakes since 1850 to relieve all the energy that had accumulated in the crust. The situation had created a quake deficit that would have to be paid back with larger or more frequent earthquakes in the future, they said (SN: 1/21/95, p. 37).

Now, two teams of researchers are disputing the evidence of a quake deficit in the region, and their studies are drawing positive reviews from other seismologists. "If you add all these things up, it's pretty clear now that this deficit can go away. It doesn't have to exist," comments Thomas L. Henyey, director of the Southern California Earthquake Center (SCEC), the Los Angeles–based consortium that issued the 1995 report.

Earthquake-producing stress in Southern California comes from the ongoing collision of the North American and Pacific tectonic plates—two large pieces of Earth's outer shell that grind past each other. The SCEC report estimated seismic hazard by looking at the rate of past earthquakes, the geology of faults, and ground movement. Written primarily by David D. Jackson of the University of California, Los Angeles, the report covered the southern 40 percent of the state.

The SCEC study concluded that magnitude 6 and larger earthquakes should have occurred once every 1.6 years—twice their actual rate. To relieve all the energy that had accumulated in the crust, earthquakes would eventually have to strike more frequently or smaller faults would have to join together to produce a giant quake, Jackson and his colleagues reported.

A reanalysis of the SCEC study has uncovered subtle flaws that combined to overestimate the amount of energy building in the crust, reported a team of scientists at a meeting of the Seismological Society of America in Boulder, Colo., earlier this week. Edward H. Field and James F. Dolan of the University of Southern California in Los Angeles and Jackson collaborated on the new study. When they constructed an improved model, no deficit emerged.

Thomas C. Hanks and Ross S. Stein of the U.S. Geological Survey in Menlo Park, Calif., questioned the SCEC study for other reasons. The report relied on a list of magnitude 6 and larger earthquakes since 1850, but Hanks and Stein argue that the list may overlook quakes from early in that period, when California's inland population was sparse. The rate of observed earthquakes since 1903 is 40 to 50 percent

Southern California's quake-producing faults.

higher than that since 1850, which reduces the apparent deficit, they report.

While the new studies will please many Californians, they will not dispel the concerns of Los Angeles residents.

Seismologists still believe that faults around metropolitan Los Angeles have a major quake deficit that will eventually lead to larger or more frequent tremors there.
—*R. Monastersky*

Racing the Waves

Seismologists try to catch quake tremors quickly enough to save lives

By RICHARD MONASTERSKY

SCIENCE NEWS, VOL. 153   MARCH 14, 1998

On June 28, 1992, a violent earthquake shattered a peaceful Sunday morning in Southern California, jerking millions of residents awake at 4:57 a.m. Centered in the Mojave Desert, some 300 kilometers east of Los Angeles, the magnitude 7.3 quake shook the region hard enough to knock out power in scattered pockets across the southern quarter of the state.

Like many of his neighbors, Los Angeles resident Ken Niles turned on the television news for details about the earthquake. "As I was watching, the camera started to shake and the newscasters voiced alarm and started to scramble."

Vibrations from a strong aftershock were rattling the television studio, but they had not yet reached Niles' home in the western part of the city. He yelled to his wife upstairs and then had enough time to dash across the room to grab a fragile clock before the floor started bucking.

Those few seconds of warning came courtesy of simple

Red spells trouble in this map of ground shaking during the 1994 Northridge earthquake. Researchers could not produce such a chart for several weeks after the disaster. In the future, they will have maps of shaking within minutes.

physics. The most damaging seismic waves ripple through Earth's crust at 3.7 kilometers per second. That's quick compared to an airplane but downright poky compared to a television signal, which moves 80,000 times faster.

Seismologists in Southern California plan to exploit that difference in speed. Over the next few years, they will test a system that provides a warning within the first few seconds of an earthquake. Although the initial goal will not be to notify the public, the prototype network could save lives and property indirectly, proponents say.

"You could broadcast a signal giving a few seconds' warning that strong ground shaking is on its way. School children could jump under their desks. Brain surgeons could pull back their knives. Workers dealing with toxic materials could possibly hit shutoff valves," says John R. Filson, chief of the earthquake hazards program at the U.S. Geological Survey (USGS) in Reston, Va.

The pilot early warning system will be the final element in an ambitious, 5-year project called TriNet, started last year by USGS, the California Division of Mines

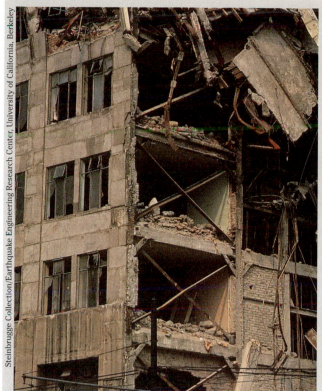

A strong earthquake took more than 10,000 lives in Mexico City in 1985, prompting authorities to set up an early warning system to provide advance notice of future shocks. Experts say that Mexico City is the ideal place for such a system because the earthquake-generating faults lie along the west coast, hundreds of kilometers from the capital. Sensors situated along the coastline can provide a minute or more of warning to residents of the city.

and Geology, and the California Institute of Technology in Pasadena. The $20-million program will wire the Southern California landscape with advanced seismic sensors, all sending their measurements into a central computer facility that quickly analyzes and disseminates the information. The initial purpose is to arm officials with vital earthquake data minutes after a disaster, providing statistics that would previously have taken hours or months to obtain.

Even before the early warning system comes online, TriNet will offer vital information, say emergency managers. "We feel this is very important. It will definitely aid us in how we carry out our response to earthquakes," says Edward J. Bortugno, chief geologist at the California Office of Emergency Services in Oakland, which coordinates relief efforts following earthquakes.

Within minutes of a sizable tremor, TriNet generates maps of shaking intensity. These maps enable emergency crews to identify quickly the sites that have suffered the strongest tremors, where buildings are most likely to have crumbled, says Bortugno.

"Having maps of shaking will help direct our response. Search and rescue, for instance, in collapsed buildings is very dicey, and it's something that has to occur awfully quickly if you're going to save a life. When you lose an hour or two, you end up with more dead people in those buildings. This is a way possibly to get at [trapped people] quicker."

The TriNet initiative represents a dramatic shift in focus for California seismologists, who once viewed earthquake prediction as their ultimate goal. The 1970s saw several successful predictions of large and small earthquakes around the world, notably the alert in 1975 that saved thousands of lives in Haicheng, China.

Hopes fizzled, however, amid some startling failures and a growing realization of the magnitude of the task. "We recognized that prediction really isn't an attainable goal," says Lucile M. Jones, a seismologist with the USGS in Pasadena and a developer of the TriNet project. "We don't want to predict every earthquake, because we have 20 a day. What we need to do is predict which of those earthquakes will grow into major events." This more difficult goal may be unreachable, she adds.

After the Northridge earthquake struck Los Angeles on Jan. 17, 1994, seismologists began to concentrate on more attainable public safety objectives, such as improving the response to earthquakes once they've struck. Jones and her colleagues knew that there was ample room for improvement.

In the early 1990s, USGS and Caltech had established a pager system for automatically disseminating quake information to key organizations within minutes. During the Northridge disaster, however, the central computer mistook the flood of incoming data for faulty communications signals. The system froze for half an hour, refusing to provide the quake's location. Even after the quake data went out,

emergency officials had little to go on. The pager system provided a quake's location and magnitude but no details on which areas suffered the worst shaking. Rescue efforts and media attention centered on the epicentral region in the San Fernando Valley, but they ignored other, more distant sites rattled hard by the quake, says Bortugno. "It was quite some time before we knew of damage south of the epicentral area, in south central Los Angeles, San Pedro, and Santa Monica."

The information blackout spread even farther following the 1989 Loma Prieta earthquake, which slammed the San Francisco area during the World Series at 5:04 p.m. on Oct. 17. Television crews that had gathered to cover the baseball game quickly captured images of the spectacular damage in San Francisco. Initial relief efforts were concentrated there, and officials heard nothing about the problems plaguing other towns and cities, says Bortugno. "It took long into the night before we knew of damage in Santa Cruz," where part of a shopping mall collapsed and killed six people.

TriNet designers hope to avoid similar breakdowns in communication by setting up sensors at 670 sites around Southern California. Unlike older networks, which included many seismometers that recorded and transmitted analog information, TriNet will use only digital equipment, thus reducing the chances of a communication malfunction like the one during Northridge, says Egill Hauksson, a seismologist at Caltech and TriNet leader.

As its cornerstone, the system will have 250 sites sending information continuously, making possible rapid processing of the measurements. TriNet organizers have designed a hardened communication system—one with special telephone lines, radios, and microwave transmitters—so that most sensors can continue transmitting data even during a strong earthquake, says Hauksson. Within 3 to 5 minutes, the central computer will produce a map showing the intensity of shaking around the region.

These so-called shakemaps are already available on the Internet (*http://www-socal.wr.usgs.gov/pga.html*) for earthquakes in Southern California above magnitude 4.0. To avoid potential traffic jams on the Internet during a disaster, critical organizations such as the Office of Emergency Services will receive the shakemaps through a direct computer link to Caltech.

In addition to speeding up rescue operations, rapid measurements of shaking provide other benefits, says James D. Goltz, manager of earthquake programs at Caltech. The state can quickly estimate the number of displaced persons needing shelter and assess financial damages. These estimates, in turn, can expedite the process leading to a presidential disaster declaration, possibly making federal funds available days earlier.

Utilities and industries can use the almost instant information to determine which facilities may require repairs and which probably escaped unharmed. In the past, some large companies set up strong-motion sensors at their facilities to obtain this information, but it took months or even years to process such recordings, making them useless for directing rapid postquake repairs.

Almost 3 dozen large electric transformers broke in the aftermath of a strong Los Angeles earthquake in 1971. "We had several spectacular fires from 6 hours to 6 months after the event," says Ron Tognazzini of the Los Angeles Department of Water and Power. With information on shaking from TriNet, he says, "we could make decisions to depower certain transformers and inspect their insides."

In another unsettling example, the 1971 earthquake damaged an earthen dam in Los Angeles. The top of the dam collapsed to within a handsbreadth of the water level in the reservoir, just shy of unleashing a major flood. There were no measurements on how hard this area had shaken in the earthquake, making it difficult to improve dam designs for future tremors. "In 1971, we had only guesses. Now we will know exactly what they were subjected to," says Tognazzini.

In the same way, engineers could use TriNet data to evaluate damage to buildings and judge the best way to retrofit structures. "When an earthquake happens and a building falls down, we need to know what the ground motions at that site were," says Jones. In the past, though, there were not enough sensors in urban sections of Los Angeles to provide that sort of information.

The new seismic system owes its existence as much to technological innovations as to past catastrophes. Improvements in seismic sensors and computers have enabled researchers to design a much speedier, hardier system than anything available before. However, it took a major earthquake to bring that system to life.

After the Northridge disaster, the Federal Emergency Management Agency (FEMA) granted close to $1 billion in aid to mitigate damage from future earthquakes in Southern California. As part of this package, the state received some $50 million of discretionary funding with few strings attached. The Office of Emergency Services decided to use this money to pay for much of the TriNet system.

Although TriNet proponents have advertised its multiple uses, state officials were sold on one feature in particular: the pilot early warning system.

At this point in the project, administrators are only beginning to explore how to set up the warning system. In certain earthquakes—a magnitude 8 on the southern San Andreas fault, for example—TriNet could provide tens of seconds of warning to downtown Los Angeles and other regions far from the fault line. For earthquakes centered below the city, however, the best warning may come only seconds before the waves or even after the initial, less damaging jitters start.

Emergency managers will try over the next several years to determine how best to use such alerts. "It will take some careful thought," says Bortugno. "There are lots of questions. Should this be like an emergency broadcast system warning? What should it be? Who would get it?"

For some industries, the benefits are obvious. Power companies, for example, suffer damages during earthquakes when overhead lines slap together and melt, shorting out the electrical grid,

Caltech

When completed, the new Southern California seismic network will include more than 200 online stations. Yellow squares denote existing stations, red indicate proposed locations.

says Tognazzini. It would take only one-twentieth of a second to de-energize lines—an act that would prevent this type of power outage.

Turning out the lights on 1.3 million customers has costly implications, however. "You have to weigh the possibilities of throwing out millions of dollars of transactions, which are going across the wires and happening in computers," says Tognazzini.

He hasn't tried calculating the impact of a false alarm yet. "We're just scared to death of what the cost would be to some of our larger customers." During the pilot phase of TriNet, the power department may try the early cutoff with only a tiny percentage of households and then evaluate it.

Emergency managers and seismologists express doubt that the public would benefit directly from just a few seconds of warning. "If the populace is not educated on what to do, it could cause panic. The thing is, the whole earthquake is over in 2 minutes. It's not the time to get in your car and drive to pick up your kids at school," says Filson.

The alerts may be most useful to certain groups that can be trained in how to respond. Workers in high-rises, which sway during quakes, could move to the center of these buildings to prevent being tossed through windows. Emergency teams could take fire trucks and ambulances out of their garages.

In general, people may expect more from a warning system than it can deliver. In a survey conducted a decade ago in California, employees of small, medium, and large businesses said they would want at least a half-minute warning, an unlikely span for any but distant quakes. They also wanted people, rather than automated systems, to decide how to respond to warnings, a step that would eat up many critical seconds.

Nonetheless, citizens may demand access to even the short warnings that will come out of TriNet. Already, some Los Angeles residents harbor the false belief that government scientist are withholding knowledge about the time and dates of future quakes, says Jones. By limiting access to quake warnings, TriNet could feed public distrust of academic and government authorities.

To move beyond the pilot network and develop a true early warning system, emergency managers would have to invest in a network far more expensive than TriNet. Southern California has so many earthquake-generating faults that an effective system would require roughly four times as many seismic stations as TriNet is slated to have, says Jones.

The TriNet project does have its skeptics.

"Just because we paid for it, doesn't mean we endorse it. It's too soon to tell what the utility of the system will be for emergency managers," says Stuart Nishenko of FEMA in Washington, D.C.

Some recall the computer problems encountered during the Northridge quake and wonder whether the new system will perform as advertised. The critical issue for potential users is reliability, says Stephanie H. Masaki-Schatz, Manager for Corporate Safety and Emergency Planning at ARCO oil company in Los Angeles. "It's the confidence that there will be redundancy, so when an event does happen, the data will be readily available and accurate."

The TriNet team plans to run the new computer system through mock earthquakes to evaluate its performance under pressure. The ultimate test will come the next time seismic waves race through Southern California, striking down buildings like so many children's toys. When that will occur, seismologists sadly admit, nobody knows.

Natural-disaster policies need shaking up 26

SCIENCE NEWS, VOL. 155   MAY 29, 1999

Twentieth-century technology has lofted people into space and conquered smallpox, but it has not made them any less susceptible to floods, hurricanes, earthquakes, and other ravages of nature, according to a comprehensive 5-year study released last week.

In fact, U.S. citizens are growing ever more vulnerable to natural disasters, says Dennis S. Mileti, a sociologist at the University of Colorado at Boulder, who led the study team of 132 academics and officials. "The U.S. has been—and still is—creating increased catastrophic disasters in its future. It's time for a paradigmatic shift in our nation's approach," says Mileti.

The federally funded report estimates that 24,000 people died in natural disasters between 1975 and 1994, and four times that number sustained injuries. These events caused damage amounting to $500 billion, not counting indirect costs such as lost business and employment, environmental harm, and emotional tolls on victims.

Severe storms caused more than 80 percent of these losses, while earthquakes and volcanoes together accounted for just 10 percent. Seven of the 10 most expensive disasters, adjusted for inflation, have happened since 1989, indicating that the cost of catastrophes is growing.

The report blames people for the increasing damage. "There is a fundamental flaw in the way our nation and culture try to solve the problems of disaster.

Many of the accepted methods of coping with disaster are based on the fantasy that people can use technology to control nature and make themselves totally safe," says Mileti.

For instance, hurricane forecasting in the past several decades has improved to the point that meteorologists can now provide many hours of warning before a storm makes landfall, perhaps fostering a false sense of security. Compounding that influence was a lull, from 1965 to 1989, in large storms that hit the U.S. coastline. "By providing advance warnings of severe storms, this country may well have encouraged more people to build in fragile coastal areas," says the report.

Indeed, between 1950 and 1991, the population of South Florida swelled from 3 million to over 13 million, with 80 percent of that expansion along the coast.

Instead of trusting in technology to eliminate disasters, Mileti and his colleagues propose that people take responsibility for reducing their risk of harm. Local communities and state and federal officials must consider the threat of natural disasters while making long-term development plans. Insurance companies can provide incentives for people to take

The 1993 Midwest floods ruined 50,000 homes.

steps mitigating future losses.

The Federal Emergency Management Agency (FEMA) has already begun to adopt some of these principles, says Jane Bullock, the agency's chief of staff. In 1997, FEMA started a program called Project Impact, designed to help communities increase their resistance to disasters. In flood-prone Darlington, Wis., for example, residents moved 15 vulnerable buildings and prevented new development near a river.

The program started out working with 7 localities and has since expanded to more than 118. The new report will help spread Project Impact to other regions, says Bullock. "It will give the program an impetus in different communities that may have been somewhat reluctant to come to the table."
—R. Monastersky

The Big Chill 27

Does dust drive Earth's ice ages?

By RICHARD MONASTERSKY

SCIENCE NEWS, VOL. 152   OCTOBER 4, 1997

Earlier this year, as millions of people gazed upward to watch Comet Hale-Bopp sail through the heavens, some observers may have caught specks of space dirt in their eyes. More than 40,000 tons of extremely fine extraterrestrial dust rains down on the planet annually, gathering imperceptibly on windowsills, furniture, poodles, people, and every other available object.

In the long run, these interplanetary motes may have profound consequences for Earth and its inhabitants. Two scientists propose, in a radical

An interplanetary dust particle.

new theory, that dust from space caused the last 10 ice ages, which have gripped the planet like recurrent cases of the flu over the last million years.

"I suspect that to really understand the climate in general, we're going to have to take into account the presence of this extraterrestrial dust, which right now is being ignored," says Richard A. Muller, a physicist at the University of California, Berkeley who collaborated with Gordon J. MacDonald of the International Institute for Applied Systems Analysis in Laxenburg, Austria.

In proposing their hypothesis, Muller and MacDonald have, in effect, picked a fight with the scientific establishment. The theory challenges the accepted explanation of the ice ages, first proposed by Scotsman James Croll in the late 1800s and then expanded by Serbian mathematician Milutin Milankovitch early in this century. The so-called Milankovitch theory holds that periodic flutters in Earth's orbit drive the ice ages by altering the strength of sunlight hitting the north.

"We have shown there is a very serious problem with Milankovitch. I think it's the most serious challenge that has ever been mounted. It's not easy for them to handle this," says Muller of those who hold to the old model.

Muller's strong rhetoric and his unorthodox views have won few converts among climate scientists who specialize in the ice ages. At the same time, opponents arguing from the standard Milankovitch theory cannot easily dismiss his points. This leaves a major gap in knowledge about both how the climate worked in the past and how it may behave in the future.

The new theory focuses on a climatic conundrum called the 100,000-year problem: Why have the last 10 ice ages struck every 100 millennia or so? Muller and MacDonald claim that the standard theory fails to account for this behavior, whereas extraterrestrial dust does. They present their argu-ment in the July 11 SCIENCE.

To understand the disagreement, it's important to go back to 1914, when strife in the Balkans dragged Europe into what would become World War I. Milankovitch, who was held as a prisoner of war by the Austro-Hungarian army that year, escaped from the surrounding chaos by immersing himself in the predictable and stately wiggles of Earth's orbit. He isolated two orbital effects that he thought had a major influence on climate.

The first was the tilt of Earth's rotation axis. Currently 23.5°, the axis bobs toward the vertical and then dips toward the horizontal every 41,000 years. During times of the greatest dip, the amount of sunlight falling on the North and South Poles increases during summer and decreases during winter, causing the seasons to become more extreme.

The second factor, orbital precession, describes the time of year in which Earth comes closest to the sun. Every 23,000 years, the planet's orbit carries it nearest the sun during the Northern Hemisphere's summer solstice, thereby strengthening the sunlight bathing the Arctic.

Milankovitch reasoned that the intensity of summer sunlight hitting the far north is critical to the creation of an ice age. When Earth's orbital cycles conspire to weaken summer sunlight near the Arctic Circle, snow can accumulate from one winter to the next. As the ice sheet spreads, its surface reflects sunlight back into space and cools the climate even further.

In 1976, a trio of oceanographers discovered support for Milankovitch's hypothesis in cores of mud pulled from the seafloor. James Hays and John Imbrie of Columbia University and Nicholas Shackleton of Cambridge University in England measured two oxygen isotopes whose ratio in the ooze reflects the amount of ice on Earth at the time the sediments fell to the ocean floor. This technique enabled them to date the ice ages over the last 500,000 years.

By analyzing the times at which the great ice sheets advanced and retreated, researchers could see hints that Earth's climate was following the beat of several long-term cycles running concurrently. The two most prominent cycles had periods of 41,000 and 23,000 years—the same periods displayed by the variation of orbital tilt and precession. Here at last was evidence supporting Milankovitch's contention that orbital processes played a major role in causing the ice ages.

At the same time, the oceanogra-phers' work pointed out a major problem. In recent ice ages—those occurring in the past million years—the strongest pattern of ice sheet contraction and expansion was one with a period of about 100,000 years, something not anticipated by Milankovitch. Imbrie and his colleagues suggested that the 100,000-year cycle may have arisen from changes in orbital eccentricity, or the amount of distortion of Earth's path around the sun.

Every 100,000 years or so, the orbit varies from completely circular to slightly oval then back to circular. In a round orbit, Earth lies slightly farther, on average, from the sun.

Milankovitch had discounted eccentricity because the variations in the shape of Earth's orbit had a minuscule effect on the amount of sunlight hitting the planet. To explain how such a small flicker can cast such a large shadow, Imbrie and others have hypothesized that the eccentricity changes are somehow amplified by Earth itself. Some factors inherent in the climate must be particularly sensitive to that 100,000-year cycle and enhance it.

Many climate scientists suspect that the amplification stems from the behavior of the largest ice sheets—the time it takes them to grow and decay. The biggest glaciations have occurred only in the last million years, beginning precisely when the 100,000-year cycle kicked in. During earlier times, between 2.5 million and 1 million years ago, the dominant cycles of the ice ages centered on periods of 41,000 and 23,000 years, just as Milankovitch had predicted.

While the more recent ice ages turned out to be more complicated than Milankovitch had envisioned, climate scientists credit his general theory with explaining the timing of the ice ages. "The Milankovitch orbital changes are the pacemaker determining when the large ice sheets grow and decay," says Imbrie, now at Brown University in Providence, R.I.

Muller and MacDonald, however, thought they heard a different drummer setting the rhythm of the ice ages. When the two analyzed the pattern of glaciation for just the last million years, they uncovered something new. By reducing the ice-age variations to a series of frequencies, the scientists found that the dominant 100,000-year cycle was surprisingly exact. It formed a single sharp peak on their graph, with no hint of other frequencies playing a substantial role in the ice ages.

The standard model cannot explain this, they say, because orbital eccentricity varies in a complicated way, involving cycles with periods of 400,000 years and 125,000 years as well as the one close to 100,000. If eccentricity changes were driving the ice ages, all three separate beats should echo through the ice ages, but they do not.

While examining astronomical records, Muller happened on a different cycle that matched exclusively the 100,000-year pattern. The plane of Earth's orbit, he found, tilts gently with respect to the orbits of the other planets every 100,000 years. At times, Earth's orbit around the sun lines up with the plane of the solar system. At other times, Earth travels in a cockeyed orbit, its plane inclined 2.5°.

Muller and MacDonald surmise that when Earth's orbit reaches a certain plane, the planet plows through an extra-thick cloud of interplanetary dust. These small particles drift through the upper atmosphere and set in motion a series of climatic effects, perhaps blocking out sunlight, stimulating cloud growth, or weakening Earth's ozone layer. Such factors cool the planet enough to produce an ice age. Later, as the planet's orbit pulls out of the dust, the climate warms enough to fight back the ice.

Records of extraterrestrial dust offer some support for this hypothesis. Kenneth A. Farley of the California Institute of Technology in Pasadena has measured the variations in helium-3—a tracer for extraterrestrial dust—in seafloor sediments. His studies show that helium-3 values swing up and down over a 100,000-year-long cycle, as Muller and MacDonald predict.

What's more, the amount of helium-3 increased markedly between 2 million and 1 million years ago. This may explain why the 100,000-year climate cycle did not dominate before then, says Farley.

One interpretation of these findings is that large asteroids collided around a million years ago, creating a rich supply of dust that filled the inner solar system. Prior to that time, the ice ages were governed by the regular Milankovitch factors of orbital tilt and precession. When the dust appeared, it overwhelmed the other factors, making ice ages larger and longer.

The dust data do not totally support Muller and MacDonald's idea, however. "These variations in dust amount are quite small; at most, they go up and down by a factor of 3 to 5. I'm skeptical that such variations could change the climate. We're only looking at a little bit of extraterrestrial dust," says Farley.

Even more troubling is the question

Planetary shivers: Oxygen isotopes in seafloor sediments track the ice ages over the last 900,000 years (left). An analysis of the frequencies present in the isotope record (right) shows a single peak corresponding to a 100,000-year-long cycle (top), which matches the pattern expected for orbital inclination (bottom), but not the triple peak of orbital eccentricity (middle).

of how the dust alters climate. While Muller and MacDonald have general ideas, they lack a solid hypothesis to explain how adding dust to the upper atmosphere could plunge the planet into an ice age. "That is what we're most severely criticized for. We do a lot of hand waving at this point. There are many possibilities, but we don't have a good mechanism," says Muller.

When Muller and MacDonald came up with their dust theory in 1993, they had a hard time getting climate scientists to listen. Last year, Wallace S. Broecker of the Lamont-Doherty Earth Observatory in Palisades, N.Y., invited Muller to present his case to the top researchers interested in the ice ages. Broecker dubbed the meeting the "Mullerfest," and the discussions continued for weeks via E-mail.

Muller scored the most points at the meeting when he attacked a standard technique, called tuning, that oceanographers use for dating layers in sediment cores. The task of dating these strata is difficult because sediments may accumulate more quickly during some eras and more slowly in others. To tell the age of layers between known benchmarks, researchers often use the Milankovitch orbital cycles to tune the sediment record: They assume that ice volume should vary with the orbital cycles, then line up the wiggles in the sediment record with ups and downs in the astronomical record.

"This whole tuning procedure, which is used extensively, has elements of circular reasoning in it," says Muller. He argues that tuning can artificially make the sediment record support the Milankovitch theory.

Muller's criticisms hit home with many researchers. "He scared the hell out of them, and they deserved it," says Broecker.

Oceanographers soon rose to the challenge. In the August PALEOCEAN-

OGRAPHY, Maureen E. Raymo of the Massachusetts Institute of Technology presents an untuned sediment record that corroborates the ice age dates determined by tuning.

To get at the causes of the ice ages, Raymo takes a different approach. Instead of doing frequency analysis to pick out the cycles in the record, she looks at the actual timing of the ice ages: their beginnings, endings, and the intervals between each. She finds that ice ages don't come like clockwork every 100,000 years. Instead, they seem to hit sporadically, sometimes ending 85,000 years apart, other times 125,000 years apart.

What is consistent is that the ice ages almost always end when Milankovitch's orbital cycles combine to warm the northern lands during summertime, as the mathematician hypothesized 80 years ago.

Still, nobody can explain satisfactorily how Earth's climate amplifies the changes in sunlight to drive the ice ages of the last million years. Until the theory is complete, it leaves the door open to challengers.

"It's not clear what is causing this cycle in the last million years. I would say it is one of the great puzzles of earth sciences," says George H. Denton, a glacial geologist at the University of Maine in Orono.

Most climate researchers do not think that extraterrestrial dust will provide the answer, but Imbrie, a long-time supporter of the Milankovitch hypothesis, says that Muller and MacDonald's concept may have some merit. It is possible that Earth's climate could respond to the unsteady orbital plane as well as to the kinds of astronomical twitches that preoccupied Milankovitch, he says. "These are not mutually exclusive ideas. Nature doesn't always follow a neat little paradigm."

Signs of unstable ice in Antarctica

SCIENCE NEWS, VOL. 154   JULY 11, 1998

The icy blanket draped over Antarctica is so stark and primordial that it seems as permanent as rock. Indeed, glaciologists have traditionally considered the ice sheet to be a stable feature that has persisted for at least the last 6 million years in West Antarctica and much longer on the eastern side of the continent. New evidence, however, melts that solid image by suggesting that the ice cap of West Antarctica collapsed and then reformed quite recently in geologic time.

Reed P. Scherer, a paleontologist at Uppsala University in Sweden, first suggested this possibility in 1991. While studying the shells of algae drilled up from beneath the ice cap, he found modern diatom species that originated no earlier than 750,000 years ago. He concluded that the ice had melted at some time since then, allowing the ocean to flood this region, which lies below sea level. The current ice cap must have formed after that event, he argued.

At the time, critics attacked Scherer's study, but now he has additional evidence. In the July 3 SCIENCE, Scherer and his colleagues report finding young diatoms in several more drill cores pulled up from beneath the West Antarctic ice cap. Further support for a recent melting comes from analysis of radioactive beryllium-10 atoms, which are created in the atmosphere by incoming cosmic rays. The concentration of these atoms in the sediments below the ice cap indicates that the region was free of ice at some time much more recent than 6 million years ago.

Richard B. Alley, a glaciologist at Pennsylvania State University in State College, says Scherer's new work answers his past critics. The finding also raises questions about the long-term stability of West Antarctica. If the ice sheet collapses, global sea levels would rise by 6 meters, enough to flood many coastal cities. "Most of us think it becomes more likely if it happened in the past," says Alley.   —R.M.

Sea Change in the Arctic

An oceanful of clues points to climatic warming in the far North

By RICHARD MONASTERSKY

SCIENCE NEWS, VOL. 155   FEBRUARY 13, 1999

The first sign that something was wrong came just a few days after the icebreaker *Des Groseilliers* left Tuktoyaktuk, a port on the Arctic coast of western Canada. Steaming north with a full load of 20 scientists and provisions for 18 months, the Canadian coast guard ship was heading for a date with the Arctic pack ice—the perennially frozen layer that covers the top of the globe like a white skullcap.

Old hands at polar research, such as oceanographer Miles G. McPhee, expected to meet the ice between 71° and 72°N. That's where the southern edge of the pack showed up in the 1970s, when McPhee made several trips far into the Arctic. Two decades later, the Canadian icebreaker cruised past 72°N at full speed, with no ice in the water to slow its progress. McPhee, who runs a research company out of Naches, Wash., wondered silently, "Oh my God, where did all the ice go?"

The case of the shrinking pack ice is only one of many climatic conundrums troubling scientists who study the Arctic. In recent years, researchers have discovered myriad signs of changes in the far North, affecting everything from the ocean currents flowing 1,000 meters beneath the ice pack to the howling winds at the top of Earth's atmosphere. "I think the changes are verging on what could be called dramatic," says John M. Wallace, an atmospheric scientist at the University of Washington in Seattle.

Some aspects of the shifting Arctic weather resemble the patterns expected from greenhouse warming, leading scientists to wonder whether they are witnessing early warning signs of conditions to come during the next century. The polar regions, especially the Arctic, are regarded as the climatic equivalent of canaries in a coal mine. The poles are the places on Earth most sensitive to global warming. With the canaries looking distinctly wan, the search is now on to determine what has caused their decline.

These are some of the scientific issues that propelled McPhee and his colleagues northward on the *Des Groseilliers* in the fall of 1997. Following the plans of the $19.5 million international research project, the Canadian icebreaker finally plowed into the pack ice, cut its way to the middle of a large floe, and then stopped its engines at 75°N.

For an entire year, the ship remained frozen amid the ice, dragged 2,800 kilometers on a roundabout course northwest across the Arctic Ocean, a landless expanse of water and ice. Over that time, planes ferried a total of 170 researchers out to the ship in shifts of several weeks to months. The scientific teams fanned out across the nearby ice to collect measurements on how heat shuttles among the ocean, the sea ice, and the atmosphere through the different seasons. They called the project SHEBA, an acronym for Surface Heat Budget of the Arctic.

From the beginning, SHEBA taught investigators to question the prevailing ideas about the Arctic. Proj-ect planners had expected to park the icebreaker amid floes measuring 3 m thick, the kind of ice seen in the 1970s during the last major U.S. Arctic initiative. The SHEBA crew, however, was dismayed by what it encountered in 1997.

"When we went up there, the first problem we had was trying to find a floe that was thick enough. The thickest ice we could find was 1.5 to 2 m," says SHE-BA chief scientist Donald K. Perovich of the U.S. Army Cold Regions Research and Engineering Laboratory in Hanover, N.H.

Once the crew set up the ice station, another startling fact came to light. During the first few weeks of work, McPhee and his colleagues found that the shallow waters of the Arctic Ocean were

At home on the Arctic Ocean: During their year frozen into the ice, scientists on the Canadian ship Des Groseilliers *saw temperatures range from a low of -42°C in late December to a high of almost 1°C in July, when substantial melting occurred around the ship.*

warmer and less salty than they had been 22 years earlier. This observation implies that a significant quantity of ice had melted during the previous summer, says McPhee's team, which published the SHEBA finding in the May 15, 1998 GEOPHYSICAL RESEARCH LETTERS.

More shocks came toward the end of the project in October 1998. "The big surprise of our work was that the ice ended up at the end of the year thinner than when we had started," says Perovich. The warm winter and long summer of 1988 had shaved off about one-third of a meter from the already thin ice.

SHEBA investigators suggest that some of the changes they saw could have stemmed from the 1997–1998 El Niño in the equatorial Pacific, although this temporary warming can't explain all

the findings. Arctic sea ice has been declining for many years.

In 1997, researchers from NASA's Goddard Space Flight Center in Greenbelt, Md., reported that the area covered by sea ice decreased by more than 5 percent between 1978 and 1996. Other satellite measurements of the sea ice have revealed evidence of increased summertime melting since 1979 (SN: 2/21/98, p. 116).

Sea ice is only a thin skin over the Arctic Ocean, but in many ways it serves as the linchpin of the region's climate and perhaps that of the whole globe. If the pack diminished substantially, climate models predict big changes in the ocean and atmosphere. The reason stems mainly from something scientists call the ice-albedo feedback.

Because ice is so bright, it reflects more than half the sunlight that hits it during summer. The dark water, by contrast, absorbs 90 percent of the incident sunlight, says Perovich. The existence of the sea ice, therefore, keeps the Arctic Ocean cool by shielding it from solar energy.

If the ice starts to disappear, according to theory, the ocean will rapidly warm and melt more ice—a potentially runaway process that could strip the Arctic of its protective cap and allow solar energy to stream into the ocean unhindered. This ice-albedo feedback would greatly amplify the effects of greenhouse gas pollution, causing the Arctic to warm much more than the rest of the globe, according to climate models. The Arctic shift would have a domino effect, rerouting ocean currents and weather patterns further south.

Until recently, scientists had little access to the Arctic Ocean, which served as the sparring ring for superpower submarines. With the end of the Cold War and the withdrawal of military forces, scientists started invading the region. Researchers who once sat on opposite sides of the Iron Curtain began sharing data and collaborating on projects. Since 1993, the U.S. Navy has invited oceanographers along on several submarine cruises beneath the ice pack.

In a strange confluence of climate and current events, the thaw in political tensions a decade ago coincided with an apparent shift toward warmer conditions in the North. Some of the biggest changes have occurred within the Arctic Ocean, an oblong body of water nearly one and a half times the size of the United States. The ocean is divided into four unequal basins separated by three roughly parallel mountain ranges on the seafloor—an arrangement that looks somewhat like an empty TV dinner tray.

Much of the water in the Arctic comes from the Atlantic Ocean, coursing north as a vestigial extension of the Gulf Stream. On the opposite side of the pole, water from the Pacific Ocean enters the Arctic via the Bering Strait. As the Pacific water passes through the shallow strait, it loses much of its heat to the atmosphere and consequently grows colder than the Atlantic water. The two distinct types of water meet midway in the Arctic, creating a front much like the kind that separates warm and cold air masses in the atmosphere.

Water measurements made by Rus-sian and Western scientists indicate that the front between these two types of water bisected the Arctic Ocean for most of the past half-century. From 1949 through the late 1980s, the front generally lined up with a submerged mountain range called the Lomonosov Ridge, which runs from northern Greenland toward eastern Siberia, passing close to the North Pole.

During a scientific submarine cruise in 1993, researchers on board the USS *Pargo* found that the front had shifted substantially. The warmer, saltier Atlantic waters had pushed further into the Arctic, causing a retreat of the Pacific waters, says James H. Morison of the University of Washington, who participated in the *Pargo* cruise.

At the same time, the Atlantic-water layer over the Lomonosov Ridge warmed by 1°C, reaching a temperature not seen in the data going back to 1949, according to a January 1998 report by Morison and his colleagues in DEEP-SEA RESEARCH PART I.

Other shifts have altered the layered arrangement of water in the Arctic. To understand these divisions, imagine lowering a thermometer and a salt meter off a ship in the middle of the Arctic. The uppermost layer of the ocean would be extremely fresh with a temperature near the freezing point. Below that, the instruments would pass

through a layer called the halocline, where the water remains cold but grows saltier with depth. Lower still would be a thick region of relatively warm water, so dense with salt that it remains stuck on the bottom, trapped below the fresher, colder layer above.

In a series of three submarine cruises during the 1990s, oceanographers witnessed the halocline weakening in the central part of the Arctic. Early in the decade, a particularly cold sheet of halocline water straddled the Lomonosov Ridge. By 1995, however, the sheet had disappeared from the European side of this range, according to Michael Steele of the University of Washington. He and colleague Timothy Boyd of Oregon State Univer-sity in Corvallis reported their finding in the May 15,

Three sets of submerged mountain ridges (red) split the Arctic Ocean into separate basins. The SHEBA project studied conditions on the Alaskan side of the Arctic.

1998 JOURNAL OF GEOPHYSICAL RESEARCH.

The retreat of this layer could have profound effects, says Steele, because the halocline helps stratify the Arctic Ocean, keeping the warm, deep waters from reaching the surface. "Because there is decreased stratification, there should be enhanced heat transfer from the warm layer up to the surface, and hence thinner ice," says Steele.

Data collected by British submarines show that ice has been thinning in this region for some time. In 1990, researchers with the Scott Polar Research Institute at Cambridge University in England reported that ice thickness on the European side of the Lomonsov Ridge had de-

creased by 15 percent between 1976 and 1987. Preliminary analysis of measurements made in 1996 now suggest that the downward trend has continued, says Cambridge's Norman Davis.

With so many signs of disruptions in the Arctic climate, scientists are looking skyward for answers. The ice pack and even the water beneath it, they suspect, are taking their cues from the pattern of winds that swirl around the edges of the Arctic. "Really, that is where it all starts," says Steele.

What's going on in the Arctic may fit into a larger picture that includes much of the globe's northern half. According to a theory proposed last year, atmospheric pressure tends to fluctuate in seesaw fashion over large parts of the Northern Hemisphere. When pressure increases over the Arctic, it decreases in a donut-shaped ring at the latitude of Washington, D.C. During the reverse part of the irregular cycle, air pressure drops in the Arctic and rebounds in the midlatitudes. Wallace and David W. J. Thompson of the University of Washington called this pattern the Arctic oscillation in the May 1, 1998 GEOPHYSICAL RESEARCH LETTERS.

Atmospheric temperatures in northern Europe and Asia dance in step with this Arctic oscillation. When the atmospheric pressure sags, air temperatures tend to rise and vice versa, says Wallace, who has tracked data going back to 1900.

For most of the century, the Arctic oscillation shifted randomly from month to month and year to year, with no distinct preference for either extreme. Recently, however, the oscillation has leaned strongly toward low Arctic air pressure and high temperatures in northern Eurasia. The shift started in the early 1970s and intensified in the late 1980s, bringing atmospheric conditions unseen in the century of measurements.

"That's the link to perhaps some of the other things going on in the Arctic," says Wallace. As the air pressure drops, the westerly winds encircling the Arctic grow stronger. The winds weaken the ice pack by pulling it apart and opening up watery channels between individual ice floes, says Wallace.

The souped-up westerlies also alter conditions in heavily populated parts of the globe. As they scream over the North Atlantic Ocean, the winds carry more heat downwind onto land. This translates into warmer winters in northern Europe and drier winters in southern Europe.

75° N

70° N

1998

1997

1996

Alaska

Canada

160° W   150° W   140° W   130° W

Frozen retreat: The edge of summertime sea ice in the Beaufort Sea pulled northward between 1996 and 1998, according to satellite data. Oceanographers are watching closely to see whether the ice withdrawal continues this summer.

"We've seen a strong warming in high latitudes over land in large areas of Russia," says Wallace. "We're seeing changes in the circulation that are big enough in winter that people in Europe are noticing them."

The key question is, What has caused all this change? In a study using computer models of the atmosphere, a team of Russian researchers has found evidence that thinning of the Arctic ozone layer could be driving the shift in the atmosphere. As chlorofluorocarbons and other chemicals eat away ozone, the polar stratosphere cools off. In the Russian model, this cooling effect enhances the westerlies that circle the Arctic and hence reduces the atmospheric pressure there.

Greenhouse gases could also have a hand in the Arctic shift. Carbon dioxide and other heat-trapping pollutants tend to warm the lowest layer of the atmosphere—the troposphere—while cooling off the stratosphere above. Researchers at NASA's Goddard Institute for Space Studies in New York City used a computer model to investigate this split effect of greenhouse gases. They found that the combination speeds up the westerly polar winds, pushing the Arctic oscillation in the direction seen during the past 30 years, says NASA's Drew T. Shindell.

Wallace thinks there is a good chance that one or both of these human influences could have precipitated the Arctic changes. Yet he cautions against drawing any conclusions. "We have to allow some possibility that this is of natural origin," as part of a long-term cycle that will eventually reverse, he says. To rule out a natural cause, the Arctic oscillation would have to keep on its present course for another 5 to 10 years, he says.

If that indeed transpires, researchers may not be able to repeat an experiment like SHEBA. With sustained warming and stronger westerlies, the ice-albedo feedback could kick in and rapidly rid the Arctic of its white cover during summer. McPhee and his colleagues raised this possibility last year in their paper in GEOPHYSICAL RESEARCH LETTERS. In the title, they asked, "Is perennial sea ice disappearing?"

Early in his career, McPhee would have found such a question unthinkable. Now, he says, "I'm starting to wonder whether we're not going to see it happen."

Limited Use License Agreement

This is the John Wiley and Sons, Inc. (Wiley) limited use License Agreement, which governs your use of any Wiley proprietary software products (Licensed Program) and User Manual (s) delivered with it.

Your use of the Licensed Program indicates your acceptance of the terms and conditions of this Agreement. If you do not accept or agree with them, you must return the Licensed Program unused within 30 days of receipt or, if purchased, within 30 days, as evidenced by a copy of your receipt, in which case, the purchase price will be fully refunded.

License: Wiley hereby grants you, and you accept, a non-exclusive and non-transferable license, to use the Licensed Program and User Manual (s) on the following terms and conditions:

a. The Licensed Program and User Manual(s) are for your personal use only.
b. You may use the Licensed Program on a single computer, or on its temporary replacement, or on a subsequent computer only.
c. The Licensed Program may be copied to a single computer hard drive for playing.
d. A backup copy or copies may be made only as provided by the User Manual(s), except as expressly permitted by this Agreement.
e. You may not use the Licensed Program on more than one computer system, make or distribute unauthorized copies of the Licensed Program or User Manual(s), create by decompilation or otherwise the source code of the Licensed Program or use, copy, modify, or transfer the Licensed Program, in whole or in part, or User Manual(s), except as expressly permitted by this Agreement. If you transfer possession of any copy or modification of the Licensed Program to any third party, your license is automatically terminated. Such termination shall be in addition to and not in lieu of any equitable, civil, or other remedies available to Wiley.

Term: This License Agreement is effective until terminated. You may terminate it at any time by destroying the Licensed Program and User Manual together with all copies made (with or without authorization).

This Agreement will also terminate upon the conditions discussed elsewhere in this Agreement, or if you fail to comply with any term or condition of this Agreement. Upon such termination, you agree to destroy the Licensed Program, User Manual (s), and any copies made (with or without authorization) of either.

Wiley's Rights: You acknowledge that all rights (including without limitation, copyrights, patents and trade secrets) in the Licensed Program (including without limitation, the structure, sequence, organization, flow, logic, source code, object code and all means and forms of operation of the Licensed Program) are the sole and exclusive property of Wiley. By accepting this Agreement, you do not become the owner of the Licensed Program, but you do have the right to use it in accordance with the provision of this Agreement. You agree to protect the Licensed Program from unauthorized use, reproduction, or distribution. You further acknowledge that the Licensed Program contains valuable trade secrets and confidential information belonging to Wiley. You may not disclose any component of the Licensed Program, whether or not in machine readable form, except as expressly provided in this Agreement.

WARRANTY: TO THE ORIGINAL LICENSEE ONLY, WILEY WARRANTS THAT THE MEDIA ON WHICH THE LICENSED PROGRAM IS FURNISHED ARE FREE FROM DEFECTS IN THE MATERIAL AND WORKMANSHIP UNDER NORMAL USE FOR A PERIOD OF NINETY (90) DAYS FROM THE DATE OF PURCHASE OR RECEIPT AS EVIDENCED BY A COPY OF YOUR RECEIPT. IF DURING THE 90 DAY PERIOD, A DEFECT IN ANY MEDIA OCCURS, YOU MAY RETURN IT. WILEY WILL REPLACE THE DEFECTIVE MEDIA WITHOUT CHARGE TO YOU. YOUR SOLE AND EXCLUSIVE REMEDY IN THE EVENT OF A DEFECT IS EXPRESSLY LIMITED TO REPLACEMENT OF THE DEFECTIVE MEDIA AT NO ADDITIONAL CHARGE. THIS WARRANTY DOES NOT APPLY TO DAMAGE OR DEFECTS DUE TO IMPROPER USE OR NEGLIGENCE.

THIS LIMITED WARRANTY IS IN LIEU OF ALL OTHER WARRANTIES, EXPRESSED OR IMPLIED, INCLUDING, WITHOUT LIMITATION, ANY WARRANTIES OF MERCHANTABILITY OR FITNESS FOR A PARTICULAR PURPOSE.

EXCEPT AS SPECIFIED ABOVE, THE LICENSED PROGRAM AND USER MANUAL(S) ARE FURNISHED BY WILEY ON AN "AS IS" BASIS AND WITHOUT WARRANTY AS TO THE PERFORMANCE OR RESULTS YOU MAY OBTAIN BY USING THE LICENSED PROGRAM AND USER MANUAL(S). THE ENTIRE RISK AS TO THE RESULTS OR PERFORMANCE, AND THE COST OF ALL NECESSARY SERVICING, REPAIR, OR CORRECTION OF THE LICENSED PROGRAM AND USER MANUAL(S) IS ASSUMED BY YOU.

IN NO EVENT WILL WILEY, THE AUTHOR, OR ALLEGIANT TECHNOLOGIES, INC. BE LIABLE TO YOU FOR ANY DAMAGES, INCLUDING LOST PROFITS, LOST SAVINGS, OR OTHER INCIDENTAL OR CONSEQUENTIAL DAMAGES ARISING OUT OF THE USE OR INABILITY TO USE THE LICENSED PROGRAM OR USER MANUAL(S), EVEN IF WILEY OR AN AUTHORIZED WILEY DEALER HAS BEEN ADVISED OF THE POSSIBILITY OF SUCH DAMAGES.

General: This Limited Warranty gives you specific legal rights. You may have others by operation of law which varies from state to state. If any of the provisions of this Agreement are invalid under any applicable statute or rule of law, they are to that extent deemed omitted.

This Agreement represents the entire agreement between us and supersedes any proposals or prior Agreements, oral or written, and any other communication between us relating to the subject matter of this Agreement.

This Agreement will be governed and construed as if wholly entered into and performed within the State of New York. You acknowledge that you have read this Agreement, and agree to be bound by its terms and conditions.

Installing Visualization 2.0

Windows 95, Windows 98 and NT 4.0

1. Start your computer. If your computer is already running, close all applications.
2. Insert the Visualization 2.0 CD-ROM in your CD-ROM drive (volume label = VSLZE2).
3. Click on the START button and then click on Run...
4. In the open box, type:
   D:\SETUP.EXE (where D is the letter of your CD-ROM drive).
   Or, click on Browse..., locate SETUP.EXE on the CD-ROM drive, and click on OK.
5. Follow the instructions in the installer.
6. Restart your computer.

MacOS

1. Start your computer.
2. Insert the Visualization 2.0 CD-ROM in your CD-ROM Drive.
3. Double-click on the Install Visualization 2.0 icon.
4. At the bottom of the opening installation screen, click on Continue.
5. Select the desired location for installation or accept the default location, then click on Install.
6. Click Continue when the dialog recommends restarting your computer after completion of installation.
7. If you already have QuickTime installed, click on Restart to ensure proper installation of fonts for Visualization 2.0. If you do not have QuickTime or Adobe Acrobat Reader installed on your computer, click Quit after completion of the installation.
8. If your computer does not have QuickTime installed or if you are running a version of QuickTime earlier than 2.5, open the QuickTime 2.5 Installer folder and double-click the QuickTime Read Me icon and follow the step-by-step instructions to install QuickTime on your hard drive.
9. To install Adobe Acrobat Reader, double-click the Install Acrobat Reader 3.01 icon on the CD-ROM and follow installation instructions.
10. Restart your computer.

RL List Font

The RL List font enables lists of media to display properly within Visualization 2.0. RL List will automatically be installed by the Visualization 2.0 installer. In order for the RL List font to display properly, you must restart your computer after installation

Getting Started

Windows 95, Windows 98 and NT 4.0
A program folder named "Visualization 2.0" will be installed during setup. Once installation is completed and you have restarted Windows, you will be ready to begin exploring Visualization 2.0. With the Visualization 2.0 CD-ROM loaded in your CD-ROM drive, click the Start button, and then point to Programs. Point to the "Visualization 2.0" program folder, then click on "Visualization 2.0" to launch the program.

MacOS
Open the newly installed "Visualization 2.0" folder on your hard drive. Double-click the "Visualization 2.0" icon to launch the program.

Online Help/User's Manual
Windows 95, Windows 98 and NT 4.0
For more information, please refer to the online user's manual. To view this document, click the "Visualization 2.0 Help" icon in the "Visualization 2.0" program folder. Help is also available from within Visualization 2.0 and can be accessed at any time by pressing the F1 key or by selecting Contents from the Help menu.

Basic Navigation and Features
4 Steps to Get Started on Visualization 2.0

Visualization 2.0 basically is a library of visual files of four types: animation, video, photo and illustration. These files can be accessed by type and/or by physical geography topic. The user, whether he or she is the instructor or the student, can organize this large library according to need, and may narrow the search for files by type or by topic. Once the user finds the files to be saved, he or she should build a "presentation" and save each presentation according to user preference. Here are some basic steps to consider and descriptions for some of the options and windows you will see on the Visualization 2.0 screen.

1 "OPEN" or "NEW"

- When you are using Visualization 2.0 for the first time, or when you are starting a new search of Visualization 2.0, click "**New**" at this screen, which is the first one to appear on your monitor after installation or start-up is successful.

- If you want to get to a Visualization file that you have already saved, choose "**Open**" from this screen.

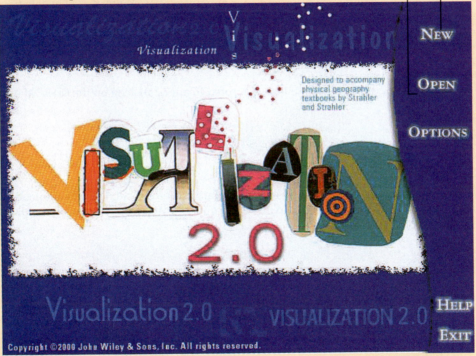

2 SEARCH

- Click on "**Resource**" from the menu bar, and select "**Search CD-ROM**."

3 CREATE

Visualization 2.0 contains over 650 resources of four types: photo, illustration, animation and video. The goal of Step 3 is to create your own files with just the resources you will want to use.

- When you highlight a resource in the Resource Window, a small preview of it appears here.

- These three fields allow you to narrow your search for resources by **Type** (animation, video, photo, illustration); by **Topics** that reflect chapter titles from the text; and by **Keyword**. Each resource is associated with three keywords.

- Movie camera and still photography camera icons indicate the type of resource: animation/video or photo/illustration.

- Resource Window

- While searching Visualization 2.0, you may "**Add**" a resource to a file, which you will create on your computer in Step 4. For now, click once on the resource, and then click once on the **Add** button. This is the way instructors can build presentations and students can build files that will be useful for studying or for a course project.

- The **Export** function allows the user to move resources from Visualization to other software, such as **Power Point**.®

- When you are finished searching and adding to your list of resources, **Close** this window and move on to Step 4.

4 SAVE

When you are ready to view your resources in full screen format, just double click on the top resource. Then use the arrow buttons on the menu bar to move through the resources in your file.

- Arrow Buttons

- Save the file you built in Step 3 by opening **File** on the menu bar and selecting **Save As**. With Visualization, save your file using the extension ".RLD", for example, WATER.RLD

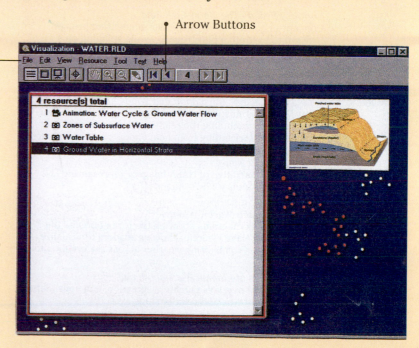

Animation and Video Contents

The animations and videos are keyed to text sections and art, as follows, by the icon shown here. 🎥

New and Improved Version 2.0!	ANIMATIONS	Chap.	Art references	Text references
Improved	*Earth-Sun Relationships* Illustrates the basic earth-sun relationships and the effect of the tilt of the earth's axis to create the seasons.	1	Figure 1.13	"The Earth's Revolution Around the Sun," p 22
NEW!	*The Sun's Path in the Sky* Captures variable noon sun angles *from the perspective of the Earth's surface* at different parts of the globe and different times of year, with the effect on the length of the daylight period.	2		"The Path of the Sun in the Sky," p 35
Improved	*Sun Angle and Length of Day* Captures variable noon sun angles *from the perspective of Space* at different parts of the globe and different times of year, with the effect on the length of the daylight period.	2	Figure 2.6	"The Path of the Sun in the Sky," p 35
Improved	*Earth's Energy Balance* Diagrams the flows of energy between the sun, the earth's surface, and the atmosphere.	2	Figure 2.11 Figure 2.12 Figure 2.13	"The Global Energy System," p 39
Improved	*Latent Heat* Prepares the reader for adiabatic processes by showing that heat energy can be tied up in a substance without raising its temperature, as the heat energy is being converted into kinetic energy during thawing or evaporation.	2		"Sensible Heat and Latent Heat Transfer," p 39
Improved	*Adiabatic Processes* Allows students to link changes in temperature, altitude, and pressure; builds on Latent Heat animation to show adiabatic cooling as a change in the form of energy.	4	Figure 4.9 Figure 4.10	"The Adiabatic Process," p 81 "Convectional Precipitation," p 96
NEW!	*Orographic Precipitation* Focuses on one parcel of moist oceanic air as it arrives at the coast, ascends up a slope and down the other side, dramatizing adiabatic and latent heat processes. A thermometer registers the changing temperature during the parcel's journey.	4	Figure 4.16	"Orographic Precipitation," p 95 "Global Precipitation," p 155
Improved	*Coriolis Effect I* Traces the diversion of objects on a rotating globe to help users understand the concept of conservation of angular momentum on the surface of a rotating sphere.	5	Figure 5.10	"The Coriolis Effect and Winds," p 110
NEW!	*Coriolis Effect II - The Merry-Go-Round* Concept of conservation of angular momentum brought down to earth to a playground. While spinning on a merry-go-round, one person tosses a ripe tomato to a mate directly opposite. Traces the diversion of the tomato in motion and the mess that results.	5	Figure 5.10	"The Coriolis Effect and Winds," p 110
Improved	*Global Atmospheric Circulation and Winds Aloft* Shows global atmospheric circulation as components and as a system, including the relationships between falling and rising air resulting from heat energy and air pressure.	5	Figure 5.17 Figure 5.18	"Surface Winds," p 112 "Global Circulation at Upper Levels,"
NEW!	*20 Months of El Nino* An animated narrative of NASA imagery representing El Nino from January 1997 - August 1998.	5	Eye on the Environment Figure	"Eye on the Environment - El Nino," p 124-125
Improved	*Development of a Wave Cyclone* Animates block diagrams showing the formation and life history of a wave cyclone, allowing readers to visualize how storm systems bring weather across the continent.	6	Figure 6.6 Figure 6.7	"Wave Cyclones," p 139
Improved	*The Interaction of Climate, Vegetation, and Soils* Sequences and overlays climate, vegetation, and soils maps to reveal patterns and relationships.	7, 8, 9	Figure 7.2 Figure S7.2 Figure 7.4 Figure 8.13 Figure 9.10	"Climate Classification," p 157 "Terrestrial Ecosystems- The Biomes," p 208 "Global Soils," p 247